GAME THEORETIC ANALYSIS

❶ Non-cooperative Games and Equilibrium Analysis
❷ Cooperative Games and Axiomatic Value

Other Related Titles from World Scientific

A Nontechnical Introduction to the Analysis of Strategy
Third Edition
by Roger A McCain
ISBN: 978-981-4578-87-5

Game Theory
Second Edition
by Leon A Petrosyan and Nikolay A Zenkevich
ISBN: 978-981-4725-38-5

Game Theory: A Comprehensive Introduction
by Hans Keiding
ISBN: 978-981-4623-65-0

GAME THEORETIC ANALYSIS

❶ Non-cooperative Games and Equilibrium Analysis
❷ Cooperative Games and Axiomatic Value

Editors

Leon A Petrosyan
Saint Petersburg State University, Russia

David Wing Kay Yeung
Hong Kong Shue Yan University, Hong Kong
Saint Petersburg State University, Russia

 **World Scientific**

NEW JERSEY · LONDON · SINGAPORE · BEIJING · SHANGHAI · HONG KONG · TAIPEI · CHENNAI · TOKYO

Published by

World Scientific Publishing Co. Pte. Ltd.

5 Toh Tuck Link, Singapore 596224

USA office: 27 Warren Street, Suite 401-402, Hackensack, NJ 07601

UK office: 57 Shelton Street, Covent Garden, London WC2H 9HE

British Library Cataloguing-in-Publication Data
A catalogue record for this book is available from the British Library.

GAME THEORETIC ANALYSIS

ISBN 978-981-120-200-1

For any available supplementary material, please visit
https://www.worldscientific.com/worldscibooks/10.1142/11326#t=suppl

Typeset by Stallion Press
Email: enquiries@stallionpress.com

Contents

Part II.   Cooperative Games and Axiomatic Values   299

List of Contributors

Giulia Bernardi, Dipartimento di Matematica, Politecnico di Milano, Italy

Jane M. Binner, Birmingham Business School, University of Birmingham, UK

Peter Borm, Department of Econometrics and Operations Research, School of Economics and Management, Tilburg University, The Netherlands

Michael R. Caputo, Department of Economics, University of Central Florida, USA

Francesco Ciardiello, Management School, University of Sheffield, UK

Marco Dall'Aglio, LUISS University, Department of Economics and Finance, Italy

Pierre Dehez, CORE, University of Louvain, Belgium

T. S. H. Driessen, Faculty of Electrical Engineering, Mathematics and Computer Science, Department of Applied Mathematics, University of Twente, The Netherlands

Gary Erickson, Department of Marketing, Michael G. Foster School of Business, University of Washington, USA

Leslie R. Fletcher, School of Computing and Mathematical Sciences, Liverpool John Moores University, UK

Vito Fragnelli, Department of Sciences and Innovative Technologies (DISIT), University of Eastern Piedmont, Italy

Josep Freixas, Departament de Matematiques, Universitat Politecnica de Catalunya, Catalunya

Mario A. García-Meza, Facultad de Economía, Contaduría y Administración, Universidad Juárez del Estado de Durango, México

Yigal Gerchak, Department of Industrial Engineering, Tel Aviv University, Israel

Dieter Grass, Department of Operations Research and Systems Theory, Vienna University of Technology, Austria

Ekaterina Viktorovna Gromova, Faculty of Applied Mathematics and Control Processes, St. Petersburg State University, Russia

Ruud Hendrickx, Department of Econometrics and Operations Research, School of Economics and Management, Tilburg University, The Netherlands

Dongshuang Hou, Department of Applied Mathematics, Northwestern Polytechnical University, P. R. China

Steffen Jørgensen, Department of Business and Economics, University of Southern Denmark, Denmark

John Kleppe, Department of Econometrics and Operations Research, School of Economics and Management, Tilburg University, The Netherlands

Eugene Khmelnitsk, Department of Industrial Engineering, Tel Aviv University, Israel

Vassili N. Kolokoltsov, Department of Statistics, University of Warwick Coventry, UK; Faculty of Applied Mathematics and Control Processes, St. Petersburg State University, Russia

Jacek Krawczyk, Victoria University of Wellington, Wellington, New Zealand

Aymeric Lardon, Université Côte d'Azur, GREDEG, CNRS, Valbonne, France

Jerzy Legut, Department of Mathematics, Wrocław University of Technology, Poland

Natividad Llorca, Center of Operations Research (CIO), Miguel Hernandez University of Elche, Spain

José Daniel López-Barrientos, Facultad de Ciencias Actuariales, Universidad Anáhuac México, México

Vladimir Mazalov, Institute of Applied Mathematical Research, Karelian Research Center, Russian Academy of Sciences, Russia

Anna Melnik, Saint-Petersburg State University, Russia

Stefano Moretti, Universite Paris Dauphine, PSL Research University, France

Pierre von Mouche, Wageningen University, Wageningen, The Netherlands

Koji Okuguchi, Tokyo Metropolitan University, Japan

Fouad El Ouardighi, Department of Operations Management, ESSEC Business School, France

Ovanes Petrosian, St. Petersburg State University, Russia; St. Petersburg School of Economics and Management; National Research University Higher School of Economics, Moscow, Russia

Leon Petrosyan, Faculty of Applied Mathematics-Control Processes, St. Petersburg State University, Russia

Hans Reijnierse, Department of Econometrics and Operations Research, School of Economics and Management, Tilburg University, The Netherlands

Anna Rettieva, Institute of Applied Mathematical Research, Karelian Research Centre of RAS, Russia

Julio Rodríguez-Segura, Facultad de Economa, UASLP, Consejo Estatal de Población (COESPO), México

Sebastien Rouillon, GREThA, University of Bordeaux, France

Joss Sánchez-Pérez, Facultad de Economía, UASLP, México

Joaquin Sanchez-Soriano, Center of Operations Research (CIO), Miguel Hernandez University of Elche, Spain

Takashi Sato, Shimonoseki City University, Japan

Atsuhiro Satoh, Faculty of Economics, Hokkai-Gakuen University, Japan

Stein Ivar Steinshamn, Department of Business and Management Science, Norwegian School of Economics, Norway

Harborne Stuart, Jr., Columbia Business School, Columbia University, USA

Yasuhito Tanaka, Faculty of Economics, Doshisha University, Japan

Mabel Tidball, Institut National de Recherche en Agronomie, LAMETA, Montpellier, France

Evangelos Toumasatos, Department of Business and Management Science, Norwegian School of Economics, Norway

Gisèle Umbhauer, Bureau d'Economie Théorique et Appliquée, University of Strasbourg, France

Thomas Weber, Operations, Economics and Strategy, Management of Technology and Entrepreneurship Institute, Ecole Polytechnique Federale de Lausanne, Switzerland.

Takeshi Yamazaki, Department of Economics, Niigata University, Japan

David Yeung, Center of Game Theory, St. Petersburg State University, Russia; SRS Consortium for Advanced Study, Shue Yan University, Hong Kong

About the Editors

David W. K. Yeung obtained his B.Soc.Sc in Economics/Statistics from Hong Kong University and PhD in Economics from York University. He studied in the DSc (Doktor Nauk — European higher doctorate above PhD) program in Applied Mathematics at Saint Petersburg State University and was awarded the University's highest degree Dr.h.c. for his outstanding contributions in differential games. He is Distinguished Research Professor at Shue Yan University and Co-director of Centre of Game Theory at Saint Petersburg State University. He is Managing Editor of *International Game Theory Review*, Guest Editor of *Annals of Operations Research*, Associate Editor of *Dynamic Games & Applications* and *Operations Research Letters*, and Editor of the *Routledge Series on Economics and Optimization*. His main research areas include dynamic game theory, control theory, stochastic processes and decision sciences. He is co-author of the world's first book on cooperative stochastic differential games and of the first text on subgame consistent economic optimization. He is also the author of the first book on dynamic consumer theory.

Leon Petrosyan obtained his MSc, PhD and DSc (Doktor Nauk) in Mathematics from Saint Petersburg State University. He is currently University Professor, Professor of Management and Director of Center of Game Theory at Saint Petersburg State University. He has been Dean of the Faculty of Applied Mathematics-Control Processes of the University for almost forty years. He is Laureate of the Isaacs Award 2014. He has been President of the International Society of Dynamic Games and the Founding President of the Russian Chapter of the Society. He is editor of the *International Game Theory Review, Mathematical Game Theory & Applications*, and *Game Theory & Applications*. His main research areas include differential games, control theory, operations research and optimization. He has published more than 170 scientific articles and over 20 books including the first ever book on cooperative stochastic differential games and the first text on subgame consistent economic optimization with David Yeung.

Chapter 1

Introduction

David W. K. Yeung

Hong Kong Shue Yan University, Hong Kong & Saint Petersburg State University, Russia

Leon A. Petrosyan

Saint Petersburg State University, Russia

Strategic behavior in the human and social world has been increasingly recognized in theory and practice. From a decision-maker's perspective, it becomes important to consider and accommodate the interdependencies and interactions of human decisions. As a result, game theory has emerged as a fundamental instrument in pure and applied research.

In game theory, a non-cooperative game is a game with competition between individual players and in which only self-enforcing are possible due to the absence of external means to enforce cooperative behavior as opposed to cooperative games. In fact, non-cooperative games are the foundation for the development of cooperative games by acting as the status quo. Non-cooperative games are generally analysed through the framework of equilibrium, which tries to predict players' individual strategies and payoffs. Indeed, equilibrium analysis is the centre of non-cooperative games. Part I is on non-cooperative games and equilibrium analysis and it contains a variety of recently developed non-cooperative games and non-cooperative game equilibria.

It is well known that non-cooperative behaviours, in general, would not lead to a Pareto optimal outcome. Highly undesirable outcomes (like that in the prisoner's dilemma) and devastating results (like the tragedy of the commons) could appear when the involved parties only care about their individual interests in a non-cooperative situation. Cooperative games offer the possibility of obtaining socially optimal and group efficient solutions to decision problems involving strategic actions. In addition, axiomatic values serve as guidance for establishing cooperative solutions. Part II presents a collection of recent cooperative games and axiomatic values.

Part I of this book — non-cooperative games and equilibrium analysis — contains a collection of thirteen articles from Chapter 2 to Chapter 14. These chapters include state-of-art non-cooperative games and original game equilibrium concepts are provided.

In Chapter 2, Jane M. Binner, Francesco Ciardiello, Leslie R. Fletcher and Vassili N. Kolokoltsov analyze strict Nash equilibria in a duopolistic market share model. They developed an advertising model where generic and brand advertising marketing effects are combined in a duopolistic market with inelastic demand, linear advertising costs and strictly positive discount factors. The nontrivial existence of investment Nash equilibrium in pure strategies is derived and the absence of cheap or free riding equilibria is proved. The relationship between predatory and generic marketing effects and the optimal timing of investments and other marketing scenarios are examined.

In Chapter 3, Atsuhiro Satoh and Yasuhito Tanaka examine maximin and minimax strategies in two-players game with two strategic variables, x and p. Two patterns of game — one is the x-game in which the strategic variables of players are x's, and the other is the p-game in which the strategic variables of players are p's — are considered. It is shown that the maximin strategy and the minimax strategy in the x-game, and the maximin strategy and the minimax strategy in the p-game are all equivalent for each player.

In Chapter 4, Sebastien Rouillon considers a non-cooperative dynamic contribution game in which a group of agents collaborates to complete a public project. The agents exert efforts over time and get rewarded upon completion of the project, once the cumulative effort has reached a pre-specified level. We explicitly derive the cooperative solution and a noncooperative Markov-perfect Nash equilibrium. The author characterizes the set of socially efficient projects, i.e., projects that cooperative groups find worth completing. Comparing with the Markov-perfect Nash equilibrium, it is found that noncooperative groups give up large socially efficient projects and take longer to complete the projects.

In Chapter 5, David Yeung and Ovanes Petrosian formulate a new approach to analyze infinite horizon dynamic games with uncertainties and unknowns in the players' future payoff structures. Many events in the considerably far future are intrinsically unknown. Information about the players' future payoffs will be revealed as the game proceeds. Making use of the newly obtained information, the players revise their strategies accordingly, and the process will continue indefinitely. This new approach for the analysis of infinite horizon dynamic games via information updating provides a practical alternative to the study of infinite horizon dynamic games.

In Chapter 6, Fouad El Ouardighi, Gary Erickson, Dieter Grass and Steffen Jørgensen study contracts and information structure in a supply chain with operations and marketing interaction. They examine how wholesale price and revenue sharing contracts affect operations and marketing decisions in a supply chain under different dynamic informational structures. A differential game framework of a supply chain consisting of a manufacturer and a single retailer is constructed. The manufacturer sets a production rate and the rate of advertising efforts while the retailer chooses a purchase rate and the consumer price. The state of the game is summarized in the firms' backlogs and the manufacturer's advertising goodwill. Depending on whether the supply chain members have and share state information, they may either make decisions contingent on the current state of the game (feedback Nash strategy), or precommit to a plan of action during the whole game (open-loop Nash strategy). Given a contract type, the impact of the availability of information regarding the state of the game on the firms' decisions and payoffs is investigated.

In Chapter 7, Vladimir Mazalov and Anna Melnik consider a non-cooperative transport game of n players on a communication graph. Here players are passenger transportation companies (carriers). Service requests form a Poisson process with an intensity rate matrix Λ. Players announce prices for their services and passengers choose an appropriate service by minimizing their individual costs (the ticket price and the expected service time). For each carrier, the authors solve the pricing problem and define the equilibrium intensity flows in the conditions of competition.

In Chapter 8, Koji Okuguchi and Takeshi Yamazaki study the existence of unique equilibrium in Cournot mixed oligopoly. They consider the properties of Cournot mixed oligopoly consisting of one public firm and one or more than one private firms. After proving the existence of a unique equilibrium in Cournot mixed oligopoly under general conditions on the market demand and each firm's cost function, the authors derive conditions ensuring the existence of a unique Nash equilibrium for the mixed oligopoly where one public firm and at least one of the private firms are active in a general model of Cournot mixed oligopoly with one public firm and several private firms.

In Chapter 9, Gisèle Umbhauer studies second-price all-pay auctions and best-reply matching equilibria. The chapter considers second-price all-pay auctions — wars of attrition — in a new way, based on classroom experiments and best-reply matching equilibrium. The behavior probability distributions in the classroom experiments are strikingly different from the mixed Nash equilibrium. They fit with best-reply matching and generalized best-reply matching. The analysis goes into the generalized best-reply matching logic, highlights the role of focal values and discusses the high or low payoffs this logic can lead to.

In Chapter 10, Pierre von Mouche and Takashi Sato discuss the issue of Cournot equilibrium uniqueness at 0 discontinuous industry revenue and decreasing price flexibility. They consider the equilibrium uniqueness problem for a large class of Cournot oligopolies with convex cost functions and proper price function with decreasing price flexibility. They also illustrate the Selten-Szidarovszky technique based on virtual backward reply correspondences. An algorithm for the calculation of the unique equilibrium is provided.

In Chapter 11, Thomas Weber considers quantifying a player's commitment in a given Nash equilibrium of a finite dynamic game. He maps the corresponding normal-form game to a canonical extension, which allows each player to adjust his or her move with a certain probability. The commitment measure relates to the average over all adjustment probabilities for which the given Nash equilibrium can be implemented as a subgame-perfect equilibrium in the canonical extension.

In Chapter 12, Anna Rettieva develops equilibria in dynamic multicriteria games. She examines dynamic games involving multiple objectives which are not comparable. These game situations are typical for game-theoretic models in economic and ecology. New approaches to construct equilibria in dynamic multicriteria games are presented. An equilibrium as a solution of a Nash bargaining scheme with the guaranteed payoffs playing the role of status quo points is set up. The obtained equilibrium, called a multicriteria Nash equilibrium, gives a possible solution concept for dynamic multicriteria games.

In Chapter 13, Jacek Krawczyk and Mabel Tidball study economic problems with constraints and how efficiency relates to equilibrium. They consider the supply of socially important goods by independent economic agents under a regulator that believes that constraining the goods delivery is desirable. The regulator can compute a constrained Pareto-efficient solution to establish optimal output levels for each agent. The chapter suggests that a coupled-constraint equilibrium (also called a "generalized" Nash or "normalized" equilibrium à la Rosen) may be more relevant for market economies than a Pareto-efficient solution. The authors illustrate their findings using a coordination problem, in which the agents' outputs depend on externalities.

In Chapter 14, Harborne Stuart, Jr. studies substitution, complementarity and stability in marketing models. He develops a duopolistic discounted marketing model with linear advertising costs and advertised prices for mature markets still in expansion. Generic and predatory advertising effects are combined together in the model. The entity of efficiency at varying of parameters of the advertising model is examined. A computational framework in which market shares can be computed at equilibrium is provided to illustrate different scenarios in practical applications.

Part II of this book — cooperative games and axiomatic values — contains a collection of fourteen articles from Chapter 15 to Chapter 28. These chapters include pioneering cooperative games and a series of axiomatic values for characterizing the solution of cooperative games.

In Chapter 15, John Kleppe, Peter Borm, Ruud Hendrickx and Hans Reijnierse analyze cost allocation problems with cooperation building structures and order problem representations. They identify associated cooperation building structures, with joint cost functions, and corresponding efficient order problem representations, with individualized cost functions. The chapter presents an approach that, when applicable, offers a way not only to adequately model a cost allocation problem by means of a cooperative cost game, but also to construct a core element of such a game by means of a generalized Bird allocation. The analysis applies the approach to both existing and new classes of cost allocation problems related to operational research problems: sequencing situations without initial ordering, maintenance problems, minimum cost spanning tree situations, permutation situations without initial allocation, public congestion network situations, traveling salesman problems, shared taxi problems and traveling repairman problems.

In Chapter 16, Evangelos Toumasatos and Stein Ivar Steinshamn study coalition formation with externalities using the case of the Northeast Atlantic mackerel fishery in a pre- and post-Brexit Context. They apply the partition function approach to study coalition formation in the presence of externalities. Atlantic mackerel is mainly exploited by the European Union (EU), the United Kingdom (UK), Norway, the Faroe Islands and Iceland. Two games are considered. First, a four-player game where the UK is still a member of the EU. Second, a five-player game where the UK is no longer a member of the union. Each game is modeled in two stages. In the first stage, players form coalitions following a predefined set of rules. In the second stage, given the coalition structure that has been formed, each coalition chooses the economic strategy that maximizes its own net present value of the fishery, given the behavior of the other coalitions.

In Chapter 17, Mario A. García-Meza, Ekaterina Viktorovna Gromova and José Daniel López-Barrientos consider stable marketing cooperation in a differential game for an oligopoly. They develop a dynamic model of an oligopoly playing an advertising game of goodwill accumulation with random terminal time. The goal is to find a cooperative solution that is time-consistent, considering a dynamic accumulation of goodwill with depreciation for a finite number of firms.

In Chapter 18, David Yeung and Leon Petrosyan present a novel cooperative dynamic environmental game of subgame consistent clean technology development. The chapter considers cooperative adoption and development of clean technology in effectively solving the continual worsening industrial pollution problem. For cooperation over time to be credible, a subgame consistency solution which requires the agreed-upon optimality principle to remain in effect throughout the collaboration duration has to hold. A subgame consistent cooperative dynamic game of collaborative environmental management with clean technology development is provided. To overcome the problem discrete choices of production techniques switching (between conventional and clean technologies), the joint optimal solutions for all the possible patterns of production techniques are computed and the pattern with the highest joint payoff is then selected.

In Chapter 19, Yigal Gerchak and Eugene Khmelnitsk present linear and nonlinear contracts for partnership's profit sharing. They provide necessary and sufficient conditions for a non-empty core in many-to-one assignment games. When players on the 'many' side (buyers) are substitutes with respect to any given player on the other side (firms), it is shown that non-emptiness requires an additional condition that limits the competition among the buyers. When buyers are complements with respect to any given firm, a sufficient condition for non-emptiness is that buyers also be complements with respect to all of the firms, collectively. A necessary condition is that no firm can be guaranteed a profit when the core is non-empty.

In Chapter 20, Dongshuang Hou, Aymeric Lardon and T. S. H. Driessen present the characterization of Stackelberg oligopoly TU-games and the nonemptiness of the core of the game. In particular, firms can be better off by forming cartels to achieve Pareto efficiency. They consider the dynamic setting of Stackelberg oligopoly TU-games in γ-characteristic function form. Any deviating coalition produces an output at a first period as a leader and then, outsiders simultaneously and independently play a quantity at a second period as followers. The inverse demand function is linear and that firms operate at constant but possibly distinct marginal costs. It is shown that the core of any Stackelberg oligopoly TU-game always coincides with the set of imputations. The analysis yields a necessary and sufficient condition, depending on the heterogeneity of firms' marginal costs, under which the core is nonempty.

In Chapter 21, Leon Petrosyan develops strong strategic support of cooperation in multistage Games. He constructs a strong Nash equilibrium with payoffs, as a new solution concept, which can be attained under cooperation for a wide class of such games; as a subset of the classical core for repeated and multistage games; this new solution concept is proved to be strongly time consistent.

In Chapter 22, Joaquin Sanchez-Soriano and Natividad Llorca discuss solution concept related to bounded rationality for some two-echelon models. They consider two-echelon models in which there are two differentiated groups of agents. Some examples of these models can be found in supply chain problems, transportation problems or two-sided markets. It deals with two-sided transportation problems which can be used to describe a wide variety of logistic and market problems. The analysis introduces a new solution concept, a core catcher, which can be motivated by a kind of bounded rationality which can arise in cooperative contexts.

In Chapter 23, Michael R. Caputo analyze intrinsic comparative statics of a Nash bargaining solution. He provides a generalization of the class of bargaining problems examined by Engwerda and Douven (2008) and studies the sensitivity matrix of the Nash bargaining solution. The generalized class consists of nonconvex bargaining problems in which the feasible set satisfies the requirement that the set of weak Pareto-optimal solutions can be described by a smooth function. The intrinsic comparative statics of the aforesaid class are derived and shown to be characterized by a symmetric and positive semidefinite matrix, and an upper bound to the rank of the matrix is established.

In Chapter 24, Jerzy Legut develops an optimal fair division for measures with piecewise linear density functions. He uses a nonlinear programming method for finding an optimal fair division of the unit interval among n players. Preferences of players are described by nonatomic probability measures with piecewise linear (PWL) density functions. The presented algorithm can be applied for obtaining "almost" optimal fair divisions for measures with arbitrary density functions approximable by PWL functions. The number of cuts needed for obtaining such divisions is given.

In Chapter 25, Julio Rodríguez-Segura and Joss Sánchez-Pérez consider an extension of the solidarity value for environments with externalities. He proposes an axiomatic extension for the solidarity value of Nowak and Radzik (1994) for n-person transferable utility games to the class of games with externalities. This value is characterized as the unique function that satisfies linearity, symmetry, efficiency and average nullity. A discussion on a key subject of how to extend the concept of average marginal contribution to settings where externalities are present is given.

In Chapter 26, Pierre Dehez considers Harsanyi dividends and asymmetric values. The concept of dividend in transferable utility games was introduced by Harsanyi (1959), offering a unifying framework for studying various valuation concepts, from the Shapley value to the different notions of values introduced by Weber. Using the decomposition of the characteristic function used by Shapley to prove uniqueness of his value, the idea of Harsanyi was to associate to each coalition a dividend to be distributed among its members to define an allocation. A synthesis on dividend distributions, starting with the seminal contributions of Vasil'ev, Hammer, Peled and Sorensen and Derks, Haller and Peters, van den Brink, van der Laan and Vasil'ev is provided.

In Chapter 27, Giulia Bernardi and Josep Freixas provide an axiomatization for two power indices in (3,2)-Simple Games. He develops a characterization of the Shapley-Shubik and the Banzhaf power indices for (3,2)-simple games. It generalizes the classical axioms for power indices on simple games, namely transfer, anonymity, null player property, and their efficiency by introducing a new axiom to support the uniqueness of the extension of the Shapley-Shubik power index in this context.

In Chapter 28, Marco Dall'Aglio, Vito Fragnelli and Stefano Moretti develop an indice of criticality in simple games. They generalize the notion of power index for simple games to different orders of criticality, where the order of criticality represents the possibility for players to gain more power over the members of a coalition. These criticality indices are used to compare the power of different players within a single voting situation, and that of the same player with varying weight across different voting situations, establishing monotonicity results á la Turnovec.

PART I

NON-COOPERATIVE GAMES AND EQUILIBRIUM ANALYSIS

Chapter 2

On Pure-Strategy Nash Equilibria in a Duopolistic Market Share Model

J. M. Binner

Birmingham Business School, University of Birmingham
Edgbaston Park Rd, Birmingham B15 2TY, UK

j.m.binner@bham.ac.uk

F. Ciardiello*

Management School, University of Sheffield
Conduit Rd, Sheffield B15 2TY, UK

f.ciardiello@shef.ac.uk

L. R. Fletcher

School of Computing and Mathematical Sciences
Liverpool John Moores University

Byrom St, Liverpool L3 3AF, UK

V. N. Kolokoltsov

Department of Statistics
University of Warwick Coventry CV4 7AL, UK

Faculty Applied Mathematics and Control Processes
St. Petersburg State University (Russia)
FRC CSC Russian Academy of Science, Russia

v.kolokoltsov@warwick.ac.uk

This chapter develops a duopolistic discounted marketing model with linear advertising costs and advertised prices for mature markets still in expansion. Generic and predatory advertising effects are combined together in the model. We characterize a class of advertising models with some lowered production costs. For such a class of models, advertising investments have a no-free-riding strict Nash equilibrium in pure strategies if discount rates are small. We discuss the entity of this efficiency at varying of parameters of our advertising model. We provide a computational framework in which market shares can be computed at *equilibrium*, too. We analyze market share dynamics for an asymmetrical numerical scenario where one of the two firms is more effective in generic and predatory

*Corresponding author.

advertising. Several numerical insights on market share dynamics are obtained. Our computational framework allows for different scenarios in practical applications and it is developed, thanks to *Mathematica* software.

Keywords: Advertising models; Nash equilibrium; generic advertising; brand advertising; computational equilibria; market shares; sticky prices; supply chains.

1. Introduction

Advertising expenditures for companies may be generally viewed as a form of investment and the main thrust of the advertising literature is to examine optimal strategies which maximize the net present value of future cash flows. When advertising is aimed at increasing product sales, then it is called generic or informative advertising, whilst when advertising is aimed at gaining market shares, it is called brand advertising. When brand advertising is devoted to stealing customers from competitors, it is called predatory advertising. Many advertising models have been built to provide solutions in monopolistic, duopolistic and oligopolistic environments [Vidale and Wolfe, 1957; Sethi, 1973; Deal, 1979; Little, 1979; Sethi, 1983] for many decision variables.[a] Distinct dynamic advertising strategies were never developed prior to the model developed in [Bass *et al.*, 2005, Table 1] for brand and generic advertising, respectively. Even the deep dynamic analysis provided in [Bass *et al.*, 2005] suggests that generic advertising expenditures must be highly resolved separately from its brand advertising ones in order to halt suboptimal advertising. However externalities from simple generic advertising may become significant and may modify brand preferences, as market demand becomes more informed [Kinnucan, 1996; Norman *et al.*, 2008; Rutz and Bucklin, 2011; Brahim *et al.*, 2014]. The main controversial *economic* issue is that generic advertising may redistribute market shares, especially in markets that have become strongly differentiated [Chakravarti and Janiszewski, 2004; Brady, 2009; Espinosa and Mariel, 2001; Piga, 1998; Friedman, 1983]. In this chapter, predatory and generic advertising expenditures are combined together.

The analysis provided in Espinosa and Mariel [2001] deals with a duopoly where generic and predatory advertising are present but separately analyzed. The authors argue that static strategies do not internalize any variations of market shares if firms play a game with time-invariant market shares. Therefore, it is expected that Nash equilibria are neither efficient if the advertising is predatory (too high expenditures), nor efficient if advertising is generic (too low expenditures). In this chapter, we face a more complex scenario than the static one analyzed in Espinosa

---

[a]For a complete review of the literature on advertising models before 1995 see Erickson [1995]. The more recent classification of advertising models consists of six categories [Huang *et al.*, 2012]; the Nerlove–Arrow model and its extensions, the Vidale–Wolfe model and its extensions, the Lanchester model and its extensions, the diffusion models, dynamic advertising-competition models with other attributes, and (vi) empirical studies for dynamic advertising problems.

and Mariel [2001]. In fact, generic and brand/predatory advertising are unified and their effect on market shares cannot be distinguished at any time for an infinite time horizon. Strategic generic advertising interactions between firms may cause short-run bubble effects but similar predatory effects may cause long-run lowering effects on market shares, especially for asymmetrical market conditions. For instance, different investment strategies may generate different market share dynamics, which yield the same discounted profit to one of the firms. To some extent, the efficiency of such equilibria is not so intuitive. For instance, in Espinosa and Mariel [2001] closed-loop equilibria are more efficient than open-loop equilibria if the advertising is only generic while open-loop are more efficient than closed-loop if the advertising is only predatory. Very little attention has been devoted to the efficiency and stability of Nash equilibria in the game theory of advertising. The simple concept of strong Nash equilibria, which incorporates a strong form of efficiency, may be useful to be analyzed in a duopolistic model such as this. By being complementary to the dynamic analysis in Espinosa and Mariel [2001], we restrict optimal analysis to static expenditure strategies. The stability we require is the following: if a firm unilaterally changes its strategy from Nash equilibrium strategies, then such a firm has to be *strictly* worse. We fill the gap in this literature and we investigate the existence of strict Nash equilibria in a context of generic and brand advertising models.

2. Results

In this chapter, we initially develop a duopolistic discounted advertising model for mature markets in expansion. Generic and predatory advertising effects are combined together. We model generic advertising by adopting a diffusion model and we model predatory advertising by adopting the Lanchester model of combat. For the sake of simplicity we allow firms to advertise prices even if we do not model prices as decision variables. In fact, prices may be announced before marketing campaigns [Jiang *et al.*, 2014; Grewal *et al.*, 1998; Tenn and Wendling, 2014; Lu *et al.*, 2016].[b] We define a class of advertising models where production costs are lowered by marketing parameters, being inspired by an economic analysis between advertising and quality of products suggested in Bagwell [2007].

We obtain payoffs in a closed-form using tools from the theory of linear differential systems. Significantly, payoffs are continuous but they are not quasi-concave on non-compact subsets. Our method provides for an algebraic analysis of best-reply correspondences. We find out that a strict Nash equilibrium in pure strategies exists for the class of advertising games with low costs. Interestingly, our equilibria are not

---

[b] We prefer to characterize our model in terms of advertised prices to make our results comparable to another literature stream, i.e. optimal pricing in markets with sticky prices [Gorodnichenko and Weber, 2016; Piga, 2000]. Our model is comparable to a limit case in this literature, i.e., the rate of price readjustment is null. Recently, static optimal solutions are also provided in advertising models subject to interferences [Baggio and Viscolani, 2014; Viscolani, 2012].

free-riding and they cannot be conceptually compared to the ones found in Krishnamurthy [2000]. The existence of equilibria is not surprising for small discount rates by taking into account Folk theorem in game theory. However, the efficiency of such equilibria is not so intuitive. Our strict equilibria are not obtained for any level of production costs, as it has been found in the static model provided in Espinosa and Mariel [2001]. Unfortunately, due to the lack of a closed form expression of pure-strategy Nash equilibria we cannot analytically discuss the influence of the parameters on optimal investments strategies. However, our algebraic method provides a computational framework in order to integrate such optimal investment strategies [Deal, 1979; Erickson, 1985].

As a further investigation, we focus on market share dynamics generated by our strict Nash equilibria in pure strategies. Since generic advertising supports the general standard of the product category, it brings advantages to firms in the market regardless of whether or not they contributed to advertising campaigns [Han *et al.*, 2017; Shapiro, 2018]. In the duopolistic model analyzed in Bass *et al.* [2005], when the asymmetries between the firms increase, there is a larger difference between their generic advertising contributions. However, the weaker firm always invests a small amount of money and, then, cheaply but not freely rides the market [Krishnamurthy, 2000]. In our numerical scenario, we assume the similar scenario adopted in Bass *et al.* [2005]. One firm is *stronger* if it is endowed with more favorable competitive and generic advertising parameters. Interestingly, we find out that the *weaker* firm is not a cheap rider. We believe that this does happen because our advertising expenditure are combined together for brand and generic advertising. As an expected result, the weaker firm enlarges its market share due to generic advertising while its market share is affected by the long run effect of predatory advertising from its competitor. Our results support those obtained in Bass *et al.* [2005]. Thanks to our numerical framework we are numerically able to integrate the time at which the weaker firm achieves its maximum market share, if the market in itself is not initially saturated. Similar insights cannot been replicated from the results in Bass *et al.* [2005] because the two models are similar but different.

The rest of the chapter is organized as follows. In Sec. 3, we provide the modeling background from the literature of advertising models. In Sec. 4, we provide details of our advertising model. In Sec. 5, we prove the existence of proper investment equilibria in pure strategies. In Sec. 6, we provide an asymmetric numerical scenario for our model and we illustrate market share dynamics if strict pure-strategy Nash equilibria investments are implemented. Our proofs are provided in Appendix A.

3. Modeling Background

Firms sell their heterogenous goods in a specific product category. Each firm satisfies a portion of demand, i.e., its market share $x_i(t) \geq 0$ for time $t \geq 0$. The sum of market shares is $x(t) = x_1(t) + x_2(t)$. The market is still in expansion and the whole demand is finite and unitary. The sum of market shares $x(t) = x_1(t) + x_2(t)$ and the remaining portion of the demand is $1 - x(t)$, i.e., the potential demand. Potential demand is the demand which has an as yet unrevealed desire for the product. However, the potential demand is not already contained within market shares of any firm and it is *uninformed* on the product category. Generic advertising is only directed to the potential demand through diffusion models. Diffusion models are used to capture the life cycle dynamics of new products or to forecast the demand on markets [Fisher and Pry, 1971]. Originally, diffusion models do not incorporate any advertising effort and one of the main challenges is to add exogenous influences, most importantly the influence of advertising efforts [Bass *et al.*, 1994]. A generic functional form for diffusion advertising models for firm i is $\dot{x}_i(t) = f(x_i(t), u_i(t), t)$ where $u_i(t)$ is the advertising effort and $\dot{x}_i(t)$ is the change in market share for firm i [Dockner and Jørgensen, 1988]. For instance, a diffusion model is $\dot{x}_i(t) = \alpha u_i(t)[1 - x_i(t)] + \gamma x_i(t)[1 - x(t)]$ for a monopolistic firm i where $\alpha > 0$ is the advertising effectiveness of the firm and γ represents the effectiveness of word-of-mouth advertising in Jørgensen *et al.* [2006]. With respect to generic advertising, we symmetrically adopt this diffusion model with $\gamma = 0$ and $\alpha = 1$ for both firms in our duopoly. The Lanchester model has often been used to model competitive advertising [Fruchter and Kalish, 1997]. The Lanchester dynamics capture the competitive market shares' shifts of firms i, j due to investments in advertising by the two market rivals [Chintagunta and Vilcassim, 1992]. The Lanchester model is $\dot{x}_i(t) = \rho_i x_j(t) u_i(t) - \rho_j x_i(t) u_j(t)$ where $\rho_i, \rho_j \in [0, 1]$ are the effectiveness of brand advertising for firms i, j [Little, 1979]. In particular, here we assume that $\rho_i, \rho_j > 0$ for both firms.

4. Our Advertising Model

We adopt some modifications to the models in the previous section. We assume that the advertising strategy variable $u_i \in [0, \infty[$ is time-invariant. Moreover we do not adopt a linear form for investment variable u_i. In spite of mathematical simplifications we add a level of complexity and we shape advertising returns as a functional form $a_i, b_i : [0, \infty[\rightarrow [0, \infty[$ for generic and brand advertising expenditures, respectively. We assume these functions satisfy a law of diminishing returns on the investment for both generic and brand advertising [Hanssens *et al.*, 2003; Freimer and Horsky, 2012]. It thus follows that

— $a_i(u_i) = \frac{u_i}{u_i + \alpha_i} : \mathbb{R}^+ \rightarrow \mathbb{R}^+$ is the generic advertising return for i. The coefficient $\alpha_i \in \mathbb{R}^+ \backslash \{0\}$ is called ineffectiveness of generic advertising return.
— $b_i(u_i) = \frac{u_i}{u_i + \beta_i} : \mathbb{R}^+ \rightarrow \mathbb{R}^+$ is the brand advertising return for i. The coefficient $\beta_i \in \mathbb{R}^+ \backslash \{0\}$ is called ineffectiveness of brand advertising return.

Functions $a_i(\cdot), b_i(\cdot)$ are increasing and concave and approximate 1 for large investments. The lower α_i and b_i's values are, the more profitable advertising investments are. Summing up all these features of the diffusion model and of the Lanchester model, the derivative in primary demand $\dot{x}_i$ is modeled by the following differential system with initial conditions

$$
\begin{cases}
\dot{x}_i(t) = [a_i(u_i) + a_j(u_j)][1 - (x_i(t) + x_j(t))] + \rho_i x_j(t) b_i(u_i) - \rho_j x_i(t) b_j(u_j) \\
x_i(0) = x_i^0
\end{cases}
\tag{1}
$$

where $u_i \in \mathbb{R}^+$ is the advertising expenditure of firm i, $x_i(t) : \mathbb{R}^+ \to [0, 1]$ is the market share of firm i at time t and $x_i^0 = x_i(0)$ are the initial market shares. We assume linear advertising costs that are, traditionally, used in the Nerlove–Arrow model [Gould, 1976]. In the literature, it is widely accepted that firms have different discount rates. Here firms adopt a unique positive discount rate $\rho \neq 0$ [Jørgensen, 1982]. Therefore, the discounted flow of profits is

$$
\pi^i(u_i, u_j) = \int_0^\infty e^{-\rho t}(r_i x_i(t) - u_i)dt.
\tag{2}
$$

where $r_i = p_i - c_i$. The quantity $p_i > 0$ is the advertised price p_i for the good i. The quantity $c_i > 0$ is the marginal cost for good i. We assume that marginal profits r_i are positive. We say that $G = ([0, +\infty[^2, \pi^i)$ is a non-cooperative advertising game.

4.1. *Advertising games with low costs*

We characterize advertising games which have lower costs (or higher profits) by considering that prices are advertised in our marketing model.

Definition 1. The following inequality is satisfied

$$
c_i + c_i^{sh} < p_i,
\tag{3}
$$

where

$$
c_i^{sh} = \frac{(\rho_j + \rho_i)}{2\rho_i}\alpha_i + \frac{\rho_j}{\rho_i}\beta_i.
\tag{4}
$$

We say that $G = ([0, +\infty[^2, \pi^i)$ is the advertising game with low costs if the above conditions are satisfied. We say that $c_i^{sh} > 0$ is the marketing incentive for firm i.

Marketing incentives need to be smaller than the advertised prices in order to satisfy inequality (3). For instance, if firm i has a brand advertising effectiveness much lower than its competitor's counterpart ρ_j, inequality (3) is not satisfied. The marketing incentive c_i^{sh} is increasing with respect to $\rho_j, \alpha_i, \beta_i$. The latter means that firm i has lower costs if the effectiveness of the brand advertising of the competitor increases or if its advertising returns are marginally worse. The marketing incentive c_i^{sh} is decreasing with respect to ρ_i. The latter means that firm i may have higher costs if the effectiveness of own brand advertising increases. If $\rho_1 = \rho_2$, then $c_i^{sh} = \alpha_i + \beta_i$. If $\alpha_i = \beta_i$, then $c_i^{sh} = \alpha_i \frac{(3\rho_j + \rho_i)}{2\rho_i}$.

Classical solution concepts are defined below.

Definition 2 (Nash equilibria, market shares at equilibrium). Let $(\hat{u}_1, \hat{u}_2)$ be a strict Nash equilibrium in pure strategies for G. Let us substitute $(\hat{u}_1, \hat{u}_2)$

in (1). A solution $\hat{x}_i(t)$ of the first-order differential system (1) is the associated market share of firm i at equilibrium. A pure-strategy profile is *free-riding* if one strategy is null and the remaining strategy is not null. A pure-strategy profile is null when both strategies are null. We say that a pure strategy profile is proper if it is neither null nor free-riding.

5. Existence of a Strict Nash Equilibrium in Pure Strategies

In this section, we find sufficient conditions which guarantee the existence of static advertising expenditures equilibria for our advertising model. Let $\widehat{U}_i :$ $[0, +\infty[\to 2^{[0, +\infty[}$ be the best-reply correspondence for firm i. The set $\widehat{U}_i(u_j)$ collects firm i's best replies to a strategy u_j. We transform system (1) through a change of variables into

$$\begin{cases} \dot{x} = 2\{a_1(u_1) + a_2(u_2)\}(1 - x), \\ \dot{w} = \rho_1 b_1(u_1)(x - w) - \rho_2 b_2(u_2)(x + w), \end{cases} \tag{5}$$

where $x = x_1 + x_2$ and $w = x_1 - x_2$.

Proposition 1. *The solution for system* (5) *is given by the following formulae*

$$\begin{cases} x(t) = 1 - (1 - x_1^0 - x_2^0)e^{-2At} \quad t \geq 0 \\ w(t) = \dfrac{B^-}{B^+} + \dfrac{B^-}{(B^+ - 2A)}(x_1^0 + x_2^0 - 1)(e^{-2At} - e^{-B^+t}) + e^{-B^+t}C_u \end{cases}$$

where C_u is a constant depending on

$$A = a_1(u_1) + a_2(u_2), \ B^+ = \rho_1 b_1(u_1) + \rho_2 b_2(u_2), \ B^- = \rho_1 b_1(u_1) - \rho_2 b_2(u_2).$$

Here, we provide the following result in which market shares and payoffs are obtained in closed formulas.

Proposition 2. *Assume $\rho \neq 0$. Then market shares and payoffs are*

$$x_i(t) = \frac{\rho_i b_i(u_i)}{B^+}(1 - e^{-B^+t}) + \frac{(x_i^0 + x_j^0 - 1)(\rho_i b_i(u_i) - A)}{(B^+ - 2A)}$$

$$\times (e^{-2At} - e^{-B^+t}) + x_i^0 e^{-B^+t}. \tag{6}$$

$$\pi^i(u_i, u_j) = \frac{r_i}{(B^+ + \rho)}\left(\frac{\rho_i b_i(u_i)}{\rho} + \frac{(x_i^0 + x_j^0 - 1)(\rho_i b_i(u_i) - A)}{(2A + \rho)} + x_i^0 \right) - \frac{u_i}{\rho} \tag{7}$$

for $i, j = 1, 2$, respectively.

Each payoff on its own variable is defined on a non-compact set $[0, \infty[$ and it may lack quasi-concavity. Therefore, we cannot apply classical results for the existence of pure-strategy Nash equilibria. We prefer to follow a different approach to the problem of whether a strict Nash equilibrium exists in pure strategies. First we identify an algebraic structure for firms' best-reply strategies, i.e., best reply strategies are zeros of a polynomial equation.

Proposition 3. *Assume $\rho \neq 0$. The payoff $\pi^1(u_1, u_2)$ is rational in u_1 and can be calculated as follows*

$$\pi^1(u_1, u_2) = -\frac{hu_1^3 + (b - r_1 d)u_1^2 + (c - r_1 f)u_1 - r_1 g}{\rho(hu_1^2 + bu_1 + c)},$$

where

$b = (\alpha_1 + \beta_1)(\rho^2 + (\theta + 2\vartheta)\rho + 2\theta\vartheta) + 2\beta_1(\rho + \theta) + \rho_1\alpha_1(\rho + 2\vartheta),$

$c = \alpha_1\beta_1(\rho^2 + (\theta + 2\vartheta)\rho + 2\vartheta\theta),$

$d = x_1^0\rho^2 + ((\vartheta + 1)(1 + x_1^0 - x_2^0) + \rho_1(x_1^0 + x_2^0))\rho + 2\rho_1(\vartheta + 1),$

$f = x_1^0(\alpha_1 + \beta_1)\rho^2 + (\alpha_1\rho_1(x_1^0 + x_2^0) + ((\alpha_1 + \beta_1)\vartheta + \beta_1)(1 + x_1^0 - x_2^0))\rho + 2\rho_1\alpha_1\vartheta,$

$g = \alpha_1\beta_1\rho(x_1^0\rho + \vartheta(1 + x_1^0 - x_2^0)),$

$h = (\rho + 2\vartheta + 2)(\rho + \rho_1 + \theta),$

$\theta = \rho_2 \dfrac{u_2}{u_2 + \beta_2}, \quad \vartheta = \dfrac{u_2}{u_2 + \alpha_2}.$

Proof. The proof follows from a direct calculation.   □

Proposition 4. *Assume $\rho \neq 0$. Let $u_2 \geq 0$ be a strategy of firm 2. Assume that a best reply to strategy u_2 exists, i.e. $u_1 > 0$. Then, u_1 is a root of*

$$h^2 u_1{}^4 + 2hbu_1{}^3 + (r_1 fh - r_1 db + 2hc + b^2)u_1{}^2$$

$$+ (2bc - 2r_1 dc + 2r_1 gh)u_1 + (r_1 gb - r_1 fc + c^2) = 0. \tag{8}$$

Proof. Let $\hat{u}_2$ be an investment of firm 2. By simple calculation, we have

$$\frac{\partial \pi^1(u_1, \hat{u}_2)}{\partial u_1} = -\frac{h^2 u_1^4 + 2hbu_1^3 + (r_1 fh - r_1 db + 2hc + b^2)u_1^2}{\rho(hu_1^2 + bu_1 + c)^2}$$

$$+ \frac{(2bc - 2r_1 dc + 2r_1 gh)u_1 + (r_1 gb - r_1 fc + c^2)}{\rho(hu_1^2 + bu_1 + c)^2}. \tag{9}$$

If there exists a best reply in u_1, trivially u_1 is a root of above equation.   □

We conventionally rank the coefficients on the left side of Eq. (8) by decreasing order. The first two lemmas are preliminary to the fundamental Lemma 3.

Lemma 1. *If $\rho \neq 0, \rho \approx 0$ and $r_1 > \frac{(\rho_2 + \rho_1)}{2\rho_1}\alpha_1 + \beta_1\frac{\rho_2}{\rho_1}$ then the fourth coefficient is strictly negative, for any $u_2 \geq 0$.*

Lemma 2. *If $\rho \neq 0, \rho \approx 0$ and $r_1 > \beta_1\frac{\rho_2}{\rho_1}$, then the fifth coefficient is strictly negative for any $u_2 \geq 0$.*

Lemma 3. *If $\rho \neq 0, \rho \approx 0$ and $r_1 > \frac{(\rho_2+\rho_1)}{2\rho_1}\alpha_1 + \beta_1\frac{\rho_2}{\rho_1}$ then there exists a unique best reply $u_1 > 0$ for any strategy $u_2 \geq 0$.*

A fundamental property of best-reply correspondences is presented below.

Lemma 4. *Under the hypothesis of Lemma 3, best-reply correspondences $\widehat{U}_i$ are continuous functions. In addition, $\lim_{u_j \to \infty} \widehat{U}_i(u_j) < \infty$.*

By a classical fixed-point argument, we prove the existence of a strict Nash equilibrium in pure strategies.

Theorem 1 (Main Existence Result). *Assume that $G = ([0, +\infty[^2, \pi^i)$ is an advertising game with low costs. Assume that the discount rate satisfies the following properties: $\rho \neq 0$ and $\rho \approx 0$. Then, a strict Nash equilibrium in proper pure strategies exists for G.*

Proof. The thesis of Lemma 4 is satisfied by hypotheses. By coercivity conditions the image of $\widehat{U}_2$ is a bounded subset in $[0, +\infty[$. Let $K_1 = Cl(\widehat{U}_2([0, +\infty[)) \subset \,]0, +\infty[$ be the closure of the image values of $\widehat{U}_2$. Let $\widehat{U}_1|_{K_1} : K_1 \to \,]0, +\infty[$ be the restricted function to the compact subset K_1. In addition $\widehat{U}_1|_{K_1}(K_1)$ is a compact subset in $]0, +\infty[$ because K_1 is compact and $\widehat{U}_1|_{K_1}$ is continuous. By definition of K_1, it follows that $\widehat{U}_1 \circ \widehat{U}_2|_{K_1} : K_1 \to K_1$. In addition, the above function is continuous because it is a composition of continuous functions and K_1 is a compact in $]0, +\infty[$. By Brower's fixed point $\widehat{U}_1 \circ \widehat{U}_2|_{K_1}$ admits a fixed point. Then there exists a strategy $\widehat{u}_1 \in K_1 \subset \,]0, +\infty[$ such that $\widehat{U}_1(\widehat{U}_2(\widehat{u}_1)) = \widehat{u}_1$. We define $\widehat{u}_2 := \widehat{U}_2(u_1)$. By Lemma 3, we have that $\widehat{u}_2 \neq 0$. It is straightforward to prove that $\widehat{U}_2(\widehat{U}_1)(\widehat{u}_2)) = \widehat{u}_2$. By definition, $(\widehat{U}_2(\widehat{U}_1))(\widehat{u}_2) = \widehat{u}_2$ is a Nash equilibrium in pure strategies. By construction we have that $\widehat{u}$ is a strategy profile with $\widehat{u}_i \neq 0$. Therefore $\widehat{u}$ is a proper strategy profile. By Lemma 3 again, $\widehat{u}$ is a strict Nash equilibrium in pure strategies. □

6. Market Shares at Equilibrium for an Asymmetric Advertising Situation

In this section, we provide a numerical example under some asymmetric conditions for two firms. We assume that firm 1 has higher investment returns for both generic and predatory advertising. Further, we assume that firm 1 is more effective at capturing demand belonging to the market share of the competitor, *ceteribus paribus* their investments. Table 1 contains the numerical parameters for the simulation of our model. Values of parameters of our model are listed in Table 1. It is straightforward to verify that our game is an advertising game with low costs. Using best reply equation (8), we obtain implicit formulae for best reply functions. We numerically integrate their intersection points, i.e., strict Nash equilibria in proper pure strategies. Interestingly, we find that a strict Nash equilibria is unique and equal to $(5.23, 7.24)$ which is not free-riding. Moreover, the weaker firm invests more than

Table 1.   Parameters of the model are listed. Then we compute marketing incentives. We set advertised marginal prices and marginal costs for both firms. In addition, the discount factor is not null and it is extremely small.

Ineffectiveness of generic advertising return	$\alpha_1 = 5, \alpha_2 = 10$
Ineffectiveness of brand advertising return	$\beta_1 = 5, \beta_2 = 15$
Effectiveness of brand advertising	$\rho_1 = 1/50, \rho_2 = 1/100$
Marketing incentive	$c_1^{sh} = 25/4, c_1^{sh} = 45$
Advertised marginal prices	$p_1 = 75, p_2 = 60$
Marginal cost	$c_1 = 25, c_2 = 10$
Discount rate	$\rho = 0.000001$

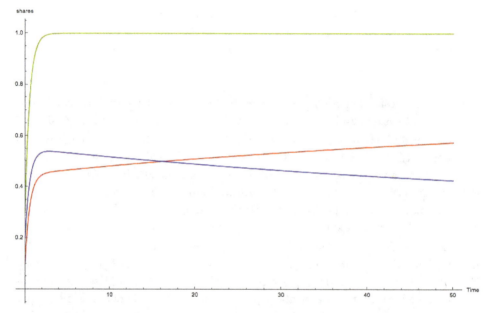

Fig. 1. $x_1^0 = \frac{1}{10}, x_2^0 = \frac{1}{5}$. Firm 1 is initially weaker in the market. The market is initially not saturated.

its opponent. Therefore, the weaker firm does not *cheaply* ride the market [Krishnamurthy, 2000].[c] By replacing Nash equilibrium strategies in formula (6), we compute firms' market shares. Figures 1–5 describe market shares dynamics. The red, blue and green lines represent the dynamics of market shares of the stronger firm (firm 1) and of weaker firm (firm 2) and of the whole market demand, respectively. If the market is initially not saturated, firms invest in marketing advertising and they improve their market positions. The *weaker* firm, i.e., firm 2, initially increases its market position (blue line) and its market share reaches its maximum at time $t \approx 2$ and, then it decreases because of predatory effects. In Fig. 1, the *stronger*

---

[c]We consider the same values of $\rho_i, \beta_i, \alpha_i, r_i$ in Table 1 and we choose higher discount rates. From our numerical implementation, it follows that pure-strategy Nash equilibria fail to exist if ρ approximatively becomes higher than a threshold value equal to 0.040821.

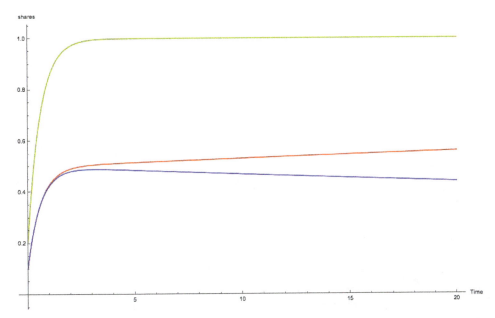

Fig. 2. $x_1^0 = x_2^0 = \frac{1}{10}$. Firms 1,2 equally share the market. The market is initially not saturated.

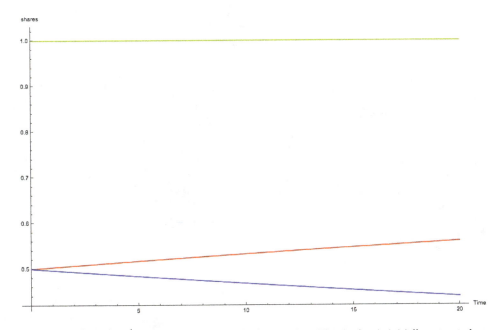

Fig. 3. $x_1^0 = x_2^0 = \frac{1}{2}$. Firms 1,2 equally share the market. The market is initially saturated.

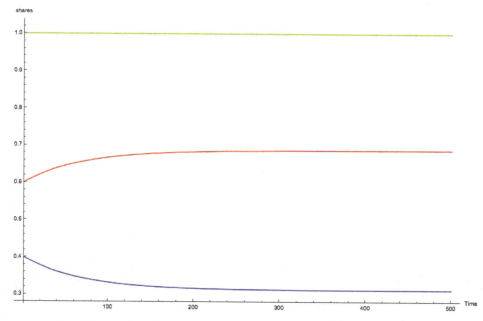

Fig. 4. $x_1^0 = \frac{9}{10}, x_2^0 = \frac{1}{10}$. Firm 1 is initially much stronger on the market. The market is initially saturated.

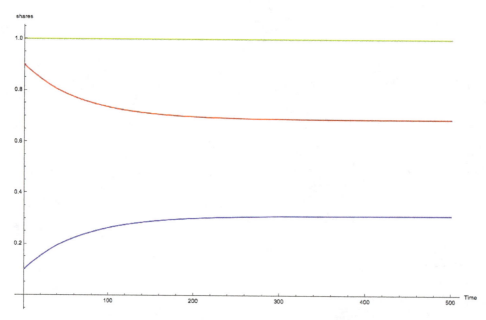

Fig. 5. $x_1^0 = \frac{3}{5}, x_2^0 = \frac{2}{5}$. Firm 1 is initially stronger on the market. The market is initially saturated.

firm's market share is initially smaller than the opponent's one and market shares become equal at $t \approx 16$. If the market size is saturated, then any generic marketing effort is uninfluential in extending the market. Although firm 1 is more effective in marketing campaigns and its initial market position is very dominant, its market share decreases (Fig. 5). The latter happens when the difference between two initial market shares is high. If this difference is not high enough, then firm 1 keeps on improving its market share by making the opponent's market position weaker (Figs. 3–4).

7. Conclusions, Limitations and Future Work

We develop a novel duopolistic advertising model describing changes to the market shares in a duopoly caused by strategic investments in generic and predatory advertising. We provide the existence of strictly optimal investment strategies for firms with lower production costs due to marketing incentives. We discuss marketing incentives by varying the parameters of the model. We provide numerical insights on market share optima in an asymmetrical marketing scenario. We provide rational insights on how competing firms might ultimately reduce the quality of manufactured goods when they publish the prices at the beginning of marketing campaigns. Our work has some limitations. Optimal advertising investments are not in a closed-form formula. Due to the lack of a closed form expression of these equilibria, we cannot study the sensitivity of optimal strategies to a change in marketing parameters. Our main existence result lies in the assumption that the unique discount rate is small. A natural question arising from this chapter is the following: if firms are given different discount rates, can we extend our main existence result? Future research should consider parameterizing this model using advanced econometric models in order to estimate marketing incentives in our model. The latter will help to measure the quality of products provided by firms competing during informative marketing campaigns.

Appendix A. Proofs

Proof of Proposition 1. Analyzing the first equation of system (5) we can see that it is possible to separate the variables and integrate it as follows

$$\frac{dx}{dt} = 2A(1 - x),$$

$$x(t) = 1 - C_x e^{-2At}.$$

If at $t = 0$ we have $x(0) = x_1^0 + x_2^0$, then $C_x = 1 - x_1^0 - x_2^0$, and hence

$$x(t) = 1 + (x_1^0 + x_2^0 - 1)e^{-2At}.$$

For the function $w(t)$ we have

$$\frac{dw}{dt} = \rho_1 b_1(u_1)(x - w) - \rho_2 b_2(u_2)(x + w) = B^- x(t) - B^+ w(t).$$

It is a linear equation of the first order. We integrate it using the standard approach and obtain that the exact solution is given as

$$w(t) = e^{-B^+t}\int e^{B^+t}B^-x(tdt + e^{-B^+t}C_u.$$

Therefore, for the function $w(t)$ we have

$$w(t) = e^{-B^+t}\int e^{B^+t}B^-x(t)dt + e^{-B^+t}C_u$$

$$= e^{-B^+t}\int e^{B^+t}B^-((1+(x_1^0+x_2^0-1)e^{-2At}))dt + e^{-B^+t}C_u$$

$$= e^{-B^+t}\left(B^-\int e^{B^+t}dt + B^-(x_1^0+x_2^0-1)\int e^{(B^+-2A)t}dt\right) + e^{-B^+t}C_u$$

$$= e^{-B^+t}\left(\frac{B^-}{B^+}e^{B^+t} + \frac{B^-(x_1^0+x_2^0-1)}{(B^+-2A)}(e^{(B^+-2A)t}-1)\right) + e^{-B^+t}C_u$$

$$= \frac{B^-}{B^+} + \frac{B^-(x_1^0+x_2^0-1)}{(B^+-2A)}(e^{-2At}-e^{-B^+t}) + e^{-B^+t}C_u. \qquad \square$$

Proof of Proposition 2. Since $x_1 = \frac{x+w}{2}$ and $x_2 = \frac{x-w}{2}$ we obtain

$$x_1(t) = \frac{x(t)+w(t)}{2} = \frac{1}{2}\left(1 + (x_1^0+x_2^0-1)e^{-2At} + \frac{B^-}{B^+}\right.$$

$$+ \frac{B^-(x_1^0+x_2^0-1)}{(B^+-2A)}(e^{-2At}-e^{-B^+t}) + \left. e^{-B^+t}C_u\right)$$

$$= \frac{1}{2}\left(\frac{B^++B^-}{B^+} + \frac{(x_1^0+x_2^0-1)(B^++B^--2A)}{(B^+-2A)}e^{-2At}\right.$$

$$+ \left(C_u - \frac{(x_1^0+x_2^0-1)(B^-)}{(B^+-2A)}\right)e^{-B^+t}\right)$$

$$= \frac{\rho_1 b_1(u_1)}{B^+} + \frac{(x_1^0+x_2^0-1)(\rho_1 b_1(u_1)-A)}{(B^+-2A)}e^{-2At} + \frac{C_u}{2}e^{-B^+t}.$$

Let us now substitute the initial condition and find the constant C_u. If $x_1(0) = x_1^0$ then

$$C_u/2 = 2\left(x_1^0 - \frac{\rho_1 b_1(u_1)}{B^+} - \frac{(x_1^0+x_2^0-1)(-2\rho_1 b_1(u_1)+2A+B^-)}{2(B^+-2A)}\right).$$

Therefore,

$$x_1(t) = \frac{\rho_1 b_1(u_1)}{B^+} + \frac{(x_1^0+x_2^0-1)(\rho_1 b_1(u_1)-A)}{(B^+-2A)}e^{-2At}$$

$$+ \left(x_1^0 - \frac{\rho_1 b_1(u_1)}{B^+} - \frac{(x_1^0+x_2^0-1)(2\rho_1 b_1(u_1)-2A)}{2(B^+-2A)}\right)e^{-B^+t}$$

$$= \frac{\rho_1 b_1(u_1)}{B^+}(1 - e^{-B^+ t})$$

$$+ \frac{(x_1^0 + x_2^0 - 1)(\rho_1 b_1(u_1) - A)}{(B^+ - 2A)}(e^{-2At} - e^{-B^+ t}) + x_1^0 e^{-B^+ t}.$$

For $x_2(t)$ we have a similar result.

$$x_2(t) = \frac{\rho_2 b_2(u_2)}{B^+} + \frac{(x_1^0 + x_2^0 - 1)(\rho_2 b_2(u_2) - A)}{(B^+ - 2A)} e^{-2At}$$

$$- \left(-x_2^0 + \frac{\rho_2 b_2(u_2)}{B^+} + \frac{(x_1^0 + x_2^0 - 1)(\rho_2 b_2(u_2) - A)}{(B^+ - 2A)} \right) e^{-B^+ t}$$

$$= \frac{\rho_2 b_2(u_2)}{B^+}(1 - e^{-B^+ t})$$

$$+ \frac{(x_1^0 + x_2^0 - 1)(\rho_2 b_2(u_2) - A)}{(B^+ - 2A)}(e^{-2At} - e^{-B^+ t}) + x_2^0 e^{-B^+ t}.$$

We therefore have obtained the formulae for the evolution of market shares given the investment rates of firms 1,2. We have

$$\pi^1(u_1, u_2) = \lim_{T \to \infty} \int_0^T e^{-\rho t}(r_1 x_1(t) - u_1) dt$$

$$= \lim_{T \to \infty} \int_0^T \left(\frac{r_1 \rho_1 b_1(u_1)}{B^+} - u_1 \right) e^{-\rho t} dt$$

$$+ \int_0^T \frac{r_1(x_1^0 + x_2^0 - 1)(\rho_1 b_1(u_1) - A)}{(B^+ - 2A)} e^{-\rho t} e^{-2At} dt$$

$$- \int_0^T r_1 \left(\frac{\rho_1 b_1(u_1)}{B^+} + \frac{(x_1^0 + x_2^0 - 1)(\rho_1 b_1(u_1) - A)}{(B^+ - 2A)} - x_1^0 \right) e^{-\rho t} e^{-B^+ t} dt$$

$$= \frac{1}{\rho} \left(\frac{r_1 \rho_1 b_1(u_1)}{B^+} - u_1 \right) + \frac{r_1(x_1^0 + x_2^0 - 1)(\rho_1 b_1(u_1) - A)}{(2A + \rho)(B^+ - 2A)}$$

$$- \frac{r_1}{(B^+ + \rho)} \left(\frac{\rho_1 b_1(u_1)}{B^+} + \frac{(x_1^0 + x_2^0 - 1)(\rho_1 b_1(u_1) - A)}{(B^+ - 2A)} - x_1^0 \right)$$

$$= \frac{r_1}{(B^+ + \rho)} \left(-\frac{\rho_1 b_1(u_1)}{\ln \beta} + \frac{(x_1^0 + x_2^0 - 1)(\rho_1 b_1(u_1) - A)}{(2A + \rho)} + x_1^0 \right) - \frac{u_1}{\rho}.$$

The expression of $\pi^2(u_1, u_2)$ can be obtained in the same way. So we have derived explicit formulae for payoffs. $\square$

Proof of Lemma 1. We assume that $\rho \neq 0$. Consider the exact form of the fourth coefficient and substitute the values for

$$\theta = \rho_2 \frac{u_2}{u_2 + \beta_2} \quad \text{and} \quad \vartheta = \frac{u_2}{u_2 + \alpha_2}.$$

Multiplying the fourth coefficient by $\frac{(u_2+\alpha_2)^2(u_2+\beta_2)^2}{\alpha_1\beta_1}$ and representing the expression $P(u_2) = \frac{(u_2+\alpha_2)^2(u_2+\beta_2)^2}{\alpha_1\beta_1}(2bc - 2r_1dc + 2r_1gh)$ as a polynomial in u_2, we find that the sign of $P(u_2)$ is equal to the sign of the fourth coefficient. We have the following polynomial expression

$$P(u_2) = 2\alpha_2^2\rho^2\beta_2^2(-2r_1\rho_1(1-x_1^0) + \bar{\bar{o}}(\rho))$$

$$+ u_2(2\alpha_2\rho\beta_2(-2\rho_1 r_1(\beta_2(1-x_1^0+x_2^0) + \alpha_2\rho_2) + \bar{\bar{o}}(\rho)))$$

$$+ u_2^2(-8r_1\rho_1\alpha_2\rho_2\beta_2 + \bar{\bar{o}}(\rho))$$

$$+ u_2^3(8\rho_2(\beta_1\alpha_2\rho_2 - r_1\rho_1\alpha_2 + \rho_1\alpha_1\beta_2 - 2r_1\rho_1\beta_2) + \bar{\bar{o}}(\rho))$$

$$+ u_2^4(8\rho_2(2\rho_2\beta_1 + \alpha_1\rho_2 + \rho_1\alpha_1 - 2r_1\rho_1) + \bar{\bar{o}}(\rho)),$$

where $f(\rho) = \bar{\bar{o}}(\rho)$ is such that $\lim_{\rho\to 0}\frac{f(\rho)}{\rho} = 0$. By hypothesis we know that $\rho \approx 0$, then $\bar{\bar{o}}(\rho)$ does not contribute to the signs of coefficients. If the following system of equations

$$\begin{cases} 8\rho_2(\beta_1\alpha_2\rho_2 - r_1\rho_1\alpha_2 + \rho_1\alpha_1\beta_2 - 2r_1\rho_1\beta_2) < 0, \\ 8\rho_2(2\rho_2\beta_1 + \alpha_1\rho_2 + \rho_1\alpha_1 - 2r_1\rho_1) < 0 \end{cases}$$

or, in equivalent way,

$$\begin{cases} r_1 > \dfrac{\beta_1\alpha_2\rho_2 + \rho_1\alpha_1\beta_2}{\rho_1(\alpha_2 + 2\beta_2)}, \\ r_1 > \dfrac{\rho_2}{\rho_1}\beta_1 + \dfrac{\rho_2 + \rho_1}{2\rho_1}\alpha_1 \end{cases} \tag{A.1}$$

is satisfied, then the number of sign alterations of polynomial equation $P(u_2) = 0$ is zero. Since

$$\frac{\rho_2}{\rho_1}\beta_1 + \frac{\rho_2 + \rho_1}{2\rho_1}\alpha_1 - \frac{\beta_1\alpha_2\rho_2 + \rho_1\alpha_1\beta_2}{\rho_1(\alpha_2 + 2\beta_2)}$$

$$= \frac{1}{2}\frac{4\beta_2\rho_2\beta_1 + \alpha_2\rho_1\alpha_1 + \alpha_2\alpha_1\rho_2 + 2\beta_2\alpha_1\rho_2}{\rho_1(\alpha_2 + 2\beta_2)} > 0$$

then we conclude that the second condition is stronger than the first in (A.1). By hypothesis the second condition in (A.1) is satisfied. Then the first condition in (A.1) is satisfied. It thus follows there are no positive roots for the equation $P(u_2) = 0$. It is straightforward to verify that $P(0) < 0$. By continuity arguments, it thus follows that $P(u_2) < 0$ for any $u_2 \geq 0$. Then the third coefficient is strictly negative for any $u_2 \geq 0$. □

Proof of Lemma 2. Arguing as in the proof of Lemma 1, we consider

$$P(u_2) = \frac{(u_2 + \alpha_2)^2(u_2 + \beta_2)^2}{\alpha_1\beta_1}(r_1gb - r_1fc + c^2).$$

Here

$$P(u_2) = \rho^3 \alpha_2^2 \beta_2^2 (-r_1(\alpha_1 \rho_1 x_2^0 - \beta_1 x_2^0 + \beta_1 - \beta_1 x_1^0) + \overline{\overline{o}}(\rho)) + u_2 \rho^2 \beta_2 \alpha_2$$

$$+ [-r_1[\alpha_2 \beta_1 \rho_2 (1 - x_1^0 - x_2^0) + \beta_2 \rho_1 \alpha_1 (1 - x_1^0 + 3 x_2^0)$$

$$+ \alpha_1 \alpha_2 \rho_2 \rho_1 (x_1^0 + x_2^0)] + \overline{\overline{o}}(\rho)]$$

$$+ u_2^2 \rho (-2 \rho_1 \alpha_1 r_1 \beta_2 (\beta_2 (1 - x_1^0 + x_2^0) + \alpha_2 \rho_2 (1 + x_1^0 + x_2^0)) + \overline{\overline{o}}(\rho))$$

$$+ u_2^3 (-4 r_1 \rho_2 \alpha_1 \rho_1 \beta_2 + \overline{\overline{o}}(\rho))$$

$$+ u_2^4 (-4 \alpha_1 \rho_2 (r_1 \rho_1 - \beta_1 \rho_2) + \overline{\overline{o}}(\rho)).$$

Therefore, we impose the set of following conditions:

$$\begin{cases} -r_1(\alpha_1 \rho_1 x_2^0 + \beta_1(1 - x_1^0 - x_2^0)) < 0, \\ -r_1(\alpha_2 \beta_1 \rho_2 (1 - x_1^0 - x_2^0) + \beta_2 \rho_1 \alpha_1 (1 - x_1^0 + 3 x_2^0) + \alpha_1 \alpha_2 \rho_2 \rho_1 (x_1^0 + x_2^0)) < 0, \\ -2 \rho_1 \alpha_1 r_1 \beta_2 (\beta_2 (1 - x_1^0 + x_2^0) + \alpha_2 \rho_2 (1 + x_1^0 + x_2^0)) < 0, \\ -4 r_1 \rho_2 \alpha_1 \rho_1 \beta_2 < 0, \\ -4 \alpha_1 \rho_2 (r_1 \rho_1 - \beta_1 \rho_2) < 0 \end{cases}$$

which is trivially equivalent to $r_1 \rho_1 - \beta_1 \rho_2 > 0$. If $r_1 > \beta_1 \frac{\rho_2}{\rho_1}$ then the fifth coefficient is negative for any non-negative value of $\hat{u}_2$. $\qquad \square$

Proof of Lemma 3. Let $\hat{u}_1$ be a best reply to $\hat{u}_2$. Then $\hat{u}_1 > 0$ is a positive root of Eq. (8). According to the Descartes's rule of sign alterations, if the number of sign alterations in the sequence of the coefficients of equations of a polynomial equation is equal to 1, then there exists exactly one positive root of the equation [Korn and Korn, 1968]. Since the first and the second coefficients in left-side of Eq. (8) are positive, we obtain the different cases in Table A.1. Cases are reduced by assuming that the third coefficient is not null in Table A.1. If the third coefficient is null, the number of sign alterations may be just lower than in the previous described case, i.e., the third coefficient is not null. This happens since the first two coefficients of (8) are not strictly negative. We clarify this ambiguity for the first four rows of Table A.1. Let us suppose that the third coefficient is null then, the second coefficient is positive. If it is strictly positive the number of alterations does not decrease. If the second coefficient is null, then the first coefficient of (8) is strictly positive, i.e., $h^2 > 0$, since $\rho_1 > 0$. Then, one of the first three coefficients is at least not null. Therefore, the four rows of Table A.1 do not present any change in the number of sign alterations if the third coefficient is null. For the remaining last four rows of Table A.1, it is straightforward to prove that the number of sign alteration may be lower if the third coefficient is null. In particular, the number of sign alterations is 0 at row 5 and it is 1 at row 6, if the third coefficient is null. Table A.1 fully represents the number of sign alterations of Eq. (8).

Taking into account our hypothesis, theses of Lemmas 1, 2 hold. Therefore, the fourth and the fifth coefficients are strictly negative. It follows from Table A.1

Table A.1.   Number of sign alterations in Eq. (8).

Third coefficient	Fourth coefficient	Fifth coefficient	Sign alterations
> 0	> 0	> 0	0
> 0	> 0	< 0	1
> 0	< 0	> 0	2
> 0	**< 0**	**< 0**	1
$< 0 (= 0)$	> 0	> 0	2 (0)
$< 0 (= 0)$	> 0	< 0	3 (1)
< 0	< 0	> 0	2
< 0	**< 0**	**< 0**	1

that $\hat{u}_1 > 0$ is the unique best reply to $\hat{u}_2$. If there is exactly one positive root of the numerator of (9), and since the denominator is positive for any $u_1 > 0$, the derivative of payoffs takes negative values for $u_1 > \hat{u}_1$ and positive values for $u_1 < \hat{u}_1$. This proves that $\hat{u}_1$ is a maximum point of $\pi^1(u_1, \hat{u}_2)$.   □

Proof of Lemma 4. By hypothesis, the thesis of Lemma 3 is satisfied. Therefore, best reply multifunctions $\widehat{U}_i$ are functions. In addition, $\widehat{U}_1$ are continuous since payoffs in (7) are continuous. The rest of the proof is just technical and we leave it to the reader. The proof is simply based on convergence properties of a_i, b_i when u converges to ∞. In fact, we have $\lim_{u_2 \to \infty} \theta(u_2) = \rho_2$ and $\lim_{u_2 \to \infty} \vartheta(u_2) = 1$.   □

Acknowledgments

This research has been partially supported by Swedish Research Council project N. 2009-20474-66896-29 and UK EPSRC project N. EP/I005765/1.

References

Baggio, A. and Viscolani, B. [2014] An advertising game with multiplicative interference, *Optimization* **63**(9), 1401–1418.

Bagwell, K. [2007] The economic analysis of advertising, *Handbook Indust. Organ.* **3**, 1701–1844.

Bass, F. M., Krishnamoorthy, A., Prasad, A. and Sethi, S. P. [2005] Generic and brand advertising strategies in a dynamic duopoly, *Market. Sci.* **24**(4), 556–568.

Bass, F. M., Krishnan, T. V. and Jain, D. C. [1994] Why the bass model fits without decision variables, *Market. Sci.* **13**(3), 203–223.

Brady, M. P. [2009] Advertising effectiveness and spillover: Simulating strategic interaction using advertising, *Syst. Dynam. Rev.* 25(4), 281–307.

Brahim, N. B. E.-B., Lahmandi-Ayed, R. and Laussel, D. [2014] Advertising spillovers and strategic interaction in media-product markets, *Louvain Econ. Rev.* **80**(2), 51–98.

Chakravarti, A. and Janiszewski, C. [2004] The influence of generic advertising on brand preferences, *J. Consumer Res.* **30**(4), 487–502.

Chintagunta, P. K. and Vilcassim, N. J. [1992] An empirical investigation of advertising strategies in a dynamic duopoly, *Manage. Sci.* **38**(9) 1230–1244.

Deal, K. R. [1979] Optimizing advertising expenditures in a dynamic duopoly, *Operations Res.* **27**(4), 682–692.

Dockner, E. and Jørgensen, S. [1988] Optimal advertising policies for diffusion models of new product innovation in monopolistic situations, *Manag. Sci.* **34**(1), 119–130.

Erickson, G. M. [1985] A model of advertising competition, *J. Market. Res.* **22**(3), 297–304.

Erickson, G. M. [1995] Differential game models of advertising competition, *Europ. J. Operat. Res.* **83**(3), 431–438.

Espinosa, M. P. and Mariel, P. [2001] A model of optimal advertising expenditures in a dynamic duopoly, *Atlant. Econ. J.* **29**(2), 135–161.

Fisher, J. C. and Pry, R. H. [1971] A simple substitution model of technological change, *Technol. Forecast. Soc. Change* **3**, 75–88.

Freimer, M. and Horsky, D. [2012] Periodic advertising pulsing in a competitive market, *Market. Sci.* **31**(4), 637–648.

Friedman, J. W. [1983] Advertising and oligopolistic equilibrium, *Bell J. Econ.* **14**(2), 464–473.

Fruchter, G. E. and Kalish, S. [1997] Closed-loop advertising strategies in a duopoly, *Manage. Sci.* **43**(1), 54–63.

Gorodnichenko, Y. and Weber, M. [2016] Are sticky prices costly? Evidence from the stock market, *Amer. Econ. Rev.* **106**(1), 165–99.

Gould, J. P. [1976] *Diffusion Processes and Optimal Advertising Policy* Lecture Notes in Economics and Mathematical Systems (Springer, Berlin), pp. 169–174.

Grewal, D., Monroe, K. B. and Krishnan, R. [1998] The effects of price-comparison advertising on buyers' perceptions of acquisition value, transaction value and behavioral intentions, *J. Market.* **62**(2), 46–59.

Han, S., Heywood, J. S. and Ye, G. [2017] Informative advertising in a mixed oligopoly, *Rev. Indust. Organ.* **51**(1), 103–125.

Hanssens, D. M., Parsons, L. J. and Schultz, R. L. [2003] *Market Response Models: Econometric and Time Series Analysis*, International Series in Quantitative Marketing, Vol. 12 (Kluwer Academic Publishers, London).

Huang, J., Leng, M. and Liang, L. [2012] Recent developments in dynamic advertising research, *Europ. J. Operat. Res.* **220**(3), 591–609.

Jiang, B., Ni, J. and Srinivasan, K. [2014] Signaling through pricing by service providers with social preferences, *Market. Sci.* **33**(5), 641–654.

Jørgensen, S. [1982] A differential games solution to a logarithmic advertising model, *J. Operat. Res. Soc.* **33**(5), 425–432.

Jørgensen, S., Kort, P. M. and Zaccour, G. [2006] Advertising an event, *Automatica* **42**(8), 1349–1355.

Kinnucan, H. [1996] A note on measuring returns to generic advertising in interrelated markets, *J. Agricult. Econ.* **47**(1–4), 261–267.

Korn, G. and Korn, T. [1968] *Mathematics Handbook for Scientist and Engineers* (McGraw-Hill, New York).

Krishnamurthy, S. [2000] Enlarging the pie vs. increasing one's slice: An analysis of the relationship between generic and brand advertising, *Market. Lett.* **11**(1), 37–48.

Little, J. D. [1979] Aggregate advertising models: The state of the art, *Operat. Res.* **27**(4), 629–667.

Lu, L., Gou, Q., Tang, W. and Zhang, J. [2016] Joint pricing and advertising strategy with reference price effect, *Int. J. Product. Res.* **54**(17), 5250–5270.

Norman, G., Pepall, L. and Richards, D. [2008] Generic product advertising, spillovers and market concentration, *Amer. J. Agricult. Econ.* **90**(3), 719–732.

Piga, C. A. [1998] A dynamic model of advertising and product differentiation, *Rev. Indust. Organ.* **13**(5), 509–522.

Piga, C. A. [2000] Competition in a duopoly with sticky price and advertising, *Int. J. Indust. Organ.* **18**(4), 595–614.

Rutz, O. J. and Bucklin, R. E. [2011] From generic to branded: A model of spillover in paid search advertising, *J. Market. Res.* **48**(1), 87–102.

Sethi, S. P. [1973] Optimal control of the vidale-wolfe advertising model, *Operat. Res.* **21**(4), 998–1013.

Sethi, S. P. [1983] Deterministic and stochastic optimization of a dynamic advertising model, *Opt. Control Appl. Methods* **4**(2), 179–184.

Shapiro, B. T. [2018] Positive spillovers and free riding in advertising of prescription pharmaceuticals: The case of antidepressants, *J. Polit. Econ.* **126**(1), 381–437.

Tenn, S. and Wendling, B. W. [2014] Entry threats and pricing in the generic drug industry, *Rev. Econ. Statist.* **96**(2), 214–228.

Vidale, M. and Wolfe, H. [1957] An operations-research study of sales response to advertising, *Operat. Res.* **5**(3), 370–381.

Viscolani, B. [2012] Pure-strategy nash equilibria in an advertising game with interference, *Europ. J. Operat. Res.* **216**(3), 605–612.

Chapter 3

Maximin and Minimax Strategies in Two-Players Game with Two Strategic Variables

Atsuhiro Satoh

Faculty of Economics, Hokkai-Gakuen University
Toyohira-ku, Sapporo, Hokkaido 062-8605, Japan

atsatoh@hgu.jp

Yasuhito Tanaka

Faculty of Economics, Doshisha University
Kamigyo-ku, Kyoto 602-8580, Japan

yasuhito@mail.doshisha.ac.jp

We examine maximin and minimax strategies for players in a two-players game with two strategic variables, x and p. We consider two patterns of game; one is the x-game in which the strategic variables of players are x's, and the other is the p-game in which the strategic variables of players are p's. We call two players Players A and B, and will show that the maximin strategy and the minimax strategy in the x-game, and the maximin strategy and the minimax strategy in the p-game are all equivalent for each player. However, the maximin strategy for Player A and that for Player B are not necessarily equivalent, and they are not necessarily equivalent to their Nash equilibrium strategies in the x-game nor the p-game. But, in a special case, where the objective function of Player B is the opposite of the objective function of Player A, the maximin strategy for Player A and that for Player B are equivalent, and they constitute the Nash equilibrium both in the x-game and the p-game.

Keywords: Two-players game; two strategic variables; maximin strategy; minimax strategy.

1. Introduction

We examine the maximin and minimax strategies for players in a two-players game with two strategic variables, x and p. We consider two patterns of game; the x-game in which the strategic variables of players are x's, and the p-game in which the strategic variables of players are p's. The maximin strategy for a player is his strategy which maximizes his objective function that is minimized by a strategy of the other player. The minimax strategy for a player is a strategy of the other player which minimizes his objective function that is maximized by his strategy. We call two players Players A and B.

A motivation for this research is to examine the relationships among maximin strategy, minimax strategy, and Nash equilibrium in a general two-players game with two alternative strategic variables, x and p. With differentiable payoff functions we have found the following results.

(1) The maximin strategy and the minimax strategy in the x-game for each player are equivalent, the maximin strategy and the minimax strategy in the p-game for each player are equivalent, and they are all equivalent.

(2) However, the maximin strategy (or the minimax strategy) for Player A and that for Player B are not necessarily equivalent (if the game is not symmetric), and they are not necessarily equivalent to their Nash equilibrium strategies in the x-game nor the p-game.[a]

(3) But in a special case, where the objective function of Player B is the opposite of the objective function of Player A, the maximin strategy (or the minimax strategy) for Player A and that for Player B are equivalent, and they constitute the Nash equilibrium both in the x-game and the p-game. Thus, in the special case the Nash equilibrium in the x-game and that in the p-game are equivalent.

(1) means that the conditions for the maximin strategy are the same as the conditions for the minimax strategy in each game, and these conditions in the x-game are the same as those in the p-game. Thus, the maximin strategy for Player A (or B) is the same as his strategy when Player B (or A) chooses the minimax strategy for Player A (or B) in each game, and these strategies in the x-game are equivalent to those in the p-game. The maximin strategy for Player A is not necessarily the same as the minimax strategy (Player B's strategy) for him. They coincide in the special case.

The special case in (3) corresponds to the relative profit maximization by firms in a duopoly with differentiated goods in which two strategic variables are the outputs and the prices. In Tanaka [2013a] it was shown that, under the assumption of linear demand and cost functions when firms in duopoly with differentiated goods maximize their relative profits, the Cournot equilibrium and the Bertrand equilibrium are equivalent. Satoh and Tanaka [2014a] extended this result to asymmetric duopoly in which the firms have different cost functions. The special case in this research is a generalization of these papers because the x-game (or the p-game) means Cournot (or Bertrand) competition.[b]

---

[a]If the game is symmetric, the maximin strategy (or the minimax strategy) for Player A and that for Player B are equivalent. But even if the game is symmetric, they are not necessarily equivalent to their Nash equilibrium strategies.
[b]We have shown in Satoh and Tanaka [2014b, 2016a] that the same result holds in symmetric oligopoly but not in asymmetric one.

In Sec. 5, we consider a mixed game in which one of the players chooses x and the other player chooses p as their strategic variables, and show that the maximin and the minimax strategies for each player in the mixed game are equivalent to those in the x-game and the p-game. In Sec. 6, we present an example of the results of this chapter in a duopoly model.

2. The Model

There are two players, Players A and B. Their strategic variables are denoted by x_A and p_A for Player A, and x_B and p_B for Player B. They are related by the following functions:

$$p_A = f_A(x_A, x_B) \quad \text{and} \quad p_B = f_B(x_A, x_B). \tag{1}$$

These functions are continuous, differentiable, and invertible. The inverses of them are written as

$$x_A = x_A(p_A, p_B), \quad x_B = x_B(p_A, p_B).$$

Differentiating (1) with respect to p_A given p_B yields

$$\frac{\partial f_A}{\partial x_A}\frac{dx_A}{dp_A} + \frac{\partial f_A}{\partial x_B}\frac{dx_B}{dp_A} = 1$$

and

$$\frac{\partial f_B}{\partial x_A}\frac{dx_A}{dp_A} + \frac{\partial f_B}{\partial x_B}\frac{dx_B}{dp_A} = 0.$$

From them we get

$$\frac{dx_A}{dp_A} = \frac{\frac{\partial f_B}{\partial x_B}}{\frac{\partial f_A}{\partial x_A}\frac{\partial f_B}{\partial x_B} - \frac{\partial f_A}{\partial x_B}\frac{\partial f_B}{\partial x_A}} \tag{2}$$

and

$$\frac{dx_B}{dp_A} = -\frac{\frac{\partial f_B}{\partial x_A}}{\frac{\partial f_A}{\partial x_A}\frac{\partial f_B}{\partial x_B} - \frac{\partial f_A}{\partial x_B}\frac{\partial f_B}{\partial x_A}}. \tag{3}$$

Symmetrically,

$$\frac{dx_B}{dp_B} = \frac{\frac{\partial f_A}{\partial x_A}}{\frac{\partial f_A}{\partial x_A}\frac{\partial f_B}{\partial x_B} - \frac{\partial f_A}{\partial x_B}\frac{\partial f_B}{\partial x_A}} \tag{4}$$

and

$$\frac{dx_A}{dp_B} = -\frac{\frac{\partial f_A}{\partial x_B}}{\frac{\partial f_A}{\partial x_A}\frac{\partial f_B}{\partial x_B} - \frac{\partial f_A}{\partial x_B}\frac{\partial f_B}{\partial x_A}}. \tag{5}$$

We assume

$$\frac{\partial f_A}{\partial x_A} \neq 0, \quad \frac{\partial f_B}{\partial x_B} \neq 0, \quad \frac{\partial f_A}{\partial x_B} \neq 0, \quad \frac{\partial f_B}{\partial x_A} \neq 0 \quad \text{and} \quad \frac{\partial f_A}{\partial x_A}\frac{\partial f_B}{\partial x_B} - \frac{\partial f_A}{\partial x_B}\frac{\partial f_B}{\partial x_A} \neq 0. \tag{6}$$

The objective functions of Players A and B are

$$\pi_A(x_A, x_B) \quad \text{and} \quad \pi_B(x_A, x_B).$$

They are continuous and differentiable. We consider two patterns of game, the x-game and the p-game. In the x-game strategic variables of the players are x_A and x_B; in the p-game their strategic variables are p_A and p_B. We do not consider simple maximization of their objective functions. Instead, we investigate the maximin strategies and minimax strategies for the players.

3. Maximin and Minimax Strategies

3.1. *x-game*

3.1.1. *Maximin strategy*

First consider the condition for minimization of π_A with respect to x_B. It is

$$\frac{\partial \pi_A}{\partial x_B} = 0. \tag{7}$$

Depending on the value of x_A we get the value of x_B which satisfies (7). Denote it by $x_B(x_A)$. From (7)

$$\frac{dx_B(x_A)}{dx_A} = -\frac{\frac{\partial^2 \pi_A}{\partial x_A \partial x_B}}{\frac{\partial^2 \pi_A}{\partial x_B^2}}.$$

We assume that it is not zero. The maximin strategy for Player A is his strategy which maximizes $\pi_A(x_A, x_B(x_A))$. The condition for maximization of $\pi_A(x_A, x_B(x_A))$ with respect to x_A is

$$\frac{\partial \pi_A}{\partial x_A} + \frac{\partial \pi_A}{\partial x_B} \frac{dx_B(x_A)}{dx_A} = 0.$$

By (7) it is reduced to

$$\frac{\partial \pi_A}{\partial x_A} = 0.$$

Thus, the conditions for the maximin strategy for Player A are

$$\frac{\partial \pi_A}{\partial x_A} = 0 \quad \text{and} \quad \frac{\partial \pi_A}{\partial x_B} = 0. \tag{8}$$

3.1.2. *Minimax strategy*

Consider the condition for maximization of π_A with respect to x_A. It is

$$\frac{\partial \pi_A}{\partial x_A} = 0. \tag{9}$$

Depending on the value of x_B we get the value of x_A which satisfies (9). Denote it by $x_A(x_B)$. From (9) we obtain

$$\frac{dx_A(x_B)}{dx_B} = -\frac{\frac{\partial^2 \pi_A}{\partial x_B \partial x_A}}{\frac{\partial^2 \pi_A}{\partial x_A^2}}.$$

We assume that it is not zero. The minimax strategy for Player A is a strategy of Player B which minimizes $\pi_A(x_A(x_B), x_B)$. The condition for minimization of $\pi_A(x_A(x_B), x_B)$ with respect to x_B is

$$\frac{\partial \pi_A}{\partial x_A} \frac{dx_A(x_B)}{dx_B} + \frac{\partial \pi_A}{\partial x_B} = 0.$$

By (9) it is reduced to

$$\frac{\partial \pi_A}{\partial x_B} = 0.$$

Thus, the conditions for the minimax strategy for Player A are

$$\frac{\partial \pi_A}{\partial x_A} = 0 \quad \text{and} \quad \frac{\partial \pi_A}{\partial x_B} = 0.$$

They are the same as conditions in (8). Similarly, we can show that the conditions for the maximin strategy and the minimax strategy for Player B are

$$\frac{\partial \pi_B}{\partial x_B} = 0 \quad \text{and} \quad \frac{\partial \pi_B}{\partial x_A} = 0. \tag{10}$$

3.2. *p-game*

The objective functions of Players A and B in the *p*-game are written as follows:

$$\pi_A(x_A(p_A, p_B), x_B(p_A, p_B)) \quad \text{and} \quad \pi_B(x_A(p_A, p_B), x_B(p_A, p_B)).$$

We can write them as

$$\pi_A(p_A, p_B) \quad \text{and} \quad \pi_B(p_A, p_B)$$

because $\pi_A(x_A(p_A, p_B), x_B(p_A, p_B))$ and $\pi_B(x_A(p_A, p_B), x_B(p_A, p_B))$ are functions of p_A and p_B. Interchanging x_A and x_B by p_A and p_B in the arguments in the previous section, we can show that the conditions for the maximin strategy and the minimax strategy for Player A in the *p*-game are

$$\frac{\partial \pi_A}{\partial p_A} = 0 \quad \text{and} \quad \frac{\partial \pi_A}{\partial p_B} = 0. \tag{11}$$

We can rewrite them as follows:

$$\frac{\partial \pi_A}{\partial x_A} \frac{dx_A}{dp_A} + \frac{\partial \pi_A}{\partial x_B} \frac{dx_B}{dp_A} = 0 \quad \text{and} \quad \frac{\partial \pi_A}{\partial x_A} \frac{dx_A}{dp_B} + \frac{\partial \pi_A}{\partial x_B} \frac{dx_B}{dp_B} = 0.$$

By (2)–(5), and the assumptions in (6), they are further rewritten as

$$\frac{\partial \pi_A}{\partial x_A}\frac{\partial f_B}{\partial x_B} - \frac{\partial \pi_A}{\partial x_B}\frac{\partial f_B}{\partial x_A} = 0 \quad \text{and} \quad \frac{\partial \pi_A}{\partial x_A}\frac{\partial f_A}{\partial x_B} - \frac{\partial \pi_A}{\partial x_B}\frac{\partial f_A}{\partial x_A} = 0.$$

Again by the assumptions in (6), we obtain

$$\frac{\partial \pi_A}{\partial x_A} = 0 \quad \text{and} \quad \frac{\partial \pi_A}{\partial x_B} = 0.$$

They are the same as conditions in (8).

The conditions for the maximin strategy and the minimax strategy for Player B in the p-game are

$$\frac{\partial \pi_B}{\partial p_B} = 0 \quad \text{and} \quad \frac{\partial \pi_B}{\partial p_A} = 0.$$

They are rewritten as

$$\frac{\partial \pi_B}{\partial x_B}\frac{dx_B}{dp_B} + \frac{\partial \pi_B}{\partial x_A}\frac{dx_A}{dp_B} = 0 \quad \text{and} \quad \frac{\partial \pi_B}{\partial x_B}\frac{dx_B}{dp_A} + \frac{\partial \pi_B}{\partial x_A}\frac{dx_A}{dp_A} = 0.$$

By (2)–(5), and the assumptions in (6), they are further rewritten as

$$\frac{\partial \pi_B}{\partial x_B}\frac{\partial f_A}{\partial x_A} - \frac{\partial \pi_B}{\partial x_A}\frac{\partial f_A}{\partial x_B} = 0 \quad \text{and} \quad \frac{\partial \pi_B}{\partial x_B}\frac{\partial f_B}{\partial x_A} - \frac{\partial \pi_B}{\partial x_A}\frac{\partial f_B}{\partial x_B} = 0.$$

Again by the assumptions in (6), we obtain

$$\frac{\partial \pi_B}{\partial x_A} = 0 \quad \text{and} \quad \frac{\partial \pi_B}{\partial x_B} = 0.$$

They are the same as conditions in (10). We have proved the following proposition.

Proposition 1.

(1) *The maximin strategy and the minimax strategy in the x-game, and the maximin strategy and the minimax strategy in the p-game for Player A are all equivalent.*
(2) *The maximin strategy and the minimax strategy in the x-game, and the maximin strategy and the minimax strategy in the p-game for Player B are all equivalent.*

As we have stated in the introduction (1) means that the conditions for the maximin strategy are the same as the conditions for the minimax strategy in each game, and these conditions in the x-game are the same as those in the p-game. Thus, the maximin strategy for Player A (or B) is the same as his strategy when Player B (or A) chooses the minimax strategy for Player A (or B) in each game, and these strategies in the x-game are equivalent to those in the p-game. The maximin strategy for Player A is not necessarily the same as the minimax strategy (Player B's strategy) for him. They coincide in the special case.

4. Special Case

The results in the previous section do not imply that the maximin strategy (or the minimax strategy) for Player A and that for Player B are equivalent (if the game is not symmetric), nor they are equivalent to their Nash equilibrium strategies in the x-game or the p-game. But in a special case the maximin strategy (or the minimax strategy) for Player A and that for Player B are equivalent, and they constitute the Nash equilibrium both in the x-game and the p-game.

The conditions for the maximin strategy and the minimax strategy for Player A are

$$\frac{\partial \pi_A}{\partial x_A} = 0 \quad \text{and} \quad \frac{\partial \pi_A}{\partial x_B} = 0. \tag{12}$$

Those for Player B are

$$\frac{\partial \pi_B}{\partial x_B} = 0 \quad \text{and} \quad \frac{\partial \pi_B}{\partial x_A} = 0. \tag{13}$$

Equations (12) and (13) are not necessarily equivalent. The conditions for Nash equilibrium in the x-game are

$$\frac{\partial \pi_A}{\partial x_A} = 0 \quad \text{and} \quad \frac{\partial \pi_B}{\partial x_B} = 0. \tag{14}$$

Equations (12) and (14) are not necessarily equivalent.

The conditions for Nash equilibrium in the p-game are

$$\frac{\partial \pi_A}{\partial p_A} = 0 \quad \text{and} \quad \frac{\partial \pi_B}{\partial p_B} = 0. \tag{15}$$

Equations (11) and (15) are not necessarily equivalent.

However, in a special case those conditions are all equivalent. We assume

$$\pi_A + \pi_B = 0, \quad \text{or} \quad \pi_B = -\pi_A. \tag{16}$$

Then, (13) is rewritten as

$$\frac{\partial \pi_A}{\partial x_B} = 0 \quad \text{and} \quad \frac{\partial \pi_A}{\partial x_A} = 0. \tag{17}$$

They are equivalent to (12). Therefore, the maximin strategy and the minimax strategy for Player A and those for Player B are equivalent. $\frac{\partial \pi_B}{\partial x_A} = 0$ and $\frac{\partial \pi_B}{\partial x_B} = 0$ in (13) mean, respectively, minimization of π_B with respect to x_A and maximization of π_B with respect to x_B. On the other hand, $\frac{\partial \pi_A}{\partial x_A} = 0$ and $\frac{\partial \pi_A}{\partial x_B} = 0$ in (12) and (17) mean, respectively, maximization of π_A with respect to x_A and minimization of π_A with respect to x_B.

Equation (14) is rewritten as

$$\frac{\partial \pi_A}{\partial x_A} = 0 \quad \text{and} \quad \frac{\partial \pi_A}{\partial x_B} = 0. \tag{18}$$

Equations (18) and (12) are equivalent. Therefore, the maximin strategy (Player A's strategy) and the minimax strategy (Player B's strategy) for Player A constitute the Nash equilibrium of the x-game. $\frac{\partial \pi_B}{\partial x_B} = 0$ in (14) means maximization of π_B with respect to x_B. On the other hand, $\frac{\partial \pi_A}{\partial x_B} = 0$ in (18) means minimization of π_A with respect to x_B.

Equation (15) is rewritten as

$$\frac{\partial \pi_A}{\partial p_A} = 0 \quad \text{and} \quad \frac{\partial \pi_A}{\partial p_B} = 0. \tag{19}$$

Equations (19) and (11) are equivalent. Therefore, the maximin strategy (Player A's strategy) and the minimax strategy (Player B's strategy) for each player in the p-game constitute the Nash equilibrium of the p-game. Since the maximin strategy and the minimax strategy for Player A in the x-game and those in the p-game are equivalent, the Nash equilibrium of the x-game and that of the p-game are equivalent.

Summarizing the results, we get the following proposition.

Proposition 2. *In the special case in which* (16) *is satisfied:*

(1) *The maximin strategy and the minimax strategy in the x-game and the p-game for Player A and the maximin strategy and the minimax strategy in the x-game and the p-game for Player B are equivalent.*

(2) *These maximin and minimax strategies constitute the Nash equilibrium both in the x-game and the p-game.*

This special case corresponds to the relative profit maximization by firms in duopoly with differentiated goods.[c] Let $\bar{\pi}_A$ and $\bar{\pi}_B$ be the absolute profits of Players A and B, and denote their relative profits by π_A and π_B. Then,

$$\pi_A = \bar{\pi}_A - \bar{\pi}_B \quad \text{and} \quad \pi_B = \bar{\pi}_B - \bar{\pi}_A.$$

From them we can see

$$\pi_B = -\pi_A.$$

5. Mixed Game

We consider a case where Player A's strategic variable is p_A, and that of Player B is x_B.

Differentiating (1) with respect to p_A given x_B yields

$$\frac{\partial f_A}{\partial x_A} \frac{dx_A}{dp_A} = 1$$

and

$$\frac{\partial f_B}{\partial x_A} \frac{dx_A}{dp_A} = \frac{dp_B}{dp_A}.$$

---

[c] About relative profit maximization under imperfect competition, please see Matsumura and Matsushima [2012], Matsumura *et al.* [2013], Satoh and Tanaka [2014a], Satoh and Tanaka [2014b], Tanaka [2013a], Tanaka [2013b] and Vega-Redondo [1997]. See also Sec. 6.

Differentiating (1) with respect to x_B given p_A yields

$$\frac{\partial f_A}{\partial x_A}\frac{dx_A}{dx_B} + \frac{\partial f_A}{\partial x_B} = 0$$

and

$$\frac{\partial f_B}{\partial x_A}\frac{dx_A}{dx_B} + \frac{\partial f_B}{\partial x_B} = \frac{dp_B}{dx_B}.$$

From them we obtain

$$\frac{dx_A}{dp_A} = \frac{1}{\frac{\partial f_A}{\partial x_A}}, \quad \frac{dp_B}{dp_A} = \frac{\frac{\partial f_B}{\partial x_A}}{\frac{\partial f_A}{\partial x_A}},$$

$$\frac{dx_A}{dx_B} = -\frac{\frac{\partial f_A}{\partial x_B}}{\frac{\partial f_A}{\partial x_A}} \quad \text{and} \quad \frac{dp_B}{dx_B} = \frac{\frac{\partial f_A}{\partial x_A}\frac{\partial f_B}{\partial x_B} - \frac{\partial f_B}{\partial x_A}\frac{\partial f_A}{\partial x_B}}{\frac{\partial f_A}{\partial x_A}}.$$

We assume $\frac{\partial f_A}{\partial x_A} \neq 0$ and $\frac{\partial f_A}{\partial x_B} \neq 0$, and so $\frac{dx_A}{dp_A} \neq 0$ and $\frac{dx_A}{dx_B} \neq 0$.

We write the objective functions of Players A and B as follows:

$$\varphi_A(p_A, x_B) = \pi_A(x_A(p_A, p_B), x_B) \quad \text{and} \quad \varphi_B(p_A, x_B) = \pi_B(x_A(p_A, p_B), x_B).$$

Then,

$$\begin{cases} \dfrac{\partial \varphi_A}{\partial p_A} = \dfrac{\partial \pi_A}{\partial x_A}\dfrac{dx_A}{dp_A}, \\[2mm] \dfrac{\partial \varphi_A}{\partial x_B} = \dfrac{\partial \pi_A}{\partial x_A}\dfrac{dx_A}{dx_B} + \dfrac{\partial \pi_A}{\partial x_B}, \\[2mm] \dfrac{\partial \varphi_B}{\partial p_A} = \dfrac{\partial \pi_B}{\partial x_A}\dfrac{dx_A}{dp_A}, \\[2mm] \dfrac{\partial \varphi_B}{\partial x_B} = \dfrac{\partial \pi_B}{\partial x_A}\dfrac{dx_A}{dx_B} + \dfrac{\partial \pi_B}{\partial x_B}. \end{cases} \tag{20}$$

By similar ways to arguments in Sec. 3, we can show that the conditions for the maximin strategy and the conditions for the minimax strategy for Player A are equivalent, and they are

$$\frac{\partial \varphi_A}{\partial p_A} = 0 \quad \text{and} \quad \frac{\partial \varphi_A}{\partial x_B} = 0. \tag{21}$$

The conditions for the maximin strategy and the minimax strategy for Player B are

$$\frac{\partial \varphi_B}{\partial p_A} = 0 \quad \text{and} \quad \frac{\partial \varphi_B}{\partial x_B} = 0. \tag{22}$$

By (20), (21) is rewritten as

$$\frac{\partial \pi_A}{\partial x_A}\frac{dx_A}{dp_A} = 0 \quad \text{and} \quad \frac{\partial \pi_A}{\partial x_A}\frac{dx_A}{dx_B} + \frac{\partial \pi_A}{\partial x_B} = 0.$$

Similarly, (22) is rewritten as follows:

$$\frac{\partial \pi_B}{\partial x_A}\frac{dx_A}{dp_A} = 0 \quad \text{and} \quad \frac{\partial \pi_B}{\partial x_A}\frac{dx_A}{dx_B} + \frac{\partial \pi_B}{\partial x_B} = 0.$$

By the assumptions $\frac{dx_A}{dp_A} \neq 0$ and $\frac{dx_A}{dx_B} \neq 0$, we obtain

$$\frac{\partial \pi_A}{\partial x_A} = 0 \quad \text{and} \quad \frac{\partial \pi_A}{\partial x_B} = 0,$$

and

$$\frac{\partial \pi_B}{\partial x_A} = 0 \quad \text{and} \quad \frac{\partial \pi_B}{\partial x_B} = 0.$$

They are the same as the conditions for the maximin and minimax strategies for Players A and B in the x-game. We have shown the following result.

Proposition 3. *The maximin strategy and the minimax strategy for each player in the mixed game are equivalent to those in the x-game and the p-game.*

6. Maximization of Weighted Average of Absolute and Relative Profits in an Duopoly

In Matsumura and Matsushima [2012] and Matsumura *et al.* [2013], firms' behavior in a duopoly is characterized as the maximization of weighted average of their absolute and relative profits. Assume that there are two firms, A and B producing differentiated goods. The outputs and the prices of the goods of the firms are x_A, x_B, p_A and p_B. The inverse demand functions are

$$p_A = a - x_A - bx_B, \quad p_B = a - x_B - bx_A, \quad 0 < b < 1.$$

Then, the demand functions are written as follows:

$$x_A = \frac{(1-b)a - p_A + bp_B}{1-b^2}, \quad x_B = \frac{(1-b)a - p_B + bp_A}{1-b^2}.$$

For simplicity we assume that cost of the firms is zero. The absolute profits of Firms A and B are

$$\bar{\pi}_A = p_A x_A = \frac{p_A[(1-b)a - p_A + bp_B]}{1-b^2},$$

and

$$\bar{\pi}_B = p_B x_B = \frac{p_B[(1-b)a - p_B + bp_A]}{1-b^2}.$$

The relative profits of Firms A and B are $\bar{\pi}_A - \bar{\pi}_B$ and $\bar{\pi}_B - \bar{\pi}_A$. The objective functions of Firms A and B are defined as the weighted averages of their absolute and relative profits as follows:

$$\pi_A = \lambda\bar{\pi}_A + (1-\lambda)(\bar{\pi}_A - \bar{\pi}_B) = \bar{\pi}_A - (1-\lambda)\bar{\pi}_B,$$

and

$$\pi_B = \lambda\bar{\pi}_B + (1-\lambda)(\bar{\pi}_B - \bar{\pi}_A) = \bar{\pi}_B - (1-\lambda)\bar{\pi}_A,$$

where $0 \le \lambda \le 1$. $\lambda = 0$ corresponds to the pure relative profit maximization, and $\lambda = 1$ corresponds to the absolute profit maximization. The smaller the value of λ is, the more competitive the market is. When the strategic variables of the firms are the outputs (Cournot competition), the maximin strategies and the minimax strategies of the firms are

$$\arg\max_{x_A}\min_{x_B} \pi_A = \arg\max_{x_B}\min_{x_A} \pi_B = \frac{a(1-\lambda)(2-b\lambda)}{b^2\lambda^2 - 4\lambda + 4},$$

$$\arg\min_{x_A}\max_{x_B} \pi_B = \arg\min_{x_B}\max_{x_A} \pi_A = \frac{a(b\lambda - 2\lambda + 2)}{b^2\lambda^2 - 4\lambda + 4}.$$

When the strategy of Firm A is $x_A = \arg\max_{x_A}\min_{x_B} \pi_A$, Firm B's strategy, which is obtained from $\frac{\partial\pi_A}{\partial x_B} = 0$, is

$$x_B = \frac{a(b\lambda - 2\lambda + 2)}{b^2\lambda^2 - 4\lambda + 4} = \arg\min_{x_B}\max_{x_A} \pi_A. \tag{23}$$

On the other hand, the Nash equilibrium strategies in the Cournot competition are

$$x_A^C = x_B^C = \frac{b}{b\lambda + 2}.$$

The differences among them are

$$\arg\max_{x_A}\min_{x_B} \pi_A - x_A^C = \frac{ab^2(\lambda - 2)\lambda^2}{(b\lambda + 2)(b^2\lambda^2 - 4\lambda + 4)},$$

$$\arg\min_{x_B}\max_{x_A} \pi_A - x_A^C = \frac{2ab(2-\lambda)\lambda}{(b\lambda + 2)(b^2\lambda^2 - 4\lambda + 4)}.$$

They are zero if and only if $\lambda = 0$.

When the strategic variables of the firms are the pries (Bertrand competition), the maximin strategies and the minimax strategies of the firms are

$$\arg\max_{p_A}\min_{p_B} \pi_A = \arg\max_{p_B}\min_{p_A} \pi_B = \frac{a(1-b)(1-\lambda)(b\lambda + 2)}{b^2\lambda^2 - 4\lambda + 4},$$

$$\arg\min_{p_A}\max_{p_B} \pi_B = \arg\min_{p_B}\max_{p_A} \pi_A = \frac{a(1-b)(2-b\lambda - 2\lambda)}{b^2\lambda^2 - 4\lambda + 4}.$$

When the strategy of Firm A is $p_A = \arg\max_{p_A}\min_{p_B} \pi_A$, Firm B's strategy, which is obtained from $\frac{\partial\pi_A}{\partial p_B} = 0$, is

$$p_B = \frac{a(1-b)(2-b\lambda - 2\lambda)}{b^2\lambda^2 - 4\lambda + 4} = \arg\min_{p_B}\max_{p_A} \pi_A.$$

Then, the outputs of the firms are

$$x_A = \frac{a(1-\lambda)(2-b\lambda)}{b^2\lambda^2 - 4\lambda + 4} = \arg\max_{x_A}\min_{x_B} \pi_A,$$

$$x_B = \frac{a(b\lambda - 2\lambda + 2)}{b^2\lambda^2 - 4\lambda + 4}. \tag{24}$$

From (23) and (24) we find that the maximin strategy in the Cournot competition and that in the Bertrand competition are equivalent. We can also show that the minimax strategy in the Cournot competition and that in the Bertrand competition are equivalent by similar procedures.

On the other hand, the Nash equilibrium strategies in the Bertrand competition are

$$p_A^B = p_B^B = \frac{(1-b)a}{2 - b\lambda}.$$

The differences among p_A^B, $\arg\max_{p_A}\min_{p_B}\pi_A$ and $\arg\min_{p_B}\max_{p_A}\pi_A$ are

$$\arg\max_{p_A}\min_{p_B}\pi_A - p_A^B = \frac{a(b-1)b^2(\lambda-2)\lambda^2}{(b\lambda - 2)(b^2\lambda^2 - 4\lambda + 4)},$$

$$\arg\min_{p_B}\max_{p_A}\pi_A - p_A^B = \frac{2a(b-1)b(\lambda-2)\lambda}{(b\lambda - 2)(b^2\lambda^2 - 4\lambda + 4)}.$$

They are zero if and only if $\lambda = 0$.

The equilibrium outputs in the Bertrand equilibrium are

$$x_A^B = x_B^B = \frac{a(1+b-b\lambda)}{(b+1)(2-b\lambda)}.$$

The difference between the equilibrium output in the Cournot competition and that in the Bertrand competition is

$$x_A^C - x_A^B = -\frac{ab^2(2-\lambda)\lambda}{(b+1)(2-b\lambda)(2+b\lambda)}.$$

This is zero if and only if $\lambda = 0$. Therefore, if and only if the objective functions of the firms are pure relative profits, the maximin strategies and the minimax strategies of the firms are the same as their Nash equilibrium strategies, and the Cournot equilibrium is equivalent to the Bertrand equilibrium.

7. Concluding Remark

We have analyzed maximin and minimax strategies in a game with two strategic variables. We assumed differentiability of objective functions of the players. In future research, we want to extend the arguments of this chapter to a case where the objective functions of players are not assumed to be differentiable[d] and to a symmetric game with more than two players. In an asymmetric game with more than two players the equivalence results of this chapter do not hold.

---

[d]One attempt along this line is Satoh and Tanaka [2016b].

Acknowledgments

We sincerely thank the constructive criticisms and valuable comments, which were of great help. This work was supported by the Japan Society for the Promotion of Science KAKENHI Grant No. 15K03481.

References

Matsumura, T. and Matsushima, N. [2012] Competitiveness and stability of collusive behavior, *Bull. Econ. Res.* **64**, S221–S31.

Matsumura, T., Matsushima, N. and Cato, S. [2013] Competitiveness and R&D competition revisited, *Econ. Model.* **31**, 541–547.

Satoh, A. and Tanaka, Y. [2014a] Relative profit maximization and equivalence of Cournot and Bertrand equilibria in asymmetric duopoly, *Econ. Bull.* **34**, 819–827.

Satoh, A. and Tanaka, Y. [2014b] Relative profit maximization in asymmetric oligopoly, *Econ. Bull.* **34**, 1653–1664.

Satoh, A. and Tanaka, Y. [2016a] Choice of strategic variables by relative profit maximizing firms in oligopoly, *Econ. Rev.* **67**, 17–25 (in Japanese).

Satoh, A. and Tanaka, Y. [2016b] Two person zero-sum game with two sets of strategic variables, MPRA Paper 73472, University Library of Munich, Germany.

Tanaka, Y. [2013a] Equivalence of Cournot and Bertrand equilibria in differentiated duopoly under relative profit maximization with linear demand, *Econ. Bull.* **33**, 1479–1486.

Tanaka, Y. [2013b] Irrelevance of the choice of strategic variables in a duopoly under relative profit maximization, *Econ. Bus. Lett.* **2**, 75–83.

Vega-Redondo, F. [1997] The evolution of Walrasian behavior, *Econometrica* **65**, 375–384.

Chapter 4

Noncooperative Dynamic Contribution to a Public Project

Sebastien Rouillon

GREThA, University of Bordeaux, Avenue Duguit
33 608 Pessac cedex, France

rouillon@u-bordeaux.fr

We consider a dynamic contribution game in which a group of agents collaborates to complete a public project. The agents exert efforts over time and get rewarded upon completion of the project, once the cumulative effort has reached a pre-specified level. We explicitly derive the cooperative solution and a noncooperative Markov-perfect Nash equilibrium. We characterize the set of socially efficient projects, i.e., projects that cooperative groups find worth completing. Comparing with the Markov-perfect Nash equilibrium, we find that noncooperative groups give up large socially efficient projects. Moreover, they take too much time to complete the projects that they undertake.

Keywords: Voluntary contribution games; differential games; free-riding; procrastination.

1. Introduction

Since the seminal work of Olson [1965], the problem of free riding in groups has been mostly analyzed in static settings. However, many voluntary contributions to public projects have a dynamic and recurring feature [Fershtman and Nitzan, 1991]. In this chapter, we consider a differential game in which players contribute to a joint project generating utility only after completion. Our model can be used to describe collaborative situations such as search teams, R&D joint ventures or funding discrete public goods. We show that the noncooperative groups fail to carry out some socially optimal projects and procrastinate on the projects which they complete.

Our model ties into the literature on free-riding in groups, showing that self-interested individual members have too little incentives to further their common interests [Olson, 1965]. In static settings, free-riding incentives are found with continuous public goods [Olson, 1965; Cornes and Sandler, 1984, 1985; Bergström *et al.*, 1986], but may vanish with discrete public goods [Palfrey and Rosenthal, 1984;

Bagnoli and Lipman, 1989; Nitzan and Romano, 1990].[a] In dynamic settings, free-riding occurs with both continuous public goods [Fershtman and Nitzan, 1991] and discrete public goods [Kessing, 2007; Yidirim, 2006; Georgiadis, 2015; Cvitanic and Georgiadis, 2016]. However, the players' contributions become strategic complements with discrete public goods, which partially mitigates the incentives to free-ride [Yidirim, 2006; Kessing, 2007; Georgiadis, 2015; Cvitanic and Georgiadis, 2016].

In this chapter, we study a game of dynamic contributions to a discrete public good in continuous time, considering a general convex technology for producing the public good. We solve the cooperative solution and a noncooperative Markov-perfect Nash equilibrium. We characterize the solutions by means of an intuitive graphical method and in the closed-form. We find that both cooperative and noncooperative groups complete larger projects,[b] when the benefit of the project is larger, when the team gathers more members and when their members are more patient. However, we show that noncooperative groups give up large projects that cooperative groups with same characteristics would bring to completion (i.e., free-riding). In addition, we prove that they take too much time to complete the projects that they undertake (i.e., procrastination).

To the best of my knowledge, the closest papers in the literature are Admati and Perry [1991], Yidirim [2006], Kessing [2007], Georgiadis [2015] and Cvitanic and Georgiadis [2016].

Admati and Perry [1991] consider a game in discrete time in which *two* players alternate in contributing a discrete public good. Using isoelastic cost functions, they can show that some socially desirable projects may not be completed in equilibrium. Besides technical differences (i.e., here, any number of players contributing simultaneously in continuous time), the present chapter extends Admati and Perry [1991] by fully characterizing the inefficiency arising in equilibrium using a *general class of convex cost functions*.

Yidirim [2006] solves a game in discrete time in which players simultaneously make binary contributions further to a joint project, which is completed only after a given number of steps where at least one player contributed. The players' costs of contributing are private information and drawn periodically from a common distribution. In equilibrium, no matter the number of steps required before completion, each player contributes with a strictly positive, but socially too small probability. The model of Yidirim [2006] differs dramatically from the one used below, in particular in the technology considered for producing the public good (respectively, the maximum versus the sum of individual contributions).

---

[a]Nitzan and Romano [1990] argue that a discrete public good induces a discontinuity in the players' payoffs which corrects the free-riding incentives.

[b]In our setting, the group is the set of agents involved in the public project. It is said cooperative when the individuals maximize the group's aggregate utility. It is said noncooperative when they follow their personal interest.

Kessing [2007] sets a dynamic game of discrete public good completed by private contributions in continuous time. Using quadratic cost functions, Kessing [2007] calculates explicitly the socially optimal and the Markov-perfect Nash equilibrium contribution paths and shows that in equilibrium, some socially profitable projects are not carried out and the completion time is too long. This chapter generalizes the results of Kessing [2007], by using a *general class of convex cost functions*.

Recently, Cvitanic and Georgiadis [2016] extended Kessing [2007] to a general class of convex cost functions. However, they characterize a Markov-perfect Nash equilibrium by a system of ordinary differential equations, which only allows to prove existence and to derive some limited properties. By contrast, this chapter explicitly solves a noncooperative Markov-perfect Nash equilibrium and fully characterizes its properties (comparative statics, closed form).

Georgiadis [2015] constructs a game in continuous time in which players contribute a discrete public project which progresses stochastically (Brownian motion). In equilibrium, no matter the cumulative effort required to complete it, the players' contributions are always strictly positive, but socially insufficient.[c] This chapter departs from Georgiadis [2015], by considering a *general class of convex cost functions* (instead of isoelastic cost functions) and by assuming that the state of the project follows a deterministic process. This allows us to solve the game explicitly and derive clearcut properties.[d]

The remainder of the chapter is organized as follows. Section 2 sets out the model. Section 3 analyzes the cooperative solution. Section 4 characterizes a Markov-perfect Nash equilibrium. Section 5 restates our results in closed form and provide an example. Section 6 discusses some normative implications of our results.

2. The Model

Consider a group composed of n agents who can collaborate to complete a project. Time is assumed continuous. Each agent i exerts an instantaneous effort $q_i(t)$ at time t. The cost of effort, denoted $c(q_i)$, is assumed strictly increasing, strictly convex and twice differentiable, with $c(0) = c'(0) = 0$. The progression of the project is represented by the state $x(t)$, which evolves according to the ordinary differential equation

$$\dot{x}(t) = -\sum_{i=1}^{n} q_i(t), \quad x(0) = \ell. \tag{1}$$

The initial state ℓ shall be interpreted as the size of the project. At any time t, the state $x(t)$ corresponds to the missing contributions to complete the project. The

---

[c]This may be surprising at first glance. It can be explained because, as the project progresses stochastically, there always is some positive probability that it goes to completion. Hence, exerting an infinitesimal effort always induce a positive marginal expected benefit.

[d]Georgiadis [2015] also addresses interesting design issues (i.e., team size, rewards), from the perspective of a residual claimant of the project. These are beyond the scope of this chapter.

project is completed at time T, which is endogenously determined when $x(T) = 0$ occurs for the first time. Each agent i then receives a reward b.[e] The agents are assumed to discount time at the common rate δ.

Each agent i's problem is to choose an individual effort path $q_i(t)$ to maximize

$$be^{-\delta T} - \int_0^T c(q_i(t))e^{-\delta t}\, dt. \tag{2}$$

The team's problem is to find a vector of individual effort paths $q_i(t)$, for all i, to maximize

$$\sum_{i=1}^n \left(be^{-\delta T} - \int_0^T c(q_i(t))e^{-\delta t}\, dt \right). \tag{3}$$

3. Optimal Policy

We determine here the socially optimal path of individual efforts and discuss its properties.

The team's problem is to find a path $q_i(t)$, for all i, and a completion time T, to maximize

$$\sum_{i=1}^n \left(be^{-\delta T} - \int_0^T c(q_i(t))e^{-\delta t}\, dt \right)$$

subject to

$$\dot{x}(t) = -\sum_{i=1}^n q_i(t), \quad x(0) = \ell \quad \text{and} \quad x(T) = 0.$$

In order to ease the comparison with the noncooperative equilibrium, we solve this problem in feedback form. This is admissible because the terminal time T is endogenous, rendering the problem autonomous. Moreover, as the cost of effort $c(q_i)$ is strictly convex, we know that the cooperative solution is symmetric. Hence, we search below an *optimal policy* $f(x)$, such that the team's problem is solved if, at any time t, the individual members contribute $q_i(t) = f(x(t))$ for all i, given that the current state of the project is $x(t)$.

In order to write the cooperative solution, define $\sigma(q) \equiv c''(q)q/c'(q)$ for all q, the elasticity of the marginal cost $c'(q)$.[f] We assume below that $\sigma(q)$ is continuous and has a strictly positive lower bound.[g]

---

[e] In general, the individual reward b may depend on the size n of the group (i.e., $b = \beta(n)$). For the sake of simplicity, we do not make this apparent in the notations.

[f] The elasticity of marginal cost $\sigma(q)$ is a measure of relative risk aversion. Its inverse gives the elasticity of intertemporal substitution of effort, that is, the sensibility of the growth rate of effort to changes in the discount rate.

[g] This assumption guaranties that condition (4) is well-defined and has a (unique) solution.

Proposition 1 (Cooperative solution). *Let $\ell^o > 0$ be the maximum size such that a cooperative team only undertakes and completes projects such that $\ell < \ell^o$.[h] The cooperative solution, given in the feedback form $q_i = f(x)$, for all i, is implicitly defined by*

$$\int_0^{f(x)} \sigma(q)\,dq = \frac{\delta}{n}\max(0, \ell^o - x), \quad \text{for all } x. \tag{4}$$

We can provide the following sketch of the proof. For all x, let $J(x)$ be the maximized discounted payoff function of the team. Assuming differentiability, it satisfies the Hamilton–Jacobi–Bellman (HJB) equation

$$\delta J(x) = \max\left\{-\sum_{i=1}^n c(q_i) - J'(x)\sum_{i=1}^n q_i; \forall i, q_i \geq 0\right\}, \quad \text{for all } x, \tag{5}$$

and the boundary condition

$$J(0) = nb. \tag{6}$$

For all x, let $q_i = f(x)$, for all i, be an optimal control. As it maximizes the right-hand side of (5), we can write after substitution

$$\delta J(x) = -nc(f(x)) - J'(x)nf(x).$$

Suppose that there exists ℓ^o such that $q_i = f(x) > 0$ if and only if $0 \leq x < \ell^o$. Then, being interior, the optimal control satisfies the first-order condition

$$-c'(f(x)) - J'(x) = 0,$$

which allows to rewrite (5) as

$$\delta J(x) = n(c'(f(x))f(x) - c(f(x))).$$

By differentiation, this yields

$$\delta J'(x) = nc''(f(x))f(x)f'(x)$$

and, using $J'(x) = -c'(f(x))$ anew,

$$\sigma(f(x))f'(x) = -\frac{\delta}{n}.$$

Integrating this between x and ℓ^o, we find

$$\int_x^{\ell^o} \sigma(f(s))f'(s)\,ds = -\frac{\delta}{n}(\ell^o - x).$$

Finally, making the substitution $q = f(s)$, $dq = f'(s)\,ds$ and $f(\ell^o) = 0$ (by definition of ℓ^o), this can be rewritten as

$$\int_0^{f(x)} \sigma(q)\,dq = \frac{\delta}{n}(\ell^o - x),$$

[h] For the sake of brevity, the calculus of ℓ^o is postponed below. Its comparative statics is expounded in Property 1.

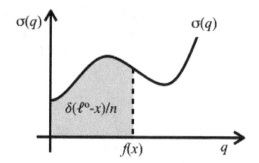

Fig. 1.   Optimal individual effort (if $x \leq \ell^o$).

which gives condition (4) in Proposition 1. The threshold ℓ^o can then be found using (6), which implies to set $f(0) = q^o$ such that

$$c'(q^o)q^o - c(q^o) = b.$$

Then, using (4), we directly get

$$\ell^o = \frac{n}{\delta} \int_0^{q^o} \sigma(q)\, dq.$$

We can use Fig. 1 to illustrate Proposition 1. As $f(x) > 0$ if and only if $x < \ell^o$, the threshold ℓ^o can indeed be interpreted as the maximum size such that a cooperative group undertakes and completes the project. For shorter projects, the optimal individual effort $f(x)$ can be constructed graphically, by finding the point on the abscissa such that the surface area below the curve representing the elasticity of marginal cost $\sigma(q)$ has a measure equal to $(\delta/n)(\ell^o - x)$. It follows that the optimal individual effort increases as the project approaches completion (i.e., $f'(x) < 0$ for all $x < \ell^o$).

Proof. The proof is constructive. We first display a value function $V(x)$, satisfying the HJB equation and the boundary condition. We show that $q_i = f(x)$, for all i, is the corresponding optimal control. We then verify that the control $q_i = f(x)$, for all i, actually induces a discounted payoff of the team equal to the value function.

Value function and optimal control:

For all x, consider the function

$$V(x) \equiv \frac{n}{\delta}(c'f(x))f(x) - c(f(x))),$$

where $f(x)$ is the policy defined by condition (4) in Proposition 1.

-HJB equation:

For all $x < \ell^o$, $f(x)$ is strictly positive and differentiable, with

$$\sigma(f(x))f'(x) = -\frac{\delta}{n}.$$

Thus, $V(x)$ is differentiable and we can calculate, after rearrangement, that

$$V'(x) = \frac{n}{\delta}\sigma(f(x))f'(x)c'f(x)).$$

After substitution, this simplifies to

$$V'(x) = -c'(f(x)).$$

It follows that

$$q_i = f(x), \quad \text{for all } i, \quad \text{maximizes} \quad \left\{-\sum_{i=1}^{n} c(q_i) - V'(x)\sum_{i=1}^{n} q_i; \forall i, q_i \geq 0\right\}$$

and therefore

$$\delta V(x) = \max\left\{-\sum_{i=1}^{n} c(q_i) - V'(x)\sum_{i=1}^{n} q_i; \forall i, q_i \geq 0\right\}.$$

This proves that $V(x)$ satisfies (5) for all $x < \ell^o$.

For all $x \geq \ell^o$, $f(x) = 0$, implying that $V(x) = V'(x) = 0$. Clearly, it follows that

$$q_i = f(x), \quad \text{for all } i, \quad \text{maximize} \quad \left\{-\sum_{i=1}^{n} c(q_i) - V'(x)\sum_{i=1}^{n} q_i; \forall i, q_i \geq 0\right\}$$

and therefore

$$\delta V(x) = \max\left\{-\sum_{i=1}^{n} c(q_i) - V'(x)\sum_{i=1}^{n} q_i; \forall i, q_i \geq 0\right\}.$$

This proves that $V(x)$ satisfies (5) for all $x \geq \ell^o$.

-Boundary condition:

The function $V(x)$ satisfies (6) if

$$V(0) = \frac{n}{\delta}(c'(f(0))f(0) - c(f(0))) = nb,$$

which requires to set $f(0) = q^o$, where q^o is implicitly defined by

$$c'(q^o)q^o - c(q^o) = \delta b. \tag{7}$$

As $f'(x) \leq 0$ for all x, it is interpreted as the maximum instantaneous effort that the agents will optimally contribute.

The upper part of Fig. 2 illustrates the determination of q^o. As $c(0) = 0$, the shaded area has a measure equal to $c'(q^o)q^o - c(q^o)$. By definition, it must be equal to δb.

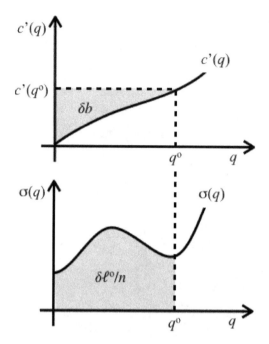

Fig. 2.   Construction of q^o and ℓ^o.

Now, by definition of $f(x)$, we have

$$\int_0^{f(0)} \sigma(s)\, ds = \frac{\delta}{n} \ell^o,$$

which allows to calculate

$$\ell^o = \frac{n}{\delta} \int_0^{q^o} \sigma(s)\, ds. \tag{8}$$

As $f(x) > 0$ if and only if $x < \ell^o$, this threshold is coherently interpreted as the maximum size of the project such that it will eventually go to completion.

The lower part of Fig. 2 describes the determination of ℓ^o. The shaded area has a measure equal to the integral of $\sigma(q)$ from 0 to q^o. By definition, it must be equal to $\delta \ell^o / n$.

Verification:

Suppose that the team implements the policy defined by condition (4) in Proposition 1. In other words, for all t, the players contribute $q_i(t) = f(x(t))$, for all i, and the state trajectory $x(t)$ satisfies $\dot{x}(t) = -nf(x(t))$, with initial condition $x(0) = \ell$.

Consider first the case where $\ell < \ell^o$. As $f(x) > 0$ and $f'(x) < 0$ for all $x < \ell^o$, $\dot{x}(t) \leq -nf(x(0)) < 0$ for all t. Hence, the state trajectory $x(t)$ reaches $x(T) = 0$

in finite time T. Accordingly, the team's payoff is equal to

$$J(\ell) = n\left(be^{-\delta T} - \int_0^T c(f(x(t))e^{-\delta t}\,dt\right).$$

Since the value function $V(x)$ satisfies (5), with $q_i = f(x)$, for all i, the corresponding optimal control, for all t, we have

$$\delta V(x(t)) = -nc(f(x(t))) - V'(x(t))nf(x(t)).$$

Using $\dot{x}(t) = -nf(x(t))$, multiplying by $e^{-\delta t}$ and rearranging, we can write

$$(-\delta V(x(t)) + V'(x(t))\dot{x}(t))e^{-\delta t} = nc(f(x(t)))e^{-\delta t}.$$

Thus, upon substitution, we obtain

$$J(\ell) = nbe^{-\delta T} - \int_0^T (-\delta V(x(t)) + V'(x(t))\dot{x}(t))e^{-\delta t}\,dt,$$

which by integration implies that

$$J(\ell) = V(x(0)) + (nb - V(x(T)))e^{-\delta T}.$$

Finally, using the boundary conditions $x(0) = \ell$, $x(T) = 0$ and $V(0) = nb$, we obtain

$$J(\ell) = V(\ell),$$

which is what we needed to show.

Consider now the case where $\ell \geq \ell^o$. As $f(x) = 0$ for all $x \geq \ell^o$, $\dot{x}(t) = -nf(x(t)) = 0$ for all t. Hence, the players never contribute and the state $x(t)$ remains equal to ℓ for all t. Accordingly, the team's payoff is equal to (using $c(0) = 0$)

$$J(\ell) = \lim_{T\to\infty} n\left(be^{-\delta T} - \int_0^T c(0)e^{-\delta t}\,dt\right) = 0,$$

which is what we needed to show. □

Property 1 below expounds our findings regarding the comparative statics of the cooperative equilibrium.

Property 1 (Comparative statics). (i) The maximum size ℓ^o such that a cooperative team completes the project is increasing in b and n, and decreasing in δ. (ii) The optimal contribution $f(x)$ is increasing in b and n for all x. There exists $\underline{\ell}^o$ such that $f(x)$ is increasing in δ if $x < \underline{\ell}^o$ and is decreasing in δ if $\underline{\ell}^o < x < \ell^o$.

Most comparative statics results are as expected. Cooperative groups undertake and complete projects with a larger size, when the projects are more valuable, when the team gathers more members and when their members are more patient. Before the project is completed, the cooperative members contribute more when the project is more valuable and when the team is larger. The comparative statics with respect to the time preferences is more surprising. We find that the cooperative members contribute more when they are more patient, if and only if the project is sufficiently far from its completion.

Proof of Property 1. (i) We have derived in Appendix A the following total derivative

$$d\ell^o = \frac{n}{c'(q^o)} \, db - \left(\ell^o - \frac{nb}{c'(q^o)}\right) \frac{d\delta}{\delta} + \ell^o \frac{dn}{n}.$$

The comparative statics with respect to b and n follows directly. To show that ℓ^o is decreasing in δ, we first remark that

$$\int_0^{q^o} c''(q)q \, dq = c'(q^o)q^o - c(q^o).$$

Hence, condition (7) is equivalent to

$$\int_0^{q^o} c''(q)q \, dq = \delta b.$$

Using $\sigma(q) = c''(q)q/c'(q)$ and rearranging, this implies that

$$\frac{nb}{c'(q^o)} = \frac{n}{\delta} \int_0^{q^o} \sigma(q) \frac{c'(q)}{c'(q^o)} \, dq.$$

Using condition (8), we directly calculate

$$\ell^o - \frac{nb}{c'(q^o)} = \frac{n}{\delta} \int_0^{q^o} \sigma(q) \left(1 - \frac{c'(q)}{c'(q^o)}\right) dq,$$

which is positive by convexity of $c(q)$ (as $c'(q) < c'(q^o)$ for all $q < q^o$).

(ii) We have derived in Appendix A the following total derivative

$$d(f(x)) = \delta \frac{\frac{n}{c'(q^o)} \, db - \left(x - \frac{nb}{c'(q^o)}\right) \frac{d\delta}{\delta} + x \frac{dn}{n}}{n\sigma(f(x))}.$$

The comparative statics with respect to b and n follow directly. We have shown previously that $\ell^o - nb/c'(q^o) > 0$, implying that $0 < \underline{\ell}^o \equiv nb/c'(q^o) < \ell^o$. Clearly, $f(x)$ is increasing in δ if $x < \underline{\ell}^o$ and is decreasing in δ if $\underline{\ell}^o < x < \ell^o$. □

4. Markov-perfect Nash Equilibrium

In this section, we derive the noncooperative solution and discuss its properties. We restrict attention to Markov-perfect Nash equilibria, such that the players condition their contribution only on the progress of the project. This assumption is standard in the literature [Admati and Perry, 1991; Yidirim, 2006; Kessing, 2007; Georgiadis, 2015; Cvitanic and Georgiadis, 2016]. It allows us to derive the solution explicitly.

A (stationary) Markovian strategy for individual i is a function s_i, associating project states x with agent i's efforts $q_i = s_i(x)$. A vector $S = (s_i)_{i=1}^n$ is called a strategic profile. It is said to be feasible if there exists a unique absolutely continuous state trajectory $x(t)$ satisfying (1), with $q_i(t) = s_i(x(t))$, for all i and t, and if the corresponding agents' objectives (2), for all i, are well defined [Dockner *et al.*, 2000].

For feasible strategic profile $S = (s_i)_{i=1}^n$ and initial state ℓ, let

$$V_i(S, \ell) = be^{-\delta T} - \int_0^T c(q_i(t))e^{-\delta t}\, dt$$

with

$$\dot{x}(t) = -\sum_{i=1}^n q_i(t),\ x(0) = \ell \quad \text{and} \quad x(T) = 0,$$

$$(q_i(t))_{i=1}^n = (s_i(x(t)))_{i=1}^n.$$

A (stationary) Markov-perfect Nash equilibrium is a feasible vector $S^* = (s_i^*)_{i=1}^n$ such that, for all i, s_i and ℓ, $V_i(S^*, \ell) \geq V_i((S^*/s_i), \ell)$, with $(S^*/s_i) = (s_1^*, \ldots, s_{i-1}^*, s_i, s_{i+1}^*, \ldots, s_n^*)$ a feasible strategic profile.

We expound our result in Proposition 2 below. We restrict attention to symmetric Markov-perfect Nash equilibria.[i]

Proposition 2 (Markov-perfect Nash equilibrium). *Let $\ell^* > 0$ be the maximum size such that a noncooperative group only undertakes and completes projects such that $\ell < \ell^*$.[j] Let each individual i's strategy $s_i^* = g(x)$, for all i, be implicitly defined by*

$$\int_0^{g(x)} \left(\sigma(q) + 1 - \frac{1}{n} \right) dq = \frac{\delta}{n} \max(0, \ell^* - x), \quad \textit{for all } x. \tag{9}$$

The strategic profile $S^* = (s_i^*)_{i=1}^n$ yields a symmetric Markov-perfect Nash equilibrium.[k]

Again, we can give a sketch of the proof. For all x, let $J_i(x)$ be equal to the maximized discounted payoff function of player i, assuming that the others adopt the strategy $s_j^* = g(x)$ for all $j \neq i$. Assuming differentiability, it satisfies the HJB equation

$$\delta J_i(x) = \max\{-c(q_i) - J_i'(x)(q_i + (n-1)g(x)); q_i \geq 0\}, \quad \text{for all } x, \tag{10}$$

and the boundary condition

$$J_i(0) = b. \tag{11}$$

<hr>

[i]Under the assumption that $\sigma(q)$ is smooth and has a strictly positive lower bound, condition (9) is well-defined and has a (unique) solution.

[j]For the sake of brevity, the calculus of ℓ^* is postponed in the proof. Its comparative statics is expounded in Property 2.

[k]The definition of ℓ^* can be found in the proof.

For all x, assume that $q_i = g(x)$ is the equilibrium strategy of player i. As it maximizes the right-hand side of (10), we can write after substitution

$$\delta J_i(x) = -c(g(x)) - J_i'(x)ng(x).$$

Suppose that there exists ℓ^* such that $q_i = g(x) > 0$ if and only if $0 \leq x < \ell^*$. Then, being interior, the equilibrium strategy satisfies the first-order condition

$$-c'(g(x)) - J_i'(x) = 0,$$

which allows to rewrite (10) as

$$\delta J_i(x) = nc'(g(x))g(x) - c(g(x)).$$

By differentiation, this yields

$$\delta J_i'(x) = (nc''(g(x))g(x) + (n-1)c'(g(x)))g'(x)$$

and, using $J_i'(x) = -c'(g(x))$ anew,

$$\left(\sigma(g(x)) + 1 - \frac{1}{n}\right)g'(x) = -\frac{\delta}{n}.$$

Integrating this between x and ℓ^*, we find

$$\int_x^{\ell^*} \left(\sigma(g(s)) + 1 + \frac{1}{n}\right)g'(s)\,\mathrm{d}s = -\frac{\delta}{n}(\ell^* - x).$$

Finally, making the substitution $q = g(s)$, $\mathrm{d}q = g'(s)\,\mathrm{d}s$ and $g(\ell^*) = 0$ (by definition of ℓ^*), this can be rewritten as

$$\int_0^{g(x)} \left(\sigma(q) + 1 + \frac{1}{n}\right)\mathrm{d}q = \frac{\delta}{n}(\ell^* - x),$$

which gives condition (9) in Proposition 2. The threshold ℓ^* can then be found using (11), which implies to set $g(0) = q^*$ such that

$$nc'(q^*)q^* - c(q^*) = b.$$

Then, using (9), we directly find

$$\ell^* = \frac{n}{\delta}\int_0^{q^*} \left(\sigma(q) + 1 - \frac{1}{n}\right)\mathrm{d}q.$$

We can use Fig. 3 to illustrate Proposition 2. As $g(x) > 0$ if and only if $x < \ell^*$, the threshold ℓ^* can indeed be interpreted as the maximum size such that a noncooperative group undertakes and completes the project. For shorter projects, the noncooperative individual effort $g(x)$ can be constructed graphically, by finding the point on the abscissa such that the surface area below the curve representing $\sigma(\cdot) + 1 - 1/n$ has a measure equal to $(\delta/n)(\ell^* - x)$. It follows that the equilibrium individual effort increases as the project approaches completion (i.e., $g'(x) < 0$ for all $x < \ell^o$).

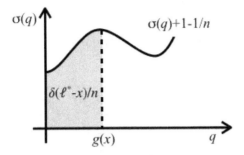

Fig. 3.   Equilibrium individual effort (if $x \leq \ell^*$).

Proof of Proposition 2. The proof is constructive. We first display a value function $v(x)$, satisfying the HJB equation and the boundary condition. We show that $q_i = g(x)$ is the corresponding equilibrium strategy. We then verify that the control $q_i = g(x)$ induces a discounted payoff of player i equal to the value function.

Value function and optimal control:

For all x, consider the function

$$v(x) = \frac{1}{\delta}(nc'(g(x))g(x) - c(g(x))),$$

where $g(x)$ is the strategy defined by condition (9) in Proposition 2.

- *HJB equation:*

For all $x < \ell^*$, $g(x)$ is strictly positive and differentiable, with

$$\left(\sigma(g(x)) + 1 - \frac{1}{n}\right)g'(x) = -\frac{\delta}{n}.$$

Thus, $v(x)$ is differentiable and we can calculate, after rearrangement, that

$$v'(x) = \frac{n}{\delta}\left(\sigma(g(x)) + 1 - \frac{1}{n}\right)g'(x)c'(g(x)).$$

After substitution, this simplifies to

$$v'(x) = -c'(g(x)).$$

It follows that

$$q_i = g(x) \text{ maximizes } \{-c(q_i) - v'(x)(q_i + (n-1)g(x)); q_i \geq 0\}$$

and therefore

$$\delta v(x) = \max\{-c(q_i) - v'(x)(q_i + (n-1)g(x)); q_i \geq 0\}.$$

This proves that $v(x)$ satisfies (10) for all $x < \ell^*$.

For all $x \geq \ell^*$, $g(x) = 0$, implying that $v(x) = v'(x) = 0$. Clearly, it follows that

$$q_i = g(x) \text{ maximizes } \{-c(q_i) - v'(x)(q_i + (n-1)g(x)); q_i \geq 0\}$$

and therefore

$$\delta v(x) = \max\{-c(q_i) - v'(x)(q_i + (n-1)g(x)); q_i \geq 0\}.$$

This proves that $v(x)$ satisfies (10) for all $x \geq \ell^*$.

- *Boundary condition*:

The function $v(x)$ satisfies (11) if

$$v(0) = (1/\delta)(nc'(g(0))g(0) - c(g(0))) = b,$$

which requires that we set $g(0) = q^*$, where q^* is implicitly defined by

$$nc'(q^*)q^* - c(q^*) = \delta b. \tag{12}$$

As $g'(x) \leq 0$ for all x, it is interpreted as the maximum instantaneous effort that player i will noncooperatively contribute.

The upper part of Fig. 4 illustrates the determination of q^*. As $c(0) = 0$, the shaded area has a measure equal to $nc'(q^*)q^* - c(q^*)$. By definition, it must be equal to δb.

Now, by definition of $g(x)$, we have

$$\int_0^{g(0)} \left(\sigma(q) + 1 - \frac{1}{n}\right) dq = \frac{\delta}{n}\ell^*,$$

which can be used to calculate

$$\ell^* = \frac{n}{\delta} \int_0^{q^*} \left(\sigma(q) + 1 - \frac{1}{n}\right) dq. \tag{13}$$

As $g(x) > 0$ if and only if $x < \ell^*$, this threshold is coherently interpreted as the maximum size of the project such that it will eventually go to completion.

The lower part of Fig. 4 describes the determination of ℓ^*. The shaded area has a measure equal to the integral of $\sigma(q) + 1 - 1/n$ from 0 to q^*. By definition of ℓ^*, it must be equal to $\delta \ell^* / n$.

Verification:

Suppose that player i adopts the strategy defined by condition (9) in Proposition 2, knowing that the others do too. In other words, for all t, the players contribute $q_i(t) = g(x(t))$, for all i, and the state trajectory $x(t)$ satisfies $\dot{x}(t) = -ng(x(t))$, with initial condition $x(0) = \ell$.

Consider first the case where $\ell < \ell^*$. As $g(x) > 0$ and $g'(x) < 0$ for all $x < \ell^*$, $\dot{x}(t) \leq -ng(x(0)) < 0$ for all t. Hence, the state trajectory $x(t)$ reaches $x(T) = 0$ in

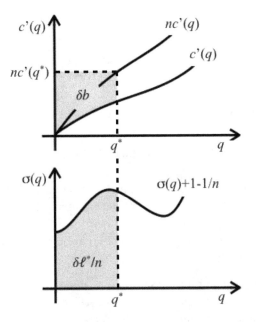

Fig. 4.   Construction of q^* and ℓ^*.

finite time T. Accordingly, player i's discounted payoff is equal to

$$J_i(\ell) = be^{-\delta T} - \int_0^T c(g(x(t)))e^{-\delta t}\,dt.$$

Since the value function $v(x)$ satisfies (10), with $q_i = g(x)$, the corresponding equilibrium strategy, for all t, we have

$$\delta v(x(t)) = -c(g(x(t))) - v'(x(t))ng(x(t)).$$

Using $\dot{x}(t) = -ng(x(t))$, multiplying by $e^{-\delta t}$ and rearranging, we can write

$$(-\delta v(x(t)) + v'(x(t))\dot{x}(t))e^{-\delta t} = c(g(x(t)))e^{-\delta t}.$$

Thus, upon substitution, we get

$$J_i(\ell) = be^{-\delta T} - \int_0^T (-\delta v(x(t)) + v'(x(t))\dot{x}(t))e^{-\delta t}\,dt,$$

which by integration implies that

$$J_i(\ell) = v(x(0)) + (b - v(x(T)))e^{-\delta T}.$$

Finally, using the boundary conditions $x(0) = \ell$, $x(T) = 0$ and $v(0) = b$, we obtain

$$J_i(\ell) = v(\ell),$$

which is what we needed to show.

Consider now the case where $\ell \geq \ell^*$. As $g(x) = 0$ for all $x \geq \ell^*$, $\dot{x}(t) = -ng(x(t)) = 0$ for all t. Hence, as the players never contribute, the state $x(t)$ will remain equal to ℓ for all t. Accordingly, the payoff of player i is equal to (using $c(0) = 0$)

$$J_i(\ell) = \lim_{T \to \infty} \left(be^{-\delta T} - \int_0^T c(0)e^{-\delta t} \, dt \right) = 0,$$

which is what we needed to show. □

Remark. Our results confirm the finding of Kessing [2007] that the individual contributions are strategic complements. Indeed, as the individual contribution is decreasing in the state, by contributing more today, a player decreases the remaining size of the project and therefore fosters a larger contribution from the other players in the future.

Property 2 below expounds our findings regarding the comparative statics of the noncooperative equilibrium.

Property 2 (Comparative statics). (i) The maximum size ℓ^* such that a non-cooperative group completes the project is increasing in b and n, and decreasing in δ. (ii) The equilibrium contribution $g(x)$ is increasing in b for all x. There exists $\underline{\ell}^*$ such that $g(x)$ is increasing in δ if $x < \underline{\ell}^*$ and is decreasing in δ if $\underline{\ell}^* < x < \ell^*$. There exists $\overline{\ell}^*$ such that $g(x)$ is decreasing in n if $x < \overline{\ell}^*$ and is increasing in n if $\overline{\ell}^* < x < \ell^*$.

Again, most comparative statics results are as expected. Noncooperative groups undertake and complete projects with a larger size, when the benefit of the project is larger, when the team gathers more members and when their members are more patient. Before its completion, the noncooperative players contribute more when the benefit of the project is larger. The comparative statics with respect to the time preferences and to the group size are more surprising, confirming the findings of Georgiadis [2015]. The noncooperative individuals contribute more when they are more patient, if and only if the project is sufficiently far from its completion. Intuitively, more patient players have more interest in long-term benefits and thus are more willing contributing in projects far from completion. Finally, we obtain that the players contribute less when the group size becomes larger, if and only if the project is sufficiently close to its completion. Intuitively, the players actually face countervailing incentives, as a larger group increases the incentive to free-ride, but also shortens the delay before the project achievement. Property 1 implies that the first effect dominates when the missing contributions to complete the project is small.

Proof of Property 2. (i) We have calculated in Appendix A the following total derivative

$$d\ell^* = \frac{1}{c'(q^*)} \, db - \left(\ell^* - \frac{b}{c'(q^*)} \right) \frac{d\delta}{\delta} + \left(\ell^* - (n-1)\frac{q^*}{\delta} \right) \frac{dn}{n}.$$

The comparative statics with respect to b follows directly. To show that ℓ^* is decreasing in δ, we first remark that

$$\int_0^{q^*} (nc''(q)q + (n-1)c'(q))\, dq = nc'(q^*)q^* - c(q^*).$$

Hence, condition (12) is equivalent to

$$\int_0^{q^*} (nc''(q)q + (n-1)c'(q))\, dq = \delta b.$$

Using $\sigma(q) = c''(q)q/c'(q)$ and rearranging, we can obtain

$$\frac{b}{c'(q^*)} = \frac{n}{\delta} \int_0^{q^*} \left(\sigma(q) + 1 - \frac{1}{n}\right) \frac{c'(q)}{c'(q^*)}\, dq.$$

Using condition (13), we directly calculate

$$\ell^* - \frac{b}{c'(q^*)} = \frac{n}{\delta} \int_0^{q^*} \left(\sigma(q) + 1 - \frac{1}{n}\right) \left(1 - \frac{c'(q)}{c'(q^*)}\right) dq,$$

which is positive by convexity of $c(q)$ (as $c'(q) < c'(q^*)$ for all $q < q^*$). To show that ℓ^* is increasing in n, it is sufficient to remark that condition (13) implies that

$$\ell^* - (n-1)\frac{q^*}{\delta} = \frac{n}{\delta} \int_0^{q^*} \sigma(q)\, dq,$$

which is positive.

(ii) We have calculated in Appendix A the following total derivative

$$dg(x) = \delta \frac{\frac{1}{c'(q^*)}\, db - \left(x - \frac{b}{c'(q^*)}\right) \frac{d\delta}{\delta} + \left(x - \frac{g(x)}{\delta} - (n-1)\frac{q^*}{\delta}\right) \frac{dn}{n}}{n\sigma(g(x)) + n - 1}.$$

The comparative statics with respect to b is immediate. We have shown above that $\ell^* - b/c'(q^*) > 0$, implying that $0 < \underline{\ell}^* \equiv nb/c'(q^*) < \ell^*$. Clearly, $g(x)$ is increasing in δ if $x < \underline{\ell}^*$ and is decreasing in δ if $\underline{\ell}^* < x < \ell^*$. Let us finally distangle the comparative statics with respect to n. If $x = 0$, then $g(0) = q^*$ and $x - g(x)/\delta - (n-1)q^*/\delta = -nq^*/\delta < 0$. If $x = \ell^*$, then $g(\ell^*) = 0$ and $x - g(x)/\delta - (n-1)q^*/\delta = \ell^* - (n-1)q^*/\delta > 0$ (see part (i) above). Finally, as $g'(x) < 0$, we can show by differentiation that $x - g(x)/\delta - (n-1)q^*/\delta$ is increasing in x. Altogether, this implies that there exists $\overline{\ell}^*$ such that $g(x)$ is decreasing in n if $x < \overline{\ell}^*$ and is increasing in δ if $\overline{\ell}^* < x < \ell^*$. □

5. Closed Forms

Above, the cooperative and noncooperative solutions have been stated in propositions 1 and 2 in such a way that makes the logic of their construction apparent and emphasize their graphical interpretation. This helped deriving and understanding some of their properties. However, in counterpart, this may obscure the fact that a derivation of the solutions in closed form follows directly from our results. In this section, we provide a restatement of Propositions 1 and 2 highlighting this. We then illustrate it by means of a standard specification in the literature (i.e., isoelastic cost).

In order to formulate the alternative statement, we need to introduce some new notations. For all $n \geq 1$, let us define $A_n(q) \equiv nc'(q)q - c(q)$ and $B_n(q) \equiv \int_0^q (\sigma(s) + 1 - 1/n)ds$. Both functions, being monotonic, admit an inverse function.[1] Proposition 3 gives the cooperative and noncooperative solutions in closed form.

Proposition 3 (Closed form). *A closed form statement of the cooperative and noncooperative solutions is respectively*

$$\ell^o = \frac{n}{\delta} B_1(A_1^{-1}(\delta b)) \quad and \quad f(x) = B_1^{-1}\left(\frac{\delta}{n}\max(0, \ell^o - x)\right)$$

and

$$\ell^* = \frac{n}{\delta} B_n(A_n^{-1}(\delta b)) \quad and \quad g(x) = B_n^{-1}\left(\frac{\delta}{n}\max(0, \ell^* - x)\right).$$

Proof. The proof is immediate, as it simply consists of rewriting our previous results using $A_n(q)$ and $B_n(q)$. For example, in the case of the cooperative solution, conditions (7) and (8) respectively write $A_1(q^o) = \delta b$ and $\ell^o = (n/\delta)B_1(q^o)$. Hence, we have $q^o = A_1^{-1}(\delta b)$ and $\ell^o = (n/\delta)B_1(A_1^{-1}(\delta b))$. Condition (4) writes $B_1(f(x)) = (\delta/n)\max(0, \ell^o - x)$. Hence, we have $f(x) = B_1^{-1}((\delta/n)\max(0, \ell^o - x))$. The proof of the noncooperative solution is not different. □

We now propose a standard specification (i.e., isoelastic cost) and write the corresponding cooperative and noncooperative solutions in closed form.

Example (Isoelastic[m]). Consider the class of cost functions $c(q) = \lambda q^{1+\mu}/(1+\mu)$, with $\lambda > 0$ and $\mu > 0$.[n] It is immediate to show that

$$A_n(q) = ((1+\mu)n - 1)\lambda\frac{q^{1+\mu}}{1+\mu}$$

and

$$B_n(q) = \mu q,$$

for all q. From this, we can write that the cooperative solution as

$$\ell^o = n\left(\frac{\mu}{\delta}\right)^{\frac{\mu}{1+\mu}}\left((1+\mu)\frac{b}{\lambda}\right)^{\frac{1}{1+\mu}} \quad and \quad f(x) = \frac{\delta}{\mu n}\max(0, \ell^o - x),$$

---

[1]We use the standard notations for the inverse functions, namely $A_n^{-1}(.)$ and $B_n^{-1}(.)$.
[m]This class of cost functions is called is elastic because, with this specification, $\sigma(q) = \mu$ for all q.
[n]Remark that the specific parametrization $\lambda = \mu = 1$ basically corresponds to the quadratic cost function case.

and the noncooperative equilibrium as

$$\ell^* = \left(\frac{(1+\mu)n - 1}{\delta} \right)^{\frac{\mu}{1+\mu}} \left((1+\mu)\frac{b}{\lambda} \right)^{\frac{1}{1+\mu}} \quad \text{and}$$

$$g(x) = \frac{\delta}{(1+\mu)n - 1} \max(0, \ell^* - x).$$

6. Comparison

We investigate here the difference between the cooperative solution and the Markov-perfect Nash equilibrium which is determined previously. We illustrate our result using the isoelastic specification.

We first need to characterize the set of socially worthwhile projects. By definition, a project, as defined by the individual reward b and the size ℓ, is said to be socially worthwhile if a cooperative team, with time preference δ and size n, would bring it to completion. Property 3 below characterizes the set of socially efficient projects.

Property 3 (Socially worthwhile projects). There exists a function $\phi(b, \delta)$, increasing in b, decreasing in δ and invariant with n, such that a group of n members finds as socially worthwhile a public project characterized by individual reward b and size ℓ, if and only if $\ell/n < \phi(b, \delta)$.°

Intuitively, Property 3 defines the largest *cumulative effort* that an agent participating in a cooperative team would be willing to contribute through time, in order to concretize a given project. Property 3 implies that it is increasing in the benefit of the project, decreasing in the agent's impatience, and does not depend on the size of the group.

Proof of Property 3. According to proposition 1, $f(x) > 0$ if and only if $x < \ell^o$. Hence, a cooperative group with time preference δ and size n initiates and completes a project characterized by parameters b and ℓ, if and only if $\ell < \ell^o$. In Property 3, we let $\phi(b, \delta) \equiv \ell^o/n$. We show here that $\phi(b, \delta)$ is increasing in b, decreasing in δ, and does not vary with n. By construction, a socially worthwhile public project satisfies $\ell/n < \phi(b, \delta)$.

We have shown in Appendix A that

$$\mathrm{d}\ell^o = \frac{n}{c'(q^o)}\, \mathrm{d}b - \left(\ell^o - \frac{nb}{c'(q^o)} \right) \frac{\mathrm{d}\delta}{\delta} + \ell^o \frac{\mathrm{d}n}{n}.$$

By total differentiation, we show that

$$\mathrm{d}\left(\frac{\ell^o}{n} \right) = \frac{1}{n}\left(\mathrm{d}\ell^o - \ell^o \frac{\mathrm{d}n}{n} \right).$$

° In Proposition 1, $\ell^o/n = \phi(b, \delta)$, for given b and δ.

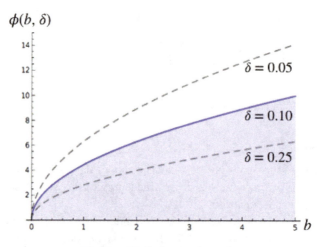

Fig. 5. $\phi(b, \delta)$ (isoelastic cost).

After substitution and simplification, we thus calculate that

$$
d\left(\frac{\ell^o}{n}\right) = \frac{1}{c'(q^o)}\,db - \frac{1}{n}\left(\ell^o - \frac{nb}{c'(q^o)}\right)\frac{d\delta}{\delta}.
$$

Knowing from the proof of Property 1 that $\ell^o - nb/c'(q^o) > 0$, this completes the proof of Property 3.   □

Example (Continued). For the class of isoelastic cost functions, it can be shown that

$$
\phi(b, \delta) = \left(\frac{\mu}{\delta}\right)^{\frac{\mu}{1+\mu}}\left((1+\mu)\frac{b}{\lambda}\right)^{\frac{1}{1+\mu}}.
$$

Assuming that $\lambda = \mu = 1$, it is depicted in Fig. 5, with the benefit of the project varying from 1 to 5. From Property 3, a project (characterized by b and ℓ) is socially worthwhile if, and only if, the effort required per individual for its completion (ℓ/n) lies below the appropriate frontier (shaded area in case where $\delta = 0.1$).[P]

Property 4 below compares the cooperative and the noncooperative paths of individual effort to complete the collective project.

Property 4 (Free-riding and procrastination). Noncooperative groups (a) give up large projects that cooperative groups with same characteristics would find worth completing and (b) take too much time completing the others. Formally, we respectively show that $\ell^* < \ell^o$ and $g(x) < f(x)$ for all $x < \ell^*$.

Property 4 extends to our setting the well-known social dilemma that self-interested individuals fail to manage collective goods optimally. On the one hand,

---

[P] Figure 5 shows three curves, with either $\delta = 0.05$, 0.1 or 0.25.

it confirms that noncooperative groups sometimes do not engage in socially worthwhile projects (free-riding). On the other hand, it explains that even when they do,
they will contribute too little effort in it and delay its completion, thus pointing
out another source of inefficiency (procrastination).

Proof of Property 4. To obtain part (a), we show that ℓ^*/ℓ^o is equal to one
when $n = 1$ and is strictly decreasing in n. Using Eqs. (8) and (13)), we verify that
ℓ^* and ℓ^o are equal when $n = 1$ and we can calculate the ratio

$$\frac{\ell^*}{\ell^o} = \frac{\int_0^{q^*}(\sigma(s) + 1 - 1/n)\,ds}{\int_0^{q^o}\sigma(s)\,ds}.$$

From Appendix A, we know that

$$\frac{dq^o}{dn} = 0 \quad \text{and} \quad \frac{dq^*}{dn} = -\frac{q^*/n}{\sigma(q^*) + 1 - 1/n}.$$

Hence, by differentiation, we can show that

$$\frac{d(\ell^*/\ell^o)}{dn} = \frac{(\sigma(q^*) + 1 - 1/n)\frac{dq^*}{dn} + q^*/n^2}{\int_0^{q^o}\sigma(s)\,ds},$$

which yields after substitution

$$\frac{d(\ell^*/\ell^o)}{dn} = -\frac{(n-1)q^*}{n^2\int_0^{q^o}\sigma(s)\,ds} < 0.$$

Part (b) follows from Propositions 1 and 2, given that $\ell^* < \ell^o$. Consider any
$x < \ell^*$ and assume, by way of contradiction, that $g(x) \geq f(x)$. Then, for all $n > 1$,
it is clear that

$$\int_0^{g(x)}(\sigma(s) + 1 - 1/n)\,ds > \int_0^{f(x)}\sigma(s)\,ds.$$

By definition of $f(x)$, it follows that

$$\int_0^{g(x)}(\sigma(s) + 1 - 1/n)\,ds > (\delta/n)(\ell^o - x).$$

Finally, as $\ell^* < \ell^o$, we obtain

$$\int_0^{g(x)}(\sigma(s) + 1 - 1/n)\,ds > (\delta/n)(\ell^* - x),$$

which contradicts the definition of $g(x)$. □

Example (Continued). For the class of isoelastic cost functions, it can be shown
that

$$\frac{\ell^*}{\ell^o} = \frac{1}{n}\left(\frac{(1+\mu)n - 1}{\mu}\right)^{\frac{\mu}{1+\mu}}.$$

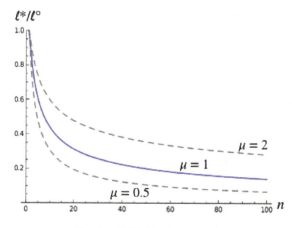

Fig. 6. ℓ^*/ℓ° (isoelastic cost).

This ratio can be used as a measure of the free-riding problem. From Property 4, we know that it is always strictly less than one. Figure 6 depicts it, with the size of the group varying from 1 to 100. We see that the maximum size of a project that a noncooperative group is able to complete, becomes quickly a small fraction of the maximum size of a socially optimal project. However, a larger elasticity of marginal cost is able slow down the phenomenon.[q]

7. Conclusion

This chapter has analyzed the problem of free-riding within groups, in case where the agents make contributions over time to complete a public good. Considering a general technology for producing the public good, we still achieve to give an explicit characterization of the noncooperative path of contributions. We show that noncooperative groups fail to carry out some socially optimal projects and procrastinate on the projects which they complete.

Acknowledgment

I am thankful for the comments given to improve this work.

Appendix A

A.1. *Comparative statics: Cooperative solution*

The cooperative solution is fully characterized by

$$c'(q^\circ)q^\circ - c(q^\circ) = \delta b,$$

$$\ell^\circ = \frac{n}{\delta} \int_0^{q^\circ} \sigma(s)\,\mathrm{d}s$$

---

[q]Figure 6 shows three curves, with either $\mu = 0.5$, 1 or 2.

and

$$\int_0^{f(x)} \sigma(s)ds = \frac{\delta}{n}(\ell^o - x).$$

By total differentiation, we obtain (using $\sigma(q) = c''(q)q/c'(q)$)

$$\sigma(q^o)\,dq^o = \frac{\delta}{c'(q^o)}\,db + \frac{b}{c'(q^o)}\,d\delta,$$

$$d\ell^o = -\ell^o\frac{d\delta}{\delta} + \ell^o\frac{dn}{n} + \frac{n}{\delta}\sigma(q^o)\,dq^o$$

and

$$\sigma(f(x))\,d(f(x)) = \frac{\delta}{n}(\ell^o - x)\frac{d\delta}{\delta} - \frac{\delta}{n}(\ell^o - x)\frac{dn}{n} + \frac{\delta}{n}\,d\ell^o.$$

Upon substitution and after rearrangement, we finally get

$$dq^o = \frac{\frac{\delta}{c'(q^o)}\,db + \frac{b}{c'(q^o)}\,d\delta}{\sigma(q^o)},$$

$$d\ell^o = \frac{n}{c'(q^o)}\,db - \left(\ell^o - n\frac{b}{c'(q^o)}\right)\frac{d\delta}{\delta} + \ell^o\frac{dn}{n}$$

and

$$d(f(x)) = \delta\frac{\frac{n}{c'(q^o)}\,db - \left(x - n\frac{b}{c'(q^o)}\right)\frac{d\delta}{\delta} + x\frac{dn}{n}}{n\sigma(f(x))}.$$

A.2. Comparative statics: Noncooperative solution

The noncooperative solution is fully characterized by

$$nc'(q^*)q^* - c(q^*) = \delta b,$$

$$\ell^* = \frac{n}{\delta}\int_0^{q^*}\left(\sigma(s) + 1 - \frac{1}{n}\right)ds$$

and

$$\int_0^{g(x)}\left(\sigma(s) + 1 - \frac{1}{n}\right)ds = \frac{\delta}{n}(\ell^* - x).$$

By total differentiation, we obtain (using $\sigma(q) = c''(q)q/c'(q)$)

$$(n\sigma(q^*) + n - 1)\,dq^* = \frac{\delta}{c'(q^*)}\,db + \frac{b}{c'(q^*)}\,d\delta - q^*\,dn,$$

$$d\ell^* = -\ell^*\frac{d\delta}{\delta} + \left(\ell^* + \frac{q^*}{\delta}\right)\frac{dn}{n} + \frac{1}{\delta}(n\sigma(q^*) + n - 1)\,dq^*$$

and

$$\left(\sigma(g(x)) + 1 - \frac{1}{n}\right)dg(x) = \frac{\delta}{n}(\ell^* - x)\frac{d\delta}{\delta} - \frac{1}{n}(g(x) + \delta(\ell^* - x))\frac{dn}{n} + \frac{\delta}{n}\,d\ell^*.$$

Upon substitution and after rearrangement, we finally get

$$dq^* = \frac{\frac{\delta}{c'(q^*)}\,db + \frac{b}{c'(q^*)}\,d\delta - q^*\,dn}{n\sigma(q^*) + n - 1},$$

$$d\ell^* = \frac{1}{c'(q^*)}\,db - \left(\ell^* - \frac{b}{c'(q^*)}\right)\frac{d\delta}{\delta} + \left(\ell^* - (n-1)\frac{q^*}{\delta}\right)\frac{dn}{n}$$

and

$$dg(x) = \delta\frac{\frac{1}{c'(q^*)}\,db - \left(x - \frac{b}{c'(q^*)}\right)\frac{d\delta}{\delta} + \left(x - \frac{g(x)}{\delta} - (n-1)\frac{q^*}{\delta}\right)\frac{dn}{n}}{n\sigma(g(x)) + n - 1}.$$

References

Admati, A. R. and Perry, M. [1991] Joint projects without commitment, *Rev. Econ. Stud.* **58**, 259–276.

Bagnoli, M. and Lipman, B. [1989] Provision of public goods: Fully implementing the core through private contributions, *Rev. Econ. Stud.* **56**(4), 583–601.

Bergström, T., Blum, L. and Varian, H. [1986] On the private provision of public goods, *J. Public Econ.* **29**, 25–49.

Bowen, T. R., Geogiardis, G. and Lambert, N. S. [2016] Collective choice in dynamic public good provision, NBER Working Paper No. 22772.

Cornes, R. and Sandler, T. [1984] Easy riders, joint production, and public goods, *Econ. J.* **94**(375), 580–598.

Cornes, R. and Sandler, T. [1985] The simple analytics of pure public good provision, *Economica* **52**(205), 103–116.

Cvitanic, J. and Georgiadis, G. [2016] Achieving efficiency in dynamic contribution games, *Amer. Econ. J. Microecon.* **8**(4), 309–342.

Dockner, E. *et al.*, [2000] *Differential Games in Economics and Management Science* (Cambridge University Press, New York, 2000).

Fershtman, C. and Nitzan, S. [1991] Dynamic voluntary provision of public goods, *Europ. Econ. Rev.* **35**, 1057–1067.

Georgiadis, G. [2015] Projects and team dynamics, *Rev. Econ. Stud.* **82**(1), 187–218.

Kessing, S. G. [2007] Strategic complementarity in the dynamic private provision of a discrete public good, *J. Public Econ. Theory* **9**(4), 699–710.

Marx, L. M. and Matthews, S. A. [2000] Dynamic voluntary contribution to a public project, *Rev. Econ. Stud.* **67**, 327–358.

Nitzan, S. and Romano, R. E. [1990] Private provision of a discrete public good with uncertain cost, *J. Public Econ.* **42**, 357–370.

Olson, M., *The Logic of Collective Action: Public Goods and the Theory of Groups* (Revised ed., 1971) (Harvard University Press, London, England).

Palfrey, T. R. and Rosenthal, H. [1984] Participation and the provision of discrete public goods: A strategic analysis, *J. Public Econ.* **24**, 171–193.

Yidirim, H. [2006] Getting the ball rolling: Voluntary contributions to a large-scale public project, *J. Public Econ. Theory* **8**(4), 503–528.

Chapter 5

Infinite Horizon Dynamic Games: A New Approach via Information Updating

David Yeung

Center of Game Theory, St. Petersburg State University
St. Petersburg 198504, Russia

SRS Consortium for Advanced Study in Cooperative Dynamic Games
Hong Kong Shue Yan University, Hong Kong

dwkyeung@hksyu.edu

Ovanes Petrosian

Faculty of Applied Mathematics and Control Process
St. Petersburg State University, St. Petersburg 198504, Russia

St. Petersburg School of Economics and Management
National Research University Higher School of Economics
Moscow 101000, Russia

petrosian.ovanes@yandex.ru

This chapter formulates a new approach to analyze infinite horizon dynamic games with uncertainties and unknowns in the players' future payoff structures. In many game situations, the game horizon would last for an indefinitely long period and one has to consider them as infinite horizon games. Existing infinite horizon dynamic games often rely on the assumption of time-invariant game structures for the derivation of equilibrium solutions. However, many events in the considerably far future are intrinsically unknown. In this chapter, information about the players' future payoffs will be revealed as the game proceeds. Making use of the newly obtained information, the players revise their strategies accordingly, and the process will continue indefinitely. This new approach for the analysis of infinite horizon dynamic games via information updating provides a more realistic and practical alternative to the study of infinite horizon dynamic games.

Keywords: Dynamic games; infinite game horizon; information updating; uncertain payoff structures.

1. Introduction

In many real-life game situations, the operation horizons are indefinite and may last for an unknown length of time. Often, we have to consider these situations as infinite horizon problems. In addition, technology development, new knowledge discovery, economic changes, scientific advancement, changes in taste and preferences often lead to uncertain future outcomes. For many events in the considerably far future, they are even intrinsically unknown. These facts make decision-making in an infinite horizon framework extremely difficult. Existing infinite horizon dynamic game studies tend to adopt time-invariant payoff structures (see Levhari and Mirman [1980], Rubioa and Ulphb [2007], Chen and Zadroznyb [2002], Gyurkovics and Takacs [2005], and Dockner and Mosburger [2007]). Similarly, existing studies of infinite horizon differential games also often assume time-invariant payoff structures (see Jorgensen and Yeung [1996], Benchekroun [2003, 2008], Fujiwara [2009a,b], Jun and Vives [2004], Reynolds [1987], Wirl [1994, 1996] and Hinloopen *et al.* [2017]). Yeung [2001] presented infinite horizon stochastic differential games with repeated intervals of identical randomly branching game structures. Although time-invariant payoff structures allow the derivation of solution in a convenient way, the assumption that there will be no change in future events may not be realistic. An essential characteristic of time — and hence decision-making over time — is that, though the individual may, through the expenditure of resources, gather past and present information, many of the far future events are inherently uncertain or intrinsically unknown. An empirically meaningful theory of infinite horizon dynamic games must therefore incorporate uncertainties and unknowns in an appropriate manner. Petrosian [2016], Gromova and Petrosian [2016], Petrosian and Barabanov [2017], Petrosian *et al.* [2017] and Yeung and Petrosyan [2016] developed new classes of finite horizon differential games in which the state dynamics and the players' payoff structures in the future are unknown. Information about the future will be revealed at some pre-determined points of time. Players would revise their strategies paths when new information is revealed at these points of time.

This chapter develops a new approach for analyzing infinite horizon dynamic games with uncertainties and unknowns in the players' future payoff structures. Information about the future will be revealed stage by stage. Adapting to the new information the players revise their strategies accordingly. The information, structure over time can be categorized as (i) events in the current stage which are known with certainty, (ii) events in future stages which are known with certainty, (iii) events in future stages with their probability distributions being known, and (iv) future events that are intrinsically unknown and only a best estimate is available. This new approach for the analysis of infinite horizon dynamic games via information updating provides a more realistic and practical alternative to the paradigm of infinite horizon dynamic games with time-invariant game structures. The chapter is organized as follows. Section 2 presents the game formulation of an infinite horizon dynamic game with the above information structure. The feedback Nash equilibrium solution in the game starting at the current stage is provided. The process of information updating is presented in Sec. 3. An illustration is given in Sec. 4 and concluding remarks are provided in Sec. 5.

2. Game Formulation and Equilibrium

We consider n-person infinite horizon dynamic games with the following information structures. In the current (initial) stage t, the players' payoff structures are known with certainty in the current stage and the next T_t^1 future stages. Only the probability distributions of the players' payoff structures are known in the following T_t^2 stages. Finally, knowledge of events in stage $t + T_t^1 + T_t^2 + 1$ and subsequent stages is intrinsically unknown in stage t. The players have only the best estimate of the players' expected values of the states at stage $t + T_t^1 + T_t^2 + 1$. Note that $T_t^1 \in \{0, \bar{T}_t^1\}$ and $T_t^2 \in \{0, \bar{T}_t^2\}$, and $\bar{T}_t^1$ and $\bar{T}_t^2$ can be very large. If $\bar{T}_t^1$ approaches infinity, the game becomes an infinite horizon dynamic game. If $\bar{T}_t^1 = 0$ and $\bar{T}_t^2$ approach infinity, the game becomes an infinite horizon version of the randomly furcating dynamic game developed by Yeung and Petrosyan [2013].

In stage $t+1$, the information not known by the players in stage t will available. In particular, the players' payoff structures are known with certainty in the current stage $t+1$ and the next T_{t+1}^1 future stages. Again, only the probability distributions of the players' payoff structures are known in the following T_{t+1}^2 stages. The players have only the best estimates of their expected values of the states of the game at stage $t + 1 + T_{t+1}^1 + T_{t+1}^2 + 1$. The information updating process will continue for stages $t+2, t+3, t+4, \ldots$, and so on. The above setting provides a characterization of the information structure of the game which includes a period with events known with certainty, a period when only the (objective) probability distributions of future events is known, and the best estimates of their expected values of the states, which are unknown, afterwards.

At stage t, player i seeks to maximize his expected payoff:

$$\sum_{\varsigma=t}^{t+T_t^1} g_\varsigma^{(t)i}\left(x_\varsigma, u_\varsigma^1, u_\varsigma^2, \ldots, u_\varsigma^n\right)\delta_t^\varsigma$$

$$+ \sum_{\varsigma=t+T_t^1+1}^{t+T_t^1+T_t^2} \sum_{\varepsilon_\varsigma^k=1}^{h_\varsigma} \gamma_\varsigma^{\varepsilon_\varsigma^k} g_\varsigma^{(t)\left(\varepsilon_\varsigma^k\right)i}\left(x_\varsigma, u_\varsigma^1, u_\varsigma^2, \ldots, u_\varsigma^n\right)\delta_t^\varsigma$$

$$+ \delta_t^{t+T_t^1+T_t^2+1} S_{t+T_t^1+T_t^2+1}^{(t)i}(x_{t+T_t^1+T_t^2+1}), \quad i \in \{1, 2, \ldots, n\} \equiv N, \quad (2.1)$$

subject to the state dynamics

$$x_{\varsigma+1} = f_\varsigma\left(x_\varsigma, u_\varsigma^1, u_\varsigma^2, \ldots, u_\varsigma^n\right), \quad x_t = x_t^0, \quad (2.2)$$

where $x_\varsigma \in X \subset R^m$ is the state variable at stage ς, $u_\varsigma^i \in U^i \subset R^{n_i}$ is the control of player i, $g_\varsigma^{(t)i}\left(x_\varsigma, u_\varsigma^{1,}, u_\varsigma^2, \ldots, u_\varsigma^n\right)$ is the payoff of player i in stage $\varsigma \in \{t, t+1, \ldots, t+T_t^1\}$, δ_t^ς is the discount factor in stage ς viewed from the initial stage t, ε_ς is a random variable with range $\{\varepsilon_\varsigma^1, \varepsilon_\varsigma^2, \ldots, \varepsilon_\varsigma^{h_\varsigma}\} = \{1, 2, \ldots, h_\varsigma\}$ and corresponding probabilities $\{\gamma_\varsigma^1, \gamma_\varsigma^2, \ldots, \gamma_\varsigma^{h_\varsigma}\}$ for $\varsigma \in \{t + T_t^1 + 1, t + T_t^1 + 2, \ldots, t + T_t^1 + T_t^2\}$, $g_\varsigma^{(t)\left(\varepsilon_\varsigma^k\right)i}\left(x_\varsigma, u_\varsigma^1, u_\varsigma^2, \ldots, u_\varsigma^n\right)$ is the payoff of player i if ε_ς^k occurs in stage

$\varsigma \in \{t + T_t^1 + 1, t + T_t^1 + 2, \ldots, t + T_t^1 + T_t^2\}$, $S_{t+T_t^1+T_t^2+1}^{(t)i}(x_{t+T_t^1+T_t^2+1})$ is the best estimate of player i's expected value of the state $x_{t+T_t^1+T_t^2+1}$ at stage $t+T_t^1+T_t^2+1$.

The functions $g_\varsigma^{(t)i}(x_\varsigma, u_\varsigma^1, u_\varsigma^2, \ldots, u_\varsigma^n)$, $g_\varsigma^{(t)(\varepsilon_\varsigma^k)i}(x_\varsigma, u_\varsigma^1, u_\varsigma^2, \ldots, u_\varsigma^n)$, $f_\varsigma(x_\varsigma, u_\varsigma^1,$ $u_\varsigma^2, \ldots, u_\varsigma^n)$, and $S_{t+T_t^1+T_t^2+1}^{(t)i}(x_{t+T_t^1+T_t^2+1})$ are continuously differentiable functions. In addition, given that events after stage $t + T_t^1 + T_t^2$ are unknown, $S_{t+T_t^1+T_t^2+1}^{(t)i}(x_{t+T_t^1+T_t^2+1})$ is the best estimate of player i's expected value at initial stage t for the stage $t + T_t^1 + T_t^2 + 1$.

A theorem characterizing a feedback Nash equilibrium solution of the game (2.1) and (2.2) is provided below.

Theorem 2.1. *A set of strategies* $\{u_\varsigma^{i*} = \phi_\varsigma^{(t)i}(x),$ *for* $i \in N$ *and* $\varsigma \in \{t, t+1, \ldots, t+$ $T_t^1\},$ *and* $u_\varsigma^{(\varepsilon_\varsigma^k)i*} = \phi_\varsigma^{(t)(\varepsilon_\varsigma^k)i}(x),$ *for* $i \in N,$ $\varepsilon_\varsigma^k \in \{1, 2, \ldots, h_\varsigma\}$ *and* $\varsigma \in \{t + T_t^1 + 1, t+T_t^1+2, \ldots, t+T_t^1+T_t^2\}\}$ *provides a feedback Nash equilibrium solution to the game (2.1) and (2.2) if there exist continuously differentiable functions* $V^{(t)i}(\tau, x),$ *for* $i \in N,$ $\tau \in \{t, t+1, \ldots, t+T_t^1\},$ *and* $V^{(t)(\varepsilon_\tau^k)i}(\tau, x),$ *for* $i \in N,$ $\varepsilon_\tau^k \in \{1, 2, \ldots, h_\tau\}$ *and* $\tau \in \{t + T_t^1 + 1, t + T_t^1 + 2, \ldots, t + T_t^1 + T_t^2\};$ *such that the following recursive relations are satisfied:*

$$V^{(t)i}(\tau, x) = \max_{u_\tau^i}\Big\{g_\tau^{(t)i}\big[x, \phi_\tau^{(t)1}(x), \phi_\tau^{(t)2}(x), \ldots, \phi_\tau^{(t)i-1}(x), u_\tau^i,$$

$$\phi_\tau^{(t)i+1}(x), \ldots, \phi_\tau^{(t)n}(x)\big]\delta_t^\tau + V^{(t)i}\big[\tau + 1, f_\tau\big[x, \phi_\tau^{(t)1}(x),$$

$$\phi_\tau^{(t)2}(x), \ldots, \phi_\tau^{(t)i-1}(x), u_\tau^i, \phi_\tau^{(t)i+1}(x), \ldots, \phi_\tau^{(t)n}(x)\big]\Big\}, \qquad (2.3)$$

for $\tau \in \{t, t+1, \ldots, t + T_t^1 - 1\};$

$$V^{(t)i}(t + T_t^1 x)$$

$$= \max_{u_{t+T_t^1}^i}\Bigg\{g_{t+T_t^1}^{(t)i}\big[x, \phi_{t+T_t^1}^{(t)1}(x), \phi_{t+T_t^1}^{(t)2}(x), \ldots$$

$$\ldots, \phi_{t+T_t^1}^{(t)i-1}(x), u_{t+T_t^1}^i, \phi_{t+T_t^1}^{(t)i+1}(x), \ldots, \phi_{t+T_t^1}^{(t)n}(x)\big]\delta_{t+T_t^1}^\tau$$

$$+ \sum_{\varepsilon_{t+T_t^1+1}^k=1}^{h_{t+T_t^1+1}} \gamma_{t+T_t^1+1}^{\varepsilon_{t+T_t^1+1}^k} V^{(t)(\varepsilon_{t+T_t^1+1}^k)i}\big[t + T_t^1 + 1, f_{t+T_t^1}\big[x, \phi_{t+T_t^1}^{(t)1}(x), \phi_{t+T_t^1}^{(t)2}(x), \ldots$$

$$\ldots, \phi_{t+T_t^1}^{(t)i-1}(x), u_{t+T_t^1}^i, \phi_{t+T_t^1}^{(t)i+1}(x), \ldots, \phi_{t+T_t^1}^{(t)n}(x)\big]\big]\Bigg\}; \qquad (2.4)$$

$$V^{(t)(\varepsilon_\tau^k)i}(\tau, x)$$

$$= \max_{u_\tau^i} \left\{ g_\tau^{(t)\left(\varepsilon_\tau^k\right)i} \left[x, \phi_\tau^{(t)\left(\varepsilon_\tau^k\right)1}(x), \phi_\tau^{(t)\left(\varepsilon_\tau^k\right)2}(x), \ldots \right. \right.$$

$$\left. \ldots, \phi_\tau^{(t)\left(\varepsilon_\tau^k\right)i-1}(x), u_\tau^i, \phi_\tau^{(t)\left(\varepsilon_\tau^k\right)i+1}(x), \ldots, \phi_\tau^{(t)\left(\varepsilon_\tau^k\right)n}(x) \right] \delta_t^\tau$$

$$+ \sum_{\varepsilon_{\tau+1}^k=1}^{h_{\tau+1}} \gamma_{\tau+1}^{\varepsilon_{\tau+1}^k} V^{(t)\left(\varepsilon_{\tau+1}^k\right)i} \left[\tau+1, f_\tau \left[x, \phi_\tau^{(t)\left(\varepsilon_\tau^k\right)1}(x), \phi_\tau^{(t)\left(\varepsilon_\tau^k\right)2}(x), \ldots \right. \right.$$

$$\left. \left. \left. \ldots, \phi_\tau^{(t)\left(\varepsilon_\tau^k\right)i-1}(x), u_\tau^i, \phi_\tau^{(t)\left(\varepsilon_\tau^k\right)i+1}(x), \ldots, \phi_\tau^{(t)(\varepsilon_\tau^k)n}(x) \right] \right] \right\}, \tag{2.5}$$

for $\varepsilon_\tau^k \in \{1, 2, \ldots, h_\tau\}$ and $\tau \in \{t + T_t^1 + 1, t + T_t^1 + 2, \ldots, t + T_t^1 + T_t^2 - 1\}$;

$$V^{(t)\left(\varepsilon_{t+T_t^1+T_t^2}^k\right)i}(t + T_t^1 + T_t^2, x)$$

$$= \max_{u_{t+T_t^1+T_t^2}^i} \left\{ g_{t+T_t^1+T_t^2}^{(t)\left(\varepsilon_{t+T_t^1+T_t^2}^k\right)i} \left[x, \phi_{t+T_t^1+T_t^2}^{(t)\left(\varepsilon_{t+T_t^1+T_t^2}^k\right)1}(x), \phi_{t+T_t^1+T_t^2}^{(t)\left(\varepsilon_{t+T_t^1+T_t^2}^k\right)2}(x), \ldots \right. \right.$$

$$\left. \ldots, \phi_{t+T_t^1+T_t^2}^{(t)\left(\varepsilon_{t+T_t^1+T_t^2}^k\right)i-1}(x), u_{t+T_t^1+T_t^2}^i, \phi_{t+T_t^1+T_t^2}^{(t)\left(\varepsilon_{t+T_t^1+T_t^2}^k\right)i+1}(x), \ldots \right.$$

$$\left. \ldots, \phi_{t+T_t^1+T_t^2}^{(t)\left(\varepsilon_{t+T_t^1+T_t^2}^k\right)n}(x) \right] \delta_t^{t+T_t^1+T_t^2} + \delta_t^{t+T_t^1+T_t^2+1}$$

$$\times S_{t+T_t^1+T_t^2+1}^{(t)i} \left(f_{t+T_t^1+T_t^2} \left[x, \phi_{t+T_t^1+T_t^2}^{(t)\left(\varepsilon_{t+T_t^1+T_t^2}^k\right)1}(x), \phi_{t+T_t^1+T_t^2}^{(t)\left(\varepsilon_{t+T_t^1+T_t^2}^k\right)2}(x), \ldots \right. \right.$$

$$\left. \ldots, \phi_{t+T_t^1+T_t^2}^{(t)\left(\varepsilon_{t+T_t^1+T_t^2}^k\right)i-1}(x), u_{t+T_t^1+T_t^2}^i, \phi_{t+T_t^1+T_t^2}^{(t)\left(\varepsilon_{t+T_t^1+T_t^2}^k\right)i+1}(x), \ldots \right.$$

$$\left. \left. \left. \ldots, \phi_{t+T_t^1+T_t^2}^{(t)\left(\varepsilon_{t+T_t^1+T_t^2}^k\right)n}(x) \right] \right) \right\}, \tag{2.6}$$

for $\varepsilon_{t+T_t^1+T_t^2}^k \in \{1, 2, \ldots, h_{t+T_t^1+T_t^2}\}$.

Proof. See Appendix A. □

For notational simplicity, we use $\phi_\tau^{(t)}(x) = (\phi_\tau^{(t)1}(x), \phi_\tau^{(t)2}(x), \ldots, \phi_\tau^{(t)n}(x))$ to denote the vector of game equilibrium strategies at stage τ.

The game equilibrium dynamics from stage t to stage $t+1$ can be obtained as

$$x_{t+1} = f_t(x_t, \phi_t^{(t)1}(x_t), \phi_t^{(t)2}(x_t), \ldots, \phi_t^{(t)n}(x_t)), \quad x_t = x_t^0. \tag{2.7}$$

3. Information Updating

In this section, we consider the updating of future information and the evolution of the game. In stage $t + 1$, the information not known in stage t becomes available. In particular, the players' payoff structures are known with certainty in the current stage $t+1$ and the next T_{t+1}^1 future stages, the probability distributions of players' payoff structures are known for the following T_{t+1}^2 stages; and the best estimates of the player's expected value of the state x at stage $(t + 1) + T_{t+1}^1 + T_{t+1}^2 + 1$ are available. In the game starting at stage $t + 1$, player $i \in N$ seeks to maximize his expected payoff:

$$\sum_{\varsigma=(t+1)}^{(t+1)+T_{t+1}^1} g_\varsigma^{(t+1)i}\left(x_\varsigma, u_\varsigma^{1}, u_\varsigma^{2}, \ldots, u_\varsigma^{n}\right)\delta_{t+1}^\varsigma$$

$$+ \sum_{\varsigma=(t+1)+T_{t+1}^1+1}^{(t+1)+T_{t+1}^1+T_{t+1}^2} \sum_{\varepsilon_\varsigma^k=1}^{h_\varsigma} \gamma_\varsigma^{\varepsilon_\varsigma^k} g_\varsigma^{(t+1)\left(\varepsilon_\varsigma^k\right)i}\left(x_\varsigma, u_\varsigma^{1}, u_\varsigma^{2}, \ldots, u_\varsigma^{n}\right)\delta_{t+1}^\varsigma$$

$$+ \delta_{(t+1)}^{(t+1)+T_{t+1}^1+T_{t+1}^2+1} S_{(t+1)+T_{t+1}^1+T_{t+1}^2+1}^{(t+1)i}\left(x_{(t+1)+T_{t+1}^1+T_{t+1}^2+1}\right) \tag{3.1}$$

subject to the state dynamics

$$x_{\varsigma+1} = f_\varsigma\left(x_\varsigma, u_\varsigma^{1}, u_\varsigma^{2}, \ldots, u_\varsigma^{n}\right), \quad x_{(t+1)} = x_{(t+1)}^0. \tag{3.2}$$

To guarantee that there is no loss of previously known information, $T_{t+1}^1 \geq T_t^1 - 1$ and $T_{t+1}^1 + T_{t+1}^2 \geq T_t^1 + T_t^2 - 1$. Given that the players' payoffs in stage $t + 1$ to stage $t + T_t^1$ are known with certainty, we have $g_\varsigma^{(t+1)i}\left(x_\varsigma, u_\varsigma^{1}, u_\varsigma^{2}, \ldots, u_\varsigma^{n}\right) = g_\varsigma^{(t)i}\left(x_\varsigma, u_\varsigma^{1}, u_\varsigma^{2}, \ldots, u_\varsigma^{n}\right)$ for $\varsigma \in \{t+1, t+2, \ldots, t+T_t^1\}$. A theorem characterizing a feedback Nash equilibrium solution of the game (3.1) and (3.2) is provided below.

Theorem 3.1. *A set of strategies* $\{u_\varsigma^{i*} = \phi_\varsigma^{(t+1)i}(x), \text{ for } i \in N \text{ and } \varsigma \in \{(t + 1), t+2, \ldots, t+T_{(t+1)}^1\}, \text{ and } u_\varsigma^{\left(\varepsilon_\varsigma^k\right)i*} = \phi_\varsigma^{(t+1)\left(\varepsilon_\varsigma^k\right)i}(x), \text{ for } i \in N, \varepsilon_\varsigma^k \in \{1, 2, \ldots, h_\varsigma\} \text{ and } \varsigma \in \{(t + 1) + T_{t+1}^1 + 1, (t+1) + T_{(t+1)}^1 + 2, \ldots, (t + 1) + T_{(t+1)}^1 + T_{(t+1)}^2\}\}$ *provides a feedback Nash equilibrium solution to the game* (3.1) *and* (3.2) *if there exist continuously differentiable functions* $V^{(t+1)i}(\tau, x)$, *for* $i \in N$, $\tau \in \{(t + 1), (t + 1) + 1, \ldots, (t + 1) + T_{(t+1)}^1\}$, *and* $V^{(t)}\left(\varepsilon_\tau^k\right)i(\tau, x)$, *for* $i \in N$, $\varepsilon_\tau^k \in \{1, 2, \ldots, h_\tau\}$ *and* $\tau \in \{(t + 1) + T_{(t+1)}^1 + 1, (t + 1) + T_{(t+1)}^1 + 2, \ldots, (t + 1) + T_{(t+1)}^1 + T_{(t+1)}^2\}$; *such that the following recursive relations are satisfied:*

$$V^{(t+1)i}(\tau, x) = \max_{u_\tau^i}\left\{g_\tau^{(t+1)i}\left[x, \phi_\tau^{(t+1)1}(x), \phi_\tau^{(t+1)2}(x), \ldots\right.\right.$$

$$\left.\ldots, \phi_\tau^{(t+1)i-1}(x), u_\tau^i, \phi_\tau^{(t+1)i+1}(x), \ldots, \phi_\tau^{(t+1)n}(x)\right]\delta_{t+1}^\tau$$

$$+ V^{(t+1)i}\left[\tau + 1, f_\tau\left[x, \phi_\tau^{(t+1)1}(x), \phi_\tau^{(t+1)2}(x), \ldots\right.\right.$$

$$\left.\left.\left.\ldots, \phi_\tau^{(t+1)i-1}(x), u_\tau^i, \phi_\tau^{(t+1)i+1}(x), \ldots, \phi_\tau^{(t+1)n}(x)\right]\right]\right\} \tag{3.3}$$

for $\tau \in \{(t+1), (t+1)+1, \ldots, (t+1)+T^1_{(t+1)} - 1\}$;

$$V^{(t+1)i}\left((t+1)+T^1_{(t+1)}, x\right)$$

$$= \max_{u^i_{(t+1)+T^1_t}} \left\{ g^{(t+1)i}_{(t+1)+T^1_t}\left[x, \phi^{(t+1)1}_{(t+1)+T^1_t}(x), \phi^{(t+1)2}_{(t+1)+T^1_t}(x), \ldots\right.\right.$$

$$\ldots, \phi^{(t+1)i-1}_{(t+1)+T^1_t}(x), u^i_{(t+1)+T^1_t}, \phi^{(t+1)i+1}_{(t+1)+T^1_t}(x), \ldots, \phi^{(t+1)n}_{(t+1)+T^1_t}(x)\right] \delta^\tau_{(t+1)}$$

$$+ \sum_{\varepsilon^k_{(t+1)+T^1_t+1}=1}^{h_{(t+1)+T^1_t+1}} \gamma^{\varepsilon^k_{(t+1)+T^1_t+1}}_{(t+1)+T^1_t+1} V^{(t+1)\left(\varepsilon^k_{(t+1)+T^1_t+1}\right)i}\left[(t+1)+T^1_{(t+1)}\right.$$

$$+ 1, f_{(t+1)+T^1_t}\left[x, \phi^{(t+1)1}_{(t+1)+T^1_{(t+1)}}(x), \phi^{(t+1)2}_{(t+1)+T^1_{(t+1)}}(x), \ldots\right.$$

$$\left.\left.\left.\ldots, \phi^{(t+1)i-1}_{(t+1')+T^1_t}(x), u^i_{(t+1)+T^1_t}, \phi^{(t+1)i+1}_{(t=1)+T^1_t}(x), \ldots, \phi^{(t+1)n}_{(t+1)+T^1_{(t+1)}}(x)\right]\right]\right\}; \quad (3.4)$$

$$V^{(t+1)(\varepsilon^k_\tau)i}(\tau, x)$$

$$= \max_{u^i_\tau} \left\{ g^{(t+1)(\varepsilon^k_\tau)i}_\tau\left[x, \phi^{(t+1)(\varepsilon^k_\tau)1}_\tau(x), \phi^{(t+1)(\varepsilon^k_\tau)2}_\tau(x), \ldots\right.\right.$$

$$\ldots, \phi^{(t+1)(\varepsilon^k_\tau)i-1}_\tau(x), u^i_\tau, \phi^{(t+1)(\varepsilon^k_\tau)i+1}_\tau(x), \ldots, \phi^{(t+1)(\varepsilon^k_\tau)n}_\tau(x)\right] \delta^\tau_{t+1}$$

$$+ \sum_{\varepsilon^k_{\tau+1}=1}^{h_{\tau+1}} \gamma^{\varepsilon^k_{\tau+1}}_{\tau+1} V^{(t+1)\left(\varepsilon^k_{\tau+1}\right)i}\left[\tau + 1, f_\tau\left[x, \phi^{(t+1)\left(\varepsilon^k_\tau\right)1}_\tau(x), \phi^{(t+1)\left(\varepsilon^k_\tau\right)2}_\tau(x), \ldots\right.\right.$$

$$\left.\left.\left.\ldots, \phi^{(t+1)\left(\varepsilon^k_\tau\right)i-1}_\tau(x), u^i_\tau, \phi^{(t+1)\left(\varepsilon^k_\tau\right)i+1}_\tau(x), \ldots, \phi^{(t+1)\left(\varepsilon^k_\tau\right)n}_\tau(x)\right]\right]\right\} \quad (3.5)$$

for $\varepsilon^k_\tau \in \{1, 2, \ldots, h_\tau\}$ *and* $\tau \in \{(t+1)+T^1_t+1, (t+1)+T^1_t+2, \ldots, (t+1)+T^1_t + T^2_t - 1\}$;

$$V^{(t+1)\left(\varepsilon^k_{(t+1)+T^1_{t+1}+T^2_{t+1}}\right)i}\left((t+1)+T^1_{(t+1)}+T^2_{(t+1)}, x\right)$$

$$= \max_{u^i_{(t+1)+T^1_{t+1}+T^2_{t+1}}} \left\{ g^{\left(\varepsilon^k_{(t+1)+T^1_{t+1}+T^2_{t+1}}\right)i}_{(t+1)+T^1_{t+1}+T^2_{t+1}}\left[x, \phi^{(t)\left(\varepsilon^k_{t+T^1_t+T^2_t}\right)1}_{(t+1)+T^1_{t+1}+T^2_{t+1}}(x),\right.\right.$$

$$\phi_{(t+1)+T_{t+1}^1+T_{t+1}^2}^{(t)\left(\varepsilon_{(t+1)+T_{t+1}^1+T_{t+1}^2}^k\right)2}(x),\dots$$

$$\dots,\phi_{(t+1)+T_t^1+T_t^2}^{(t+!)\left(\varepsilon_{(t+1)+T_{t+1}^1+T_{t+1}^2}^k\right)i-1}(x),u_{(t+1)+T_{t+1}^1+T_{t+1}^2}^i,$$

$$\phi_{(t+1)+T_{t+1}^1+T_{t+1}^2}^{(t+1)\left(\varepsilon_{(t+1)+T_t^1+T_t^2}^k\right)i+1}(x),\dots$$

$$\dots,\phi_{(t+1)+T_{t+1}^1+T_{t+1}^2}^{(t+1)\left(\varepsilon_{(t+1)+T_t^1+T_t^2}^k\right)n}(x)\Big]\delta_{t+1}^{(t+1)+T_{t+1}^1+T_{t+1}^2}+\delta_{t+1}^{(t+1)+T_{t+1}^1+T_{t+1}^2+1}$$

$$\times S_{(t+1)+T_{t+1}^1+T_{t+1}^2+1}^{(t+1)i}\left(f_{(t+1)+T_{t+1}^1+T_{t+1}^2}\left[x,\phi_{(t+1)+T_{t+1}^1+T_t^2}^{(t+1)\left(\varepsilon_{(t+1)+T_t^1+T_t^2}^k\right)1}(x),\right.\right.$$

$$\phi_{(t+1)+T_{t+1}^1+T_{t+1}^2}^{(t+1)\left(\varepsilon_{(t+1)+T_t^1+T_t^2}^k\right)2}(x),\dots,\phi_{(t+1)+T_{t+1}^1+T_{t+1}^2}^{(t+1)\left(\varepsilon_{(t+1)+T_t^1+T_t^2}^k\right)i-1}(x),u_{(t+1)+T_{t+1}^1+T_{t+1}^2}^i,$$

$$\phi_{(t+1)+T_{t+1}^1+T_{t+1}^2}^{(t+1)(\varepsilon_{(t+1)+T_t^1+T_t^2}^k)i-1}(x),\dots,\phi_{(t+1)+T_{t+1}^1+T_{t+1}^2}^{(t+1)\left(\varepsilon_{(t+1)+T_t^1+T_t^2}^k\right)n}(x)\Big]\Big)\Big\}\qquad(3.6)$$

for $\varepsilon_{(t+1)+T_{t+1}^1+T_{t+1}^2}^k\in\{1,2,\dots,h_{(t+1)+T_{t+1}^1+T_{t+1}^2}\}$.

Proof. Follow the proof of Theorem 2.1.   □

If there is no new information gained from stage t to stage $t+1$, we have $T_{t+1}^1 = T_t^1 - 1$ and $T_{t+1}^1 + T_{t+1}^2 = T_t^1 + T_t^2 - 1$, and $S_{t+1+T_{t+1}^1+T_{t+1}^2+1}^{(t+1)i}(x_{t+1+T_{t+1}^1+T_{t+1}^2+1})$ equals $S_{t+T_t^1+T_t^2+1}^{(t)i}(x_{t+T_t^1+T_t^2+1})$. Under this situation, the game equilibrium strategies $\phi_{t+1}^{(t)}(x)$ obtained in Theorem 2.1 would equal the game equilibrium strategies $\phi_{(t+1)}^{(t+1)}(x)$ in Theorem 3.1. If there is new information gained from stage t to stage $t+1$, the game equilibrium strategies at stage $t+1$ are $\phi_{t+1}^{(t+1)}(x)$, which will be different from $\phi_{t+1}^{(t)}(x)$. If there is new information gained in successive stages for k times, the game equilibrium strategies executed from stage t to stage $t+k$ will be $\phi_{(t)}^{(t)}(x)$, $\phi_{(t+1)}^{(t+1)}(x)$, $\phi_{(t+2)}^{(t+2)}(x)$, $\dots,\phi_{(t+k)}^{(t+k)}(x)$. The resultant payoffs of the players after the change in information are characterized by the value functions in Theorem 3.1. The game equilibrium dynamics from stage $t+1$ to stage $t+2$ can be obtained as

$$x_{t+2} = f_{t+1}\big(x_{t+1},\phi_{(t+1)}^{(t+1)1}(x_{t+1}),\phi_{(t+1)}^{(t+1)2}(x_{t+1}),\dots,\phi_{(t+1)}^{(t+1)n}(x_t)\big),\qquad(3.7)$$

where x_{t+1} is obtained from (2.7).

In similar manners, one can solve the games with the initial stage being stage $t+2$, $t+3$, $t+4$, and so on. By re-labeling the initial stage of the game as stage t, an infinite horizon game framework with information updating is constructed for an infinite number of appearances of initial stage t.

4. An Illustration in Resource Extraction

Consider an infinite-horizon version of the renewable resource game in Yeung and Petrosyan [2010, 2011] in which there are two extraction firms. The operation horizons of the firms may last for an indefinite length of time. Hence, one has to consider the game as an infinite horizon problem. In the current (initial) stage t, the players' payoff structures are known with certainty in the current stage and the next $T_t^1 = 4$ future stages, the probability distributions of the firms' payoff structures are known in the following $T_t^2 = 10$ stages, and the firms' best estimates of their expected values of the states of the game at stage $t + T_t^1 + T_t^2 + 1 = t + 15$ are available. Let $u_\zeta^i \in U^i \subset R^+$ denote the amount of resource extracted by firm i at stage ζ, for $i \in \{1, 2\}$, and $x_\zeta \in X \subset R^+$ the size of the resource stock. In stage t to stage $t + 4$, it is known with certainty that the extraction cost for firm i is $c_\zeta^i (u_\zeta^i)^2 / x_\zeta$ and the price of the resource is P_ζ, for $\zeta \in \{t, t+1, \ldots, t+4\}$. In stage $t + 5$ to stage $t + 14$, it is known that the probability for the extraction cost for firm i being $c_\varsigma^{(\varepsilon_\varsigma^k)i} (u_\varsigma^i)^2 / x_\varsigma$ and resource price being $P_\varsigma^{(\varepsilon_\varsigma^k)}$ is $\gamma_\varsigma^{\varepsilon_\varsigma^k}$ for $\varepsilon_\varsigma^k \in \{1, 2, \ldots, h_\varsigma\}$. The best estimate of player i's expected value of the state x_{t+15} at stage $t + 15$ is $(\bar{A}_{t+15}^{(t)i} x_{t15} + \bar{C}_{t+15}^{(t)i})(1 + r)^{-15}$. Firm $i \in \{1, 2\}$ would maximize its expected profits

$$
\sum_{\zeta=t}^{t+4} \left[P_\zeta u_\zeta^i - \frac{c_\zeta^i}{x_\zeta} (u_\zeta^i)^2 \right] (1+r)^{-(\zeta-t)} + \sum_{\varsigma=t+5}^{t+14} \sum_{\varepsilon_\varsigma^k=1}^{h_\varsigma} \gamma_\varsigma^{\varepsilon_\varsigma^k} \left[P_\varsigma^{(\varepsilon_\varsigma^k)} u_\varsigma^i - \frac{c_\varsigma^{(\varepsilon_\varsigma^k)i}}{x_\varsigma} (u_\varsigma^i)^2 \right]
$$

$$
\times (1+r)^{-(\varsigma-t)} + \left(\bar{A}_{t+15}^{(t)i} x_{t+15} + \bar{C}_{t+15}^{(t)i} \right)(1+r)^{-15} \tag{4.1}
$$

subject to the growth dynamics of the resource stock

$$
x_{\zeta+1} = x_\zeta + a - b x_\zeta - \sum_{j=1}^{2} u_\zeta^j, \quad x_t = x_t^0, \tag{4.2}
$$

where a is the natural increment of the resource and b is the decay rate.

4.1. *Game equilibrium*

Invoking Theorem 2.1, one can characterize the feedback Nash equilibrium for the game (4.1) and (4.2). In particular, a set of strategies $\{u_\zeta^{i*} = \phi_\zeta^{(t)i}(x)$, for $i \in \{1, 2\}$ and $\zeta \in \{t, t+1, \ldots, t+4\}$, and $u_\varsigma^{(\varepsilon_\varsigma^k)i*} = \phi_\varsigma^{(t)(\varepsilon_\varsigma^k)i}(x)$, for $i \in \{1, 2\}$, $\varepsilon_\varsigma^k \in \{1, 2, \ldots, h_\varsigma\}$ and $\varsigma \in \{t+5, t+6, \ldots, t+14\}\}$ provides a feedback Nash equilibrium solution to the game (4.1) and (4.2) if there exist continuously differentiable functions $V^{(t)i}(\tau, x)$, for $i \in \{1, 2\}$, $\tau \in \{t, t+1, \ldots, t+4\}$, and $V^{(t)(\varepsilon_\tau^k)i}(\tau, x)$, for $\varepsilon_\tau^k \in \{1, 2, \ldots, h_\tau\}$ and $\tau \in \{t+5, t+6, \ldots, t+14\}$; such that the following recursive relations are satisfied:

$$V^{(t)i}(\tau, x) = \max_{u_\tau^i} \left\{ \left[P_\tau u_\tau^i - \frac{c_\tau^i}{x}(u_\tau^i)^2 \right](1+r)^{-(\tau-t)} \right.$$

$$\left. + V^{(t)i}[\tau+1, x+a-bx-u_\tau^i - \phi_\tau^{(t)j}(x)] \right\} \tag{4.3}$$

for $\tau \in \{t, t+1, \ldots, t+3\}$, $i, j \in \{1, 2\}$ and $i \neq j$;

$$V^{(t)i}(t+4, x) = \max_{u_{t+4}^i} \left\{ \left[P_{t+4}u_{t+4}^i - \frac{c_{t+4}^i}{x}(u_{t+4}^i)^2 \right](1+r)^{-4} \right.$$

$$\left. + \sum_{\varepsilon_{t+5}^k=1}^{h_{t+5}} \gamma_{t+5}^{\varepsilon_{t+5}^k} V^{(t)} \left(\varepsilon_{t+5}^k\right)^i [t+5, x+a-bx-u_{t+4}^i - \phi_{t+4}^{(t)j}(x)] \right\}; \tag{4.4}$$

$$V^{(t)\left(\varepsilon_\tau^k\right)i}(\tau, x) = \max_{u_\tau^i} \left\{ \left[P_\tau^{\left(\varepsilon_\tau^k\right)} u_\tau^i - \frac{c_\tau^{\left(\varepsilon_\tau^k\right)i}}{x}(u_\tau^i)^2 \right](1+r)^{-(\tau-t)} \right.$$

$$\left. + \sum_{\varepsilon_{\tau+1}^k=1}^{h_{\tau+1}} \gamma_{\tau+1}^{\varepsilon_{\tau+1}^k} V^{(t)}\left(\varepsilon_{\tau+1}^k\right)^i [\tau+1, x+a-bx-u_\tau^i - \phi_\tau^{(t)j}(x)] \right\}, \tag{4.5}$$

$\varepsilon_\tau^k \in \{1, 2, \ldots, h_\tau\}$ and $\tau \in \{t+5, t+6, \ldots, t+13\}$;

$$V^{(t)\left(\varepsilon_{t+14}^k\right)i}(t+14, x)$$

$$= \max_{u_{t+14}^i} \left\{ \left[P_{t+14}^{\left(\varepsilon_{t+14}^k\right)} u_{t+14}^i - \frac{c_{t+14}^{\left(\varepsilon_{t+14}^k\right)i}}{x}(u_{t+14}^i)^2 \right](1+r)^{-14} \right.$$

$$\left. + \left(\bar{A}_{t+15}^{(t)i}[x+a-bx-u_{t+14}^i - \phi_{t+14}^{(t)j}(x)] + \bar{C}_{t+15}^{(t)i} \right)(1+r)^{-15} \right\} \tag{4.6}$$

for $\varepsilon_{t+14}^k \in \{1, 2, \ldots, h_{t+14}\}$.

The value functions in (4.3)–(4.6) can be obtained as follows.

Proposition 4.1. *The game equilibrium value functions in* (4.3)–(4.6) *are:*

$$V^{(t)\left(\varepsilon_{t+14}^k\right)i}(t+14, x) = \left[A_{t+14}^{(t)\left(\varepsilon_{t+14}^k\right)i} x + C_{t+14}^{(t)\left(\varepsilon_{t+14}^k\right)i} \right](1+r)^{-14},$$

$$i \in \{1, 2\} \quad and \quad \varepsilon_{t+14}^k \in \{1, 2, \ldots, h_{t+14}\},$$

where

$$A_{t+14}^{(t)\left(\varepsilon_{t+14}^k\right)i} = \left[P_{t+14}^{\left(\varepsilon_{t+14}^k\right)} - (1+r)^{-1}\bar{A}_{t+15}^{(t)i}\right]^2 \frac{1}{4c_{t+14}^{\left(\varepsilon_{t+14}^k\right)i}}$$

$$+\bar{A}_{t+15}^{(t)i}(1+r)^{-1}\left(1 - b - \sum_{j=1}^{2}\left[P_{t+14}^{\left(\varepsilon_{t+14}^k\right)} - (1+r)^{-1}\bar{A}_{t+15}^{(t)j}\right]\frac{1}{2c_{t+14}^{\left(\varepsilon_{t+14}^k\right)j}}\right),$$

and

$$C_{t+14}^{(t)\left(\varepsilon_{t+14}^k\right)i} = \bar{A}_{t+15}^{(t)i}(1+r)^{-1}a + \bar{C}_{t+15}^{(t)i}(1+r)^{-1};$$

$$V^{(t)(\varepsilon_\tau^k)i}(\tau, x) = \left[A_\tau^{(t)(\varepsilon_\tau^k)i}x + C_\tau^{(t)(\varepsilon_\tau^k)i}\right](1+r)^{-(\tau-t)}, \quad \varepsilon_\tau^k \in \{1, 2, \ldots, h_\tau\} \quad and$$

$$\tau \in \{t+5, t+6, \ldots, t+13\};$$

where

$$A_\tau^{(t)(\varepsilon_\tau^k)i} = \left(P_\tau^{\left(\varepsilon_\tau^k\right)} - (1+r)^{-1}\sum_{\varepsilon_{\tau+1}^k=1}^{h_{\tau+1}}\gamma_{\tau+1}^{\varepsilon_{\tau+1}^k}A_{\tau+1}^{(t)\left(\varepsilon_{\tau+1}^k\right)i}\right)^2 \frac{1}{4c_\tau^{\left(\varepsilon_\tau^k\right)i}}$$

$$+ \sum_{\varepsilon_{\tau+1}^k=1}^{h_{\tau+1}}\gamma_{\tau+1}^{\varepsilon_{\tau+1}^k}A_{\tau+1}^{(t)\left(\varepsilon_{\tau+1}^k\right)i}(1+r)^{-1}\left[1 - b\right.$$

$$\left. - \sum_{j=1}^{2}\left(P_\tau^{\left(\varepsilon_\tau^k\right)} - (1+r)^{-1}\sum_{\varepsilon_{\tau+1}^k=1}^{h_{\tau+1}}\gamma_{\tau+1}^{\varepsilon_{\tau+1}^k}A_{\tau+1}^{(t)\left(\varepsilon_{\tau+1}^k\right)j}\right)\frac{1}{2c_\tau^{\left(\varepsilon_\tau^k\right)j}}\right], \quad and$$

$$C_\tau^{(t)\left(\varepsilon_\tau^k\right)i} = (1+r)^{-1}\sum_{\varepsilon_{\tau+1}^k=1}^{h_{\tau+1}}\gamma_{\tau+1}^{\varepsilon_{\tau+1}^k}\left(A_{\tau+1}^{(t)\left(\varepsilon_{\tau+1}^k\right)i}a + C_{\tau+1}^{(t)\left(\varepsilon_{\tau+1}^k\right)i}\right);$$

$$V^{(t)i}(t+4, x) = \left[A_{t+4}^{(t)i}x + C_{t+4}^{(t)i}\right](1+r)^{-4},$$

where

$$A_{t+4}^{(t)i} = \left(P_{t+4} - (1+r)^{-1}\sum_{\varepsilon_{t+5}^k=1}^{h_{t+5}}\gamma_{t+5}^{\varepsilon_{t+5}^k}A_{t+5}^{(t)\left(\varepsilon_{t+5}^k\right)i}\right)^2 \frac{1}{4c_{t+4}^i}$$

$$+ \sum_{\varepsilon_{t+5}^k=1}^{h_{t+5}}\gamma_{t+5}^{\varepsilon_{t+5}^k}A_{t+5}^{(t)\left(\varepsilon_{t+5}^k\right)i}(1+r)^{-1}\left[1 - b\right.$$

$$\left. - \sum_{j=1}^{2}\left(P_{t+4} - (1+r)^{-1}\sum_{\varepsilon_{t+5}^k=1}^{h_{t+5}}\gamma_{t+5}^{\varepsilon_{t+5}^k}A_{t+5}^{(t)\left(\varepsilon_{t+5}^k\right)j}\right)\frac{1}{2c_{t+4}^j}\right], \quad and$$

$$C_{t+4}^{(t)i} = (1+r)^{-1} \sum_{\varepsilon_{t+5}^k=1}^{h_{t+5}} \gamma_{t+5}^{\varepsilon_{t+5}^k} \left(A_{t+5}^{(t)\left(\varepsilon_{t+5}^k\right)i} a + C_{t+5}^{(t)\left(\varepsilon_{t+5}^k\right)i} \right);$$

$$V^{(t)i}(\tau, x) = \left[A_\tau^{(t)i} x + C_\tau^{(t)i} \right] (1+r)^{-(\tau-t)}, \quad \tau \in \{t, t+1, \ldots, t+3\},$$

where

$$A_\tau^{(t)i} = \left(P_\tau - (1+r)^{-1} A_{\tau+1}^{(t)i} \right)^2 \frac{1}{4c_\tau^i}$$

$$+ A_{\tau+1}^{(t)i}(1+r)^{-1} \left[1 - b - \sum_{j=1}^{2} \left(P_\tau - (1+r)^{-1} A_\tau^{(t)j} \right) \frac{1}{2c_\tau^j} \right], \quad and$$

$$C_\tau^{(t)i} = (1+r)^{-1} [A_{\tau+1}^{(t)i} a + C_{\tau+1}^{(t)i}]. \tag{4.7}$$

Proof. See Appendix B.   □

The game equilibrium strategies are

$$\phi_\tau^{(t)i}(x) = \left(P_\tau - (1+r)^{-1} A_{\tau+1}^{(t)i} \right) \frac{x}{2c_\tau^i},$$

for $i \in \{1, 2\}$ and $\tau \in \{t, t+1, \ldots, t+3\}$,

$$\phi_{t+4}^{(t)i}(x) = \left(P_{t+4} - (1+r)^{-1} \sum_{\varepsilon_{t+5}^k=1}^{h_{t+5}} \gamma_{t+5}^{\varepsilon_{t+5}^k} A_{t+5}^{(t)\left(\varepsilon_{t+5}^k\right)i} \right) \frac{x}{2c_{t+4}^i},$$

for $i \in \{1, 2\}$,

$$\phi_\tau^{(t)(\varepsilon_\tau^k)i}(x) = \left(P_\tau^{(\varepsilon_\tau^k)} - (1+r)^{-1} \sum_{\varepsilon_{\tau+1}^k=1}^{h_{\tau+1}} \gamma_{\tau+1}^{\varepsilon_{\tau+1}^k} A_{\tau+1}^{(t)\left(\varepsilon_{\tau+1}^k\right)i} \right) \frac{x}{2c_\tau^{(\varepsilon_\tau^k)i}},$$

for $i \in \{1, 2\}$ and $\varepsilon_\tau^k \in \{1, 2, \ldots, h_\tau\}$, and $\tau \in \{t+5, t+6, \ldots, t+13\}$,

$$\phi_{t+14}^{(t)(\varepsilon_{t+14}^k)i}(x) = \left[P_{t+14}^{(\varepsilon_{t+14}^k)} - (1+r)^{-1} \bar{A}_{t+15}^{(t)i} \right] \frac{x}{2c_{t+14}^{(\varepsilon_{t+14}^k)i}},$$

for $i \in \{1, 2\}$ and $\varepsilon_{t+14}^k \in \{1, 2, \ldots, h_{t+14}\}$.

When the game proceeds to stage $t+1$, the players' payoff structures in stages $t+5 = (t+1)+4$ and $t+6 = (t+1)+5$ become known with certainty, the probability distributions of the firms' payoff structures in stages $t+15 = (t+1)+14$, $t+16 = (t+1)+15$ and $t+17 = (t+1)+16$ are also known, and the best estimates of the firms' expected values of the states of the game at stage $t + 18 = (t+1) + 17$ are available. Hence, $T^1_{(t+1)} = 5$ and $T^2_{(t+1)} = 11$.

At stage $t + 1$, firm $i \in \{1, 2\}$ would maximize its expected profits:

$$\sum_{\varsigma=(t+1)}^{(t+1)+6} \left[P_\varsigma u^i_\varsigma - \frac{c^i_\varsigma}{x_\varsigma}(u^i_\varsigma)^2 \right] (1+r)^{-(\varsigma-(t+1))}$$

$$+ \sum_{\varsigma=(t+1)+6}^{(t+1)+13} \sum_{\varepsilon^k_\varsigma=1}^{h_\varsigma} \gamma^{\varepsilon^k_\varsigma}_\varsigma \left[P_\varsigma^{\left(\varepsilon^k_\varsigma\right)} u^i_\varsigma - \frac{c_\varsigma^{\left(\varepsilon^k_\varsigma\right)i}}{x_\varsigma}(u^i_\varsigma)^2 \right] (1+r)^{-(\varsigma-(t+1))}$$

$$+ \sum_{\varsigma=(t+1)+14}^{(t+1)+16} \sum_{\varepsilon^k_\varsigma=1}^{h_\varsigma} \gamma^{\varepsilon^k_\varsigma}_\varsigma \left[P_\varsigma^{\left(\varepsilon^k_\varsigma\right)} u^i_\varsigma - \frac{c_\varsigma^{\left(\varepsilon^k_\varsigma\right)i}}{x_\varsigma}(u^i_\varsigma)^2 + h_\varsigma^{\left(\varepsilon^k_\varsigma\right)i} x_\varsigma \right] (1+r)^{-(\varsigma-(t+1))}$$

$$+ \left(\bar{A}^{(t)i}_{(t+1)+17} x_{(t+1)+17} + C^{(t)i}_{(t+1)+17} \right)(1+r)^{-17}, \tag{4.8}$$

subject to the growth dynamics of the resource

$$x_{\varsigma+1} = x_\varsigma + a - b x_\varsigma - \sum_{j=1}^{2} u^j_\varsigma, \quad x_{t+1} = x^0_{t+1}. \tag{4.9}$$

Note that the payoff structures in stages $(t + 1) + 14$, $(t + 1) + 15$ and $(t + 1) + 16$ are new (and unknown in stage t). In particular, new technology or government environmental subsidies enable the resource stock x_ς to contribute to the revenues of the firms. Following the proof of Proposition 4.1, one can obtain

$$V^{(t+1)\left(\varepsilon^k_{t+17}\right)i}[(t+1)+16, x] = \left[A^{(t+1)\left(\varepsilon^k_{t+17}\right)i}_{(t+1)+16} x + C^{(t+1)\left(\varepsilon^k_{t+17}\right)i}_{(t+1)+16} \right](1+r)^{-16},$$

$$i \in \{1, 2\} \quad \text{and} \quad \varepsilon^k_{(t+1)+16} \in \{1, 2, \ldots, h_{(t+1)+16}\},$$

where

$$A^{(t+1)\left(\varepsilon^k_{t+17}\right)i}_{(t+1)+16} = \left[P_{(t+1)+16}^{\left(\varepsilon^k_{t+17}\right)} - (1+r)^{-1} \bar{A}^{(t+1)i}_{(t+1)+17} \right]^2 \frac{1}{4c_{(t+1)+16}^{\left(\varepsilon^k_{t+17}\right)i}} + h_{(t+1)+16}^{\left(\varepsilon^k_{t+17}\right)i}$$

$$+ \bar{A}^{(t+1)i}_{(t+1)+17}(1+r)^{-1} \left(1 - b - \sum_{j=1}^{2} \left[P_{(t+1)+16}^{\left(\varepsilon^k_{t+17}\right)} \right. \right.$$

$$\left. \left. - (1+r)^{-1} \bar{A}^{(t+1)j}_{(t+1)+17} \right] \frac{1}{2c_{(t+1)+16}^{\left(\varepsilon^k_{t+17}\right)j}} \right), \quad \text{and}$$

$$C^{(t+1)\left(\varepsilon^k_{t+17}\right)i}_{(t+1)+16} = (1+r)^{-1} \left[\bar{A}^{(t+1)i}_{(t+1)+17} a + \bar{C}^{(t+1)i}_{(t+1)+17} \right];$$

$$V^{(t+1)\left(\varepsilon_\tau^k\right)i}(\tau,x) = \left[A_\tau^{(t+1)\left(\varepsilon_\tau^k\right)i}x + C_\tau^{(t+1)\left(\varepsilon_\tau^k\right)i}\right](1+r)^{-(\tau-(t+1))},$$

$$\varepsilon_\tau^k \in \{1,2,\ldots,h_\tau\} \quad \text{and} \quad \tau \in \{(t+1)+14,(t+1)+15\},$$

where

$$A_\tau^{(t+1)\left(\varepsilon_\tau^k\right)i} = \left(P_\tau^{\left(\varepsilon_\tau^k\right)} - (1+r)^{-1}\sum_{\varepsilon_{\tau+1}^k=1}^{h_{\tau+1}}\gamma_{\tau+1}^{\varepsilon_{\tau+1}^k}A_{\tau+1}^{(t+1)\left(\varepsilon_{\tau+1}^k\right)i}\right)^2 \frac{1}{4c_\tau^{\left(\varepsilon_\tau^k\right)i}}$$

$$+ h_\tau^{\left(\varepsilon_\tau^k\right)i} + \sum_{\varepsilon_{\tau+1}^k=1}^{h_{\tau+1}}\gamma_{\tau+1}^{\varepsilon_{\tau+1}^k}A_{\tau+1}^{(t+1)\left(\varepsilon_{\tau+1}^k\right)i}(1+r)^{-1}$$

$$\times\left[1-b-\sum_{j=1}^2\left(P_\tau^{\left(\varepsilon_\tau^k\right)} - (1+r)^{-1}\right.\right.$$

$$\left.\left.\times\sum_{\varepsilon_{\tau+1}^k=1}^{h_{\tau+1}}\gamma_{\tau+1}^{\varepsilon_{\tau+1}^k}A_{\tau+1}^{(t+1)\left(\varepsilon_{\tau+1}^k\right)j}\right)\frac{1}{2c_\tau^{\left(\varepsilon_\tau^k\right)j}}\right], \quad \text{and}$$

$$C_\tau^{(t+1)\left(\varepsilon_\tau^k\right)i} = (1+r)^{-1}\sum_{\varepsilon_{\tau+1}^k=1}^{h_{\tau+1}}\gamma_{\tau+1}^{\varepsilon_{\tau+1}^k}\left(A_{\tau+1}^{(t+1)\left(\varepsilon_{\tau+1}^k\right)i}a + C_{\tau+1}^{(t+1)\left(\varepsilon_{\tau+1}^k\right)i}\right).$$

$$V^{(t+1)\left(\varepsilon_{t+14}^k\right)i}(t+14,x) = \left[A_{t+14}^{(t+1)\left(\varepsilon_{t+14}^k\right)i}x + C_{t+14}^{(t+1)\left(\varepsilon_{t+14}^k\right)i}\right](1+r)^{-13},$$

$$\varepsilon_{t+14}^k \in \{1,2,\ldots,h_{t+14}\},$$

where

$$A_{t+14}^{(t+1)\left(\varepsilon_{t+14}^k\right)i} = \left(P_{t+14}^{\left(\varepsilon_{t+14}^k\right)} - (1+r)^{-1}\sum_{\varepsilon_{t+15}^k=1}^{h_{t+15}}\gamma_{t+15}^{\varepsilon_{t+15}^k}A_{t+15}^{(t+1)\left(\varepsilon_{t+15}^k\right)i}\right)^2 \frac{1}{4c_{t+14}^{\left(\varepsilon_{t+14}^k\right)i}}$$

$$+ \sum_{\varepsilon_{t+15}^k=1}^{h_{t+15}}\gamma_{t+15}^{\varepsilon_{t+15}^k}A_{t+15}^{(t+1)\left(\varepsilon_{t+15}^k\right)i}(1+r)^{-1}\left[1-b-\sum_{j=1}^2\left(P_{t+14}^{\left(\varepsilon_{t+14}^k\right)}\right.\right.$$

$$\left.\left.- (1+r)^{-1}\sum_{\varepsilon_{t+15}^k=1}^{h_{t+15}}\gamma_{t+15}^{\varepsilon_{t+15}^k}A_{t+15}^{(t+1)\left(\varepsilon_{t+15}^k\right)j}\right)\frac{1}{2c_{t+14}^{\left(\varepsilon_{t+14}^k\right)j}}\right],$$

and

$$C_{t+14}^{(t+1)\left(\varepsilon_{t+14}^k\right)i} = (1+r)^{-1}\sum_{\varepsilon_{t+15}^k=1}^{h_t=15}\gamma_{t+15}^{\varepsilon_{t+15}^k}\left(A_{t+15}^{(t+1)\left(\varepsilon_{t+15}^k\right)i}a + C_{t+15}^{(t+1)\left(\varepsilon_{t+15}^k\right)i}\right). \quad (4.10)$$

The value functions,

$$V^{(t+1)\left(\varepsilon_{t+14}^k\right)i}[(t+1)+13,x] = \left[A_{(t+1)+13}^{(t+1)\left(\varepsilon_{t+14}^k\right)i}x + C_{(t+1)+13}^{(t+1)\left(\varepsilon_{t+14}^k\right)i}\right](1+r)^{-13},$$

$$V^{(t+1)\left(\varepsilon_\tau^k\right)i}(\tau,x) = \left[A_\tau^{(t+1)\left(\varepsilon_\tau^k\right)i}x + C_\tau^{(t+1)\left(\varepsilon_\tau^k\right)i}\right](1+r)^{-(\tau-t+1)} \quad \text{for}$$

$$\tau \in \{t+7,t+8,\ldots,t+13\},$$

$$V^{(t+1)i}[(t+1)+5,x] = \left[A_{(t+1)+5}^{(t+1)i}x + C_{(t+1)+5}^{(t+1)i}\right](1+r)^{-5}, \quad \text{and}$$

$$V^{(t+1)i}(\tau,x) = \left[A_\tau^{(t+1)i}x + C_\tau^{(t+1)i}\right](1+r)^{-(\tau-t+1)},$$

$$\tau \in \{t+1,(t+1)+1,\ldots,(t+1)+4\}, \qquad (4.11)$$

have similar analytical solutions to those of the corresponding value functions in Proposition 4.1.

In particular, the game equilibrium strategies in stage $t+1$ with the information revealed in stage $t+1$ are

$$\phi_{(t+1)}^{(t+1)i}(x) = \left(P_{(t+1)} - (1+r)^{-1}A_{(t+1)+1}^{(t+1)i}\right)\frac{x}{2c_{(t+1)}^i}, \quad i \in \{1,2\},$$

which are different from

$$\phi_{t+1}^{(t)i}(x) = \left(P_{t+1} - (1+r)^{-1}A_{t+2}^{(t)i}\right)\frac{x}{2c_{t+1}^i}, \quad i \in \{1,2\}. \qquad (4.12)$$

Hence, revision in the players' strategies occurs as new information is revealed. One can label stage $t+1$ as stage t and move on so that stage t is the starting time over an infinite time horizon. Continuing with the information updating processes for subsequent stages enters into a framework of infinite horizon dynamic games.

4.2. *Comparison with standard infinite horizon games*

First consider the case of a standard infinite horizon dynamic game in which $T_t^1 = \infty$ and $T_t^2 = 0$. The game becomes

$$\max_{u_\zeta^i,\zeta \geq t} \sum_{\zeta=t}^{\infty} \left\{\left[P_\zeta u_\zeta^i - \frac{c_\zeta^i}{x_\zeta}(u_\zeta^i)^2\right](1+r)^{-(\zeta-t)}\right\} \qquad (4.13)$$

subject to the growth dynamics of the resource stock (4.2),
where

$$c_t^i = c_\zeta^i = c^i \quad \text{for all} \quad \zeta \geq t.$$

Invoking the standard technique of Bellman [1957] for solving an infinite horizon dynamic, one can characterize the solution of the game (4.1) and (4.13) as follows.

A set of strategies $\{u^{i*} = \phi^i(x)$ for $i \in \{1,2\}$ and $\varsigma \geq t\}$ constitutes a feedback Nash equilibrium for the game (4.1) and (4.13), if there exist continuously differentiable functions $V^i(x)$ such that the following Hamilton–Jacobi–Bellman equation is satisfied:

$$V^i(x) = \max_{u^i}\left\{\left[Pu^i - \frac{c^i}{x}(u^i)^2\right] + (1+r)^{-1}V^i[x + a - bx - u^i - \phi^j(x)]\right\}.$$

$$(4.14)$$

Solving (4.14) yields

$$V^i(x) = (A^i x + C^i),$$

where

$$A^i = (P - (1+r)^{-1}A^i)^2 \frac{1}{4c^i} + A^i(1+r)^{-1}\left[1 - b - \sum_{j=1}^{2}(P - (1+r)^{-1}A^j)\frac{1}{2c^j}\right],$$

and

$$C^i = (1+r)^{-1}[A^i a + C^i], \quad i \in \{1,2\}. \tag{4.15}$$

The game equilibrium strategies are

$$\phi^i(x) = (P - (1+r)^{-1}A^i)\frac{x}{2c^i}, \quad i \in \{1,2\}. \tag{4.16}$$

The differences between the solutions of the standard infinite horizon game and the information updating game include (i) the removal of the possible changes in future prices and costs, (ii) the removal of possible variations in value of the expected payoffs, and (iii) the requirement of a constant discount rate. However, if it is shown that the game parameters in (4.1) closely resemble those in (4.13), the term $V^i(x) = (A^i x + C^i)$ can serve as a proxy of the best estimates $\left(\bar{A}_{t+15}^{(t)i}x_{t+15} + \bar{C}_{t+15}^{(t)i}\right)$ in the information updating game.

Then, we consider the case where $T_t^1 = 0$ and $T_t^2 = \infty$. The game becomes

$$\max_{u_\varsigma^i, \varsigma \geq t}\left\{\left[P_t^{\left(\varepsilon_t^k\right)}u_t^i - \frac{c_\varsigma^{\left(\varepsilon_t^k\right)i}}{x_t}(u_\varsigma^i)^2\right](1+r)^{-(\varsigma-t)}\right.$$

$$\left. + \sum_{\varsigma=t+1}^{\infty}\sum_{\varepsilon^\ell=1}^{h}\gamma_\varsigma^{\varepsilon_\varsigma^\ell}\left[P_\varsigma^{\left(\varepsilon_\varsigma^\ell\right)}u_\varsigma^i - \frac{c_\varsigma^{\left(\varepsilon_\varsigma^\ell\right)i}}{x_\varsigma}(u_\varsigma^i)^2\right](1+r)^{-(\varsigma-t)}\right\} \tag{4.17}$$

subject to the growth dynamics of the resource stock (4.2), if ε^k happens in stage t, where $\varepsilon_\varsigma = \varepsilon\{\varepsilon^1, \varepsilon^2, \ldots, \varepsilon^h\}$ are identical and independent random variables for $\varsigma \geq t$.

Again invoking Bellman [1957], the corresponding Hamilton–Jacobi–Bellman equation becomes

$$
V^{(\varepsilon^k)i}(x) = \max_{u^i} \left\{ \left[P_t^{(\varepsilon^k)} u_t^i - \frac{c_\zeta^{(\varepsilon^k)i}}{x_t}(u^i)^2 \right] \right.
$$

$$
\left. + (1+r)^{-1} \sum_{\varepsilon^\ell=1}^{h} \gamma^{\varepsilon^\ell} V^{(\varepsilon^\ell)i}[x + a - bx - u^i - \phi^j(x)] \right\}, \quad i \in \{1,2\}.
$$

$$(4.18)$$

Performing the indicated maximization in (4.18) yields

$$
\phi^{(\varepsilon^k)i}(x) = \left(P^{(\varepsilon^k_t)} - (1+r)^{-1} \sum_{\varepsilon^\ell=1}^{h} \gamma^{\varepsilon^\ell} V_x^{(\varepsilon^\ell)i}[x + a - bx - u^i - \phi^j(x)] \right)
$$

$$
\times \frac{x}{2c^{(\varepsilon^k)i}}, \quad i \in \{1,2\}.
$$

$$(4.19)$$

Substituting (4.19) into (4.18) yields

$$
V^{(\varepsilon^k)i}(x) = \left(A^{(\varepsilon^k)i} x + C^{(\varepsilon^k)i} \right), \quad \text{where}
$$

$$
A^{(\varepsilon^k)i} = \left(P^{(\varepsilon^k)} - (1+r)^{-1} \sum_{j=1}^{2} \gamma^{\varepsilon^\ell} A^{(\varepsilon^\ell)j} \right)^2 \frac{1}{4c^{(\varepsilon^k)i}} + A^{(\varepsilon^k)i}(1+r)^{-1}
$$

$$
\times \left[1 - b - \sum_{j=1}^{2} \left(P^{(\varepsilon^k)} - (1+r)^{-1} \sum_{j=1}^{2} \gamma^{\varepsilon^\ell} A^{(\varepsilon^\ell)j} \right) \frac{1}{2c^{(\varepsilon^k)j}} \right], \quad \text{and}
$$

$$
C^{(\varepsilon^k)i} = (1+r)^{-1} [A^{(\varepsilon^k)i} a + C^{(\varepsilon^k)i}], \quad i \in \{1,2\}.
$$

$$(4.20)$$

The differences between the solutions of the stochastic infinite horizon game and the information updating game include (i) while possible changes in future prices and costs are allowed, the probability distributions of prices and costs remain the same, and (ii) possible variations in value of the expected payoffs. If the random fluctuations in (4.17) closely resemble those in (4.1), the term $\sum_{\varepsilon^k=1}^{h} \gamma^{(\varepsilon^k)i} V^{(\varepsilon^k)i}(x) = \sum_{\varepsilon^k=1}^{h} \gamma^{(\varepsilon^k)i} (A^{(\varepsilon^k)i} x + C^{(\varepsilon^k)i})$ can serve as a proxy of the best estimates $(\bar{A}_{t+15}^{(t)i} x_{t+15} + \bar{C}_{t+15}^{(t)i})$ in the information updating game.

5. Conclusions

In many real-life game situations, the game horizons may last for an indefinite length of time. Hence, the problem has to be considered as an infinite horizon game. Existing infinite horizon dynamic games often resorted to the assumption of time-invariant game structures to obtain a solution. While uncertainties and unknowns in future events are inevitable, time-invariant game structures can hardly reflect the real situations. This chapter presents a new approach to analyze the infinite horizon dynamic games via information updating. The analysis preserves the infinite horizon nature of the game while allowing the adaptation of the game to the new environments as the game evolves. This approach will certainly be a more realistic and applicable alternative to the paradigm of infinite horizon dynamic games with time-invariant game structures. Further research along this line is expected.

Appendix A. Proof of Theorem 2.1

We first consider the last game stage, that is stage $t + T_t^1 + T_t^2$. Given the other players' game equilibrium strategies and the occurrence of $\varepsilon_{t+T_t^1+T_t^2}^k$, the game problem facing player i becomes the maximization of

$$
g_{t+T_t^1+T_t^2}^{\left(\varepsilon_{t+T_t^1+T_t^2}^k\right)i} \left[x, \phi_{t+T_t^1+T_t^2}^{(t)\left(\varepsilon_{t+T_t^1+T_t^2}^k\right)1}(x), \phi_{t+T_t^1+T_t^2}^{(t)\left(\varepsilon_{t+T_t^1+T_t^2}^k\right)2}(x), \ldots \right.
$$

$$
\ldots, \phi_{t+T_t^1+T_t^2}^{(t)\left(\varepsilon_{t+T_t^1+T_t^2}^k\right)i-1}(x), u_{t+T_t^1+T_t^2}^i, \phi_{t+T_t^1+T_t^2}^{(t)\left(\varepsilon_{t+T_t^1+T_t^2}^k\right)i+1}(x), \ldots
$$

$$
\left. \ldots, \phi_{t+T_t^1+T_t^2}^{(t)\left(\varepsilon_{t+T_t^1+T_t^2}^k\right)n}(x) \right] \delta_t^{t+T_t^1+T_t^2}
$$

$$
+ \delta_t^{t+T_t^1+T_t^2+1} S_{t+T_t^1+T_t^2+1}^{(t)i}\left(x_{t+T_t^1+T_t^2+1}\right), \tag{A.1}
$$

subject to

$$
x_{t+T_t^1+T_t^2+1} = f_{t+T_t^1+T_t^2}\left[x, \phi_{t+T_t^1+T_t^2}^{(t)\left(\varepsilon_{t+T_t^1+T_t^2}^k\right)1}(x), \phi_{t+T_t^1+T_t^2}^{(t)\left(\varepsilon_{t+T_t^1+T_t^2}^k\right)2}(x), \ldots \right.
$$

$$
\ldots, \phi_{t+T_t^1+T_t^2}^{(t)\left(\varepsilon_{t+T_t^1+T_t^2}^k\right)i-1}(x), u_{t+T_t^1+T_t^2}^i, \phi_{t+T_t^1+T_t^2}^{(t)\left(\varepsilon_{t+T_t^1+T_t^2}^k\right)i+1}(x), \ldots
$$

$$
\left. \ldots, \phi_{t+T_t^1+T_t^2}^{(t)\left(\varepsilon_{t+T_t^1+T_t^2}^k\right)n}(x) \right], \quad x_{t+T_t^1+T_t^2} = x. \tag{A.2}
$$

Problem (A.1)–(A.2) can be expressed as a one-stage problem which is the right-hand sign of the Bellman equation (2.6).

Then, we consider the subgame starting at stage $\tau \in \{t + T_t^1 + 1, t + T_t^1 + 2, \ldots, t + T_t^1 + T_t^2 - 1\}$. Given the other players' game equilibrium strategies and the occurrence of ε_τ^k, the game problem facing player i becomes the maximization of

$$
g_\tau^{\left(\varepsilon_\tau^k\right)i} \left[x, \phi_\tau^{(t)\left(\varepsilon_\tau^k\right)1}(x), \phi_\tau^{(t)\left(\varepsilon_\tau^k\right)2}(x), \ldots, \phi_\tau^{(t)\left(\varepsilon_\tau^k\right)i-1}(x), u_\tau^i, \phi_\tau^{(t)\left(\varepsilon_\tau^k\right)i+1}(x), \ldots \right.
$$

$$
\left. \ldots, \phi_\tau^{(t)\left(\varepsilon_\tau^k\right)n}(x) \right] \delta_t^\tau + \sum_{\varsigma=\tau+1}^{t+T_t^1+T_t^2} \sum_{\varepsilon_\varsigma^k=1}^{h_\varsigma} \gamma_\varsigma^{\varepsilon_\varsigma^k} g_\varsigma^{\left(\varepsilon_\varsigma^k\right)i}\left[x_\varsigma, \phi_\varsigma^{(t)\left(\varepsilon_\varsigma^k\right)1}(x_\varsigma), \right.
$$

$$\phi_\varsigma^{(t)\left(\varepsilon_\varsigma^k\right)2}(x_\varsigma),\dots,\phi_\varsigma^{(t)(\varepsilon_\varsigma^k)i-1}(x_\varsigma),u_\varsigma^i,\phi_\varsigma^{(t)\left(\varepsilon_\varsigma^k\right)i+1}(x_\varsigma),\dots,\phi_\varsigma^{(t)\left(\varepsilon_\varsigma^k\right)n}(x_\varsigma)\Big]\delta_t^\varsigma$$

$$+\,\delta_t^{t+T_t^1+T_t^2+1}S_{t+T_t^1+T_t^2+1}^{(t)i}\left(x_{t+T_t^1+T_t^2+1}\right),\tag{A.3}$$

subject to

$$x_{\varsigma+1}=f_\varsigma\Big[x_\varsigma,\phi_\varsigma^{(t)\left(\varepsilon_\varsigma^k\right)1}(x_\varsigma),\phi_\varsigma^{(t)\left(\varepsilon_\varsigma^k\right)2}(x_\varsigma),\dots,$$

$$\times\,\phi_\varsigma^{(t)\left(\varepsilon_\varsigma^k\right)i-1}(x_\varsigma),u_\varsigma^i,\phi_\varsigma^{(t)\left(\varepsilon_\varsigma^k\right)i+1}(x_\varsigma),\dots,\phi_\varsigma^{(t)(\varepsilon_\varsigma^k)n}(x_\varsigma)\Big],\quad x_\tau=x.\tag{A.4}$$

Invoking the results in stage $t+T_t^1+T_t^2$ and following Yeung and Petrosyan [2013] we use $\sum_{\varepsilon_{t+T_t^1+T_t^2}^k=1}^{h_{t+T_t^1+T_t^2}}\gamma_{t+T_t^1+T_t^2}^k V^{(t)\left(\varepsilon_{t+T_t^1+T_t^2}^k\right)i}(t+T_t^1+T_t^2,x_{t+T_t^1+T_t^2})$ as the terminal conditions at stage $t+T_t^1+T_t^2$ for the subgame problem starting in stage $t+T_t^1+T_t^2-1$. Similarly, we use $\sum_{\varepsilon_{\tau+1}^k=1}^{h_{\tau+1}}\gamma_{\tau+1}^k V^{(t)\left(\varepsilon_{\tau+1}^k\right)i}(\tau+1,x_\tau)$ as the terminal conditions at stage $\tau+1$ for the subgame problem starting in stage τ. The problem (A.3)–(A.4) then can be expressed as a one-stage problem which is the right-hand sign of the Bellman equation (2.5).

We proceed to consider the subgame starting at stage $t+T_t^1$. Following the above analysis, the game problem facing player i becomes the maximization of

$$g_{t+T_t^1}^i\Big[x,\phi_{t+T_t^1}^{(t)1}(x),\phi_{t+T_t^1}^{(t)2}(x),\dots,\phi_{t+T_t^1}^{(t)i-1}(x),u_{t+T_t^1}^i,\phi_{t+T_t^1}^{(t)i+1}(x),\dots,\phi_{t+T_t^1}^{(t)n}(x)\Big]\delta_{t+T_t^1}^\tau$$

$$+\sum_{\varepsilon_{t+T_t^1+1}^k=1}^{h_{t+T_t^1+1}}\gamma_{t+T_t^1+1}^k V^{(t)\left(\varepsilon_{t+T_t^1+1}^k\right)i}\Big[t+T_t^1+1,f_{t+T_t^1}\big[x,\phi_{t+T_t^1}^{(t)1}(x),$$

$$\phi_{t+T_t^1}^{(t)2}(x),\dots,\phi_{t+T_t^1}^{(t)i-1}(x),u_{t+T_t^1}^i,\phi_{t+T_t^1}^{(t)i+1}(x),\dots,\phi_{t+T_t^1}^{(t)n}(x)\big]\Big]\Bigg\}.\tag{A.5}$$

The problem (A.5) is the right-hand side of the Bellman equation (2.4).

Finally, we consider the subgame starting at stage $\tau\in\{t,t+1,\dots,t+T_t^1-1\}$. The game problem facing player i becomes the maximization of

$$g_\tau^i\big[x,\phi_\tau^{(t)1}(x),\phi_\tau^{(t)2}(x),\dots,\phi_\tau^{(t)i-1}(x),u_\tau^i,\phi_\tau^{(t)i+1}(x),\dots,\phi_\tau^{(t)n}(x)\big]\delta_t^\tau$$

$$+V^{(t)i}[\tau+1,f_\tau[x,\phi_\tau^{(t)1}(x),\phi_\tau^{(t)2}(x),\dots,\phi_\tau^{(t)i-1}(x),u_\tau^i,\phi_\tau^{(t)i+1}(x),\dots$$

$$\dots,\phi_\tau^{(t)n}(x)]\tag{A.6}$$

for $\tau\in\{t,t+1,\dots,t+T_t^1-1\}$.

Problem (A.6) is the right-hand side of the Bellman equation (2.3). Hence, Theorem 2.1 follows.

Appendix B. Proof of Proposition 4.1

We first consider stage $t+14$. Performing the indicated maximization in (4.6) yields

$$\phi_{t+14}^{(t)\left(\varepsilon_{t+14}^k\right)i}(x) = \left[P_{t+14}^{\left(\varepsilon_{t+14}^k\right)} - (1+r)^{-1}\bar{A}_{t+15}^{(t)i}\right]\frac{x}{2c_{t+14}^{\left(\varepsilon_{t+14}^k\right)i}} \quad \text{for} \quad i \in \{1,2\}.$$

(B.1)

Invoking Proposition 4.1 and substituting (B.1) into the Bellman equations (4.6) yields

$$\left[A_{t+14}^{(t)\left(\varepsilon_{t+14}^k\right)i}x + C_{t+14}^{(t)\left(\varepsilon_{t+14}^k\right)i}\right]$$

$$= \left[P_{t+14}^{\left(\varepsilon_{t+14}^k\right)} - (1+r)^{-1}\bar{A}_{t+15}^{(t)i}\right]^2\frac{x}{4c_{t+14}^{\left(\varepsilon_{t+14}^k\right)i}}$$

$$+ \bar{A}_{t+15}^{(t)i}(1+r)^{-1}\left(x - bx - \left[P_{t+14}^{\left(\varepsilon_{t+14}^k\right)} - (1+r)^{-1}\bar{A}_{t+15}^{(t)j}\right]\frac{x}{2c_{t+14}^{\left(\varepsilon_{t+14}^k\right)j}}\right)$$

$$+ \bar{A}_{t+15}^{(t)i}(1+r)^{-1}a + \bar{C}_{t+15}^{(t)i}(1+r)^{-1}.$$

Hence, one can obtain

$$A_{t+14}^{(t)\left(\varepsilon_{t+14}^k\right)i} = \left[P_{t+14}^{\left(\varepsilon_{t+14}^k\right)} - (1+r)^{-1}\bar{A}_{t+15}^{(t)i}\right]^2\frac{1}{4c_{t+14}^{\left(\varepsilon_{t+14}^k\right)i}} + \bar{A}_{t+15}^{(t)i}(1+r)^{-1}$$

$$\times\left(1 - b - \left[P_{t+14}^{\left(\varepsilon_{t+14}^k\right)} - (1+r)^{-1}\bar{A}_{t+15}^{(t)j}\right]\frac{1}{2c_{t+14}^{\left(\varepsilon_{t+14}^k\right)j}}\right), \quad \text{and}$$

$$C_{t+14}^{(t)\left(\varepsilon_{t+14}^k\right)i} = \bar{A}_{t+15}^{(t)i}(1+r)^{-1}a + \bar{C}_{t+15}^{(t)i}(1+r)^{-1}.$$

(B.2)

For stage $\tau \in \{t+5, t+6, \ldots, t+13\}$, performing the indicated maximization in (4.5) yields

$$\phi_\tau^{(t)\left(\varepsilon_\tau^k\right)i}(x) = \left(P_\tau^{\left(\varepsilon_\tau^k\right)} - (1+r)^{-1}\sum_{\varepsilon_{\tau+1}^k=1}^{h_{\tau+1}}\gamma_{\tau+1}^{\varepsilon_{\tau+1}^k}A_{\tau+1}^{(t)\left(\varepsilon_{\tau+1}^k\right)i}\right)\frac{x}{2c_\tau^{\left(\varepsilon_\tau^k\right)i}},$$

$$i \in \{1,2\} \quad \text{and} \quad \varepsilon_\tau^k \in \{1,2,\ldots,h_\tau\}. \quad \text{(B.3)}$$

Invoking Proposition 4.1 and substituting (B.3) into the Bellman equations (4.5) yields

$$V^{(t)\left(\varepsilon_\tau^k\right)i}(\tau,x) = \left[A_\tau^{(t)\left(\varepsilon_\tau^k\right)i}x + C_\tau^{(t)\left(\varepsilon_\tau^k\right)i}\right](1+r)^{-(\tau-t)},$$

where

$$
A_\tau^{(t)\left(\varepsilon_\tau^k\right)i} = \left(P_\tau^{\left(\varepsilon_\tau^k\right)} - (1+r)^{-1} \sum_{\varepsilon_{\tau+1}^k=1}^{h_{\tau+1}} \gamma_{\tau+1}^{\varepsilon_{\tau+1}^k} A_{\tau+1}^{(t)\left(\varepsilon_{\tau+1}^k\right)i}\right)^2 \frac{1}{4c_\tau^{\left(\varepsilon_\tau^k\right)i}}
$$

$$
+ \sum_{\varepsilon_{\tau+1}^k=1}^{h_{\tau+1}} \gamma_{\tau+1}^{\varepsilon_{\tau+1}^k} A_{\tau+1}^{(t)\left(\varepsilon_{\tau+1}^k\right)i}(1+r)^{-1}\left[1-b\right.
$$

$$
\left. - \left(P_\tau^{\left(\varepsilon_\tau^k\right)} - (1+r)^{-1} \sum_{\varepsilon_{\tau+1}^k=1}^{h_{\tau+1}} \gamma_{\tau+1}^{\varepsilon_{\tau+1}^k} A_{\tau+1}^{(t)\left(\varepsilon_{\tau+1}^k\right)j}\right) \frac{1}{2c_\tau^{\left(\varepsilon_\tau^k\right)j}}\right] \quad \text{and}
$$

$$
C_\tau^{(t)\left(\varepsilon_\tau^k\right)i} = (1+r)^{-1} \sum_{\varepsilon_{\tau+1}^k=1}^{h_{\tau+1}} \gamma_{\tau+1}^{\varepsilon_{\tau+1}^k} \left(A_{\tau+1}^{(t)\left(\varepsilon_{\tau+1}^k\right)i} a + C_{\tau+1}^{(t)\left(\varepsilon_{\tau+1}^k\right)i}\right). \tag{B.4}
$$

For stage $t+4$, performing the indicated maximization in (4.4) yields

$$
\phi_{t+4}^{(t)i}(x) = \left(P_{t+4} - (1+r)^{-1} \sum_{\varepsilon_{t+5}^k=1}^{h_{t+5}} \gamma_{t+5}^{\varepsilon_{t+5}^k} A_{t+5}^{(t)\left(\varepsilon_{t+5}^k\right)i}\right) \frac{x}{2c_{t+4}^i}, \quad i \in \{1,2\}. \tag{B.5}
$$

Invoking Proposition 4.1 and substituting (B.5) into the Bellman equations (4.4) yields

$$
V^{(t)i}(t+4, x) = \left[A_{t+4}^{(t)i} x + C_{t+4}^{(t)i}\right](1+r)^{-4},
$$

where

$$
A_{t+4}^{(t)i} = \left(P_{t+4} - (1+r)^{-1} \sum_{\varepsilon_{t+5}^k=1}^{h_{t+5}} \gamma_{t+5}^{\varepsilon_{t+5}^k} A_{t+5}^{(t)\left(\varepsilon_{t+5}^k\right)i}\right)^2 \frac{1}{4c_{t+4}^i}
$$

$$
+ \sum_{\varepsilon_{t+5}^k=1}^{h_{t+5}} \gamma_{t+5}^{\varepsilon_{t+5}^k} A_{t+5}^{(t)\left(\varepsilon_{t+5}^k\right)i}(1+r)^{-1}\left[1-b\right.
$$

$$
\left. - \left(P_{t+4} - (1+r)^{-1} \sum_{\varepsilon_{t+5}^k=1}^{h_{t+5}} \gamma_{t+5}^{\varepsilon_{t+5}^k} A_{t+5}^{(t)\left(\varepsilon_{t+5}^k\right)j}\right) \frac{1}{2c_{t+4}^j}\right] \quad \text{and}
$$

$$
C_{t+4}^{(t)i} = (1+r)^{-1} \sum_{\varepsilon_{t+5}^k=1}^{h_{t+5}} \gamma_{t+5}^{\varepsilon_{t+5}^k} \left(A_{t+5}^{(t)\left(\varepsilon_{t+5}^k\right)i} a + C_{t+5}^{(t)\left(\varepsilon_{t+5}^k\right)i}\right). \tag{B.6}
$$

Finally, for stage $\tau \in \{t, t+1, \ldots, t+3\}$, performing the indicated maximization in (4.3) yields

$$
\phi_\tau^{(t)i}(x) = \left(P_\tau - (1+r)^{-1} A_{\tau+1}^{(t)i}\right) \frac{x}{2c_\tau^i}, \quad i \in \{1,2\}. \tag{B.7}
$$

Invoking Proposition 4.1 and substituting (B.7) into the Bellman equations (4.3) yields

$$V^{(t)i}(\tau, x) = \left[A_\tau^{(t)i} x + C_\tau^{(t)i}\right](1+r)^{-(\tau-t)},$$

where

$$A_\tau^{(t)i} = \left(P_\tau - (1+r)^{-1} A_{\tau+1}^{(t)i}\right)^2 \frac{1}{4c_\tau^i}$$

$$+ \sum A_{\tau+1}^{(t)i}(1+r)^{-1}\left[1 - b - \left(P_\tau - (1+r)^{-1} A_\tau^{(t)j}\right)\frac{1}{2c_\tau^j}\right], \quad \text{and}$$

$$C_\tau^{(t)i} = (1+r)^{-1}[A_{\tau+1}^{(t)i} a + C_{\tau+1}^{(t)i}].$$

Hence Proposition 4.1 follows.

References

Bellman, R. [1957] *Dynamic Programming* (Princeton University Press, Princeton).

Benchekroun, H. [2003] Unilateral production restrictions in a dynamic duopoly, *J. Econ. Theory* **111**(2), 214–239.

Benchekroun, H. [2008] Comparative dynamics in a productive asset oligopoly, *J. Econo. Theory* **138**, 237–261.

Chen, B. and Zadrozny, P. A. [2002] An anticipative feedback solution for the infinite-horizon, linear-quadratic, dynamic, Stackelberg game, *J. Econ. Dyn. Control* **26**, 1397–1416.

Dockner, E. J. and Mosburger, G. [2007] Capital accumulation, asset values and imperfect product market competition, *J. Differ. Equ. Appl.* **13**(2–3), 197–215.

Fujiwara, K. [2009a] Gains from trade in a differential game model of asymmetric oligopoly, *Rev. Int. Econ.* **17**(5), 1066–1073.

Fujiwara, K. [2009b] A dynamic reciprocal dumping model of international trade, *Asia-Pacific J. Account. Econ.* **16**(3), 255–270.

Gromova, E. and Petrosian, O. [2016] Control of information horizon for cooperative differential game of pollution control, 2016, *Int. Conf. Stability and Oscillations of Nonlinear Control Systems (Pyatnitskiy's Conf.)*, IEEE, pp. 1–4, doi:10.1109/STAB.2016.7541187.

Gyurkovics, E. and Takacs, T. [2005] Guaranteeing cost strategies for infinite horizon difference games with uncertain dynamics, *Int. J. Control* **78**(8), 587–599.

Hinloopen, J., Smrkolj, G. and Wagener, F. [2017] Research and development cooperatives and market collusion: A global dynamic approach, *J. Optim. Theory Appl.* **174**, 567–612, doi:10.1007/s10957-017-1133-0.

Jorgensen, S. and Yeung, D. W. K. [1996] Stochastic differential game model of a common property fishery, *J. Optim. Theory Appl.* **90**, 381–403.

Jun, B. and Vives, X. [2004] Strategic incentives in dynamic duopoly, *J. Econ. Theory* **116**, 249–281.

Levhari, D. and Mirman, L. [1980] The great fish wars: An example of using a dynamic Nash–Cournot solution, *Bell J. Econ.* **11**, 322–334.

Petrosian, O. [2016] Looking forward approach in cooperative differential games, *Int. Game Theo. Rev.* **18**, 1640007.

Petrosian, O. and Barabanov, A. [2017] Looking forward approach in cooperative differential games with uncertain stochastic dynamics, *J. Optim. Theory Appl.* **172**, 328–347.

Petrosian, O., Nastych, M. and Volf, D. [2017] Differential game of oil market with moving informational horizon and non-transferable utility, 2017 *Constructive Nonsmooth Analysis and Related Topics (dedicated to the memory of V. F. Demyanov) (CNSA)*, IEEE, pp. 1–4, doi:10.1109/CNSA.2017.7974002.

Reynolds, S. [1987] Capacity investment, preemption, and commitment in an infinite horizon model, *Int. Econ. Rev.* **28**, 69–88.

Rubioa, S. J. and Ulphb, A. [2007] An infinite-horizon model of dynamic membership of international environmental agreements, *J. Environ. Econ. Manag.* **54**, 296–310.

Wirl, F. [1994] Pigouvian taxation of energy for flow and stock externalities and strategic, non-competitive energy pricing, *J. Environ. Econ. Manag.* **26**, 1–18.

Wirl, F. [1996] Dynamic voluntary provision of public goods: Extension to non-linear strategies, *Eur. J. Political Econ.* **12**, 555–560.

Yeung, D. W. K. [2001] Infinite horizon stochastic differential games with branching payoffs, *J. Optim. Theory Appl.* **111**(2), 445–460.

Yeung, D. W. K. and Petrosyan, L. A. [2010] Subgame consistent solutions for cooperative stochastic dynamic games, *J. Optim. Theory Appl.* **145**(3), 579–596.

Yeung, D. W. K. and Petrosyan, L. A. [2011] Subgame consistent cooperative solution of dynamic games with random horizon, *J. Optim. Theory Appl.* **150**, 78–97.

Yeung, D. W. K. and Petrosyan, L. A. [2013] Subgame-consistent cooperative solutions in randomly furcating stochastic dynamic games, *Math. Comput. Model.* **57**(3–4), 976–991.

Yeung, D. W. K. and Petrosyan, O. [2016] Cooperative stochastic differential games with information adaptation, in *Advances in Engineering Research*, Vol. 116, eds. Kim, H. and Hwang, L.-C. pp. 375–381, Atlantis Press, Paris.

Chapter 6

Contracts and Information Structure in a Supply Chain with Operations and Marketing Interaction

Fouad El Ouardighi

Department of Operations Management
ESSEC Business School, France
elouardighi@essec.fr

Gary Erickson

Department of Marketing, Michael G. Foster School of Business
University of Washington, USA

Dieter Grass

Department of Operations Research and Systems Theory
Vienna University of Technology, Austria

Steffen Jørgensen

Department of Business and Economics
University of Southern Denmark, Denmark

The objective of this chapter is to study how wholesale price and revenue sharing contracts affect operations and marketing decisions in a supply chain under different dynamic informational structures. We suggest a differential game model of a supply chain consisting of a manufacturer and a single retailer that agree on the contract parameters at the outset of the game. The model includes key operational and marketing activities related to a single product in the supply chain. The manufacturer sets a production rate and the rate of advertising efforts while the retailer chooses a purchase rate and the consumer price. The state of the game is summarized in the firms' backlogs and the manufacturer's advertising goodwill. Depending on whether the supply chain members have and share state information, they may either make decisions contingent on the current state of the game (feedback Nash strategy), or precommit to a plan of action during the whole game (open-loop Nash strategy). Given a contract type, the impact of the availability of information regarding the state of the game on the firms' decisions and payoffs is investigated. It is shown that double marginalization can be better mitigated if the supply chain members adopt a contingent strategy under a wholesale price contract and a commitment strategy under a revenue sharing contract.

Keywords: Supply chain management; wholesale price contract; revenue sharing contract; information structure; operations; marketing.

1. Introduction

Operations management and marketing are critical functions for not only individual manufacturing firms but also, with increased global competition, entire supply chains. Also, the two functions are strategically interlinked, since marketing's role is to create and manage demand, and operations are responsible for efficiently meeting the demand. Conflicts between marketing and operations management arise, since the two functions have differing objectives, marketing to enhance demand, and operations to minimize manufacturing and inventory/backlog costs, so management of the interface between the areas is critical. The importance of effective management of the interface has been recognized for some time [Malhotra and Sharma, 2002], and a review of models of the interface is provided by Tang [2010].

An important research area of operations management and marketing literature dealing with supply chains/marketing channels is the study of inefficiencies caused by lack of coordination. A classic example is the double marginalization problem that occurs when a retailer pays a supplier a fixed transfer price per unit ordered. The markups applied by both chain members lead to lower ordering quantities, higher consumer prices, and smaller overall profits than if decisions were centralized [Spengler, 1950; Bresnahan and Reiss, 1985]. To avoid this inefficiency decision making in the supply chain must be coordinated. This often entails using an appropriate contract to regulate the flow of payments between members of the supply chain.

The coordination problem has been addressed in a sizable body of literature in operations management as well as marketing. This literature has, however, tended to disregard the interactions between the two functional areas: *"Operations management has a wealth of literature that deals with the inventory aspects of the supply chain, but ignores marketing expenses... At the same time, the marketing literature tends to deal with customer acquisition costs in the form of advertising [...], but ignores operational issues"* [Simchi-Levi et al., 2004, p. 612].

The current research focuses on two types of contracts, the wholesale price contract (WPC) and the revenue sharing contract (RSC). A RSC lowers the transfer price to the retailer's benefit. The retailer then pays the supplier a part of its revenue [Dana and Spier, 2001]. The relative merits of the two contracts in static settings are well known [e.g., Cachon and Lariviere, 2005]. Mortimer [2008] provides an empirical study of WPC and RSC in the video rental industry. RSCs have been used in other industries, for example, in telecommunication services [Qin, 2008; Chakravarty and Werner, 2011] and cell phone manufacturing [Linh and Hong, 2009].

An operations management approach to coordination with WPC and RSC has often employed a static setup [see, e.g., Lariviere, 1999; Cachon, 2003]. Less is known, however, about how the two contracts work in a supply chain involving operational and marketing functional areas with repeated interaction. Thus, a primary aim of this chapter is to study the relative merits of the two contracts in a game setting where a retailer and its supplier first agree on a contractual scheme and then make operations and marketing decisions over time in the supply chain.

In this setting, firms' strategy depends on the extent to which chain members have, and share, state information [Başar and Olsder, 1999; Dockner *et al.*, 2000]. If a chain member has no such information, it cannot condition its actions on the state vector, and the firm's actions can be based only on time. Although decisions are dynamically optimized, strategic interaction takes place at the initial instant of time only; each player makes a commitment to execute a predetermined plan of action. This is known as an open-loop strategy. Such strategy applies in inventory management when there exist information delays [Bensoussan *et al.*, 2007], inaccurate records [DeHoratius and Raman, 2008], or hidden information. A good example of the latter is when a retailer is unwilling to reveal consumer sales data to its supplier. Conversely, if chain members have, and share, state information a firm's actions can also be based on the current state. This is known as a feedback strategy. Strategic interaction takes place throughout the game because players make decisions that are contingent upon the current state and time. In a supply chain, this situation is plausible whenever mutual trust is prevalent.

Both open-loop and feedback strategies have been applied in the supply chain [e.g., Gaimon, 1998; Kogan and Tapiero, 2007] and marketing channel [e.g., Jørgensen and Zaccour, 2004] literature dealing with intertemporal decisions.

A major aim of this chapter is to compare the performance of different informational structures under a WPC and a RSC, respectively. For example, given a contract, a comparison will reveal the relative merits of commitment and contingent strategies and quantify how availability of information on the current state of key operational and marketing variables affect supply chain members' decisions and payoffs.

The chapter focuses on production and sales of a single product/brand and suggests a dynamic model that includes key operational and marketing activities in the supply chain. The operations-marketing model that we use in the current research is a variant of the model in El Ouardighi *et al.* [2008] and Jørgensen [2011] (see also Jørgensen, 1986; Eliashberg and Steinberg, 1987) and an advertising goodwill model; see, e.g., Jørgensen and Zaccour [2004]. None of these papers, however, were concerned with the availability of state information in a supply chain.

The setup is one in which a single manufacturer sets her production rate and the national advertising rate for her product. Manufacturer advertising supposedly builds up a stock of consumer goodwill. There is a single retailer who sets the purchase rate and the consumer price. The state of the system is represented by the stock of consumer goodwill as well as the manufacturer's and retailer's respective backlogs of the product.

The following research issues are addressed:

— How are operations and marketing decisions in the supply chain, under a WPC and an RSC, respectively, affected by the availability of state information?
— Depending on the availability of state information, how do the two kinds of contracts compare in the long run?
— How do the two kinds of strategies mitigate channel inefficiency (double marginalization) under each contract?
— How are supply chain members' payoffs affected by the type of strategy and contract?

The main contribution of the current research lies in the fact that we look simultaneously at two important aspects of supply chain management: the choice of a contract that specifies how payments flow in the supply chain, and the role of information for the design of operations and marketing strategies in the supply chain. We show that with a WPC, feedback strategies mitigate the double marginalization problem because an increase in price does not deter growth of demand. Conversely, with a RSC, open-loop strategies provide results that are closest to those of a cooperative strategy.

The chapter proceeds as follows. Section 2 develops a differential game model where a manufacturer and a retailer agree on a supply chain contract at the outset. Sections 3 and 4 study the operations and marketing decisions in the supply chain under the two contracts in the context of open-loop and feedback Nash equilibria. Section 5 compares the contracts and the strategies. Section 6 concludes.

2. Differential Game Model

A manufacturer's product is ordered by an exclusive retailer who resells the product to final consumers Each chain member controls two decisions: one in marketing and one in operations. The manufacturer determines its production rate and investment in advertising goodwill whereas the retailer controls its procurement rate and the consumer price.

Time t is continuous and the game starts at time zero. State variables are the manufacturer's and retailer's backlogs as well as the manufacturer's advertising goodwill. Backlogging means that there is a delay in the delivery of some of the product quantity the retailer ordered from the manufacturer, as well as from retailer to customers, with associated costs to the manufacturer and retailer. Backlogging reflects an on-time fulfillment rate of less than 100%, which is not unusual in practice. Feichtinger and Hartl [1985] and Erickson [2011, 2012] show that backlogging may be profitable in the long run because production and purchase costs can be deferred. Sapra *et al.* [2010] argue that backlogging can be optimal because it adds to the allure and sense of exclusivity of a product and stimulates its demand.

The manufacturer's backlog at time t is denoted by $X(t)$ and evolves over time according to

$$\dot{X}(t) = v(t) - u(t), \quad X(0) = X_0 \geq 0, \tag{1}$$

where $u(t) \geq 0$ is the manufacturer's production rate and $v(t) \geq 0$ is the retailer's purchase rate. We assume that the manufacturer has a safety stock that can be used to fill backlogged products if there is backlogging [see, e.g., Maimon *et al.*, 1998]. The reason is that whenever the production rate is lower than the purchasing rate, the difference has to be filled, though with delay, thanks to the safety stock.

The retailer's backlog is $Y(t)$ and evolves according to

$$\dot{Y}(t) = S(t) - v(t), \quad Y(0) = Y_0 \geq 0, \tag{2}$$

where $S(t)$ is the consumer demand rate.

Consumer demand is affected negatively by the retail price $p(t)$ and positively by advertising goodwill $G(t)$ in a simple linear fashion

$$S(t) = \alpha - \beta p(t) + G(t), \tag{3}$$

where $\alpha, \beta > 0$ and constant. The market potential is time-dependent and equals $\alpha + G(t)$.

The manufacturer's goodwill evolves according to the Nerlove and Arrow [1962] model

$$\dot{G}(t) = w(t) - \delta G(t), \quad G(0) = G_0 \geq 0, \tag{4}$$

where $w(t) \geq 0$ is the manufacturer's national advertising effort and $\delta > 0$ a constant decay rate.

If advertising goodwill is omitted in the demand function, Eqs. (1)–(4) reduce to a variant of the model developed by Jørgensen [1986], Eliashberg and Steinberg [1987], Desai [1992, 1996], Kogan [2012] and others. If supply chain members adopt a zero-stock policy, they produce and order to meet retailer and consumer demand, respectively, at any time and Eqs. (1)–(4) reduce to the model studied in Jørgensen and Zaccour [2004].

Our next task is to define a payoff functional for each firm. Operations management literature usually assumes short planning horizons to evaluate the performance of operational strategies [Kogan and Tapiero, 2007]. Marketing literature has often adopted an infinite planning horizon, partly to disclose longrun effects of marketing strategies, and partly for mathematical convenience [Erickson, 2003; Jørgensen and Zaccour, 2004]. This chapter assumes an infinite planning horizon, that is $t \in [0, +\infty[$, which enables us to study the long-run stability of game equilibria. The use of an infinite horizon is also motivated by the inclusion of advertising goodwill in the model. Building a stock of goodwill takes time and assuming a finite (and possibly short) horizon might leave out interesting aspects of goodwill evolution.

We assume that an agreement on (i) the type of contract and (ii) its parameter(s) has been reached before playing the game. Here supply chain members can choose among a WPC or an RSC. The WPC has one parameter, the transfer price w, while the RSC has two parameters, the transfer price and the share, ϕ, of the retailer's revenue that the manufacturer gets from the retailer. The parameters w and ϕ are exogenously determined and remain constant over time.

- The transfer price $w \geq 0$ is paid by the retailer to the manufacturer for each unit purchased and we denote by $w|_{\text{WPC}}$ and $w|_{\text{RSC}}$ the transfer prices that apply under a WPC and an RSC, respectively. We require a transfer price to be positive and $w|_{\text{RSC}}$ be lower than $w|_{\text{WPC}}$, that is, the retailer would agree to an RSC only if the manufacturer lowers the transfer price that applies in the WPC: *quid pro quo*.
- The second parameter is the manufacturer's share ϕ of the retailer's revenue and we require this share to be nonnegative and less than one. The WPC is a special case of an RSC with $\phi = 0$ and a larger transfer price. The rationale for an RSC is to decrease the retailer's unit procurement cost in order to induce the retailer to buy more units.

It remains to formulate the objective functions of the two firms. As to the manufacturer's costs, we suppose that the manufacturing cost increases with the production rate according to the quadratic function $au(t)^2/2$, $a > 0$ and constant. The manufacturer incurs a cost of advertising effort, expressed by the quadratic function $bw(t)^2/2$, $b > 0$ and constant. Finally, the manufacturer's cost of backlogging is $cX(t)^2/2$, $c > 0$ and constant. Note that if the backlogging turns to inventory, i.e., $X < 0$, the manufacturer then incurs an inventory cost. Finally, if the manufacturer uses a safety stock, its cost is sunk and be disregarded.

Assume that both firms employ a constant discounting rate, denoted by $r > 0$. Then the manufacturer's objective is

$$\underset{u(t),w(t)\geq 0}{\text{Max}} \ \Pi^M = \int_0^\infty e^{-rt}[\phi p(t)S(t) + wv(t) - au(t)^2/2 - bw(t)^2/2 - cX(t)^2/2]dt.$$

(5)

The retailer's gross revenue is $p(t)S(t)$ under a WPC, and $\bar{\phi}p(t)S(t) - w|_{\text{RSC}}v(t)$, $\bar{\phi} \equiv 1 - \phi$, under an RSC. The retailer's ordering/processing cost is increasing and is convex in the purchase rate: $dv(t)^2/2$, $d > 0$ and constant Finally, the retailer's backlogging cost is $eY(t)^2/2$, $e > 0$ and constant.

The retailer's objective is

$$\underset{v(t),p(t)\geq 0}{\text{Max}} \ \Pi^R = \int_0^\infty e^{-rt}[\bar{\phi}p(t)S(t) - wv(t) - dv(t)^2/2 - eY(t)^2/2]dt \quad (6)$$

We note that all costs for the manufacturer and retailer, production, advertising, retailer ordering, and backlog, are modeled as being quadratic. This is to be consistent with existing literatures, which typically assume strictly convex costs for advertising [Jørgensen and Zaccour, 2004; Erickson, 2003], production [Jørgensen et al., 1999], ordering [Jørgensen, 1986], and backlog [Feichtinger and Hartl, 1985; Erickson, 2011].

To avoid being entangled in mathematical subtleties we confine our interest to equilibrium outcomes for which the objective integrals in (5) and (6) converge for all admissible states, controls, and parameter values.

3. Open-loop Nash Equilibrium Strategies

This section identifies an open-loop Nash equilibrium (OLNE). The assumption here is that information on the backlogs as well as the goodwill stock is not available and the firms' strategies depend on time only.

Omitting from now on the time argument when no confusion can arise, the manufacturer's (current-value) Hamiltonian is

$$H^M = \phi p(\alpha - \beta p + G) + wv - au^2/2 - bw^2/2 - cX^2/2$$
$$+ \lambda_1(v - u) + \lambda_2(\alpha - \beta p + G - v) + \lambda_3(w - \delta G), \tag{7}$$

where $\lambda_1(t)$, $\lambda_2(t)$, and $\lambda_3(t)$ are the manufacturer's (current-value) costate variables, associated with state variables, X, Y, and G, respectively.

The costate equations are given by

$$\dot{\lambda}_1 = r\lambda_1 + cX, \tag{8}$$

$$\dot{\lambda}_2 = r\lambda_2, \tag{9}$$

$$\dot{\lambda}_3 = (r + \delta)\lambda_3 - \lambda_2 - \phi p. \tag{10}$$

The retailer's (current-value) Hamiltonian is

$$H^R = \bar{\phi} p(\alpha - \beta p + G) - wv - dv^2/2 - eY^2/2$$
$$+ \mu_1(v - u) + \mu_2(\alpha - \beta p + G - v) + \mu_3(w - \delta G), \tag{11}$$

where $\mu_1(t)$, $\mu_2(t)$, and $\mu_3(t)$ are (current-value) costates.

The costate equations are

$$\dot{\mu}_1 = r\mu_1, \tag{12}$$

$$\dot{\mu}_2 = r\mu_2 + eY, \tag{13}$$

$$\dot{\mu}_3 = (r + \delta)\mu_3 - \bar{\phi} p - \mu_2. \tag{14}$$

The state equations are as in (1), (2), and (4). For the case where optimal controls are positive, necessary optimality conditions are (1), (2), (4), (8)–(10), (12)–(14), and

$$H_u^M = 0 \Rightarrow u^{\text{ol}} = -\lambda_1/a, \tag{15}$$

$$H_w^M = 0 \Rightarrow w^{\text{ol}} = \lambda_3/b \tag{16}$$

for the manufacturer, and

$$H_v^R = 0 \Rightarrow v^{\text{ol}} = -(\omega + \mu_2)/d, \tag{17}$$

$$H_p^R = 0 \Rightarrow p^{\text{ol}} = [(\alpha + G^{\text{ol}})/\beta + (dv^{\text{ol}} + w)/\bar{\phi}]/2 \tag{18}$$

for the retailer. The superscript "ol" refers to "open-loop". It is readily shown that the Hamiltonians are strictly concave in the decision variables, which guarantees unique maxima. Analysis of the necessary conditions provides the following results.

Lemma 1. *For the case where optimal controls are positive, the manufacturer's OLNE production rate and advertising effort rate satisfy*

$$\dot{u}^{ol} = ru^{ol} - cX^{ol}/a, \qquad u^{ol}_\infty = cX^{ol}_\infty/ra, \tag{19}$$

$$\dot{w}^{ol} = (r+\delta)w^{ol} - \phi p^{ol}/b, \quad w^{ol}_\infty = \phi p^{ol}_\infty/b(r+\delta), \tag{20}$$

where the subscript ∞ indicates that controls are computed in steady state.

Proof. See Appendix A.1. □

The second equation in (19) shows — as expected — that the larger the backlog, the larger the production rate in steady state. The second equation in (20) shows that, in steady state, the larger the consumer price, the larger the manufacturer's advertising effort. This is because price and goodwill have opposite effects on demand. The second equation in (20) also shows that a larger discounting rate implies a smaller steady state advertising effort. The intuition is that shortsighted manufacturers should not invest very much in long-run advertising goodwill.

Proposition 1. *Under a WPC, the manufacturer's advertising effort rate is always zero. Under an RSC, the effort rate is always positive.*

Proof. See Appendix A.2. □

The first part of Proposition 1 tells a manufacturer operating under a WPC that it is not worthwhile to invest in advertising (to increase consumer demand) because the manufacturer's gross revenue is independent of retail sales. Consequently, brand goodwill decreases steadily over time. With an RSC, the manufacturer has an incentive to raise brand goodwill — and hence consumer demand — because the manufacturer receives a positive share of the retailer's revenue.

Lemma 2. *The retailer's OLNE purchase rate and consumer price satisfy*

$$\dot{v}^{ol} = rv^{ol} - (eY^{ol} - rw)/d, \quad v^{ol}_\infty = (eY^{ol}_\infty - rw)/rd, \tag{21}$$

$$\dot{p}^{ol} = \frac{1}{2}(\dot{G}^{ol}/\beta + d\dot{v}^{ol}/\bar{\phi}), \quad p^{ol}_\infty = [(\alpha + G^{ol}_\infty)/\beta + eY^{ol}_\infty/r\bar{\phi}]/2. \tag{22}$$

Proof. See Appendix A.3. □

Equation (20) shows that in steady state, the larger the retailer's backlog the higher the purchase rate. Given that the transfer price is higher under a WPC than an RSC, the retailer has a lower steady state purchase rate under a WPC. This is intuitive. Using (21) shows that (in steady state) the larger the retailer's backlog, the higher the consumer price. Also this is intuitive: if the retailer is understocked, she should decrease demand by raising the consumer price.

Consider an RSC in steady state and use (3), (20), and (22). A higher consumer price leads (by (3)) to lower consumer demand but it also implies a higher level of manufacturer advertising (by (20)). More advertising implies a higher stock of goodwill and (by (3)) greater consumer demand. By (22), a higher goodwill stock implies a higher consumer price. The managerial implication of these effects is that should — for any reason — the retailer increase the consumer price, the resulting decrease in demand/sales will be mitigated by the manufacturer, who responds by increasing advertising to make the goodwill stock larger and thereby stimulate consumer demand.

The next proposition shows that an equilibrium under a WPC exhibits nice structural properties.

Proposition 2. *Under a WPC, the steady state is unique and the equilibrium path converging to the steady state is monotonic.*

Proof. See Appendix A.4. □

Under a WPC, steady state values for production and backlogging, as well as procurement and backlogging, are all strictly positive if the transfer price is not larger than the steady state consumer price, i.e., $p^{ol}_{\infty|WPC} \geq \omega_{|WPC}$. This requirement is likely to be satisfied in practice; one would not expect a retailer to have a negative gross margin.

Technically, long-run feasibility and local stability of the equilibrium under a WPC did not require additional assumptions. We conclude that in steady state, backlogs are constant and consumer demand equals production, which in turn equals the purchase rate. Advertising effort as well as the goodwill stock is zero.

Proposition 3. *Under an RSC, the supply chain has a unique steady state. This steady state is feasible and is a saddle point in the control-state space if the following conditions are satisfied*

$$r < 1/\beta\delta b - \delta \quad and \quad \phi < \tilde{\phi},$$

where $\tilde{\phi}$ is an upper threshold of ϕ that is strictly lower than 1. Under the two conditions, the equilibrium path converging to the steady state is monotonic.

Proof. See Appendix A.5. □

According to Proposition 3, the equilibrium path to the steady state under an RSC is monotonic if the sharing parameter is sufficiently small, $\phi \leq \tilde{\phi}$. The parameter ϕ may be seen as a proxy for the bargaining position for the manufacturer. If ϕ is higher than $\tilde{\phi}$, the steady state cannot be reached from some or all initial states [Engwerda, 1998]. The implication is that the RSC is not superior to a WPC as a long-term contract in an OLNE because it does not ensure stability while WPC does. This interesting feature of the RSC was not envisioned in the literature on supply chain contracting and coordination in static or shortterm setups [Cachon, 2003; Cachon and Lariviere, 2005; El Ouardighi *et al.*, 2008; El Ouardighi and Kim, 2010]. Long-run feasibility and stability of an equilibrium under an RSC can be ensured only by contracts that have a sharing parameter being lower than the threshold $\tilde{\phi}$. In addition, the inequality $r < 1/\beta\delta b - \delta$ must hold. The interpretation of this inequality is that the discount rate must not be "too large", that is, supply chain members must be relatively farsighted.

In the case where the threshold for the sharing parameter is close to 1, the manufacturer's production rate and the retailer's purchase and sales rates are close to zero. This situation is sustainable, however, because the retailer is able to compensate the high manufacturer share in the sales revenue by charging a high consumer price, which is backed up by a substantial manufacturer advertising effort. (Technically, the local monotonicity of the equilibrium path can still be assured, though globally the paths may be nonmonotonic).

It can be shown that whatever the compensation scheme, the manufacturer's steady state backlog rate, relative to the retailer's steady state purchase rate, i.e., $X_\infty^{\text{ol}}/v_\infty^{\text{ol}}$, equals ra/c (see Appendix A.5). However, the retailer's steady state backlog rate, relative to consumer sales, i.e., $Y_\infty^{\text{ol}}/S_\infty^{\text{ol}}$, is lower under an RSC than a WPC if $\phi \leq (\omega_{|\text{WPC}} - \omega_{|\text{RSC}})/\omega_{|\text{WPC}}$. If this inequality is satisfied, an RSC is more effective than a WPC in meeting demand in the sense that relative understocking at the retailer's outlet is reduced. This result confirms what happens in practice when an RSC is introduced. The main idea of an RSC is to enable the retailer to buy more units by lowering the transfer price, compared to a WPC. This reduces the likelihood of the retailer being understocked.

The condition $\phi \leq (\omega_{|\text{WPC}} - \omega_{|\text{RSC}})/\omega_{|\text{WPC}}$ can be illustrated by the data from the Blockbuster case in Mortimer [2008]. We rewrite the condition as follows: $\bar{\phi} \geq \omega_{|\text{RSC}}/\omega_{|\text{WPC}}$. The left-hand side of this inequality is the fraction of retail revenue that the retailer keeps for herself while the right-hand side is a positive number, strictly less than 1. In the Blockbuster example, $\bar{\phi}$ is approximately 60% while the fraction on the right-hand side is $\$8/\$65 \approx 0.12$. Hence the condition is satisfied.

4. Feedback Nash Equilibrium Strategies

If information on backlogs and advertising goodwill is available to both firms throughout the game, they may condition their current actions on the current values of these stocks, as well as on time In this case one can look for a feedback Nash equilibrium (FBNE). To identify an FBNE, we derive the Hamilton–Jacobi–Bellman (HJB) equations of the firms and then characterize the equilibrium strategies. The model we have formulated has an infinite horizon and parameters are constant. In such a case, one can look for strategies that are stationary in the sense that they depend on the state but not on time (explicitly).

Let $V^M(X, Y, G)$ be the manufacturer's value function, which represents the optimal profit of the manufacturer in a game that starts out at time t in state $(X(t), Y(t), G(t))$. The manufacturer's HJB equation then is

$$rV^M = \underset{u,w}{\text{Max}}\{\phi p(\alpha - \beta p + G) + wv - au^2/2 - bw^2/2 - cX^2/2$$

$$+ V_X^M(v - u) + V_Y^M(\alpha - \beta p + G - v) + V_G^M(w - \delta G)\}, \qquad (23)$$

where subscripts on V^M denote partial differentiation. Necessary conditions for a maximum on the right-hand side of (22) are, if production and advertising rates are positive,

$$u^{\text{fb}} = -V_X^M/a, \qquad (24)$$

$$w^{\text{fb}} = V_G^M/b, \qquad (25)$$

where the superscript "*fb*" refers to "feedback".

The retailer's HJB equation is

$$rV^R = \underset{v,p}{\text{Max}}\{\bar{\phi} p(\alpha - \beta p + G) - wv - dv^2/2 - eY^2/2$$

$$+ V_X^R(v - u) + V_Y^R(\alpha - \beta p + G - v) + V_G^R(w - \delta G)\}, \qquad (26)$$

where $V^R(X, Y, G)$ is the retailer's value function. Necessary conditions for a maximum are, if the purchase rate and the consumer price are positive,

$$v^{\text{fb}} = (V_X^R - V_Y^R - w)/d, \qquad (27)$$

$$p^{\text{fb}} = [(\alpha + G)/\beta - V_Y^R/\bar{\phi}]/2. \qquad (28)$$

Proposition 4. *Whenever all decision variables are positive, FBNE strategies for production, advertising effort, procurement, and consumer price are as follows:*

$$u^{\text{fb}} = -(B^M + 2C^M X + I^M Y + J^M G)/a, \qquad (29)$$

$$w^{\text{fb}} = (F^M + J^M X + K^M Y + 2H^M G)/b, \qquad (30)$$

$$v^{\text{fb}} = [B^R - D^R - w + (2C^R - I^R)X + (I^R - 2E^R)Y + (J^R - K^R)G]/d, \qquad (31)$$

$$p^{\text{fb}} = \frac{1}{2}\left[\frac{\alpha}{\beta} - \frac{D^R}{\bar{\phi}} - \frac{I^R}{\bar{\phi}}X - \frac{2E^R}{\bar{\phi}}Y + \left(\frac{1}{\beta} - \frac{K^R}{\bar{\phi}}\right)G\right] \qquad (32)$$

in which B^M, C^M, F^M, H^M, I^M, J^M, K^M, B^R, C^R, D^R, E^R, I^R, J^R, K^R *are time-independent parameters of the two value functions*

$$V^M(X, Y, G) = A^M + B^M X + C^M X^2 + D^M Y + E^M Y^2$$

$$+ F^M G + H^M G^2 + I^M XY + J^M XG + K^M YG, \qquad (33)$$

$$V^R(X, Y, G) = A^R + B^R X + C^R X^2 + D^R Y + E^R Y^2$$

$$+ F^R G + H^R G^2 + I^R XY + J^R XG + K^R YG. \qquad (34)$$

Proof. See Appendix A.6. □

The FBNE strategies are determined analytically by (29)–(32) as linear functions of the three state variables. To determine the value function coefficients one has to solve a system of 14 nonlinear algebraic equations which does not seem to be possible. Therefore we use a numerical procedure. From the set of numerical solutions we select those that ensure a globally asymptotically stable steady state solution (if it exists).

Technical Remark. See Appendix A.7.

For the numerical solutions we use the baseline parameters given in Table 1.

In the supply chain management literature, it is usual to set the value of r between 0.05 [e.g., El Ouardighi and Erickson, 2015] and 0.1 [e.g., El Ouardighi, 2014]. Here, we choose the upper value $r = 0.1$ to emphasize the relative importance of short time operational performances. Note that the wholesale price is set equal to zero in the RSC. In sensitivity analyses, a broad range of values is used for all parameters except w and ϕ. For each parameter, the solutions are calculated for values deviating 25% and 50% above and below the baseline value. Neither the feasibility nor the qualitative pattern of the solutions generated was affected by variations up to 50% from the baseline values.

Technical Remark. See Appendix A.8.

We use the optimal values of the contract parameters in an FBNE that are derived in the next section (see Sec. 5.1). For an optimal WPC with transfer price $w = 73.98$, the eigenvalues of the Jacobian in Appendix A.7. are all real and negative which ensures monotonic convergence to the stable steady state. The coefficients of the value functions for this solution are given in Table 2.

Using (29)–(32) and the values in Table 2, FBNE strategies with the optimal WPC are

$$u^{\text{fb}}_{|\text{WPC}} = 91.36123 + 0.32733X + 0.050286Y + 0.34015G, \tag{35}$$

Table 1.   Baseline parameters.

Parameter	α	β	a	b	c	d	e	δ	r	w	ϕ
Value	1000	10	0.1	2	0.05	0.01	0.1	0.1	0.1	$w_{|\text{WPC}} = 100,$ $w_{|\text{RSC}} = 0$	$\phi_{|\text{WPC}} = 0,$ $\phi_{|\text{RSC}} = 0.8$

Table 2.   Value function coefficients in a WPC.

A^M	B^M	C^M	D^M	E^M	F^M	H^M	I^M	J^M	K^M
112,829.5367	−9.1361	−0.0327	59.244	−0.0215	127.5785	−0.0302	−0.0502	−0.034	−0.032
A^R	B^R	C^R	D^R	E^R	F^R	H^R	I^R	J^R	K^R
74,405.9114	−2.199571	0.00003	−75.1883	−0.0151	88.6002	0.0651	0.0007	−0.00249	−0.0068

$$w^{\text{fb}}_{|\text{WPC}} = 63.78926 - 0.0170075X - 0.016004Y - 0.030274G, \tag{36}$$

$$v^{\text{fb}}_{|\text{WPC}} = -99.1623 - 0.0019X + 3.0371Y + 0.4402G, \tag{37}$$

$$p^{\text{fb}}_{|\text{WPC}} = 87.59417 - 0.0000395X + 0.015146Y + 0.053449G. \tag{38}$$

According to (35), the production rate increases if the manufacturer or retailer backlog grows. This is to avoid understocking and is expected. A higher level of goodwill increases consumer demand which motivates increasing production.

Equation (36) says that the manufacturer should use less advertising effort to promote consumer demand when there are backlogs or when the goodwill stock is already large. These results confirm intuition.

Using (37) shows that a larger manufacturer backlog (i.e., poorer availability) deters retailer procurement, as expected. The retailer buys more if her own backlog increases. Realizing that a higher level of goodwill increases consumer demand induces the retailer to purchase more.

According to (38), the retailer sets a lower consumer price when there is a backlog at the manufacturer level (due to insufficient supply). This is consistent with the manufacturer's decreasing advertising effort in such situations. In contrast, the retailer sets a higher price to lower demand when there is a backlog at the retail level. Finally, a higher level of goodwill increases consumer demand and justifies an increase in the consumer price.

For an optimal RSC with parameters $\omega = 0$, $\phi = 0.7$ (see Sec. 5.1), the eigenvalues associated with the stable steady state are real and negative and hence convergence to the steady state is monotonic. The values of the coefficients of the value functions under an RSC are given in Table 3.

Using (29)–(32) and the values in Table 3, the strategies in an optimal RSC are

$$u^{\text{fb}}_{|\text{RSC}} = 369.96161 + 0.6551X + 0.45885Y + 0.32778G, \tag{39}$$

$$w^{\text{fb}}_{|\text{RSC}} = 50.98738 - 0.016389X - 0.014757Y + 0.032101G, \tag{40}$$

$$v^{\text{fb}}_{|\text{RSC}} = 402.262 - 0.0013X + 2.8889Y + 0.3985G, \tag{41}$$

$$p^{\text{fb}}_{|\text{RSC}} = 60.22474 - 0.000065X + 0.048083Y + 0.058467G. \tag{42}$$

Comparing the signs of the value function coefficients under a WPC in (35)–(38) with those of an RSC in (39)–(42) shows no differences, except for manufacturer advertising effort which depend negatively on goodwill under a WPC, but positively under an RSC. There is no straightforward explanation to this, noting that goodwill

Table 3.   Value function coefficients in an RSC.

A^M	B^M	C^M	D^M	E^M	F^M	H^M	I^M	J^M	K^M
112,599.0816	−36.9961	−0.0327	33.6214	−0.0201	101.9747	0.0321	−0.0458	−0.0327	−0.0295
A^R	B^R	C^R	D^R	E^R	F^R	H^R	I^R	J^R	K^R
113,011.2873	−2.1122	0.00001	−6.1348	−0.0144	93.9565	0.0275	0.0003	−0.001	−0.005

enters in a complicated manner in the value functions given by (33)–(34). With this exception we conclude that there are no qualitative differences between the way in which the state variables influence the operations and marketing decisions under the two contracts.

5. Comparisons of Contracts and Strategies

We assume, as would be the case in practice, that the choice of a contract precedes that of strategy type. We wish to assess the "coordination ability" of the two contracts, under open-loop and feedback strategies, respectively, and characterize the evolution of operations and marketing variables over time. For this purpose we use the analytical solutions from Secs. 3 and 4. To see how efficient a given contract is in coordinating the supply chain, we need a benchmark which is taken to be optimal cooperative solution which is defined as the solution which maximizes the overall profit of the supply chain.

Payoffs are investigated when contract parameters and strategies vary. For each strategy, an optimal contract is first defined. Then, contracts and strategies are compared with the cooperative solution in terms of their payoffs and policies.

To determine the cumulative payoffs and the operations and marketing policies under each contract and each strategy, we use the baseline parameters of Table 1. Solution paths are calculated for initial state values $(X_0, Y_0, G_0) = (0, 0, 50)$. Thus the assumption is that firms start out with zero backlogs. For the numerical solution, a boundary value approach and a continuation algorithm, described in Grass *et al.* [2008] and Grass [2012] are used.

5.1. *Optimal contracts and strategies*

The reader should be aware that now we abandon the assumption that contract parameters have fixed values that are exogenously given. In this section, we wish to determine an optimal contract for each type of strategy (open-loop and feedback). The meaning of "optimal" will be made clear below.

Starting with a WPC (that is, given the sharing parameter ϕ is zero) we vary the transfer price ω to find out what happens to manufacturer and retailer profits in an OLNE and a FBNE, respectively. Then, considering an RSC, we vary both the transfer price ω and the sharing parameter ϕ to answer the same question. As in El Ouardighi [2014], the manufacturer determines an optimal WPC by choosing a transfer price that maximizes her individual profit. The retailer can accept or reject the contract. Given that the retailer accepts the contract, Fig. 1 shows the firms' cumulative profits under WPC and both strategy types. Figure 1 depicts the manufacturer and retailer overall payoffs in a WPC and an OLNE, $\Pi^{M^{ol}}_{|WPC}$ and $\Pi^{R^{ol}}_{|WPC}$, respectively, as well as $\Pi^{M^{fb}}_{|WPC}$ and $\Pi^{R^{fb}}_{|WPC}$ in a WPC and a FBNE, respectively, as functions of the transfer price ω. For comparison, an uniform division of the cooperative profits, $\Pi^c/2$, is also depicted.

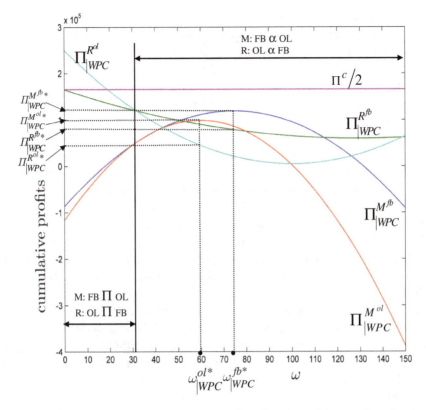

Fig. 1. OLNE and FBNE payoffs in a WPC, as functions of the transfer price.

An optimal WPC is characterized both by a higher transfer price ($\omega^{\text{fb}*}_{|\text{WPC}} = 73.98 > \omega^{\text{ol}*}_{|\text{WPC}} = 60.15$) and payoff in a FBNE than in an OLNE, for both manufacturer, $\Pi^{M^{\text{fb}}*}_{|\text{WPC}} = 1.19125 \times 10^5 > \Pi^{M^{\text{ol}}*}_{|\text{WPC}} = 9.8865 \times 10^4$ and retailer, $\Pi^{R^{\text{fb}}*}_{|\text{WPC}} = 7.8993 \times 10^4 > \Pi^{R^{\text{ol}}*}_{|\text{WPC}} = 4.4626 \times 10^4$. Therefore, if the supply chain members are able to have and share state information, both firms prefer an optimal WPC and FBNE to an optimal WPC and OLNE because they get higher profits although it creates a greater double marginalization effect The reason is that a higher transfer price in a FBNE leads to a higher consumer price, but — as a countermeasure to stimulate consumer demand — also induces the manufacturer to develop goodwill through increased advertising effort. A higher transfer price is desirable in a FBNE because it increases supply chain members' payoffs due to greater sales. This suggests that advertising in a FBNE could mitigate the effects of double marginalization on profits by stimulating the consumer demand. If the supply chain members do not share state information, or if the cost of getting and sharing state information is too large, that is, greater than $\Pi^{M^{\text{fb}}*}_{|\text{WPC}} - \Pi^{M^{\text{ol}}*}_{|\text{WPC}}$ for the manufacturer and $\Pi^{R^{\text{fb}}*}_{|\text{WPC}} - \Pi^{R^{\text{ol}}*}_{|\text{WPC}}$ for the retailer, an optimal WPC in an OLNE is the only option for both firms. Although the transfer price in this case is smaller than in a FBNE, double marginalization is stronger because the manufacturer cannot use advertising to mitigate its effects on sales and profits The rationale behind this result is suggested by Eq. (10), which represents the rate of change of the manufacturer's marginal incentive to invest in goodwill advertising. Under WPC and OLNE, the manufacturer has no incentive to invest in advertising because the retailer's backlog and marginal sales revenue are both payoff-irrelevant for the manufacturer (i.e., both λ_2 and ϕ are zero). In an OLNE, the retailer's marginal sales revenue is payoff-relevant for the manufacturer if ϕ is positive, that is, under a RSC. Under a WPC, the retailer's backlog is payoff-relevant for the manufacturer if λ_2 is nonzero. In particular, if the retailer's backlog is implicitly beneficial for the manufacturer, it will increase sales through greater goodwill advertising. In return, greater sales will induce the retailer to increase its purchase rate from the manufacturer to incur lower backlog costs. Overall, the manufacturer invests in advertising under WPC if it has an interest in raising the retailer's backlog cost. Conversely, if the manufacturer cannot observe the retailer's backlog, advertising efforts are not beneficial.

The search for a mutually beneficial alternative to an optimal WPC should aim at minimizing the difference between the supply chain's cooperative and the sum of noncooperative payoffs. The reason is that an optimal RSC is acceptable by both firms if it is profit Pareto-improving, that is, no firm gets a lower profit under an optimal RSC than under an optimal WPC. The joint cooperative profits are the upper bound on firms' profits and the Nash bargaining scheme [Dockner *et al.*, 2000] is used to determine the RSC parameters The retailer computes and selects the value of the sharing parameter that minimizes the difference between the cooperative profit and the sum of noncooperative payoffs that is, $\Pi^c - (\Pi^M_{|\text{RSC}} + \Pi^R_{|\text{RSC}})$.

To motivate the above we note that if the retailer sets the sharing parameter to maximize her own profit we can end up in a solution which is not Pareto-improving. That is, for a fixed transfer price being lower than that under a WPC, the best noncooperative solution for the retailer might be to share as little revenue as possible with the manufacturer. For the manufacturer to agree to charge a transfer price which is lower than that under a WPC, the RSC should be profit Pareto-improving. Given that the manufacturer charges such a transfer price, the retailer has to choose a sharing parameter between 0 and 1 as neither 0 nor 1 are Pareto-improving, but there is an interval of values of the sharing parameter that is profit Pareto-improving. We know the upper bounds on the firms' profits and to get the best Pareto-improving RSC (given the value of the transfer price) we must find the value of the sharing parameter that makes individual profits as close as possible to their respective upper bounds. The adoption of a RSC is not to maximize the retailer's profits but to coordinate the SC. This is the way in which we compute the curves in Fig. 2.

Under an RSC, one contract parameter needs to be exogenously specified [Cachon and Lariviere, 2005; Jørgensen, 2011]. We assume that the manufacturer

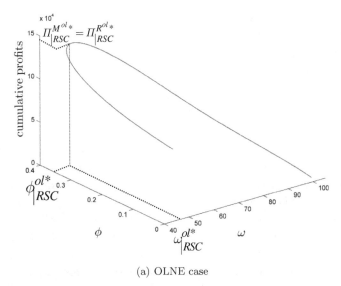

(a) OLNE case

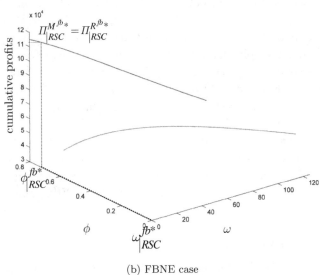

(b) FBNE case

Fig. 2.   Optimal RSC in OLNE and FBNE.

sets a more advantageous transfer price to the retailer than that under an optimal WPC. The retailer, in turn, computes the value of the sharing parameter that minimizes the difference between the supply chain's cooperative and the sum of noncooperative payoffs. To identify an optimal RSC we compute for each strategy the intersection points between the firms' payoffs curves that are at the shortest distance from the cooperative payoffs curve. Figure 2 represents the curves of these intersection points as functions of transfer price and sharing parameter in an OLNE (a) and an FBNE (b).

As expected, an optimal RSC is characterized by a strictly lower transfer price than an optimal WPC in OLNE or in FBNE. However, while the transfer price should be strictly positive in an OLNE ($\omega^{\text{ol}*}_{|\text{RSC}} = 45.5$), it is zero in an FBNE ($\omega^{\text{fb}*}_{|\text{RSC}} = 0$). The optimal sharing parameter is considerably larger in an FNBE than in an OLNE, $\phi^{\text{fb}*}_{|\text{RSC}} = 0.7 > \phi^{\text{ol}*}_{|\text{RSC}} = 0.35$. The firms' payoffs are signifi-cantly greater in an OLNE than in an FBNE, $\Pi^{M^{\text{ol}}*}_{|\text{RSC}} = \Pi^{R^{\text{ol}}*}_{|\text{RSC}} = 1.40689 \times 10^5 > \Pi^{M^{\text{fb}}*}_{|\text{RSC}} = \Pi^{R^{\text{fb}}*}_{|\text{RSC}} = 1.18 \times 10^5$. Both firms prefer an optimal RSC in an OLNE to an optimal RSC in an FBNE. Sensitivity analysis shows that this holds also true for smaller discounting rates and greater demand price-sensitivity. This suggests that farsightedness under an RSC increases the value of the commitment which is inherent in the OLNE. However, because $\Pi^{M^{\text{fb}}*}_{|\text{WPC}} > \Pi^{M^{\text{fb}}*}_{|\text{RSC}}$, an RSC in an FBNE is not profit Pareto-improving because the manufacturer prefers a WPC in an FBNE. Actually, an RSC in an FBNE is a dominated option for both firms. Under an RSC and OLNE, the state information is disregarded which makes it not only more profitable but also less constraining in terms of state information collecting and sharing than RSC and FBNE for both firms. Because both firms' profits are strictly greater under optimal RSC in an OLNE than under optimal WPC in an FBNE, $\Pi^{M^{\text{ol}}*}_{|\text{RSC}} > \Pi^{M^{\text{fb}}*}_{|\text{WPC}}$ and $\Pi^{R^{\text{ol}}*}_{|\text{RSC}} > \Pi^{R^{\text{fb}}*}_{|\text{WPC}}$, both firms are willing to adopt an RSC in an OLNE. Finally, if supply chain members are unable to condition their decisions upon the current state, they have a strong incentive to select an RSC because the increase in profits entails no additional constraints in terms of collecting and shar-ing state information. This result is notable, although it may be counterintuitive. Therefore, a switch from a WPC to an RSC may require a change in the strategy type, from a contingent to a commitment strategy. Each contract type enables the supply chain members to effectively "align their interests" so that they both prefer the same type of equilibrium — either a commitment or a contingent strategy — depending on the state information availability under a WPC, and a commitment strategy under an RSC.

5.2. *Optimal operations and marketing policies*

Supply chain members play for any given contract, the Nash strategies identified above. To analyze the transient behavior of operations and marketing policies, we generate the time paths of the control variables under each contract (WPC and RSC) and strategy (open-loop and feedback).

In Figs. 3(a) and 3(b), the time paths of the manufacturer's production rate, the retailer's purchase rate and the sales rate are graphed on a logarithmic scale for the cooperative and noncooperative equilibria. The time axis is also logarithmically scaled to show the paths in more detail on an initial interval of time where the significant dynamic changes take place.

In the cooperative setting (Figs. 3(a) and 3(b)), operations decisions variables and sales have S-shaped time paths. The production and the purchase rates are initially low and increase steadily over time until the steady state is reached in finite

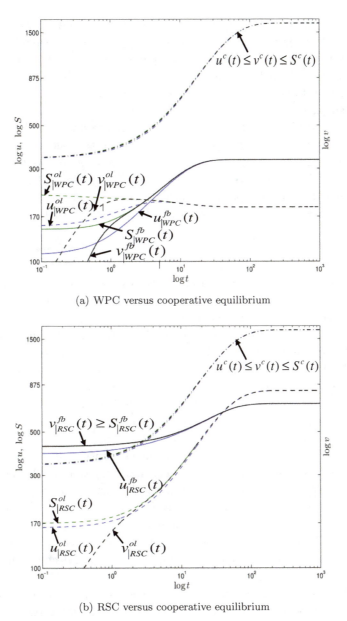

(a) WPC versus cooperative equilibrium

(b) RSC versus cooperative equilibrium

Fig. 3.   Operations policies and sales under optimal WPC and RSC in OLNE and FBNE versus cooperative case.

time. Though not quite visible, the transient sales are greater than the purchase rate, which in turn exceeds the production rate. This makes the backlogs of both firms increase. Except during an initial phase, the cooperative production, purchase and sales rates are greater than in the noncooperative case.

Under a WPC and OLNE (Fig. 3(a)), operations policies and sales have quite different patterns than in the cooperative setting. Sales are (slightly) monotonically decreasing while the production and purchase rates first increase and then (slightly) decrease. Initially, the manufacturer holds a positive inventory because the production rate is greater than the retailer's purchase rate. Later on, the purchase rate surpasses the production rate for a finite time interval until they both equalize at the steady state. However, the purchase rate starts at a lower level than sales until the steady state. The production, purchase and sales rates in an OLNE are significantly lower than in FBNE during an initial period. With a WPC and FBNE (Fig. 3(a)), convergence of these three variables is similar than to that of an OLNE, except that all time paths are increasing. The dynamic adjustment of the purchase and sales rates toward the steady state is similar to that in the cooperative solution.

Under an RSC (Fig. 3(b)), operations policies in OLNE and FBNE, by and large, follow the same pattern as the cooperative solution. Yet they differ in that the OLNE strategy displays (inferior) parallel paths to the cooperative paths, while the FBNE paths are flatter with greater initial values than the cooperative equilibrium and lower steady state values than the OLNE.

Figures 4(a) and 4(b) show the time paths of the manufacturer's advertising effort and the consumer price for the cooperative solution and the noncooperative equilibria. In the cooperative equilibrium, advertising effort and consumer price also have S-shaped time paths with increasing values until the steady state is reached. The marketing instruments are used in a way such that an increase in the consumer price (which decreases demand) goes along with an increase in advertising effort (which increases demand).

Under a WPC in an OLNE (Fig. 4(a)), the advertising effort is zero while the consumer price decreases (slightly) over time. According to Eq. (22), the decreasing time path of the consumer price results from the decrease of both the manufacturer's goodwill and the retailer's purchase rate (see Fig. 3(a)). The decreasing time path of sales (Fig. 3(a)) is due to the fact that goodwill decreases faster than the consumer price. Under a WPC and FBNE, marketing instruments affect demand in the same way: An increase in the consumer price goes along with a decrease in advertising effort (both decrease demand).

Marketing instruments are strategic complements under an RSC in both OLNE and FBNE, as in the cooperative solution. For a given strategy type, the time paths of marketing strategies are affected by the contract type. With an RSC and OLNE, the marketing policy prescribes under-advertising and successively over-pricing over a (brief) initial time interval and under-pricing thereafter. Cooperation is the more effective way of increasing demand because cooperative advertising effort is greater.

A WPC and OLNE or FBNE will in general lead to lower consumer prices than an RSC and OLNE. In a static setting, Cachon and Lariviere [2005] obtain that the consumer price is lower with an RSC than a WPC. In a dynamic setting, the retailer can set a higher consumer price under an RSC by rewarding the manufacturer for

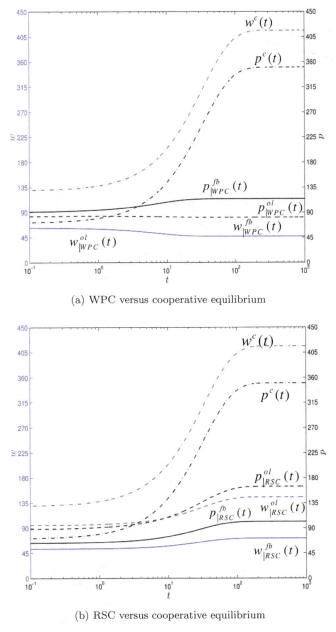

(a) WPC versus cooperative equilibrium

(b) RSC versus cooperative equilibrium

Fig. 4.   Marketing policies under optimal WPC and RSC in OLNE and FBNE versus cooperative equilibrium.

her efforts to develop the market potential through increased goodwill. The strategic interaction between the two marketing instruments eliminates the need to reduce the consumer price, a standard argument for the use of an RSC. Finally, the strategy type with the highest consumer price is more profitable.

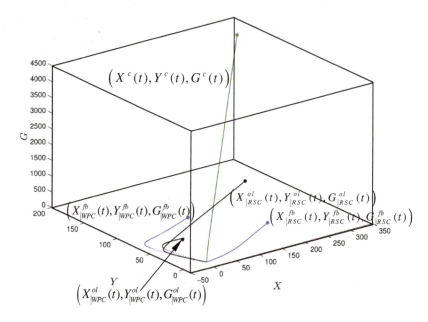

Fig. 5.   Phase diagram in the state space for cooperative and noncooperative equilibria.

In Fig. 5, the phase diagram of the manufacturer's backlog and advertising goodwill and the retailer's backlog is depicted for the cooperative solution and the noncooperative equilibria.

Although cooperation is more effective than noncooperation in developing demand (owing to higher advertising goodwill), it also leads to larger backlogs for the manufacturer and retailer. In the noncooperative game, the case of a WPC and OLNE can be described in a way opposite to that of the cooperative solution because it leads to a lower sales rate, due to absence of advertising effort, as well as smaller backlogs for both firms. An RSC and OLNE have more similarities with the cooperative solution than a WPC and OLNE. Sensitivity analysis verifies this for low values of the production and advertising cost coefficients and high price-sensitivity of demand. An RSC and FBNE differ from a WPC and FBNE because the former results in a lower retailer backlog and greater manufacturer backlog and goodwill. This holds for low values of the production cost coefficient and the retailer's backlogging cost, high values of the advertising cost coefficient, and high price-sensitivity of demand.

Due to the differences between the cooperative and noncooperative values of the retailer's purchase rate and sales volume, it is appropriate to assess understocking at the manufacturing and retailing levels in relative instead of absolute terms. To do so, we determine the transient paths of the relative backlog for each firm, that is, $X(t)/v(t)$ for the retailer, defined as the percentage of backlogged retailer's order at the manufacturer's plant, and $Y(t)/S(t)$ for the manufacturer, representing the percentage of backlogged consumer demand by the retailer. These time paths are illustrated in Figs. 6(a) and 6(b).

As suggested in Sec. 3, in an OLNE the contract type does not affect the manufacturer's steady state relative backlog. Figure 6(a) suggests that the manufacturer's

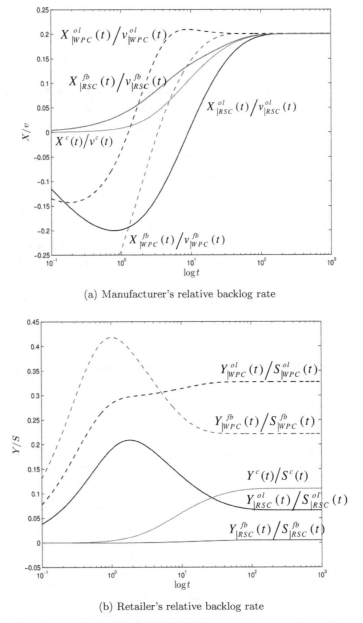

(a) Manufacturer's relative backlog rate

(b) Retailer's relative backlog rate

Fig. 6.   Manufacturer's and retailer's relative backlog rates in cooperative and noncooperative equilibria.

steady state relative backlog, i.e., $X_\infty/v_\infty = ra/c$, is also independent of the strategy type in both the cooperative solution and the noncooperative equilibria. However, the manufacturer's transient relative backlog shows significant differences between cooperation and noncooperation. In a cooperative solution and under an RSC and FBNE, the manufacturer has no initial inventory and the convergence of the relative unavailability of the product to the steady state is slower in the cooperative case. In contrast, in an OLNE under both contracts and in an FBNE and RSC, the manufacturer holds a positive inventory during an initial time interval. The inventory gradually turns into backlogging as time passes.

In Fig. 6(b), the retailer's steady state relative backlog is affected by both the compensation scheme and the type of game equilibrium. As suggested in Sec. 3, in an OLNE, the retailer's steady state relative backlog is greater under a WPC than an RSC if the manufacturer's relative share of the retailer's revenue does not exceed the relative reduction of the transfer price that the retailer obtains from the manufacturer under an RSC versus a WPC, i.e., if $\phi \leq (\omega_{|\text{WPC}} - \omega_{|\text{RSC}})/\omega_{|\text{WPC}}$; this requirement is fulfilled in our case both for OLNE and FBNE. Figure 6(b) suggests that the retailer's relative backlog also is greater under a WPC and FBNE than an RSC and FBNE. A WPC and OLNE is the least effective in meeting demand, followed by a WPC and FBNE. At the steady state, relative backlogging at the retailer is even lower under an RSC and both strategy types than in the cooperative equilibrium. The lowest (relative) unavailability level at the retailer is observed for an RSC and FBNE. Overall, the relative availability is better under an RSC than under a WPC for both strategy types. However, the retailer's transient relative backlog evolves conversely to the manufacturer's rate under RSC and OLNE, as under WPC with both strategy types.

The key results of our analysis are summarized in Table 4. The cooperative outcome reflects a vertically integrated supply chain that sells a heavily advertised, high-priced product where (relative) availability is of high concern. In contrast, under a WPC and OLNE, the supply chain sells an unadvertised, lower-priced

Table 4.   Supply chain outcomes under WPC and RSC with open-loop and feedback strategies.

		Compensation scheme	
		WPC	RSC
Decision rule	Open-loop strategy	Unadvertised, lower priced product with poorer marketing and operational performance	Higher advertised and priced product with greater marketing performance and lower operational performance
	Feedback strategy	Advertised and higher-priced product with greater marketing and operational performance	Lower advertised and priced product with lower marketing performance and greater operational performance

product with poorer availability. As for a WPC and FBNE, the RSC and OLNE lead to higher-priced and advertised products. Marketing is important in both cases, and availability has a significantly higher priority under RSC and OLNE than under WPC and FBNE. Finally, an RSC and FBNE have the most effective operational performance in terms of relative availability.

6. Conclusions

In this chapter, we analyze how the performances of WPC and RSC are affected by the information structure and the resulting decision rules in a supply chain in terms of operational and marketing decisions, and payoffs.

We point out an important, yet undisclosed merit of the WPC, that it is always a stable contract in the long run, regardless of whether state information is available or not in the supply chain. In contrast, when state information is unavailable in the supply chain, employing a RSC is not optimal in the long term if the manufacturer's bargaining power is excessively large. Our results also suggest that the unavailability of information on the current state of key operational and marketing variables is more detrimental under WPC than under RSC. An important implication of this is that it serves supply chain members better to share information on their respective inventories under a WPC than a RSC.

The analysis of the operations/marketing game has provided a series of observations that should be useful for operations, marketing, and supply chain managers:

- Integrating marketing and operations decisions is strongly believed to be beneficial to the supply chain. We have demonstrated that using advertising to create goodwill leads to higher consumer prices and stimulates sales while maintaining high product availability.
- Noncooperative behavior leads to under-investment in goodwill and a lower consumer price. If supply chain firms play a noncooperative game, the type of strategy that better enhances individual profits depends on the type of contract chosen. With a WPC, feedback strategies provide better outcomes than open-loop strategies. Under a RSC, open-loop strategies provide preferable results for all firms.
- With a WPC, feedback strategies mitigate the double marginalization problem because an increase in price does not deter growth of demand. While such strategies are effective in expanding sales, their effectiveness in meeting demand is limited. In such a situation, supply chain members should try to improve, throughout the game, the mutual availability of information on the current state of key operational and marketing variables of the supply chain.
- With a RSC, open-loop strategies provide results that are closest to those of a cooperative strategy. Such strategies are effective both in expanding and meeting demand. Supply chain members could profitably precommit to a plan of action during the entire game if the manufacturer's bargaining power is not excessively large and supply chain members are farsighted.

The current research contributes to the understanding of the role of contracts, the use of information in a supply chain, and the implications for supply chain members' operations and marketing strategies. The importance of the availability of information and the strategies used by supply chain members under dynamic conditions has received little attention in the literature. Moreover, the literature seems to have followed two disparate streams, one in marketing and another in operations management. This chapter has demonstrated the importance of the informational basis on which marketing and operations strategies are designed and the role of the contact under which decisions are made. The research also contributes to the stream of literature which views marketing and operational decision-making from an integrated point of view.

Appendix A

A.1 *Appendix*

First, note that the manufacturer's backlog (for both positive and negative values) has a negative influence on its objective function because backlogging is costly and hence we expect that its costate λ_1 is negative. Next, because the manufacturer's goodwill has no (positive) influence on its objective under a WPC (RSC), λ_3 is expected to be zero (positive). For λ_2, Eq. (9) can be explicitly solved and has the solutions $\lambda_2 = 0$ and $\lambda_2 = Ce^{rt}$, where C is an arbitrary, nonzero constant of integration. Equations (15) and (16) show the following. A positive value of the manufacturer's production rate requires a negative value of the costate λ_1. A positive value of the manufacturer's advertising effort requires a positive value of the costate variable λ_3. The manufacturer's optimal decisions in (15)–(16) do not depend on λ_2 and henceforth we choose the zero solution for λ_2. Differentiating (15) with respect to time and using (8) and (15) give (19). Differentiating (16) with respect to time and using (10) and (16) yield (20).

A.2 *Appendix*

Regarding the first part of the proposition, note that WPC implies $\phi = 0$ and, in (20), $\dot{w}^{ol} = (r + \delta)w^{ol}$ and $w^{ol}_\infty = 0$. The equation $\dot{w}^{ol} = (r + \delta)w^{ol}$ has the general solution $w^{ol}(t) = De^{(r+\delta)t}$, where D is a constant of integration. For $D = 0$, the solution is $w^{ol} = 0$, while $D > 0$ makes $w^{ol} \mapsto \infty$ when $t \mapsto \infty$. Since $w^{ol}(0) \geq 0$, $D < 0$ is not possible, which proves that $w^{ol} = 0$. For the second part of the proposition, we note that in RSC the term ϕp in (10) is positive when we exclude the possibility that $p^{ol} = 0$. We know that λ_3 nonnegative is necessary for $w^{ol} \geq 0$. Suppose $\lambda_3 = 0$. Then the optimality condition (10) cannot hold. Hence λ_3 must be positive and the second part of the proposition follows from (16).

A.3 *Appendix*

A similar argument as the one used for costate λ_2 leads to choose the solution $\mu_1 = 0$ in (12). The retailer's backlog has a negative influence on its objective, which suggests that the costate μ_2 should be negative. Finally, because goodwill positively influences the retailer's objective, whatever the compensation scheme, μ_3 is expected to be positive. According to (17), a positive value of the retailer's purchase rate requires $-\mu_2 \geq w$, which means that the retailer's unit purchase cost is lower than the imputed cost of backlogging one unit at the retail level. (The consumer price stated in (18) is clearly positive.) Differentiating (17) with respect to time and using (13) and (17) give (21). Differentiating (18) with respect to time and using (4), (20) and (21) yield (22).

A.4 *Appendix*

To compute the steady state under WPC, we form the associated canonical system in the state-control space. To do so, we plug the right-hand side of (18) for p in the right-hand side of (20) and (2) to eliminate the equation of p. We obtain the following system of six linear equations

$$
\begin{bmatrix} \dot{X}^{\text{ol}}_{|\text{WPC}} \\ \dot{Y}^{\text{ol}}_{|\text{WPC}} \\ \dot{G}^{\text{ol}}_{|\text{WPC}} \\ \dot{u}^{\text{ol}}_{|\text{WPC}} \\ \dot{w}^{\text{ol}}_{|\text{WPC}} \\ \dot{v}^{\text{ol}}_{|\text{WPC}} \end{bmatrix} = \begin{bmatrix} 0 & 0 & 0 & -1 & 0 & 1 \\ 0 & 0 & \frac{1}{2} & 0 & 0 & -1 - \frac{\beta d}{2} \\ 0 & 0 & -\delta & 0 & 1 & 0 \\ -\frac{c}{a} & 0 & 0 & r & 0 & 0 \\ 0 & 0 & 0 & 0 & r+\delta & 0 \\ 0 & -\frac{e}{d} & 0 & 0 & 0 & r \end{bmatrix} \begin{bmatrix} X^{\text{ol}}_{|\text{WPC}} \\ Y^{\text{ol}}_{|\text{WPC}} \\ G^{\text{ol}}_{|\text{WPC}} \\ u^{\text{ol}}_{|\text{WPC}} \\ w^{\text{ol}}_{|\text{WPC}} \\ v^{\text{ol}}_{|\text{WPC}} \end{bmatrix} + \begin{bmatrix} 0 \\ \frac{\alpha + \beta w}{2} \\ 0 \\ 0 \\ 0 \\ \frac{rw}{d} \end{bmatrix}.
$$

$$(\text{A.1})$$

The resolution of (A.1) along with (18) gives

$$
\left[u^{\text{ol}}_{\infty|\text{WPC}} \quad w^{\text{ol}}_{\infty|\text{WPC}} \quad v^{\text{ol}}_{\infty|\text{WPC}} \quad p^{\text{ol}}_{\infty|\text{WPC}} \quad X^{\text{ol}}_{\infty|\text{WPC}} \quad Y^{\text{ol}}_{\infty|\text{WPC}} \quad G^{\text{ol}}_{\infty|\text{WPC}} \right]^T
$$

$$
= \left[\frac{\alpha - \beta w_{|\text{WPC}}}{2 + \beta d} \quad 0 \quad \frac{\alpha - \beta w_{|\text{WPC}}}{2 + \beta d} \quad \frac{(1 + \beta d)\,\alpha + \beta w_{|\text{WPC}}}{\beta(2 + \beta d)} \right.
$$

$$
\left. \times \frac{ra\,(\alpha - \beta w_{|\text{WPC}})}{c\,(2 + \beta d)} \quad \frac{r\,(d\alpha + 2w_{|\text{WPC}})}{e\,(2 + \beta d)} \quad 0 \right]^T
$$

$$(\text{A.2})$$

To check the stability of the system, we find the roots of the characteristic equation associated with the Jacobian in (A.1), that is [Grass *et al.*, 2008],

$$
\xi^{\text{ol}}_{1|\text{WPC}} = -\delta, \quad \xi^{\text{ol}}_{2|\text{WPC}} = r + \delta, \quad \xi^{\text{ol}}_{3,4|\text{WPC}} = r/2 \pm \sqrt{r^2/4 + e(\beta d + 2)/2d},
$$

$$
\xi^{\text{ol}}_{5,6|\text{WPC}} = r/2 \pm \sqrt{r^2/4 + c/a}.
$$

$$(\text{A.3})$$

All the eigenvalues are real, three having a positive sign and three having a negative sign. The saddle-point property of the steady state is thus granted and the path converging to it is monotonic.

A.5 *Appendix*

Under RSC, the canonical system in the state-control space is

$$
\begin{bmatrix} \dot{X}^{ol}_{|RSC} \\ \dot{Y}^{ol}_{|RSC} \\ \dot{G}^{ol}_{|RSC} \\ \dot{u}^{ol}_{|RSC} \\ \dot{w}^{ol}_{|RSC} \\ \dot{v}^{ol}_{|RSC} \end{bmatrix} =
\begin{bmatrix}
0 & 0 & 0 & -1 & 0 & 1 \\
0 & 0 & \dfrac{1}{2} & 0 & 0 & -1 - \dfrac{\beta d}{2\phi} \\
0 & 0 & -\delta & 0 & 1 & 0 \\
-\dfrac{c}{a} & 0 & 0 & r & 0 & 0 \\
0 & 0 & -\dfrac{\phi}{2\beta b} & 0 & r+\delta & -\dfrac{\phi d}{2\bar\phi b} \\
0 & -\dfrac{e}{d} & 0 & 0 & 0 & r
\end{bmatrix}
$$

$$
\times
\begin{bmatrix} X^{ol}_{|RSC} \\ Y^{ol}_{|RSC} \\ G^{ol}_{|RSC} \\ u^{ol}_{|RSC} \\ w^{ol}_{|RSC} \\ v^{ol}_{|RSC} \end{bmatrix}
+
\begin{bmatrix}
0 \\
\dfrac{1}{2}\left(\alpha + \dfrac{\beta w_{|RSC}}{\bar\phi}\right) \\
0 \\
0 \\
-\dfrac{\phi}{2b}\left(\dfrac{\alpha}{\beta} - \dfrac{w_{|RSC}}{\phi}\right) \\
\dfrac{r w_{|RSC}}{d}
\end{bmatrix}. \tag{A.4}
$$

The solution of the system, along with (18), gives

$$
u^{ol}_{\infty|RSC} = \frac{\beta\delta b(r+\delta)\bar\phi\alpha + [\phi - \beta\delta b(r+\delta)]\beta w_{|RSC}}{\beta\delta b(r+\delta)(2\bar\phi + \beta d) - \phi(\bar\phi + \beta d)}, \tag{A.5}
$$

$$
w^{ol}_{\infty|RSC} = \frac{\phi\delta[(\bar\phi + \beta d)\alpha + \beta w_{|RSC}]}{\beta\delta b(2\bar\phi + \beta ds)(r+\delta) - \phi(\bar\phi + \beta d)}, \tag{A.6}
$$

$$
v^{ol}_{\infty|RSC} = \frac{\beta\delta b(r+\delta)\bar\phi\alpha + [\phi - \beta\delta b(r+\delta)]\beta w_{|RSC}}{\beta\delta b(r+\delta)(2\bar\phi + \beta d) - \phi(\bar\phi + \beta d)}, \tag{A.7}
$$

$$
p^{ol}_{\infty|RSC} = \frac{\delta b(r+\delta)[(\bar\phi + \beta d)\alpha + \beta w_{|RSC}]}{\beta\delta b(r+\delta)(2\bar\phi + \beta d) - \phi(\bar\phi + \beta d)}, \tag{A.8}
$$

$$
X^{ol}_{\infty|RSC} = \frac{ra\{\beta\delta b(r+\delta)\bar\phi\alpha + [\phi - \beta\delta b(r+\delta)]\beta w_{|RSC}\}}{c[\beta\delta b(r+\delta)(2\bar\phi + \beta d) - \phi(\bar\phi + \beta d)]}, \tag{A.9}
$$

$$Y^{\text{ol}}_{\infty|\text{RSC}} = \frac{r\bar{\phi}\{\beta\delta b(r+\delta)d\alpha + [2\beta\delta b(r+\delta) - \phi]\omega_{|\text{RSC}}\}}{e[\beta\delta b(r+\delta)(2\bar{\phi} + \beta d) - \phi(\bar{\phi} + \beta d)]}, \tag{A.10}$$

$$G^{\text{ol}}_{\infty|\text{RSC}} = \frac{\phi[(\bar{\phi} + \beta d)\alpha + \beta\omega_{|\text{RSC}}]}{\beta\delta b(r+\delta)(2\bar{\phi} + \beta d) - \phi(\bar{\phi} + \beta d)}. \tag{A.11}$$

We check the local stability of the steady state. Because the number of negative eigenvalues is at most three, a change in signs implies that the number of negative eigenvalues is reduced and therefore the number of stable eigendirections of the steady state declines, making the locally stable steady state unstable. Because the determinant of the Jacobian is given as the product of the eigenvalues, and three negative eigenvalues exist for $\phi = 0$ (see Appendix A.4), the determinant becomes zero if one of the eigenvalues becomes zero for $\phi > 0$. Due to the sparsity of the Jacobian in (A.4), the determinant is given as

$$|J_{\text{RSC}}| = ce[\phi(\bar{\phi} + \beta d) - \beta\delta b(2\bar{\phi} + \beta d)(r+\delta)]/2\bar{\phi}\beta abd. \tag{A.12}$$

It then suffices to find the zero values of (A.12). For ϕ this leads us to find the roots of

$$-\phi^2 + \phi[1 + \beta d + 2\beta\delta b(r+\delta)] - \beta\delta b(2 + \beta d)(r+\delta) = 0 \tag{A.13}$$

that is,

$$\phi_{1,2} = \beta\delta b(r+\delta) + (1 + \beta d)/2 \pm \sqrt{\beta\delta b(r+\delta)[\beta\delta b(r+\delta) - 1] + (1 + \beta d)^2/4}. \tag{A.14}$$

Five scenarios are then possible, depending on the value of the discriminant $D \equiv \beta\delta b(r+\delta)[\beta\delta b(r+\delta) - 1] + (1 + \beta d)^2/4$ in (A.14), that is,

— $D < 0$, which implies that the number of negative and positive eigenvalues does not change;
— $D > 0$ so that either $0 < \phi_1 < \phi_2 < 1$ or $\phi_1 < 0 < \phi_2 < 1$ exists;
— $D > 0$ so that either $\phi_1 < 0 < 1 < \phi_2$ or $0 < \phi_1 < 1 < \phi_2$ exists.

If $\beta\delta b(r+\delta) \geq 1$, it is obvious that $D > 0$. Conversely, if $\beta\delta b(r+\delta) < 1$, we have: $\beta\delta b(r+\delta)[\beta\delta b(r+\delta) - 1] \geq -1/4$, which also implies that $D > 0$. Therefore, the first scenario is invalid. Regarding the second scenario, it can be shown that $0 < \phi_1 < 1$ if $\beta\delta b(r+\delta) < 1$, which invalidates the third and fourth scenarios, that is, $\phi_1 < 0 < \phi_2 < 1$ and $\phi_1 < 0 < 1 < \phi_2$. However, if $\beta\delta b(r+\delta) \geq 1$, we get $\phi_2 > 1$. Conversely, if $\beta\delta b(r+\delta) < 1$, we have

$$\phi_2 \geq \beta\delta b(r+\delta) + 1/2 + \sqrt{1/4 - \beta\delta b(r+\delta)[1 - \beta\delta b(r+\delta)]}$$

$$= \beta\delta b(r+\delta) + 1/2 + \sqrt{(1/2 - \beta\delta b(r+\delta))^2}$$

$$= \beta\delta b(r+\delta) + 1/2 + \sqrt{(\beta\delta b(r+\delta) - 1/2)^2}. \tag{A.15}$$

If $\beta\delta b(r+\delta) > 1/2$, then we get

$$\phi_2 \geq \beta\delta b(r+\delta) + 1/2 + \beta\delta b(r+\delta) - 1/2 = 2\beta\delta b(r+\delta) > 1. \qquad (A.16)$$

Conversely, if $\beta\delta b(r+\delta) \leq 1/2$, we obtain

$$\phi_2 \geq \beta\delta b(r+\delta) + 1/2 + 1/2 - \beta\delta b(r+\delta) = 1. \qquad (A.17)$$

Therefore, the second scenario according to which $0 < \phi_1 < \phi_2 < 1$ is also invalid. Finally, only the smallest root in (A.14), denoted as $\tilde{\phi} \equiv \phi_1$, where $\phi_1 < 1$ requires that $\beta\delta b(r+\delta) < 1$, should be considered as a bifurcation threshold, which validates the fifth scenario whereby $0 < \phi_1 < 1 < \phi_2$, so that for $0 < \phi < \tilde{\phi}$, the number of negative eigenvalues cannot change. Next, requiring nonnegativity of the manufacturer's production and the retailer's purchase rate under RSC, we get $\phi \leq \frac{\delta b(r+\delta)(\alpha-\beta\omega_{|\text{RSC}})}{\delta b(r+\delta)\alpha-\omega_{|\text{RSC}}} < 1$. It can be shown that for $\frac{\delta b(r+\delta)(\alpha-\beta\omega_{|\text{RSC}})}{\delta b(r+\delta)\alpha-\omega_{|\text{RSC}}} < \tilde{\phi}$ to hold, it is necessary that $\beta\delta b(r+\delta) > 1$. This contradicts the condition that $\beta\delta b(r+\delta) < 1$, and proves that $\phi < \tilde{\phi} < 1$ is a necessary and sufficient condition for the feasibility of the steady state.

A.6 Appendix

Substituting from (24)–(25) and (27)–(28) into (23) and (26) yields:

$$rV^M = -\frac{cX^2}{2} + \frac{\phi G^2}{4\beta} + \left(\frac{\phi\alpha}{2\beta} + \frac{V_Y^M}{2} - \delta V_G^M\right)G + \frac{V_X^{M2}}{2a} + \frac{V_G^{M2}}{2b}$$

$$-\frac{\phi\beta V_Y^{R2}}{4\tilde{\phi}^2} + \frac{V_X^M V_X^R}{d} - \frac{V_X^M V_Y^R}{d} - \frac{V_Y^M V_X^R}{d} + \left(\frac{1}{d} + \frac{\beta}{2\tilde{\phi}}\right)V_Y^M V_Y^R$$

$$-\frac{\omega V_X^M}{d} + \left(\frac{\alpha}{2} + \frac{\omega}{d}\right)V_Y^M + \frac{\omega V_X^R}{d} - \frac{\omega V_Y^R}{d} + \frac{\phi\alpha^2}{4\beta} - \frac{\omega^2}{d}. \qquad (A.18)$$

$$rV^R = -\frac{eY^2}{2} + \frac{\tilde{\phi}G^2}{4\beta} + \left(\frac{\tilde{\phi}\alpha}{2\beta} + \frac{V_Y^R}{2} - \delta V_G^R\right)G + \frac{V_X^{R2}}{2d} + \left(\frac{\beta}{4\tilde{\phi}} + \frac{1}{2d}\right)V_Y^{R2}$$

$$+\frac{V_X^M V_X^R}{a} - \frac{V_X^R V_Y^R}{d} + \frac{V_G^R V_G^M}{b} - \frac{\omega V_X^R}{d} + \left(\frac{\alpha}{2} + \frac{\omega}{d}\right)V_Y^R + \frac{\tilde{\phi}\alpha^2}{4\beta} + \frac{\omega^2}{2d}. \qquad (A.19)$$

We need to establish the existence of bounded and continuously differentiable value functions V^M and V^S that solve the HJB equations as well as unique and nonnegative solutions X, Y, and G to the state equations. To do so, we make the following conjectures

$$V^M = A^M + B^M X + C^M X^2 + D^M Y + E^M Y^2 + F^M G$$

$$+ H^M G^2 + I^M XY + J^M XG + K^M YG, \qquad (A.20)$$

$$V^R = A^R + B^R X + C^R X^2 + D^R Y + E^R Y^2$$
$$+ F^R G + H^R G^2 + I^R XY + J^R XG + K^R YG \tag{A.21}$$

from which we obtain

$$V_X^M = B^M + 2C^M X + I^M Y + J^M G, \tag{A.22}$$

$$V_Y^M = D^M + 2E^M Y + I^M X + K^M G, \tag{A.23}$$

$$V_G^M = F^M + 2H^M G + J^M X + K^M Y, \tag{A.24}$$

$$V_X^R = B^R + 2C^R X + I^R Y + J^R G, \tag{A.25}$$

$$V_Y^R = D^R + 2E^R Y + I^R X + K^R G, \tag{A.26}$$

$$V_G^R = F^R + 2H^R G + J^R X + K^R Y. \tag{A.27}$$

Plugging the right-hand side of (A.22) and (A.24) in (24) and (25), respectively, gives (29) and (30). Plugging the right-hand side of (A.25) and (A.26) in (27), and the right-hand side of (A.26) in (28) and rearranging, respectively, yields (31) and (32). Inserting the right-hand side of (A.22)–(A.27) into (A.18)–(A.19), and equating coefficients, we develop a system of 20 connected and quadratic equations in $A^M, \ldots, K^M, A^R, \ldots, K^R$, which can be solved numerically with given values for the parameters. This leads to the following coupled quadratic equations

$$rA^M = B^{M2}/2a + F^{M2}/2b - \phi\beta D^{R2}/4\bar{\phi}^2$$
$$- [\omega(\omega + B^M - B^R + D^R) + B^M D^R + D^M B^R - B^M B^R]/d$$
$$+ (1/d + \beta/2\bar{\phi})D^M D^R + (\alpha/2 + \omega/d)D^M + \phi\alpha^2/4\beta, \tag{A.28}$$

$$rB^M = [\omega(2C^R - 2C^M - I^R) + 2(B^M C^R + C^M B^R)$$
$$- B^M I^R - 2C^M D^R - 2D^M C^R - I^M B^R]/d + F^M J^M/b + (1/d + \beta/2\bar{\phi})$$
$$\times (D^M I^R + I^M D^R) + 2B^M C^M/a - \phi\beta D^R I^R/2\bar{\phi}^2 + (\alpha/2 + \omega/d)I^M, \tag{A.29}$$

$$rC^M = 2C^{M2}/a + J^{M2}/2b - \phi\beta I^{R2}/4\bar{\phi}^2 + (1/d + \beta/2\bar{\phi})I^M I^R$$
$$- [2(C^M I^R + I^M C^R) - 4C^M C^R]/d - c/2, \tag{A.30}$$

$$rD^M = [\omega(I^R - I^M - 2E^R) + (B^M I^R + I^M B^R) - 2B^M E^R - D^R I^M$$
$$- D^M I^R - 2E^M B^R]/d + F^M K^M/b + 2(1/d + \beta/2\bar{\phi})$$
$$\times (E^M D^R + D^M E^R) + B^M I^M/a - \phi\beta D^R E^R/\bar{\phi}^2 + 2(\alpha/2 + \omega/d)E^M, \tag{A.31}$$

$$rE^M = I^{M2}/2a + K^{M2}/2b - \phi\beta E^{R2}/\bar{\phi}^2 + (4/d + 2\beta/\bar{\phi})E^M E^R$$
$$- [2(I^M E^R + E^M I^R) - I^M I^R]/d, \tag{A.32}$$

$$rF^M = [\omega(J^R - J^M - K^R) + (B^M J^R + J^M B^R) - B^M K^R - D^R J^M - D^M J^R$$
$$- K^M B^R]/d + (2H^M/b - \delta)F^M + (1/d + \beta/2\bar{\phi})(D^M K^R + K^M D^R)$$
$$+ B^M J^M/a - \phi\beta D^R K^R/2\bar{\phi}^2 + D^M/2 + (\alpha/2 + \omega/d)K^M + \phi\alpha/2\beta,$$
$$\tag{A.33}$$

$$rH^M = 2H^{M2}/b + J^{M2}/2a - \phi\beta K^{R2}/4\bar{\phi}^2 + (1/d + \beta/2\bar{\phi})K^M K^R$$
$$- (K^R J^M + K^M J^R - J^M J^R)/d + K^M/2 - 2\delta H^M + \phi/4\beta, \tag{A.34}$$

$$rI^M = 2C^M I^M/a + J^M K^M/b - \phi\beta E^R I^R/\bar{\phi}^2 - 2(2C^M E^R + 2E^M C^R + I^M I^R$$
$$- C^M I^R - I^M C^R)/d + (2/d + \beta/\bar{\phi})(E^M I^R + I^M E^R), \tag{A.35}$$

$$rJ^M = 2C^M J^M/a + (1/d + \beta/2\bar{\phi})(I^M K^R + K^M I^R)$$
$$- \phi\beta I^R K^R/2\bar{\phi}^2 + I^M/2 + (2H^M/b - \delta)J^M - [I^R J^M + I^M J^R$$
$$- 2(C^M J^R + J^M C^R) + 2(C^M K^R + 2K^M C^R)]/d, \tag{A.36}$$

$$rK^M = I^M J^M/a + 2(1/d + \beta/2\bar{\phi})(E^M K^R + K^M E^R)$$
$$- \phi\beta E^R K^R/\bar{\phi}^2 + E^M + (2H^M/b - \delta)K^M$$
$$- [K^R I^M + K^M I^R + 2(E^R J^M + E^M J^R) - (I^M J^R + J^M I^R)]/d, \tag{A.37}$$

$$rA^R = \bar{\phi}\alpha^2/4\beta + \omega^2/2d + (B^R)^2/2d - \omega B^R/d + (\omega/d + \alpha/2)D^R$$
$$+ (2\bar{\phi} + \beta d)(D^R)^2/4\bar{\phi}d + B^M B^R/a + F^M F^R/b - B^R D^R/d, \tag{A.38}$$

$$rB^R = -2\omega^2 C^R/d + (\omega/d + \alpha/2)I^R + (2\bar{\phi} + \beta d)D^R I^R/2\bar{\phi}d$$
$$+ 2(B^M C^R + C^M B^R)/a + (F^M J^R + J^M F^R)/b$$
$$+ [2B^R C^R - B^R I^R - 2C^R D^R]/d, \tag{A.39}$$

$$rC^R = 2(C^R)^2/d + (2\bar{\phi} + \beta d)(I^R)^2/4\bar{\phi}d + 4C^M C^R/a + J^M J^R/b - 2C^R I^R/d, \tag{A.40}$$

$$rD^R = -\omega I^R/d + 2(\omega/d + \alpha/2)E^R + (2\bar{\phi} + \beta d)D^R E^R/\bar{\phi}d + (B^M I^R + I^M B^R)/a$$
$$+ (F^M K^R + K^M F^R)/b + [B^R I^R - 2B^R E^R - D^R I^R]/d, \tag{A.41}$$

$$rE^R = -e/2 + (I^R)^2/2d + (2\bar{\phi} + \beta d)(E^R)^2/\bar{\phi}d + I^M I^R/a + K^M K^R/b - 2E^R I^R/d, \tag{A.42}$$

$$rF^R = \bar{\phi}\alpha/2\beta - \omega J^R/d + (\omega/d + \alpha/2)K^R + D^R/2 - \delta F^R$$
$$+ (2\bar{\phi} + \beta d)D^R K^R/2\bar{\phi}d + (B^M J^R + J^M B^R)/a$$
$$+ 2(F^M H^R + H^M F^R)/b + (B^R J^R - B^R K^R - D^R J^R)/d, \tag{A.43}$$

$$rH^R = \bar{\phi}/4\beta + (J^R)^2/2d + K^R/2 - 2\delta H^R$$
$$+ (2\bar{\phi} + \beta d)(K^R)^2/4\bar{\phi}d + J^M J^R/a + K^M K^R/b - J^R K^R/d, \tag{A.44}$$

$$rI^R = -(I^R)^2/d + 2(C^M I^R + I^M C^R)/a + (J^M K^R + K^M J^R)/b$$
$$+ 2(C^R I^R - 2C^R E^R)/d, \tag{A.45}$$

$$rJ^R = I^R/2 - \delta J^R + (2\bar{\phi} + \beta d)I^R K^R/2\bar{\phi}d + 2(C^M J^R + J^M C^R)/a$$
$$+ 2(J^M H^R + H^M J^R)/b + (2C^R J^R - 2C^R K^R - I^R J^R)/d, \tag{A.46}$$

$$rK^R = E^R - \delta K^R + (2\bar{\phi} + \beta d)E^R K^R/\bar{\phi}d + (I^M J^R + J^M I^R)/a$$
$$+ 2(K^M H^R + H^M K^R)/b + (I^R J^R - 2E^R J^R - I^R K^R)/d. \tag{A.47}$$

A.7 Appendix

Plugging the right-hand side of (32) into (3) gives the equilibrium sales rate

$$S = \alpha/2 + [\beta(D^R + I^R X + 2E^R Y) + (\bar{\phi} + \beta K^R)G]/2\bar{\phi}. \tag{A.48}$$

Plugging the right-hand side of (29)–(31), and (35) into (1)–(2), and (4), respectively, the system is

$$
\begin{bmatrix} \dot{X} \\ \dot{Y} \\ \dot{G} \end{bmatrix} =
\begin{bmatrix}
\dfrac{2C^R - I^R}{d} + \dfrac{2C^M}{a} & \dfrac{I^R - 2E^R}{d} + \dfrac{I^M}{a} & \dfrac{J^R - K^R}{d} + \dfrac{J^M}{a} \\[2ex]
\dfrac{I^R - 2C^R}{d} + \dfrac{\beta I^R}{2\bar{\phi}} & \dfrac{2E^R - I^R}{d} + \dfrac{\beta E^R}{\bar{\phi}} & \dfrac{K^R - J^R}{d} + \dfrac{1}{2} + \dfrac{\beta K^R}{2\bar{\phi}} \\[2ex]
\dfrac{J^M}{b} & \dfrac{K^M}{b} & \dfrac{2H^M}{b} - \delta
\end{bmatrix}
$$
$$
\times \begin{bmatrix} X \\ Y \\ G \end{bmatrix} +
\begin{bmatrix}
\dfrac{B^M}{a} - \dfrac{\omega - B^R + D^R}{d} \\[2ex]
\dfrac{\beta D^R}{2\bar{\phi}} + \dfrac{\omega - B^R + D^R}{d} + \dfrac{\alpha}{2} \\[2ex]
\dfrac{F^M}{b}
\end{bmatrix}. \tag{A.49}
$$

The FBNE is globally asymptotically stable if the eigenvalues of the Jacobian matrix associated with the dynamic system in (A.49) all are negative.

A.8 Appendix

The roots of the system of 20 nonlinear equations in Appendix A.6 are computed with *Matlab* 7.5.0.342 (R2007b). To ensure that all equilibria are identified, a two-step approach is necessary. First we randomly search for equilibria and we use an algorithm (*Matcont*) to continue these equilibria by changing a parameter value of the model. If the branch of any equilibrium undergoes a limit point bifurcation, this is detected by the algorithm and further equilibria can be detected. For both contracts, 10,000 initial vectors were chosen randomly for the coefficients and used to solve the equation system. Only a single steady state satisfying the criterion of global asymptotic stability was found for each contract. *Matcont* was used to detect any other equilibria due to bifurcation. However, even after this step, a single and stable steady state remained in all cases.

Acknowledgments

This research was financially supported by ESSEC Business School Research Centre (France) and the Austrian Science Fund (FWF) under grant No. P 23084-N13 (A MATLAB Package for Analyzing Optimal Control Problems). This work was partly written while the first author was visiting the University of Washington in Seattle, whose hospitality is gratefully acknowledged. The authors acknowledge helpful comments. The usual disclaimer applies.

References

Başar, T. and Olsder, G. J. [1999] *Dynamic Noncooperative Game Theory*, 3rd edition, Classics in Applied Mathematics (SIAM, Philadelphia).

Bensoussan, A., Cakanyildirim, M. and Sethi, S. P. [2007] Optimal ordering policies for inventory problems with dynamic information delays, *Prod. Oper. Manag.* **16**(2), 241–256.

Bresnahan, T. F. and Reiss, P. C. [1985] Dealer and manufacturer margins, *Rand J. Econ.* **16**(2), 253–268.

Cachon, G. P. [2003] Supply chain coordination with contracts, in *Handbooks in Operations Research and Management Science: Supply Chain Management*, eds. Graves, S. and de Kok, T. (Elsevier, Amsterdam).

Cachon, G. P. and Lariviere, M. A. [2005] Supply chain coordination with revenue sharing contracts: Strengths and limitations, *Manag. Sci.* **51**(1), 30–44.

Chakravarty, A. K. and Werner, A. S. [2011] Telecom service provider portal: Revenue sharing and outsourcing, *Eur. J. Oper. Res.* **215**(1), 289–300.

Dana, J. and Spier, K. [2001] Revenue sharing and vertical control in the video rental industry, *J. Ind. Econ.* **59**(3), 223–245.

DeHoratius, N. and Raman, A. [2008] Inventory record inaccuracy: An empirical analysis, *Manag. Sci.* **54**(4), 627–641.

Desai, V. S. [1992] Marketing-production decisions under independent and integrated channel structure, *Ann. Oper. Res.* **34**(1), 275–306.

Desai, V. S. [1996] Interactions between members of a marketing-production channel under seasonal demand, *Eur. J. Oper. Res.* **90**(1), 115–141.

Dockner, E., Jørgensen, S., Long, N. V. and Sorger, G. [2000] *Differential Games in Economics and Management Science* (Cambridge University Press, Cambridge).

Eliashberg, J. and Steinberg, R. [1987] Marketing-production decisions in an industrial channel of distribution, *Manag. Sci.* **33**(8), 981–1000.

El Ouardighi, F. [2014] Supply quality management with optimal wholesale price and revenue sharing contracts: A two-stage game approach, *Int. J. Prod. Econ.* **156**(1), 260–268.

El Ouardighi, F. and Erickson, G. M. [2015] Production capacity buildup and double marginalization mitigation in a dynamic supply chain, *J. Oper. Res. Soc.* **66**(1), 1281–1296.

El Ouardighi, F., Jørgensen, S. and Pasin, F. [2008] A dynamic game model of operations and marketing management in a supply chain, *Int. Game Theory Rev.* **10**(4), 373–397.

El Ouardighi, F. and Kim, B. [2010] Supply quality management with wholesale price and revenue-sharing contracts under horizontal competition, *Eur. J. Oper. Res.* **206**(2), 329–340.

Engwerda, J. C. [1998] On the open-loop Nash equilibrium in LQ-games, *J. Econ. Dyn. Control* **22**(5), 729–762.

Erickson, G. M. [2003] *Dynamic Models of Advertising Competition*, 2nd edition (Kluwer Academic Publishers, Boston).

Erickson, G. M. [2011] A differential game model of the marketing-operations interface, *Eur. J. Oper. Res.* **211**(2), 394–402.

Erickson, G. M. [2012] Transfer pricing in a dynamic marketing-operations interface, *Eur. J. Oper. Res.* **216**(2), 326–333.

Feichtinger, G. and Hartl, R. [1985] Optimal pricing and production in an inventory model, *Eur. J. Oper. Res.* **19**(1), 45–56.

Gaimon, C. [1998] The price-production problem: An operations and marketing interface, in *Operations Research: Methods, Models, and Applications*, eds. Aronson, J. E. and Zionts, S. (Quorum Books, Westport).

Grass, D. [2012] Numerical computation of the optimal vector field: Exemplified by a fishery model, *J. Econ. Dyn. Control* **36**(10), 1626–1658.

Grass, D., Caulkins, J. P., Feichtinger, G., Tragler, G. and Behrens, D. A. [2008] *Optimal Control of Nonlinear Processes with Applications in Drugs, Corruption, and Terror* (Springer, Heidelberg).

Jørgensen, S. [1986] Optimal production, purchasing and pricing: A differential games approach, *Eur. J. Oper. Res.* **24**(1), 64–76.

Jørgensen, S. [2011] Intertemporal contracting in a supply chain, *Dyn. Games Appl.* **1**(2), 280–300.

Jørgensen, S., Kort, P. M. and Zaccour, G. [1999] Production, inventory, and pricing under cost and demand learning effects, *Eur. J. Oper. Res.* **117**(1), 382–395.

Jørgensen, S. and Zaccour, G. [2004] *Differential Games in Marketing* (Kluwer, Dordrecht).

Kogan, K. [2012] Ship-to-order supplies: Contract breachability and the effect of outlet sales, *Eur. J. Oper. Res.* **218**(1), 113–123.

Kogan, K. and Tapiero, C. S. [2007] *Supply Chain Games: Operations Management and Risk Valuation* (Springer, New York).

Lariviere, M. A. [1999] Supply chain contracting and coordination with stochastic demand, in *Quantitative Models for Supply Chain Management*, eds. Tayur, S., Magazine, M. and Ganeshan, R. (Kluwer, Dordrecht).

Linh, C. T. and Hong, Y. [2009] Channel coordination through a revenue sharing contract in a two-period newsboy problem, *Eur. J. Oper. Res.* **198**(3), 822–829.

Maimon, O., Khmelnitsky, E. and Kogan, K. [1998] *Optimal Flow Control in Manufacturing Systems: Production Planning and Scheduling* (Kluwer Academic Publishers, Boston).

Malhotra, M. K. and Sharma, S. [2002] Spanning the continuum between marketing and operations, *J. Oper. Manag.* **20**(3), 209–219.

Mortimer, J. H. [2008] Vertical contracts in the video rental industry, *Rev. Econ. Stud.* **75**(1), 165–199.

Nerlove, M. and Arrow, K. J. [1962] Optimal advertising policy under dynamic conditions, *Economica* **29**(114), 129–142.

Qin, Z. [2008] Towards integration: A revenue-sharing contract in a supply chain, *IMA J. Manag. Math.* **19**(1), 3–15.

Sapra, A., Truong, V. A. and Zhang, R. Q. [2010] How much demand should be fulfilled? *Oper. Res.* **58**(3), 719–733.

Simchi-Levi, D., Wu, S. D. and Shen, Z. J. [2004] *Handbook of Quantitative Supply Chain Analysis: Modeling in the E-Business Era* (Springer, New York).

Spengler, J. [1950] Vertical integration and antitrust policy, *J. Political Economy* **4**(58), 347–352.

Tang, C. [2010] A review of marketing-operations interface models: From co-existence to coordination and collaboration, *Int. J. Prod. Econ.* **125**(1), 22–40.

Chapter 7

Equilibrium Prices and Flows in the Passenger Traffic Problem

V. V. Mazalov[*]

Institute of Applied Mathematical Research
Karelian Research Center
Russian Academy of Sciences
Pushkinskaya str. 11
Petrozavodsk 185910, Russia

vmazalov@krc.karelia.ru

A. V. Melnik

Saint-Petersburg State University
Universitetskii pr. 35
Saint Petersburg 198504, Russia

a.melnik@spbu.ru

This chapter considers a noncooperative transport game of n players on a communication graph. Here players are passenger transportation companies (carriers). Service requests form a Poisson process with an intensity rate matrix Λ. Players announce prices for their services and passengers choose an appropriate service by minimizing their individual costs (the ticket price and the expected service time). For each carrier, we solve the pricing problem and define the equilibrium intensity flows in the conditions of competition. A special emphasis is placed on polynomial latency functions.

Keywords: Traffic flow; equilibrium prices; Wardrop equilibrium; BPR latency function.

1. Introduction

As a matter of fact, the traffic flow distribution problem possesses a rich history. Starting from the 1950s, this field of research employed models with different optimality principles and corresponding numerical methods, both for estimating transport system parameters and constructing equilibria in such systems. For instance, in 1952 Wardrop hypothesized that any transport system reaches an equilibrium

---

[*]Corresponding author.

state after some period of time, as well as formulated two principles of equilibrium traffic flows distribution [Wardrop, 1952]. According to the first principle, the trip time on all existing routes is same for all road users and smaller than the trip time of any road user in the case of its route diversion. The second principle claimed that the average trip time gets minimized. The rigorous mathematical statement of these principles was first suggested by Beckmann. The concept of Wardrop equilibria gradually become generally accepted; nowadays, it represents a major tool in theory of traffic flows [Spiess and Florian, 1989; Yang and Huang, 2004]. In many cases, Wardrop equilibria coincide with Nash equilibria as a basic solution concept in noncooperative game theory. Wardrop's ideas can be further developed by assuming that not only trip time, but also the total costs of road users on all routes are same and minimal. This agrees with recent trends in operations research: investigators incorporate the behavioral aspects guiding agents into their mathematical models [Hammalainen *et al.*, 2013]. The cost function may include service price, the average trip time, risks and so on. A series of publications adhered to this approach within the framework of queueing theory in the following way: Luski [1976], Levhari and Luski [1978], Altman and Shimkin [1998], Hassin and Haviv [2003], Chen and Wan [2003], Chen and Wan [2005] and Bure *et al.* [2015]. For a transport flow of intensity λ, the latency was defined as the average service time $1/(\mu - \lambda)$, i.e., the cited works expressed the expected sojourn time of a user in a queueing system $M/M/n$.

In this chapter, we introduce a general analysis procedure for such problems and illustrate its applicability to problems with the BPR (Bureau of Public Road) latency functions [U.S. Bureau of Public Roads, 1964]. The BPR latency function on edge e is defined by

$$f_e(\lambda_e) = t_e \left(1 + h \left(\frac{\lambda_e}{d_e} \right)^{\beta} \right),$$

where the transport costs on edge e depend on the flow λ_e on this edge, the specific transport costs t_e on an empty edge and the capacity d_e of edge e. The quantities h and β are some positive constants, t_e specifies the trip time on a free route (i.e., in the case of zero flow on edge e). Without loss of generality, set $h = 1$, as far as this coefficient can be included in the parameter d_e.

Section 2 gives a general formalization of the game-theoretic pricing model on a communication graph. In Secs. 3 and 4 this approach is illustrated using networks with the linear latency functions. Next, Sec. 5 provides modeling results for networks with parallel routes and the nonlinear latency functions. And finally, in Sec. 6 we evaluate equilibria on a more complex network represented by a Euler graph.

2. Game-Theoretic Model

Consider a noncooperative nonzero-sum game of n players, which is associated with queue system operation on a communication graph. Denote a transport n-player game by $\Gamma = \langle N, G, Z_{i,i \in N}, H_{i,i \in N} \rangle$, where $N = \{1, \ldots, n\}$ means the set of players (carriers) serving passengers on a graph $G = \langle V, E \rangle$ with a vertex set V and an edge set E. Suppose that all vertices are numbered: $V = \{v_1, \ldots, v_k\}$. For each player i, there exists a set of routes Z_i from vertex $v_s \in V$ to vertex $v_t \in V$ served by player i. Therefore, $Z_i = (R_1^i, R_2^i, \ldots, R_{n_i}^i)$, $i = 1, \ldots, n$. Each route represents a path, i.e., a sequence of vertices connected by edges $R = (v_s, v_{s+1}, \ldots, v_t)$ so that the end of one edge is the beginning of another edge: $(v_s, v_{s+1}), \ldots, (v_{t-1}, v_t) \in E$. In the sequel, routes and subpaths are designated by capital letters R and lowercase letters r, respectively. The notation R_{st} or r_{st} emphasizes that vertex v_s is the beginning of a route or path and vertex v_t is its end. We say that path $r_{s't'}$ is a subpath of path r_{st} and write $r_{s't'} \subset r_{st}$ if path $r_{s't'}$ makes a subsequence of vertices belonging to r_{st}.

Assume that passenger flows running between graph vertices form Poisson processes of some intensities. Introduce the matrix of flow intensities $\{\lambda_{st}\}$ from vertex v_s to vertex v_t for different $s, t = 1, \ldots, k$.

The game is organized as follows. Player i establishes prices p_i^R, p_i^r for its services on each route $R \in Z_i$ and all its subpaths $r \subset R$. As a result, a strategy profile $\{p_i^{Z_i}\} = \{p_i^r\}, r \subset R \in Z_i, i = 1, \ldots, n$ is constructed. We believe that passengers minimize their costs expressed as the sum of ticket price and the expected sojourn time (trip time) and choose the cheapest service. The expected sojourn time is a nondecreasing function of traffic intensity and on each edge e it takes the form $f_e(\lambda)$.

Then the incoming Poisson flow of intensity λ_{st} is decomposed into n subflows of intensities λ_{st}^i, where $\sum_{i=1}^{n} \lambda_{st}^i = \lambda_{st}$. We obtain $\lambda_{st}^i = 0$ whenever none of the route in the set Z_i of player i has subpath r_{st}.

The costs of a passenger preferring service i on subpath r of some route $R \in Z_i$ make up

$$p_i^r + \sum_{e \in r} f_e(\lambda_e^i),$$

where

$$\lambda_e^i = \sum_{r_{st}: e \in r_{st} \subset R} \lambda_{st}^i, \quad i = 1, 2, \ldots, n.$$

Therefore, the equilibrium costs of all passengers on competing directions coincide for all services. This feature allows evaluating intensities λ_{st}^i for all services $i = 1, \ldots, m$ and subpaths r_{st}. That is,

$$p_i^r + \sum_{e \in r} f_e(\lambda_e^i) = p_j^r + \sum_{e \in r} f_e(\lambda_e^j) \tag{1}$$

for all i, j such that $r \subset R \in Z_i$ and $r \subset R' \in Z_j$. If the price on the route of a certain service appears too high, the passenger flow is distributed among other services, and the former service does not compete. In other words, equilibrium prices should be searched among balanced prices.

The payoff of player $i \in N$ can be defined by its income per unit time from serving all flows on all routes of this player, i.e.,

$$H_i(\{p_i^{Z_i}\}_{i \in N}) = \sum_{r_{st}:r_{st} \subset C \subset R \in Z_i} \lambda_{st}^i p_i^r.$$

This chapter focuses on the transport game with the BPR latency functions [Chen and Wan, 2003]

$$f_e(\lambda_e) = t_e \left(1 + \left(\frac{\lambda_e}{d_e}\right)^\beta\right), \tag{2}$$

where λ_e gives the flow intensity on edge e, t_e specifies the trip time on a free route (i.e., in the case of zero flow on edge e), d_e means the capacity of edge e and β is some positive constant.

Our analysis begins with the transport game involving two carriers rendering their services between two points v_1 and v_2.

3. Transport Game on Two Parallel Routes with the Linear Latency Function

Consider the transport game on a network composed of two parallel routes (see Fig. 1) with the linear *BPR* latency functions:

$$t_1 \left(1 + \frac{\lambda_1}{d_1}\right), \quad t_2 \left(1 + \frac{\lambda_2}{d_2}\right).$$

Suppose that each route is served by a carrier. Without loss of generality, set $t_1 \leq t_2$, i.e., route 1 is faster than route 2. Depending on the incoming flow intensity λ, the flow runs through (faster) route 1 or gets distributed between both routes.

First, assume that the incoming flow possesses a small intensity satisfying the condition

$$\lambda \leq \frac{t_2 - t_1}{2\frac{t_1}{d_1} + \frac{t_2}{d_2}}. \tag{3}$$

In this case, all traffic λ runs through route 1 and carrier 1 naturally establishes some price p_1 for its services. Passengers prefer service 1 under the inequality

$$p_1 + t_1 \left(1 + \frac{\lambda}{d_1}\right) \leq t_2.$$

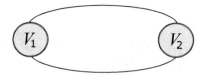

Fig. 1.   Two parallel routes.

The optimal price for player 1 service acquires the form

$$p_1^* = t_2 - t_1\left(1 + \frac{\lambda}{d_1}\right),$$

yielding its payoff

$$H_1 = \left(t_2 - t_1 - \frac{\lambda t_1}{d_1}\right)\lambda.$$

Player 2 gains nothing: $H_2 = 0$.

Now, let

$$\lambda > \frac{t_2 - t_1}{2\frac{t_1}{d_1} + \frac{t_2}{d_2}}. \tag{4}$$

In this case, the flow runs through both routes and player 2 also assigns some price for its service. There arises a two-player game on the described transport network, denote it by $\Gamma_2(t_1, t_2)$. Two players establish prices p_1 and p_2 for their services. The incoming passenger flow of the intensity λ is decomposed into two subflows of intensities λ_1 and λ_2 so that $\lambda_1 + \lambda_2 = \lambda$ and

$$p_1 + t_1\left(1 + \frac{\lambda_1}{d_1}\right) = p_2 + t_2\left(1 + \frac{\lambda_2}{d_2}\right). \tag{5}$$

The payoffs of the players have the form

$$H_1 = p_1\lambda_1, \quad H_2 = p_2\lambda_2.$$

Find an equilibrium by solving the pricing game. Rewrite Eq. (5) as

$$p_1 + t_1\left(1 + \frac{\lambda_1}{d_1}\right) = p_2 + t_2\left(1 + \frac{\lambda - \lambda_1}{d_2}\right).$$

Express λ_1 from this equation. Then the payoff function of player 1 represents the concave parabola

$$H_1 = p_1\left(\frac{p_2 - p_1 + t_2\left(1 + \frac{\lambda}{d_2}\right) - t_1}{\frac{t_1}{d_1} + \frac{t_2}{d_2}}\right),$$

and the maximum point is

$$p_1 = \frac{p_2 + t_2\left(1 + \frac{\lambda}{d_2}\right) - t_1}{2}.$$

This is the best response of player 1 to the price p_2. Similarly, we calculate the best response of player 2 to the price p_1. And the equilibrium in the game acquires the form

$$p_1^* = \lambda_1\left(\frac{t_1}{d_1} + \frac{t_2}{d_2}\right),$$

$$p_2^* = \lambda_2\left(\frac{t_1}{d_1} + \frac{t_2}{d_2}\right),$$

where the flow intensities λ_1 and λ_2 meet the balance condition (5).

Consequently, the pricing game admits the equilibrium

$$p_1^* = \frac{1}{3}\left(t_1\left(\frac{\lambda}{d_1}-1\right)+t_2\left(1+2\frac{\lambda}{d_2}\right)\right), \tag{6}$$

$$p_2^* = \frac{1}{3}\left(t_2\left(\frac{\lambda}{d_2}-1\right)+t_1\left(1+2\frac{\lambda}{d_1}\right)\right). \tag{7}$$

Interestingly, under the conditions (4) the expressions (6)–(7) are nonnegative and the payoff of player 1,

$$H_1 = p_1^*\lambda_1 = \lambda_1^2\left(\frac{t_1}{d_1}+\frac{t_2}{d_2}\right),$$

is not smaller than its payoff $H_1 = \left(t_2 - t_1 - \frac{\lambda t_1}{d_1}\right)\lambda$ in the case of no competition with player 2.

We summarize the above considerations. Player 1 enjoys an advantage for incoming flows of small intensities. Actually, it establishes the service price so that player 2 does not enter the market regardless of the latter's offer. As passenger traffic grows, such strategy becomes nonoptimal, and player 2 joins the game. This principle will be employed below for games with three and more players.

Theorem 1. *Under the condition* (3), *all traffic runs through route 1 and the optimal price of player 1 equals $p_1^* = t_2 - t_1 - \frac{\lambda t_1}{d_1}$. Otherwise, the incoming flow is distributed between both routes and the equilibrium prices have the form* (6)–(7).

The symmetrical case. Consider the symmetrical case when both players have identical trip times and edge capacities: $t_1 = t_2 = t$, $d_1 = d_2 = d$. Obviously, the equilibrium prices coincide and the incoming flow is equally shared by both players. The equilibrium prices make up

$$p_1^* = p_2^* = t\frac{\lambda}{d}.$$

Thus and so, the equilibrium prices are proportional to the trip time on an empty edge. Higher capacities reduce the prices.

Table 1.   The equilibrium prices (p_1^*, p_2^*) under $\lambda=1$, $d_1 = d_2 = 2$.

t_2	Prices	t_1			
		1	2	3	4
1	$(p_1^*; p_2^*)$	(0.5;0.5)			
	$(\lambda_1; \lambda_2)$	(0.5;0.5)			
2	$(p_1^*; p_2^*)$	(1.16;0.33)	(1;1)		
	$(\lambda_1; \lambda_2)$	(0.78;0.22)	(0.5;0.5)		
3	$(p_1^*; p_2^*)$	(1.83;0.16)	(1.67;0.83)	(1.5;1.5)	
	$(\lambda_1; \lambda_2)$	(0.08;0.92)	(0.33;0.67)	(0.5;0.5)	
4	$(p_1^*; p_2^*)$	(2.5;0)	(2.33;0.67)	(2.16;1.33)	(2;2)
	$(\lambda_1; \lambda_2)$	(1;0)	(0.78;0.22)	(0.62;0.38)	(0.5;0.5)

The asymmetrical case. Imagine that $t_1 \neq t_2$, $d_1 = d_2 = d$. The corresponding calculations are presented by Table 1.

According to this table, increasing the trip time t_2 for player 2 enlarges the equilibrium price of player 1 and the flow of passengers preferring this service.

4. Transport Game on N Parallel Channels with the Linear Latency Function

Consider competition of N carriers on N parallel channels, see Fig. 2.

Denote this game by $\Gamma_N(\{t_i, i \in N\})$. Each carrier i serves passengers on its edge and assigns some price p_i, $i = 1, \ldots, N$. For convenience, renumber channels so that latencies in them form a nondecreasing sequence: $t_1 \leq t_2 \leq \cdots \leq t_N$.

Depending on traffic intensity, the incoming passenger flow runs through some n routes, $1 \leq n \leq N - 1$. Let us study two scenarios as follows. In the first scenario, players equalize their costs and maximize their payoffs. Here we establish the existence conditions of a Nash equilibrium. In the second scenario, players $\{1, 2, \ldots, n\}$ also equalize their costs but maintain them at the level of t_{n+1} trying not to admit a new player to the market.

Our analysis begins with the first scenario. Under the announced service prices $\{p_1, \ldots, p_n\}$, the incoming passenger flow λ is decomposed into n subflows λ_i, $\sum_{i=1}^{n} \lambda_i = \lambda$, according to the balance equations

$$p_1 + t_1 \left(1 + \frac{\lambda_1}{d_1}\right) = p_2 + t_2 \left(1 + \frac{\lambda_2}{d_2}\right) = \cdots = p_n + t_n \left(1 + \frac{\lambda_n}{d_n}\right). \quad (8)$$

For Nash equilibrium evaluation find the best response of each player using Lagrange's method of multipliers. Take player i and construct an associated Lagrange function for maximization of $p_i \lambda_i$ under the fixed strategies $\{p_j, j \neq i\}$ of all other players and the conditions (8):

$$L_i(p, \lambda, k, \gamma) = p_1 \lambda_1 + \sum_{j=1, j \neq i}^{n} k_j \left(p_i + t_i \left(1 + \frac{\lambda_i}{d_i}\right) - p_j - t_i \left(1 + \frac{\lambda_j}{d_j}\right)\right)$$

$$+ \gamma \left(\sum_{j=1}^{n} \lambda_j - \lambda\right).$$

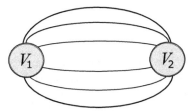

Fig. 2. N parallel routes.

The first-order necessary optimality conditions have the form

$$\frac{\partial L_i}{\partial p_i} = \lambda_i + \sum_{j=1, j\neq i}^{n} k_j = 0,$$

$$\frac{\partial L_i}{\partial \lambda_i} = p_i + t_i \frac{\sum_{j=1, j\neq i}^{n} k_j}{d_i} + \gamma = 0,$$

$$\frac{\partial L_i}{\partial \lambda_j} = -t_j \frac{k_j}{d_j} + \gamma = 0.$$

By solving the above system of equations for all $i = 1, \ldots, n$, we derive the following equilibrium solution in this pricing game:

$$p_i^* = \lambda_i^{(n)} \left(\frac{1}{\frac{d_i}{t_i}} + \frac{1}{\sum_{j=1, j\neq i}^{n} \frac{d_i}{t_i}} \right), \quad i = 1, \ldots, n, \tag{9}$$

$$\lambda_i^{(n)} = \frac{\lambda - \sum_{j=1}^{n} \frac{t_i - t_j}{b_j^{(n)}}}{\sum_{j=1}^{n} \frac{b_i^{(n)}}{b_j^{(n)}}}, \quad i = 1, \ldots, n, \tag{10}$$

where

$$b_i^{(n)} = \frac{2}{\frac{d_i}{t_i}} + \frac{1}{\sum_{j=1, j\neq i}^{n} \frac{d_j}{t_j}}, \quad i = 1, \ldots, n.$$

For the equilibrium to exist, it suffices that

$$\lambda \geq \sum_{j=1}^{n-1} \frac{t_n - t_j}{b_j^{(n)}}. \tag{11}$$

Note that, as we increase λ, the passenger costs in the conditions (11) gradually reach t_{n+1}. This happens under

$$\lambda = \sum_{j=1}^{n} \frac{t_{n+1} - t_j}{b_j^{(n)}}.$$

The following result is true.

Theorem 2. *Suppose that for some $n = 1, \ldots, N - 1$:*

$$\sum_{j=1}^{n-1} \frac{t_n - t_j}{b_j^{(n)}} < \lambda \leq \sum_{j=1}^{n} \frac{t_{n+1} - t_j}{b_j^{(n)}}.$$

Then the incoming flow is distributed among first n routes and the equilibrium prices have the form (9). The corresponding equilibrium payoffs of the players make up

$$H_i = p_i \lambda_i^{(n)} = p_i^2 \left(\frac{1}{\frac{d_i}{t_i}} + \frac{1}{\sum_{j=1, j\neq i}^{n} \frac{d_i}{t_i}} \right)^{-1}, \quad i = 1, \ldots, n. \tag{12}$$

Now, study the second scenario:

$$\sum_{j=1}^{n} \frac{t_{n+1} - t_j}{b_j^{(n)}} < \lambda \le \sum_{j=1}^{n} \frac{t_{n+1} - t_j}{b_j^{(n+1)}}. \tag{13}$$

Assume that players choose prices so that the balance condition (8) remains in force but

$$p_i + t_i \left(1 + \frac{\lambda_i}{d_i}\right) = t_{n+1}, \quad i = 1, \ldots, n. \tag{14}$$

Equalities (14) guarantee that passengers prefer service $(n+1)$. Let p_i be such that λ_i become linear in λ:

$$\lambda_i = k_i \lambda + c_i, \quad i = 1, \ldots, n,$$

and coincide with $\lambda_i^{(n)}$ and $\lambda_i^{(n+1)}$ (see (10)) on the left and right boundaries of the interval (13), respectively. In other words,

$$k_i \sum_{j=1}^{n} \frac{t_{n+1} - t_j}{b_j^{(n)}} + c_i = \frac{t_{n+1} - t_i}{b_i^{(n)}}, \quad i = 1, \ldots, n,$$

$$k_i \sum_{j=1}^{n} \frac{t_{n+1} - t_j}{b_j^{(n+1)}} + c_i = \frac{t_{n+1} - t_i}{b_i^{(n+1)}}, \quad i = 1, \ldots, n. \tag{15}$$

These conditions and linearity of the function λ_i are guaranteed by $\sum_{i=1}^{n} \lambda_i = \lambda$. We underline that in the conditions (13) the quantities λ_i meet the inequalities

$$\frac{t_{n+1} - t_i}{b_i^{(n)}} < \lambda_i \le \frac{t_{n+1} - t_i}{b_i^{(n+1)}}, \quad i = 1, \ldots, n. \tag{16}$$

Theorem 3. *Suppose that the conditions (12) take place for some* $n = 1, \ldots, N-1$. *Then the incoming flow is distributed among first* n *routes and the equilibrium prices acquire the linear form*

$$p_i = t_{n+1} - t_i \left(1 + \frac{k_i \lambda + c_i}{d_i}\right), \quad i = 1, \ldots, n, \tag{17}$$

where k_i, c_i *are uniquely defined by (15).*

Proof. Assume that all players use strategies $p_i, i = 1, \ldots, n$ of the form (17) and a certain player (e.g., player 1) deviates from the equilibrium strategy. If player i decreases the price p_i, after equalization the costs of the players become smaller than t_{n+1}. Hence, the conditions of Theorem 2 are satisfied, and the payoff of player i is described by (12), *ergo*, goes down.

Now, imagine that player i increases its price. Then after equalization the costs of the players exceed t_{n+1}. And player $n + 1$ joins the game. Let us demonstrate that, in this case, the payoff of player 1 does not grow. Really, the payoff of player i obeys formula (12), where the flow intensity λ_i' is smaller than λ_i:

$$H_i'(\lambda_i') = (\lambda_i')^2 \left(\frac{1}{\frac{d_i}{t_i}} + \frac{1}{\sum_{j=1, j\neq i}^{n+1} \frac{d_i}{t_i}} \right). \tag{18}$$

Under the price (17), its payoff constitutes

$$H_i(\lambda_i) = \lambda_i \left(t_{n+1} - t_i - \frac{\lambda_i t_i}{d_i} \right). \tag{19}$$

The function (19) represents a parabola and λ_i in (16) lies to the left from the parabola maximum $\frac{t_{n+1}-t_i}{2t_i/d_i}$. Hence, we obtain that $H_i(\lambda_i) \geq H_i(\lambda_i')$. At the same time, the conditions (16) imply that $H_i(\lambda_i') \geq H_i'(\lambda_i')$. The above inequalities argue that such prices form an equilibrium in the game. □

Figure 3 illustrates the competition of five carriers under $d_i = 2$, $t_i = i$, $i = 1, 2, \ldots, 5$. For $0 < \lambda \leq 0.5$, no competition takes place, and the route is served by carrier 1 only. As we increase incoming traffic ($0.5 < \lambda \leq 1.4$), player 2 joins the game according to Theorem 2. In the case of $1.4 < \lambda \leq 1.67$ (formula (13)), players 1 and 2 avoid competition from player 3 by reducing their ticket prices and maintaining the passenger costs at the level of t_3. Player 3 joins the game when $1.67 < \lambda \leq 3.01$ (see Theorem 2). And the game continues by analogy. If $3.01 < \lambda \leq 3.22$, players 1, 2 and 3 avoid competition from player 4 by reducing

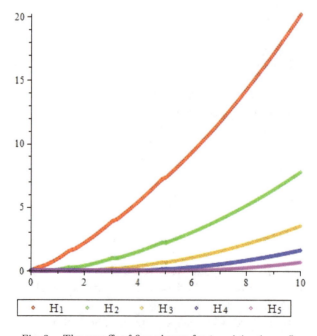

Fig. 3.   The payoffs of five players for $t_i = i, i = 1, \ldots, 5$.

their ticket prices and maintaining the passenger costs at the level of t_4. Player 4 joins the game when $3.22 < \lambda \leq 4.875$. Under $4.875 < \lambda \leq 5.04$, the four carriers avoid competition from player 5, but as soon as $5.04 < \lambda$ all players compete at the transportation market.

Numerical examples. Consider the competition of 10 players each serving its own route. The equilibrium prices and flow intensities are provided by Tables 2 and 3.

Table 2 demonstrates that, the higher is the latency on player's empty road in comparison with competitors, the lower is the service price established by this carrier. As the capacities of all players increase, the price goes down but the flow of passengers preferring carriers with small t grows (the flow of passengers choosing carriers with high t is accordingly reduced). Therefore, trip time has a higher priority for passengers than service price.

Table 2.   The values of p_i and λ_i $i = 1, 2, \ldots, 10$ under $\lambda=1$, $t_i = i$, $i = 1, 2, \ldots, 10$.

d	p_1 λ_1	p_2 λ_2	p_3 λ_3	p_4 λ_4	p_5 λ_5	p_6 λ_6	p_7 λ_7	p_8 λ_8	p_9 λ_9	p_{10} λ_{10}
1	47.54	42.56	40.74	39.61	38.75	38	37.33	36.71	36.11	35.53
	31.31	17.65	12.04	9.06	7.22	5.97	5.07	4.39	3.86	3.43
2	24.55	21.72	20.53	19.7	19.01	18.39	17.8	17.23	16.68	16.14
	32.34	18.01	12.13	9.01	7.09	5.78	4.84	4.12	3.57	3.12
3	16.89	14.77	13.79	13.07	12.43	11.85	11.28	10.74	10.2	9.67
	33.37	18.37	12.22	8.96	6.95	5.59	4.6	3.85	3.27	2.8
4	13.06	11.3	10.42	9.75	9.14	8.58	8.03	7.49	6.96	6.44
	34.41	18.74	12.32	8.92	6.82	5.39	4.36	3.59	2.98	2.49
5	10.76	9.21	8.4	7.76	7.17	6.61	6.07	5.54	5.02	4.5
	35.44	19.1	12.41	8.87	6.68	5.2	4.13	3.32	2.68	2.17
6	9.23	7.82	7.06	6.43	5.85	5.31	4.77	4.25	3.73	3.21
	36.47	19.46	12.51	8.82	6.55	5	3.89	3.05	2.39	1.86
7	8.14	6.83	6.09	5.48	4.91	4.37	3.84	3.32	2.8	2.29
	37.51	19.82	12.6	8.77	6.41	4.81	3.65	2.78	2.1	1.55

Table 3.   The values of (p_1^*, p_2^*) under $\lambda=10$, $d_1 = d_2 = 2$, $\beta = 2$.

t_2	Prices	t_1 1	2	3	4
1	$(p_1^*; p_2^*)$ $(\lambda_1; \lambda_2)$	(25;25) (5;5)			
2	$(p_1^*; p_2^*)$ $(\lambda_1; \lambda_2)$	(41.62;40.95) (5.91;4.09)	(50;50) (5;5)		
3	$(p_1^*; p_2^*)$ $(\lambda_1; \lambda_2)$	(55.08;53.74) (6.42;3.58)	(67.68;67.01) (5.53;4.47)	(75;75) (5;5)	
4	$(p_1^*; p_2^*)$ $(\lambda_1; \lambda_2)$	(66.65;64.65) (6.77;3.23)	(83.23;81.9) (5.91;4.09)	(93.1;92.44) (5.38;4.62)	(100;100) (5;5)

5. Transport Game with the Nonlinear Latency Function

Consider the nonlinear latency in a network composed of N parallel routes served by carriers with the trip times

$$t_i \left(1 + \left(\frac{\lambda}{d_i} \right)^\beta \right), \quad i = 1, \ldots, N.$$

The same method as in the case of the linear latency function is applicable here, but it fails to yield analytical expressions.

5.1. *Transport game on two parallel routes*

Imagine two carriers serving two parallel routes with prices p_1 and p_2, respectively. Write down the balance equation:

$$p_1 + t_1 \left(1 + \left(\frac{\lambda_1}{d_1} \right)^\beta \right) = p_2 + t_2 \left(1 + \left(\frac{\lambda_2}{d_2} \right)^\beta \right). \tag{20}$$

If $t_2 > t_1$ (channel 2 has a higher latency than channel 1), passengers do not choose carrier 2 even under zero price in the case of small incoming flow; therefore, player 2 is eliminated from competition.

That is, under the condition

$$p_1 + t_1 \left(1 + \left(\frac{\lambda}{d_1} \right)^\beta \right) \le t_2,$$

the incoming flow runs through route 1 only. The optimal price of player 1 makes up

$$p_1^* = t_2 - t_1 \left(1 + \left(\frac{\lambda}{d_1} \right)^\beta \right).$$

Yet, a sufficiently large flow such that

$$\lambda > d_1 \left(\frac{t_2 - t_1}{t_1 (1 + \beta)} \right)^{\frac{1}{\beta}} \tag{21}$$

runs through both routes. As before, the payoffs of the players have the form

$$H_1 = p_1 \lambda_1, \quad H_2 = p_2 \lambda_2.$$

Fix the strategy p_2 of player 2 and evaluate the best response of its opponent:

$$\frac{\partial H_1}{\partial p_1} = \lambda_1 + p_1 \frac{\partial \lambda_1}{\partial p_1} = 0.$$

Hence, it follows that

$$p_1 = -\frac{\lambda_1}{\frac{\partial \lambda_1}{\partial p_1}}.$$

Perform differentiation of (17) and take into account that $\lambda_1 + \lambda_2 = \lambda$ to obtain

$$\frac{\partial \lambda_1}{\partial p_1} = -\frac{1}{\beta} \left(\frac{t_1 \lambda_1^{\beta-1}}{d_1^\beta} + \frac{t_2 \lambda_2^{\beta-1}}{d_2^\beta} \right)^{-1},$$

and so

$$p_1^* = \lambda_1 \beta \left(\frac{t_1}{d_1} \left(\frac{\lambda_1}{d_1} \right)^{\beta-1} + \frac{t_2}{d_2} \left(\frac{\lambda_2}{d_2} \right)^{\beta-1} \right).$$

Similar reasoning for player 2 gives

$$p_2^* = \lambda_2 \beta \left(\frac{t_1}{d_1} \left(\frac{\lambda_1}{d_1} \right)^{\beta-1} + \frac{t_2}{d_2} \left(\frac{\lambda_2}{d_2} \right)^{\beta-1} \right).$$

By substituting p_1^*, p_2^* into formula (17), we calculate the equilibrium flow intensities λ_1, λ_2. Due to the assumption (18), these quantities are nonnegative. Interestingly, for $\lambda = d_1 \left(\frac{t_2-t_1}{t_1(1+\beta)} \right)^{\frac{1}{\beta}}$ the result is $\lambda_2 = 0$.

The derived expressions define the equilibrium prices and flows.

Numerical examples. In the symmetrical case of $t_1 = t_2 = t$, $d_1 = d_2 = d$, the system possesses the solution

$$\lambda_1 = \lambda_2 = \frac{\lambda}{2},$$

$$p_1^* = p_2^* = 2t\beta \left(\frac{\lambda}{2d} \right)^\beta.$$

Thus, the equilibrium prices grow exponentially with traffic intensity. At the same time, higher capacity appreciably reduces the prices.

Now, suppose that $t_1 \neq t_2$, $d_1 = d_2 = d$. The corresponding calculations are demonstrated by Table 3.

For instance, take $d = 4$ and $t_1 = t_2 = 4$; then the equilibrium prices are $p_1^* = p_2^* = 25$. Clearly, the higher is the trip time on an empty route, the higher is the service price. But as the capacity d increases, the trip time goes down and the price also drops appreciably.

5.2. *Transport game on n parallel routes with the nonlinear latency function*

When n carriers compete on n parallel routes, the incoming passenger flow of intensity λ is decomposed into n subflows of intensities λ_i, $\sum_{i=1}^{n} \lambda_i = \lambda$ depending on λ and the announced service prices $\{p_1, \ldots, p_n\}$, just like in the previous subsection.

The quantities λ_i can be calculated using the conditions

$$p_1 + t_1\left(1 + \left(\frac{\lambda_1}{d_1}\right)^\beta\right) = p_2 + t_2\left(1 + \left(\frac{\lambda_2}{d_2}\right)^\beta\right)$$

$$= \cdots = p_n + t_n\left(1 + \left(\frac{\lambda_n}{d_n}\right)^\beta\right). \tag{22}$$

Again, Lagrange's method of multipliers for best response evaluation yields the equilibrium prices:

$$p_i^* = \beta\lambda_i\left(\frac{t_i}{d_i}\left(\frac{\lambda_i}{d_i}\right)^{\beta-1} + \frac{1}{\sum_{j=1,j\neq i}^n \frac{d_j}{t_j}\left(\frac{d_j}{\lambda_j}\right)^{\beta-1}}\right), \quad i = 1,\ldots,n.$$

The corresponding equilibrium flows are uniquely defined by the conditions (22).

Numerical examples. Within the framework of the linear latency model, we have studied an example with 10 players. Consider this example with the quadratic latency function. Table 4 contains the values of the equilibrium prices and flow intensities.

Note that players with small channel latencies (i.e., players 1, 2, 3 and 4) have more passengers as the capacity increases. For the rest of the players, the opposite picture takes place: higher capacity reduces passenger flows preferring their services.

Table 4.   The values of p_i and λ_i, $i = 1, 2, \ldots, 10$ under $\lambda{=}100$, $t_i = i$, $i = 1, 2, \ldots, 10$, $\beta = 2$.

d	p_1 λ_1	p_2 λ_2	p_3 λ_3	p_4 λ_4	p_5 λ_5	p_6 λ_6	p_7 λ_7	p_8 λ_8	p_9 λ_9	p_{10} λ_{10}
1	930.8	909.2	899.9	894.2	890.3	887.2	884.6	882.5	880.6	878.9
	19.22	13.97	11.52	10.04	9.01	8.25	7.65	7.17	6.76	6.42
2	234.5	228.6	225.7	223.8	222.3	221	219.9	218.8	217.9	216.9
	19.3	14	11.54	10.04	9	8.23	7.63	7.14	6.73	6.38
3	105.6	102.5	100.9	99.7	98.6	97.7	96.8	95.9	95.1	94.4
	19.44	14.08	11.58	10.05	9	8.21	7.59	7.08	6.67	6.31
4	60.42	58.41	57.19	56.2	55.32	54.49	53.7	52.94	52.19	51.45
	19.64	14.18	11.62	10.07	8.98	8.17	7.53	7.01	6.58	6.21
5	39.54	38	36.98	36.1	35.29	34.52	33.77	33.04	32.32	31.6
	19.89	14.3	11.69	10.09	8.97	8.13	7.47	6.92	6.47	6.08
6	28.21	26.92	26	25.18	24.4	23.67	22.95	22.23	21.53	20.83
	20.2	14.46	11.77	10.11	8.95	8.08	7.38	6.81	6.33	5.91
7	21.37	20.25	19.39	18.61	17.86	17.14	16.43	15.72	15.03	14.33
	20.56	14.65	11.86	10.14	8.93	8.01	7.28	6.68	6.16	5.72

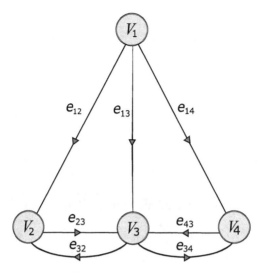

Fig. 4.   A Euler graph.

6. Transport Game on Euler Graph

This section demonstrates application of the suggested methods on a well-known Euler graph, see Fig. 4. The graph arises in Euler's solution to the Königsberg Bridge problem. Further exposition focuses on the linear latency functions.

Consider the following noncooperative game. Three carriers serve passengers on the above graph: player 1 serves two edges e_{12} and e_{23}, player 2 serves three edges e_{13}, e_{32} and e_{34}, whereas player 3 serves the rest of the edges e_{14} and e_{43}. Passengers have to move from vertex v_1 to vertex v_2, v_3 or v_4. And so, there exist three passenger flows λ_{12}, λ_{13} and λ_{14}. Each carrier announces a ticket price p^i_{1j} on its route, $i = 1, 2, 3$, $j = 2, 3, 4$, and the oncoming flow is decomposed into passenger subflows preferring the services of the cheapest carrier. The balance equations acquire the form

$$p_{12}^{(1)} + t_{12}\left(1 + \frac{\lambda_{12}^{(1)} + \lambda_{13}^{(1)}}{d_{12}}\right)$$

$$= p_{12}^{(2)} + t_{13}\left(1 + \frac{\lambda_{12}^{(2)} + \lambda_{13}^{(2)} + \lambda_{14}^{(2)}}{d_{13}}\right) + t_{32}\left(1 + \frac{\lambda_{12}^{(2)}}{d_{32}}\right),$$

$$p_{14}^{(3)} + t_{14}\left(1 + \frac{\lambda_{14}^{(3)} + \lambda_{13}^{(3)}}{d_{14}}\right)$$

$$= p_{14}^{(2)} + t_{13}\left(1 + \frac{\lambda_{12}^{(2)} + \lambda_{13}^{(2)} + \lambda_{14}^{(2)}}{d_{13}}\right) + t_{34}\left(1 + \frac{\lambda_{14}^{(2)}}{d_{32}}\right),$$

$$p_{13}^{(1)} + t_{12} \left(1 + \frac{\lambda_{12}^{(1)} + \lambda_{13}^{(1)}}{d_{12}} \right) + t_{23} \left(1 + \frac{\lambda_{13}^{(1)}}{d_{23}} \right)$$

$$= p_{13}^{(2)} + t_{13} \left(1 + \frac{\lambda_{12}^{(2)} + \lambda_{13}^{(2)} + \lambda_{14}^{(2)}}{d_{13}} \right)$$

$$= p_{13}^{(3)} + t_{14} \left(1 + \frac{\lambda_{13}^{(3)} + \lambda_{14}^{(3)}}{d_{14}} \right) + t_{43} \left(1 + \frac{\lambda_{13}^{(3)}}{d_{43}} \right),$$

$$\lambda_{12}^{(1)} + \lambda_{12}^{(2)} = \lambda_{12}, \quad \lambda_{13}^{(1)} + \lambda_{13}^{(2)} + \lambda_{13}^{(3)} = \lambda_{13}, \quad \lambda_{14}^{(5)} + \lambda_{14}^{(2)} = \lambda_{14}.$$

Using Lagrange's method of multipliers, find the best response of player 1:

$$p_{12}^{(1)} = \lambda_{12}^{(1)} \left(\frac{t_{12}}{d_{12}} + \frac{t_{13}}{d_{13}} + \frac{t_{32}}{d_{32}} \right) + \lambda_{13}^{(1)} \left(\frac{t_{12}}{d_{12}} + \frac{t_{13}}{d_{13}} \right)$$

$$- \left(\frac{t_{13}}{d_{13}} \right)^2 \left(\frac{\lambda_{12}^{(1)} + \lambda_{13}^{(1)}}{\frac{t_{13}}{d_{13}} + \frac{t_{14}}{d_{14}} + \frac{t_{43}}{d_{43}}} \right).$$

$$p_{13}^{(1)} = \lambda_{12}^{(1)} \left(\frac{t_{12}}{d_{12}} \right) + \lambda_{13}^{(1)} \left(\frac{t_{12}}{d_{12}} + \frac{t_{23}}{d_{23}} \right) + \left(\frac{\lambda_{12}^{(1)} + \lambda_{13}^{(1)}}{\frac{1}{\frac{t_{14}}{d_{14}} + \frac{t_{43}}{d_{43}}} + \frac{d_{13}}{t_{13}}} \right).$$

Owing to the problem's symmetry, the best response of player 3 is defined by

$$p_{14}^{(3)} = \lambda_{14}^{(3)} \left(\frac{t_{14}}{d_{14}} + \frac{t_{13}}{d_{13}} + \frac{t_{34}}{d_{34}} \right) + \lambda_{13}^{(3)} \left(\frac{t_{14}}{d_{14}} + \frac{t_{13}}{d_{13}} \right)$$

$$- \left(\frac{t_{13}}{d_{13}} \right)^2 \left(\frac{\lambda_{13}^{(3)} + \lambda_{14}^{(3)}}{\frac{t_{13}}{d_{13}} + \frac{t_{12}}{d_{12}} + \frac{t_{23}}{d_{23}}} \right).$$

$$p_{13}^{(3)} = \lambda_{14}^{(3)} \left(\frac{t_{14}}{d_{14}} \right) + \lambda_{13}^{(3)} \left(\frac{t_{14}}{d_{14}} + \frac{t_{43}}{d_{43}} \right) + \left(\frac{\lambda_{13}^{(3)} + \lambda_{14}^{(3)}}{\frac{1}{\frac{t_{12}}{d_{12}} + \frac{t_{23}}{d_{23}}} + \frac{d_{13}}{t_{13}}} \right).$$

It remains to derive the best response conditions for player 2. Again, Lagrange's method of multipliers yields

$$p_{12}^{(2)} = \lambda_{12}^{(2)} \left(\frac{t_{12}}{d_{12}} + \frac{t_{13}}{d_{13}} + \frac{t_{32}}{d_{32}} \right) + \lambda_{13}^{(2)} \left(\frac{t_{12}}{d_{12}} + \frac{t_{13}}{d_{13}} \right) + \lambda_{14}^{(2)} \left(\frac{t_{13}}{d_{13}} \right)$$

$$- \left(\frac{t_{12}}{d_{12}} \right) \left(\frac{\lambda_{13}^{(2)} \left(\frac{t_{12}}{d_{12}} + \frac{t_{23}}{d_{23}} \right) + \lambda_{12}^{(2)} \frac{t_{12}}{d_{12}} - \lambda_{14}^{(2)} \frac{t_{14}}{d_{14}}}{\frac{t_{12}}{d_{12}} + \frac{t_{23}}{d_{23}} + \frac{t_{14}}{d_{14}} + \frac{t_{43}}{d_{43}}} \right),$$

$$p_{13}^{(2)} = (\lambda_{12}^{(2)} + \lambda_{13}^{(2)} + \lambda_{14}^{(2)})\frac{t_{13}}{d_{13}}$$

$$+ \frac{\left(\frac{t_{14}}{d_{14}} + \frac{t_{43}}{d_{43}}\right)\left(\lambda_{12}^{(2)}\frac{t_{12}}{d_{12}} + \lambda_{13}^{(2)}\left(\frac{t_{12}}{d_{12}} + \frac{t_{13}}{d_{13}}\right)\right) + \lambda_{14}^{(2)}\frac{t_{14}}{d_{14}}\left(\frac{t_{12}}{d_{12}} + \frac{t_{23}}{d_{23}}\right)}{\frac{t_{12}}{d_{12}} + \frac{t_{23}}{d_{23}} + \frac{t_{14}}{d_{14}} + \frac{t_{43}}{d_{43}}},$$

$$p_{14}^{(2)} = \lambda_{12}^{(2)}\left(\frac{t_{13}}{d_{13}}\right) + \lambda_{13}^{(2)}\left(\frac{t_{13}}{d_{13}}\right) + \lambda_{14}^{(2)}\left(\frac{t_{13}}{d_{13}} + \frac{t_{14}}{d_{14}} + \frac{t_{34}}{d_{34}}\right)$$

$$+ \left(\frac{t_{14}}{d_{14}}\right)\left(\frac{\lambda_{13}^{(2)}\left(\frac{t_{12}}{d_{12}} + \frac{t_{23}}{d_{23}}\right) + \lambda_{12}^{(2)}\frac{t_{12}}{d_{12}} - \lambda_{14}^{(2)}\frac{t_{14}}{d_{14}}}{\frac{t_{12}}{d_{12}} + \frac{t_{23}}{d_{23}} + \frac{t_{14}}{d_{14}} + \frac{t_{43}}{d_{43}}}\right).$$

The constructed equations can be used for equilibrium evaluation in this communication network.

Numerical examples. Let $t_{12} = t_{14}$, $d_{12} = d_{14}$, $t_{23} = t_{43}$, $d_{23} = d_{43}$. Obviously, the following equalities take place in an equilibrium: $p_{12}^{(1)} = p_{14}^{(3)}$, $\lambda_{12}^{(1)} = \lambda_{14}^{(3)}$ and $p_{13}^{(1)} = p_{13}^{(3)}$, $\lambda_{13}^{(1)} = \lambda_{13}^{(3)}$. The corresponding calculations are presented in Table 5, where d_{13} varies and the rest of the parameters have fixed values: $\lambda_{12}=20$, $\lambda_{13}=30$, $\lambda_{14}=10$, $t_{12} = t_{14} = 4$, $d_{12} = d_{14} = 2$, $t_{23} = t_{43} = 2$, $d_{23} = d_{43} = 1$, $t_{32} = t_{34} = 2$, $d_{32} = d_{34} = 1$.

Direct analysis of Table 5 brings to the following observation. Passenger subflows depend on latencies rather than on ticket prices. For instance, under $d_{13} = 1$ player 2 has a smaller ticket price (p_{13}^2) than the opponents do and a higher subflow; in the

Table 5. The values of $(p_{12}^{(1)}, p_{12}^{(2)})$ and $(p_{13}^{(1)}, p_{13}^{(2)})$ in the transport game on the Euler graph.

| d_{13} | Equilibrium | t_{13} | | | |
		4	5	6	7
1	$(p_{12}^{(1)};p_{12}^{(2)})$	(135.9;101.9)	(147.2;107)	(156.5;111.3)	(164.4;114.4)
	$(\lambda_{12}^{(1)};\lambda_{12}^{(2)})$	(17.19;2.81)	(17.73;2.27)	(18.16;1.84)	(18.51;1.49)
	$(p_{13}^{(1)};p_{13}^{(2)})$	(117.9;109.9)	(129.2;115.1)	(138.5;119.3)	(146.4;122.7)
	$(\lambda_{13}^{(1)};\lambda_{13}^{(2)})$	(8.19;13.62)	(8.73;12.54)	(9.16;11.68)	(9.5;11)
2	$(p_{12}^{(1)};p_{12}^{(2)})$	(104.7;86.7)	(114.2;90.6)	(122.5;94.3)	(129.9;97.5)
	$(\lambda_{12}^{(1)};\lambda_{12}^{(2)})$	(15.55;4.45)	(16.1;3.9)	(16.54;3.46)	(16.93;3.07)
	$(p_{13}^{(1)};p_{13}^{(2)})$	(86.7;94.4)	(96.2;98.6)	(104.5;102.3)	(111.9;105.5)
	$(\lambda_{13}^{(1)};\lambda_{13}^{(2)})$	(6.55;16.9)	(7.1;15.8)	(7.54;14.92)	(7.93;14.14)
4	$(p_{12}^{(1)};p_{12}^{(2)})$	(82;75)	(88.6;77.8)	(94.7;80.3)	(100.4;82.7)
	$(\lambda_{12}^{(1)};\lambda_{12}^{(2)})$	(14.1;5.9)	(14.57;5.43)	(14.97;5.03)	(15.33;4.67)
	$(p_{13}^{(1)};p_{13}^{(2)})$	(64;83)	(70.6;85.8)	(76.7;88.3)	(82.4;90.7)
	$(\lambda_{13}^{(1)};\lambda_{13}^{(2)})$	(5.1;19.8)	(5.57;18.86)	(5.97;18.06)	(6.33;17.34)

case of $d_{13} = 4$, the price p_{13}^2 exceeds those of both opponents, but the subflow is still higher.

7. Conclusion

This chapter has proposed a pricing model in a competitive transport network described by a communication graph. We have evaluated an equilibrium in the pricing game under the BPR latency functions on all routes. In the linear case, we have established the competitiveness conditions of all carriers. The introduced model has been illustrated using a Euler graph, including equilibrium construction in the associated pricing problem. According to the modeling results, the greater the trip latencies, the higher the equilibrium prices. The parameter β answers for trip time and, whenever jamming occurs, the latter grows proportionally to the former. An increase in the parameter t (trip time on an empty edge) enlarges an equilibrium price, but passengers prefer the fastest service under competing carriers with different values of this parameter. If the capacity d of some player goes up, it may happen that the equilibrium price of its service is higher against the opponents.

And finally, note that the suggested approach is applicable only when competition exists at least on one edge of the communication graph. Future investigations will be dedicated to passenger traffic problems with price constraints on certain network edges and more complicated network topologies.

Acknowledgments

The authors acknowledge the support of the Mathematical Sciences Division of the Russian Academy of Sciences and the Russian Foundation for Basic Research (Project No. 16-51-55006), and the Russian Foundation for the Humanities (project no. 15-02-00352), and Saint-Petersburg State University (project no. 9.38.245.2014).

References

Altman, E. and Shimkin, N. [1998] Individual equilibrium and learning in processor sharing systems, *Oper. Res.* **46**(6), 776–784.

Bure, V. M., Mazalov, V. V., Melnik, A. V. and Plaksina, N. V. [2015] Passenger traffic evaluation and price formation on the transportation services market, *Adv. Oper. Res.* 2015, 1–10.

Chen, H. and Wan, Y. [2003] Price competition of make-to-order firms, *IIE Trans.* **35**(9), 817–832.

Chen, H. and Wan, Y. [2005] Capacity competition of make-to-order firms, *Oper. Res. Lett.* **33**(2), 187–194.

Hammalainen, R., Luoma, J. and Saarinen, E. [2013] On the importance of behavioral operational research: The case of understanding and communicating about dynamic systems, *Eur. J. Oper. Res.* **228**(3), 623–634.

Hassin, R. and Haviv, M. [2003] *To Queue or Not to Queue. Equilibrium Behavior in Queueing Systems*, Kluwer Academic Publishers (Boston/Dordrecht/London).

Levhari, D. and Luski, I. [1978] Duopoly pricing and waiting lines, *Eur. Econ. Rev.* **11**, 17–35.

Luski, I. [1976] On partial equilibrium in a queueing system with two services, *Rev. Econ. Studies* **43**(3), 519–525.

Spiess, H. and Florian, M. [1989] Optimal strategies: A new assignment model for transit networks, *Transp. Res. B* **23**, 83–102.

U.S. Bureau of Public Roads [1964] Traffic Assignment Manual, U.S. Department of Commerce, Washington, DC.

Wardrop, J. G. [1952] Some theoretical aspects of road traffic research, *Proc. Institution of Civil Engineers II*, pp. 325–378.

Yang, H. and Huang, H.-J. [2004] The multi-class, multi-criteria traffic network equilibrium and systems optimum problem, *Transport. Res. B* **38**, 1–15.

Chapter 8

Existence of Unique Equilibrium in Cournot Mixed Oligopoly

Koji Okuguchi[*,‡] and Takeshi Yamazaki[†,§]

*Tokyo Metropolitan University, Japan
†Department of Economics, Niigata University, Japan
‡kokuguchi@jcom.zaq.ne.jp

§tyamazak@econ.niigata-u.ac.jp

The properties of Cournot mixed oligopoly consisting of one public firm and one or more than one private firms have mostly been analyzed for simple cases on the basis of numerical calculations of the equilibrium values for a linear market demand function and linear or quadratic cost functions. In this chapter, after proving the existence of a unique equilibrium in Cournot mixed oligopoly under general conditions on the market demand and each firm's cost function, we derive conditions ensuring the existence of a unique Nash equilibrium for the mixed oligopoly where one public firm and at least one of the private firms are active in a general model of Cournot mixed oligopoly with one public firm and several private firms.

Keywords: Mixed oligopoly; Cournot; equilibrium; existence.

1. Introduction

Cournot oligopoly has been much analyzed from the point of view of existence and stability of its equilibrium since Theocharis [1960] and McManus [1962, 1964]. Bamón and Fraysse [1985], Gaudet and Salant [1991], Kolstad and Mathiesen [1987], Okuguchi [1993], and Szidarovszky and Yakowitz [1977] study the existence or the unique existence of the Cournot equilibrium, whereas Al-Nowaihi and Levine [1985], Dastidar [2000], Furth [1986], Hahn [1962], Okuguchi [1964], Okuguchi and Yamazaki [2004, 2008, 2014], Seade [1980], and Theocharis [1960] study the local or global stability of the Cournot equilibrium with or without product differentiation. The model of Cournot oligopoly has been extended along many directions. For

---

§Corresponding author.

example, Colombo and Labrecciosa [2013a, 2013b, 2015], Grilli and Bisceglia [2017], Lambertini [2014], Lambertini and Mantovani [2014], Possajennikov [2003], and Vega-Redondo [1997] study various economic issues in dynamic models of Cournot oligopoly or more general oligopoly.

Many researchers have also analyzed the traditional rent-seeking game, which can be interpreted as a model of Cournot oligopoly with a specific demand function. For example, Szidarovszky and Okuguchi [1997], Yamazaki [2008], and Hirai and Szidarovszky [2013] prove the unique existence of the pure-strategy Nash equilibrium in a fundamental model, an asymmetric model, and an asymmetric model with endogenous prizes of the rent-seeking game, respectively. Another variant of Cournot oligopoly is the Cournot mixed oligopoly in which one social welfare-maximizing public firm and several profit-maximizing private firms compete one another in a single market, while only private profit-maximizing firms compete in Cournot oligopoly. There are many papers on Cournot mixed oligopoly as well as on Stackelberg one. Almost all recent papers on mixed oligopoly, however, have derived their results basing on numerical calculations of the equilibrium values for simple cases of a linear market demand function, and linear or quadratic cost functions, which casts a serious doubt on the robustness of their results. Matsumura [1998] studies the effect of privatization of a public firm in a general model of mixed duopoly with one public firm and only one private firm. Myles [2002] and Matsumura and Kanda [2005] study a general model of mixed oligopoly with one public firm and several private firms. However, to derive their interesting results, they analyze the equilibrium condition without proving the existence of Nash equilibrium. Okuguchi [2012] derives some conditions ensuring the unique existence of the Nash equilibrium in a general model of the mixed oligopoly, but one public firm or all private firms may not produce in the equilibrium under the conditions of Okuguchi [2012]. Matsumura [1998], Myles [2002], Matsumura and Kanda [2005], and Okuguchi [2012] do not show any condition ensuring that one public firm and at least one of the private firms produce in the equilibrium. Okuguchi [1985] pays some attention to the positivity of each firm's output in a general model of oligopoly with a competitive fringe, which can be interpreted as a general model of mixed oligopoly with one public firm and several private firms. However, as we will see later, one additional condition is needed to ensure that at least one of oligopolistic firms and the competitive fringe produce in the equilibrium of the general model of oligopoly with a competitive fringe. The aim of this chapter is to derive a set of general conditions ensuring the existence of a unique Nash equilibrium for the mixed oligopoly where one public firm and at least one of the private firms are active in a general model of Cournot mixed oligopoly with one public firm and several private firms. To ensure that one public firm and at least one of the private firms produce in the equilibrium, we treat the public firm's output and the total output of the private firms as strategic analytical variables, examine whether these two variables are positive in the equilibrium, and derive the set of general conditions ensuring the existence of a unique Nash equilibrium for the mixed oligopoly where one public firm and at least one of the private firms are active.

2. Model and Analysis

Let firm 0 be a public firm and firm i, $i = 1, 2, \ldots, n$, private one. All firms produce a homogeneous good to sell in a single market. Denote firm i's output as x_i, $i = 0, 1, \ldots, n$. First we impose the following assumption.

Assumption 1. Firm i's cost function $C_i(x_i)$ is twice continuously differentiable and

$$C_i'(x_i) > 0 \quad \text{for any } x_i \geq 0, \ i = 0, 1, \ldots, n.^{\text{a}}$$

Next we describe the inverse demand function $p = f(X)$ in detail, where $X \equiv \sum_{i=0}^{n} x_i$ and p is the market price of the goods produced by all firms.

Definition 1. If $f(X) = 0$ for some $X > 0$, $\bar{X} \equiv \min\{X | f(X) = 0\}$. Otherwise, define $\bar{X} \equiv \infty$.

We add the following assumptions.

Assumption 2. $f(X)$ is twice continuously differentiable for any $X \in \mathbb{R}_{++}\backslash\{\bar{X}\}$, $f'(X) < 0$ for any $X \in (0, \bar{X})$ and $f'(X) = 0$ for any $X \geq \bar{X}$.

Assumption 3. $f(0) > \max\{C_i'(0)\}_{i=0}^{n}.^{\text{b}}$

Assumption 3 together with Assumptions 1 and 2 implies that $f(X) > 0$ for any $X \in [0, \bar{X})$ and $f(X) = 0$ for any $X \geq \bar{X}$. Under Assumptions 1–3, $f(X) - C_i'(0)$ can be negative for a sufficiently large X.

Definition 2. If $f(X) - C_i'(0) \leq 0$ for some $X > 0$, $\bar{X}^i \equiv \min\{X | f(X) - C_i'(0) \leq 0\}$. Otherwise, define $\bar{X}^i \equiv \infty$.

Now, we are ready to prove the following lemma.

Lemma 1. *Under Assumptions 2 and 3, if $\bar{X}^i < \infty$, $f(X) - C_i'(0) > 0$ for any $X \in [0, \bar{X}^i)$, $f(\bar{X}^i) - C_i'(0) = 0$ and $f(X) - C_i'(0) < 0$ for any $X > \bar{X}^i$. If $\bar{X}^i = \infty$, $f(X) - C_i'(0) > 0$ for any $X \geq 0$.*

Proof. Suppose $\bar{X} < \infty$. Since $f(X) > 0$ and $f'(X) < 0$ for any $X \in [0, \bar{X})$ and $f(X) = 0$ for any $X \geq \bar{X}$, it is clear that the solution to the equation $f(X) = C_i'(0) > 0$, $\bar{X}^i$, is uniquely determined and the solution $\bar{X}^i$ should be less than $\bar{X}$, that is, $\bar{X}^i < \bar{X}$. Since $\bar{X}^i < \bar{X}$, $f(\bar{X}^i) - C_i'(0) = 0$, $f'(X) < 0$ for any

---

[a] As usual, $C_i'(0) \equiv \lim_{x_i \to 0+} C_i'(x_i)$.

[b] As in Note 1, $f'(0) \equiv \lim_{X \to 0+} f'(X)$. By Assumption 2, $f'(0) \leq 0$.

$X \in (0, \bar{X})$, and $f'(X) = 0$ for any $X \geq \bar{X}$, $f(X) - C_i'(0) > 0$ for any $X \in [0, \bar{X}^i)$, $f(\bar{X}^i) - C_i'(0) = 0$ and $f(X) - C_i'(0) < 0$ for any $X > \bar{X}^i$.[c]

Suppose $\bar{X} = \infty$. If $\bar{X}^i < \infty$, $\bar{X}^i < \bar{X} = \infty$ and $f(\bar{X}^i) - C_i'(0) \leq 0$ by the definition of $\bar{X}^i$, i.e., $\bar{X}^i \equiv \min\{X | f(X) - C_i'(0) \leq 0\}$. To show $f(\bar{X}^i) - C_i'(0) = 0$, assume $f(\bar{X}^i) - C_i'(0) < 0$ on the contrary. Since $f'(X) < 0$ for any $X \in \mathbb{R}_{++}$, $f(\bar{X}^i - \varepsilon) - C_i'(0) \leq 0$ for any $\varepsilon > 0$ small enough, which contradicts with the definition of $\bar{X}^i$. Hence, $f(\bar{X}^i) - C_i'(0) = 0$. Since $\bar{X}^i < \bar{X}$, $f(\bar{X}^i) - C_i'(0) = 0$, $f'(X) < 0$ for any $X \in \mathbb{R}_{++}$, $f(X) - C_i'(0) > 0$ for any $X \in [0, \bar{X}^i)$, $f(\bar{X}^i) - C_i'(0) = 0$ and $f(X) - C_i'(0) < 0$ for any $X > \bar{X}^i$. If $\bar{X}^i = \infty$, by Definition 2, $f(X) - C_i'(0) > 0$ for any $X \geq 0$. $\square$

The social welfare W is the sum of the consumers' surplus and profits of all firms.

$$W \equiv \int_0^X f(x)dx - \sum_{i=0}^n C_i(x_i). \tag{1}$$

Firm i's profits π_i are

$$\pi_i \equiv x_i f(X) - C_i(x_i), \quad i = 0, 1, \ldots, n. \tag{2}$$

Under the Cournot behavioristic assumption, the first-order conditions for the interior maximum of the public and private firm's outputs are given by Eqs. (3) and (4), respectively:

$$\frac{\partial W}{\partial x_0} = f(x_0 + X_{-0}) - C_0'(x_0) = 0, \tag{3}$$

$$\frac{\partial \pi_i}{\partial x_i} = f(X) + x_i f'(X) - C_i'(x_i) = 0, \quad i = 1, 2, \ldots, n, \tag{4}$$

where $X_{-i} \equiv \sum_{j \neq i} x_j$. According to Eq. (3), the price of the produced goods is equal to the public firm's marginal cost, whereas the first-order condition (4) shows that the price is higher than the private firm's marginal costs.

Before going any further, let us now introduce the following fundamental assumptions on the market demand and firm's cost functions.

Assumption 4. $f'(X) < C_i''(x_i)$ for any $x_i > 0$ and $X \in [x_i, \bar{X})$, $i = 0, 1, \ldots, n$.

Assumption 5. $f'(X) + x f''(X) \leq 0$ for any $x > 0$ and $X \geq x$.

Assumption 6. $f(x) < C_i'(x)$ for some $x > 0$, $i = 0, 1, \ldots, n$.

Note that Assumption 4 is imposed only for $X < \bar{X}$. For $X \geq \bar{X}$, Assumptions 1 and 2 ensure that $0 = f'(X) \leq C_i''(x_i)$ holds for any $x_i > 0$, $i = 0, 1, \ldots, n$. Okuguchi [1985, 2012] and Matsumura [1998] impose Assumptions 4 and 5 on private firms' cost functions in the mixed duopoly and oligopoly, respectively. Okuguchi [1985]

---

[c]In this case, $f(X) - C_i'(0) = -C_i'(0) < 0$ for any $X \geq \bar{X}$.

assumes that $C_0''(x_0) > 0$ for a competitive fringe, which can be interpreted as the public firm. Matsumura and Kanda [2005] impose an assumption that $C_i(x) = C(x)$ and $C''(x) > 0$ for any $x > 0$, $i = 1, 2, \ldots, n$. The assumption is much more restrictive than Assumptions 4 and 5.

It is easy to prove that the following lemma holds under Assumptions 3, 4 and 6.

Lemma 2. *Under Assumptions 3, 4, and 6, for each $i = 0, 1, \ldots, n$, there exists a unique positive number $\hat{x}_i$ such that $f(x) > C_i'(x)$ for any $x \in [0, \hat{x}_i)$, $f(\hat{x}_i) = C_i'(\hat{x}_i)$, and $f(x) < C_i'(x)$ for any $x > \hat{x}_i$.*

Proof. By Assumption 3, $f(0) - C_i'(0) > 0$. By Assumption 6, $f(x) - C_i'(x) < 0$ for some x. As $f(x)$ and $C_i'(x)$ are differentiable, $f(x) - C_i'(x)$ is continuous in x. Hence, the intermediate value theorem proves that there exists *a real number* $\hat{x}_i \in (0, x)$ such that $f(\hat{x}_i) - C_i'(\hat{x}_i) = 0$. By Assumption 4, $f'(x) - C_i''(x) < 0$ for any $x > 0$, which implies that the positive number $\hat{x}_i$ is unique and that $f(x) - C_i'(x) > 0$ for any $x \in [0, \hat{x}_i)$, $f(\hat{x}_i) - C_i'(\hat{x}_i) = 0$ and $f(x) - C_i'(x) < 0$ for any $x > \hat{x}_i$. $\square$

We can also prove Lemma 3 below by making use of Lemma 2.

Lemma 3. *Under Assumptions 2, 3, 4, and 6, for each $i = 0, 1, \ldots, n$,*

$$f(x_i + X_{-i}) < C_i'(x_i) \quad \text{for any } x_i > \hat{x}_i \quad \text{and} \quad X_{-i} \geq 0, \tag{5}$$

$$f(x_i + X_{-i}) + x_i f'(x_i + X_{-i}) < C_i'(x_i) \quad \text{for any } x_i > \hat{x}_i \quad \text{and} \quad X_{-i} \geq 0. \tag{6}$$

Proof. Since $f(x) < C_i'(x)$ for any $x > \hat{x}_i$ by Lemma 2 and $f'(X) \leq 0$ by Assumption 2, the inequality (5) must hold. Since $f'(X) \leq 0$, the inequality (5) implies the inequality (6). $\square$

Now we are ready to derive the best reply function of the public firm.

Lemma 4. *The best reply function of the public firm, $x_0 = \phi_0(X_{-0})$, for any $X_{-0} \in \mathbb{R}_+$ can be written as*

$$\phi_0(X_{-0}) = \begin{cases} \tilde{\phi}_0(X_{-0}) & \text{for } X_{-0} \in [0, \bar{X}^0], \\ 0 & \text{for } X_{-0} \geq \bar{X}^0, \end{cases} \tag{7}$$

which satisfies $\phi_0(0) = \tilde{\phi}_0(0) = \hat{x}_0$, where $\tilde{\phi}_0(X_{-0})$ is the unique solution x_0 to Eq. (3) for some $X_{-0} \geq 0$ and for any $X_{-0} \in (0, \bar{X}^0)$

$$\phi_0'(X_{-0}) = \tilde{\phi}_0'(X_{-0}) = -\frac{f'(X)}{f'(X) - C_0''(x_0)} < 0. \tag{8}$$

Proof. If $x_0 = 0$, Eq. (3) becomes $f(X_{-0}) - C_0'(0) = 0$. It should be clear that the solution X_{-0} to this equation is $\bar{X}^0$ in Definition 2. By Lemma 1, for any $X_{-0} \in [0, \bar{X}^0)$

$$f(X_{-0}) - C_0'(0) > 0. \tag{9}$$

Since (5) and (9) hold, Assumption 4 ensures that there exists a unique solution $x_0 \in (0, \hat{x}_0)$ to Eq. (3), $x_0 = \tilde{\phi}_0(X_{-0})$, which is the best reply x_0 to $X_{-0} \in [0, \bar{X}^0]$. Under Assumption 4, we can apply the implicit function theorem to (3) to get (8) for any $X_{-0} \in (0, \bar{X}^0)$.

If $X_{-0} = 0$, Eq. (3) becomes $f(x_0) - C_0'(x_0) = 0$. The solution to this equation is the unique positive number $\hat{x}_0$ in Lemma 2, that is, $\phi_0(0) = \tilde{\phi}_0(0) = \hat{x}_0$. By Lemma 2, $f(\bar{X}^0) - C_0'(0) = 0$ so that Eq. (3) holds with $x_0 = 0$ and $X_{-0} = \bar{X}^0$. Hence, $\phi_0(\bar{X}^0) = \tilde{\phi}_0(\bar{X}^0) = 0$. By Lemma 1, for any $X_{-0} > \bar{X}^0$ and $x_0 \geq 0$

$$f(x_0 + X_{-0}) - C_0'(x_0) < 0. \tag{10}$$

This inequality (10) ensures that the best reply x_0 to any $X_{-0} > \bar{X}^0$ is 0, that is, $\phi_0(X_{-0}) = 0$ for any $X_{-0} > \bar{X}^0$. Since $\phi_0(\bar{X}^0) = \tilde{\phi}_0(\bar{X}^0) = 0$, $\phi_0(X_{-0}) = 0$ for any $X_{-0} \geq \bar{X}^0$. □

Next, we want to derive each private firm's cumulative best reply function in the sense of Vives [1999, Sec. 2.3.2].

Lemma 5. *The cumulative best reply function of private firm i, $x_i = \Phi_i(X)$, can be written as*

$$\Phi_i(X) = \begin{cases} \tilde{\Phi}_i(X) & \text{for } X \in [0, \bar{X}^i], \\ 0 & \text{for } X \geq \bar{X}^i, \end{cases} \tag{11}$$

where $\tilde{\Phi}_i(X)$ is the unique solution x_i to Eq. (4) for some $X \geq 0$ and for any $X \in (0, \bar{X}^i)$

$$\Phi_i'(X) = \tilde{\Phi}_i'(X) = -\frac{f'(X) + f''(X)x_i}{f'(X) - C_i''(x_i)} < 0, \quad i = 1, 2, \ldots, n. \tag{12}$$

Proof. If $x_i = 0$ solves the first-order condition (4),

$$f(X_{-i}) - C_i'(0) = f(X) - C_i'(0) = 0, \quad i = 1, 2, \ldots, n. \tag{13}$$

It should be clear that the solution X to Eq. (13) is $\bar{X}^i$ in Definition 2. By Lemma 1, for any $X \in [0, \bar{X}^i)$

$$f(X) - C_i'(0) > 0. \tag{14}$$

Next, note that $C_i'(x_i) > 0$ for any $x_i \geq 0$ by Assumption 1. For any fixed $X \geq 0$, $f(X) + x_i f'(X)$ is linear and strictly decreasing in $x_i \geq 0$. Hence,

$$f(X) + x_i f'(X) - C_i'(x_i) < 0 \tag{15}$$

for a sufficiently large $x_i \geq 0$. Since Eqs. (14) and (15) hold, Assumptions 4 and 5 ensure that there exists a unique solution $x_i > 0$ to Eq. (4), $x_i = \tilde{\Phi}_i(X)$, which is the cumulative best reply x_i to $X \in [0, \bar{X}^i)$. Under Assumption 5, we can apply the implicit function theorem to Eq. (4) to get Eq. (12) for any $X \in (0, \bar{X}^i)$.

By Lemma 1, $f(\bar{X}^i) - C_i'(0) = 0$ so that Eq. (4) holds with $x_i = 0$ and $X = \bar{X}^i$. Hence, $\Phi_i(\bar{X}^i) = \tilde{\Phi}_i(\bar{X}^i) = 0$. By Lemma 1, for any $X_{-i} > \bar{X}^i$ and $x_i \geq 0$

$$f(x_i + X_{-i}) - C_i'(x_i) = f(X) - C_i'(x_i) < 0. \tag{16}$$

The inequality (16) ensures that the best reply x_i to any $X > \bar{X}^i$ is 0, that is, $\Phi_i(X) = 0$ for any $X > \bar{X}^i$. Since $\Phi_i(\bar{X}^i) = \tilde{\Phi}_i(\bar{X}^i) = 0$, $\Phi_i(X) = 0$ for any $X \geq \bar{X}^i$. $\qquad \square$

Define

$$\Phi_{-0}(X) \equiv \sum_{i=1}^{n} \Phi_i(X). \tag{17}$$

Note that $\Phi_{-0}(X) > 0$ for any $X \in [0, \bar{X}^{-0})$ and $\Phi_{-0}(X) = 0$ for any $X > \bar{X}^{-0}$, where

$$\bar{X}^{-0} \equiv \max\{\bar{X}^i\}_{i=1}^{n}. \tag{18}$$

Since $\Phi_i'(X) = \tilde{\Phi}_i'(X) < 0$ for any $X \in (0, \bar{X}^i)$ and $\Phi_i'(X) = 0$ for any $X > \bar{X}^i$, $\Phi_{-0}(X)$ is strictly decreasing in X for any $X \in [0, \bar{X}^{-0})$.

Denote the Nash equilibrium combination of (x, X_{-0}) as (x_0^*, X_{-0}^*). The equilibrium combination (x_0^*, X_{-0}^*) is the solution to the following two equations:

$$x_0 = \phi_0(X_{-0}), \tag{19}$$

$$X_{-0} = \Phi_{-0}(x_0 + X_{-0}). \tag{20}$$

To derive some conditions for the unique existence of the Nash equilibrium, we examine the properties of the solution X_{-0} to Eq. (20) for some $x_0 \geq 0$.

Lemma 6. *There exists a unique solution X_{-0} to Eq. (20) for some x_0. The solution X_{-0} can be written as a function of x_0, $X_{-0} = \psi_{-0}(x_0)$, where*

$$\psi_{-0}(x_0) = \begin{cases} \tilde{\psi}_{-0}(x_0) & \text{for } x_0 \in [0, \bar{X}^{-0}), \\ 0 & \text{for } x_0 \geq \bar{X}^{-0}. \end{cases} \tag{21}$$

Furthermore, $\psi_{-0}(x_0)$ satisfies that $\psi_{-0}(0) = \hat{X}_{-0}$, $\psi_{-0}(x_0) > 0$ for any $x_0 \in [0, \bar{X}^{-0})$, $\psi_{-0}(x_0) = 0$ for any $x_0 \geq \bar{X}^{-0}$, and for any $x_0 \in (0, \bar{X}^{-0})$

$$\psi_{-0}'(x_0) = \tilde{\psi}_{-0}'(x_0) = \frac{\Phi_{-0}'(x_0 + \psi_{-0}(x_0))}{1 - \Phi_{-0}'(x_0 + \psi_{-0}(x_0))} \in (-1, 0), \tag{22}$$

where $\hat{X}_{-0}$ stands for the total output of the Cournot equilibrium in the market with only n private firms.

Proof. Since $\Phi_{-0}(X) > 0$ for any $X \in [0, \bar{X}^{-0})$, $\Phi_{-0}(X) = 0$ for any $X > \bar{X}^{-0}$, and $\Phi_{-0}(X)$ is strictly decreasing in X for any $X \in [0, \bar{X}^{-0})$, for any $x_0 \in [0, \bar{X}^{-0})$, $\Phi_{-0}(x_0 + X_{-0}) > 0$ for any $X_{-0} \in [0, \bar{X}^{-0} - x_0)$, $\Phi_{-0}(x_0 + X_{-0}) = 0$ for any

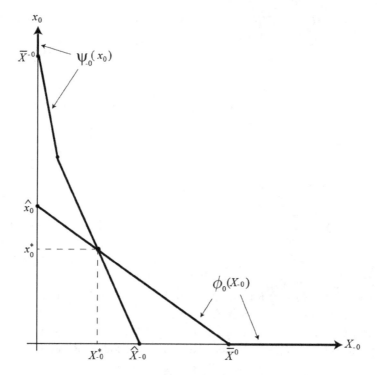

Fig. 1.   The unique intersection of two curves of $X_{-0} = \psi_{-0}(x_0)$ and $x_0 = \phi_0(X_{-0})$.

$X_{-0} \geq \bar{X}^{-0} - x_0$, and $\Phi_{-0}(x_0 + X_{-0})$ is strictly decreasing in X_{-0} for any $X_{-0} \in [0, \bar{X}^{-0} - x_0)$. These properties of $\Phi_{-0}(x_0 + X_{-0})$ ensure that for any $x_0 \in [0, \bar{X}^{-0})$, there exists a unique solution $X_{-0} \in (0, \bar{X}^{-0} - x_0)$ to Eq. (20). Since $1 - \Phi'_{-0}(x_0 + X_{-0}) \neq 0$, the implicit function theorem proves that the unique solution can be written as $X_{-0} = \tilde{\psi}_{-0}(x_0)$ in Eq. (21) and that the derivative of $\tilde{\psi}_{-0}(x_0)$ can be written as in Eq. (22). In particular, for $x_0 = 0$, Eq. (20) becomes $X_{-0} = \Phi_{-0}(X_{-0})$, which is the condition for the Cournot equilibrium in the market with only n private firms. Hence, the solution X_{-0} to Eq. (20) for $x_0 = 0$ is $\hat{X}_{-0}$, the total output of the Cournot equilibrium in the market with only n private firms, that is, $\psi_{-0}(0) = \tilde{\psi}_{-0}(0) = \hat{X}_{-0}$.

For $x_0 \geq \bar{X}^{-0}$, $\Phi_{-0}(x_0 + X_{-0}) = 0$ for any $X_{-0} \geq 0$ so that the solution to Eq. (20) is 0 for any $x_0 \geq \bar{X}^{-0}$, that is, $\psi_{-0}(x_0) = 0$ for any $x_0 \geq \bar{X}^{-0}$.   □

Now we can state that the Nash equilibrium combination (x_0^*, X_{-0}^*) is the solution to two equations (19) and (23) below:

$$X_{-0} = \psi_{-0}(x_0). \tag{23}$$

As in Fig. 1, the curve of $X_{-0} = \psi_{-0}(x_0)$ always intersects with that of $x_0 = \phi_0(X_{-0})$ only once. Hence, the following proposition holds.

Proposition 1. *Under Assumptions* 1–6, *there exists a unique Nash equilibrium.*

In Fig. 1, as $X_{-0}^* = \psi_{-0}(x_0^*) > 0$ and $x_0^* = \phi_0(X_{-0}^*) > 0$, the public firm and at least one of the private firms produce the good in the unique equilibrium. However, since one of $\psi_{-0}(x_0)$ or $\phi_0(X_{-0})$ can be zero at the unique intersection of two curves of $X_{-0} = \psi_{-0}(x_0)$ and $x_0 = \phi_0(X_{-0})$, the public firm or all private firms may not produce the good in the unique equilibrium. Now we are ready to prove the main proposition.

Proposition 2. *Under Assumptions* 1–6, *if*

(C1) $f(\hat{x}_0) = C_0'(\hat{x}_0) > \min\{C_i'(0)\}_{i=1}^n$,
(C2) $f(\hat{X}_{-0}) > C_0'(0)$,

there exists a unique Nash equilibrium where the public firm and at least one private firm produce the good of the market.

Proof. If $\hat{x}_0 < \bar{X}^{-0}$ and $\hat{X}_{-0} < \bar{X}^0$, there exists a unique Nash equilibrium where the public firm and at least one private firm produce the good of the market. It is clear that the condition (C1) in Proposition 2 is satisfied if and only if $\hat{x}_0 < \bar{X}^{-0}$ is satisfied. Similarly, the condition (C2) in Proposition 2 is satisfied if and only if $\hat{X}_{-0} < \bar{X}^0$ is satisfied. Hence, Proposition 2 holds. $\qquad\square$

As we can see in the proof above, the condition (C1) ensures that at least one of the private firms produces in the equilibrium, $X_{-0}^* > 0$, whereas the condition (C2) ensures that the public firm produces in the equilibrium, $x_0^* > 0$. Since $f(\cdot)$ and $\hat{x}_0$ in (C1), which is defined in Lemma 2, do not depend on $C_i(\cdot)$, $i = 1, 2, \ldots, n$, and since $f(\cdot)$ and $\hat{X}_{-0}$ in (C2), which is the total output of the Cournot equilibrium in the market with only n private firms, do not depend on $C_0(\cdot)$, two conditions (C1) and (C2) are independent of each other. As we have already mentioned in Sec. 1, previous works including Matsumura [1998], Myles [2002], Matsumura and Kanda [2005], and Okuguchi [2012] do not show any condition ensuring that one public firm and at least one of the private firms produce in the equilibrium. Okuguchi [1985] pays some attention to the positivity of each firm's output and assumes the condition (C2) in a general model of oligopoly with a competitive fringe, which can be interpreted as a general model of mixed oligopoly with one public firm and several private firms. Now we know that the condition (C1) is also needed to ensure that at least one of the oligopolistic firms and the competitive fringe produce in the equilibrium of the Cournot oligopoly with a competitive fringe. Proposition 2 is the first complete proposition in the literature which clarifies the conditions ensuring that one public firm and at least one of the private firms produce in the equilibrium of the Cournot mixed oligopoly.

Note that $f(\hat{x}_0) = C_0'(\hat{x}_0) > 0$ and $f(\hat{X}_{-0}) = C_i'(\Phi_i(\hat{X}_{-0})) - \Phi_i(\hat{X}_{-0}) \times f'(\hat{X}_{-0}) > 0$ for some $i \neq 0$. Hence, if both $C_0'(0)$ and $\min\{C_i'(0)\}_{i=1}^n$ are small enough, the public firm and at least one of the private firms are active in the unique

Nash equilibrium, provided that Assumptions 1–6 are satisfied. We can show in a concrete example how small $C_0'(0)$ and $\min\{C_i'(0)\}_{i=1}^n$ should be for the public firm and at least one private firm to be active in the unique Nash equilibrium. Yamazaki [2017] considers a special case in which the inverse demand function and the cost functions of firms have a specific functional form as follows:

$$f(X) = \begin{cases} a - bX & \text{for any } X \in [0, \bar{X}] \\ 0 & \text{for any } X > \bar{X} \end{cases}, \quad a > 0, \quad b > 0,$$

$$C_i(x_i) = c_0^i + c_1^i x_i + c_2^i (x_i)^2, \quad c_0^i \geq 0, \quad c_1^i > 0, \quad c_2^i \geq 0, \quad i = 0, 1, \ldots, n,$$

$$c_1^n \geq c_1^{n-1} \geq \cdots \geq c_1^1 \geq 0,$$

where $\bar{X} \equiv a/b$. This example satisfies all assumptions of this chapter, if we assume $a > \max\{c_1^0, c_1^n\} = \max\{c_1^i\}_{i=0}^n$, which is equivalent to Assumption 3. In this case, we can show that the conditions (C1) and (C2) are equivalent to

(C1') $c_1^0 > \dfrac{2c_2^0 + b}{b} c_1^1 - \dfrac{2c_2^0}{b} a,$

(C2') $c_1^0 < Aa - B(c_1^1)c_1^1,$

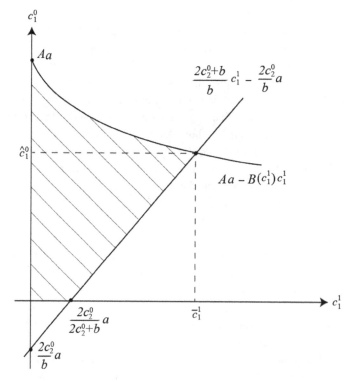

Fig. 2.   The region of (c_1^0, c_1^1) satisfying (C1') and (C2').

where

$$A = \frac{\left(\sum_{i\in I(X)} \frac{b}{2c_2^i+b}\right)^{-1}}{1 + \left(\sum_{i\in I(X)} \frac{b}{2c_2^i+b}\right)^{-1}} \in (0,1),$$

$$B(c_1^1) \equiv \frac{(\bar{c}_1/c_1^1)}{1 + \left(\sum_{i\in I(X)} \frac{b}{2c_2^i+b}\right)^{-1}} > 0,$$

$$I(X) \equiv \{i \mid X \le \bar{X}^i\}$$

and $\bar{c}_1 \in [c_1^1, c_1^n]$ is a positive real number which uniquely solves the equation

$$\sum_{i\in I(X)} \frac{a - \bar{c}_1}{2c_2^i + b} = \sum_{i\in I(X)} \frac{a - c_1^i}{2c_2^i + b}. \tag{24}$$

The unique number $\bar{c}_1$ can be interpreted as the "average" of $\{c_1^i\}_{i=1}^n$. Yamazaki [2017] shows that $B(c_1^1)$ is strictly decreasing in c_1^1 and that the region of (c_1^0, c_1^1) satisfying (C1′) and (C2′) can be depicted as in Fig. 2. This example confirms to us that two conditions (C1) and (C2) are compatible with each other and that if both $C_0'(0)$ and $\min\{C_i'(0)\}_{i=1}^n$ are small enough, which ensures that the two conditions (C1) and (C2) are satisfied, then the public firm and at least one private firm are active in the unique Nash equilibrium, provided that Assumptions 1–6 are satisfied.

3. Conclusion

In this chapter, we have proved that under Assumptions 1–6, there exists a unique Nash equilibrium in Cournot mixed oligopoly with one public and one or more private firms. Previous works including Matsumura [1998], Myles [2002], Matsumura and Kanda [2005], and Okuguchi [2012] do not show any condition ensuring that one public firm and at least one of the private firms produce in the equilibrium. We have derived two conditions ensuring that the public firm and at least one of the private firms produce the good in the unique Nash equilibrium, under Assumptions 1–6. In some papers including Matsumura [1998], Matsumura and Kanda [2005], and Okuguchi [2012], a public firm can be partially or fully privatized in the models, while we do not take into account the possibility of privatization in this chapter. Our next task in progress is to derive conditions ensuring the unique existence of the Nash equilibrium with a public firm and at least one of the private firms active in a general model of Cournot mixed oligopoly where the public firm can be partially or fully privatized.

Acknowledgment

We are thankful for the comments which help us to improve this work. All remaining errors are of course ours.

References

Al-Nowaihi, A. and Levine, P. L. [1985] The stability of the Cournot oligopoly model: A reassessment, *J. Econ. Theory* **35**, 307–321.

Bamón, R. and Frayssé, J. [1985] Existence of Cournot equilibrium in large markets, *Econometrica* **53**, 587–597.

Colombo, L. and Labrecciosa, P. [2013a] Oligopoly exploitation of a private property productive asset, *J. Econ. Dyn. Control* **37**, 838–853.

Colombo, L. and Labrecciosa, P. [2013b] On the convergence to the Cournot equilibrium in a productive asset oligopoly, *J. Math. Econ.* **49**, 441–445.

Colombo, L. and Labrecciosa, P. [2015] On the Markovian efficiency of Bertrand and Cournot equilibria, *J. Econ. Theory* **155**, 332–358.

Dastidar, K. G. [2000] Is a unique Cournot equilibrium locally stable?, *Games Econ. Behav.* **32**, 206–218.

Furth, D. [1986] Stability and instability in oligopoly, *J. Econ. Theory* **40**, 197–228.

Gaudet, G. and Salant, S. W. [1991] Uniqueness of Cournot equilibrium: New results from old methods, *Rev. Econ. Stud.* **58**, 399–404.

Grilli, L. and Bisceglia, M. [2017] A duopoly with common renewable resource and incentives, *Int. Game Theory Rev.* **19**(4) 1750018.

Hahn, F. H. [1962] The stability of the Cournot oligopoly solution, *Rev. Econ. Stud.* **29**, 329–333.

Hirai, S. and Szidarovszky, F. [2013] Existence and uniqueness of equilibrium in asymmetric contests with endogenous prizes, *Int. Game Theory Rev.* **15**(1) 1350005.

Kolstad, C. D. and Mathiesen, L. [1987] Necessary and sufficient conditions for uniqueness of a Cournot equilibrium, *Rev. Econ. Stud.* **54**, 681–690.

Lambertini, L. [2014] Exploration for nonrenewable resources in a dynamic oligopoly: An Arrovian result, *Int. Game Theory Rev.* **16**(2) 1440011.

Lambertini, L. and Mantovani, P. [2014] Feedback equilibria in a dynamic renewable resource oligopoly: Pre-emption, voracity and exhaustion, *J. Econ. Dyn. Control* **47**, 115–122.

Matsumura, T. [1998] Partial privatization in mixed duopoly, *J. Public Econ.* **70**, 473–483.

Matsumura, T. and Kanda, O. [2005] Mixed oligopoly at free entry markets, *J. Econ.* **84**, 27–48.

McManus, M. [1962] Numbers and size in Cournot oligopoly, *Yorkshire Bull. Soc. Economic Res.* **14**, 14–22.

McManus, M. [1964] Equilibrium, numbers and size in Cournot oligopoly, *Yorkshire Bull. Soc. Econ. Res.* **16**, 68–75.

Myles, G. [2002] Mixed oligopoly, subsidization and the order of firms' moves: An irrelevance result for the general case, *Econ. Bull.* **12**(1), 1–6.

Okuguchi, K. [1964] The stability of the Cournot oligopoly solution: A generalization, *Rev. Econ. Stud.* **31**, 143–146.

Okuguchi, K. [1985] Nash–Cournot equilibrium for an industry with oligopoly and a competitive fringe, *Keio Econ. Stud.* **22**(2), 51–56.

Okuguchi, K. [1993] Unified approach to Cournot models: Oligopoly, taxation and aggregate provision of a pure public good, *Eur. J. Polit. Econ.* **9**, 233–245.

Okuguchi, K. [2012] General analysis of Cournot mixed oligopoly with partial privatization, *Euras. Econ. Rev.* **2**(1), 48–62.

Okuguchi, K. and Yamazaki, T. [2004] Stability of equilibrium in Bertrand and Cournot duopolies, *Int. Game Theory Rev.* **6**(3), 381–390.

Okuguchi, K. and Yamazaki, T. [2008] Global stability of unique Nash equilibirum in Cournot oligopoly and rent-seeking game, *J. Econ. Dyn. Control* **32**, 1204–1211.

Okuguchi, K. and Yamazaki, T. [2014] Global stability of Nash equilibrium in aggregative games, *Int. Game Theory Rev.* **16**(4) 1450014.

Possajennikov, A. [2003] Imitation dynamic and Nash equilibrium in Cournot oligopoly with capacities, *Int. Game Theory Rev.* **5**(3), 291–305.

Seade, J. [1980] The stability of Cournot revisited, *J. Econ. Theory* **23**, 15–27.

Szidarovszky, F. and Okuguchi, K. [1997] On the existence and uniqueness of pure Nash equilibrium in rent-seeking games, *Games Econ. Behav.* **18**, 135–140.

Szidarovszky, F. and Yakowitz, S. [1977] A new proof of the existence and uniqueness of the Cournot equilibrium, *Int. Econ. Rev.* **18**, 787–789.

Theocharis, R. D. [1960] On the stability of the Cournot solution on the oligopoly problem, *Rev. Econ. Stud.* **27**, 133–134.

Vives, X. [1999] *Oligopoly Pricing, Old Ideas and New Tools* (MIT Press, Cambridge, USA and London).

Vega-Redondo, F. [1997] The evolution of Walrasian behavior, *Econometrica* **65**, 375–384.

Yamazaki, T. [2008] On the existence and uniqueness of pure-strategy Nash equilibrium in asymmetric rent-seeking contests, *J. Public Econ. Theory* **10**, 317–327.

Yamazaki, T. [2017] On the unique existence of the Nash equilibrium in Cournot mixed oligopoly with linear demand and quadratic cost functions, *Econ. J.* **102**, 51–61. http://dspace.lib.niigata-u.ac.jp/dspace/bitstream/10191/47154/1/102 51-61.pdf.

Chapter 9

Second-Price All-Pay Auctions and Best-Reply Matching Equilibria

Gisèle Umbhauer

Bureau d'Economie Théorique et Appliquée
University of Strasbourg, Strasbourg, France
umbhauer@unistra.fr

This chapter studies second-price all-pay auctions — wars of attrition — in a new way, based on classroom experiments and Kosfeld *et al.*'s best-reply matching (BRM) equilibrium. Two players fight over a prize of value V, and submit bids not exceeding a budget M; both pay the lowest bid and the prize goes to the highest bidder. The behavior probability distributions in the classroom experiments are strikingly different from the mixed Nash equilibrium (NE). They fit with BRM and generalized best-reply matching (GBRM), an ordinal logic according to which, if bid A is the best response to bid B, then A is played as often as B. The chapter goes into the GBRM logic, highlights the role of focal values and discusses the high or low payoffs this logic can lead to.

Keywords: Second-price all-pay auction; war of attrition; best-reply matching; Nash equilibrium; classroom experiment.

1. Introduction

This chapter studies second-price all-pay auctions — wars of attrition — in a new way, based on classroom experiments and Kosfeld *et al.*'s [2002] best-reply matching (BRM) equilibrium. A lot has been said about first-price all-pay auctions, but there are only few papers with experiments on second-price all-pay auctions (see Hörisch and Kirchkamp [2010] and Dechenaux *et al.* [2015] for experiments with this class of games). The second-price all-pay auction studied in this chapter is the most standard one: two players fight over a prize of value V, have a budget M, and simultaneously submit bids not exceeding M. Both pay the lowest bid and the prize goes to the highest bidder; in case of a tie, each player gets the prize with probability $1/2$.

The mixed Nash equilibrium (NE) is strikingly different from the behavior observed in the classroom experiments (an experiment with 116 students, another with 109 students). The distributions are so different that we cannot conclude on overbidding or underbidding in comparison with the NE behavior. This observation is partly shared by Hörisch and Kirchkamp [2010]: whereas overbidding is regularly observed in first-price all-pay auction experiments (see, for example, Gneezy and Smorodinsky [2006] and Lugovskyy *et al.* [2010]), Hörisch and Kirchkamp [2010] establish that underbidding prevails in sequential war of attrition experiments.

In our classroom experiments, the students' behavior better fits with best-reply (and generalized best-reply) matching, a behavioral concept developed by Kosfeld *et al.* [2002]. According to BRM, if bid A is the best response to bid B, then A is played as often as B. Mixed NE and BRM equilibria follow a different logic: whereas NE probabilities are calculated so as to equalize the payoffs of the bids played at equilibrium, BRM probabilities just aim to match best responses, each bid being played as often as the bid to which it is a best reply.

In the second-price all-pay auction, there are multiple best responses to a bid, when the bid does not exceed the value of the prize. In case of multiple best responses, BRM requires that each best response be played with the same probability, whereas generalized best-reply matching (GBRM) (Umbhauer [2016]) allows any probability distribution over the set of best responses. So we mostly work with GBRM, given that real players may spontaneously select some best responses more than others. We say that some best responses are more focal than others and we exploit this characteristic to come closer to the students' behavior.

We also call attention to the payoffs of the bidders. In the mixed NE, the payoffs are slightly positive in the discrete setting (they are null in a continuous setting). This is no longer true with best-reply and generalized best-reply matching: the players can lose or win a lot of money.

Section 2 recalls the pure and mixed NE in the discrete second-price all-pay auctions, when M, V and the bids are integers. It also presents the two classroom experiments and compares the student's behavior with the mixed NE. In Sec. 3, we present the BRM equilibrium and the GBRM equilibrium, and we compare the mixed NE philosophy with the BRM philosophy. In Sec. 4, we establish the BRM equilibria and compare them to the students' behavior. Section 5 is the main part. We turn to GBRM and focal behavior. Given that there are multiple best responses to a given bid, players may only play some focal best responses. Focusing on the best responses 0, V and M leads to a GBRM equilibrium which is close to the students' behavior. And focusing on cautious best responses leads to a GBRM equilibrium that is also, surprisingly, a GBRM equilibrium of the first-price all-pay auction. Section 6 is on the mixed NE payoffs and the (generalized) BRM equilibrium payoffs. Section 7 concludes on the role of M in second-price all-pay auctions.

2. Classroom Experiments, NE and Students' Behavior

Two players have a budget M. They fight over a prize of value V. Each player i submits a bid b_i, $i = 1, 2$ lower than or equal to M. The prize goes to the highest bidder but both bidders pay the lowest bid. In case of a tie, the prize goes to each bidder with probability $1/2$. Throughout the chapter, we suppose $M \geq V$. V, M and the rules of the game are common knowledge. Moreover, in the classroom experiments, M, V and the bids are integers. So we restrict attention to discrete games.

This game is known to have a lot of intuitive asymmetric NE, one player bidding nothing (0), the other bidding V or more. In some way, if the first player is cautious and afraid of losing money, whereas the second player is a hothead, the first player bids 0, whereas the second can afford to bid M (even if M is much larger than V), getting the prize without paying anything, thanks to the cautious behavior of the first player. Second-price all-pay auctions are rightly seen as dangerous games with great opportunities: when a hothead meets another one, both lose a lot of money, but if he meets a cautious player, then he wins V and pays nothing. And if a cautious player meets another one, each player gets the expected amount $V/2$ without paying anything.

Things become less intuitive when turning to the unique mixed NE of this game, which is a symmetric mixed NE. We call q_i the probability of bidding i.

Result 1 (Umbhauer [2017]).[a] *V, M and the bids are integers, $M \geq V$ and the bids go from 0 to M; we assume that V is odd. The unique mixed NE is symmetric. We note $\bar{i} = M - V/2 - 5/2$. The main recurrence equation that defines the probabilities is*

$q_i = 2q_{i+1}/V + q_{i+2}$ *i from 0 to $\bar{i}$ (for $\bar{i} \geq 0$).*
The additional equations are:
$q_{\bar{i}+1} = q_M(1/V + 2/V^2)$ *(for $\bar{i} + 1 \geq 0$)*
$q_{\bar{i}+2} = q_M/V$
$q_i = 0$ *for i from to $\bar{i} + 3$ to $M - 1$ (for $V \geq 3$)*
and $\sum_{i=0}^{\bar{i}+2} q_i + q_M = 1$.

Proof. See Appendix A. □

We study the mixed NE for $(V = 3, M = 5)$, $(V = 3, M = 6)$ (these equilibria are necessary in the classroom experiments) and $(V = 9, M = 12)$. The NE are given in Table 1 and illustrated in Figs. 1(a)–1(c).

The shape of the three distributions is close to that obtained in the continuous setting, illustrated in Fig. 1(b) (the decreasing curve, the segment and the point)

---

[a] Result 1 only holds for odd values of V (V and M being integers and the bids being all integers from 0 to M). When V is even (V and M being integers, the bids being all integers from 0 to M), a different result applies, due to a different gap between $M - V/2$ and the highest played bid lower than M in case of V odd and V even. This has a strong impact on the probability distribution, that is much less regular in shape for V even than for V odd (see Umbhauer [2017]).

Table 1. Mixed NE for $(V = 3, M = 5)$, $(V = 3, M = 6)$, $(V = 9, M = 12)$.

	q_0	q_1	q_2	q_3	q_4	q_5	q_6	q_7	$q_8, q_9,$ q_{10}, q_{11}	q_{12}
$V = 3$ $M = 5$	$83/$ $293 =$ 0.283	$57/$ $293 =$ 0.195	$45/$ $293 =$ 0.154	$27/$ $293 =$ 0.092	0	$81/$ $293 =$ 0.276				
$V = 3$ $M = 6$	$337/$ $1,216 =$ 0.277	$249/$ $1,216 =$ 0.205	$171/$ $1,216 =$ 0.141	$135/$ $1,216 =$ 0.111	$81/$ $1,216 =$ 0.066	0	$243/$ $1,216 =$ 0.20			
$V = 9$ $M = 12$	0.106	0.093	0.085	0.074	0.069	0.058	0.056	0.046	0	0.413

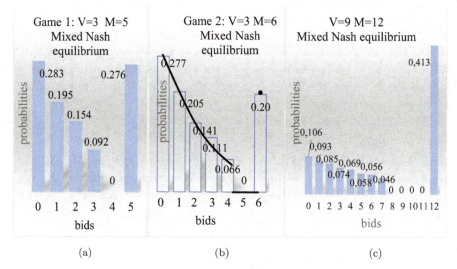

Fig. 1. (a) Mixed NE $(V = 3, M = 5)$, (b) Mixed NE $(V = 3, M = 6)$, (c) Mixed NE $(V = 9, M = 12)$.

and briefly recalled below (Result 2). This similarity is always observed when V, M and the bids are integers and V is odd (see Umbhauer [2017]).

Result 2. *We call b a bid in $[0, M]$. The unique mixed NE in the continuous game is symmetric and given by:*

— *The support of the equilibrium is $[0, M - V/2] \cup \{M\}$.*
— *The cumulative probability distribution on $[0, M - V/2]$ is given by $F(b) = 1 - e^{-b/V}$.*
— *M is a mass point played with probability $1 - F(M - V/2) = e^{1/2 - M/V}$.*

The net payoff (payoff minus M) is equal to 0 at equilibrium.

Proof. See Appendix B.   □

It derives from the Figs. 1(a)–1(c) that the NE probabilities decrease from 0 to $M-V/2-1/2$ ($M-V/2$ in the continuous setting), and are null from $M-V/2+1/2$ to $M-1$ (in] $M-V/2$, $M[$ in the continuous setting), whereas bid M is played with a positive probability. In the continuous setting, M is a mass point; this is rather intuitive, in that M is a focal point with a special property. Given that nobody can play more than M and given that many players bid less, a player, when he bids M, is sure to get the prize with a high probability and he often pays less than V, especially if $M - V/2 < V$. In this latter case, at equilibrium, bidding M leads to a negative payoff only if the opponent bids M too.

0 is more often played than the bids from 1 to $M - V/2 - 1/2$, which is rather intuitive, because 0 is the only best reply to M and never leads to losing money. Yet the probability of playing 0 seems unrelated to the probability of playing M (q_0 is very close to, larger than, much smaller than q_M, respectively in Figs. 1(a)–1(c)) and this nonintuitive fact is confirmed in the continuous setting: 0 is not a mass point and the cumulative probability on $[0, \mathrm{db}]$ is db/V ($\mathrm{db} \to 0$) so does not depend on M, in contrast to the probability assigned to M. It namely follows that when V and M become large, but M/V remains constant, the cumulative probability on $[0, \mathrm{db}]$ tends towards 0 whereas $f(M)$ remains constant. This is not intuitive.

What about risk aversion? In the above equilibria, we implicitly suppose that the players are risk neutral, given that the utility is assumed to be equal to the payoff. We could follow Hörisch and Kirchkamp [2010], and opt for the utility function $U(x) = e^{-rM} - e^{-rx}$ to express risk aversion (r being the player's degree of risk aversion). This would lead to steeper density probability distributions, with a stronger probability on low bids, but it would not change the nature of the mixed NE.

We now present the two classroom experiments.

In the first classroom experiment,[b] 116 L3 students, i.e., undergraduate students in their third year of training, played the second-price all-pay auction game in matrix 1 (Game 1), with $V = 3$, $M = 5$ and the possible bids 0, 1, 2, 3, 4 and 5. In the second classroom experiment, 109 L3 students played the second-price all-pay auction game in matrix 2 (Game 2), with $V = 30$, $M = 60$ and the possible bids 0, 10, 20, 30, 40, 50 and 60. This second game has the same best responses and the same NE as the game with $V = 3$, $M = 6$, and bids 0, 1, 2, 3, 4, 5 and 6 (to get the payoffs in the game with $V = 3$, $M = 6$, it is sufficient to divide the payoffs of Game 2 by 10). So, to get the mixed NE of Game 2, we can apply result 1 to the game with $V = 3$, $M = 6$ and bids $0, 1, 2, 3, 4, 5, 6$.[c]

---

[b]This experiment has also been partly studied in Umbhauer [2016].
[c]For reasons of presentation homogeneity, it would be logical to give the students' results by referring to the game with $V = 3$, $M = 6$ and the bids 0, 1, 2, 3, 4 and 5. But of course this cannot be done from an experimental point of view.

Player 2

		0	1	2	3	4	5
	0	(6.5,6.5)	(5, 8)	(5, 8)	(5, 8)	(5, 8)	(5, 8)
	1	(8, 5)	(5.5,5.5)	(4, 7)	(4, 7)	(4, 7)	(4, 7)
Player 1	2	(8, 5)	(7, 4)	(4.5,4.5)	(3, 6)	(3, 6)	(3, 6)
	3	(8, 5)	(7, 4)	(6, 3)	(3.5,3.5)	(2, 5)	(2, 5)
	4	(8, 5)	(7, 4)	(6, 3)	(5, 2)	(2.5,2.5)	(1, 4)
	5	(8, 5)	(7, 4)	(6, 3)	(5, 2)	(4, 1)	(1.5,1.5)

Matrix 1 : Game 1, V=3, M=5.

Player 2

		0	10	20	30	40	50	60
	0	(75, 75)	(60, 90)	(60, 90)	(60, 90)	(60, 90)	(60, 90)	(60, 90)
	10	(90, 60)	(65, 65)	(50, 80)	(50, 80)	(50, 80)	(50, 80)	(50, 80)
	20	(90, 60)	(80, 50)	(55, 55)	(40, 70)	(40, 70)	(40, 70)	(40, 70)
Player 1	30	(90, 60)	(80, 50)	(70, 40)	(45, 45)	(30, 60)	(30, 60)	(30, 60)
	40	(90, 60)	(80, 50)	(70, 40)	(60, 30)	(35, 35)	(20, 50)	(20, 50)
	50	(90, 60)	(80, 50)	(70, 40)	(60, 30)	(50, 20)	(25, 25)	(10, 40)
	60	(90, 60)	(80, 50)	(70, 40)	(60, 30)	(50, 20)	(40, 10)	(15, 15)

Matrix 2: Game 2, V=30, M=60.

The games were played during game theory lectures and the students knew what a normal-form game is. So they had no difficulty in understanding the two games, and the meaning of the normal forms in Matrix 1 and Matrix 2. Several couples of payoffs were explained in detail to the students, to ensure that they understood the rules of the game. The students had the matrices, as well as the explanation of the couples of payoffs, in front of them while choosing their bid. We add that the first game was played by students trained in NE and dominance. In contrast, the second game was played by novice students with no training in NE and dominance. Students were also allowed, but not compelled to, add comments on their way of playing.

The students' way of playing is given in Table 2.

Table 2.   Comparison of the mixed NE and the students' behavior.

Game 1 $V = 3$, $M = 5$ Bids	NE probabilities (%)	Students: Frequencies of the bids(%)	Game 2 $V = 30$, $M = 60$ bids	NE probabilities (%)	Students: Frequencies of the bids(%)
0	28.3	37.9	0	27.7	33
1	19.5	9.5	10	20.48	5.5
2	15.4	1.7	20	14.06	2.8
3	9.2	20.7	30	11.1	21.1
4	0	15.5	40	6.66	4.6
5	27.6	14.7	50	0	5.5
			60	20	27.5

The students' distributions do not fit with the mixed NE distributions. The main difference concerns low bids different from 0. The probabilities assigned to low bids (1 and 2 in Game 1, 10 and 20 in Game 2) by the NE are much higher than the frequencies with which the students play these bids: bids 1 and 2 in Game 1 are played with probability 34.9% in the NE, with probability 11.2% by the students; bids 10 and 20 in Game 2 are played with probability 34.54% in the NE, with probability 8.3% by the students. According to the comments added by the students, they fear that if they play low bids different from 0, the opponent will bid more and make money whereas they lose money. In some way, bidding a low amount is seen as a way to encourage the opponent to bid more, even if there is no sequentiality in this game.

Another difference concerns the probability assigned to the value of the prize. Students bid this value much more often than in the NE (20.7% versus 9.2% in the first game, 21.1% versus 11.1% in the second game).

We also observe that the students bid 0 more often than in the NE (37.9% versus 28.3% in Game 1, 33% versus 27.7% in Game 2), yet the difference in the probabilities in the second game is less significant.

The way students play bids higher than V is different in the two games. Whereas 30.2% of them play bids 4 and 5 almost with the same probabilities in Game 1 (in contrast to the NE that assigns probability 0 to bid 4 and 27.6% to bid 5), students, like the NE, assign a small probability to bids 40 and 50 in Game 2 (10.1% for the students, 6.66% in the NE), and a large probability to bid 60 (27.5% for the students, 20% in the NE).

It derives from these facts that the students' probabilities are different from the NE ones, and — this matters more — that the shapes of the students' distributions are quite different from the mixed NE one.

We claim that these differences simply highlight the fact that the philosophy of a mixed NE does not fit with the way of playing of real players. We justify this point of view by turning to BRM.

3. Philosophy of Mixed NE and Philosophy of BRM

We first recall Kosfeld *et al.*'s [2002] BRM equilibrium.

Definition 1 (Kosfeld *et al.* [2002]: Normal-form BRM equilibrium). Let $G = (N, S_i, \succ_i, i \in N)$ be a game in normal form (N is the set of players, $K = \mathrm{Card}N$, S_i is player i's set of pure strategies, and $\succ_i$ is player i's preference relation on $X_{i=1}^K S_i$). A mixed strategy p is a BRM equilibrium if for every player $i \in N$ and for every pure strategy $s_i \in S_i$:

$$p_i(s_i) = \sum_{s_{-i} \in B_i^{-1}(s_i)} \frac{1}{\mathrm{Card}B_i(s_{-i})} p_{-i}(s_{-i}),$$

where $B_i(s_{-i})$ is the set of player i's best replies to the strategies s_{-i} played by the other agents.

In a BRM equilibrium (BRME), the probability assigned to a pure strategy by player i is linked to the probability assigned to the opponents' strategies to which this pure strategy is a best reply: if player i's opponents play s_{-i} with probability $p_{-i}(s_{-i})$, and if the set of player i's best responses to s_{-i} is the subset of pure strategies $B_i(s_{-i})$, then each strategy of this subset is played with probability $p_{-i}(s_{-i})$ divided by the cardinal of $B_i(s_{-i})$.

This criterion builds on the notion of rationalizability developed by Bernheim [1984] and Pearce [1984], a strategy s_i being rationalizable if it is a best response to at least one profile s_{-i} played by the other players. Kosfeld *et al.* [2002] simply observe that, if the opponents often play s_{-i}, then s_i often becomes the best response, so should often be played. In some way, they rationalize the probabilities of a player by the other players' probabilities.

We illustrate the concept on the normal-form game in matrix 3.

Player 2

		A_2	B_2
Player 1	A_1	(2.9, 1)	(5, 5)
	B_1	(3, 3)	(1, 2.9)

Matrix 3.

We call p and $1 - p$, respectively q and $1 - q$, the probabilities assigned to A_1 and B_1, respectively to A_2 and B_2. In the mixed NE, we equalize player 1's payoffs[d] obtained with A_1 and B_1, so we set $2.9q + 5(1 - q) = 3q + 1 - q$, and we get $q = 40/41$, i.e., a condition on player 2's probabilities. We also equalize player 2's payoffs obtained with A_2 and B_2, $p + 3(1 - p)$ and $5p + 2.9(1 - p)$, so we get $p = 1/41$, a condition on player 1's probabilities. This may seem quite strange from a behavioral point of view: a player's probabilities have no impact on his own payoff, they only ensure that the opponent is indifferent between his actions in the NE support. So, when player 1 plays A_1 with probability $1/41$ and B_1 with probability $40/41$, these probabilities mean nothing for herself. She could play A_1 and B_1 with any probabilities given that, due to player 2's probabilities, she is indifferent between A_1 and B_1. She chooses probabilities $1/41$ and $40/41$ only to help player 2 to become indifferent between his two actions. This explains why she plays B_1 with a probability close to 1, even though B_1 is not interesting for her when considering the range of payoffs (2.9 and 5 for A_1, 3 and 1 for B_1). As a matter of fact, player 2's payoffs — when he plays A_2 and B_2 — are close when she plays B_1 (he gets 3 with A_2 and 2.9 with B_2) whereas they are quite different when she plays A_1 (he gets 1 with A_2 and 5 with B_2). And vice versa for player 2.

We think that real players do not choose probabilities in this way. Especially, if the support of the mixed NE is the whole set of pure strategies, have you ever seen

---

[d]By 'payoff' we mean 'expected payoff', for all concepts of equilibria. For ease of notations, we will proceed so throughout.

a player who says: "let's try to put probabilities on my pure strategies so that the opponent gets the same payoff with all his pure strategies"?[e]

In real life, behavior is less sophisticated (and less strange). When somebody plays A with probability $1/41$ and B with probability $40/41$, it is because he thinks that B is much more often his best response than A, 40 times more often, which justifies that he plays B with probability $40/41$. In real life, probabilities (often) simply translate the frequency with which an action is supposed to be a best response, and, as a consequence, the frequency with which a player is ready to play it. And this is what is done in the BRME. Players simply try to be consistent with the way other players are playing, adapting their probability of playing an action to the probabilities with which the others play the actions to which this action is a best reply. This way of dealing with probabilities has no link with the mixed NE way of dealing with probabilities.

We define the BRME for the game in matrix 3:

A_1 is player 1's best response to B_2, so has to be played as often as B_2, i.e., $p = 1 - q$.

B_1 is player 1's best response to A_2, so has to be played as often as A_2, i.e., $1 - p = q$.

A_2 is player 2's best response to B_1, so has to be played as often as B_1, i.e., $q = 1 - p$.

B_2 is player 2's best response to A_1, so has to be played as often as A_1, i.e., $1 - q = p$.

And $0 \leq p \leq 1$, $0 \leq q \leq 1$. So, for the game studied, we get an infinite number of BRME, characterized by the fact that player 1 plays A_1 as often as player 2 plays B_2 and plays B_1 as often as player 2 plays A_2 and vice versa.

Three remarks derive from these results:

- First, the BRM way of defining probabilities allows us to cope with the asymmetric pure strategy NE. Given that A_1 is the best response to B_2 and B_2 is the best response to A_1, the BRME allows player 1 to play A_1 with probability 1 and player 2 to play B_2 with probability 1, given that A_1 is player 1's best response as often as player 2 plays B_2 and vice versa. A similar reasoning holds for the profile (B_1, A_2). So, in this game, the pure strategy NE are also BRME.
- Second, the payoff in the mixed NE (here $121/41$) may be higher or lower than the payoff in the BRME. In the game studied, as long as we choose p between $1/41$ and $20/41$, the players get more with the mixed NE than with the BRME, but for p lower than $1/41$ and p larger than $20/41$, the players get more with the BRME.

[e]We do not say that this way of playing is always meaningless. If, by doing so, the payoff of the opponent is always low regardless of what he plays, and if the game is a zero-sum game (so a player is better off when his opponent is worse off) then behaving in such a way may be strategically meaningful. But, in a usual game like the one in matrix 3, this behavior is quite strange.

- Third, in the game studied, the mixed NE is also a BRME (because $p = 1 - q = 1/41$). Most often, mixed NE are not BRME. Observe that the justification of this special BRME is not the mixed NE one. In the BRME, player 1 plays A_1 with probability $1/41$ because it is her best response to B_2 which is also played with probability $1/41$, and she plays B_1 with probability $40/41$ because it is her best response to A_2, which is played with the same probability (and the symmetric explanation holds for player 2). So both actions are played because each is a best response, and not because they lead to the same payoff. In some degree, even when the BRME is a mixed equilibrium, players reason in a pure strategy way: in our example the aim is to play A when the other plays B, and to play B when he plays A. This is not the case in a mixed NE, where each player best reacts to the mixed strategies of the others.

In this chapter, we mainly focus on a generalization of the BRM concept. As a matter of fact, when there are several best replies to a profile s_{-i}, there is no reason to demand that each best reply be played with the same probability, so it is reasonable to generalize Kosfeld *et al.*'s criterion by allowing players to play the different best replies with different probabilities.

Definition 2 (Umbhauer [2016]: Normal-form GBRM equilibrium). Let $G = (N, S_i, \succ_i, i \in N)$ be a game in normal form. A mixed strategy p is a GBRM equilibrium if for every player $i \in N$ and for every pure strategy $s_i \in S_i$:

$$p_i(s_i) = \sum_{s_{-i} \in B_i^{-1}(s_i)} \delta_{s_i s_{-i}} p_{-i}(s_{-i})$$

with $\delta_{s_i s_{-i}} \in [0, 1]$ for any s_i belonging to $B_i(s_{-i})$ and $\sum_{s_i \in B_i(s_{-i})} \delta_{s_i s_{-i}} = 1$.

Pure NE, in contrast to mixed ones, are automatically GBRM equilibria (GBRME): if player 1 plays A and player 2 plays B in a pure strategy NE — so they play the actions with probability 1 — player 1 plays A as often as the opponent plays the action B to which A is a — perhaps among several — best reply, and player 2 plays B as often as player 1 plays the action A to which B is a — perhaps among several — best reply (Umbhauer [2016]).

We make two additional remarks. Firstly, BRME and GBRME are ordinal concepts. This may prevent a good strategy, for example, a cautious strategy that leads to high payoffs but is never a best response to the pure strategy profiles of the opponents, from being played in a BRME or in a GBRME. This is not intuitive and allows us to conjecture that BRM behavior will not always be observed in real life (in real life, players like playing cautious strategies that lead to high payoffs even when these strategies are neither best responses to the pure strategy profiles of the opponents, nor belong to a NE). Yet in the second-price all-pay auction, there do not exist strategies that lead to high payoffs without being a best response. Bidding 0 is a cautious way of playing but it leads to a positive net payoff only if the other player bids 0 too. Secondly, the second-price all-pay auction has a special structure as regards best responses: there are many best responses to a given bid (when it is not higher than V), so that GBRM becomes very useful and powerful; this will be developed in the following sections.

4. BRM and Students' Way of Playing

We establish the BRME in the second-price all-pay auctions.

We start with the games played by the students. To do so, we write the best-reply matrices 4a (Game 1) and 4b (Game 2), where b_i means that player i's action is a best reply to the opponent's action, $i = 1, 2$. For example, the bold b_1 in italics in matrix 4a means that bid 4 is one of player 1's best replies to player 2's bid 1.

Player 2

Pl.1		q_0 0	q_1 1	q_2 2	q_3 3	q_4 4	q_5 5
p_0	0		b_2	b_2	$b_1 b_2$	$b_1 b_2$	$b_1 b_2$
p_1	1	b_1		b_2	b_2	b_2	b_2
p_2	2	b_1	b_1		b_2	b_2	b_2
p_3	3	$b_1 b_2$	b_1	b_1		b_2	b_2
p_4	4	$b_1 b_2$	$\boldsymbol{b_1}$	b_1	b_1		
p_5	5	$b_1 b_2$	b_1	b_1	b_1		

Matrix4a. Best-reply Matrix Game 1.

Player 2

Pl.1		q_0 0	q_{10} 10	q_{20} 20	q_{30} 30	q_{40} 40	q_{50} 50	q_{60} 60
p_0	0		b_2	b_2	$b_1 b_2$	$b_1 b_2$	$b_1 b_2$	$b_1 b_2$
p_{10}	10	b_1		b_2	b_2	b_2	b_2	b_2
p_{20}	20	b_1	b_1		b_2	b_2	b_2	b_2
p_{30}	30	$b_1 b_2$	b_1	b_1		b_2	b_2	b_2
p_{40}	40	$b_1 b_2$	b_1	b_1	b_1			
p_{50}	50	$b_1 b_2$	b_1	b_1	b_1			
p_{60}	60	$b_1 b_2$	b_1	b_1	b_1			

Matrix 4b. Best-reply Matrix Game 2.

It derives from the matrices 4a and 4b that the best responses in the second-price all-pay auction match a special structure: each bid x higher than 0 and up to V is a best response to each bid lower than x, each bid higher than V is a best response to all bids from 0 to V, and 0 is a best response to each bid from V to M. This fact is illustrated for player 1 in both matrices: the diagrams contain player 1's best responses.

In all the games studied, we write p_i, respectively q_i, the probability assigned to bid i by player 1, respectively by player 2.

Table 3.   BRME for Game 1, Game 2 and $(V = 9, M = 12)$[a,b,c].

	q_0	q_1	q_2	q_3	q_4	q_5	q_6	q_7	q_8	q_9	q_{10}, q_{11}, q_{12}
$V = 3$	180/	36/	45/	60/	80/	80/					
$M = 5$	481 =	481 =	481 =	481 =	481 =	481 =					
*	0.374	0.075	0.094	0.125	0.166	0.166					
$V = 30$	240/	40/	48/	60/	75/	75/	75/				
$M = 60$	613 =	613 =	613 =	613 =	613 =	613	613				
*	0.3915	0.065	0.078	0.098	0.1225						
$V = 9$	55,440/	4,620/	5,040/	5,544/	6,160/	6,930/	7,920/	9,240/	11,088/	13,860/	17,325/
$M = 12$	$D =$	$D =$	$D =$	$D =$	$D =$	$D =$	$D =$	$D =$	$D =$	$D =$	$D =$
*	0.312	0.026	0.0285	0.031	0.035	0.039	0.045	0.052	0.0625	0.078	0.097

[a]The notations q_0, q_1, q_2, q_3, q_4, q_5, q_6 should be read q_0, q_{10}, q_{20}, q_{30}, q_{40}, q_{50}, q_{60} for $V = 30$ and $M = 60$.

[b]* = unique BRME.

[c]$D = 177,817$.

In Game 1,[f] i goes from 0 to 5. The symmetry of the game leads to the symmetric set of equations:

$$p_0 = q_3/3 + q_4 + q_5, \qquad q_0 = p_3/3 + p_4 + p_5,$$
$$p_1 = q_0/5, \qquad q_1 = p_0/5,$$
$$p_2 = q_0/5 + q_1/4, \qquad q_2 = p_0/5 + p_1/4,$$
$$p_3 = q_0/5 + q_1/4 + q_2/3, \qquad q_3 = p_0/5 + p_1/4 + p_2/3,$$
$$p_4 = q_0/5 + q_1/4 + q_2/3 + q_3/3, \qquad q_4 = p_0/5 + p_1/4 + p_2/3 + p_3/3,$$
$$p_5 = q_0/5 + q_1/4 + q_2/3 + q_3/3 = p_4, \quad q_5 = p_0/5 + p_1/4 + p_2/3 + p_3/3 = q_4,$$
$$p_0 + p_1 + p_2 + p_3 + p_4 + p_5 = 1, \qquad q_0 + q_1 + q_2 + q_3 + q_4 + q_5 = 1.$$

This system of equations has a unique solution: $p_0 = q_0 = 180/481 = 37.4\%$, $p_1 = q_1 = p_0/5 = 7.5\%$, $p_2 = q_2 = p_0/4 = 9.4\%$, $p_3 = q_3 = p_0/3 = 12.5\%$, $p_4 = p_5 = q_4 = q_5 = 4p_0/9 = 16.6\%$. These results are reproduced in Table 3.

Game 2 leads to the equations:

$$p_0 = q_{30}/4 + q_{40} + q_{50} + q_{60}, \qquad q_0 = p_{30}/4 + p_{40} + p_{50} + p_{60},$$
$$p_{10} = q_0/6, \qquad q_{10} = p_0/6,$$
$$p_{20} = q_0/6 + q_{10}/5, \qquad q_{20} = p_0/6 + p_{10}/5,$$
$$p_{30} = q_0/6 + q_{10}/5 + q_{20}/4, \qquad q_{30} = p_0/6 + p_{10}/5 + p_{20}/4,$$
$$p_{40} = q_0/6 + q_{10}/5 + q_{20}/4 + q_{30}/4, \qquad q_{40} = p_0/6 + p_{10}/5 + p_{20}/4 + p_{30}/4,$$
$$p_{50} = q_0/6 + q_{10}/5 + q_{20}/4$$
$$+ q_{30}/4 = p_{40}, \qquad q_{50} = p_0/6 + p_{10}/5 + p_{20}/4$$
$$+ p_{30}/4 = q_{40},$$

[f]For some results of Game 1, see also Umbhauer [2016].

$$p_{60} = q_0/6 + q_{10}/5 + q_{20}/4$$
$$+ q_{30}/4 = p_{40},$$
$$p_0 + p_{10} + p_{20} + p_{30} + p_{40}$$
$$+ p_{50} + p_{60} = 1,$$

$$q_{60} = p_0/6 + p_{10}/5 + p_{20}/4$$
$$+ p_{30}/4 = q_{40},$$
$$q_0 + q_{10} + q_{20} + q_{30} + q_{40}$$
$$+ q_{50} + q_{60} = 1.$$

The unique solution, reproduced in Table 3, is $p_0 = q_0 = 240/613 = 39.15\%$, $p_{10} = q_{10} = p_0/6 = 40/613 = 6.5\%$, $p_{20} = q_{20} = p_0/5 = 48/613 = 7.8\%$, $p_{30} = q_{30} = p_0/4 = 60/613 = 9.8\%$, $p_{40} = p_{50} = p_{60} = q_{40} = q_{50} = q_{60} = 5p_0/16 = 75/613 = 12.25\%$.

More generally, due to the structure of the second-price all-pay auction, there is a unique BRME, given in proposition 1.

Proposition 1. *For $M \geq V$ and $M > 1$ the unique BRME is symmetric and given by*

$$q_0 = \frac{1}{2 + \sum_{i=0}^{V-1} \frac{1}{M-i} - \frac{1}{(M-V+1)^2}},$$

$$q_i = q_0/(M - i + 1), \quad i \text{ from } 1 \text{ to } V,$$

$$q_i = q_0(M - V + 2)/(M - V + 1)^2 \quad i \text{ from } V + 1 \text{ to } M \text{ (when } M > V),$$

where q_i is the probability of playing bid i, i from 0 to M.

We get: $\sum_{i=V+1}^{M} q_i < q_0 < \sum_{i=V}^{M} q_i$ if $M > V$ and $q_0 = q_V$ if $M = V$.

For $M = V = 1$ there exist an infinite number of BRME defined by: $q_1 = p_0$, $p_0 + p_1 = 1$, $q_0 + q_1 = 1$.

Proof. See Appendix C. □

Proposition 1 allows us to establish the BRME for $V = 9$ and $M = 12$. The probabilities are given in Table 3. The BRME are given in Figs. 2(a)–2(c).

We first compare the shape of the BRME distribution and the shape of the mixed NE distribution.

In the BRME, q_i is increasing in i for i from 1 to $V + 1$, and is constant from $V + 1$ to M, a result in sharp contrast with the mixed NE probabilities that are decreasing from 1 to $M - V/2 - 1/2$ and null from $M - V/2 + 1/2$ to $M - 1$. BRM clearly takes into account that a higher bid is more often a best reply than a lower one (different from 0), in that each bid x (different from 0) is a best reply to all the bids lower than x, if $x \leq V$, and a best-reply to all bids from 0 to V if $x > V$. And bid 0, in contrast to the other low bids, has a special status in that it is a best reply to all bids from V to M.

Clearly, the Nash and BRM distributions have no common points, except the fact that the probability on bid 0 is higher than the probabilities assigned to the bids higher than 0 and lower than or equal to V, both in the BRME and in the NE

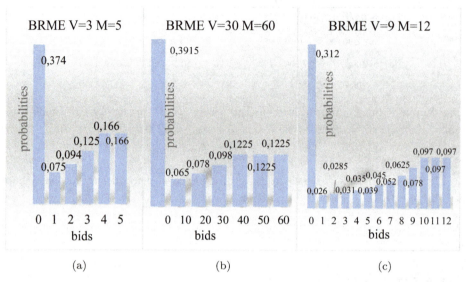

Fig. 2.   (a) BRME ($V = 3$, $M = 5$), (b) BRME ($V = 30$, $M = 60$), (c) BRME ($V = 9$, $M = 12$).

(when $M > V$). The strong differences and the few similarities are highlighted by comparing two by two Figs. 1(a) and 2(a), 1(b) and 2(b), 1(c) and 2(c).

But what about BRM and the student's behavior?

Figure 3(a)–3(c) give the students' behavior, the NE and the BRME in Game 1.

In Game 1, the BRME probabilities fit much better with the students' probabilities, except p_2 (higher) and p_3 (lower). This proximity is due to the fact that

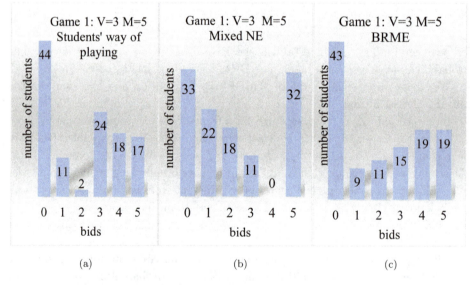

Fig. 3.   (a) Game 1, Students' way of playing, (b) NE Game 1, (c) BRME Game 1.

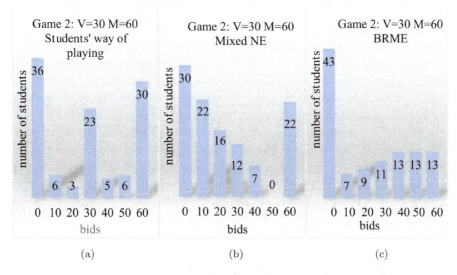

Fig. 4. (a) Game 2, Students' way of playing, (b) NE Game 2, (c) BRME Game 2.

BRM exploits some observations made by the students in their comments, namely that bids 1 and 2 are seldom best responses. In fact, bid 1 is a best response only if the opponent bids 0 (and in this case, bids 2, 3, 4, 5 are also best responses), bid 2 is a best response only if the opponent bids 0 or 1 and in these two cases, bids 3, 4 and 5 are also best responses. By contrast, 0, 3, 4 and 5 are often best responses (bid 0 is the unique best response to bids 4 and 5 and one best response to bid 3, bid 3 is a best response to bids 0, 1 and 2, bids 4 and 5 are best responses to bids 0, 1, 2 and 3).

We turn to Game 2: Figs. 4(a)–4(c) give the students' behavior, the NE and the BRME.

This time, both the BRME and the NE are strongly different from the students' behavior. This can be explained by the fact that the students, especially when the number of bids grows and when there are several best replies, do not play all best replies with the same probability, and may even choose to play only some of them, as allowed by GBRM.

5. GBRM, Focal Values and Focal Behavior, a Way to Bring Closer First-Price and Second-Price All-Pay Auctions

In a second-price all-pay auction, all the bids higher than x and up to M are best responses to x, when x is not exceeding V. It is very reasonable to expect that players will not play each best response with the same probability but may prefer playing best responses that are focal, that is to say that have something special that make them prominent. That is why GBRM fits better with real behavior. So, for example, in Game 2, bids 40, 50 and 60 are best replies to bids 0, 10, 20 and 30, which explains that they are each played with the same probability 0.1225 in the BRME. Yet 40 and 50 are much less focal than 60, given that 60, besides being a best response, has focal properties not shared by 40 and 50: 60 is the largest possible bid, so a player is sure to get the prize at least with probability 1/2 by bidding 60. This may lead a player to only (mostly) playing 60, instead of playing 40, 50 and 60 with the same probability 0.1225 (this does not change the other probabilities in the system of equations). So, according to GBRM, the player can play 40 and 50 with a small probability, and 60 with the probability complementary to $3 \times 0.1225 = 0.3675$. And this is exactly what is done by the students; they play 40 with probability 0.046, 50 with probability 0.055 and 60 with probability 0.275, which leads to the total amount 0.376, a probability which is close to 0.3675.

In Game 2, 3 bids are more focal than others: 0, because it is the cautious bid,[g] V, because it is the value of the prize, hence the fair price to pay, and M, which is the highest possible bid and which leads to getting the prize at least with probability 1/2.

So suppose that the students prefer playing focal values each time they belong to the best responses. For example, when player 1 bids 10, player 2 only best replies with bid 30 and bid 60, even though bids 20, 40 and 50 are also best responses. This leads to the GBRM matrix 5 (consider only the b_1 and b_2, the B_1 and B_2 are used in a further study).

			Player 2						
			q_0	q_{10}	q_{20}	q_{30}	q_{40}	q_{50}	q_{60}
			0	10	20	30	40	50	60
	p_0	0		B_2		b_1b_2	b_1	b_1	b_1b_2
	p_{10}	10	B_1		B_2	b_2			b_2
Player 1	p_{20}	20		B_1		b_2			b_2
	p_{30}	30	b_1b_2	b_1	b_1		B_2	B_2	b_2
	p_{40}	40	b_2			B_1			
	p_{50}	50	b_2			B_1			
	p_{60}	60	b_1b_2	b_1	b_1	b_1			

Matrix 5. Selected best responses in Game 2.

The system of equations becomes:

$$p_0 = q_{30}/2 + q_{40} + q_{50} + q_{60}, \qquad q_0 = p_{30}/2 + p_{40} + p_{50} + p_{60},$$

$$p_{10} = 0, \qquad q_{10} = 0,$$

[g]0 is sometimes also viewed as a cooperative bid, because bidding 0 allows a player to get the prize half of the time without paying anything, when both players bid 0.

$$p_{20} = 0, \qquad\qquad\qquad\qquad q_{20} = 0,$$

$$p_{30} = q_0/2 + q_{10}/2 + q_{20}/2, \qquad q_{30} = p_0/2 + p_{10}/2 + p_{20}/2,$$

$$p_{40} = 0, \qquad\qquad\qquad\qquad q_{40} = 0,$$

$$p_{50} = 0, \qquad\qquad\qquad\qquad q_{50} = 0,$$

$$p_{60} = q_0/2 + q_{10}/2 + q_{20}/2 + q_{30}/2, \qquad q_{60} = p_0/2 + p_{10}/2 + p_{20}/2 + p_{30}/2,$$

$$p_0 + p_{10} + p_{20} + p_{30} + p_{40} + p_{50} + p_{60} = 1, \quad q_0 + q_{10} + q_{20} + q_{30} + q_{40} + q_{50} + q_{60} = 1.$$

The unique GBRME, solution of this system of equations, is: $p_0 = q_0 = 4/9 = 44.4\%$, $p_{30} = q_{30} = p_0/2 = 22.2\%$, $p_{60} = q_{60} = 3p_0/4 = 33.3\%$, $p_{10} = q_{10} = p_{20} = q_{20} = p_{40} = q_{40} = p_{50} = q_{50} = 0$.

We get a 3 peak distribution which is similar to that of the students as regards the shape (highest peak on 0, second highest peak on 60 and lowest peak on 30) (see Table 4 and Fig. 5(a)).

To focal bids we can add focal behavior. We suggest that the players, among the multiple best responses, may prefer choosing responses that fit with a type of behavior. For example, they may add cautiousness to best replying. So, when the opponent bids 0, the bids 10, 20, 30, 40, 50 and 60 are best responses, but they choose 10 because 10 is the most cautious bid in the set of best responses (in that it maximizes the minimum payoff in this set, the minimum payoffs being respectively 50, 40, 30, 20, 10 and 15 for bids 10, 20, 30, 40, 50 and 60). In the same way, 20 is the most cautious best reply to 10 and 30 is the most cautious best response to 20. So cautiousness in some groups of players (and other types of behavior in other groups of players) may help them to select actions in the set of best responses.

To come closer to the students' distribution, we add some cautious behavior (in addition to the play of the three focal values 0, V and M), by assuming that players also focus on 10 as the cautious best response to 0, and on 20 as the cautious best response to 10. And we also add 40 and 50 as best responses to 30 (perhaps because some players like outbidding the fair price bidders, without necessarily focusing exclusively on 60). So we add the B_1 and B_2 in matrix 5 and we get the

Table 4.   NE, Students' behavior and GBRME for $V = 30$ and $M = 60$.

$V = 30$, $M = 60$ bids	NE(%)	Students' behavior(%)	GBRME with bids 0, 30 and 60(%)	GBRME with weighted focal values(%)
0	27.7	33	44.5	35.5
10	20.48	5.5	0	11.8
20	14.06	2.8	0	3.9
30	11.1	21.1	22.2	17.7
40	6.66	4.6	0	4.45
50	0	5.5	0	4.45
60	20	27.5	33.3	22.2

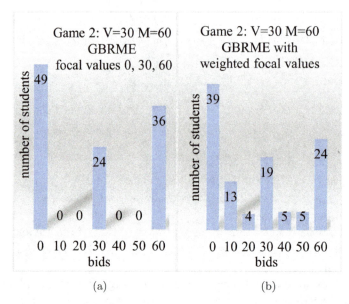

Fig. 5.   (a) GBRME Game 2 focal values 0, V and M, (b) GBRME Game 2 weighted focal values.

equations:

$$p_0 = q_{30}/4 + q_{40} + q_{50} + q_{60},$$
$$p_{10} = q_0/3,$$
$$p_{20} = q_{10}/3,$$
$$p_{30} = q_0/3 + q_{10}/3 + q_{20}/2,$$
$$p_{40} = q_{30}/4 = p_{50},$$
$$p_{60} = q_0/3 + q_{10}/3 + q_{20}/2 + q_{30}/4,$$
$$p_0 + p_{10} + p_{20} + p_{30} + p_{40} + p_{50} + p_{60} = 1,$$

$$q_0 = p_{30}/4 + p_{40} + p_{50} + p_{60},$$
$$q_{10} = p_0/3,$$
$$q_{20} = p_{10}/3,$$
$$q_{30} = p_0/3 + p_{10}/3 + p_{20}/2,$$
$$q_{40} = p_{30}/4 = q_{50},$$
$$q_{60} = p_0/3 + p_{10}/3 + p_{20}/2 + p_{30}/4,$$
$$q_0 + q_{10} + q_{20} + q_{30} + q_{40} + q_{50} + q_{60} = 1.$$

The unique solution of this system of equations is: $p_0 = q_0 = 72/203 = 35.5\%$, $p_{10} = q_{10} = 24/203 = 11.8\%$, $p_{20} = q_{20} = 8/203 = 3.9\%$, $p_{30} = q_{30} = 36/203 = 17.7\%$, $p_{40} = p_{50} = q_{40} = q_{50} = 9/203 = 4.45\%$ and $p_{60} = q_{60} = 45/203 = 22.2\%$.

These values are illustrated in Fig. 5(b) and in Table 4. It derives from this that we can come close to the students' distribution both in probabilities (see Table 4) and in shape (see Figs. 5(b) and 4(a)).

We now turn to the general discrete second-price all-pay auction (V, M are integers and the bids are the integers from 0 to M) and show that we get a GBRME of special interest, in that it is also a GBRME of the first-price all-pay auction.

We recall that the only — but crucial — difference between a first-price all-pay auction and a second-price all-pay auction is that in the first-price all-pay auction each player pays *his* bid. So the structure of the best-reply matrix is different.

We show the first-price all-pay auction for $V = 3$ and $M = 6$ in matrix 6a.

Player 2

	0	1	2	3	4	5	6
0	(7.5,7.5)	(6, 8)	(6, 7)	(6, 6)	(6, 5)	(6, 4)	(6, 3)
1	(8, 6)	(6.5,6.5)	(5, 7)	(5, 6)	(5, 5)	(5, 4)	(5, 3)
2	(7, 6)	(7, 5)	(5.5,5.5)	(4, 6)	(4, 5)	(4, 4)	(4, 3)
3	(6, 6)	(6, 5)	(6, 4)	(4.5,4.5)	(3, 5)	(3, 4)	(3, 3)
4	(5, 6)	(5, 5)	(5, 4)	(5, 3)	(3.5,3.5)	(2, 4)	(2, 3)
5	(4, 6)	(4, 5)	(4, 4)	(4, 3)	(4, 2)	(2.5,2.5)	(1, 3)
6	(3, 6)	(3, 5)	(3, 4)	(3, 3)	(3, 2)	(3, 1)	(1.5,1.5)

Player 1 labels rows 0–6.

Matrix 6a. First–price all-pay auction (V=3, M=6).

We recall in matrix 6b the best-reply matrix for the second-price all-pay auction with $V = 3$ and $M = 6$; matrix 6c is the best-reply matrix for the first-price all-pay auction with $V = 3$ and $M = 6$.

It derives from matrices 6b and 6c that, for any bid, the intersection of the sets of best responses to this bid, in the second-price all-pay auction, and in the first-price all-pay auction, is never empty. It follows from this that, by selecting the same best responses in the intersection, we necessarily get the same GBRME for both games.

So suppose that the players, in the second-price all-pay auction, among the best responses, only choose the most cautious one. It follows that each player best replies to the opponent's bid x by playing the lowest possible best reply (because it is the bid that yields the max-min payoff in the set of best responses). So bid $x + 1$ is chosen as the best response to bid x, for x from 0 to $V - 1$, and bid 0 is chosen as the best response to all bids higher than or equal to V.

Player 2

		q_0 0	q_1 1	q_2 2	q_3 3	q_4 4	q_5 5	q_6 6
p_0	0		**b₂**	b₂	**b₁**b₂	**b₁**b₂	**b₁**b₂	**b₁**b₂
p_1	1	**b₁**		**b₂**	b₂	b₂	b₂	b₂
p_2	2	b₁	**b₁**		**b₂**	b₂	b₂	b₂
p_3	3	b₁**b₂**	b₁	**b₁**		b₂	b₂	b₂
p_4	4	b₁**b₂**	b₁	b₁	b₁			
p_5	5	b₁**b₂**	b₁	b₁	b₁			
p_6	6	b₁**b₂**	b₁	b₁	b₁			

Player 1 labels rows p_0–p_6.

Matrix 6b. Best-reply matrix, second-price all-pay auction (V=3, M=6).

Player 2

		q_0	q_1	q_2	q_3	q_4	q_5	q_6
		0	1	2	3	4	5	6
p_0	0		$\mathbf{b_2}$	b_1	$\mathbf{b_1}$	$\mathbf{b_1}$	$\mathbf{b_1}$	$\mathbf{b_1}$
p_1	1	$\mathbf{b_1}$		$\mathbf{b_2}$				
p_2	2	b_2	$\mathbf{b_1}$		$\mathbf{b_2}$			
p_3	3	$\mathbf{b_2}$		$\mathbf{b_1}$				
p_4	4	$\mathbf{b_2}$						
p_5	5	$\mathbf{b_2}$						
p_6	6	$\mathbf{b_2}$						

(Player 1 labels the rows p_0–p_6.)

Matrix 6c. Best-reply matrix, first-price all-pay auction (V=3, M=6).

In the first-price all-pay auction, bid $x + 1$ is the only best response to bid x, x from 0 to $V - 2$, bid 0 is the only best reply to all bids higher than or equal to V, and bid 0 and bid V are the only best responses to bid $V - 1$. Suppose that the players choose V as the best response to $V - 1$, i.e., that they select the *less cautious* best response.

By so doing, in both games, the players choose for any bid the same best response (the bold and underlined best responses in matrices 6b and 6c). So they play the same GBRME.

Given that $p_i = q_i = 0$, for i from $V + 1$ to M, the GBRM equations become

$$p_0 = q_V \quad q_0 = p_V$$

$$p_i = q_{i-1} \quad q_i = p_{i-1} \quad i \text{ from 1 to } V \qquad \sum_{i=0}^{V} p_i = \sum_{i=0}^{V} q_i = 1.$$

It follows that the unique symmetric GBRME assigns the same probability $1/(V + 1)$ to each bid from 0 to V and a null probability to the higher bids.

So playing in a cautious (best-replying) way in the second-price all-pay auction leads to the same strategy profile as playing in a noncautious (best-replying) way in the first-price all-pay auction.

This result is of interest, both from a game theory point of view and from an economic point of view.

Firstly, in game theory, getting a same equilibrium behavior for two different games means that the equilibrium is robust to important changes in the game. And this is interesting given that players do often not perfectly understand the rules of a game. Especially, as regards all-pay auctions, which are special nonobvious auctions, some players may be unsure about what they will have to pay: even if one takes time explaining the rules of a second-price all-pay auction, one can reasonably anticipate that some players keep on thinking that they pay what they bid. So some players play without knowing what they have to pay, i.e., they play a game with incomplete information on the payoffs. Hence, it is fine to get an equilibrium that is robust to different payoff structures. In our approach, getting the same equilibrium in two different games is possible when the intersection of the sets of best responses to each action in both games is not empty, which restricts the ways in which the payoff structures are allowed to differ. This observation contributes to the literature on the robustness of equilibrium behavior in games with incomplete information (see, for example, Bergemann and Morris [2013]).

Secondly, it is sometimes difficult to distinguish a first-price all-pay auction from a second-price all-pay auction. Consider for example Shubik's[h] [1971] dollar auction game, where two bidders make bids and can outbid each other (in multiple of five cents) in order to get a dollar. The game stops when one bidder drops out; the dollar goes to the highest bidder, and both bidders pay the largest bid they made. This game is close to a first-price all-pay auction, in that each bidder pays his largest bid. But, because the game is sequential, one may observe a sequential overbidding process — the escalation process Shubik [1971] analyses in his paper — such that each player systematically outbids the opponent by 5 cents, till one player stops; if so, the player who stops loses his last bid, and the opponent, who gets the dollar, just pays 5 cents more than the loser: so he (almost) pays the second-price, that is to say the game looks like a second-price all-pay auction.

To give a more economic (political) example, consider an electoral campaign that leads a candidate to visiting a given town several times in order to persuade the inhabitants of the city to vote for him. Suppose that the citizens vote for the candidate that most visited their city. At first sight, this game is a first-price all-pay auction, given that each candidate loses the money and the time he invests in each visit. Yet it is also close to a war of attrition because the visits are sequential. So suppose that, at a given time t, both candidates visited the town a same number of times. But assume that one of the candidates, for possibly different reasons — he has not enough money, he prefers visiting other towns. . . — stops visiting the town. Then one (and only one) additional visit is enough for the opponent to be voted by the inhabitants of the city. So the winner just pays one more visit than the loser. This looks like a second-price all-pay auction. This example is linked to vote-buying literature (see, for example, Dekel *et al.* [2008]).

---

[h]Shubik's game is in fact neither exactly a first-price all-pay auction nor a second price all-pay auction. It is more complex. There are potentially more than two bidders (a large crowd around the auctioneer) and only the two highest bidders pay their bid. The way ties are resolved is quite original, given that the dollar goes to the bidder closest to the auctioneer (so, at the start of the game, especially if the crowd forms a disk around the auctioneer, a potential bidder does not know if he will be the lucky closest person). Shubik expects an escalation process that lasts a long time, players overbidding each other by 5 cents, but he is not even sure that the game will start. And Shubik tells us nothing about what happens if nobody makes a bid. Does the auctioneer keep the dollar? If so, there is a discontinuity between what happens for no bid (i.e., a bid of 0 cents) and what happens for a single bid of the lowest amount (5 cents). For the game and the escalation process to start, Shubik expects a kind of bounded rationality in that he hopes that 'the propensity to calculate does not settle in until at least two bids have been made'. . . . This easily explains why Shubik's game is more than an all-pay auction and that it has been studied in many different ways.

To summarize, when a first-price all-pay auction gives rise to a dynamic process of overbidding, then it can become close to a war of attrition, i.e., a second-price all-pay auction.

6. BRME, GBRME, NE and Payoffs

What about payoffs? In the second-price all-pay auction, the NE always leads to a null payoff (in the continuous setting) or a small positive payoff in the discrete setting; this (slight) positivity is due to the fact that bid 0 always yields a nonnegative payoff, regardless of the bids chosen by the opponent.

With BRM and GBRM, we get different results.

We first come back to the BRME.

For M not too far from V, the BRME payoffs can be higher than the NE ones. For example, for $V = 3$ and $M = 5$, the NE payoff is $1589.5/293 = 5.425$, whereas the BRME payoff is $5.457(= 180(6.5 \times 180 + 5(481 - 180)) + 36(8 \times 180 + 5.5 \times 36 + 4(481 - 180 - 36)) + 45(8 \times 180 + 7 \times 36 + 4.5 \times 45 + 3 \times 60 + 3 \times 80 + 3X80) + 60(8 \times 180 + 7 \times 36 + 6 \times 45 + 3.5 \times 60 + 2 \times 80 + 2 \times 80) + 80(8 \times 180 + 7 \times 36 + 6 \times 45 + 5 \times 60 + 2.5 \times 80 + 1 \times 80) + 80(8 \times 180 + 7 \times 36 + 6 \times 45 + 5 \times 60 + 4 \times 80 + 1.5 \times 80)/481^2 = 1262620.5/481^2)$.

For $V = 9$ and $M = 12$, the BRME is 13.528, whereas the NE payoff is only 12.477.

For $V = 30$ and $M = 60$ (and bids in increments of ten) the results are reversed: the NE payoff is 64.157 and the BRME payoff is slightly lower $(24082745/613^2 = 64.089)$.

In the general setting (V, M are integers and the bids are the integers from 0 to M), we get the result in proposition 2.

Proposition 2. *We call $Eg(i)$ the payoff obtained with bid i in the BRME. We get*
$Eg(i + 1) - Eg(i) = V q_i/2 + V q_{i+1}/2 - \sum_{j=i+1}^{M} q_j$ *i from 0 to $M - 1$.*
$Eg(i + 1) - Eg(i)$ *is increasing in i, i from 1 to $M - 1$.*
For $M > 2V$, $Eg(i)$ decreases from bid 1 to bid $M - V$, then increases up to M.

Proof. See Appendix D.   □

It derives from proposition 2 that the shape of the payoff function goes as follows. Often $Eg(1) > Eg(0)$ due to the fact that $\frac{V q_0}{2} + \frac{V q_1}{2} - \sum_{j=1}^{M} q_j > 0$ (because of the large value of q_0). Then, generally, the payoff function decreases (due to the fact that $\frac{V q_1}{2} + \frac{V q_2}{2} - \sum_{j=2}^{M} q_j < 0$) for a while; and, when $Eg(i + 1) - Eg(i)$ becomes positive, the payoff function increases up to bid M. We represent the net payoff function for $V = 9$ and $M = 12$ in Fig. 6.

We comment on these payoffs. If we compare the payoffs in Fig. 6 and the probabilities assigned to the bids in Fig. 2(c), we may feel uncomfortable in that the evolution of the payoffs does not follow the evolution of the probabilities. For example, q_1 is the smallest probability but bid 1 yields one of the highest net payoffs; and q_i increases from 1 to 7 whereas the net payoffs decrease at the same time. Yet

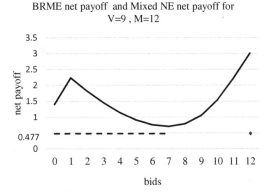

Fig. 6.   BRME net payoffs (full curve) and mixed NE net payoffs (dashed line).

the player, regardless of his chosen bid, never gets a negative net payoff and he always gets a net payoff larger than the NE one: his lowest net payoff, obtained for bid 7, is equal to 0.704 which is higher than 0 and higher than the NE net payoff 0.477. So the BRME is of course not stabilized as regards the Nash logic, but its own consistency ensures large payoffs.

Yet the payoffs may become negative, notably when M is much larger than V. In that case q_0 tends towards $1/2$, and all the other probabilities tend towards 0 (but are still increasing in i from 1 to $V+1$ and constant from $V+1$ to M) because $\sum_{i=1}^{V} q_i < Vq_V = Vq_0/(M-V+1) \to 0$, and $\sum_{i=V+1}^{M} q_i = q_0(M-V+2)(M-V)/(M-V+1)^2 \to q_0$, so q_0 and $\sum_{i=V+1}^{M} q_i \to 1/2$.

Accordingly, when M is large, the BRM probabilities are shared on bid 0 (probability $1/2$) and homogenously shared over the set of bids from $V+1$ to M (probability $1/2$ on this set). We have a kind of bimodal distribution, $1/2$ on bid 0 and $1/2$ on a set (each bid in the set being played with the same probability).

It follows that for large values of M, a player often suffers from the winner's curse. Playing high bids leads him to often win the prize, but he pays too much given that he often wins against a player who bids too much.

Proposition 3. *For large values of M, M large in comparison to V ($M/V \to +\infty$, V is a constant), the mean net loss of a player, at the BRME, is equal to $1/12$th of his budget M. The main net loss is obtained for the bid $M-V$: the player loses $1/4$th of M.*

Proof. See Appendix E. □

We now turn to GBRM. The degree of liberty in the way of weighting the best responses — i.e., in choosing the probability distributions over the sets of best responses — gives rise to a large set of payoffs: players can get payoffs much larger than the NE payoff, but they may also get lower, negative payoffs, depending on the best responses they focus on.

Before talking about some specific payoffs, we specify the set of GBRME and their associated payoffs.

Proposition 4. *For any probability distribution over each set of best responses, there exists a symmetric GBRME. These equilibria lead to a closed interval of net payoffs that includes all the payoffs in $[V/2 - M/4, V/4]$. The maximal symmetric equilibrium net payoff will not be higher than $V/2 - 1/4$ and the minimal one will not be lower than $-V/4 - M/4$.*

Proof. See Appendix F. □

Given that all the net payoffs in the interval $[V/2 - M/4, V/4]$ are reachable, players can get large positive net payoffs ($V/4$ is far from a mixed NE null payoff) but they may also get very negative ones, notably if $M \gg 2V$. So the first reaction may be to say that the degree of liberty allowed by the GBRME concept has the defects of its qualities. On the one hand, it allows us to give a consistent foundation to different potential observed bidding behaviors and their associated payoffs, but on the other hand, it is not very useful if used to predict the payoffs in a second-price all-pay auction game. The second reaction, which involves two steps, is more positive and is linked to behavioral economics. Firstly, true facts often reduce the acceptable probability distributions on the sets of best responses; for example, if we work with senior bidders, usually known as being cautious, we automatically weight the best responses in a given way and so we reduce the set of achievable equilibria and associated payoffs. Secondly, a large range of possible payoffs should encourage players to try to reach the largest positive payoffs, so should lead them to learning to select best responses in a way to get nice payoffs.

We now give some examples. We first observe that by selecting cautious best responses as in matrix 6*b*, which leads both players to putting the same probability $1/(V + 1)$ on each bid from 0 to V, the players get a nice positive payoff.

Proposition 5. *The GBRME that assigns the probability $1/(V + 1)$ to each bid from 0 to V leads to a null net payoff in the first-price all-pay auction. But it leads to the net payoff $(V^2 + 2V)/(6V + 6)$ in the second-price all-pay auction. Moreover, in the second-price all-pay auction, the net payoff obtained with each played bid is positive.*

Proof. See Appendix G. □

It follows from proposition 5 that by playing this GBRME in the second-price all-pay auction, the players are as safe as in the NE, given that they get a positive net payoff with each played bid. And the net payoff is much larger than the NE one, especially if V is large (for $V = 3$ and $M = 6$, respectively $V = 9$ and $M = 12$, the NE net payoff is 0.4157, respectively 0.477, whereas the GBRME net payoff is 0.625, respectively 1.65). More generally, the GBRME payoff is close to $V/6$ when V is large.

We now come back, in the general setting (M, V and the bids are integers), to our students who focus on 0, V and M (more than 4/5 of the students only play these bids in Game 2); observe that for any bid from 0 to M, at least one of these three bids is a best response to it. So we get the game in matrix 7a, and the best-reply matrix 7b ('...' represents the other bids and probabilities).

Player 2

		0	...	V	...	M
	0	(M+V/2, M+V/2)		(M, M+V)		(M, M+V)
	...					
Player 1	V	(M+V, M)		(M-V/2, M-V/2)		(M-V, M)
	...					
	M	(M+V, M)		(M, M-V)		(V/2, V/2)

Matrix 7a. Payoffs with focal values 0, V and M.

Player 2

			q_0	...	q_V	...	q_M
			0	...	V	...	M
p_0		0			b_1b_2	b_1	b_1b_2
	...	...			b_2		b_2
Player 1	p_V	V	b_1b_2	b_1			b_2
	...	...	b_2				
	p_M	M	b_1b_2	b_1	b_1		

Matrix 7b. Best-reply Matrix with 0, V and M.

For any values of V and M, given that p_i and q_i, i from 1 to $M-1$, $i \neq V$, are equal to 0 (the associated bids being never chosen as best responses), the GBRM equations become:

$$p_0 = q_V/2 + q_M, \qquad q_0 = p_V/2 + p_M,$$
$$p_V = q_0/2, \qquad q_V = p_0/2,$$
$$p_M = q_0/2 + q_V/2, \qquad q_M = p_0/2 + p_V/2,$$
$$p_0 + p_V + p_M = 1, \qquad q_0 + q_V + q_M = 1.$$

The unique solution of this system of equations is $p_0 = q_0 = 4/9$, $p_V = q_V = 2/9$, $p_M = q_M = 3/9$.

These probabilities exploit true facts: bid 0 is played more than M because it is the unique best response to M and is also one among the 2 best responses to V. M is more played than V because each time V is a best response, M is a best response too, and M is also a best response to V. These facts may easily be observed by a real player, so he can play in accordance with the hierarchy $p_0 > p_M > p_V$ (like the students in Game 2).

An important fact is that the probabilities do not depend on the values of V and M (what matters is that $M + V > M + V/2$, $M > M - V/2$, $M > V/2$ and $M > M - V$, which is true for all values of M and V, given that $M \geq V$). This derives from the fact that, contrary to the mixed NE concept, BRME and GBRME are ordinal concepts, so only take into account the sign of the differences in payoffs. Yet this may imply bad payoffs. As a matter of fact, the net payoff is equal to: $4/9 \times 4V/18 + 2/9(4V/9 - V/9 - 3V/9) + 3/9(4V/9 + 3(V/2 - M)/9) = 24.5V/81 - M/9$. It follows that the net payoff is negative as soon as $M > 24.5V/9$.

Our students do not lose money even if their net payoff is barely positive (their payoff is $60.026 > 60$), because M/V is not too large: in the experiment, $M = 2V < 24.5V/9$. So it is possible that the students (mainly) focus on 0, $V = 30$ and $M = 60$, because they estimate that the possible loss with M is not large enough to prevent them from bidding M. In other terms, M is not felt as being risky. We can reasonably conjecture that the students would behave differently for other values of V, M, and the ratio M/V.[i] There may exist a kind of bifurcation in the focal bids chosen as best replies when the values of M, V and M/V change and exceed threshold values.

An extreme way of playing, notably useful when M/V becomes large, allowed by GBRM, consists in selecting only the focal bids 0 and V as best responses; this is possible because 0 or V belong to the best responses to each possible bid (0 is a best response to all bids higher than or equal to V, and V is a best response to each bid lower than V). In that case, the GBRM equations, after deleting all the null probabilities, reduce to $q_0 = p_V$ and $q_V = p_0$, each player bidding 0 as often as the opponent bids V and vice versa. The symmetric GBRME, which consists in bidding 0 and V with probability $1/2$, leads to the large net positive payoff $V/4$, which is much larger than the mixed NE net payoff when V is large (for $V = 9$ and $M = 12$, $V/4 = 2.25$ whereas the mixed NE net payoff is only 0.477).

In other terms, if real players behave in accordance with GBRM, then it is crucial to know what is focal for a player, and if a change in the values of V and M is able to induce an appropriate change in the focal bids and in the way of selecting best responses, in order always to get nice positive payoffs.

7. Concluding Remarks

We have shown in this chapter that real behavior, in second-price all-pay auctions, fits much better with GBRM than with mixed NE. Second-price all-pay auctions are games with many best responses to a given strategy. In such games, players can adopt a BRM behavior and combine it with additional characteristics, like cautiousness for example. Depending on the way players select their best responses, the equilibrium behavior and payoffs will be different.

Whereas the mixed NE behavior always leads to a small positive payoff (in the discrete setting) or null payoff (in the continuous setting), a best-reply (and

---

[i]Other experiments with other students give support to this conjecture. The percentage of players playing M shrinks when M becomes much larger than V, and new focal values emerge, like $2V$.

a generalized best-reply) matching behavior can lead to lose or to get a lot of money.

With BRM, when M is close to V, players get a nice payoff, but when M is far from V, the bidders lose money. With GBRM, all depends on the way players choose among their best responses. Some best responses may be more focal, hence be more chosen than others, like 0, V and M. In that case, the payoff is positive when M/V is lower than a given threshold, negative otherwise. When players choose cautious bids in the best response set, then they usually make money. For example, when the bidders focus on the lowest possible best responses, they get a net payoff close to $V/6$ (when V is large). And if they only focus on 0 or V as best responses, each bidder gets a net payoff $V/4$, even if M/V is large.

M seems to be a dangerous focal point, when it becomes large relative to V, because bidding M when the opponent also bids M leads to losing a lot of money. Yet other experiments with other students not reproduced in this chapter highlight a more subtle fact. When M is large in comparison to V (for example $M \geq 3V$) students focus less on M. As a consequence, M stops acting as a bad focal point. Students prefer focusing on 0, V and $2V$ for example. So the negative payoffs the bidders can get by playing M too often, may be observed more when M is larger than $2V$ but not too large. This raises the question of where M comes from, and, as a consequence, the question of the real number of players in the game.

For example, when the second-price all-pay auction is a true auction, then of course there is a third player, the auctioneer, who takes the bids and offers the prize, and the true game can be viewed as a three-player zero-sum game. In that case, in the mixed NE, the payoffs of the auctioneer and the two bidders are slightly negative and positive (are equal to 0 in the continuous setting). With GBRM, the auctioneer and the bidders may alternately be losers and winners, depending on the value of M/V and on the best responses the bidders focus on. Observe also that M can be the budget of the bidders, but it may also be arbitrarily fixed by the rules of the auction.

We may also view the second-price all-pay auction as a "casino" game, like Shubik's [1971] dollar auction game; in this game, there is an auctioneer and he can, if he wants, set an upper bound M to invest in the game (if so, he surely chooses the amount that leads to the largest bids). But he may also fix no upper bound (M does not exist). Shubik [1971] observed that, with no upper bound, the game is usually highly beneficial to the auctioneer, in that it is not uncommon for him to get a total payoff between 3 and 5 dollars. To avoid such a situation, in a casino game, a game authority regulation could impose the value of M, so that the players, even when focusing on M (and 0 and V), do not lose money or at least not too much.

But what happens in real life? A seller rarely uses a second-price all-pay auction when he wants to sell an object, so second-price all-pay auctions are seldom true auctions. Usually, a second-price all-pay auction is used to study an economic war of attrition.

For example, consider the theory of exit from a duopoly (see, for example, Fudenberg and Tirole [1986]), where two firms incur a loss by staying in the duopoly, but where the remaining firm (if the other firm leaves the market) gets a positive monopoly payoff. In this game, V is the monopoly payoff, the bids are the losses incurred each period by staying in the market, and M is the maximal total loss a firm can incur before being constrained to cease trading. M is partly fixed by law, so could be changed; in France for example, a firm has to declare the state of insolvency at most 45 days after having observed it, so M is linked to these 45 days. Observe also that, in addition to the law player, there is an additional player, who would like a more generous deadline (so that M is larger): the consumer, who prefers the duopoly to the monopoly, because he pays less for a same amount.

More generally, there are many economic or political contexts where M is arbitrarily fixed and could be changed. Consider, as a last example, the presidential electoral process in France. For reasons of equality, there are campaign spending limits. In 2017 (the limit amount M can change from one campaign to the other), each candidate could at most invest 16.851 Million euros if eliminated in the first round of the presidential electoral process, 22.509 Million euros if he reached the second round. These amounts are fixed by law. So, if we assume that the result of the electoral process is positively linked to the amount invested in the campaign, we get a war of attrition[j] with an arbitrary limit M, V being the value of winning the election. And one may wonder if changing M, or even deleting M, may change the electoral campaign and its result (see, for example, Che and Gale [1998] and Pastine and Pastine [2012]).

To conclude, M, in economic or political second-price all-pay auctions, when it exists, is not always a natural limit. Given that the value of the ratio M/V may have an impact on the focal values of the bidders, and hence on their behavior and payoff, this is not a neutral observation.

Acknowledgments

I am grateful for helpful suggestions and comments and I wish to thank the L3 students (year's class 2014/2015 and year's class 2016/2017) at the Faculté des Sciences Economiques et de Gestion (Faculty of Economic and Management Sciences) of the University of Strasbourg who played the games with an endless supply of good humor.

Appendix A: Proof of Result 1 (Umbhauer [2017])

Consider player 1. If $V = 1$, there is no dominated strategy. If $V > 1$, the bids from $M - V/2 + 1/2$ to $M - 1$ are weakly dominated by M and played with probability 0.

---

[j]This game can be seen as a first-price or a second-price all-pay auction, depending on the way the candidates are able to adapt their spending to that of the others.

$M - V/2 - 1/2$ and M lead to the same payoff except if player 2 bids $M - V/2 - 1/2$ or M. So player 1 gets the same payoff with both bids if and only if:

$$q_{M-V/2-1/2}(V + 1/2) + q_M(V/2 + 1/2) = q_{M-V/2-1/2}(3V/2 + 1/2) + q_M(V/2),$$

we get $q_{M-V/2-1/2} = q_M/V$.

We now compare the bids $M - V/2 - 3/2$ and $M - V/2 - 1/2$. Both lead to the same payoff, except if player 2 bids $M - V/2 - 3/2$, $M - V/2 - 1/2$ or M. We need

$$q_{M-V/2-3/2}(V + 3/2) + q_{M-V/2-1/2}(V/2 + 3/2) + q_M(V/2 + 3/2)$$

$$= q_{M-V/2-3/2}(3V/2 + 3/2) + q_{M-V/2-1/2}(V + 1/2) + q_M(V/2 + 1/2)$$

hence, $q_{M-V/2-3/2}(-V/2) + q_{M-V/2-1/2}(-V/2 + 1) + q_M = 0$,

we get, $q_{M-V/2-3/2} = q_M(1/V + 2/V^2)$.

More generally, $M - V/2 - 1/2 - j$ and $M - V/2 - 1/2 - j - 1$, j from 1 to $M - V/2 - 3/2$, lead to the same payoff, except if player 2 bids $M - V/2 - 1/2 - k$ or M, with k going from 0 to $j + 1$.

We need

$$q_{M-\frac{V}{2}-\frac{1}{2}-j-1}\left(V + \frac{3}{2} + j\right) + q_{M-\frac{V}{2}-\frac{1}{2}-j}\left(\frac{V}{2} + \frac{3}{2} + j\right)$$

$$+ \sum_{k=0}^{j-1} q_{M-\frac{V}{2}-\frac{1}{2}-k}\left(\frac{V}{2} + \frac{3}{2} + j\right) + q_M\left(\frac{V}{2} + \frac{3}{2} + j\right)$$

$$= q_{M-\frac{V}{2}-\frac{1}{2}-j-1}\left(\frac{3V}{2} + \frac{3}{2} + j\right) + q_{M-\frac{V}{2}-\frac{1}{2}-j}\left(V + \frac{1}{2} + j\right)$$

$$+ \sum_{k=0}^{j-1} q_{M-\frac{V}{2}-\frac{1}{2}-k}\left(\frac{V}{2} + \frac{1}{2} + j\right) + q_M\left(\frac{V}{2} + \frac{1}{2} + j\right).$$

Hence,

$$q_{M-\frac{V}{2}-\frac{3}{2}-j}\left(\frac{-V}{2}\right) + q_{M-\frac{V}{2}-\frac{1}{2}-j}\left(\frac{-V}{2} + 1\right) + \sum_{k=0}^{j-1} q_{M-\frac{V}{2}-\frac{1}{2}-k} + q_M = 0.$$

We also get

$$q_{M-\frac{V}{2}-\frac{3}{2}-(j+1)}\left(\frac{-V}{2}\right) + q_{M-\frac{V}{2}-\frac{1}{2}-(j+1)}\left(\frac{-V}{2} + 1\right)$$

$$+ q_{M-\frac{V}{2}-\frac{1}{2}-j} + \sum_{k=0}^{j-1} q_{M-\frac{V}{2}-\frac{1}{2}-k} + q_M = 0.$$

It follows $q_{M-V/2-j-5/2} = 2q_{M-V/2-j-3/2}/V + q_{M-V/2-j-1/2}$ for any j from 0 to $M - V/2 - 5/2$.

i.e., $q_i = 2q_{i+1}/V + q_{i+2}$ for i from 0 to $M - V/2 - 5/2$.

Observe that if $M = (V + 1)/2$, which requires $V = M = 1$ (because $V \leq M$), then the proof stops with $q_{M-V/2-1/2} = q_0 = q_M/V = q_1 = 1/2$. And if

$M = (V + 3)/2$, which requires $M \leq 3$, then the proof stops with $q_1 = q_M/V$ and $q_0 = q_M(1/V + 2/V^2)$. $M = (V+3)/2$ implies $M = 3$ and $V = 3$ (and $q_0 = 5/17$, $q_1 = 3/17$ and $q_3 = 9/17$), or $M = 2$ and $V = 1$ (and $q_0 = 3/5$, $q_1 = q_2 = 1/5$).

The analysis is the same for player 2, so leads to the same equilibrium probabilities for player 1.

Appendix B: Proof of Result 2

A first proof of this (almost folk) result goes back to Hendricks *et al.* [1988]. The following proof is from Umbhauer [2016].

All bids from $M - V/2$ to M (excluded) are weakly dominated by M, so it is conjectured that the NE strategy is a density function $f(\cdot)$ on $[0, M - V/2]$ with a mass point on M.

Call $f_2(\cdot)$ player 2's equilibrium strategy. Suppose that player 1 plays b. She wins the auction each time player 2 bids less than b. Her payoff $Eg(b)$ is equal to:

$$Eg(b) = M + \int_0^b (V - x)f_2(x)dx - b\left(\int_b^{M-V/2} f_2(x)dx + f_2(M)\right)$$

$Eg(b)$ has to be constant for each b in $[0, M - V/2]\bigcup\{M\}$.

We get $Eg(M) = M + \int_0^{M-V/2}(V - x)f_2(x)dx + (\frac{V}{2} - M)f_2(M) = Eg(M - \frac{V}{2})$.

We need $Eg'(b) = 0$ for b in $[0, M - V/2]$.

We get $(V - b)f_2(b) - F_2(M - V/2) + F_2(b) - f_2(M) + bf_2(b) = 0$,

where $F_2(\cdot)$ is the cumulative distribution of the density function $f_2(\cdot)$.

By construction $f_2(M) = 1 - F_2(M - V/2)$, and the differential equation becomes $Vf_2(b) - 1 + F_2(b) = 0$.

The solution is $F_2(b) = 1 + Ke^{-b/V}$, where K is a constant determined as follows:

$F_2(0) = 0$ because there is no mass point on 0, so $K = -1$.

It follows $F_2(b) = 1 - e^{-b/V}$ for b in $[0, M - V/2]$, $f_2(M) = 1 - F_2(M - V/2) = e^{1/2 - M/V}$ (< 1), $f_2(b) = e^{-b/V}/V$ for b in $[0, M - V/2]$ (and $f_2(b) = 0$ for b in $]M - V/2, M[$).

By symmetry, we get $f_1(b) = e^{-b/V}/V$ for b in $[0, M - V/2]$, $f_1(M) = e^{1/2 - M/V}$ (and $f_1(b) = 0$ for b in $]M - V/2, M[$).

Given that $Eg(0)$ is equal to M, given that bid 0 is played at equilibrium, and given that a player gets the same payoff with each played bid, each player gets the payoff M, hence a net payoff equal to 0 at equilibrium.

Appendix C: Proof of Proposition 1

Each bid i, i from 1 to V is a best reply to all bids j, j from 0 to $i - 1$.

Each bid i, i from $V + 1$ to M is a best reply to all bids j, j from 0 to V (if $M > V$).

Bid 0 is a best reply to V and the only best reply to bid j, j from $V + 1$ to M, if $M > V$, and bid 0 is the only best reply to V if $M = V$.

We get

$$p_0 = \frac{q_V}{M-V+1} + \sum_{i=1}^{M-V} q_{V+i}, \quad q_0 = \frac{p_V}{M-V+1} + \sum_{i=1}^{M-V} p_{V+i}.$$

These two equations reduce to $p_0 = q_V$ and $q_0 = p_V$ if $M = V$.

$$
\begin{aligned}
p_1 &= q_0/M, & q_1 &= p_0/M, \\
p_2 &= q_0/M + q_1/(M-1), \quad \text{(if } V > 1\text{)}, & q_2 &= p_0/M + p_1/(M-1), \quad \text{(if } V > 1\text{)}, \\
p_i &= \sum_{j=0}^{i-1} q_j/(M-j), \quad i \text{ from 1 to } V, & q_i &= \sum_{j=0}^{i-1} p_j/(M-j), \quad i \text{ from 1 to } V.
\end{aligned}
$$

$$\text{(C.1)}$$

$$p_i = \sum_{j=0}^{V-1} \frac{q_j}{M-j} + q_V/(M-V+1), \quad i \text{ from } V+1 \text{ to } M,$$

$$q_i = \sum_{j=0}^{V-1} \frac{p_j}{M-j} + p_V/(M-V+1), \quad i \text{ from } V+1 \text{ to } M,$$

$$\text{and } \sum_{i=0}^{M} p_i = \sum_{i=0}^{M} q_i = 1. \qquad \text{(C.2)}$$

It can easily be checked that the equations defining p_0 and q_0 are redundant.

We first look for a symmetric BRME.

We start with $M > V$.

We get,

$$q_0 = \frac{q_V}{M-V+1} + \sum_{i=1}^{M-V} q_{V+i},$$

$$q_1 = q_0/M,$$

$$q_2 = q_0/M + q_1/(M-1), \qquad \text{(if } V > 1\text{)},$$

$$q_i = \sum_{j=0}^{i-1} q_j/(M-j), \qquad i \text{ from 1 to } V,$$

$$q_i = \sum_{j=0}^{V-1} \frac{q_j}{M-j} + q_V/(M-V+1), \quad i \text{ from } V+1 \text{ to } M.$$

We observe that $\sum_{i=V+1}^{M} q_i < q_0 < \sum_{i=V}^{M} q_i$.

Suppose $V > 1$. Given that $q_1 = q_0/M$, we get $q_2 = q_0/M + q_0/((M-1)M) = q_0/(M-1)$.

$q_i = q_0/(M-i+1)$ implies $q_{i+1} = q_i + q_i/(M-i) = q_0/(M-i+1) + q_0/((M-i+1)(M-i)) = q_0/(M-i)$.

It derives by induction that $q_i = q_0/(M-i+1)$, for i from 1 to V.

And $q_i = q_0/(M-V+1) + q_V/(M-V+1) = q_0/(M-V+1) + q_0/(M-V+1)^2 = q_0(M-V+2)/(M-V+1)^2$ for i from $V+1$ to M.

It follows: $q_0 + q_0/M + q_0/(M-1) + \cdots + q_0/(M-V+1) + q_0 - q_0/(M-V+1)^2 = 1$.

So $q_0(2 + \sum_{i=0}^{V-1} 1/(M-i) - 1/(M-V+1)^2 = 1$.

It can be checked that the result is the same for $V = 1$.

We now suppose that $M = V$.

In that case the bids from $V+1$ to M vanish and we only focus on the boxed equations (C.1) and (C.2) (and the redundant equations $p_0 = q_V$ and $q_0 = p_V$). It easily follows that $q_0/(1 + \sum_{i=0}^{V-1} 1/(M-i)) = 1$.

Given that $1 + \sum_{i=0}^{V-1} \frac{1}{M-i} = 2 + \sum_{i=0}^{V-1} 1/(M-i) - 1/(M-V+1)^2$ (for $M = V$), we keep the same notation for q_0 for $M > V$ and $M = V$.

It can be checked that the result also holds for $M = V = 1$. It reduces to $q_0 = q_1 = 1/2$.

We now show that the equilibrium is unique, by showing that a BRME is necessarily symmetric (for $M \neq 1$).

We observe that $q_0 = p_0$ implies $q_i = p_i$ for i from 1 to M.

Now we set $p_o = aq_0$ where a is a constant, and we show that $a = 1$.

We start with $M > V$.

We have

$$
\boxed{
\begin{array}{ll}
p_1 = q_0/M, & q_1 = p_0/M, \\
p_2 = p_1 + q_1/(M-1), \quad \text{(if } V > 1\text{)}, & q_2 = q_1 + p_1/(M-1), \quad \text{(if } V > 1\text{)}, \\
p_i = p_{i-1} + q_{i-1}/(M-i+1), & q_i = q_{i-1} + p_{i-1}/(M-i+1), \\
\quad i \text{ from 1 to } V, & \quad i \text{ from 1 to } V
\end{array}
}
$$

$$\text{(C.3)}$$

$$p_i = p_V + q_V/(M-V+1), \quad i \text{ from } V+1 \text{ to } M,$$

$$q_i = q_V + p_V/(M-V+1), \quad i \text{ from } V+1 \text{ to } M$$

$$p_0 = \frac{q_V}{M-V+1} + \sum_{i=1}^{M-V} q_{V+i}, \qquad q_0 = \frac{p_V}{M-V+1} + \sum_{i=1}^{M-V} p_{V+i}.$$

Suppose $V > 1$.

It follows:

$$
\begin{aligned}
p_1 - q_1 &= (q_0 - p_0)/M = (1-a)q_0/M, \qquad \text{i.e., } p_1 = q_1 + (1-a)q_0/M \\
p_2 - q_2 &= (p_1 - q_1)(1 - 1/(M-1)) \\
&= (1-a)(M-2)q_0/[M(M-1)] \\
p_i - q_i &= (p_{i-1} - q_{i-1})(1 - 1/(M-i+1)) \\
&= (1-a)(M-i)q_0/[M(M-1)], \qquad i \text{ from 2 to } V
\end{aligned}
$$

(C.4)

$$
\begin{aligned}
p_{V+i} - q_{V+i} &= (p_V - q_V)(1 - 1/(M-V+1)) \\
&= (1-a)(M-V)^2 q_0/[M(M-1)(M-V+1)]
\end{aligned}
$$

$$ i \text{ from 1 to } M-V. $$

$$
\begin{aligned}
p_0 &= \frac{q_V}{M-V+1} + \sum_{i=1}^{M-V} q_{V+i} \quad \Rightarrow \\
aq_0 &= q_V/(M-V+1) + (M-V)q_{V+1}.
\end{aligned}
$$

(C.5)

$$
\begin{aligned}
q_0 &= \frac{p_V}{M-V+1} + \sum_{i=1}^{M-V} p_{V+i} \quad \Rightarrow \\
q_0 &= (q_V + (1-a)(M-V)q_0/[M(M-1)])/(M-V+1) \\
&\quad + (M-V)(q_{V+1} + (1-a)(M-V)^2 q_0/ \\
&\quad [M(M-1)(M-V+1)]).
\end{aligned}
$$

$$
\begin{aligned}
q_0 &= q_V/(M-V+1) + (M-V)q_{V+1} + (1-a)q_0((M-V) \\
&\quad + (M-V)^3)/[M(M-1)(M-V+1)].
\end{aligned}
$$

(C.6)

It follows from Eq. (C.5) and (C.6):

$aq_0 = q_0 - (1-a)q_0((M-V) + (M-V)^3)/[M(M-1)(M-V+1)]$ i.e., $(1-a)(1-((M-V)+(M-V)^3)/[M(M-1)(M-V+1)]) = 0$ i.e., $a = 1$ except if $((M-V)+(M-V)^3)/[M(M-1)(M-V+1)] = 1$, which is not possible for $M > V$. By developing $((M-V)+(M-V)^3)/[M(M-1)(M-V+1)] - 1$, we get the expression $3MV^2 - 2M^2V - V^3 + 2M - VM - V = V(V-M)(2M-V) + M(2-V) - V$ which is always negative.

Hence $a = 1$, and the symmetric BRME is the only BRME of the game.

Easier calculations lead to the same result for $V = 1$.

We now switch to $M = V$.

If $V = M = 1$, it is obvious to get the infinite set of BRME, defined by: $p_1 = q_0$, $q_0 + q_1 = 1$, $p_0 + p_1 = 1$.

Suppose $M > 1$; given that $V = M$, we only focus on the boxed equations (C.3) and (C.4) and the equations $p_0 = q_V$ and $q_0 = p_V$. Given the expression of $p_i - q_i$, for i from 1 to V, we have $p_V = q_V$. Given that $p_0 = q_V$ and $q_0 = p_V$, we get $p_0 = q_0$. So the symmetric BRME is the only BRME of the game.

Appendix D: Proof of Proposition 2

We have $Eg(0) = M + q_0 V/2$, $Eg(1) = M + q_0 V + q_1(V/2 - 1) - 1(1 - q_0 - q_1) = M + q_0 V + q_1 V/2 - 1(1 - q_0)$,

$$Eg(i) = M + \sum_{j=0}^{i-1}(V-j)q_j + q_i V/2 - i\left(1 - \sum_{j=0}^{i-1} q_j\right), \quad i \text{ from 1 to } M.$$

It follows: $Eg(i+1) - Eg(i) = (V-i)q_i + \frac{q_{i+1}V}{2} - \frac{q_i V}{2} + iq_i - (1 - \sum_{j=0}^{i} q_j) = \frac{Vq_i}{2} + \frac{Vq_{i+1}}{2} - \sum_{j=i+1}^{M} q_j$.

And $Eg(i+2) - Eg(i+1) = \frac{Vq_{i+1}}{2} + \frac{Vq_{i+2}}{2} - \sum_{j=i+2}^{M} q_j$.

Given that $q_{i+2} \geq q_i$ for i from 1 to $M-2$, and given that $-\sum_{j=i+2}^{M} q_j > -\sum_{j=i+1}^{M} q_j$ for i from 0 to $M-2$, we get $Eg(i+2) - Eg(i+1) > Eg(i+1) - Eg(i)$ for i from 1 to $M-2$. So $Eg(i+1) - Eg(i)$ is increasing in i, for i from 1 to $M-1$.

We now suppose $M > 2V$.

We get, for i from 1 to $M - V - 1$,

$$Eg(V+i+1) - Eg(V+i) = \frac{Vq_{V+i}}{2} + \frac{Vq_{V+i+1}}{2} - \sum_{j=i+1}^{M-V} q_{V+j}$$

$$= q_0(M - V + 2)(2V + i - M)/(M - V + 1)^2$$

given that $q_{V+j} = q_0(M - V + 2)/(M - V + 1)^2$ for i from 1 to $M - V$. It follows that $Eg(V+i+1) - Eg(V+i)$ becomes positive only for $i > M - 2V$.

Hence $Eg(j+1) - Eg(j)$ becomes positive only for $j > M - V$.

Putting the results together, it derives that, for $M > 2V$, $Eg(b)$ decreases for b from 1 to $M - V$ and increases from $M - V + 1$ to M.

Appendix E: Proof of Proposition 3

E.1. *First part of the proposition*

We calculate the mean payoff for each player when M is large. We know that, when M is large ($M \to +\infty$) and V is a constant, then $q_0 \to 1/2$, $\sum_{i=1}^{V} q_i \to 0$, $\sum_{i=V+1}^{M} q_i \to 1/2$, so $q_i = a = 1/(2(M - V))$ for each i from $V + 1$ to M.

We look for the payoff obtained with each played bid. We omit the bids from 1 to V, given that they lead to a payoff which is a constant that will be multiplied by a probability (of playing the bid) so close to 0 that the sum of the payoffs obtained with these bids also tends towards 0 (each $q_i \to 0$ and $\sum_{i=1}^{V} q_i \to 0$). For similar reasons, we omit the payoff a player gets when he meets a player who plays a bid from 1 to V. So we calculate the payoff obtained with bid 0, the payoff obtained with bid $V + i$, i from 1 to $M - V$, and the expected net payoff.

Net payoff obtained with bid $0 = q_0 V/2 = V/4$.

Net payoff obtained with bid $V + 1 = q_0 V + a(V/2 - V - 1) - (V + 1)(M - V - 1)a = q_0 V + a(V/2) - a(V + 1)(M - V)$.

Net payoff obtained with bid $V + 2 = q_0 V - a + a(V/2 - V - 2) - (V + 2)(M - V - 2)a = q_0 V - a + a(V/2) - a(V + 2)(M - V - 1)$.

More generally, the net payoff obtained with bid $V + i$ is equal to:
$q_0 V - a - \cdots - (i - 1)a + a(V/2) - a(V + i)(M - V - i + 1)$, i from 2 to $M - V$.
So, for i from 1 to $M - V$, the net payoff obtained with $V + i$ is equal to:
$V/2 + ai^2/2 - aV(M - V + 1/2) - ia(M - 2V + 1/2)$.

To calculate the expected net payoff, we multiply $V/4$ by q_0, the net payoff for bid $V + i$ by a, i from 1 to $M - V$, and we sum these payoffs. We get:

$$\frac{V}{8} + \frac{a(M - V)V}{2} - a^2(M - V)V\left(M - V + \frac{1}{2}\right)$$

$$+ a\sum_{i=1}^{M-V}\left(\frac{ai^2}{2} - ia\left(M - 2V + \frac{1}{2}\right)\right).$$

$$a\sum_{i=1}^{M-V}\left(\frac{ai^2}{2} - ia\left(M - 2V + \frac{1}{2}\right)\right)$$

$$= a^2(M - V)(M - V + 1)(2M - 2V + 1)/12$$

$$- a^2(M - V)(M - V + 1)(M - 2V + 1/2)/2,$$

which tends towards $2M/48 - M/8 = -M/12$ because $a = 1/(2(M - V))$ and because V and the other constants are small in comparison to M.

$V/8 + a(M - V)V/2 - a^2(M - V)V(M - V + 1/2) \to V/8 + V/4 - V/4$ because $a = 1/(2(M - V))$.

So the BRME net payoff tends towards $-M/12$.

E.2. *Second part of the proposition*

Given that $M \gg 2V$, we know from proposition 2 that the lowest payoff is obtained for bid $M - V$, i.e., bid $V + (M - 2V)$. This net payoff is equal to:
$V/2 + ai^2/2 - aV(M - V + 1/2) - ia(M - 2V + 1/2)$ with $i = M - 2V$ and $a = 1/(2(M - V))$.

So it is equal to: $V/2 + a(M - 2V)^2/2 - aV(M - V + 1/2) - (M - 2V)a(M - 2V + 1/2)$ which tends towards $-M/4$ (we can omit V and the others constants given that only $M \to +\infty$).

Appendix F: Proof of Proposition 4

We only look for symmetric GBRME and we only study the case $M > V$ (there is no difficulty in adapting the proof to $M = V$). We get the equations (F.1):

$$\left.\begin{aligned}
q_0 &= \delta_{0V} q_V + \sum_{i=V+1}^{M} q_i, \\[2mm]
q_i &= \sum_{j=0}^{i-1} \delta_{ij} q_j, &\quad i \text{ from 1 to } V, \\[2mm]
q_i &= \sum_{j=0}^{V} \delta_{ij} q_j, &\quad i \text{ from } V+1 \text{ to } M \quad \text{and} \quad \sum_{i=0}^{M} q_i = 1.
\end{aligned}\right\} \quad \text{(F.1)}$$

For ease of notation we will write "δ" for all "the constants/variables δ_{0V} and δ_{ij}, j from 0 to V and i from $j+1$ to M". We call D the set of δ that check the constraints (F.2):

$$\left.\begin{aligned}
\sum_{i=j+1}^{M} \delta_{ij} &= 1 \quad j \text{ from 0 to } V-1. \\[2mm]
\delta_{0V} + \sum_{i=V+1}^{M} \delta_{iV} &= 1. \\[2mm]
\delta_{0V} \geq 0, \quad \delta_{ij} \geq 0, &\quad j \text{ from 0 to } V \quad \text{and} \quad i \text{ from } j+1 \text{ to } M.
\end{aligned}\right\} \quad \text{(F.2)}$$

There is a redundant equation in (F.1):

$q_i = \sum_{j=0}^{i-1} \delta_{ij} q_j$, i from 1 to V, $q_i = \sum_{j=0}^{V} \delta_{ij} q_j$, i from $V+1$ to M, and $\sum_{i=0}^{M} q_i = 1 \Rightarrow$

$$q_0 = 1 - \sum_{i=1}^{M} q_i = 1 - \sum_{j=0}^{V-1} \sum_{i=j+1}^{M} \delta_{ij} q_j - \sum_{j=V+1}^{M} \delta_{jV} q_V = 1 - \sum_{j=0}^{V-1} q_j - (1 - \delta_{0V}) q_V$$

$$= \sum_{i=V}^{M} q_i - (1 - \delta_{0V}) q_V = \sum_{i=V+1}^{M} q_i + \delta_{0V} q_V.$$

So the first equation is redundant and we suppress it.

The existence of a unique symmetric GBRME for all δ in D follows by construction.

$$q_1 = \delta_{10} q_0 = a_1 q_0 \quad \text{with } a_1 \geq 0$$

$$q_2 = \delta_{20} q_0 + \delta_{21} a_1 q_0 = a_2 q_0 \quad \text{with } a_2 \geq 0.$$

Suppose that $q_i = a_i q_0$ with $a_i \geq 0$ for $i < V$.
It follows $q_{i+1} = \sum_{j=0}^{i} \delta_{(i+1)j} q_j = \sum_{j=0}^{i} \delta_{(i+1)j} a_j q_0 = a_{i+1} q_0$ with $a_{i+1} \geq 0$.
So $q_i = a_i q_0$ with $a_i \geq 0$ i from 1 to V.
And $q_i = \sum_{j=0}^{V} \delta_{ij} q_j = \sum_{j=0}^{V} \delta_{ij} a_j q_0 = a_i q_0$ $\quad a_i \geq 0$, i from $V+1$ to M.

We get, $\sum_{i=0}^{M} q_i = 1 \Leftrightarrow q_0 = 1/(1 + \sum_{i=1}^{M} a_i)$ with at least one strictly positive a_i, i from 1 to M.

So q_0 is unique, positive and lower than 1. Given that $q_i = a_i q_0$ with $a_i \geq 0$, and given that $\sum_{i=0}^{M} q_i = 1$, it follows that all the q_i, i from 1 to M are unique, positive or null and lower than 1. So, for all δ in D, there exists a unique symmetric GBRME.

More precisions can be given on q_0. Given that $q_i = \delta_{i0} q_0 + b_i$, with $b_i \geq 0$, i from 1 to M, we get $\sum_{i=1}^{M} q_i = \sum_{i=1}^{M} (\delta_{i0} q_0 + b_i) = q_0 + \sum_{i=1}^{M} b_i$. It follows that $\sum_{i=0}^{M} q_i = 1 \Leftrightarrow 2q_0 + \sum_{i=1}^{M} b_i = 1$ so $q_0 < 1/2$ given that each b_i is positive or null (i from 1 to M), and at least one of them is strictly positive if $V > 1$ (if $V = 1$, q_0 can be equal to $1/2$).

We now look at player 1's net payoff (the analysis is symmetric for player 2). It is equal to

$(q_0, q_1, q_2, \ldots, q_{M-1}, q_M)$

$$
\times
\begin{pmatrix}
\frac{V}{2} & 0 & 0 & 0 & 0 & \cdots & 0 & 0 \\
V & \frac{V}{2} - 1 & -1 & -1 & -1 & \cdots & -1 & -1 \\
V & V-1 & \frac{V}{2} - 2 & -2 & -2 & \cdots & -2 & -2 \\
\cdots & \cdots & \cdots & \cdots & \cdots & \cdots & \cdots & \cdots \\
V & V-1 & V-2 & V-3 & V-4 & \cdots & \frac{V}{2} - (M-1) & -(M-1) \\
V & V-1 & V-2 & V-3 & V-4 & \cdots & V-(M-1) & \frac{V}{2} - M
\end{pmatrix}
\cdot
\begin{pmatrix}
q_0 \\ q_1 \\ q_2 \\ \cdots \\ q_{M-1} \\ q_M
\end{pmatrix}
$$

$$
= \frac{V}{2} q_0^2 + \sum_{i=1}^{M-1} q_i \left(\sum_{j=0}^{i-1} (V-j) q_j + \left(\frac{V}{2} - i \right) q_i - i \sum_{j=i+1}^{M} q_j \right)
$$

$$
+ q_M \left(\sum_{j=0}^{M-1} (V-j) q_j + \left(\frac{V}{2} - M \right) q_M \right)
$$

$$
= \frac{V}{2} \sum_{i=0}^{M} q_i^2 + V \sum_{j=i+1}^{M} \sum_{i=0}^{M-1} q_i q_j - \sum_{i=1}^{M-1} i q_i \left(q_i + 2 \sum_{j=i+1}^{M} q_j \right) - M q_M^2
$$

$$
= \frac{V}{2} \left[\sum_{i=0}^{M} q_i \right]^2 - \sum_{i=1}^{M} \left[\sum_{j=i}^{M} q_j \right]^2 = \frac{V}{2} - \sum_{i=1}^{M} \left[\sum_{j=i}^{M} q_j \right]^2 .
$$

When $M = V$, we get the same result.

We write $g(\cdot) = \frac{V}{2} - \sum_{i=1}^{M} [\sum_{j=i}^{M} q_j]^2$ with $q_i = a_i/(1 + \sum_{i=1}^{M} a_i)$ where a_i is defined as above, i from 1 to M, in function of δ. So $g(\cdot)$ is a function of δ and can be maximized and minimized on δ subject to $\delta \in D$.

Given that $g(\delta)$ is continuous on D, given that D is closed and bounded, Weierstrass's theorem ensures that $g(\delta)$ has a maximum and a minimum on D.

We give a trivial upper bound for the maximum. Given that $(\sum_{j=1}^{M} q_j)^2 = (1 - q_0)^2$ and given that $q_0 < 1/2$, we get $\frac{V}{2} - \sum_{i=1}^{M}[\sum_{j=i}^{M} q_j]^2 < \frac{V}{2} - \frac{1}{4}$.

This result also holds when $M = V$.

We provide a lower bound:

Given that $q_0 \geq \sum_{i=V+1}^{M} q_i$

$$\frac{V}{2} - \sum_{i=1}^{M}\left[\sum_{j=i}^{M} q_j\right]^2 = \frac{V}{2} - \sum_{i=1}^{V}\left[\sum_{j=i}^{M} q_j\right]^2 - \sum_{i=V+1}^{M}\left[\sum_{j=i}^{M} q_j\right]^2$$

$$\sum_{i=V+1}^{M}\left[\sum_{j=i}^{M} q_j\right]^2 \leq (M - V)\left(\sum_{i=V+1}^{M} q_i\right)^2$$

$$\sum_{i=1}^{V}\left[\sum_{j=i}^{M} q_j\right]^2 \leq V(1 - q_0)^2.$$

$$\frac{V}{2} - \sum_{i=1}^{V}[\sum_{j=i}^{M} q_j]^2 - \sum_{i=V+1}^{M}[\sum_{j=i}^{M} q_j]^2 \geq \frac{V}{2} - V(1-q_0)^2 - (M-V)(\sum_{i=V+1}^{M} q_i)^2$$

$$\geq \frac{V}{2} - V - (M - V)q_0^2$$

$$\geq \frac{V}{2} - V - (M - V)/4$$

$$\geq -V/4 - M/4.$$

This result also holds for $M = V$.

Given that $(q_0 = 1/2, q_V = 1/2, q_i = 0, i \neq 0, V)$ satisfies the equations (F.1) when $\delta_{0V} = 1$ and $\delta_{Vi} = 1$, i from 0 to $V - 1$, the other δ variables being equal to 0, the maximum is necessarily higher than or equal to $V/4$. And given that $(q_0 = 1/2, q_M = 1/2, q_i = 0, i \neq 0, M)$ satisfies the equations (F.1) when $\delta_{Mi} = 1$, i from 0 to V, the other δ variables being equal to 0, the minimum is necessarily lower than or equal to $V/2 - M/4$.

D is convex and $g(\delta)$ is continuous on D. So the intermediate value theorem ensures that all the payoffs between the minimum and the maximum, and *a fortiori*, any payoff between $V/2 - M/4$ and $V/4$, can be observed for some specific δ. Of course, this last result is of interest only if $M > V$.

Appendix G: Proof of Proposition 5

We first calculate the net payoff in the first-price all-pay auction. It is well-known that the discrete NE leads to playing each bid from 0 to $V - 1$ with probability $1/V$ and that the NE payoff is 0.5 (so each bid from 0 to $V - 1$ leads to the payoff 0.5 in the NE).

It follows that, in the GBRME:

Bid 0 leads to the net payoff $0.5V/(V+1)$.

Bid 1 leads to the net payoff $0.5V/(V+1) - 1/(V+1)$ (because we have to add the payoff obtained by bid 1 when confronted to bid V).

Bid i leads to the net payoff $0.5V/(V+1) - i/(V+1)$ i from 2 to $V-1$.

Bid V leads to the net payoff $-0.5V/(V+1)$.

And the GBRME net payoff is equal to:

$0.5V^2/(V+1)^2 - (1+2+\cdots+V-1)/(V+1)^2 - 0.5V/(V+1)^2 = [0.5V^2 - 0.5V(V-1) - 0.5V]/(V+1)^2 = 0$.

We now focus on the second-price all-pay auction.

Bid 0 leads to the net payoff $0.5V/(V+1)$.

Bid 1 leads to the net payoff $[V + V/2 - V]/(V+1)$.

Bid i leads to the net payoff $[\sum_{j=0}^{i-1}(V-j) + V/2 - i(V-i+1)]/(V+1)$ i from 2 to V.

We get $[\sum_{j=0}^{i-1}(V-j) + V/2 - i(V-i+1)]/(V+1) = [V/2 + i(i-1)/2]/(V+1)$ for i from 1 to V.

And the GBRME net payoff becomes:

$[0.5V(V+1) + 0.5\sum_{i=1}^{V} i^2 - 0.5\sum_{i=1}^{V} i]/(V+1)^2 = 0.5[V(V+1) + V(V+1)(2V+1)/6 - 0.5V(V+1)]/(V+1)^2 = 0.5V[0.5 + (2V+1)/6]/(V+1) = (V^2 + 2V)/(6(V+1))$.

Regardless of the played bid, the player gets a positive net payoff, given that bid 0 leads to the net payoff $0.5V/(V+1)$ and that bid i leads to the net payoff $[V/2 + i(i-1)/2]/(V+1)$ for i from 1 to V.

References

Bergemann, D. and Morris, S. [2013] Robust predictions in games with incomplete information, *Econometrica* **81**, 1251–1308.

Bernheim, D. B. [1984] Rationalizable strategic behaviour, *Econometrica* **52**, 1007–1028.

Che, Y. K. and Gale, I. L. [1998] Caps on political lobbying, *Am. Econ. Rev.* **88**, 643–651.

Dechenaux, E., Kovenock, D. and Sheremeta, R. M. [2015] A survey of experimental research on contests, all-pay auctions and tournaments, *Exp. Econ.* **18**, 609–669.

Dekel, E., Jackson, M. O. and Wolinsky, A. [2008] Vote buying: General elections, *J. Political Econ.* **116**, 351–380.

Fudenberg, D. and Tirole, J. [1986] A theory of exit in duopoly, *Econometrica* **54**, 943–960.

Gneezy, U. and Smorodinsky, R. [2006] All-pay auctions — an experimental study, *J. Econ. Behav. Org.* **61**, 255–275.

Hendricks, K., Weiss, A. and Wilson, C. [1988] The war of attrition in continuous time with complete information, *Int. Econ. Rev.* **29**, 663–680.

Hörisch, H. and Kirchkamp, O. [2010] Less fighting than expected, experiments with wars of attrition and all-pay auctions, *Public Choice* **144**, 347–367.

Kosfeld, M., Droste, E. and Voorneveld, M. [2002] A myopic adjustment process leading to best-reply matching, *Games Econ. Behav.* **40**, 270–298.

Lugovskyy, V., Puzzello, D. and Tucker, S. [2010] An experimental investigation of overdissipation in the all-pay auction, *Eur. Econ. Rev.* **54**, 974–997.

Pastine, I. and Pastine, T. [2012] Incumbency advantage and political campaign spending limits, *J. Public Econ.* **96**, 20–32.

Pearce, D. [1984] Rationalizable strategic behavior and the problem of perfection, *Econometrica* **52**, 1029–1050.

Shubik, M. [1971] The dollar auction game: A paradox in noncooperative behaviour and escalation, *J. Conflict Resol.* **15**, 109–111.

Umbhauer, G. [2016] *Game Theory and Exercises*, Routledge Advanced Texts in Economics and Finance (Routledge Editors, London).

Umbhauer, G. [2017] Equilibria in discrete and continuous second-price all-pay auctions, convergence or yoyo phenomena, BETA Working Paper 2017–14, University of Strasbourg, France.

Chapter 10

Cournot Equilibrium Uniqueness: At 0 Discontinuous Industry Revenue and Decreasing Price Flexibility

Pierre Von Mouche*

Wageningen University, Wageningen, The Netherlands

pvmouche@deds.nl

Takashi Sato

Shimonoseki City University, Shimonoseki, Japan

sato@shimonoseki-cu.ac.jp

We consider the equilibrium uniqueness problem for a large class of Cournot oligopolies with convex cost functions and a proper price function $\tilde{p}$ with decreasing price flexibility. This class allows for (at 0) discontinuous industry revenue and in particular for $\tilde{p}(y) = y^{-\alpha}$. This chapter illustrates in an exemplary way the Selten–Szidarovszky technique based on virtual backward reply functions. An algorithm for the calculation of the unique equilibrium is provided.

Keywords: Cournot oligopoly; decreasing price flexibility; discontinuous payoff functions; equilibrium uniqueness; pseudo-concavity; Selten–Szidarovszky technique.

1. Introduction

Nowadays, dozens of equilibrium results exist for Cournot oligopolies,[a] but only a few for the case where industry revenue is discontinuous at 0.[b] Besides its intrinsic relevance for Cournot games,[c] understanding situations with at 0 discontinuous industry revenue is important for analyzing rent-seeking contests, innovation

---

*Corresponding author.

[a]For some overview the reader may like to consult Vives [2001], Ewerhart [2014] and Von Mouche and Quartieri [2016] and references in these papers.

[b]See Szidarovszky and Okuguchi [1997], Yamazaki [2008], Von Mouche and Quartieri [2012], Hirai and Szidarovszky [2013], Cornes and Sato [2016] and Matsushima and Yamazaki [2016] and references in these papers.

[c]Note that various consumer utility functions, e.g., that of Cobb–Douglas type, lead to an inverse demand function for which the associated revenue function is discontinuous at 0.

tournaments and patent-race games, because various instances of these games are 'structurally equivalent' to Cournot oligopolies [Szidarovszky and Okuguchi, 1997; Baye and Hoppe, 2003]. The first equilibrium result dealing with this type of discontinuity is in Szidarovszky and Okuguchi [1997] for the proper price function $\tilde{p}(y) = y^{-1}$. The problem of at 0 discontinuous industry revenue also is theoretically interesting as (assuming continuous cost functions) profit functions are discontinuous (at 0). This indicates, as far as we know, that no known equilibrium result guarantees the existence of an equilibrium for this case.

A recent result dealing with at 0 discontinuous industry revenue is the following result in Cornes and Sato [2016][d]:

Theorem 1. *Consider a homogeneous Cournot oligopoly without proper capacity constraints. Suppose the proper price function equals $\tilde{p}(y) = y^{-\alpha}$ with $\alpha > 0$ and each cost function c_i is convex with $c_i(0) = 0$ and twice continuously differentiable with $Dc_i > 0$. Then, with n the number of firms, the game has an equilibrium if and only if $n > \alpha$. In this case, the equilibrium even is unique.*

In Theorem 1, the proper price function $\tilde{p}$ has constant price flexibility (i.e., elasticity) $-\alpha$. In this chapter, we improve with Theorem 2 upon Theorem 1 by allowing for a proper price function with decreasing price flexibility.[e] An important class of (positive decreasing differentiable) proper price functions with decreasing price flexibility are the log-concave ones.

An interesting aspect in Theorem 1 is that there is an equilibrium if and only if the number n of firms is larger than α. Theorem 2(5a) contains a generalization of this result. This also sheds some new light on the well-known result that a monopoly with proper price function $\tilde{p}(y) = y^{-1}$ and a strictly increasing continuous cost function does not have an equilibrium.

Theorem 1 was obtained by the Selten–Szidarovszky technique. This technique is very powerful for games in strategic form with an aggregative structure. The technique transforms the n-dimensional fixed point problem for the joint best-reply correspondence into a 1-dimensional one. The technique is not limited to situations where each best-reply correspondence has a decreasing single-valued selection or where each best-reply correspondence has an increasing single-valued selection. Another advantage of this technique is that it is at the base of constructive algorithms for equilibrium calculations. (See Von Mouche [2016] and Jensen [2018] for a discussion of this technique and further references.)

---

[d]We note that Theorem 1 is not exactly stated as in Cornes and Sato [2016]. Therein it is for example stated that for $n \geq \alpha$ with $n \geq 2$ the game has a unique equilibrium. However, this statement is not true as Theorem 2(1,5a) below will show; also see the second example in Sec. 6. The present article repairs and also presents various omitted details.
[e]Theorem 2(1,5a) implies Theorem 1. In order to see this note that Proposition 1 shows that in Theorem 1 each firm (indeed) has an effective capacity constraint.

The proof of Theorem 2 is based on ideas in Cornes and Sato [2016] and Von Mouche and Yamazaki [2015]. It illustrates in an exemplary way the variant of the Selten–Szidarovszky technique based on virtual backward reply functions. Such functions find their origin in Szidarovszky and Yakowitz [1977].

The chapter is organized as follows. Section 2 fixes the setting and defines the virtual backward reply functions. Section 3 contains the generalization of Theorem 1 and Sec. 5 its proof. In Sec. 4, Proposition 3 presents a basic result for the Selten–Szidarovszky technique. An algorithm implied by Theorem 2 is presented and illustrated in Sec. 6. Section 7 contains some concluding remarks.

2. Setting and Various Fundamental Objects

2.1. *Setting*

A (*homogeneous*) *Cournot oligopoly* is a game in strategic form with a player set $N := \{1, \ldots, n\}$ whose elements are called *firms*. We assume that each firm i has a strategy set X_i that is a proper interval of $\mathbb{R}_+$ containing 0. This implies that $X_i = [0, m_i]$ or $X_i = [0, m_i[$ (with $m_i > 0$) or $X_i = \mathbb{R}_+$. The elements of X_i, i.e., the strategies, are also called *production levels*, and those of (the Minkowski-sum)

$$Y := \sum_{l \in N} X_l$$

as *industry production levels*. Given a production level profile $\mathbf{x}$, a firm i is called *active* (in $\mathbf{x}$) if $x_i \neq 0$ and *inactive* (in $\mathbf{x}$) if $x_i = 0$. And $\mathbf{x}$ is called *active* if at least one firm is active in $\mathbf{x}$; otherwise said if $\mathbf{x} \neq \mathbf{0}$. With $\mathbf{X} := X_1 \times \cdots \times X_n$, each firm i has a payoff function, called *profit function*, $f_i : \mathbf{X} \to \mathbb{R}$ given by[f]

$$f_i(\mathbf{x}) := p(x_N)x_i - c_i(x_i).$$

Here, $p : Y \to \mathbb{R}$ is called *price function* (also called *inverse demand function*) and $c_i : X_i \to \mathbb{R}$ is called *firm i's* (*net*) *cost function*. A Nash equilibrium of a Cournot oligopoly is called (*Cournot*) *equilibrium*. We denote the set of equilibria by

$$E.$$

Each conditional profit function of firm i (i.e., its profit function as a function of the own production level, given the strategies of the other firms) only depends on the aggregate of the strategies of the other firms. Having observed this, denote for $i \in N$ the Minkowski-sum $\sum_{l \in N \setminus \{i\}} X_l$ by[g]

$$T_i$$

and define for $i \in N$ and $z \in T_i$, the function $\tilde{f}_i^{(z)} : \mathbb{R}_+ \to \mathbb{R}$ by

$$\tilde{f}_i^{(z)}(x_i) = p(x_i + z)x_i - c_i(x_i). \tag{1}$$

---

[f]Notation: for $\mathbf{x} \in \mathbf{X}$ and $A \subseteq N$, let $x_A := \sum_{l \in A} x_l$.
[g]Note that for $n = 1$, we have $T_i = \{0\}$.

We refer to $\tilde{f}_i^{(z)}$ as a *reduced conditional profit function* of firm i and to $\tilde{f}_i^{(0)}$ as its *monopoly profit function*. Note that $f_i(\mathbf{x}) = \tilde{f}_i^{(x_N - x_i)}(x_i)$.

The profit functions (and therefore equilibria too) do not depend on the value of p at 0. Thus, with $Y^\oplus := Y \backslash \{0\}$, only the *proper price function*

$$\tilde{p} := p \upharpoonright Y^\oplus$$

matters.

When $X_i = [0, m_i]$ or $X_i = [0, m_i[$, firm i is said to have (m_i as) a *proper capacity constraint*. Also Y is a proper interval of $\mathbb{R}$ with $0 \in Y \subseteq \mathbb{R}_+$. We say $\bar{x}_i \in X_i \backslash \{0\}$ is an *effective capacity constraint* of firm i if every $x_i \in X_i$ with $x_i > \bar{x}_i$ is strongly dominated by some element in $[0, \bar{x}_i]$.

Proposition 1. *If firm i does not have a proper capacity constraint (i.e., $X_i = \mathbb{R}_+$) and $c_i(0) = 0$, then sufficient for this firm to have an effective capacity constraint is that the proper price function is decreasing and the firm makes for large enough production levels a negative monopoly profit.*

Proof. By assumption there exists $\bar{x}_i \in X_i \backslash \{0\}$ such that $\tilde{f}_i^{(0)}(x_i) < 0$ for every $x_i \in X_i$ with $x_i > \bar{x}_i$. Such an $\bar{x}_i$ is an effective capacity constraint as every $x_i \in X_i$ with $x_i > \bar{x}_i$ is strongly dominated by 0: $\tilde{f}_i^{(z)}(0) = -c_i(0) = 0 > \tilde{f}_i^{(0)}(x_i) = p(x_i)x_i - c_i(x_i) \geq p(x_i + z)x_i - c_i(x_i) = \tilde{f}_i^{(z)}(x_i)$. $\qquad\square$

The function $r_p : Y \to \mathbb{R}$ defined by

$$r_p(y) := p(y)y$$

is called the *industry revenue* function. When $\tilde{p}$ is positive and differentiable, then the function $\epsilon_{\tilde{p}} : Y^\oplus \to \mathbb{R}$, defined by[h]

$$\epsilon_{\tilde{p}}(y) := D\tilde{p}(y)\frac{y}{\tilde{p}(y)},$$

is called the *price flexibility* of $\tilde{p}$. If $\epsilon_{\tilde{p}}$ is decreasing, then

$$\overline{\epsilon_{\tilde{p}}} := \lim_{y \downarrow 0} \epsilon_{\tilde{p}}(y)$$

is well-defined as element of $\mathbb{R} \cup \{+\infty\}$.

If $\lim_{y \downarrow 0} \tilde{p}(y)$ exists in $\overline{\mathbb{R}} := \mathbb{R} \cup \{-\infty, +\infty\}$, then

$$\overline{p}(0) := \lim_{y \downarrow 0} \tilde{p}(y).$$

In this case, the derivative of the function r_p at 0 exists in $\overline{\mathbb{R}}$ and equals

$$Dr_p(0) = \overline{p}(0).$$

Also: if $\overline{p}(0) \in \mathbb{R}$, then r_p is continuous at 0.

---

[h] Using Euler's notation D for derivatives.

If $\bar{p}(0)$ is well-defined (for example if $\tilde{p}$ is decreasing) and c_i is differentiable, then the derivative of $\tilde{f}_i^{(0)}$ at 0 exists in $\overline{\mathbb{R}}$ and equals

$$D\tilde{f}_i^{(0)}(0) = \bar{p}(0) - Dc_i(0). \tag{2}$$

If $\bar{p}(0)$ is well-defined and every c_i is differentiable, then we define

$$N_> := \{k \in N \mid \bar{p}(0) > Dc_k(0)\}$$

and then for every $k \in N_>$ the formula (2) implies

$$\tilde{f}_k^{(0)}(x_k) > \tilde{f}_k^{(0)}(0) \text{ for some } x_k \in X_k. \tag{3}$$

Interpretation of (3) for the case where $c_k(0) = 0$: firm $k \in N_>$ can make a positive monopoly profit.

Below, if not otherwise stated, we consider a Cournot oligopoly.

2.2. *Marginal reductions*

For $i \in N$, define the correspondence[i] $\tilde{R}_i : T_i \multimap X_i$ by

$$\tilde{R}_i(z) := \operatorname{argmax} \tilde{f}_i^{(z)}. \tag{4}$$

We refer to $\tilde{R}_i$ as a *reduced best reply correspondence* of firm i.

If firm i has an effective capacity constraint $\bar{x}_i$, then for every $z \in T_i$ it is easy to see that

$$\tilde{R}_i(z) = \operatorname{argmax} \tilde{f}_i^{(z)} \upharpoonright [0, \bar{x}_i]. \tag{5}$$

If $\tilde{p}$ and c_i are differentiable, then we define the function $\tilde{t}_i : X_i \times Y^\oplus \to \mathbb{R}$ by

$$\tilde{t}_i(x_i, y) := D\tilde{p}(y)x_i + \tilde{p}(y) - Dc_i(x_i).$$

Then

$$D_1\tilde{t}_i(x_i, y) = D\tilde{p}(y) - D^2c_i(x_i)$$

and for all $x_i \in X_i$ and $z \in T_i$ with $x_i + z \neq 0$

$$D\tilde{f}_i^{(z)}(x_i) = \tilde{t}_i(x_i, x_i + z). \tag{6}$$

If $\tilde{p}$ and c_i are two times differentiable, then for such x_i and z also

$$D^2\tilde{f}_i^{(z)}(x_i) = (D_1 + D_2)\tilde{t}_i(x_i, x_i + z). \tag{7}$$

Following Folmer and von Mouche [2004], we refer to the $\tilde{t}_i$ as *marginal reductions*.

Using the notation in footnote f, define the *equilibrium aggregator* as the function $\sigma : E \to \mathbb{R}$ given by

$$\sigma(\mathbf{e}) := e_N.$$

Proposition 2. *If each $\tilde{t}_i$ is strictly decreasing in its first variable, then the equilibrium aggregator σ is injective.*

---

[i]For correspondences the symbol $\multimap$ will be used.

Proof. By contradiction suppose $\mathbf{a}, \mathbf{b} \in E$ with $\mathbf{a} \neq \mathbf{b}$ and $\sigma(\mathbf{a}) = \sigma(\mathbf{b}) =: y$. This implies $y \neq 0$. Fix $i \in N$ with $b_i > a_i$. As $\mathbf{a}, \mathbf{b} \in E$, we have $\tilde{t}_i(b_i, b_N) = 0 \geq \tilde{t}_i(a_i, a_N)$, so $\tilde{t}_i(b_i, y) \geq \tilde{t}_i(a_i, y)$. But, by assumption, $\tilde{t}_i(b_i, y) < \tilde{t}_i(a_i, y)$. $\qquad\square$

Assuming that $\tilde{t}_i$ is strictly decreasing in its first variable, substantially simplifies the analysis. In Theorem 2 it even holds that $D_1\tilde{t}_i < 0$ (see subsection 5.1).

2.3. *Virtual backward reply functions*

Given a Cournot oligopoly the *backward reply correspondence for firm i* is the correspondence $B_i : Y \multimap X_i$

$$B_i(y) := \{x_i \in X_i \mid x_i \leq y \wedge x_i \in \tilde{R}_i(y - x_i)\}.$$

In the case where the marginal reductions $\tilde{t}_i$ are well-defined working with so-called virtual backward reply correspondences has advantages.[j] In order to define these correspondences, we suppose in this subsection that

- there are no proper capacity constraints, i.e., $X_i = \mathbb{R}_+ (i \in N)$;
- the proper price function $\tilde{p}$ is decreasing and differentiable and each cost function c_i is differentiable;
- for every $y \in Y^\oplus$ each function $\tilde{t}_i(\cdot, y) : X_i \to \mathbb{R}$ is strictly decreasing.

Now define, as follows, for every $i \in N$ a function

$$\tilde{b}_i : Y^\oplus \to X_i.$$

Well, $\tilde{b}_i(y) := 0 (i \notin N_>)$. And for $i \in N_>$, let

$$\tilde{Y}_i^{(\mathrm{ess})} := \{y \in Y^\oplus \mid \text{there exists an } x_i \in X_i \text{ with } \tilde{t}_i(x_i, y) = 0\}$$

and define $\tilde{b}_i$ as follows:

— if $y \in \tilde{Y}_i^{(\mathrm{ess})}$, then with $x_i \in X_i$ such that $\tilde{t}_i(x_i, y) = 0,$[k] let $\tilde{b}_i(y) := x_i$;

— if $y \notin \tilde{Y}_i^{(\mathrm{ess})}$, then $\tilde{b}_i(y) := 0$. We refer to the $\tilde{b}_i$ as *virtual backward reply function* of firm i.[l] Also let

$$\tilde{b} := \sum_{k \in N} \tilde{b}_k.$$

So for $i \in N_>$ and $y \in Y^\oplus$, we have $\tilde{b}_i(y) > 0 \Rightarrow y \in \tilde{Y}_i^{(\mathrm{ess})} \Rightarrow \tilde{t}_i(\tilde{b}_i(y), y) = 0$. Finally, for $i \in N$, define the *virtual share function* $s_i : Y^\oplus \to \mathbb{R}$ by

$$s_i := \frac{\tilde{b}_i(y)}{y} \quad \text{and let} \quad s := \sum_{i \in N} s_i.$$

---

[j] For example, mastering the restriction $x_i \leq y$ in analyzing B_i gives serious technical difficulties; but for $\tilde{b}_i$ there is not such a restriction.

[k] Note that the strict decreasingness of $\tilde{t}_i(\cdot, y)$ makes that such an x_i is unique.

[l] Our assumptions make that the virtual backward reply correspondences are single-valued, and so can be interpreted as functions.

Although we assumed that there are no proper capacity constraints, $\tilde{b}_i$ and s_i are in the above also well-defined in the case firm i has a proper capacity constraint m_i. However, for this case, the definition of these objects should be modified in order to have a result similar to the below Proposition 3.

3. Main Result

The next theorem is our main result.

Theorem 2. *Consider a homogeneous Cournot oligopoly where*

- *no firm has a proper capacity constraint;*
- *every cost function c_i is increasing and twice continuously differentiable with $D^2 c_i \geq 0$;*
- *the proper price function $\tilde{p}$ is positive, decreasing, twice continuously differentiable with decreasing flexibility $\epsilon_{\tilde{p}}$.*

In addition, with $\overline{p}(0) = \lim_{y\downarrow 0} \tilde{p}(y) \in \,]0, +\infty]$, assume that at least one of the following properties hold:

- *$D\epsilon_{\tilde{p}} < 0$;*
- *$D^2 c_i > 0 (i \in N)$ and $\lim_{x_i \to \infty} Dc_i(x_i) > \overline{p}(0)(i \in N)$;*
- *$\tilde{p}$ is strictly decreasing and every c_i is strictly increasing.*

Then, with $N_> = \{k \in N \,|\, \overline{p}(0) > Dc_k(0)\}$ and with $\tilde{b}_i$ the virtual backward reply function of firm i

(1) *if $N_> \neq \emptyset$, then each equilibrium is active;*
(2) *in each equilibrium each firm $i \in N \backslash N_>$ is inactive;*
(3) *$\#E \leq 1$;*
(4) *$E \backslash \{\mathbf{0}\} = \{(\tilde{b}_1(y), \ldots, \tilde{b}_n(y)) \,|\, y \in \mathrm{fix}(\tilde{b})\}$.*

Further assume each firm in $N_>$ has an effective capacity constraint. Then with $\overline{\epsilon_{\tilde{p}}} = \lim_{y\downarrow 0} \epsilon_{\tilde{p}}(y) \in \,]-\infty, 0]$

(5a) *if $\overline{\epsilon_{\tilde{p}}} < 0$, then $n > -\overline{\epsilon_{\tilde{p}}} \Leftrightarrow [\#E = 1$ and each equilibrium is active];*
(5b) *if $\overline{\epsilon_{\tilde{p}}} = 0$ and $N_> \neq \emptyset$, then $\#E = 1$ and each equilibrium is active.*

Remarks:[m]

(1) Industry revenue r_p is continuous on $\mathbb{R}_{++}$, lower-semi-continuous at 0, but may be not upper-semi-continuous at 0. If $\overline{p}(0) < +\infty$, then r_p is continuous.
(2) As $D\tilde{p} \leq 0$, we have $\epsilon_{\tilde{p}} \leq 0$. If $\tilde{p}$ is strictly decreasing or $D\epsilon_{\tilde{p}} < 0$, then $D\tilde{p} < 0$ and therefore $\epsilon_{\tilde{p}} < 0$.

---

[m] Assuming the first three bullets hold for remarks 1–6 and that also at least one of the additional bullets holds for remarks 7–9.

(3) p is differentiable $\Rightarrow \bar{\epsilon}_{\tilde{p}} = 0$.[n]

(4) One easily verifies the identity $\tilde{p}(y) = \tilde{p}(1)e^{\int_y^1 -\frac{\epsilon_{\tilde{p}}(\xi)}{\xi}d\xi}$. It implies: if $\bar{\epsilon}_{\tilde{p}} < 0$, then $\bar{p}(0) = +\infty$ and thus $N_> = N$. And also: if $\bar{\epsilon}_{\tilde{p}} > -1$, then r_p is continuous.

(5) If (as may be in the fifth bullet) $\tilde{p}$ is not strictly decreasing, then there exists $y_1 \in Y^\oplus$ such that $\tilde{p}(y) = \bar{p}(0)(0 < y \leq y_1)$ and thus $\tilde{p}$ allows for a differentiable p.

(6) Each reduced conditional profit function $\tilde{f}_i^{(z)}$ (see (1)) is twice continuously differentiable for $z \neq 0$. Each $\tilde{f}_i^{(0)}$ is twice continuously differentiable on $X_i \backslash \{0\}$ and may be discontinuous at 0. Also see (2).

(7) Step 2 in Sec. 5 shows that each $\tilde{f}_i^{(z)}$ is strictly pseudo-concave[o] for $z \neq 0$ and that each $\tilde{f}_i^{(0)}$ is strictly pseudo-concave on $X_i \backslash \{0\}$.

(8) In Step 4a, in Sec. 5, we shall see that for $i \in N_>$ the reduced best-reply correspondence $\tilde{R}_i$ (see (4)) is at most single-valued and that in the case where firm i has an effective capacity constraint it is single-valued on $T_i \backslash \{0\}$. (Also see Lemma 1.)

(9) The second example in Sec. 6 shows that $\tilde{b}_i$ has not to be monotone.

4. The Selten–Szidarovszky Technique: A Basic Result

In this section, we suppose that the same assumptions as in subsection 2.3 (stated in the three bullets there) hold.

The following proposition provides a basic result of the Selten–Szidarovszky technique. Its proof shows how pseudo-concavity of reduced conditional profit functions $\tilde{f}_i^{(z)}$ is going to play a role in the proof of the main theorem.

Lemma 1. *Further assume* $i \notin N_>$ *and* $z \in T_i$. *Then* $D\tilde{f}_i^{(z)}(0) \leq 0, D\tilde{f}_i^{(z)}(x_i) < 0$ $(x_i \neq 0)$ *and* $\tilde{R}_i(z) \subseteq \{0\}$.

Proof. First statement: for $z = 0$ this follows by (2). Now suppose $z \neq 0$. Then with (6), as $\tilde{p}$ is decreasing, $D\tilde{f}_i^{(z)}(0) = \tilde{t}_i(0, z) = \tilde{p}(z) - Dc_i(0) \leq \bar{p}(0) - Dc_i(0) \leq 0$.

Second statement: we obtain with (6), as $\tilde{p}$ is decreasing, $D\tilde{f}_i^{(z)}(x_i) = \tilde{t}_i(x_i, x_i + z) < \tilde{t}_i(0, x_i + z) = \tilde{p}(x_i + z) - Dc_i(0) \leq \bar{p}(0) - Dc_i(0) \leq 0$.

Third statement: by contradiction suppose $x_i \in \tilde{R}_i(z)$ with $x_i \neq 0$. As $\tilde{f}_i^{(z)}$ is differentiable on $X_i \backslash \{0\}$ and x_i is an interior point of the domain of $\tilde{f}_i^{(z)}$, by Fermat's theorem gives $D\tilde{f}_i^{(z)}(x_i) = 0$. But, by the second statement, $D\tilde{f}_i^{(z)}(x_i) < 0$. $\square$

---

[n] Indeed: with $\epsilon_p : Y \to \mathbb{R}$ the price flexibility of p, we have that $\epsilon_p(0) = 0$, that ϵ_p is decreasing, and that ϵ_p is continuous on $Y^\oplus$. We now show, by contradiction that ϵ_p is continuous at 0 and then, as desired $\bar{\epsilon}_{\tilde{p}} = \lim_{y \downarrow 0} \epsilon_{\tilde{p}}(y) = \lim_{y \downarrow 0} \epsilon_p(y) = \epsilon_p(0) = 0$. So suppose ϵ_p is not continuous at 0. As ϵ_p is decreasing, there exists $A > 0$ such that $yDp(y)/p(y) < -A(y > 0)$. So $Dp(y) < -Ap(y)/y < -Ap(1)/y(0 < y < 1)$. This implies that for $y > 0$ small enough $Dp(y) < Dp(0) - 137$. This is a contradiction with (as p is differentiable) Dp being a Darboux function.
[o] Remember: A differentiable function $g : I \to \mathbb{R}$ where I is a proper real interval is (strictly) pseudo-concave if for all $x, y \in I$: $Dg(x)(y - x) \leq 0 \Rightarrow g(y)(<) \leq g(x)$.

For $i \in N_>$ let

$$\tilde{Y}_i^{(\text{ess}+)} := \{y \in Y^\oplus \mid \tilde{b}_i(y) > 0\}.$$

Proposition 3. *Further assume*

(I) *for $i \in N$ and $z \in T_i \backslash \{0\}$ the function $\tilde{f}_i^{(z)}$ is pseudo-concave and for $i \in N_>$ the function $\tilde{f}_i^{(0)}$ is pseudo-concave on $X_i \backslash \{0\}$;*

(II) *for $i \in N_>$: $\tilde{Y}_i^{(\text{ess}+)} = \{y \in Y^\oplus \mid \tilde{t}_i(0, y) > 0\}$.*

Then

(1) *If $y \in \text{fix}(\tilde{b})$, then $(\tilde{b}_1(y), \ldots, \tilde{b}_n(y)) \in E \backslash \{\mathbf{0}\}$.*
(2) *If $\mathbf{e} \in E \backslash \{\mathbf{0}\}$, then $e_i = \tilde{b}_i(e_N)(i \in N)$.*
(3) *$E \backslash \{\mathbf{0}\} = \{(\tilde{b}_1(y), \ldots, \tilde{b}_n(y)) \mid y \in \text{fix}(\tilde{b})\}$.*
(4) *$\sigma(E \backslash \{\mathbf{0}\}) = \text{fix}(\tilde{b}) = \{y \in Y^\oplus \mid s(y) = 1\}$.*
(5) *$\#(E \backslash \{\mathbf{0}\}) = \#\text{fix}(\tilde{b}) = \#\{y \in Y^\oplus \mid s(y) = 1\}$.*

Proof. Suppose $y \in \text{fix}(\tilde{b})$. So $y \neq 0$ Let $e_i = \tilde{b}_i(y)(i \in N)$. We have $e_N = \sum_{k \in N} e_k = \sum_{k \in N} \tilde{b}_k(y) = \tilde{b}(y) = y \neq 0$. Thus, $(e_1, \ldots, e_n) \neq \mathbf{0}$. In order to prove that $\mathbf{e} \in E$, we fix $i \in N$ and show that e_i is a maximizer of $\tilde{f}_i^{(e_N \backslash \{i\})}$. We distinguish between various cases.

(1) First consider the case where $e_i = 0$: now $e_i = \tilde{b}_i(y) = 0$. As $e_N \neq 0$, also $e_{N \backslash \{i\}} \neq 0$. By, Assumption II, $i \notin N_>$ or $[i \in N_>$ and $t_i(0, e_{N \backslash \{i\}}) \leq 0]$. If $i \notin N_>$, then by Lemma 1, $D\tilde{f}_i^{(e_N \backslash \{i\})}(e_i) \leq 0$ and in the other case also $D\tilde{f}_i^{(e_N \backslash \{i\})}(e_i) = \tilde{t}_i(e_i, e_N) = \tilde{t}_i(0, e_N) \leq 0$. As, by Assumption I, the function $\tilde{f}_i^{(e_N \backslash \{i\})}$ is pseudo-concave, $e_i = 0$ is a maximizer of this function.

Further suppose $e_i \neq 0$. As $e_i = \tilde{b}_i(y)$, this implies $i \in N_>$.

Case $e_{N \backslash \{i\}} \neq 0$. Noting that $0 < e_i = \tilde{b}_i(e_N)$, we have $\tilde{t}_i(\tilde{b}_i(e_N), e_N) = 0$. So $\tilde{t}_i(e_i, e_N) = 0$. Now $D\tilde{f}_i^{(e_N \backslash \{i\})}(e_i) = \tilde{t}_i(e_i, e_N) = 0$. As $\tilde{f}_i^{(e_N \backslash \{i\})}$ is pseudo-concave, e_i is a maximizer of this function.

Case $e_{N \backslash \{i\}} = 0$: noting that $e_i = \tilde{b}_i(e_N) > 0$, we have $\tilde{t}_i(e_i, e_N) = \tilde{t}_i(\tilde{b}_i(e_N), e_N) = 0$. Now $D\tilde{f}_i^{(0)}(e_i) = D\tilde{f}_i^{(e_N \backslash \{i\})}(e_i) = \tilde{t}_i(e_i, e_N) = 0$. As $\tilde{f}_i^{(0)}$ is pseudo-concave on $X_i \backslash \{0\}$, e_i is a maximizer of $\tilde{f}_i^{(0)} \upharpoonright X_i \backslash \{0\}$. It follows that e_i also is a maximizer of $\tilde{f}_i^{(0)}$. Indeed: if not, then there exists $a_i \in X_i$ with $\tilde{f}_i^{(0)}(a_i) > \tilde{f}_i^{(0)}(e_i)$. It then follows that $a_i = 0$ and $\tilde{f}_i^{(0)}(0) > \tilde{f}_i^{(0)}(e_i)$. Also now 0 is a maximizer of $\tilde{f}_i^{(0)}$. Thus, $\tilde{f}_i^{(0)}(x_i) \leq \tilde{f}_i^{(0)}(0)(x_i \in X_i)$ which is a contradiction with (3).

(2) Suppose $\mathbf{e} \in E \backslash \{\mathbf{0}\}$. We have $e_i \in \tilde{R}_i(e_{N \backslash \{i\}})(i \in N)$. If $i \notin N_>$, then $e_i = 0$ by Lemma 1. Further suppose $i \in N_>$.

Case where $e_i > 0$: as $\mathbf{e} \in E$ and $X_i = \mathbb{R}_+$, we have $\tilde{t}_i(e_i, e_N) = D\tilde{f}_i^{(e_N \backslash \{i\})}(e_i) = 0$ and therefore $e_i = \tilde{b}_i(e_N)$.

Case where $e_i = 0$: as $\mathbf{e} \in E$ we have $\tilde{t}_i(e_i, e_N) = D\tilde{f}_i^{(e_N \backslash \{i\})}(e_i) \leq 0$. Thus, by Assumption II, $\tilde{b}_i(e_N) = 0 = e_i$.

(3) '⊇': by part 1. '⊆': by part 2.

(4) The second equality is clear. Now we prove the first one.
'⊇': by part 1. '⊆': suppose $y = \sigma(\mathbf{e})$ with $\mathbf{e} \in E \backslash \{\mathbf{0}\}$. Part 2 gives $y = \sum_{i \in N} e_i = \sum_{i \in N} \tilde{b}_i(e_N) = \tilde{b}(e_N) = \tilde{b}(y)$. Thus $y \in \text{fix}(\tilde{b})$.

(5) By part 4 and Proposition 2.   □

For the analysis of the $\tilde{b}_i$ it is important to know where $\tilde{b}_i$ is positive. Concerning this, we present in the next lemma an assumption that guarantees that each $\tilde{Y}_i^{(\text{ess}+)}$ is a proper interval of the form $]0, \lambda_i[$. This substantially will simplify the analysis.

Lemma 2. *Suppose $i \in N_>$. Further assume that for every $y \in Y^{\oplus}$ there exists $x_i \in X_i$ such that $\tilde{t}_i(x_i, y) < 0$.*

(1) $\tilde{Y}_i^{(\text{ess})} = \{y \in Y^{\oplus} \mid \tilde{t}_i(0, y) \geq 0\}$ *and* $\tilde{Y}_i^{(\text{ess}+)} = \{y \in Y^{\oplus} \mid \tilde{t}_i(0, y) > 0\}$.

(2) *There exists $\lambda_i \in]0, +\infty]$ such that $\tilde{Y}_i^{(\text{ess}+)} =]0, \lambda_i[$. Also $\tilde{Y}_i^{(\text{ess})} = \mathbb{R}_{++}$ or $\tilde{Y}_i^{(\text{ess})} =]0, \tau_i]$ with $\tau_i \in [\lambda_i, +\infty[$.*

(3) *In the case where $\tilde{Y}_i^{(\text{ess})} =]0, \tau_i]$, we have $\tilde{b}_i(y) = 0 (y \geq \tau_i)$.*

Proof. (1) First statement: if $y \in \tilde{Y}_i^{(\text{ess})}$, then as $\tilde{t}_i$ is strictly decreasing in its first variable, we have $\tilde{t}_i(0, y) \geq \tilde{t}_i(\tilde{b}_i(y), y) = 0$. Now suppose $y \in Y^{\oplus}$ with $\tilde{t}_i(0, y) \geq 0$. We note that $\tilde{t}_i(\cdot, y)$ is continuous (see subsection 5.1). As $\tilde{t}_i(\cdot, y)$ assumes, by assumption, a nonpositive value, it follows that there exists, $x_i \in X_i$ with $\tilde{t}_i(x_i, y) = 0$; thus $y \in \tilde{Y}_i^{(\text{ess})}$.

Second statement. First suppose $y \in \tilde{Y}_i^{(\text{ess}+)}$. Now $\tilde{b}_i(y) > 0$. Therefore, $\tilde{t}_i(\tilde{b}_i(y), y) = 0$. It follows that $\tilde{t}_i(0, y) > \tilde{t}_i(\tilde{b}_i(y), y) = 0$. Next, suppose $\tilde{t}_i(0, y) > 0$. We prove by contradiction that $y \in \tilde{Y}_i^{(\text{ess}+)}$. So suppose $y \notin \tilde{Y}_i^{(\text{ess}+)}$, i.e., $\tilde{b}_i(y) = 0$. By assumption there exists $x_i \in X_i$ such that $\tilde{t}_i(x_i, y) < 0$. Of course, $x_i > 0$. As $\tilde{t}_i(\cdot, y)$ is a derivative, it is a Darboux function. This implies the existence of $x_i' \in]0, x_i[$ such that $\tilde{t}_i(x_i', y) = 0$. So $\tilde{b}_i(y) = x_i' > 0$, a contradiction.

(2) By part 1, $\tilde{Y}_i^{(\text{ess})} = \{y \in Y^{\oplus} \mid \tilde{p}(y) \geq Dc_i(0)\}$ and $\tilde{Y}_i^{(\text{ess}+)} = \{y \in Y^{\oplus} \mid \tilde{p}(y) > Dc_i(0)\}$. As $\tilde{p}$ is decreasing, $\tilde{Y}_i^{(\text{ess})}$ and $\tilde{Y}_i^{(\text{ess}+)}$ are intervals. As $i \in N_>$, we have $\tilde{p}(0) > Dc_i(0)$. As $\tilde{p}$ is continuous, this implies that $y \in \tilde{Y}_i^{(\text{ess}+)}$ for $y > 0$ small enough. Now, as $\tilde{p}$ is continuous, all statements follow.

(3) This is clear if $y > \tau_i$. As, by part 1, $\tilde{t}_i(0, y) < 0 (y > \tau_i)$ and $\tilde{t}_i(0, y) \geq 0 (0 < y < \tau_i)$, the continuity of $\tilde{t}_i(0, .)$ implies $\tilde{t}_i(0, \tau_i) = 0$ and thus $\tilde{b}_i(\tau_i) = 0$.   □

5. Proof of Theorem 2

5.1. *Observations on assumptions*

The general assumptions in Theorem 2 imply the following.

- $D_1 \tilde{t}_i < 0$ ('first Fisher–Hahn condition'). However, $D_2 \tilde{t}_i \leq 0$ ('second Fisher–Hahn condition') may not hold. For example, it does not hold in Theorem 1.

- As c_i is differentiable and convex, c_i is continuously differentiable and thus $\tilde{t}_i$ is continuous in its first variable.
- $\lim_{x_i \to \infty} Dc_i(x_i)$ exists in $\mathbb{R} \cup \{+\infty\}$. If c_i is strictly increasing, then $Dc_i(x_i) > 0(x_i \neq 0)$.
- For every $y \in Y^\oplus$ there exists $x_i \in X_i$ such that $\tilde{t}_i(x_i, y) < 0$ and therefore Lemma 2 applies.[P]

Below we shall use these results without explicitly referring to them.

5.2. *Proof*

We now deliver the proof of Theorem 2 in 14 steps. In these steps, we prove that Proposition 3 can be applied and further analyze the virtual backward reply functions $\tilde{b}_i$. From now on we assume in this subsection that the general assumptions in Theorem 2 hold and that in parts 5a and 5b the extra assumptions hold.

Step 0. Theorem 2(1,2) holds

Proof of Theorem 2(1). Suppose $k \in N_>$. By (3), there exists $x_k \in X_k \backslash \{0\}$ with $\tilde{f}_k^{(0)}(x_k) > \tilde{f}_k^{(0)}(0)$. Therefore, $\mathbf{0}$ cannot be an equilibrium. $\qquad\square$

Proof of Theorem 2(2). Suppose $e \in E$ and $i \notin N_>$. As $e \in E$, we have $e_i \in \tilde{R}_i(\sum_{l \neq i} e_l)$. Lemma 1 applies and implies $e_i = 0$. $\qquad\square$

Step 1a. $D^2\tilde{p} \cdot \mathrm{Id} + D\tilde{p} \leq \mathrm{Id} \cdot \frac{(D\tilde{p})^2}{\tilde{p}}$.[q] And the strict inequality holds if $D\epsilon_{\tilde{p}} < 0$.

Proof. $D\epsilon_{\tilde{p}} = D(D\tilde{p} \cdot \frac{\mathrm{Id}}{\tilde{p}}) = \frac{D^2\tilde{p} \cdot \mathrm{Id} \cdot \tilde{p} + D\tilde{p} \cdot \tilde{p} - \mathrm{Id} \cdot (D\tilde{p})^2}{\tilde{p}^2}$. Thus, the desired result follows as $[D\epsilon_{\tilde{p}} \leq 0$ or $D\epsilon_{\tilde{p}} < 0]$ and $\tilde{p} > 0$. $\qquad\square$

Step 1b. $\tilde{t}_i(x_i, y) = 0 \Rightarrow (x_i D_1 + y D_2)\tilde{t}_i(x_i, y) \leq \epsilon_{\tilde{p}}(y)Dc_i(x_i) - x_i D^2 c_i(x_i)$. And the strict inequality holds if $D\epsilon_{\tilde{p}} < 0$ and $x_i \neq 0$.

Proof. Suppose $\tilde{t}_i(x_i, y) = 0$. With Step 1a, we obtain

$$(x_i D_1 + y D_2)\tilde{t}_i(x_i, y) = x_i(D\tilde{p}(y) - D^2 c_i(x_i)) + y(D^2\tilde{p}(y)x_i + D\tilde{p}(y))$$

$$= x_i(y D^2\tilde{p}(y) + D\tilde{p}(y)) + y D\tilde{p}(y) - x_i D^2 c_i(x_i)$$

$$\leq x_i y \frac{(D\tilde{p}(y))^2}{\tilde{p}(y)} + y D\tilde{p}(y) - x_i D^2 c_i(x_i)$$

$$= \frac{y D\tilde{p}(y)}{\tilde{p}(y)}(x_i D\tilde{p}(y) + \tilde{p}(y)) - x_i D^2 c_i(x_i)$$

[P]Indeed: If the fourth or sixth bullet in Theorem 2 holds, then $D\tilde{p} < 0$ by Remark 2, and if the fifth bullet holds, then note that $\tilde{t}_i(x_i, y) = D\tilde{p}(y)x_i + \tilde{p}(y) - Dc_i(x_i) \leq \tilde{p}(y) - Dc_i(x_i) \leq \overline{p}(0) - Dc_i(x_i)$.
[q]'Id' denotes the identity mapping.

$$= \frac{yD\tilde{p}(y)}{\tilde{p}(y)}(\tilde{t}_i(x_i, y) + Dc_i(x_i)) - x_iD^2c_i(x_i)$$

$$= \epsilon_{\tilde{p}}(y)Dc_i(x_i) - x_iD^2c_i(x_i).$$

Further, the inequality in this calculation is strict if $D\epsilon_{\tilde{p}} < 0$ and $x_i \neq 0$.   □

Step 1c. For all $i \in N$, $x_i \in X_i$ and $y \in Y^{\oplus}$

$$\tilde{t}_i(x_i, y) = 0 \Rightarrow (x_iD_1 + yD_2)\tilde{t}_i(x_i, y) \begin{cases} < 0, & \text{if } x_i \neq 0, \\ \leq 0, & \text{if } x_i = 0. \end{cases}$$

Proof. By Step 1b. If $x_i = 0$, then note that $\epsilon_{\tilde{p}}(y) \leq 0, Dc_i(0) \geq 0$ and $D^2c_i(x_i) \geq 0$. Now suppose $x_i > 0$. If the sixth bullet in Theorem 2 holds, then note that $Dc_i(x_i) > 0$, $\epsilon_{\tilde{p}} < 0$ and $D^2c_i(x_i) \geq 0$. If the fifth bullet holds, then note $Dc_i(x_i) \geq 0, \epsilon_{\tilde{p}} \leq 0$ and $D^2c_i(x_i) > 0$. And if the fourth bullet holds, then note that the strict inequality holds in Step 1b.   □

Step 2. Suppose $i \in N$. For $z \in T_i \backslash \{0\}$ the reduced conditional profit function $\tilde{f}_i^{(z)}$ is strictly pseudo-concave. And $\tilde{f}_i^{(0)}$ is strictly pseudo-concave on $X_i \backslash \{0\}$.

Proof. We shall prove this by using the following general result for a two times continuously differentiable real-valued function f on a proper real interval [Truchon, 1987, Théorème 9.2.6]: if for all $x \in I$ the implication $Df(x) = 0 \Rightarrow D^2f(x) < 0$ holds, then f is strictly pseudo-concave.

First statement: Suppose $x_i \in X_i$ and $z \in T_i \backslash \{0\}$ are such that $D\tilde{f}_i^{(z)}(x_i) = 0$. So, writing $y = x_i + z$, we have $\tilde{t}_i(x_i, y) = 0$. We have to prove that $D^2\tilde{f}_i^{(z)}(x_i) < 0$. As, by (7),

$$D^2\tilde{f}_i^{(z)}(x_i) = \frac{(zD_1 + x_iD_1 + (x_i + z)D_2)\tilde{t}_i(x_i, x_i + z)}{x_i + z},$$

we have to prove that $(zD_1 + x_iD_1 + yD_2)\tilde{t}_i(x_i, y) < 0$. By virtue of the first Fisher–Hahn condition $zD_1\tilde{t}_i(x_i, y) < 0$ holds. Therefore, it is sufficient to prove that $(x_iD_1 + yD_2)\tilde{t}_i(x_i, y) \leq 0$. Well, this is guaranteed by Step 1c.

Second statement: Suppose $x_i \in X_i \backslash \{0\}$ is such that $D\tilde{f}_i^{(0)}(x_i) = 0$, i.e., $\tilde{t}_i(x_i, x_i) = 0$. We have to prove that $D^2\tilde{f}_i^{(0)}(x_i) < 0$, so by (7) that $\frac{1}{x_i}(x_iD_1 + x_iD_2)\tilde{t}_i(x_i, x_i) < 0$. Well, this again is guaranteed by Step 1c.   □

Step 3a. The assumptions in Proposition 3 hold and therefore also its statements.

Proof. By Step 2 and Lemma 2(1).   □

Step 3b. Theorem 2(4) holds

Proof. By Step 3a, Proposition 3(3) applies.   □

Step 4a. Suppose $i \in N_>$. Then: $\#\tilde{R}_i(z) \leq 1$. If firm i has an effective capacity constraint $\overline{x}_i$, then $\tilde{R}_i(0) \subseteq]0, \overline{x}_i]$ and for $z \in T_i \backslash \{0\}$ it holds that $\tilde{R}_i(z) \subseteq [0, \overline{x}_i]$ and $\#\tilde{R}_i(z) = 1$.

Proof of the first statement. First suppose $z = 0$. (3) implies that $0 \notin \tilde{R}_i(0)$. By contradiction now suppose that $\#\tilde{R}_i(0) \geq 2$. By Step 2, the function $\tilde{f}_i^{(0)} \lceil X_i \backslash \{0\}$ is strictly quasi-concave. Therefore this function has at most one maximizer. As $\tilde{f}_i^{(0)}$ has at least two maximizers, it follows that 0 is a maximizer of $\tilde{f}_i^{(0)}$. So $0 \in \tilde{R}_i(0)$, a contradiction. Next, suppose $z \neq 0$. Step 2 guarantees that $\tilde{f}_i^{(z)}$ is strictly quasi-concave and therefore $\#\tilde{R}_i(z) \leq 1$. □

Proof of the second statement. (5) implies $\tilde{R}_i(0) \subseteq [0, \overline{x}_i]$. (3) implies that $0 \notin \tilde{R}_i(0)$. Thus, $\tilde{R}_i(0) \subseteq]0, \overline{x}_i]$. □

Proof of the third statement. (5) implies $\tilde{R}_i(z) \subseteq [0, \overline{x}_i]$. □

Proof of the fourth statement. By (5), we have $\tilde{R}_i(z) = \operatorname{argmax} \tilde{f}_i^{(z)} \lceil [0, \overline{x}_i]$. As $z \neq 0$, the function $\tilde{f}_i^{(z)}$ is continuous. The Weierstrass' theorem implies $\#\tilde{R}_i(z) \geq 1$. With the first statement $\#\tilde{R}_i(z) = 1$ follows. □

Step 4b. Suppose $i \in N_>$. If firm i has an effective capacity constraint $\overline{x}_i$, then for $x_i > \overline{x}_i$ it holds that $D\tilde{f}_i^{(0)}(x_i) < 0$.

Proof. Let $g := \tilde{f}_i^{(0)} \lceil \mathbb{R}_{++}$. Note that g is differentiable, strictly pseudo-concave (Step 2) and that, by (3), g assumes a value greater than $\tilde{f}_i^{(0)}(0)$. We distinguish between two cases.

Case where $\tilde{R}_i(0) \neq \emptyset$. By Step 4a, $\tilde{f}_i^{(0)}$ has a unique maximizer, say c, and $c \neq 0$. By Fermat's theorem, $D\tilde{f}_i^{(0)}(c) = 0$; so $Dg(c) = 0$. As g is strictly pseudo-concave, we have $Dg(x_i) < 0$ for every $x_i > c$.[r] So $D\tilde{f}_i^{(0)}(x_i) < 0$ for every $x_i > c$. As $c \leq \overline{x}_i$, it also follows that $D\tilde{f}_i^{(0)}(x_i) < 0$ for every $x_i > \overline{x}_i$.

Case where $\tilde{R}_i(0) = \emptyset$. So $\tilde{f}_i^{(0)} : \mathbb{R}_+ \to \mathbb{R}$ does not have a maximizer. We now first prove that g also does not have a maximizer. Well, if q would be a maximizer of g, then as q is not a maximizer of $\tilde{f}_i^{(0)}$ we would have $g(q) < \tilde{f}_i^{(0)}(0)$. So then $g < \tilde{f}_i^{(0)}(0)$, a contradiction with the above. Next, we prove by contradiction that g is not strictly increasing: so suppose g is strictly increasing. Let $b > \overline{x}_i$. There exists $a \in [0, \overline{x}_i]$ with $\tilde{f}_i^{(0)}(a) > \tilde{f}_i^{(0)}(b)$. As g is strictly increasing, $a = 0$ follows. So $\tilde{f}_i^{(0)}(0) > g(b)$. As g is strictly increasing and b is arbitrary, it follows that $g < \tilde{f}_i^{(0)}(0)$, a contradiction. Now we are ready to prove the second case. Well, as g is strictly quasi-concave, there are two disjunct real intervals I_1, I_2 such that

---

[r]We use here: Suppose $h : I \to \mathbb{R}$ is a function on a proper real interval I and $c \in I$. If h is differentiable, strictly pseudo-concave and $Dh(c) = 0$, then $Dh(x) < 0$ for every $x \in I$ with $x > c$.

$\mathbb{R}_{++} = I_1 \cup I_2$, $I_1 \le I_2$, $g\restriction_{I_1}$ is strictly increasing and $g\restriction_{I_2}$ is strictly decreasing. As g is not strictly increasing, g is continuous and g does not have a maximizer, it follows that g is strictly decreasing. So $g : \mathbb{R}_{++} \to \mathbb{R}$ is a strictly decreasing differentiable function. As g also is strictly pseudo-concave, it follows that $Dg < 0$ and so the proof is complete. □

Step 4c. Suppose firm $i \in N_>$ has an effective capacity constraint $\overline{x}_i$ and $y > \overline{x}_i$. Then $\tilde{b}_i(y) < y$ and $[\tilde{b}_i(y) > 0 \Rightarrow \tilde{b}_i(y) \in \tilde{R}_i(y - \tilde{b}_i(y))]$.

Proof of the first statement. If $\tilde{b}_i(y) = 0$, then this statement trivially holds. Now suppose $\tilde{b}_i(y) > 0$. This implies $\tilde{t}_i(\tilde{b}_i(y), y) = 0$. By Step 4b, $D\tilde{f}_i^{(0)}(y) < 0$. So $\tilde{t}_i(y, y) = D\tilde{f}_i^{(0)}(y) < 0 = \tilde{t}_i(\tilde{b}_i(y), y)$. As $\tilde{t}_i$ is strictly decreasing in its first variable, $\tilde{b}_i(y) < y$ follows. □

Proof of the second statement. As $\tilde{b}_i(y) > 0$, $\tilde{t}_i(\tilde{b}_i(y), y) = 0$ holds. By the first statement $y - \tilde{b}_i(y) \in T_i \setminus \{0\}$. So we obtain $D\tilde{f}_i^{(y - \tilde{b}_i(y))}(\tilde{b}_i(y)) = \tilde{t}_i(\tilde{b}_i(y), y) = 0$. As $\tilde{f}_i^{(y - \tilde{b}_i(y))}$ is pseudo-concave, $\tilde{b}_i(y)$ is a maximizer of $\tilde{f}_i^{(y - \tilde{b}_i(y))}$, i.e., $\tilde{b}_i(y) \in \tilde{R}_i(y - \tilde{b}_i(y))$. □

Step 5. If firm $i \in N_>$ has an effective capacity constraint $\overline{x}_i$, then $\tilde{b}_i(y) \le \overline{x}_i$ for $y > \overline{x}_i$. If each firm $i \in N_>$ has an effective capacity constraint, then $\lim_{y \to \infty} s(y) = 0$.

Proof of the first statement. If $\tilde{b}_i(y) = 0$, then $\tilde{b}_i(y) \le \overline{x}_i$. Now suppose $\tilde{b}_i(y) > 0$. By Steps 4c and 4a, $\tilde{b}_i(y) \in \tilde{R}_i(y - \tilde{b}_i(y)) \subseteq [0, \overline{x}_i]$. □

Proof of the second statement. The statement holds if $N_> = \emptyset$. Now suppose $N_> \ne \emptyset$. Next note that, with $\overline{x} := \max_{l \in N_>} \overline{x}_l$, the first statement implies $s(y) \le (\sum_{l \in N_>} \overline{x}_l)/y$ for $y > \overline{x}$. □

Step 6. Every $\tilde{b}_i$ and s_i is continuous.

Proof. We are done if prove continuity of $\tilde{b}_i$. Well, if $i \notin N_>$, then $\tilde{b}_i = 0$. Further suppose $i \in N_>$. Consider the (restricted) function $\tilde{t}_i : X_i \times \tilde{Y}_i^{(\text{ess})} \to \mathbb{R}$. As $\tilde{p}$ and c_i are continuously differentiable, $\tilde{t}_i$ is continuous. For $y \in \tilde{Y}_i^{(\text{ess})}$ it holds that $\tilde{b}_i(y)$ is the unique element x_i of X_i with $\tilde{t}_i(x_i, y) = 0$. By Lemma 2(2), $\tilde{Y}_i^{(\text{ess})}$ is a proper interval of $\mathbb{R}$. As $D_1 \tilde{t}_i < 0$, we have $\tilde{t}_i(x_i, y) < 0$ for every $y \in \tilde{Y}_i^{(\text{ess})}$ and $x_i > \tilde{b}_i(y)$. Theorem 7.1 in Von Mouche and Yamazaki [2015] applies and implies that $\tilde{b}_i$ is continuous on $\tilde{Y}_i^{(\text{ess})}$. So the proof is complete if $\tilde{Y}_i^{(\text{ess})} = \mathbb{R}_{++}$. Now suppose $\tilde{Y}_i^{(\text{ess})} = \,]0, \tau_i]$. With Lemma 2(3) we see that $\tilde{b}_i$ is continuous. □

Step 7. For $i \in N_>$, $\tilde{b}_i$ is differentiable at every $y_0 \in \tilde{Y}_i^{(\text{ess}+)}$ and the formula $D\tilde{b}_i(y_0) = -\frac{D_2 \tilde{t}_i}{D_1 \tilde{t}_i}(\tilde{b}_i(y_0), y_0)$ holds.

Proof. Let $W_i := \mathbb{R}_{++} \times \tilde{Y}_i^{(\text{ess}+)}$. Note that $\tilde{b}_i(y_0) > 0$, that $0 = t_i(\tilde{b}_i(y_0), y_0)$ and that, by Lemma 2(2), W_i is open in $\mathbb{R}^2$. The function $\tilde{t}_i : W_i \to \mathbb{R}$ is continuously differentiable and, by the first Fisher–Hahn condition, we have $D_1\tilde{t}_i(\tilde{b}_i(y_0), y_0) \neq 0$. The implicit function theorem guarantees that there exists an open neighborhood U_i of $\tilde{b}_i(y_0)$ in $\mathbb{R}$, an open neighborhood V_i of y_0 in $\mathbb{R}$ such that $U_i \times V_i \subseteq W_i$ and a unique function $\Psi_i : V_i \to \mathbb{R}$ with $\Psi_i(V_i) \subseteq U_i$ such that

$$\{(\Psi_i(y), y) \mid y \in V_i\} = \{(x_i, y) \in U_i \times V_i \mid \tilde{t}_i(x_i, y) = 0\}.$$

In addition: This function Ψ_i is continuously differentiable. So $\tilde{t}_i(\Psi_i(y), y) = 0 (y \in V_i)$. By definition of $\tilde{b}_i$ it follows that $\tilde{b}_i = \Psi_i$ on V_i and so $\tilde{b}_i$ is differentiable at y_0. Differentiating the identity $\tilde{t}_i(\tilde{b}_i(y), y) = 0 (y \in V_i)$ gives $D\tilde{b}_i(y) = -\frac{D_2\tilde{t}_i}{D_1\tilde{t}_i}(\tilde{b}_i(y), y))$. □

Step 8. For all $i \in N_>$ and $y \in \tilde{Y}_i^{(\text{ess})}$ with $\epsilon_{\tilde{p}}(y) \neq 0$, we have

$$s_i(y) = \frac{-1}{\epsilon_{\tilde{p}}(y)}\left(1 - \frac{Dc_i(\tilde{b}_i(y))}{\tilde{p}(y)}\right) \leq \frac{-1}{\epsilon_{\tilde{p}}(y)}\left(1 - \frac{Dc_i(0)}{\overline{p}(0)}\right) \leq \frac{-1}{\epsilon_{\tilde{p}}(y)}.$$

Proof. We have $\tilde{t}_i(\tilde{b}_i(y), y) = 0$, so $D\tilde{p}(y)\tilde{b}_i(y) + \tilde{p}(y) - Dc_i(\tilde{b}_i(y)) = 0$. This in turn implies the desired results as $Dc_i(\tilde{b}_i(y)) \geq Dc_i(0)$, $\tilde{p}(y) \leq \overline{p}(0)$, $\epsilon_{\tilde{p}} < 0$ and $i \in N_>$. □

Step 9. For $i \in N_>$ and $y \in \tilde{Y}_i^{(\text{ess}+)}$ (using shortened notations)

$$Ds_i = -\frac{\tilde{b}_i \cdot D_1\tilde{t}_i + \text{Id} \cdot D_2\tilde{t}_i}{\text{Id}^2 \cdot D_1\tilde{t}_i}.$$

Proof. By Step 7, $\tilde{b}_i$ is differentiable at every $y \in \tilde{Y}_i^{(\text{ess}+)}$. It follows that $Ds_i = \frac{D\tilde{b}_i \cdot \text{Id} - \tilde{b}_i}{\text{Id}^2}$. With the formula in Step 7 this leads to the desired identity. □

Step 10. For $i \in N_>$ the function s_i is decreasing and continuous, the set $\{y \in Y^\oplus \mid s_i(y) > 0\}$ is an interval of the form $]0, \lambda_i[$ with $\lambda_i \in]0, +\infty]$ and $Ds_i < 0$ on this interval. The function $s = \sum_{i \in N} s_i$ is decreasing and continuous, $\{y \in Y^\oplus \mid s(y) > 0\}$ is not empty if and only if $N_> \neq \emptyset$ and in this case equals $]0, \max_{i \in N_>} \lambda_i[$ and s is strictly decreasing on this interval.

Proof of statement about s_i. By Step 6, s_i is continuous. As $\{y \in Y^\oplus \mid s_i(y) > 0\} = \tilde{Y}_i^{(\text{ess}+)}$, this set equals $]0, \lambda_i[$ by Lemma 2(2). By Step 9, for $y \in \tilde{Y}_i^{(\text{ess}+)}$

$$Ds_i(y) = \frac{\tilde{b}_i(y) \cdot D_1\tilde{t}_i(\tilde{b}_i(y), y) + y \cdot D_2\tilde{t}_i(\tilde{b}_i(y), y)}{-y^2 \cdot D_1\tilde{t}_i(\tilde{b}_i(y), y)}.$$

As the denominator of this fraction is positive and, by Step 1c, the nominator is negative, we have that $Ds_i < 0$ on $]0, \lambda_i[$. As $s_i > 0$ on $]0, \lambda_i[$ and $s_i(y) = 0 (y \geq \lambda_i)$, it follows that s_i is decreasing. □

Proof of statement about s. First note that $s = \sum_{i \in N_>} s_i$. So by the above s is decreasing and continuous. We have $\{y \in Y^\oplus \mid s(y) > 0\} = \bigcup_{i \in N_>} \tilde{Y}_i^{(\text{ess}+)} = \bigcup_{i \in N_>} \,]0, \lambda_i[$. So this set is not empty if and only if $N_> \neq \emptyset$ and in this case, this equals $]0, \max_{i \in N_>} \lambda_i[$. Of course, s_i is strictly decreasing on this interval. $\qquad\square$

Step 11. Theorem 2(3) holds

Proof. By contradiction suppose $\#E \geq 2$. Fix $\mathbf{a}, \mathbf{b} \in E$ with $\mathbf{a} \neq \mathbf{b}$. By Proposition 2, $a_N \neq b_N$. Theorem 2(1,2) implies $\mathbf{0} \notin E$. So, as by Step 3a Proposition 3(4) applies, we have $s(a_N) = s(b_N) = 1$. Thus, we have a contradiction with the strict monotonicity property of s in Step 10. $\qquad\square$

Step 12. For every $i \in N$, the limit

$$\bar{s}_i := \lim_{y \downarrow 0} s_i(y)$$

exists in $\mathbb{R}_+ \cup \{+\infty\}$. If $i \in N_>$, then $\bar{s}_i \in \mathbb{R}_{++} \cup \{+\infty\}$.

Proof. As s_i is decreasing by Step 10. $\qquad\square$

Step 13. (1) If $\overline{\epsilon_{\tilde{p}}} < 0$, then $\bar{s}_i = \frac{-1}{\overline{\epsilon_{\tilde{p}}}} \in \mathbb{R}_{++} (i \in N)$.
(2) If $\overline{\epsilon_{\tilde{p}}} = 0$ and $i \in N_>$, then $\bar{s}_i = +\infty$.

Proof. (1) Suppose $\overline{\epsilon_{\tilde{p}}} < 0$. By Remark 4, $N_> = N$. As $\epsilon_{\tilde{p}} \leq \overline{\epsilon_{\tilde{p}}} < 0$, we have by Step 8 for every $y \in \tilde{Y}_i^{(\text{ess})}$ that $0 \leq \tilde{b}_i(y) \leq \frac{-y}{\epsilon_{\tilde{p}}(y)} \leq \frac{-y}{\overline{\epsilon_{\tilde{p}}}}$. So $\lim_{y \downarrow 0} \tilde{b}_i(y) = 0$. Also by Step 8, $s_i(y) = \frac{-1}{\epsilon_{\tilde{p}}(y)}(1 - \frac{Dc_i(\tilde{b}_i(y))}{\tilde{p}(y)})$ for all $y \in \tilde{Y}_i^{(\text{ess})}$. As $\lim_{y \downarrow 0} \epsilon_{\tilde{p}}(y) = \overline{\epsilon_{\tilde{p}}} < 0$, $\lim_{y \downarrow 0} \tilde{p}(y) = \tilde{p}(0) = +\infty$, Dc_i is continuous, $\lim_{y \downarrow 0} \tilde{b}_i(y) = 0$ and $i \in N_>$, we obtain $\bar{s}_i = \frac{-1}{\overline{\epsilon_{\tilde{p}}}} \in \mathbb{R}_{++}$.

(2) Suppose $\overline{\epsilon_{\tilde{p}}} = 0$ and $i \in N_>$. First consider the case where $\tilde{p}$ strictly decreasing. This implies that $\epsilon_{\tilde{p}} < 0$. By contradiction suppose that $\bar{s}_i \neq +\infty$. By Step 13a, $\bar{s}_i \in \mathbb{R}_{++}$. It follows that $\lim_{y \downarrow 0} \tilde{b}_i(y) = \lim_{y \downarrow 0} y s_i(y) = 0$. By Lemma 2(2), we have for $y > 0$ small enough that $\tilde{b}_i(y) > 0$ and further by Step 8 that $s_i(y) = \frac{-1}{\epsilon_{\tilde{p}}(y)}(1 - \frac{Dc_i(\tilde{b}_i(y))}{\tilde{p}(y)})$ and therefore $\epsilon_{\tilde{p}}(y) = \frac{-1}{s_i(y)}(1 - \frac{Dc_i(\tilde{b}_i(y))}{\tilde{p}(y)})$. As $\bar{s}_i \in \mathbb{R}_{++}$ and $i \in N_>$, we obtain $\overline{\epsilon_{\tilde{p}}} = \lim_{y \downarrow 0} \epsilon_{\tilde{p}}(y) = \frac{-1}{\bar{s}_i}(1 - \frac{Dc_i(0)}{\tilde{p}(0)}) < 0$, which is a contradiction.

Next, suppose that $\tilde{p}$ is not strictly decreasing. We already know that there exists $y_1 \in Y^\oplus$ such that $\tilde{p} = \tilde{p}(0)(0 < y \leq y_1)$. So for such y we have $\tilde{t}_i(x_i, y) = \tilde{p}(0) - Dc_i(x_i)$. Also $\tilde{t}_i(0, y) = \tilde{p}(0) - Dc_i(0) > 0$. As $\lim_{x_i \to \infty} Dc_i(x_i) > \tilde{p}(0)$ and $\tilde{t}_i(\cdot, y)$ is strictly decreasing and continuous there exists a unique $\bar{x} \in \mathbb{R}_{++}$ such that $\tilde{t}_i(\bar{x}, y) = 0$. So $\tilde{b}_i(y) = \bar{x}(0 < y < y_1)$ and thus $\bar{s}_i = +\infty$ follows. $\qquad\square$

Step 14a. Theorem 2(5a) holds

Proof. Suppose $\overline{\epsilon_{\tilde{p}}} < 0$. Note that $N_> = N$ (by Remark 4) and that by Step 3a, Proposition 3 applies. $\qquad\square$

Proof of '⇒'. Suppose $n > -\overline{\epsilon_{\tilde{p}}}$. By Theorem 2(1), $0 \notin E$. Next, we prove that there exists $y_\star \in Y^\oplus$ with $s(y_\star) = 1$. Then (using Step 3b) Proposition 3(5) implies $\#E \geq 1$ and by Theorem 2(3) the proof is complete. Well, with Step 13(1), we have $\lim_{y \downarrow 0} s(y) = \sum_{i \in N} \overline{s}_i = \frac{-n}{\overline{\epsilon_{\tilde{p}}}} > 1$. As, $\lim_{y \to \infty} s(y) = 0$ (Step 5) and s is continuous (Step 10), it follows that there exists an $y_\star$ as desired. $\quad\square$

Proof of '⇐'. Assume $\#E = 1$ and $0 \notin E$. By Proposition 3(5) there exists $y_\star$ with $s(y_\star) = 1$. By Step 10, the function s is strictly decreasing on the proper interval $]0, \max_{i \in N_>} \lambda_i[$. This interval contains $y_\star$. By Step 13(1), $\lim_{y \downarrow 0} s(y) = \frac{-n}{\overline{\epsilon_{\tilde{p}}}}$. So $s(y) < \frac{-n}{\overline{\epsilon_{\tilde{p}}}}$ for all $y \in]0, \max_{i \in N_>} \lambda_i[$. In particular $1 = s(y_\star) < \frac{-n}{\overline{\epsilon_{\tilde{p}}}}$. Thus, $n > -\overline{\epsilon_{\tilde{p}}}$. $\quad\square$

Step 14b. Theorem 2(5b) holds

Proof. Suppose $\overline{\epsilon_{\tilde{p}}} = 0$ and $N_> \neq \emptyset$. By Theorem 2(1), $0 \notin E$. We prove that there exists $y_\star \in Y^\oplus$ with $s(y_\star) = 1$. Then (using Step 3b) Proposition 3(5) implies $\#E \geq 1$ and by Theorem 2(3) the proof is complete. Well, with Step 13(2), we have $\lim_{y \downarrow 0} s(y) = \sum_{i \in N_>} \overline{s}_i = +\infty$. As, $\lim_{y \to \infty} s(y) = 0$ (Step 5) and s is continuous (Step 10), it follows that there exists an $y_\star$ as desired. $\quad\square$

6. Algorithm

Suppose the general assumptions of Theorem 2 hold, that $N_> \neq \emptyset$, that $n > -\overline{\epsilon_{\tilde{p}}}$ and that each firm $i \in N_>$ has an effective capacity constraint. By Theorem 2(5a) and Proposition 3(3,5), we know that the game has a unique equilibrium $\mathbf{e}$, that $\mathbf{e} \neq \mathbf{0}$, that the function $\tilde{b} : Y^\oplus \to \mathbb{R}$ has a unique fixed point $y_\star$ and that

$$\mathbf{e} = (\tilde{b}_1(y_\star), \ldots, \tilde{b}_n(y_\star)).$$

So we have the following algorithm for the calculation of $\mathbf{e}$:

$$f_i \to \tilde{t}_i \to \tilde{b}_i \to \tilde{b} \to \text{fix}(\tilde{b}) \to \mathbf{e}.$$

We illustrate with three examples.

First example: $p(y) = e^{-y}$, $c_i(x_i) = x_i^2/2$. Then one quickly verifies that $\overline{p}(0) = 1, \overline{\epsilon_{\tilde{p}}} = 0, N_> = N, \tilde{t}_i(x_i, y) = e^{-y}(1 - x_i) - x_i, \tilde{b}_i(y) = \frac{e^{-y}}{1+e^{-y}}, \tilde{b}(y) = n\frac{e^{-y}}{1+e^{-y}}$ and that (with Proposition 1) each firm has an effective capacity constraint.

Second example: $\tilde{p}(y) = y^{-\alpha}$, where $0 < \alpha < n$, $c_i(x_i) = x_i^2/2$. Then one quickly verifies that $\overline{p}(0) = +\infty, \overline{\epsilon_{\tilde{p}}} = -\alpha, N_> = N, \tilde{t}_i(x_i, y) = y^{-\alpha}(1 - \alpha\frac{x_i}{y}) - x_i, \tilde{b}_i(y) = \frac{y}{\alpha + y^{1+\alpha}}, \tilde{b}(y) = n\frac{y}{\alpha + y^{1+\alpha}}$ and that each firm has an effective capacity constraint. Now $\text{fix}(\tilde{b}) = \{(n - \alpha)^{1/(1+\alpha)}\}$ and $\mathbf{e} = \{((n - \alpha)^{1/(1+\alpha)}/n, \ldots, (n - \alpha)^{1/(1+\alpha)}/n)\}$.

Third example: $p(y) = e^{-y}$, $c_i(x_i) = \gamma_i x_i$ with $\gamma_i > 0$ and $\gamma_i < 1$ for at least one i. Then one quickly verifies that $\overline{p}(0) = 1, \overline{\epsilon_{\tilde{p}}} = 0, N_> = \{i \in N \mid \gamma_i < 1\}, \tilde{t}_i(x_i, y) = e^{-y}(1 - x_i) - \gamma_i, \tilde{b}_i(y) = \max{(1 - \gamma_i e^y, 0)}, \tilde{b}(y) = \sum_{i \in N_>} \max{(1 - \gamma_i e^y, 0)}$ and that each firm has an effective capacity constraint.

7. Concluding Remarks

(1) Our proof of Theorem 2 only uses elementary mathematics. This is typical for the Selten–Szidarovszky technique. The most sophisticated mathematical result was the implicit function theorem (in Step 7 of the proof). Our proof is a bit long. This is due to the fact that we allow for at 0 discontinuous industry revenue and for equilibria with inactive firms.

(2) In Gaudet and Salant [1991] sufficient and necessary conditions are given for a certain class of oligopolies to have a unique equilibrium. A generalization of this result to aggregative games was given in Von Mouche and Yamazaki [2015]. However, the results in both these articles assume that payoff functions are continuous (everywhere). Having the approach of the present article it should be possible to further generalize the results in Von Mouche and Yamazaki [2015] such that Theorem 2 just is a special case.

(3) Concerning equilibrium results it may be good to distinguish between equilibrium existence (i.e., the existence of at least one equilibrium) and equilibrium semi-uniqueness (i.e., the existence of at most one equilibrium). So proving an equilibrium uniqueness result boils down to proving equilibrium existence and semi-uniqueness. Various equilibrium semi-uniqueness results can be obtained by a technique which is, contrary to the Selten–Szidarovszky technique, local in the sense that it only relies on an analysis of first order conditions. This technique finds its origin in Corchón [1996] and Folmer and von Mouche [2004] and was for Cournot oligopolies substantially improved in Quartieri [2008] and Von Mouche and Quartieri [2017]. As shown in for example Von Mouche and Quartieri [2015], this local technique can lead to simple proofs for the equilibrium semi-uniqueness part. For this technique continuity of the industry revenue is not important.

(4) The assumption $\lim_{x_i \to \infty} Dc_i(x_i) > \overline{p}(0) (i \in N)$ in the fifth bullet of the theorem does not look nice. However, we made this assumption in order that the simplifying assumption in Lemma 2 holds. Further note that this implies (by Remark 1) the comfortable situation of continuous industry revenue. As Theorem 3 in Von Mouche and Quartieri [2013] shows, Theorem 2(3), even holds without supposing $\lim_{x_i \to \infty} Dc_i(x_i) > \overline{p}(0) (i \in N)$.

Acknowledgment

P.v.M. likes to thank Jun-Ichi Itaya for a valuable discussion.

References

Baye, M. R. and Hoppe, H. C. [2003] The strategic equivalence of rent-seeking, innovation, and patent-race games, *Games Economic Behavior* **44**, 217–226.

Corchón, L. C. [1996] *Theories of Imperfectly Competitive Markets*, Volume 442, Lecture Notes in Economics and Mathematical Systems, 2nd Edition (Springer-Verlag, Berlin).

Cornes, R. and Sato, T. [2016] Existence and uniqueness of Nash equilibrium in aggregative games: An expository treatment, *Equilibrium Theory for Cournot Oligopolies and Related Games: Essays in Honour of Koji Okuguchi*, von Mouche, P. H. M. and Quartieri, F. (eds.) (Springer, Cham), pp. 47–61.

Ewerhart, C. [2014] Cournot games with biconcave demand, *Games Econ. Behav.* **85**, 37–47.

Folmer, H. and von Mouche, P. H. M. [2004] On a less known Nash equilibrium uniqueness result, *J. Math. Sociol.* **28**, 67–80.

Gaudet, G. and Salant, S. W. [1991] Uniqueness of Cournot equilibrium: New results from old methods, *Rev. Econ. Stud.* **58**(2), 399–404.

Hirai, S. and Szidarovszky, F. [2013] Existence and uniqueness of equilibrium in asymmetric contests with endogenous prizes, *Int. Game Theory Rev.* **15**(1).

Jensen, M. K. [2018] Aggregative games, *Handbook of Game Theory and Industrial Organization*, Corchón, L. and Marini, M. A. (eds.), Volume I (Edward Elgar, New York).

Matsushima, N. and Yamazaki, T. [2016] Heterogeneity and number of players in rent-seeking, innovation, and patent-race games, *Equilibrium Theory for Cournot Oligopolies and Related Games: Essays in Honour of Koji Okuguchi*, von Mouche, P. H. M. and Quartieri, F. (eds.) (Springer, Cham), pp. 281–294.

Quartieri, F. [2008] *Necessary and Sufficient Conditions for the Existence of a Unique Cournot Equilibrium*, Ph.D. thesis, Siena-Università di Siena, Italy.

Szidarovszky, F. and Okuguchi, K. [1997] On the existence and uniqueness of pure Nash equilibrium in rent-seeking games, *Games Econ. Behav.* **18** 135–140.

Szidarovszky, F. and Yakowitz, S. [1977] A new proof of the existence and uniqueness of the Cournot equilibrium, *Int. Econ. Rev.* **18**, 787–789.

Truchon, M. [1987] *Théorie de l'Optimisation Statique et Différentiable* (Gaëtan morin, Cicoutimi).

Vives, X. [2001] *Oligopoly Pricing: Old Ideas and New Tools* (MIT Press, Cambridge).

Von Mouche, P. and Quartieri, F. [2017] Cournot equilibrium uniqueness via demi-concavity, *Optimization* **67**(4), 441–455.

Von Mouche, P. H. M. [2016] The Selten–Szidarovszky technique: she transformation part, Petrosyan, L. A. and Mazalov, V. V. (eds.) *Recent Advances in Game Theory Applications* (Birkhäuser, Cham), pp. 147–164.

Von Mouche, P. H. M. and Quartieri, F. [2012] Existence of equilibria in Cournotian games with utility functions that are discontinuous at the origin, Technical report, 2528435.

Von Mouche, P. H. M. and Quartieri, F. [2013] on the uniqueness of Cournot equilibrium in case of concave integrated price flexibility, *J. Glob. Optim.* **57**(3), 707–718.

Von Mouche, P. H. M. and Quartieri, F. [2015] Cournot equilibrium uniqueness in case of concave industry revenue: A simple proof, *Econ. Bull.* **35**(2), 1299–1305.

Von Mouche, P. H. M. and Quartieri, F. [2016] *Equilibrium Theory for Cournot Oligopolies and Related Games: Essays in Honour of Koji Okuguchi* (Springer, Cham).

Von Mouche, P. H. M. and Yamazaki, T. [2015] Sufficient and necessary conditions for equilibrium uniqueness in aggregative games, *J. Nonlin. Convex Anal.* **16**(2), 353–364.

Yamazaki, T. [2008] On the existence and uniqueness of pure-strategy Nash equilibrium in asymmetric rent-seeking contests, *J. Public Econ. Theory* **10**(2), 317–327.

Chapter 11

Quantifying Commitment in Nash Equilibria

Thomas A. Weber

Chair of Operations, Economics and Strategy
Ecole Polytechnique Fédérale de Lausanne
Station 5, CH-1015 Lausanne, Switzerland

thomas.weber@epfl.ch

To quantify a player's commitment in a given Nash equilibrium of a finite dynamic game, we map the corresponding normal-form game to a "canonical extension," which allows each player to adjust his or her move with a certain probability. The commitment measure relates to the average overall adjustment probabilities for which the given Nash equilibrium can be implemented as a subgame-perfect equilibrium in the canonical extension.

Keywords: First-mover advantage; second-mover advantage; partial commitment; subgame perfection.

1. Introduction

How much commitment ability is needed to credibly deter market entry? How much adjustment flexibility is enough to guarantee a second-mover advantage in a pricing game? What lock-in is required to reach an advantageous outcome in a coordination game? In what follows, we develop a measure of commitment for Nash equilibria of finite-horizon dynamic games. Even in the absence of informational imperfections (assumed throughout), dynamic games can have many Nash equilibria, but almost always only one of them relies exclusively on credible threats and is therefore singled out as a subgame-perfect Nash equilibrium.[a] Which of the Nash equilibria are

---

[a] A subgame-perfect Nash equilibrium [Selten, 1965] is generically unique in the sense that only payoff indifference at a terminal node would give rise to multiplicity (see, e.g., Mas-Colell *et al.* [1995], Proposition 9.B.2). Since any small extension of the payoffs would break such indifference, the set of (finite) games with multiple subgame-perfect equilibria is of measure zero in the set of all (finite) games. Put differently, the probability that a randomly selected dynamic game has multiple subgame-perfect Nash equilibria is zero.

subgame-perfect depends on the structure of the dynamic game, in terms of the players' move order. Changing this structure does not change the set of Nash equilibria, only the specific one thought of as "credible." And any Nash equilibrium of a given normal-form representation can become subgame-perfect for a nontrivial set of corresponding dynamic games. The question we address here is how to quantify the credibility (or "commitment ability") expected from each player for a given Nash equilibrium of a finite game to be subgame-perfect.

To arrive at a measure of player-specific credibility inherent in a Nash equilibrium, independent of the structure of a particular dynamic game, we first note that a normal-form game presents an equivalence class of dynamic games which all have that same normal-form representation. For this equivalence class, we then ask: given a Nash equilibrium, *how much* commitment ability does each player expect to have in a "generic" subgame-perfect implementation of this equilibrium in extensive form? Because there is no manageable representation of all possible dynamic games belonging to a given normal form, we consider a smaller but fairly generic class of dynamic games, which we term "canonical extensions." The class of canonical extensions for a given normal-form game is a family of sequential-move games where players move once quasi-simultaneously and then, with positive probability, are subject to a termination move by one of the players. At the end of each turn, there is a positive probability for the game to end. The family of canonical extensions is fully parametrized by the vector of the players' individual continuation probabilities.

We show that for any Nash equilibrium of a finite game the set of canonical extensions for which a given Nash equilibrium is subgame-perfect has positive measure. This enables us to quantify a player's commitment in terms of a lack of expected flexibility, where the latter is given by the (expected) probability that a randomly selected canonical extension grants him an extra turn, conditional on a subgame-perfect implementation of the Nash equilibrium at hand. Our commitment measure allows for baseline comparisons of the requirements various Nash equilibria impose on each player's credibility.

1.1. *Literature*

Heinrich Freiherr von Stackelberg [1934] pointed out the importance of move order in a duopoly. GalOr [1985] characterizes when, in such a duopoly setting, it is better to move first or second in terms of the firms' actions being either strategic substitutes or strategic complements. Henkel [2002] finds a possible advantage of partial commitment in duopoly games resulting from adjustment costs which may be self-imposed. Caruana and Einav [2008] endogenize commitment by making a switching cost dependent on a player's choice to wait. Weber [2014] considers the optimal choice of commitment in terms of a continuously varying adjustment cost. In the last two approaches, players have choices available for changing their commitment levels. By contrast, the question addressed here is about *how much* commitment ability is intrinsically available in a given Nash equilibrium.

Commitment also plays a role in bargaining games. For example, demand commitment in coalitional bargaining [Bennett and van Damme, 1990; Selten, 1992; Winter, 1994] refers to the notion of players' sequentially announcing a reservation payoff for joining a coalition, usually in random order.[b] The players' commitment ability is thereby exogenous and lasts for one round of the coalition-formation game, with the last player of any given round having no commitment power at all and the first player enjoying arguably the largest commitment power [Montero and Vidal-Puga, 2007]. Path-dependent asymmetries, such as the random appointment of a coalition orchestrator (or "formateur") in each round with hold-up power also skews the ex-post payoff distribution [Breitmoser, 2009].

The commitment measure we construct here can be used *ex post* for any Nash equilibrium in a finite-horizon dynamic game with endogenous commitment. We remain unconcerned with the question of how precisely the parties' commitment might arise, which commitment devices they might be using, and so forth. More specifically, we think of commitment as a lack of flexibility, whereby a player's flexibility in a given Nash equilibrium is measured as an "expected continuation probability" in a randomly selected canonical extension, conditional on the fact that this dynamic game may serve to implement the said Nash equilibrium.

1.2. *Outline*

In Sec. 2, we assume that a finite dynamic game is represented in (reduced) normal form [Fudenberg and Tirole, 1991] and use the primitives of this representation to introduce a "canonical extension," which contains a flexibility parameter for each player. Section 3 shows how a pure-strategy Nash equilibrium of the normal-form game can be mapped to an equilibrium of the canonical extension for suitable values of the flexibility parameter. In Sec. 4, we argue that the flexibility parameter can be used to measure the players' commitment levels inherent in any given Nash equilibrium of the normal-form game. Section 5 concludes.

2. Models

2.1. *Normal-form game*

Consider the normal-form game $\Gamma = \{\mathcal{N}, \{\mathcal{A}_i\}_{i\in\mathcal{N}}, \{u_i\}_{i\in\mathcal{N}}\}$ for $N \geq 2$ players in the player set $\mathcal{N} = \{1, \ldots, N\}$. Each player $i \in \mathcal{N}$ has a nonempty finite action set $\mathcal{A}_i$ and a real-valued payoff function $u_i : \mathcal{A} \to \mathbb{R}$, where $\mathcal{A} = \mathcal{A}_1 \times \cdots \times \mathcal{A}_N$ is the set of all strategy profiles. A pure-strategy profile $a^* \in \mathcal{A}$ is a Nash equilibrium of Γ if

$$a^* \in \mathcal{R}(a^*), \tag{1}$$

---

[b]The idea of the demand commitment game originated with Reinhard Selten [Bennett and van Damme, 1990, p. 2].

where, for any strategy profile $a \in \mathcal{A}$, the (static) best-response correspondence is

$$\mathcal{R}(a) \triangleq \prod_{i \in \mathcal{N}} \mathcal{R}_i(a_{-i}) = \prod_{i \in \mathcal{N}} \arg \max_{a_i \in \mathcal{A}_i} u_i(a_i, a_{-i}), \tag{2}$$

using the standard notational convention that $a = (a_i, a_{-i})$ for any $i \in \mathcal{N}$. The definition of a mixed-strategy Nash equilibrium is more general and is obtained by replacing each action set $\mathcal{A}_i$ with the corresponding simplex $\Delta(\mathcal{A}_i)$ of probability distributions over actions. To simplify the exposition, we first assume that there exists at least one pure-strategy Nash equilibrium a^* of Γ, postponing the discussion of more general situations to Sec. 5.[c]

Example 1. Consider the standard battle-of-the-sexes game [Luce and Raiffa, 1957, pp. 90–91; Osborne and Rubinstein, 1984, pp. 15–16], where each of two players, whom we call Ann (player 1) and Bert (player 2), can choose to either go dancing ('*D*') or go to the movies ('*M*'). Whenever the players choose different actions, their payoffs vanish. Otherwise the players obtain the payoffs $u_1(D, D) = u_2(M, M) = 2$ and $u_1(M, M) = u_2(D, D) = 1$, respectively. Since each player i's static best-response correspondence is $\mathcal{R}_i(a_{-i}) \equiv \{a_{-i}\}$, this 'coordination game' has two pure-strategy Nash equilibria $a^* \in \{(D, D), (M, M)\}$; it also has a mixed-strategy Nash equilibrium where each player chooses his or her preferred action with probability $2/3$.

2.2. *Canonical extension*

Let $p = (p_1, \ldots, p_N) \in \Delta_N$ be a vector of continuation probabilities in the N-dimensional simplex

$$\Delta_N \triangleq \{(p_1, \ldots, p_N) \in \mathbb{R}_+^N : p_1 + \cdots + p_N \leq 1\}. \tag{3}$$

The *canonical extension* $\hat{\Gamma}(p)$ of the normal-form game Γ is a two-stage dynamic game which extends over the time periods $t \in \{0, 1\}$. At time $t = 0$, all players play a simultaneous-move stage game of the form Γ. At the end of that first period, the game ends with probability

$$p_0 \triangleq 1 - (p_1 + \cdots + p_N). \tag{4}$$

Otherwise, with probability p_i, player i is allowed to adjust his move in the second period, at time $t = 1$; all other players are stuck with their first-period actions. At the end of the second period, the game ends and all players obtain their (undiscounted) payoffs, based on the strategy profile $a = (a_i, a_{-i})$ for player i's second-period choice a_i and the profile a_{-i} of the other players' first-period choices.

The canonical extension can be viewed as a dynamic perturbation of Γ where each player's influence may spill into a second period, thus potentially limiting

---

[c]The Kakutani fixed-point theorem guarantees that Γ has a (mixed-strategy) Nash equilibrium [Nash, 1950].

the commitment power for any given player. For $p = 0$, the game terminates with probability 1 at the end of the first period and we obtain the original game. That is

$$\hat{\Gamma}(0) = \Gamma. \tag{5}$$

A subgame-perfect equilibrium of the canonical extension $\hat{\Gamma}(p)$ is obtained by backward-induction. At time $t = 1$, conditional on the first-period strategy profile $a = (a_i, a_{-i}) \in \mathcal{A}$, player i—upon being able to adjust his move—chooses an element of his best response

$$r_i(a_{-i}) \in \mathcal{R}_i(a_{-i}) \triangleq \arg \max_{\hat{a}_i \in \mathcal{A}_i} u_i(\hat{a}_i, a_{-i}). \tag{6}$$

Hence, at time $t = 0$, each player i obtains the expected utility

$$U_i(a_i, a_{-i}; p) \triangleq p_0 u_i(a_i, a_{-i}) + p_i u_i(r_i(a_{-i}), a_{-i})$$

$$+ \sum_{j \in \mathcal{N} \setminus \{i\}} p_j u_i(a_i, r_j(a_i, (a_{-i})_{-j}), (a_{-i})_{-j}),$$

where

$$(a_{-i})_{-j} \triangleq (a_l)_{l \in \mathcal{N} \setminus \{i,j\}}, \quad i, j \in \mathcal{N}, \ i \neq j,$$

denotes the strategy profile of the players other than players i and j. With probability p_0 player i obtains his standard stage-game payoff, and with probability p_i he gets to observe the other players' strategy profile and choose a best response; otherwise, player i can anticipate the expected payoff impact of other players' reacting to the strategy profile they may observe upon being given the opportunity to adjust their actions. Figure 1 illustrates the payoffs in the canonical extension $\hat{\Gamma}(p)$ for a two-player game where each player has two possible actions. After playing a simultaneous-move game, with probability p_i player i gets to move again, thus instead of the previously played a_i choosing a best response $r_i(a_{-i})$ to the other player's action a_{-i}.

Remark 1 (Stackelberg Variants). A pure Stackelberg variant of $\hat{\Gamma}(p)$ with player i moving last is obtained for $p = (1, \ldots, 1) - e_i$, where e_i is the ith unit vector in the standard Euclidean base of $\mathbb{R}^N$. Similarly, a Stackelberg variant of $\hat{\Gamma}(p)$ where player i moves first is obtained for $p = (p_i, p_{-i})$ where $p_i = 0$ and $p_{-i} \gg 0$ is such that $\sum_{j \in \mathcal{N} \setminus \{i\}} p_j = 1$, so the probability of stopping at the end of the first period vanishes ($p_0 = 0$).[d] One of the other players will surely terminate the game. For $N = 2$, the dynamic game $\hat{\Gamma}(p)$ can therefore embed any pure player order. For $N > 2$, the dimensionality of the simplex Δ_N is not high enough for p to be able to continuously vary among all possible player orders (of which there are $N! > N$ for $N > 2$). Still, it is possible to analyze within $\hat{\Gamma}(p)$ the comparative

---

[d]We use standard notation for vector inequalities, where for $p = (p_1, \ldots, p_N)$ and $\hat{p} = (\hat{p}_1, \ldots, \hat{p}_N)$ it is $p \geq \hat{p} \Leftrightarrow (p_i \geq \hat{p}_i$ for all $i \in \mathcal{N})$, $p > \hat{p} \Leftrightarrow (p \geq \hat{p}$ and $p_i > \hat{p}$ for at least one $i \in \mathcal{N})$, and $p \gg \hat{p} \Leftrightarrow (p_i > \hat{p}_i$ for all $i \in \mathcal{N})$.

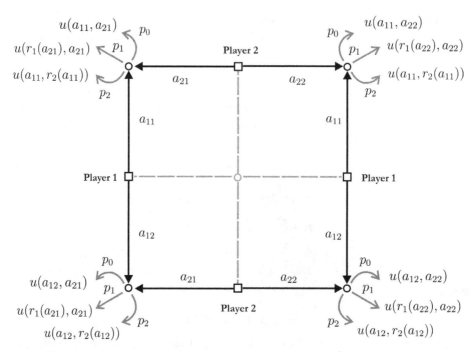

Fig. 1. Extensive-form representation of the canonical extension $\hat{\Gamma}(p)$ for $N = 2$ and $\mathcal{A} = \{a_{11}, a_{12}\} \times \{a_{21}, a_{22}\}$, where $u = (u_1, u_2)$ and $p_0 = 1 - p_1 - p_2$.

statics when any given player's move position is varied from first to last or vice versa.

Given any $p \in \Delta_N$, the canonical extension $\hat{\Gamma}(p)$ can be viewed as the original normal-form game Γ with the payoff functions $u_i(\cdot)$ replaced by $U_i(\cdot; p)$, for all $i \in \mathcal{N}$. Thus, we consider the parametrized family of *canonical normal-form games*

$$\Gamma(p) \triangleq \{\mathcal{N}, \{\mathcal{A}_i\}_{i \in \mathcal{N}}, \{U_i(\cdot; p)\}_{i \in \mathcal{N}}\}, \quad p \in \Delta_N.$$

In particular, a strategy profile $a^* = (a_1^*, \ldots, a_N^*)$ is a Nash equilibrium of the normal-form game $\Gamma(p)$ if and only if it is a (subgame-perfect) Nash equilibrium of the canonical extension $\hat{\Gamma}(p)$ for each player $i \in \mathcal{N}$ to play a_i^* in the first period and to play $a_i^* = r_i(a_{-i}^*)$ provided he gets to move again in the second period.

Lemma 1. *For any $p \in \Delta_N$, there is a one-to-one correspondence between the Nash equilibria of the canonical normal-form game $\Gamma(p)$ and the subgame-perfect Nash equilibria of the canonical extension $\hat{\Gamma}(p)$.*

The preceding result allows us to think of the canonical extension $\hat{\Gamma}(p)$ as equivalent to the canonical normal-form game $\Gamma(p)$. Each one also presents an embedding of the original normal-form game, as $\Gamma = \Gamma(0) = \hat{\Gamma}(0)$.

Example 2. In the canonical extension $\hat{\Gamma}(p)$ of the battle-of-the-sexes game in Example 1, Ann and Bert effectively choose their activities in turn. As noted in

Remark 1, the value of $p = (p_1, p_2) \in \Delta_2$ determines who starts and who terminates the game. For example, when $p = (1, 0)$ Ann effectively starts the game, while Bert's move concludes their interaction. Consider now the strategy profile $a^* = (D, D)$ in the canonical normal-form game $\Gamma(p)$. Ann would never want to deviate because

$$U_1(a^*; p) = 2p_0 + 2p_1 + 2p_2 \equiv 2 \geq 2p_1 + p_2 = U_1(M, D; p),$$

for all $p \in \Delta_2$. However, Bert would prefer a deviation, provided that Ann's continuation probability is large enough. Indeed,

$$U_2(M, D; p) = 0 + p_2 + 2p_1 \geq 1 = p_0 + p_2 + p_1 = U_2(a^*; p),$$

if and only if $p_1 \geq (1 - p_2)/2$; this condition is automatically satisfied if $p_1 \geq 1/2$. In other words, high flexibility for Ann destroys her commitment ability.

3. Sequential Implementation

For any given $p \in \Delta_N$, we say that a Nash equilibrium $a^* = (a_i^*)_{i \in N}$ of the simultaneous-move game Γ can be *implemented sequentially* in the canonical extension $\hat{\Gamma}(p)$ if and only if a^* is a Nash equilibrium of the canonical normal-form game $\Gamma(p)$. As discussed in the last section, this Nash equilibrium then induces a subgame-perfect equilibrium in $\hat{\Gamma}(p)$.

Remark 2 (Comparison with Repeated Games). The idea of dynamically implementing static Nash equilibria is not new. Indeed, one can implement the payoffs of any pure-strategy Nash equilibrium of Γ as average payoffs of a repeated game Γ^∞ where in each period the players play the simultaneous-move stage game Γ and payoffs are discounted from period to period. In such a repeated game, a subgame-perfect strategy profile is for players to ignore the history of past actions and simply play the Nash equilibrium a^* in each period. However, this approach affords no insights into the relationship of the normal-form game Γ to its sequential-move variants.[e]

For any player $i \in N$, let

$$\mathcal{P}_i(a^*) \triangleq \{\hat{p} \in \Delta_N : \max\{U_i(a_i, a_{-i}^*; \hat{p}) : a_i \in \mathcal{A}_i\} \leq u_i(a^*)\} \tag{7}$$

be the set of all continuation probabilities p such that player i's deviation in $\Gamma(p)$ is *not* profitable relative to the equilibrium payoff at the Nash equilibrium a^* in Γ.

---

[e]Infinitely repeated games feature generic equilibrium multiplicities due to well-known folk theorems (see, e.g., Friedman [1971] and Fudenberg and Maskin [1986]) that rely on a variety of credible out-of-equilibrium threats. To bypass such complications we opt here for simple sequential implementations in two periods.

Proposition 1. *A Nash equilibrium a^* of Γ can be implemented sequentially in $\hat{\Gamma}(p)$ if and only if*

$$p \in \mathcal{P}(a^*), \tag{8}$$

where $\mathcal{P}(a^*) \triangleq \bigcap_{i \in \mathcal{N}} \mathcal{P}_i(a^*)$.

Proof. Fix $p \in \Delta_N$, and let a^* be a Nash equilibrium of Γ. In the canonical extension $\hat{\Gamma}(p)$, any given player $i \in \mathcal{N}$ prefers the Nash equilibrium action a_i^* to any other available action $a_i \in \mathcal{A}_i$ (at least weakly) if and only if

$$U_i(a_i, a_{-i}^*; p) \leq U_i(a_i^*, a_{-i}^*; p), \quad a_i \in \mathcal{A}_i. \tag{9}$$

Since $U_i(a_i^*, a_{-i}^*; p) = u_i(a^*)$, the last relation is equivalent to p being in the set $\mathcal{P}_i(a^*)$ as specified in Eq. (7). In a sequential implementation of a^*, no player can have an incentive to deviate from the equilibrium strategy profile, restricting the corresponding probability vectors p to lie in the intersection $\mathcal{P}(a^*) = \bigcap_{i \in \mathcal{N}} \mathcal{P}_i(a^*)$, completing our proof. □

Consider now the measure of the set of continuation probabilities,

$$\|\mathcal{P}(a^*)\| \triangleq \int_{\mathcal{P}(a^*)} dp.$$

The sequential implementation of a^* is called *strong* if the set of continuation probabilities has positive measure. Our main result is that any Nash equilibrium of Γ has a (strong) sequential implementation.

Corollary 1. *Any Nash equilibrium a^* of Γ has a strong sequential implementation in $\hat{\Gamma}(p)$ for $p \in \mathcal{P}(a^*)$; and $\|\mathcal{P}(a^*)\| > 0$.*

Proof. Let a^* be a Nash equilibrium of Γ. By Eq. (5) this Nash equilibrium is trivially implementable in $\hat{\Gamma}(0)$ (i.e., $0 \in \mathcal{P}(a^*)$). Consider now the possible deviation a_i for player $i \in \mathcal{N}$, such that his action is *not* in his best response to the equilibrium strategy profile a_{-i}^*; that is, let[f]

$$a_i \in \bar{\mathcal{R}}_i(a_{-i}^*) \triangleq \mathcal{A}_i \backslash \mathcal{R}_i(a^*).$$

Thus, because $\mathcal{A}_i$ is finite, there exists an $\varepsilon_i > 0$ such that

$$\max_{a_i \in \mathcal{A}_i} U_i(a_i, a_{-i}^*; 0) + \varepsilon_i \leq U_i(a_i^*, a_{-i}^*; 0).$$

---

[f] If $\bar{\mathcal{R}}_i(a^*) = \emptyset$, then $\mathcal{R}_i(a^*) = \mathcal{A}_i$, which implies that $\mathcal{P}_i(a^*) = \Delta_N$, so that player i has no part in restricting $\mathcal{P}(a^*)$ and can therefore be neglected. If $\bar{\mathcal{R}}_i(a^*) = \emptyset$ for all $i \in \mathcal{N}$, then $\|\mathcal{P}(a^*)\| = \|\Delta_N\| > 0$.

By the maximum theorem [Berge, 1959] the envelope $m_i(p) \overset{\triangle}{=} \max_{a_i \in \mathcal{A}_i} U_i(a_i, a^*_{-i}; p)$ is continuous in p, so that there exists a $\delta_i > 0$ for which:[g]

$$\|p\|_\infty \in [0, \delta_i] \Rightarrow |m(0) - m(p)| \leq \varepsilon_i.$$

This implies that $\mathcal{P}_i(a^*) \subset [0, \delta_i]^N$. If we set $\delta \overset{\triangle}{=} \min_{i \in \mathcal{N}} \delta_i > 0$ and repeat the preceding arguments for all other players, one obtains that $\mathcal{P}(a^*) \subset [0, \delta]^N$, which implies that $\|\mathcal{P}(a^*)\| \geq \delta^N > 0$, completing the proof.   □

For any $i, j \in \mathcal{N}$ and any Nash equilibrium $a^* = (a^*_i, a^*_{-i}) \in \mathcal{A}$ of Γ, consider the mapping $q_{i,j}(\cdot; a^*) : \mathcal{A}_i \to \mathbb{R}$, with

$$q_{i,j}(a_i; a^*) \overset{\triangle}{=} \begin{cases} \dfrac{u_i(a_i, r_j(a_i, (a^*_{-i})_{-j}), (a^*_{-i})_{-j}) - u_i(a_i, a^*_{-i})}{u_i(a^*) - u_i(a_i, a^*_{-i})}, \\[3mm] \qquad\qquad \text{if } a_i \in \bar{\mathcal{R}}_i(a^*_{-i}) \quad \text{and} \quad i \neq j, \\[3mm] 1, \quad \text{otherwise}, \end{cases}$$

for all $a_i \in \mathcal{A}_i$. The quotient $q_{i,j}(a_i; a^*)$ quantifies player i's payoff variation if player j is able to adjust to i's deviation a_i (resulting in j's choosing the best response $r_j(a_i, (a^*_{-i})_{-j})$ instead of a^*_j), relative to player i's benefit of playing the Nash equilibrium action a^*_i instead of unilaterally deviating to a_i. The underlying assumption is that all agents other than i and j are playing according to the Nash equilibrium strategy profile $(a^*_{-i})_{-j}$. Intuitively, the quotient $q_{i,j}(a_i; a^*)$ quantifies the ratio of payoff variations between bilateral deviations (first i, then j) and unilateral deviations (for i only).

Proposition 2. *The set of continuation probabilities for which a sequential implementation of the Nash equilibrium a^* is possible can be represented in the form:*

$$\mathcal{P}(a^*) = \{p \in \Delta_N : f(p; a^*) \leq 1\}, \tag{10}$$

where $f(p; a^*)$ is piecewise linear and convex in p with $f(0; a^*) = 0$, and

$$f(p; a^*) = \max_{i \in \mathcal{N}} \left\{ \max_{a_i \in \bar{\mathcal{R}}_i(a^*_{-i})} \sum_{j \in \mathcal{N}} q_{i,j}(a_i; a^*) p_j \right\}, \quad p \in \Delta_N. \tag{11}$$

Proof. Given any Nash equilibrium a^* of Γ and any $p \in \Delta_N$, the implementability condition in Eq. (9) is equivalent to

$$p_0 u_i(a_i, a^*_{-i}) + p_i u_i(a^*) + \sum_{j \in \mathcal{N} \setminus \{i\}} p_j u_i(a_i, r_j(a_i, (a^*_{-i})_{-j}), (a^*_{-i})_{-j}) \leq u_i(a^*),$$

$$a_i \in \bar{\mathcal{R}}_i(a^*_{-i}).$$

---

[g]Because of the norm-equivalence in finite-dimensional Euclidean spaces, the precise choice of the norm is qualitatively unimportant. For simplicity, we here make use of the maximum-norm, defined as $\|p\|_\infty \overset{\triangle}{=} \max\{p_1, \ldots, p_N\}$, for all $p = (p_1, \ldots, p_N) \in \Delta_N$.

Using the fact that $p_0 = 1 - \sum_{i \in \mathcal{N}} p_i$ (see Eq. (4)) and that $u_i(a^*) - u_i(a_i, a^*_{-i}) > 0$ for all $a_i \in \bar{\mathcal{R}}_i(a^*_{-i})$, the preceding inequality can be rewritten in the form

$$p_i + \sum_{j \in \mathcal{N} \setminus \{i\}} p_j \frac{u_i(a_i, r_j(a_i, (a^*_{-i})_{-j}), (a^*_{-i})_{-j}) - u_i(a_i, a^*_{-i})}{u_i(a^*) - u_i(a_i, a^*_{-i})} \leq 1, \quad a_i \in \bar{\mathcal{R}}_i(a^*_{-i}).$$

Hence, player i's expected equilibrium payoff $u_i(a^*) = U_i(a^*; p)$ exceeds the deviation payoff $U_i(a_i, a^*_{-i}; p)$ (at least weakly) if and only if

$$\sum_{j \in \mathcal{N}} q_{i,j}(a_i; a^*) p_j \leq 1. \tag{12}$$

Maximizing the left-hand side of the last condition over all $a_i \in \bar{\mathcal{R}}_i(a^*_{-i})$ and then over all $i \in \mathcal{N}$ yields the expression for $f(p; a^*)$ and the claimed representation of $\mathcal{P}(a^*)$ in Eq. (10). Finally, as an upper envelope of linear functions, $f(\cdot; a^*)$ is naturally convex. This completes the proof. □

As long as player i's possible deviations a_i are not in his best response $\mathcal{R}_i(a^*_{-i})$ (i.e., they should lie in the complement $\bar{\mathcal{R}}_i(a^*_{-i})$), the denominator of $q_{i,j}(a_i; a^*)$ is strictly positive. However, the numerator may well be negative, indicating an additional disbenefit from player j's ability to also deviate. This in turn points to a first-mover advantage and facilitates the sequential implementation of a^* in $\hat{\Gamma}(p)$ for a particular vector of continuation probabilities p. In other words, the more negative the $q_{i,j}(a_i; a^*)$ (for $j \neq i$), the more likely the Nash equilibrium survives a perturbation from Γ to $\hat{\Gamma}(p)$.

Lottery Interpretation. If we introduce the discrete lottery

$$L_i(a_i; a^*, p) = \left[p_0, 0; [p_j, q_{i,j}(a_i; a^*)]_{j=1}^N \right],$$

with the $N + 1$ payoffs $0, q_{i,1}(a_i; a^*), \ldots, q_{i,N}(a_i; a^*)$ that occur with probabilities $p_0, p_1, \ldots, p_N$, respectively,[h] then

$$U_i(a_i, a^*_{-i}; p) \leq U_i(a^*; p) \Leftrightarrow \mathbb{E}[L_i(a_i; a^*, p)] \leq 1.$$

In other words, for the Nash equilibrium a^* to be sequentially implementable, any player i's expected relative payoff variations from bilateral reactions to unilateral deviations cannot exceed unity.

---

[h] We recall that $p_0 = 1 - (p_1 + \cdots + p_N)$ by Eq. (4), so that there is no additional dependence on p_0.

Corollary 2. *For any $i \in \mathcal{N}$, the set $\mathcal{P}_i(a^*)$ is convex.*

Proof. Consider $p, \hat{p} \in \mathcal{P}_i(a^*)$ and $\theta \in (0, 1)$. Then for any $a_i \in \bar{\mathcal{R}}_i(a_{-i}^*)$, it is

$$\sum_{j \in \mathcal{N}} q_{i,j}(a_i; a_{-i}^*)(\theta p_j + (1 - \theta)\hat{p}_j)$$

$$= \theta \sum_{j \in \mathcal{N}} q_{i,j}(a_i; a_{-i}^*) \, p_j + (1 - \theta) \sum_{j \in \mathcal{N}} q_{i,j}(a_i; a_{-i}^*)\hat{p}_j \leq 0,$$

which by inequality (12) implies that

$$U_i(a^*; \theta p + (1 - \theta)\hat{p}) \geq U_i(a_i, a_{-i}^*; \theta p + (1 - \theta)\hat{p}), \quad a_i \in \bar{\mathcal{R}}_i(a_{-i}^*),$$

so that, by definition, $\theta p + (1 - \theta)\hat{p} \in \mathcal{P}_i(a^*)$, which establishes the claim.   □

Example 3. Using Proposition 2, the Nash equilibria of the simultaneous-move battle-of-the-sexes game in Example 1 can be implemented sequentially. From the discussion in Example 2 we obtain that the Nash equilibrium $a^* = (D, D)$ of Γ is also a Nash equilibrium of the canonical extension $\hat{\Gamma}(p)$ if and only if p lies in the set $\mathcal{P}(a^*) = \mathcal{P}_1(a^*) \cap \mathcal{P}_2(a^*)$, where $\mathcal{P}_1(a^*) = \Delta_2$ and $\mathcal{P}_2(a^*) = \{(\hat{p}_1, \hat{p}_2) \in \Delta_2 : \hat{p}_1 \leq (1 - \hat{p}_2)/2\}$. Hence, $\mathcal{P}(a^*) = \mathcal{P}_2(a^*)$. By symmetry, for $\hat{a}^* = (M, M)$ we obtain that $\mathcal{P}(\hat{a}^*) = \{(\hat{p}_1, \hat{p}_2) \in \Delta_2 : \hat{p}_2 \leq (1 - \hat{p}_1)/2\}$. Thus, both pure-strategy Nash equilibria in the battle-of-the-sexes game of Example 1 can be implemented sequentially in $\hat{\Gamma}(p)$ as long as $p = (p_1, p_2)$ lies in the region $AECF$; see Fig. 2. More specifically, to sequentially implement Ann's preferred Nash equilibrium (D, D) it is enough if p lies in the region $AE(C)D$, while Bert's preferred equilibrium (M, M) can be sequentially implemented if p is in $AB(C)F$. By embedding the

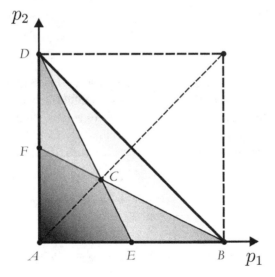

Fig. 2.   Regions of sequential implementability (as subsets of Δ_2) in Example 3.

normal-form game into the dynamic framework, one learns that in order for Ann to obtain her favorite choice (D), her probability (p_1) of being able to adjust her strategy cannot exceed half of Bert's probability of not being able to adjust his choice $(1 - p_2)$. Thus, it is *not* necessary that player 2's flexibility is smaller than player 1's because for any $p_1 \in (0,1)$, there exists a $p_2 < p_1$ that still allows for a dynamic implementation of player 1's preferred equilibrium.[i] By continuously varying p over the parameter space Δ_2, we obtain the implementable equilibria in the form of an upper semicontinuous correspondence, and thus a conceptually richer picture of commitment than can be obtained by just comparing the two Stackelberg versions. The parameter space captures the underlying continuity of commitment levels [Weber, 2014].

4. A Measure of Commitment

The nonempty set of continuation-probability vectors $\mathcal{P}(a^*)$ for any given Nash equilibrium a^* of the normal-form game Γ serves as the basis for a measure of the players' commitment ability.

Expected Flexibility. For each player i, the higher the continuation probability p_i that sustains a^*, the higher his *expected flexibility* in this Nash equilibrium, defined as

$$\varphi_i(a^*) \triangleq \frac{\int_{\mathcal{P}(a^*)} p_i \, dp}{\|\mathcal{P}(a^*)\|}, \tag{13}$$

where $\|\mathcal{P}(a^*)\| = \int_{\mathcal{P}(a^*)} dp > 0$ by Corollary 1. The flexibility measure $\varphi_i(a^*) \in [0,1]$ represents player i's expected continuation probability in a random instance of the canonical extension $\hat{\Gamma}(p)$ conditional on being in a sequential implementation of the Nash equilibrium a^* of Γ.

Lemma 2. *For any Nash equilibrium a^* of Γ and any player $i \in \mathcal{N}$: $\varphi_i(a^*) \in [0, 1/2]$.*

Proof. Fix $i \in \mathcal{N}$. Given that $p \in \Delta_N$, player i's flexibility p_i can be largest (i.e., $p_i = 1$) if $p_0 = 0$ and all other players' flexibilities vanish (i.e., $p_j = 0$ for all $j \in \mathcal{N}\backslash\{i\}$). By Cor. 2, the set $\mathcal{P}_i(a^*)$ is convex, so that if $p_i = \hat{p}_i \in (0,1)$ is feasible, then $p_i \in [0, \hat{p}_i)$ must also be feasible. By Cor. 1, the set $\mathcal{P}(a^*)$ has a positive measure, which implies there exists an $\varepsilon \in (0, 1/N)$, so that

$$p \in \mathcal{B}_{N,\varepsilon} \triangleq \{\hat{p} \in \Delta_N : \|\hat{p}\|_\infty \le \varepsilon\} \Rightarrow p \in \mathcal{P}(a^*).$$

Since $\mathcal{B}_{N,\varepsilon} \subset \mathcal{P}(a^*)$, the largest possible value for p_i is $\bar{p}_{i,\varepsilon} = 1 - (N-1)\varepsilon$. If we now set $\hat{p}_i$ equal to the largest value in the ith direction of elements in the set $\mathcal{P}(a^*)$,

$$\hat{p}_i = \sup\{p_i \in [0,1] : (p_i, p_{-i}) \in \mathcal{P}(a^*)\},$$

[i] Any player i's ability to commit decreases in p_i.

then $\hat{p}_i < 1$. By choosing ε such that $0 < \varepsilon < (1 - \hat{p}_i)/(N-1)$, we obtain that $\hat{p}_i < \bar{p}_{i,\varepsilon}$, which in turn implies that

$$\varphi_i(a^*) < \frac{\int_{[0,\varepsilon]^{N-1} \times [0,\bar{p}_{i,\varepsilon}]} p_i \, dp}{\|[0,\varepsilon]^{N-1} \times [0,\bar{p}_{i,\varepsilon}]\|} = \frac{1 - (N-1)\varepsilon}{2} \uparrow \frac{1}{2} \quad (\text{for } \varepsilon \downarrow 0^+),$$

concluding our proof. □

The proof of Lemma 2 shows that $1/2$ is in fact a *tight* upper bound for any player i's expected flexibility. It turns out that the sum of all players' expected flexibilities cannot exceed twice the expected flexibility of any one player.

Lemma 3. *For any Nash equilibrium a^* of Γ: $\sum_{i \in \mathcal{N}} \varphi_i(a^*) \le 1$.*

Proof. Let a^* be a Nash equilibrium of Γ. By the definition of the players' expected flexibility $\varphi_i(a^*)$ in Eq. (13) and the definition of the set Δ_N of admissible probability vectors in Eq. (3), it is

$$\sum_{i \in \mathcal{N}} \varphi_i(a^*) = \frac{\int_{\mathcal{P}(a^*)} \left(\sum_{i \in \mathcal{N}} p_i\right) dp}{\|\mathcal{P}(a^*)\|} \le \frac{\int_{\mathcal{P}(a^*)} dp}{\|\mathcal{P}(a^*)\|} = 1,$$

which establishes the claim. □

Commitment Measure. We are now ready to introduce player i's *measure of commitment (in the Nash equilibrium a^*)* as follows:

$$\kappa_i(a^*) \triangleq \max\{0, 1 - N\varphi_i(a^*)\}. \tag{14}$$

This measure of commitment takes on values between the tight bounds 0 and 1. It can be interpreted as the renormalized probability with which player i's decision cannot be revised in a random canonical extension that allows for an implementation of the Nash equilibrium a^*. The renormalization is such that player i's commitment vanishes whenever his expected flexibility exceeds $1/N$, which for $N = 2$ cannot happen at all because φ_i is, by Lemma 2, limited to values in $[0, 1/2]$. The reason for the normalization becomes clear with the next result and the example thereafter.

Lemma 4. *For any Nash equilibrium a^* of Γ, it is $(1/N)\sum_{i \in \mathcal{N}} \kappa_i(a^*) \in [0,1]$.*

Proof. The claim follows immediately when applying Lemma 3 to Eq. (14). □

The players' average commitment is bounded tightly by 0 and 1, and therefore the measure retains full informativeness as the number of players goes up.[j]

Example 4 (Full Sequential Implementability and Reference Commitment). Consider a *trivial* game (e.g., with constant payoffs for each player) where any Nash equilibrium a^* can be implemented sequentially in the dynamic extension $\hat{\Gamma}(p)$ for *any* $p \in \Delta_N$. Given that $\|\Delta_N\| = 1/N!$, we find that

$$\kappa_i(a^*) = \max\{0, 1 - N\varphi_i(a^*)\}$$

$$= 1 - \frac{N\int_{\Delta_N} p_i \, dp}{\|\Delta_N\|} = 1 - \frac{N/(N+1)!}{(1/N!)} = \frac{1}{N+1} = \varphi_i(a^*) > 0,$$

for any $i \in \{1, \ldots, N\}$ and for any $N \geq 2$. Hence in a game with *full sequential implementability*, the players' commitment levels are $\bar{\kappa} \triangleq 1/(N+1)$; this can be viewed as a *reference commitment*, against which one can compare the commitments achieved in nontrivial games. In a two-player game, for instance, the reference commitment is $\bar{\kappa} = 1/3$.

Lemma 5. *The commitment measure is invariant with respect to a positive-linear transformation of any agent's utility function.*

Proof. For any player $i \in \mathcal{N}$, let $\hat{u}_i = \alpha_i u_i + \beta_i$, where $(\alpha_i, \beta_i) \in \mathbb{R}_{++} \times \mathbb{R}$. Then for any Nash equilibrium a^* of Γ the modified payoff ratios in Proposition 2 are the same as before:

$$\hat{q}_{ij}(a_i; a^*) = q_{ij}(a_i; a^*), \quad a_i \in \mathcal{A}_i, \quad i, j \in \mathcal{N},$$

so that Eqs. (10) and (11) together imply that the set $\mathcal{P}(a^*)$ remains unchanged. The claim now follows from Eqs. (13) and (14), concluding our proof. $\square$

Invariance with respect to probabilistically equivalent representations of the players' payoffs is important because such rescaling does not affect any choice in the game. In particular, it does not affect the set of Nash equilibria and—as the preceding result shows—leaves the commitment measure unchanged as well.

Example 5. In the setting of the battle-of-the-sexes coordination game, let $a^* = (D, D)$ and $\hat{a}^* = (M, M)$ denote the two pure-strategy Nash equilibria. Using the results of Example 3 it is $\|\mathcal{P}(a^*)\| = \|\mathcal{P}(\hat{a}^*)\| = 1/4$. The expected flexibility is

$$\varphi_i(a^*) \triangleq \frac{\int_{\mathcal{P}(a^*)} p_i dp}{\|\mathcal{P}(a^*)\|},$$

---

[j]If instead one considered $\max\{0, 1 - \alpha\varphi_i\}$ for any fixed $\alpha > 0$ as commitment measure, then the players' average commitment level would lie in the interval $[1 - (\alpha/N), 1]$ and would therefore tend to 1 as $N \to \infty$, completely eroding the informativeness of the commitment measure for large games.

so that

$$\varphi_1(a^*) = 4 \int_0^{1/2} (1 - 2p_1)p_1 dp_1 = \frac{1}{6}$$

for Ann, and

$$\varphi_2(a^*) = 2 \int_0^{1/2} (1 - 2p_1)^2 dp_1 = \frac{1}{3}$$

for Bert. Consequently, in the Nash equilibrium a^* favorable for Ann, by Eq. (14) her commitment measure is twice as large as Bert's (who merely achieves the reference commitment level of $\bar{\kappa} = 1/(N+1) = 1/3$ introduced in Example 4):

$$\kappa_1(a^*) = \frac{2}{3} > \frac{1}{3} = \kappa_2(a^*).$$

By symmetry, in the Nash equilibrium $\hat{a}^*$ favorable for Bert, one obtains $\kappa_i(\hat{a}^*) = \kappa_{-i}(a^*)$ for player $i \in \{1, 2\}$, so the expected commitment levels are reversed compared to a^*; see Fig. 3.

Example 6 (Entry Deterrence). Consider the classical two-player entry-deterrence game as, for example, in Rasmusen [2001, p. 94]. The first player can either "enter" (E) or "not enter" ($\bar{E}$) the market. The second player, the incumbent monopolist, can either "accommodate" (A) or "fight" (F), e.g., by starting a price war. The extensive-form representation of the game and the players' payoffs are given in Fig. 4(a); the corresponding normal-form game Γ, depicted in Fig. 4(b), has two Nash equilibria: $a^* = (E, A)$ and $\hat{a}^* = (\bar{E}, F)$. Consider first $a^* = (E, A)$ and $p \in \Delta_2$. From the entrant's perspective this equilibrium can be sequentially implemented if and only if

$$U_1(\bar{E}, A; p) = 0 + 2p_1 + 0 \le 2 = u_1(a^*) = U_1(a^*; p),$$

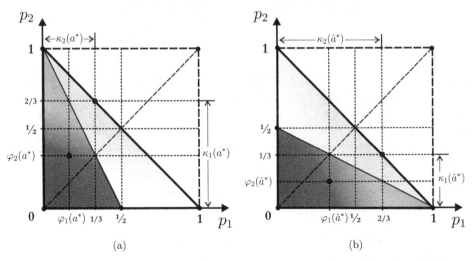

Fig. 3. Measures of commitment and flexibility in Example 5: (a) for $a^* = (D, D)$; (b) for $\hat{a}^* = (M, M)$.

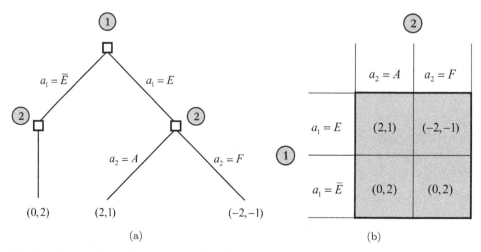

Fig. 4.   Extensive-form (a) and normal-form (b) representations of the dynamic entry-deterrence game in Example 6.

so $P_1(a^*) = \Delta_2$. On the other hand,

$$U_2(E, F; p) = -p_0 + p_2 + 2p_1 = -1 + 3p_1 + 2p_2 \leq 1 = u_2(a^*) = U_2(a^*; p),$$

whence $P_2(a^*) = \{p \in \Delta_2 : p_2 \leq 1 - (3/2)p_1\} = P(a^*)$ and

$$\|P(a^*)\| = \int_0^1 \left(\int_0^{\max\{0,1-(3/2)p_1\}} dp_2 \right) dp_1 = \int_0^{2/3} (1 - (3/2)p_1) dp_1 = \frac{1}{3}.$$

Determine now the expectations:

$$\int_{P(a^*)} p_1 dp = \int_0^{2/3} (1 - (3/2)p_1) p_1 dp_1 = \frac{2}{27},$$

$$\int_{P(a^*)} p_2 dp = \frac{1}{2} \int_0^{2/3} (1 - (3/2)p_1)^2 dp_1 = \frac{1}{9}.$$

The expected flexibilities are therefore $\varphi_1(a^*) = 2/9$ and $\varphi_2(a^*) = 1/3$, resulting in the commitment levels of $\kappa_1(a^*) = 5/9$ and $\kappa_2(a^*) = 1/3$. Hence, in the entry-accommodation Nash equilibrium, the entrant is 66.67% more committed than the incumbent.

Consider now the Nash equilibrium $\hat{a}^* = (\bar{E}, F)$, which contains a noncredible threat. For this, note that

$$U_2(A, \hat{a}_1^*; p) = 2p_0 + 2p_2 + p_1 = 2 - p_1 \leq 2 = u_2(\hat{a}^*) = U_2(\hat{a}^*; p),$$

for all $p \in P_2(\hat{a}^*) = \Delta_2$. On the other hand,

$$U_1(E, \hat{a}_2^*) = -2p_0 + 0 + p_2 = -2 + 2p_1 + 3p_2 \leq 0 = u_1(\hat{a}^*) = U_1(\hat{a}^*; p),$$

if and only if $p \in P_1(\hat{a}^*) = \{p : p_2 \leq (2/3)(1 - p_1)\}$. Therefore, $P(\hat{a}^*) = P_1(\hat{a}^*) = P_2(a^*) = P(a^*)$. Hence, we obtain the expected flexibilities, $\varphi_1(a^*) = 1/3$

and $\varphi_2(a^*) = 2/9$, implying the commitments of $\kappa_1(a^*) = 1/3$ and $\kappa_2(a^*) = 5/9$. As a result, in the entry-deterrence Nash equilibrium $\hat{a}^*$ the situation is reversed, with the incumbent being 66.67% more committed than the entrant.

5. Conclusion

In the physical world, two events are considered "simultaneous" when they take place at exactly the same time in the same frame of reference. In a game, players' actions are called "simultaneous moves" when, at the time one player chooses his action, he has not yet observed any of the other players' actions. Simultaneity in a game is therefore usually viewed as an informational imperfection, rather than a condition brought about by the timing of actions. Yet, it is the timing of actions just as much as the resulting information structure that creates advantages through commitment (or lack thereof) when players move sequentially, each being able to observe earlier moves.

By viewing the Nash equilibria of a (finite) dynamic game as equilibria of the corresponding normal-form game Γ, we abstract from the particular structure and from any generically unique subgame-perfect equilibrium. Each Nash equilibrium a^* of Γ can be characterized by a set $\mathcal{P}(a^*)$ which contains vectors $p = (p_i)$ in a richer canonical extension $\hat{\Gamma}(p)$. The results in this chapter embed the normal-form game Γ in the richer dynamic game $\hat{\Gamma}(p)$.[k] To determine player i's commitment ability in a given Nash equilibrium a^* we first determine his expected flexibility $\varphi_i(a^*)$ as the expected value of p_i in a random instance $\hat{\Gamma}(p)$, conditional on the fact that a^* can be sequentially implemented (i.e., $p \in \mathcal{P}(a^*)$). That player's commitment $\kappa_i(a^*) = \max\{0, 1 - N\varphi_i(a^*)\}$ is then the renormalized lack of flexibility, conditional on implementing the equilibrium in the canonical extension. Since κ_i can take on values between 0 and 1, and retains that range even on average across all players, for any size game, the measure retains informativeness. It is invariant with respect to probabilistically equivalent representations of the players' utility functions. In addition, the proposed measure of commitment is naturally adapted to the continuous nature of credibility (and commitment) in dynamic games, thus providing an indication about how small changes in the structure of the initial dynamic game can lead to a different subgame-perfect Nash equilibrium, without changing the underlying set of Nash equilibria.

---

[k]The somewhat unsurprising consequence of this embedding is that when the game $\hat{\Gamma}(p)$ is "sufficiently close" to Γ, that is, for $\|p\|$ small enough, any Nash equilibrium of Γ can be implemented sequentially in $\hat{\Gamma}(p)$. The implementation can be strong if all deviations from the best-responses in equilibrium lead to payoff differences that are uniformly bounded from below by some $\varepsilon > 0$. It is straightforward, and therefore asserted without proof, that the lack of "smooth" indifference around the equilibrium is not only sufficient but also necessary for a strong sequential implementation. The discreteness of choice around the equilibrium implies a certain robustness with respect to extensions of the game Γ, in the sense that the sets of possible equilibrium outcomes in Γ and $\hat{\Gamma}(p)$ coincide as long as $\|p\|$ is small enough.

Acknowledgments

The author is grateful for comments and suggestions by Bernhard von Stengel and others, as well as participants of the 2017 Conference on Game Theory and Management (GTM) at the Saint Petersburg State University and the 2017 European Meeting on Game Theory (SING13) at the Université Paris-Dauphine.

References

Bennett, E. and van Damme, E. [1990] Demand Commitment Bargaining — The Case of Apex Games, CentER Discussion Paper No. 9062, Tilburg University, Tilburg, Netherlands.

Berge, C. [1959] *Espaces Topologiques et Fonctions Multivoques* (Dunod, Paris), [English translation: [1963] *Topological Spaces*, Oliver and Boyd (Edinburgh, UK), reprinted: [1997] Dover Publications, Mineola, NY.

Breitmoser, Y. [2009] Demand commitments in majority bargaining or how formateurs get their way, *Int. J. Game Theory* **38**(2), 183–191.

Caruana, G. and Einav, L. [2008] A theory of endogenous commitment, *Rev. Econ. Stud.* **75**(1), 99–116.

Friedman, J. W. [1971] A non-cooperative equilibrium for supergames, *Rev. Econ. Stud.* **38**(1), 1–12.

Fudenberg, D. and Maskin, E. [1986] The folk theorem in repeated games with discounting or with incomplete information, *Econometrica* **54**(3), 533–554.

Fudenberg, D. and Tirole, J. [1991] *Game Theory* (MIT Press, Cambridge, MA).

Gal-Or, E. [1985] First mover and second mover advantages, *Int. Econ. Rev.* **26**(3), 649–653.

Henkel, J. [2002] The 1.5th mover advantage, *RAND J. Econ.* **33**(1), 156–170.

Luce, R. D. and Raiffa, H. [1957] *Games and Decisions* (Wiley, New York, NY).

Mas-Colell, A., Whinston, M. D. and Green, J. D. [1995] *Microeconomic Theory* (Oxford University Press, Oxford, UK).

Montero, M. and Vidal-Puga, J. J. [2007] "Demand Commitment in Legislative Bargaining, *Am. Political Sci. Rev.* **101**(4), 847–850.

Nash, J. F. [1950] Equilibrium points in n-person games, *Proc. Natl. Acad. Sci.* **36**(1), 48–49.

Osborne, M. J. and Rubinstein, A. [1984] *A Course in Game Theory* (MIT Press, Cambridge, MA).

Selten, R. [1965] Spieltheoretische behandlung eines oligopolmodells mit Nachfrageträgheit, *Zeitschrift für die Gesamte Staatswissenschaft* **121**, 301–324, 667–689.

Selten, R. [1992] "A demand commitment model of coalition bargaining, *Rational Interaction* Selten, R. (ed.) (Springer, Berlin, Germany), pp. 245–282.

von Stackelberg, H. [1934] *Marktform und Gleichgewicht* (Springer, Vienna, Austria).

Weber, T. A. [2014] A continuum of commitment, *Econ. Lett.* **124**(1), 67–73.

Winter, E. [1994] The demand commitment bargaining and snowballing cooperation, *Econ. Theory* **4**(2), 255–273.

Chapter 12

Equilibria in Dynamic Multicriteria Games

Anna Rettieva

Institute of Applied Mathematical Research
Karelian Research Centre of RAS
Pushkinskaya str., 11, Petrozavodsk 185910, Russia
annaret@krc.karelia.ru

Mathematical models involving more than one objective seem more adherent to real problems. Often players have more than one goal which are often not comparable. These situations are typical for game-theoretic models in economic and ecology. In this chapter, new approaches to construct equilibria in dynamic multicriteria games are constructed. We consider a dynamic, discrete-time, game model where the players use a common resource and have different criteria to optimize. First, we construct the guaranteed payoffs in a several ways. Then, we find an equilibrium as a solution of a Nash bargaining scheme with the guaranteed payoffs playing the role of *status quo* points. The obtained equilibrium, called a multicriteria Nash equilibrium, gives a possible solution concept for dynamic multicriteria games.

Keywords: Dynamic games; multicriteria games; Nash bargaining solution.

1. Introduction

Mathematical models involving more than one objective seem more adherent to real problems. Often players have more than one goal which are often not comparable. These situations are typical for game-theoretic models in economics and ecology. For example, in management problems the decision maker wants to maximize her profit and to minimize her production costs, in environmental agreement problems the players wish to maximize their production and to minimize their pollution reduction costs and so on. Hence, a multicriteria game approach helps to make decisions in multi-objective problems.

Shapley [1959] introduced the concept of multicriteria games that are games with vector payoffs, and gave a generalization of the classical Nash equilibrium to Pareto equilibrium for such games. In recent years, many authors have studied game problem with vector payoffs. Some concepts have been suggested to solve multicriteria games. Voorneveld *et al.* [2000] presented the notion of ideal Nash equilibrium, Patrone *et al.* [2007] connected multicriteria games with potential games, Pusillo and Tijs [2013] suggested the E-equilibrium concept and Pieri and Pusillo [2015] considered coalition formation processes in multicriteria games. The classical approach for solving multicriteria problems is to scalarize it by optimizing a weighted sum of the criteria. This approach draws just criticism: it is impossible to apply when the players' criteria are not comparable. However, the notion of Pareto equilibrium is the most studied concept in multicriteria game theory.

For cooperative multicriteria games the natural generalization of the Shapley value is used to distribute the cooperative payoff among the players.

Traditionally, equilibrium analysis in multicriteria problems is based on the static or steady-state variant. For dynamic multicriteria games the proposed equilibrium concepts do not assist in evaluating the players' behavior. As a matter of fact, the equilibrium design problem is underinvestigated in this case, in spite of the fact that dynamics appear to be widespread in real problems.

Furthermore, the situation with incomplete information, namely when the players' planning horizons are different and random, is considered. Models with random planning horizons in bioresource exploitation processes are most appropriate for describing reality: external random factors can cause a game breach and the participants know nothing about them *a priori*. As a matter of fact, the equilibrium behavior of the participants has not yet been examined in this case.

According to the aforesaid, equilibrium behavior design in dynamic models with many objectives calls for elaborating new methods. Thus, this chapter is dedicated to linking multicriteria games with dynamic games. A new approach to construct the equilibrium in dynamic games with many objectives is proposed.

2. Multicriteria Games and Solution Concepts

A multicriteria noncooperative game is

$$G = \langle (X_i)_{i \in N}, (u_i)_{i \in N} \rangle,$$

where $N = \{1, \ldots, n\}$ gives the set of players, X_i is the set of strategies of player i, and u_i denotes the payoff function of player i, $u_i \colon \prod_{i=1}^n X_i \to \mathbb{R}^m$, $i = 1, \ldots, n$.

Shapley [1959] gave a generalization of the classical Nash equilibrium to Pareto equilibrium for such games.

Definition 1. A strategy profile $x \in X = \prod_{i=1}^n X_i$ is a

(1) weak Pareto equilibrium if $\forall i \in N$

$$\neg \exists y_i \in X_i : u_i(y_i, x_{-i}) > u_i(x),$$

(2) strong Pareto equilibrium if $\forall\, i \in N$

$$\neg \exists y_i \in X_i : u_i(y_i, x_{-i}) \geqq u_i(x).$$

Here $a > b \Leftrightarrow a_i > b_i$, $a \geqq b \Leftrightarrow a_i \geq b_i$, $\forall\, i = 1, \ldots, m$.

Other solution concepts for multicriteria games, namely ideal Nash equilibrium and E-equilibrium, were introduced by Voorneveld *et al.* [2000] and Pusillo and Tijs [2013], respectively.

Patrone *et al.* [2007] connected multicriteria games with potential games, and Pieri and Pusillo [2015] considered coalition formation processes in multicriteria games.

Multicriteria games can be considered under cooperation also. Multicriteria cooperative game is defined as

$$\langle N, v \rangle,$$

where $N = \{1, \ldots, n\}$ is the set of players, $v : 2^N \to \mathbb{R}^m$ denotes the characteristic function, $v(\emptyset) = 0$ and

$$v(S) = \begin{pmatrix} v^1(S) \\ v^2(S) \\ \cdots \\ v^m(S) \end{pmatrix}, \quad \forall\, S \in 2^N.$$

For cooperative multicriteria games the natural generalization of the Shapley value is applied to distribute the cooperative payoff among the players.

Definition 2. The Shapley value $\phi(v)$ of the multicriteria game $\langle N, v \rangle$ is

$$\phi_i(v) = \sum_{S \subset N, i \in S} \frac{(s-1)!(n-s)!}{n!} \begin{pmatrix} v^1(S) - v^1(S\ \{i\}) \\ v^2(S) - v^2(S\ \{i\}) \\ \cdots \\ v^m(S) - v^m(S\ \{i\}) \end{pmatrix}.$$

3. Multicriteria Dynamic Game and Multicriteria Nash Equilibrium

Consider a bicriteria dynamic game with two participants in discrete time. The players exploit a common resource and both wish to optimize two different criteria. The state dynamics is in the form

$$x_{t+1} = f(x_t, u_{1t}, u_{2t}), \quad x_0 = x, \tag{1}$$

where $x_t \geq 0$ is the resource size at time $t \geq 0$ and $u_{it} \in U_i$ denotes the strategy of player i at time $t \geq 0$, $i = 1, 2$.

The payoff functions of the players over the infinite time horizon are defined by

$$
J_1 = \begin{pmatrix} J_1^1 = \sum_{t=0}^{\infty} \delta^t g_1^1(u_{1t}, u_{2t}) \\ J_1^2 = \sum_{t=0}^{\infty} \delta^t g_1^2(u_{1t}, u_{2t}) \end{pmatrix}, \quad J_2 = \begin{pmatrix} J_2^1 = \sum_{t=0}^{\infty} \delta^t g_2^1(u_{1t}, u_{2t}) \\ J_2^2 = \sum_{t=0}^{\infty} \delta^t g_2^2(u_{1t}, u_{2t}) \end{pmatrix}, \quad (2)
$$

where $g_i^j(u_{1t}, u_{2t}) \geq 0$ gives the instantaneous utility, $i, j = 1, 2$, and $\delta \in (0, 1)$ denotes a common discount factor.

In this chapter, we design the equilibrium in a multicriteria game using the Nash bargaining solution (Rettieva [2014]). Therefore, we begin with the construction of guaranteed payoffs which play the role of *status quo* points.

There are three possible concepts to determine the guaranteed payoffs. In the first one four guaranteed payoff points are obtained as the solutions of zero-sum games. In particular, the first guaranteed payoff point is a solution of zero-sum game where player 1 wishes to maximize her first criterion and player 2 wants to minimize it. Other points are obtained by analogy. Namely,

G_1^1 is the solution of zero-sum game $\langle I, II, U_1, U_2, J_1^1 \rangle$,

G_1^2 is the solution of zero-sum game $\langle I, II, U_1, U_2, J_1^2 \rangle$,

G_2^1 is the solution of zero-sum game $\langle I, II, U_1, U_2, J_2^1 \rangle$,

G_2^2 is the solution of zero-sum game $\langle I, II, U_1, U_2, J_2^2 \rangle$.

The second approach can be applied when the players' objectives are comparable. Consequently, the guaranteed payoff points for player 1 are obtained as the solution of a zero-sum game where she wants to maximize the sum of her criteria and player 2 wishes to minimize it. And, by analogy, for player 2. Namely,

G_1^1 and G_1^2 are the solutions of zero-sum game $\langle I, II, U_1, U_2, J_1^1 + J_1^2 \rangle$,

G_2^1 and G_2^2 are the solutions of zero-sum game $\langle I, II, U_1, U_2, J_2^1 + J_2^2 \rangle$.

In the third approach the guaranteed payoff points are constructed as the Nash equilibrium with the first and the second criteria of both players, respectively. Namely,

G_1^1 and G_2^1 are the Nash equilibrium payoffs in the game $\langle I, II, U_1, U_2, J_1^1, J_2^1 \rangle$,

G_1^2 and G_2^2 are the Nash equilibrium payoffs in the game $\langle I, II, U_1, U_2, J_1^2, J_2^2 \rangle$.

To construct multicriteria payoff functions we adopt the Nash products. The role of the *status quo* points belongs to the guaranteed payoffs of the players:

$$
H_1(u_{1t}, u_{2t}) = (J_1^1(u_{1t}, u_{2t}) - G_1^1)(J_1^2(u_{1t}, u_{2t}) - G_1^2), \quad (3)
$$

$$
H_2(u_{1t}, u_{2t}) = (J_2^1(u_{1t}, u_{2t}) - G_2^1)(J_2^2(u_{1t}, u_{2t}) - G_2^2). \quad (4)
$$

The next definition presents the suggested solution concept.

Definition 3. A strategy profile (u_{1t}^*, u_{2t}^*) is called a multicriteria Nash equilibrium of the problem (1), (2) if

$$H_1(u_{1t}^*, u_{2t}^*) \geq H_1(u_{1t}, u_{2t}^*), \quad \forall u_{1t} \in U_1, \tag{5}$$

$$H_2(u_{1t}^*, u_{2t}^*) \geq H_2(u_{1t}^*, u_{2t}), \quad \forall u_{2t} \in U_2. \tag{6}$$

Just like in the classical Nash equilibrium approach it is not profitable for both players to deviate from their equilibrium strategies. However, under the presented equilibrium concept players maximize the product of the differences between the optimal and guaranteed payoffs (3), (4).

Now we pass to a simple dynamic multicriteria model related to a bioresource management problem (fish catching) to show how the suggested concept works.

4. Dynamic Multicriteria Model with Infinite Horizon

Consider a two-player discrete-time game-theoretic bioresource management model with an identical planning horizon and discount factor for both players. Suppose that the two players (countries or fishing firms) harvest a fish stock over an infinite planning horizon. The fish population evolves according to the equation

$$x_{t+1} = \varepsilon x_t - u_{1t} - u_{2t}, \quad x_0 = x, \tag{7}$$

where $x_t \geq 0$ is the population size at time $t \geq 0$, $\varepsilon \geq 1$ denotes the natural birth rate, and $u_{it} \geq 0$ gives the catch of player i at time t, $i = 1, 2$. We assume that $\delta\varepsilon \geq 1$, where $\delta \in (0, 1)$ denotes the common discount factor.

Each player has two goals to optimize, they wish to maximize their profit from selling fish and minimize their catching cost. Suppose that the market price of the resource differs for both players, but their costs are identical and depend on both of players catches. Specifically, the payoff functions of the players over the infinite time horizon are defined by

$$J_1 = \begin{pmatrix} J_1^1 = \sum_{t=0}^{\infty} \delta^t p_1 u_{1t} \\ J_1^2 = -\sum_{t=0}^{\infty} \delta^t m u_{1t} u_{2t} \end{pmatrix}, \quad J_2 = \begin{pmatrix} J_2^1 = \sum_{t=0}^{\infty} \delta^t p_2 u_{2t} \\ J_2^2 = -\sum_{t=0}^{\infty} \delta^t m u_{1t} u_{2t} \end{pmatrix}, \tag{8}$$

where, for $i = 1, 2$, $p_i \geq 0$ is the market price of the resource for player i, $m \geq 0$ indicates the catching cost and $\delta \in (0, 1)$ denotes the discount factor.

4.1. *The guaranteed payoffs and multicriteria Nash equilibrium: Variant 1*

We begin with the construction of guaranteed payoffs using the Bellman optimality principle.

To determine the first guaranteed payoff point G_1^1 it is required to solve zero-sum game $\langle I, II, U_1, U_2, J_1^1 \rangle$. Let $V_1(x)$ be a value function for player 1, and $V_2(x)$ — for player 2.

According to the Bellman principle these functions satisfy

$$V_1(x) = \max_{u_1}\{p_1 u_1 + \delta V_1(\varepsilon x - u_1 - u_2)\},$$

$$V_2(x) = \max_{u_2}\{-p_1 u_1 + \delta V_2(\varepsilon x - u_1 - u_2)\}.$$

Assuming the value functions and the strategies have the linear forms $V_i(x) = A_i x + B_i$ and $u_i = a_i x + b_i$, $i = 1, 2$ we get the solution

$$u_{1t} = 0, \quad t = 1, 2, \ldots, \quad u_{21} = \varepsilon x_0, \quad u_{2t} = 0, \quad t = 2, \ldots,$$

so that the game ends in one time step.

Therefore, the guaranteed payoff is equal to zero $G_1^1 = 0$ and, by analogy, for player 2 — $G_2^1 = 0$.

The guaranteed payoff for the second criterion G_1^2: according to Bellman's optimality principle the value functions satisfy

$$V_1(x) = \max_{u_1}\{-m u_1 u_2 + \delta V_1(\varepsilon x - u_1 - u_2)\},$$

$$V_2(x) = \max_{u_2}\{m u_1 u_2 + \delta V_2(\varepsilon x - u_1 - u_2)\}.$$

Again assuming that the strategies are linear forms and the value functions are quadratic forms, say $V_i(x) = A_i x^2 + B_i x + D_i$ and $u_i = a_i x + b_i$, $i = 1, 2$, we get the solution

$$u_{1t} = u_{2t} = \frac{\delta \varepsilon^2 - 1}{2 \delta \varepsilon} x_t,$$

and the dynamics becomes

$$x_{t+1} = \frac{1}{\delta \varepsilon} x_t.$$

Hence $x_t = x_0 / (\delta \varepsilon)^t$ and the guaranteed payoff takes the form

$$G_1^2 = -\sum_{t=0}^{\infty} \delta^t m u_{1t} u_{2t} = -m \frac{\delta \varepsilon^2 - 1}{4 \delta} x_0^2.$$

By analogy we get

$$G_2^2 = G_1^2 = G = -m \frac{\delta \varepsilon^2 - 1}{4 \delta} x_0^2. \tag{9}$$

Since the first guaranteed payoffs are equal to zero then to define multicriteria Nash equilibrium of the problem (7), (8) it is required to solve the problem

$$J_1^1(u_{1t}, u_{2t})(J_1^2(u_{1t}, u_{2t}) - G) \to \max_{u_{1t}},$$

$$J_2^1(u_{1t}, u_{2t})(J_2^2(u_{1t}, u_{2t}) - G) \to \max_{u_{2t}}$$

or

$$\sum_{t=0}^{\infty} \delta^t p_1 u_{1t} \left(-\sum_{t=0}^{\infty} \delta^t m u_{1t} u_{2t} + m \frac{\delta \varepsilon^2 - 1}{4\delta} x_0^2 \right) \to \max_{u_{1t}},$$

$$\sum_{t=0}^{\infty} \delta^t p_2 u_{2t} \left(-\sum_{t=0}^{\infty} \delta^t m u_{1t} u_{2t} + m \frac{\delta \varepsilon^2 - 1}{4\delta} x_0^2 \right) \to \max_{u_{2t}}.$$

By analogy to the guaranteed payoff points' construction, using Bellman's optimality principle we get the multicriteria Nash optimal strategies

$$u_{1t}^N = u_{2t}^N = \frac{(\delta \varepsilon^2 - 1)(\varepsilon - 1)}{\varepsilon \delta^2 - 1 + \delta \varepsilon(\varepsilon - 1)} x_t, \tag{10}$$

and the dynamics under the multicriteria Nash equilibrium becomes

$$x_t = \left[\frac{\varepsilon^2 \delta + \varepsilon - 2}{\varepsilon \delta^2 - 1 + \delta \varepsilon(\varepsilon - 1)} \right]^t x_0. \tag{11}$$

4.2. *The guaranteed payoffs and multicriteria Nash equilibrium: Variant 2*

G_1^1 and G_1^2:

In the second variant of the guaranteed points' construction it is required to solve the zero-sum game $\langle I, II, U_1, U_2, J_1^1 + J_1^2 \rangle$, hence the players' objectives are

$$\sum_{t=0}^{\infty} \delta^t p_1 u_{1t} - \sum_{t=0}^{\infty} \delta^t m u_{1t} u_{2t} \to \max_{u_{1t}},$$

$$-\sum_{t=0}^{\infty} \delta^t p_1 u_{1t} + \sum_{t=0}^{\infty} \delta^t m u_{1t} u_{2t} \to \max_{u_{2t}}.$$

By analogy to Sec. 4.1 using Bellman's optimality principle we get the solution

$$u_{1t} = \frac{\delta \varepsilon^2 - 1}{2\delta \varepsilon} x_t - \frac{p_1(\delta \varepsilon^2 - 1)}{2m\delta \varepsilon(\varepsilon - 1)}, \quad u_{2t} = u_{1t} + \frac{p_1}{m},$$

and the dynamics becomes

$$x_t = \frac{x_0}{(\delta \varepsilon)^t} + \frac{p_1((\delta \varepsilon)^t - 1)}{m(\delta \varepsilon)^t(\varepsilon - 1)}.$$

Hence the guaranteed payoffs take the forms

$$G_1^1 = \sum_{t=0}^{\infty} \delta^t p_1 u_{1t} = \frac{p_1(\delta \varepsilon^2 - 1)}{2\delta(\varepsilon - 1)} x_0 - \frac{p_1^2(\delta \varepsilon^2 - 1)}{2m\delta(\varepsilon - 1)^2},$$

$$G_1^2 = -\sum_{t=0}^{\infty} \delta^t m u_{1t} u_{2t} = -\frac{m(\delta \varepsilon^2 - 1)}{4\delta} x_0^2 + \frac{p_1^2(\delta \varepsilon^2 - 1)}{4m\delta(\varepsilon - 1)^2}.$$

(12)

By analogy we get the guaranteed payoffs for player 2

$$G_2^1 = \sum_{t=0}^{\infty} \delta^t p_2 u_{2t} = \frac{p_2(\delta\varepsilon^2 - 1)}{2\delta(\varepsilon - 1)} x_0 - \frac{p_2^2(\delta\varepsilon^2 - 1)}{2m\delta(\varepsilon - 1)^2},$$

$$G_2^2 = -\sum_{t=0}^{\infty} \delta^t m u_{1t} u_{2t} = -\frac{m(\delta\varepsilon^2 - 1)}{4\delta} x_0^2 + \frac{p_2^2(\delta\varepsilon^2 - 1)}{4m\delta(\varepsilon - 1)^2}. \tag{13}$$

To determine the multicriteria Nash equilibrium of the problem (7), (8) it is required to solve the problem

$$\left(\sum_{t=0}^{\infty} \delta^t p_1 u_{1t} - G_1^1 \right) \left(-\sum_{t=0}^{\infty} \delta^t m u_{1t} u_{2t} - G_1^2 \right) \to \max_{u_{1t}},$$

$$\left(\sum_{t=0}^{\infty} \delta^t p_2 u_{2t} - G_2^1 \right) \left(-\sum_{t=0}^{\infty} \delta^t m u_{1t} u_{2t} - G_2^2 \right) \to \max_{u_{2t}}.$$

Again using Bellman's optimality principle and seeking the linear strategies we get the multicriteria Nash equilibrium

$$u_{1t}^N = \frac{(\delta\varepsilon^2 - 1)(\varepsilon - 1)}{\delta\varepsilon^2 - 1 + \delta\varepsilon(\varepsilon - 1)} x_t + \frac{\delta\varepsilon p_2 G_1^1(\varepsilon - 1) - G_2^1 p_1(\delta\varepsilon^2 - 1)}{2p_1 p_2(\delta\varepsilon^2 - 1 + \delta\varepsilon(\varepsilon - 1))},$$

$$u_{2t}^N = \frac{(\delta\varepsilon^2 - 1)(\varepsilon - 1)}{\delta\varepsilon^2 - 1 + \delta\varepsilon(\varepsilon - 1)} x_t + \frac{\delta\varepsilon p_1 G_2^1(\varepsilon - 1) - G_1^1 p_2(\delta\varepsilon^2 - 1)}{2p_1 p_2(\delta\varepsilon^2 - 1 + \delta\varepsilon(\varepsilon - 1))}, \tag{14}$$

and the dynamics under multicriteria Nash equilibrium is

$$x_t = \left[\frac{\delta\varepsilon^2 + \varepsilon - 2}{\delta\varepsilon^2 - 1 + \delta\varepsilon(\varepsilon - 1)} \right]^t x_0 + \frac{p_2 G_1^1 + p_1 G_2^1}{2p_1 p_2(\varepsilon - 1)}. \tag{15}$$

4.3. *The guaranteed payoffs and multicriteria Nash equilibrium: Variant 3*

In the third variant the guaranteed payoff points G_1^1 and G_2^1 are defined as Nash equilibrium in the game $\langle I, II, U_1, U_2, J_1^1, J_2^1 \rangle$, hence the value functions satisfy

$$V_1(x) = \max_{u_1}\{p_1 u_1 + \delta V_1(\varepsilon x - u_1 - u_2)\},$$

$$V_2(x) = \max_{u_2}\{p_2 u_2 + \delta V_2(\varepsilon x - u_1 - u_2)\}.$$

As usual, we seek for the players' value functions and strategies as linear forms. In this case we get the solution

$$u_{1t} = u_{2t} = \frac{\varepsilon\delta - 1}{\delta} x_t,$$

and the dynamics becomes

$$x_t = \left[\frac{2 - \varepsilon\delta}{\delta} \right]^2 x_0.$$

Hence, the guaranteed payoffs take the forms

$$G_1^1 = \frac{p_1}{\delta} x_0, \quad G_2^1 = \frac{p_2}{\delta} x_0. \tag{16}$$

By analogy, determining the Nash equilibrium in the game with the second criteria of both players $J_1^2(u_{1t}, u_{2t})$ and $J_2^2(u_{1t}, u_{2t})$ we get two more guaranteed payoff points

$$G_1^2 = G_2^2 = -m\frac{\delta\varepsilon^2 - 1}{4\delta} x_0^2. \tag{17}$$

The multicriteria Nash equilibrium in this case has the same form as in variant 2 (14), (15), but with proper guaranteed payoffs (16), (17).

4.4. *Comparison and numerical experiments*

Next we compare players' strategies and the size of the population for different variants of the guaranteed points' construction. There are two possible cases depending on the parameters of the model: in the first one the worst variant for the environment is the second one (when we determine the guaranteed payoffs as the solutions of a zero-sum games with the sum of the criteria), in the second case — the worst variant for the environment is the first one (when we determine guaranteed payoffs as the solutions of zero-sum games).

If $p_2 \geq p_1$, $x_0 \leq \frac{p_1+p_2}{2}$ or $p_2 < p_1$, $x_0 < \frac{p_2(\delta\varepsilon^2-1)-p_1\varepsilon\delta(\varepsilon-1)}{\varepsilon\delta-1}$, then

$$x_t^{\text{var2}} \leq x_t^{\text{var1}} \leq x_t^{\text{var3}},$$

$$u_{it}^{\text{var2}} \leq u_{it}^{\text{var1}} \leq u_{it}^{\text{var3}}, \quad i = 1, 2.$$

If $x_0 > \frac{\max\{p_1, p_2\}}{m(\varepsilon-1)}$, then

$$x_t^{\text{var1}} \leq x_t^{\text{var2}} \leq x_t^{\text{var3}},$$

$$u_{it}^{\text{var1}} \leq u_{it}^{\text{var2}} \leq u_{it}^{\text{var3}}, \quad i = 1, 2.$$

We have performed numerical simulation for a 50-step game with the following parameters:

$$\varepsilon = 1.3, \quad p_1 = 100, \quad p_2 = 150, \quad m = 50, \quad \delta = 0.8.$$

Figure 1 shows the dynamics of the population size, whereas Fig. 2 shows the catch of player 1 for different variants of the guaranteed payoffs' construction. As one can notice the worst variant for the environment is the first one since it leads to overfishing. The variant where the guaranteed payoffs are determined as Nash equilibrium is beneficial to both players and, moreover, improves the ecological situation.

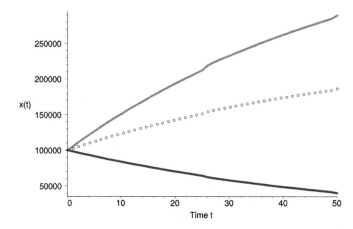

Fig. 1. Population size: dark — 1 variant, dotted — 2 variant, light — 3 variant.

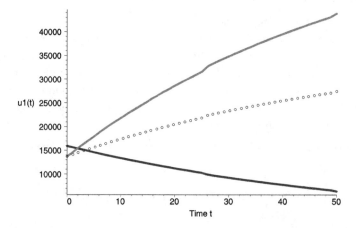

Fig. 2. The catch of player 1: dark — 1 variant, dotted — 2 variant, light — 3 variant.

5. Dynamic Multicriteria Model with Random Harvesting Times

A meaningful applied problem is to find equilibrium payoffs in the case of different planning horizons. When one player exploits a bioresource for a shorter period than the other, the former joins the exploitation process (in our case, fishing) for a fixed time. But this player has a smaller planning horizon than his/her partner; and so, the player under consideration is interested in gaining more from exploitation than the player who continues harvesting.

A model with random planning horizons in the bioresource exploitation process is most appropriate to describe reality: external random factors can cause a game breach without *a priori* knowledge of the participants. For instance, fishing firms can go bankrupt, their fleet can be damaged, etc. In the case of countries, negative factors include an economic crisis, abrupt variations in the rate of inflation, international or national economic and political situations, and so on. All these processes can possibly interrupt the game process, and the equilibrium behavior of the participants of such games has yet to be examined.

We now explore a model in which the players possess heterogeneous planning horizons (Rettieva [2015]). By assumption, the players stop joint exploitation at random time steps: external stochastic processes can cause a game breach.

Suppose that players 1 and 2 harvest the fish stock during n_1 and n_2 steps, respectively. Here n_1 represents a discrete random variable taking values $\{1, \ldots, n\}$ with the corresponding probabilities $\{\theta_1, \ldots, \theta_n\}$. Similarly, n_2 is a discrete random variable with the value set and the probabilities $\{\omega_1, \ldots, \omega_n\}$. We assume that the planning horizons are independent. Therefore, during the time period $[0, n_1]$ or $[0, n_2]$ the players harvest the same stock, and the problem consists in evaluating their optimal strategies.

The payoffs of the players are determined via the expectation operator:

$$
J_1^1 = E\left\{ \sum_{t=1}^{n_1} \delta^t p_1 u_{1t} I_{\{n_1 \le n_2\}} + \left(\sum_{t=1}^{n_2} \delta^t p_1 u_{1t} + \sum_{t=n_2+1}^{n_1} \delta^t p_1 u_{1t}^a \right) I_{\{n_1 > n_2\}} \right\}
$$

$$
= \sum_{n_1=1}^{n} \theta_{n_1} \left[\sum_{n_2=n_1}^{n} \omega_{n_2} \sum_{t=1}^{n_1} \delta^t p_1 u_{1t} + \sum_{n_2=1}^{n_1-1} \omega_{n_2} \left(\sum_{t=1}^{n_2} \delta^t p_1 u_{1t} + \sum_{t=n_2+1}^{n_1} \delta^t p_1 u_{1t}^a \right) \right],
$$
(18)

$$
J_2^1 = E\left\{ \sum_{t=1}^{n_2} \delta^t p_2 u_{2t} I_{\{n_2 \le n_1\}} + \left(\sum_{t=1}^{n_1} \delta^t p_2 u_{2t} + \sum_{t=n_1+1}^{n_2} \delta^t p_2 u_{2t}^a \right) I_{\{n_2 > n_1\}} \right\}
$$

$$
= \sum_{n_2=1}^{n} \omega_{n_2} \left[\sum_{n_1=n_2}^{n} \theta_{n_1} \sum_{t=1}^{n_2} \delta^t p_2 u_{2t} + \sum_{n_1=1}^{n_2-1} \theta_{n_1} \left(\sum_{t=1}^{n_1} \delta^t p_2 u_{2t} + \sum_{t=n_1+1}^{n_2} \delta^t p_2 u_{2t}^a \right) \right],
$$
(19)

where u_{it}^a specifies the strategy of player i when its partner leaves the game, $i = 1, 2$.

And for the second criteria we suppose that if the player harvests the stock alone, then there is no cost, hence

$$
J_1^2 = E\left\{ -\sum_{t=1}^{n_1} \delta^t m u_{1t} u_{2t} I_{\{n_1 \le n_2\}} - \sum_{t=1}^{n_2} \delta^t m u_{1t} u_{2t} I_{\{n_1 > n_2\}} \right\}
$$

$$
= -\sum_{n_1=1}^{n} \theta_{n_1} \left[\sum_{n_2=n_1}^{n} \omega_{n_2} \sum_{t=1}^{n_1} \delta^t m u_{1t} u_{2t} + \sum_{n_2=1}^{n_1-1} \omega_{n_2} \sum_{t=1}^{n_2} \delta^t m u_{1t} u_{2t} \right], \quad (20)
$$

$$
J_2^2 = E\left\{ -\sum_{t=1}^{n_2} \delta^t m u_{1t} u_{2t} I_{\{n_2 \le n_1\}} - \sum_{t=1}^{n_1} \delta^t m u_{1t} u_{2t} I_{\{n_2 > n_1\}} \right\}
$$

$$
= -\sum_{n_2=1}^{n} \omega_{n_2} \left[\sum_{n_1=n_2}^{n} \theta_{n_1} \sum_{t=1}^{n_2} \delta^t m u_{1t} u_{2t} + \sum_{n_1=1}^{n_2-1} \theta_{n_1} \sum_{t=1}^{n_1} \delta^t m u_{1t} u_{2t} \right]. \quad (21)
$$

For the duration of the game, the payoffs (Bellman functions) of the players have the form

$$V_1^{1N}(1,x) = \max_{u_{11}^N,\ldots,u_{1n}^N} \left\{ \sum_{n_1=1}^{n} \theta_{n_1} \left[\sum_{n_2=n_1}^{n} \omega_{n_2} \sum_{t=1}^{n_1} \delta^t p_1 u_{1t}^N \right. \right.$$

$$+ \sum_{n_2=1}^{n_1-1} \omega_{n_2} \left(\sum_{t=1}^{n_2} \delta^t p_1 u_{1t}^N + \sum_{t=n_2+1}^{n_1} \delta^t p_1 u_{1t}^a \right) \left. \right] \left. \right\},$$

$$V_2^{1N}(1,x) = \max_{u_{21}^N,\ldots,u_{1n}^N} \left\{ \sum_{n_2=1}^{n} \omega_{n_2} \left[\sum_{n_1=n_2}^{n} \theta_{n_1} \sum_{t=1}^{n_2} \delta^t p_2 u_{2t}^N \right. \right.$$

$$+ \sum_{n_1=1}^{n_2-1} \theta_{n_1} \left(\sum_{t=1}^{n_1} \delta^t p_2 u_{2t}^N + \sum_{t=n_1+1}^{n_2} \delta^t p_2 u_{2t}^a \right) \left. \right] \left. \right\}.$$

$$V_1^{2N}(1,x) = \max_{u_{11}^N,\ldots,u_{1n}^N} \left\{ -\sum_{n_1=1}^{n} \theta_{n_1} \left[\sum_{n_2=n_1}^{n} \omega_{n_2} \sum_{t=1}^{n_1} \delta^t m u_{1t}^N u_{2t}^N \right. \right.$$

$$+ \sum_{n_2=1}^{n_1-1} \omega_{n_2} \sum_{t=1}^{n_2} \delta^t m u_{1t}^N u_{2t}^N \left. \right] \left. \right\},$$

$$V_2^{2N}(1,x) = \max_{u_{21}^N,\ldots,u_{2n}^N} \left\{ -\sum_{n_2=1}^{n} \omega_{n_2} \left[\sum_{n_1=n_2}^{n} \theta_{n_1} \sum_{t=1}^{n_2} \delta^t m u_{1t}^N u_{2t}^N \right. \right.$$

$$+ \sum_{n_1=1}^{n_2-1} \theta_{n_1} \sum_{t=1}^{n_1} \delta^t m u_{1t}^N u_{2t}^N \left. \right] \left. \right\}.$$

Further exposition operates payoffs gained by the players as the game reaches step τ, $\tau = 1, 2, \ldots$. Note that the probabilities that player 1 continues harvesting for $\tau, \tau+1, \ldots, n$ steps constitute

$$\frac{\theta_\tau}{\sum_{l=\tau}^n \theta_l}, \quad \frac{\theta_{\tau+1}}{\sum_{l=\tau}^n \theta_l}, \ldots, \frac{\theta_n}{\sum_{l=\tau}^n \theta_l}.$$

Hence, as step τ occurs, the Bellman functions $V_i^{jN}(\tau, x)$, $i, j = 1, 2$ of the players have the form

$$V_1^{1N}(\tau, x) = \max_{u_{1\tau}^N,\ldots,u_{1n}^N} \left\{ \sum_{n_1=\tau}^{n} \frac{\theta_{n_1}}{\sum_{l=\tau}^n \theta_l} \left[\sum_{n_2=n_1}^{n} \frac{\omega_{n_2}}{\sum_{l=\tau}^n \omega_l} \sum_{t=\tau}^{n_1} \delta^t p_1 u_{1t}^N \right. \right.$$

$$+ \sum_{n_2=\tau}^{n_1-1} \frac{\omega_{n_2}}{\sum_{l=\tau}^n \omega_l} \sum_{t=\tau}^{n_2} \delta^t p_1 u_{1t}^N + V_1^a(\tau, n_1) \left. \right] \left. \right\}, \tag{22}$$

$$V_2^{1N}(\tau, x) = \max_{u_{2\tau}^N, \ldots, u_{1n}^N} \left\{ \sum_{n_2=\tau}^{n} \frac{\omega_{n_2}}{\sum_{l=\tau}^{n} \omega_l} \left[\sum_{n_1=n_2}^{n} \frac{\theta_{n_1}}{\sum_{l=\tau}^{n} \theta_l} \sum_{t=\tau}^{n_2} \delta^t p_2 u_{2t}^N \right. \right.$$

$$\left. \left. + \sum_{n_1=\tau}^{n_2-1} \frac{\theta_{n_1}}{\sum_{l=\tau}^{n} \omega_l} \sum_{t=\tau}^{n_1} \delta^t p_2 u_{2t}^N + V_2^a(\tau, n_2) \right] \right\}, \tag{23}$$

$$V_1^{2N}(\tau, x) = \max_{u_{1\tau}^N, \ldots, u_{1n}^N} \left\{ -\sum_{n_1=\tau}^{n} \frac{\theta_{n_1}}{\sum_{l=\tau}^{n} \theta_l} \left[\sum_{n_2=n_1}^{n} \frac{\omega_{n_2}}{\sum_{l=\tau}^{n} \omega_l} \sum_{t=\tau}^{n_1} \delta^t m u_{1t}^N u_{2t}^N \right. \right.$$

$$\left. \left. + \sum_{n_2=\tau}^{n_1-1} \frac{\omega_{n_2}}{\sum_{l=\tau}^{n} \omega_l} \sum_{t=\tau}^{n_2} \delta^t m u_{1t}^N u_{2t}^N \right] \right\}, \tag{24}$$

$$V_2^{2N}(\tau, x) = \max_{u_{2\tau}^N, \ldots, u_{1n}^N} \left\{ -\sum_{n_2=\tau}^{n} \frac{\omega_{n_2}}{\sum_{l=\tau}^{n} \omega_l} \left[\sum_{n_1=n_2}^{n} \frac{\theta_{n_1}}{\sum_{l=\tau}^{n} \theta_l} \sum_{t=\tau}^{n_2} \delta^t m u_{1t}^N u_{2t}^N \right. \right.$$

$$\left. \left. + \sum_{n_1=\tau}^{n_2-1} \frac{\theta_{n_1}}{\sum_{l=\tau}^{n} \omega_l} \sum_{t=\tau}^{n_1} \delta^t m u_{1t}^N u_{2t}^N \right] \right\}, \tag{25}$$

where

$$V_1^a(\tau, n_1) = \sum_{n_2=\tau}^{n_1-1} \frac{\omega_{n_2}}{\sum_{l=\tau}^{n} \omega_l} \sum_{t=n_2+1}^{n_1} \delta^t p_1 u_{1t}^a,$$

$$V_2^a(\tau, n_2) = \sum_{n_1=\tau}^{n_2-1} \frac{\theta_{n_1}}{\sum_{l=\tau}^{n} \theta_l} \sum_{t=n_1+1}^{n_2} \delta^t p_2 u_{2t}^a$$

are the profits when player i exploits the stock alone. As we will show later these profits can be estimated easily as (30).

Appendix A establishes a relationship between $V_i^{jN}(\tau, x)$ and $V_i^{jN}(\tau+1, x)$ of the form

$$V_1^{1N}(\tau, x) = \delta^\tau p_1 u_{1\tau}^N + P_\tau^{\tau+1} V_1^{1N}(\tau+1, x) + C_{1\tau} \sum_{n_1=\tau+1}^{n} \theta_{n_1} \sum_{t=\tau}^{n_1} \delta^t p_1 u_{1t}^a, \tag{26}$$

$$V_2^{1N}(\tau, x) = \delta^\tau p_2 u_{2\tau}^N + P_\tau^{\tau+1} V_2^{1N}(\tau+1, x) + C_{2\tau} \sum_{n_2=\tau+1}^{n} \omega_{n_2} \sum_{t=\tau}^{n_2} \delta^t p_2 u_{2t}^a, \tag{27}$$

$$V_1^{2N}(\tau, x) = -\delta^\tau m u_{1\tau}^N u_{2\tau}^N + P_\tau^{\tau+1} V_1^{2N}(\tau+1, x), \tag{28}$$

$$V_2^{2N}(\tau, x) = -\delta^\tau m u_{1\tau}^N u_{2\tau}^N + P_\tau^{\tau+1} V_2^{2N}(\tau+1, x), \tag{29}$$

where

$$P_\tau^{\tau+1} = \frac{\sum_{l=\tau+1}^{n} \omega_l}{\sum_{l=\tau}^{n} \omega_l} \frac{\sum_{l=\tau+1}^{n} \theta_l}{\sum_{l=\tau}^{n} \theta_l}, \quad C_{1\tau} = \frac{\omega_\tau}{\sum_{l=\tau}^{n} \omega_l} \frac{1}{\sum_{l=\tau}^{n} \theta_l},$$

$$C_{2\tau} = \frac{\theta_\tau}{\sum_{l=\tau}^{n} \theta_l} \frac{1}{\sum_{l=\tau}^{n} \omega_l}.$$

Now, it is necessary to find a player's strategy in the case when its opponent leaves the game. Suppose that player 1 has a smaller planning horizon than player 2. Consider the time period $[n_1 + 1, n_2]$, where only player 2 harvests the fish stock. First, we analyze the one-shot game and assume that at the end step the player receives the entire residual bioresource. We emphasize that such an assumption means a certain compensation for the unexploited bioresource (not its complete exhaustion). Let the initial size of the population be x. Then the second player's profit is

$$H_{21}(u_{21}^a) = p_2 u_{21}^a + \delta(\varepsilon x - u_{21}^a).$$

As $p_2 > \delta$ then $u_{21}^a = \varepsilon x$ and the payoff is in the form

$$H_{21}(u_{21}^a) = p_2 \varepsilon x.$$

Hence the objective function of player 2 for the two-step game is

$$H_{22}(u_{22}^a) = p_2 u_{22}^a + \delta p_2 \varepsilon (\varepsilon x - u_{22}^a).$$

Since we assumed that $\delta \varepsilon > 1$ one has $u_{22}^a = 0$ and

$$H_{22}(u_{22}^a) = \delta p_2 \varepsilon^2 x.$$

Hence the objective function of player 2 for the three-step game is

$$H_{23}(u_{23}^a) = p_2 u_{23}^a + \delta p_2 \varepsilon^2 (\varepsilon x - u_{23}^a)$$

and again $u_{23}^a = 0$ as $\delta \varepsilon^2 > 1$.

Hence we conclude that player 2's optimal strategy is to harvest the stock only at the last step, thus

$$u_{2n_2}^a = \varepsilon x_{n_2}, \quad u_{2t}^a = 0, \quad t = n_1 + 1, \ldots, n_2 - 1.$$

By analogy for player 1, we get the profits when player i, $i = 1, 2$ exploits the stock alone in the form

$$V_1^a(\tau, n_1) = \sum_{n_2=\tau}^{n_1-1} \frac{\omega_{n_2}}{\sum_{l=\tau}^{n} \omega_l} \delta^{n_1} p_1 u_{1n_1}^a, \quad V_2^a(\tau, n_2) = \sum_{n_1=\tau}^{n_2-1} \frac{\theta_{n_1}}{\sum_{l=\tau}^{n} \theta_l} \delta^{n_2} p_2 u_{2n_2}^a, \quad (30)$$

and the relations (26), (27) become

$$V_1^{1N}(\tau, x) = \delta^\tau p_1 u_{1\tau}^N + P_\tau^{\tau+1} V_1^{1N}(\tau + 1, x) + C_{1\tau} \sum_{n_1=\tau+1}^{n} \theta_{n_1} \delta^{n_1} p_1 u_{1n_1}^a, \quad (31)$$

$$V_2^{1N}(\tau, x) = \delta^\tau p_2 u_{2\tau}^N + P_\tau^{\tau+1} V_2^{1N}(\tau + 1, x) + C_{2\tau} \sum_{n_2=\tau+1}^{n} \omega_{n_2} \delta^{n_2} p_2 u_{2n_2}^a. \quad (32)$$

To construct the multicriteria Nash equilibrium it is required to determine the guaranteed payoffs (variant 1).

$G_1^1(\tau, x)$: Using the relations (31), (32) and searching for linear value functions $V_i(\tau, x) = A_i^\tau x + B_i^\tau$, $i = 1, 2$ we get

$$A_1^\tau x + B_1^\tau = \delta^\tau p_1 u_1 + P_\tau^{\tau+1}(A_1^\tau(\varepsilon x - u_1 - u_2) + B_1^\tau)$$

$$+ C_{1\tau} \sum_{n_1=\tau+1}^{n} \theta_{n_1} \delta^{n_1} p_1 u_{1n_1}^a,$$

$$A_2^\tau x + B_2^\tau = -\delta^\tau p_1 u_1 - P_\tau^{\tau+1}(A_2^\tau(\varepsilon x - u_1 - u_2) + B_2^\tau)$$

$$- C_{2\tau} \sum_{n_2=\tau+1}^{n} \omega_{n_2} \delta^{n_2} p_2 u_{2n_2}^a$$

and as in Sec. 4.1 we conclude that the game ends in one step, hence

$$G_1^1(\tau, x) = 0.$$

$G_1^2(\tau, x)$: Using the relations (28), (29) and searching for quadratic value functions $V_i(\tau, x) = A_i^\tau x^2 + B_i^\tau x + D_i^\tau$, $i = 1, 2$ we get

$$A_1^\tau x^2 + B_1^\tau x + D_1^\tau = -\delta^\tau m u_1 u_2 + P_\tau^{\tau+1}(A_1^\tau(\varepsilon x - u_1 - u_2)^2$$

$$+ B_1^\tau(\varepsilon x - u_1 - u_2) + D_1^\tau),$$

$$A_2^\tau x^2 + B_2^\tau x + D_2^\tau = \delta^\tau m u_1 u_2 - P_\tau^{\tau+1}(A_2^\tau(\varepsilon x - u_1 - u_2)^2$$

$$+ B_2^\tau(\varepsilon x - u_1 - u_2) + D_2^\tau),$$

hence

$$G_1^2(\tau, x) = G(\tau, x) = -\frac{m\delta^\tau(P_\tau^{\tau+1}\varepsilon^2 - 1)}{4P_\tau^{\tau+1}} x^2.$$

By analogy for player 2

$$G_2^1(\tau, x) = 0, \quad G_2^2(\tau, x) = G(\tau, x) = -\frac{m\delta^\tau(P_\tau^{\tau+1}\varepsilon^2 - 1)}{4P_\tau^{\tau+1}} x^2.$$

Therefore, to construct the multicriteria Nash equilibrium strategies it is required to solve the problem

$$V_1^{1N}(1, x)(V_1^{2N}(1, x) - G(1, x)) \to \max_{u_1}, \qquad (33)$$

$$V_2^{1N}(1, x)(V_2^{2N}(1, x) - G(1, x)) \to \max_{u_2}, \qquad (34)$$

where $V_i^{jN}(1, x)$ have the forms (22)–(25), $i, j = 1, 2$.

Theorem 1. *The multicriteria Nash equilibrium payoffs in the problem* (7), (18)–(21) *with random planning horizons have the form*

$$V_i^{1N}(n-k,x) = \delta^{n-k}p_i u_{in-k}^N + p_i P_{n-k}^{n-k+1}(\varepsilon x - u_{1n-k}^N - u_{2n-k}^N)$$

$$\cdot\,[\delta^{n-k+1}\gamma_{in-k+1} + \sum_{j=n-k+2}^{n-1}\delta^j\gamma_{ij}\prod_{l=n-k+2}^{j-2}P_l^{l+1}(\varepsilon - \gamma_{1l} - \gamma_{2l})]$$

$$+\sum_{l=1}^{k}P_{n-k}^{n-l}C_{in-l}V_i^l(n_i),\quad i=1,2,\tag{35}$$

where

$$V_1^l(n_1) = \sum_{n_1=n-l+1}^{n}\theta_{n_1}\delta^{n_1}p_1 u_{1n_1}^a,\quad V_2^l(n_2) = \sum_{n_2=n-l+1}^{n}\omega_{n_2}\delta^{n_2}p_2 u_{2n_2}^a,$$

$$V_i^{2N}(n-k,x) = -\delta^{n-k}mu_{1n-k}^N u_{2n-k}^N + P_{n-k}^{n-k+1}(\varepsilon x - u_{1n-k}^N - u_{2n-k}^N)^2$$

$$\cdot\,[-\delta^{n-k+1}m\gamma_{1n-k+1}\gamma_{2n-k+1}$$

$$-\sum_{j=n-k+1}^{n-2}\delta^{j+1}m\gamma_{1j+1}\gamma_{2j+1}(\varepsilon - \gamma_{1j} - \gamma_{2j})^2$$

$$\cdot\,\prod_{l=n-k+1}^{j}P_l^{l+1} + A\prod_{l=n-2}^{j}P_l^{l+1}(\varepsilon - \gamma_{1l} - \gamma_{2j})^2$$

$$\times\,P_{n-1}^n(\varepsilon - \gamma_{1n_1} - \gamma_{2n_1})],\quad i=1,2.\tag{36}$$

The multicriteria Nash equilibrium strategies are related by

$$\gamma_{2n-k}^N = \frac{2P_{n-k}^{n-k+1}\delta^{n-k+1}m\gamma_{1n-k+1}\gamma_{2n-k+1}(p_1 - p_2) + \varepsilon(K_{n-k}^1 - K_{n-k}^2)}{-\delta^{n-k}mp_2 - K_{n-k}^1 - K_{n-k}^2}$$

$$+\,\gamma_{1n-k}^N\frac{-\delta^{n-k}mp_1 - K_{n-k}^1 - K_{n-k}^2}{-\delta^{n-k}mp_2 - K_{n-k}^1 - K_{n-k}^2},\tag{37}$$

where

$$K_{n-k}^i = 2Ap_i\prod_{l=k}^{n}P_{n-l}^l\left(-\delta^{n-k}m\gamma_{in-k+1}\right.$$

$$-\sum_{l=k+1}^{n-1}\prod_{j=n-l}^{l}P_{j-1}^j m\delta^{n-l}\gamma_{il+1}(\varepsilon - \gamma_{1l} - \gamma_{2l})$$

$$\left.-\prod_{j=n-k+1}^{n-1}P_j^{j+1}A(\varepsilon - \gamma_{1j} - \gamma_{2j})\right),\quad i=1,2,\quad A = -\frac{m\delta^n(\varepsilon^2 - 1)}{4}.$$

The strategy of player 1 at the last step (the quantity γ_{1n-1}^N) is evaluated through one of the first-order optimality conditions.

Proof. The proof of this result is given in Appendix B. □

Appendix A

We give a proof for player 1's first criterion (i.e., find a relationship between $V_1^{1N}(\tau, x)$ and $V_1^{1N}(\tau + 1, x)$). In the case of player 2 and both players' second criteria, the line of reasoning is the same. Using (22), construct the Bellman function of player 1 as the game reaches step τ:

$$
V_1^N(\tau, x) = \max_{u_{1\tau}^N, \ldots, u_{1n}^N} \left\{ \frac{\theta_\tau}{\sum_{l=\tau}^n \theta_l} \sum_{n_2=\tau}^n \frac{\omega_{n_2}}{\sum_{l=\tau}^n \omega_l} \delta^\tau p_1 u_{1\tau}^N \right.
$$

$$
+ \sum_{n_1=\tau+1}^n \frac{\theta_{n_1}}{\sum_{l=\tau}^n \theta_l} \left[\sum_{n_2=n_1}^n \frac{\omega_{n_2}}{\sum_{l=\tau}^n \omega_l} \sum_{t=\tau}^{n_1} \delta^t p_1 u_{1t}^N \right.
$$

$$
\left. \left. + \sum_{n_2=\tau}^{n_1-1} \frac{\omega_{n_2}}{\sum_{l=\tau}^n \omega_l} \sum_{t=\tau}^{n_2} \delta^t p_1 u_{1t}^N + V_1^a(\tau, n_1) \right] \right\}
$$

$$
= \frac{\theta_\tau}{\sum_{l=\tau}^n \theta_l} \delta^\tau p_1 u_{1\tau}^N + \sum_{n_1=\tau+1}^n \frac{\theta_{n_1}}{\sum_{l=\tau}^n \theta_l}
$$

$$
\times \left[\sum_{n_2=n_1}^n \frac{\omega_{n_2}}{\sum_{l=\tau}^n \omega_l} \left(\sum_{t=\tau+1}^{n_1} \delta^t p_1 u_{1t}^N + \delta^\tau p_1 u_{1\tau}^N \right) \right.
$$

$$
\left. + \sum_{n_2=\tau}^{n_1-1} \frac{\omega_{n_2}}{\sum_{l=\tau}^n \omega_l} \left(\sum_{t=\tau+1}^{n_2} \delta^t p_1 u_{1t}^N + \delta^\tau p_1 u_{1\tau}^N \right) + V_1^a(\tau, n_1) \right]
$$

$$
= \delta^\tau p_1 u_{1\tau}^N + \sum_{n_1=\tau+1}^n \frac{\theta_{n_1}}{\sum_{l=\tau}^n \theta_l}
$$

$$
\cdot \left[\sum_{n_2=n_1}^n \frac{\omega_{n_2}}{\sum_{l=\tau}^n \omega_l} \sum_{t=\tau+1}^{n_1} \delta^t p_1 u_{1t}^N \right.
$$

$$
\left. + \sum_{n_2=\tau+1}^{n_1-1} \frac{\omega_{n_2}}{\sum_{l=\tau}^n \omega_l} \sum_{t=\tau+1}^{n_2} \delta^t p_1 u_{1t}^N + V_1^a(\tau, n_1) \right]
$$

$$
= \delta^\tau p_1 u_{1\tau}^N + \sum_{n_1=\tau+1}^n \frac{\theta_{n_1}}{\sum_{l=\tau+1}^n \theta_l} \frac{\sum_{l=\tau+1}^n \theta_l}{\sum_{l=\tau}^n \theta_l}
$$

$$
\times \left[\sum_{n_2=n_1}^n \frac{\omega_{n_2}}{\sum_{l=\tau+1}^n \omega_l} \frac{\sum_{l=\tau+1}^n \omega_l}{\sum_{l=\tau}^n \omega_l} \sum_{t=\tau+1}^{n_1} \delta^t p_1 u_{1t}^N \right.
$$

$$+ \sum_{n_2=\tau+1}^{n_1-1} \frac{\omega_{n_2}}{\sum_{l=\tau+1}^{n} \omega_l} \frac{\sum_{l=\tau+1}^{n} \omega_l}{\sum_{l=\tau}^{n} \omega_l} \sum_{t=\tau+1}^{n_2} \delta^t p_1 u_{1t}^N + \frac{\omega_\tau}{\sum_{l=\tau}^{n} \omega_l} \sum_{t=\tau}^{n_1} \delta^t p_1 u_{1t}^a$$

$$+ \frac{\sum_{l=\tau+1}^{n} \omega_l}{\sum_{l=\tau}^{n} \omega_l} V_1^a (\tau+1, n_1) \Bigg]$$

$$= \delta^\tau p_1 u_{1\tau}^N + P_\tau^{\tau+1} V_1^{1N}(\tau+1, x) + C_{1\tau} \sum_{n_1=\tau+1}^{n} \theta_{n_1} \sum_{t=\tau}^{n_1} \delta^t p_1 u_{1t}^a,$$

where

$$P_\tau^{\tau+1} = \frac{\sum_{l=\tau+1}^{n} \omega_l}{\sum_{l=\tau}^{n} \omega_l} \frac{\sum_{l=\tau+1}^{n} \theta_l}{\sum_{l=\tau}^{n} \theta_l}, \quad C_{1\tau} = \frac{\omega_\tau}{\sum_{l=\tau}^{n} \omega_l \sum_{l=\tau}^{n} \theta_l}.$$

Similarly, we establish a relationship between $V_2^{1N}(\tau, x)$ and $V_2^{1N}(\tau+1, x)$ in the form

$$V_2^{1N}(\tau, x) = \delta^\tau p_2 u_{2\tau}^N + P_\tau^{\tau+1} V_2^{1N}(\tau+1, x) + C_{2\tau} \sum_{n_2=\tau+1}^{n} \omega_{n_2} \sum_{t=\tau}^{n_2} \delta^t p_2 u_{2t}^a,$$

where

$$C_{2\tau} = \frac{\theta_\tau}{\sum_{l=\tau}^{n} \theta_l \sum_{l=\tau}^{n} \omega_l}.$$

According to the above, by analogy we get the relations for the second criteria.

Appendix B. Proof of Theorem 1

Our analysis begins with the occurrence of step n. Both players have zero payoffs at the next step $n+1$. Hence, the optimal strategies coincide with the guaranteed ones, and the payoffs have the form

$$V_1^{1N}(n, x) = V_2^{1N}(n, x) = 0, \quad V_1^{2N}(n, x) = V_2^{2N}(n, x) = Ax^2, \tag{B.1}$$

where

$$A = -\frac{m\delta^n(\varepsilon^2 - 1)}{4}.$$

Now, suppose that the game reaches step $n-1$. In this case, the problem (33), (34) reduces to

$$V_1^{1N}(n-1, x)(V_1^{2N}(n-1, x) - G(n-1, x)) \to \max_{u_{1n-1}},$$

$$V_2^{1N}(n-1, x)(V_2^{2N}(n-1, x) - G(n-1, x)) \to \max_{u_{2n-1}}, \tag{B.2}$$

where

$$V_1^{1N}(n-1,x) = \delta^{n-1}p_1 u_{1n-1}^N + C_{1n-1}\theta_n\delta^n p_1 u_{1n}^a,$$

$$V_2^{1N}(n-1,x) = \delta^{n-1}p_2 u_{2n-1}^N + C_{2n-1}\omega_n\delta^n p_2 u_{2n}^a,$$

$$V_1^{2N}(n-1,x) = -\delta^{n-1}m u_{1n-1}^N u_{2n-1}^N + P_{n-1}^n V_1^{2N}(n,\varepsilon x - u_{1n-1}^N - u_{2n-1}^N),$$

$$V_2^{2N}(n-1,x) = -\delta^{n-1}m u_{1n-1}^N u_{2n-1}^N + P_{n-1}^n V_2^{2N}(n,\varepsilon x - u_{1n-1}^N - u_{2n-1}^N).$$

Then the first-order optimality conditions take the form

$$\delta^{n-1}p_1(-\delta^{n-1}m u_{1n-1}^N u_{2n-1}^N + P_{n-1}^n A(\varepsilon x - u_{1n-1}^N - u_{2n-1}^N)^2 - G(n-1,x))$$

$$+(\delta^{n-1}p_1 u_{1n-1}^N + C_{1n-1}\theta_n\delta^n p_1 u_{1n}^a)$$

$$\cdot(-\delta^{n-1}m u_{2n-1}^N - 2P_{n-1}^n A(\varepsilon x - u_{1n-1}^N - u_{2n-1}^N)) = 0, \tag{B.3}$$

$$\delta^{n-1}p_2(-\delta^{n-1}m u_{1n-1}^N u_{2n-1}^N + P_{n-1}^n A(\varepsilon x - u_{1n-1}^N - u_{2n-1}^N)^2 - G(n-1,x))$$

$$+(\delta^{n-1}p_2 u_{2n-1}^N + C_{2n-1}\omega_n\delta^n p_2 u_{2n}^a)$$

$$\cdot(-\delta^{n-1}m u_{1n-1}^N - 2P_{n-1}^n A(\varepsilon x - u_{1n-1}^N - u_{2n-1}^N)) = 0. \tag{B.4}$$

By subtracting (B.4) multiplied by p_1 from (B.3) multiplied by p_2, we get the expression

$$2AP_{n-1}^n(\varepsilon x - u_{1n-1}^N - u_{2n-1}^N)$$

$$= \frac{\delta^{n-1}m[p_2 C_{2n-1}\omega_n\delta^n p_2 u_{2n}^a u_{2n-1}^N - p_1 C_{1n-1}\theta_n\delta^n p_1 u_{1n}^a u_{1n-1}^N]}{\delta^{n-1}p_1 p_2(u_{2n-1}^N - u_{1n-1}^N) + p_1 C_{2n-1}\omega_n\delta^n p_2 u_{2n}^a - p_2 C_{1n-1}\theta_n\delta^n p_1 u_{1n}^a}. \tag{B.5}$$

As usual, we seek for the linear strategies of the form $u_{in-1}^N = \gamma_{in-1}^N x$. Substituting these strategies into (B.5) brings us to the following relationship between the equilibrium strategies of the players:

$$\gamma_{2n-1}^N = \frac{2A\varepsilon P_{n-1}^n(K_{n-1}^1 - K_{n-1}^2)}{2AP_{n-1}^n(K_{n-1}^1 - K_{n-1}^2) - \delta^{n-1}mK_{n-1}^2}$$

$$- \gamma_{1n-1}^N \frac{2AP_{n-1}^n(K_{n-1}^1 - K_{n-1}^2) + \delta^{n-1}mK_{n-1}^1}{2AP_{n-1}^n(K_{n-1}^1 - K_{n-1}^2) - \delta^{n-1}mK_{n-1}^2}, \tag{B.6}$$

where

$$K_{n-1}^1 = p_1 C_{2n-1}\omega_n\delta^n p_2 u_{2n}^a, \quad K_{n-1}^2 = p_2 C_{1n-1}\theta_n\delta^n p_1 u_{1n}^a.$$

Now, we study the situation when step $n-2$ occurs in the game. Then the problem (33), (34) takes the form

$$V_1^{1N}(n-2,x)(V_1^{2N}(n-2,x) - G(n-2,x)) \to \max_{u_{1n-2}},$$

$$V_2^{1N}(n-2,x)(V_2^{2N}(n-2,x) - G(n-2,x)) \to \max_{u_{2n-2}}, \tag{B.7}$$

where

$$V_1^{1N}(n-2, x) = \delta^{n-2} p_1 u_{1n-2}^N + P_{n-2}^{n-1} V_1^{1N}(n-1, \varepsilon x - u_{1n-2}^N - u_{2n-2}^N)$$

$$+ C_{1n-2} \sum_{t=n-1}^{n} \theta_t \delta^t p_1 u_{1t}^a,$$

$$V_2^{1N}(n-2, x) = \delta^{n-2} p_2 u_{2n-2}^N + P_{n-2}^{n-1} V_2^{1N}(n-1, \varepsilon x - u_{1n-2}^N - u_{2n-2}^N)$$

$$+ C_{2n-2} \sum_{t=n-1}^{n} \omega_t \delta^t p_2 u_{2t}^a,$$

$$V_1^{2N}(n-2, x) = -\delta^{n-2} m u_{1n-2}^N u_{2n-2}^N + P_{n-2}^{n-1} V_1^{2N}(n-1, \varepsilon x - u_{1n-2}^N - u_{2n-2}^N),$$

$$V_2^{2N}(n-2, x) = -\delta^{n-2} m u_{1n-2}^N u_{2n-2}^N + P_{n-2}^{n-1} V_2^{2N}(n-1, \varepsilon x - u_{1n-2}^N - u_{2n-2}^N).$$

Searching for linear strategies $u_{in-2}^N = \gamma_{in-2}^N x$, from the first-order optimality conditions we obtain the following relationship between equilibrium strategies of the players:

$$\gamma_{2n-2}^N = \frac{2 P_{n-2}^{n-1} \delta^{n-1} m \gamma_{1n-1} \gamma_{2n-1}(p_1 - p_2) + \varepsilon(K_{n-2}^1 - K_{n-2}^2)}{-\delta^{n-2} m p_2 - K_{n-2}^1 - K_{n-2}^2}$$

$$+ \gamma_{1n-2}^N \frac{-\delta^{n-2} m p_1 - K_{n-2}^1 - K_{n-2}^2}{-\delta^{n-2} m p_2 - K_{n-2}^1 - K_{n-2}^2}, \tag{B.8}$$

where

$$K_{n-2}^i = 2 A p_i P_{n-2}^{n-1} P_{n-1}^n (-\delta^{n-1} m \gamma_{in-1} - P_{n-1}^n A), \quad i = 1, 2.$$

By continuing the described process until the game reaches step k, we easily obtain the payoffs (35), (36) and the cooperative strategies of the form (37).

Acknowledgments

This work was supported by the Russian Foundation for Basic Research, project nos. 16 01 00183 a and 16-41-100062 p_a.

References

Patrone, F., Pusillo, L. and Tijs, S. H. [2007] Multicriteria games and potentials, *Top* **15**, 138–145.

Pieri, G. and Pusillo, L. [2015] Multicriteria partial cooperative games, *Appl. Math.* **6**, 2125–2131.

Pusillo, L. and Tijs, S. [2013] E-equilibria for multicriteria games, *Ann. ISDG* **12**, 217–228.

Rettieva, A. N. [2014] A discrete-time bioresource management problem with asymmetric players, *Autom. Remote Control* **75**(9), 1665–1676.

Rettieva, A. N. [2015] A bioresource management problem with different planning horizons, *Autom. Remote Control* **76**(5), 919–934.

Shapley, L. S. [1959] Equilibrium points in games with vector payoffs, *Nav. Res. Logis. Q.* **6**, 57–61.

Voorneveld, M., Grahn, S. and Dufwenberg, M. [2000] Ideal equilibria in noncooperative multicriteria games, *Math. Methods Oper. Res.* **52**, 65–77.

Chapter 13

Economic Problems with Constraints:
How Efficiency Relates to Equilibrium

Jacek B. Krawczyk

Victoria University of Wellington
Wellington, New Zealand

J.Krawczyk@vuw.ac.nz

Mabel Tidball

Institut National de Recherche en Agronomie
LAMETA, Montpellier, France

tidball@supagro.inra.fr

We consider situations, in which socially important goods (like transportation capacity or hospital beds) are supplied by independent economic agents. There is also a regulator that believes that constraining the goods delivery is desirable. The regulator can compute a constrained Pareto-efficient solution to establish optimal output levels for each agent. We suggest that a *coupled-constraint* equilibrium (also called a "generalized" Nash or "normalized" equilibrium à la Rosen) may be more relevant for market economies than a Pareto-efficient solution. We examine under which conditions the latter can equal the former. We illustrate our findings using a coordination problem, in which the agents' outputs depend on externalities. It becomes evident that the correspondence between an efficient and equilibrium solutions cannot be complete if the agents' activities generate both negative and positive externalities at the same time.

Keywords: Coupled constraints; generalized Nash equilibrium; Pareto-efficient solution; game engineering.

1. Introduction

This chapter looks at equilibria that are Pareto-efficient and are such that the players' joint actions satisfy a coupling constraint. The aim of this chapter is to show when such equilibria exist — that is, when a correspondence exists between a constrained Pareto-efficient maximization problem (such as utilitarian welfare maximization) and a coupled-constraint equilibrium (CCE). We will show that the joint presence of both positive and negative externalities may preclude a correspondence.

Such a correspondence, if it exists, should be of interest to regulators who want to control competitive agents so that their jointly created outputs and externalities satisfy constraints and are *efficient*. In a market context, it is unlikely that players will internalize these constraints while choosing their optimal strategies. The correspondence which we will identify can help in establishing a tax scheme which decentralizes the constrained utilitarian solution and assists in enforcing it.

Coupling constraints will typically represent exogenous standards formulated by politicians and restrict actions in the combined strategy space of all players (they could also correspond to bargaining solutions, compare Dávila and Eeckhout [2008]). These constraints can be imposed as upper limits e.g., on pollution emitted by a cluster of pulp mills, as in Haurie and Krawczyk [1997] or Krawczyk [2005], or by thermal generators, as in Contreras *et al.* [2013]. Or, these constraints could be imposed as lower limits, e.g., on the amount of some public good like transportation capacity or hospital beds available to the local population. To this group belong problems involving competition for a scarce resource demanded by independent operators subjected to some legislation, like regions in a country as in Boucekkine *et al.* [2010], or concerning internet users logging onto a server as in Kesselman *et al.* [2005]. All above problems were solved using a game solution-concept due to Rosen [1965], called (by him) CCE.[a]

However, little attempt was made to confront a constrained equilibrium with a corresponding constrained efficient solution. Conditions for a match between these two solutions have been studied in Tidball and Zaccour [2005] — albeit in an stylistically "environmental" context. That is, their agents' utility functions contained an explicit damage term and externalities were always negative. As a result they obtained a bijection between Pareto and game solutions. Here we abstract from those assumptions and obtain conditions under which an efficient solution can be represented as an equilibrium. The bijection forms a particular case.

We illustrate our results by analyzing the two-player stylized real-life coordination game introduced in the seminal Rosen [1965]. In this game, each player's output depends on its rival's externality and the joint output should exceed a required value.

To label what a regulator needs to do to achieve a prescribed outcome of a game we use the term *game engineering*, which we have borrowed from Aumann [2008].

---

[a] Also known as *generalized Nash equilibrium* (see e.g., Arrow and Debreu [1954], Harker [1991] or Pang and Fukushima [2005]). Rosen [1965] calls a *coupled-constraint set* the convex, closed and bounded set $X \in \mathbf{R}^m$ to which all n-players' strategies belong ($m \geq n$). Then, a *coupled-constraint game* is the game in which all players' strategies belong to this set, rather than to the Cartesian product $X_1 \times X_2 \times \cdots \times X_n$ where each $X_j \subset \mathbf{R}^{m_n}$ and $m_1 + m_2 + \cdots + m_n = m$, of which X is a subset. A CCE is a solution to a coupled-constraint game, which we formulate in Sec. 2. For the history of this game-solution concept and examples of use see Rosen [1965] and e.g., Haurie [1994], Haurie and Krawczyk [1997], Krawczyk [2007], Tidball and Zaccour [2005], Drouet *et al.* [2008]; also Pang and Fukushima [2005].

This meaning may occasionally be close to what some authors call mechanism design. However, conventionally, a mechanism's design concerns conditions under which a dynamic process converges to an unconstrained equilibrium allocation (e.g., Walrasian or Lindahl).[b]

We want to remark that there exists a link between our results and welfare economics. The welfare theorems (see e.g., Groves and Ledyard [1977] or the seminal Arrow and Debreu [1954]) assert that under suitable conditions, which *exclude* externalities and action-coupling constraints[c]: (i) every competitive allocation is Pareto-efficient and (ii) every Pareto-efficient allocation can be attained through lump-sum transfers. As known, such lump-sum transfers are difficult to arrange. Furthermore, real-life economies are often composed of agents that impose externalities on each other, and their outputs are subjected to constraints (as said, imposed exogenously, perhaps by social or environmental pressure groups). Our results overcome these shortcomings: (1) we show how to construct a game whose equilibrium coincides with a given Pareto-efficient solution for agents that are allowed to interact through externalities, including negative; (2) we let the regulator's Pareto-optimization problem be constrained. In addition, the efficient outcome is obtained through a *threat* of taxes that are never collected in equilibrium and thus the need to rebate them is overcome in our engineered equilibrium.

What follows is a brief outline of what this chapter contains. In Secs. 2 and 3, we speak about the solution concepts for a coordination problem with externalities and constraints, and review the mathematics needed for existence and uniqueness of a relevant equilibrium solution. We develop sufficient conditions for existence of a map between a constrained Pareto solution and a game equilibrium in Sec. 4. We apply these conditions to a motivating example in Sec. 5. The concluding remarks summarize our findings, which include a socio-economic interpretation.

2. A Constrained Efficient Solution or Game Equilibrium?

Suppose a regulator is interested in agents' output levels x_i such that

$$(x_i, x_{-i}) \in X \equiv \{(x_i, x_{-i}) : h(x_i, x_{-i}) \geq 0, \ x_i, \ x_{-i} \geq 0\}, \tag{1}$$

where $i = 1, 2, \ldots, n$ and $h(x_i, x_{-i}) \geq 0$ is a (vector) restriction that the regulator wants to enforce on all outputs and where for notational convention, "$-i$" denotes all agents but i. We will assume that the function $h(\cdot, \cdot)$ is such that X is a convex subset of a space of real numbers whose dimensionality depends on both the number of players and the number of controls each player has. Then, given agents' payoff functions $\phi_i(x_i, x_{-i})$, differentiable and concave in x_i, a satisfactory level of outputs

[b]See a representative paper of Healy and Mathevet [2012]. Also notice that Nguyen and Vojnovic [2011] speak about an allocation mechanism of *constrained* resources in yet another context, which is an auction where *providers* allot scarce resources that are bid for by *users*.
[c]Dávila and Eeckhout [2008] allow for quantity constrains but do not deal with externalities.

$\hat{x} = (\hat{x}_1, \ldots, \hat{x}_n)$ can be established as a *Pareto-efficient* solution as follows

$$\hat{x}(\alpha) = \arg\max_{x \in X} \left\{ \sum_{i=1}^{n} \alpha_i \phi_i(x) \right\}, \tag{2}$$

where $\alpha_i \in [0,1]$, $i = 1, 2, \ldots, n$ and $\sum_{i=1}^{n} \alpha_i = 1$ are weights with which the regulator can appraise a player's payoff. Obviously, the larger α_i, the more important the regulator assigns to the payoff of the ith player. To highlight that an optimal solution depends on the weights α imposed by the regulator, we will write $\hat{x}(\alpha)$ where $\alpha \equiv [\alpha_1, \alpha_2, \ldots, \alpha_n]$.

Rosen [1965] introduced in 1965 a (then) novel solution concept for games with constraints in the combined strategy space of all agents under the name of *coupled-constraint equilibrium* (CCE).[d] In these games, the regulator may seek a solution that can be adopted by competitive players such that it guarantees fulfilment of the coupled, or shared, constraints whose satisfaction depends on the actions undertaken by all agents.

Formally, let all agents $i = 1, 2, \ldots, n$ maximize their payoff functions $\phi(x_i, x_{-i})$, differentiable and concave in x_i. Assume that they are willing to coordinate their individual solutions so that their combined choices satisfy (1). This may be achieved if they use x^*, which is a coupled constrained equilibrium, computed as follows:

$$\phi_i(x^*) = \max_{y_i | x^*_{-i} \in X} \phi_i(y_i \mid x^*_{-i}), \tag{3}$$

where $y_i | x_{-i} \equiv (y_i, x_{-i})$. At x^* no player can improve their own payoff by a unilateral change in his strategy, which keeps the combined strategy vector in X. We will refer to (3) as a coupled constraint game. In the special case where $X = X_i \times X_{-i}$ i.e., each player's action is individually constrained, the game is said to have *uncoupled* constraints.

However, unlike self-enforcing Nash equilibrium in standard games, CCE will not, in general, be "played" by selfish agents unless they are compelled to do so. This can be achieved through the imposition of individualized taxes on players' payoffs.

We do not intent to delve now into the details of how to design taxes to decentralize the utilitarian solution. Still, we will briefly comment in the next section on how a shared constraint's Karush–Kuhn–Tucker (KKT) multiplier can be used to create a necessary threat under which the agent's output will be efficient.

Here, we want to remark that CCE may have a politico-economic appeal superior to a Pareto optimum. This may be because, after the imposition of the taxes, CCE has all properties of a Nash equilibrium. In particular, it is self-enforcing, hence easy to monitor (usually more cheaply than a Pareto-efficient solution, see Contreras *et al.* [2012, 2013]). If so, the regulator may ask whether there exist tax

---

[d] Also see footnote a in page 3.

rates that would enforce a desired Pareto-efficient solution $\hat{x}(\alpha)$ as a CCE, which is the question we endeavor to answer in this chapter.

3. Existence and Uniqueness of CCE

Rosen [1965] allows for a discriminatory treatment of players through the introduction of *weights* $r_i > 0$, $i = 1, 2, \ldots, n$. Similarly to α_i, these weights have a role in controlling agents' behavior. However, the mechanisms are different: while α_i helps the regulator to directly appraise the ith agent's payoff, r_i will modify the agent's KKT multipliers and make an impact on his marginal cost. By choosing a particular set of r_i, the regulator can alter social marginal costs of the externalities and steer the players to select actions leading to a desired equilibrium outcome.

Unlike in *uncoupled* games, equilibria in coupled-constraint games change when the utility functions are subjected to increasing or decreasing transformations. So, if each payoff function ϕ_i is multiplied by weight $r_i > 0$, then x^* is an equilibrium or a *fixed point* of the *global reaction function* $\rho : X \times X \times \mathbf{R}_+^n \mapsto \mathbf{R}$ (see Rosen [1965] or Haurie *et al.* [2012])

$$\rho(x, y, r) \equiv \sum_{i=1}^{n} r_i \phi_i(y_i, x_{-i}) \tag{4}$$

as follows

$$x^* = \arg\max_{y \in X}\{\rho(x^*, y, r)\}. \tag{5}$$

To highlight that the fixed point depends on r will refer to it as $x^*(r)$.

We will define a special type of equilibrium from possibly many solutions to (5). Assume that the set X is defined by $h(x) \geq 0$ where $h(\cdot)$ is a vector of K differentiable constraint functions and such that the qualification assumptions are satisfied.

Definition 1. Denote the constraint shadow-price vector[e] for player i by $\lambda_i \geq 0$. Then, $(x^*, \lambda_i^*) \in X \times \mathbf{R}_+^K$, where K is the dimension of the (vector) constraint function $h(\cdot)$ and $\mathbf{R}_+^K$ is the closed half-space of $\mathbf{R}^K$, is a CCE point if and only if it satisfies the following KKT conditions:

$$h(x^*) \geq 0, \tag{6}$$

$$\lambda_i^* h(x^*) = 0, \tag{7}$$

$$\phi_i(x^*) \geq \phi_i(y_i|x_{-i}^*) + \lambda_i^* h(y_i|x_{-i}^*) \tag{8}$$

for all $i = 1, \ldots n$.

Conditions (6)–(8) define x^* as a vector of nonimprovable strategies when $x^* \in X$.

---

[e]If $h(x)$ is a scalar function, then λ_i is also a scalar. If not, the left-hand side of (7) defines a scalar product $\sum_{\ell=1}^{K} \lambda_{i\ell}^* h^\ell(x^*)$ where ℓ is the constraint index.

In general, the multipliers λ_i and λ_{-i} will not be related to each other. However, in a particular kind of the above equilibrium, called *normalized*, they can be connected and reflect the different levels of agent responsibility for the constraint satisfaction as in the following definition.

Definition 2. An equilibrium point x^* is a Rosen (Nash normalized[f] equilibrium point if, for some vector $r > 0$ and $\lambda^* \geq 0$, conditions (6)–(8) determine x^* and are satisfied for

$$\lambda_i^* = \frac{\lambda^*}{r_i} \tag{9}$$

for each i.[g]

For shortness we have dropped *coupled-constraint* from the equilibrium definition.

Now, we can better understand the role of the weighting vector $r = [r_1, r_2, \ldots, r_n]'$. Substitute λ_i in (8) for (9). It is then evident that if an agent's weight r_i is greater than those of his competitors, then his KKT multipliers are lessened, relative to the competitors'. This means that the marginal cost of the constraint's violation is lower for this agent than for his competitors. Paraphrasing, the vector r tells us how the regulator has distributed the responsibility for the constraints' satisfaction among the agents.

The following theorem is due to Rosen [1965]. It gives conditions of existence and uniqueness of CCE. The proof of the theorem in Rosen [1965] (or Haurie *et al.* [2012]) relies on *diagonal strict concavity* (DSC) of the *joint payoff function* $f(x, r) \equiv \sum_{i=1}^{n} r_i \phi_i(x)$. A "smooth" game is DSC if the pseudo-Hessian[h] of $f(x, r)$ is negative definite.

Theorem 1. *If $f(x, \bar{r})$ is diagonally strictly concave for some $\bar{r} > 0$ on the convex set X and such that the KKT multipliers exist, then the equilibrium point $x^*(\bar{r})$ satisfying (5) is unique.*

In other words, if a game is DSC for a feasible weight distribution r, then the game possesses a unique CCE for each such distribution. Varying r will generate different equilibria and, presumably, one of them can equal $\hat{x}(\alpha)$ — the desired outcome selected by the regulator. This outcome can be obtained as a decoupled equilibrium "played" by the competitive agents, once they include penalty threats

$$T(\lambda^*, r_i, x) \equiv -\sum_{\ell=1}^{K} \frac{\lambda_\ell^*}{r_i} \max(0, -h_\ell(x)) \tag{10}$$

---

[f]In this context "normalization" means that every player faces the same constraint shadow price λ if for all $i = 1, \ldots, n, r_i = 1$.
[g]We could say that λ^* are the "objective" shadow prices while λ_i^* are the "subjective" ones.
[h]That is Jacobian of the pseudo-gradient of $f(x, r)$.

in their payoffs. So, the penalty kicks in only when the constraint is violated. The game in which the agents have included the penalties in their payoffs will have a (possibly unique) equilibrium.[i] This will coincide with the utilitarian maximum at which the players will *not* pay any penalty fee and $T(\lambda^*, r_i, x)$ will remain a taxation threat. For a discussion and examples see Krawczyk and Uryasev [2000] and Krawczyk [2005, 2007].

4. The Relationship Between Pareto-Efficient Solutions and Rosen's Equilibria

Here, we present the main result of this chapter that consists of two theorems on a correspondence between α_i — the weights in the social optimum and r_i — the weights in the normalized equilibrium. We concentrate on the case when there is no slack on constraints, i.e., the joint restrictions are binding ($h(x) = 0$). For example, if water is scarce, we assume that the agents will use all available water. If there was abundance of water, there would be no restriction for the regulator to enforce. Mathematically, this is when a concave utility function is increasing outside the feasibility region.

4.1. *A two-player game*

First, we analyze this relationship for a duopoly model; later, we obtain an n-player extension.

4.1.1. *Pareto-efficiency and Rosen's equilibrium first-order conditions*

Consider problem (2) of a regulator dealing with two agents whose outputs need to be controlled.

Given our interest in active constraints only, let $\overline{X}$ be the set of x for which $h(x_1, x_2) = 0$ and let $x = (x_1, x_2) \in \overline{X}$. The mathematical model for a welfare maximization (regulator's) problem is:

$$\max_{x \in \overline{X}} \{\alpha\phi_1(x_1, x_2) + (1 - \alpha)\phi_2(x_1, x_2)\}. \tag{11}$$

As before, the functions $\phi_i(\cdot, \cdot)$, $i = 1, 2$ are differentiable payoff functions concave in the player's own decision variables. We will use $P(\cdot, \cdot)$ or simply P to refer to the contents of the curly brackets above.

The Lagrangian of this problem is:

$$L^P = \alpha\phi_1(x_1, x_2) + (1 - \alpha)\phi_2(x_1, x_2) + \mu h(x_1, x_2). \tag{12}$$

---

[i]If players select off-equilibrium actions, by mistake or else, the marginal costs increase and the players have incentives to return to equilibrium.

The first-order conditions for a Pareto optimal solution (when $h(x_1, x_2) = 0$) are:

$$
\left.
\begin{aligned}
\frac{\partial L^P}{\partial x_1} &= \alpha \frac{\partial \phi_1(x_1, x_2)}{\partial x_1} + (1 - \alpha) \frac{\partial \phi_2(x_1, x_2)}{\partial x_1} + \mu \frac{\partial h(x_1, x_2)}{\partial x_1} = 0, \\
\frac{\partial L^P}{\partial x_2} &= \alpha \frac{\partial \phi_1(x_1, x_2)}{\partial x_2} + (1 - \alpha) \frac{\partial \phi_2(x_1, x_2)}{\partial x_2} + \mu \frac{\partial h(x_1, x_2)}{\partial x_2} = 0,
\end{aligned}
\right\}
\tag{13}
$$

Equations (13) together with $h(x_1, x_2) = 0$ constitute a system of nonlinear equations with a parameter. Given concavity of P and constraints on the arguments, a maximum exists. However, for some values of α, $P(x_1, x_2)$ for x_1, x_2 such that $h(x_1, x_2) > 0$ might dominate this maximum. If so, this maximum will not be relevant for the regulator (we will illustrate this case in Sec. 5.2). Here we assume that there are $\alpha \in [0, 1]$, for which a maximum with $h(x_1, x_2) = 0$ maximizes welfare.

It is well known that a Pareto optimal (efficient) solution (so, the pair $\hat{x}_1(\alpha), \hat{x}_2(\alpha)$) that solves problem (11) need not be a Nash equilibrium. Consequently, it does not have the self-enforcing properties that the latter solution concept enjoys.

On the other hand, the regulator knows (e.g., from Sec. 3) that it is possible to control competitive agents, to satisfy a common constraint. The regulator may then seek $x_1^*(r), x_2^*(r)$ that satisfy[j]:

$$
\left.
\begin{aligned}
&\max_{x_1} r\, \phi_1(x_1, x_2) \\
&\max_{x_2} \phi_2(x_1, x_2) \\
&h(x_1, x_2) = 0.
\end{aligned}
\right\}
\tag{14}
$$

and select such r for which the unique (because of Theorem 1) equilibrium of (14) matches the utilitarian solution $\hat{x}_1(\alpha), \hat{x}_2(\alpha)$. Here $r > 0$ is the weight, which the regulator attaches to the first player's payoff relative[k] to the second player's payoff.

The player Lagrangians are:

$$
\begin{aligned}
L_1^R &= r\phi_1(x_1, x_2) + \lambda h(x_1, x_2), \\
L_2^R &= \phi_2(x_1, x_2) + \lambda h(x_1, x_2).
\end{aligned}
\tag{15}
$$

Following (6)–(9) and when $h(x_1, x_2) = 0$, a pair $x_1(r), x_2(r)$ is a *normalized equilibrium*, called Rosen's, of game (14) if it satisfies the following first-order

---

[j] We deal with the case when there is no slack on constraints and will want to claim uniqueness of equilibrium $x_1^*(r), x_2^*(r)$, defined by (14). We will use Theorem 1 for that purpose. This is because the KKT conditions (6)–(8) that determine a fixed point of (5) are the same as for problem (14) if $h(x_1, x_2) \geq 0$.

[k] So, we have scaled $r_2 = 1$ and set $r_1 = r$. See Appendix A for evidence that Theorem 1 is true when the regulator uses $r_n = 1$ (here $n = 2$).

conditions:

$$\left.\begin{aligned}\frac{\partial L_1^R}{\partial x_1} &= r\frac{\partial \phi_1(x_1, x_2)}{\partial x_1} + \lambda\,\frac{\partial h(x_1, x_2)}{\partial x_1} = 0, \\[2mm] \frac{\partial L_2^R}{\partial x_2} &= \frac{\partial \phi_2(x_1, x_2)}{\partial x_2} + \lambda\,\frac{\partial h(x_1, x_2)}{\partial x_2} = 0.\end{aligned}\right\} \tag{16}$$

We will assume that a solution to (16) is relevant for the regulator in that an unconstrained equilibrium lies in the "scarcity" region.

4.1.2. *Relations between* α *and* r

We want to find a relationship between α and r such that the solutions for the two problems (Pareto and Rosen) are identical i.e., $x^*(r) = \hat{x}(\alpha)$.

Assume that

$$\mu = K\lambda. \tag{17}$$

If we find $K > 0$ that satisfies this equation then the regulator will be able to use a Rosen's equilibrium (CCE) to enforce a Pareto optimal solution. In fact we are able to prove the following theorem.

Theorem 2. *If an unconstrained equilibrium (possibly multiple) of the two-player game defined by* $\phi_1(\cdot, \cdot)$ *and* $\phi_2(\cdot, \cdot)$ *generates more welfare than a constrained equilibrium and if for* $\alpha \in [0, 1]$ *and* x_1, x_2 *such that* $h(x_1, x_2) \geq 0$ P *is maximized on the constraint* $h(x_1, x_2) = 0$, *then the regulator can implement a Pareto-efficient solution as a Rosen (Nash-normalized) equilibrium. In particular, formula (20) determines* $r(\alpha)$, *the level of responsibility of the first player for the constraint's satisfaction relative to that of the second player.*

Proof. If solutions $x_1(\alpha), x_2(\alpha)$ and $x_1(r), x_2(r)$ are to be the same, then (13) and (16) imply:

$$Kr\frac{\partial \phi_1(x_1, x_2)}{\partial x_1} = \alpha\frac{\partial \phi_1(x_1, x_2)}{\partial x_1} + (1 - \alpha)\frac{\partial \phi_2(x_1, x_2)}{\partial x_1}, \tag{18}$$

$$K\frac{\partial \phi_2(x_1, x_2)}{\partial x_2} = \alpha\frac{\partial \phi_1(x_1, x_2)}{\partial x_2} + (1 - \alpha)\frac{\partial \phi_2(x_1, x_2)}{\partial x_2}. \tag{19}$$

Conditions (18) and (19) give two equations for the two unknown K and r. Solving these equations yields

$$r(\alpha) = \frac{\frac{\partial \phi_2}{\partial x_2}}{\frac{\partial \phi_1}{\partial x_1}}\,\frac{\alpha\frac{\partial \phi_1}{\partial x_1} + (1 - \alpha)\frac{\partial \phi_2}{\partial x_1}}{\alpha\frac{\partial \phi_1}{\partial x_2} + (1 - \alpha)\frac{\partial \phi_2}{\partial x_2}} \tag{20}$$

and

$$K(\alpha) = \frac{\alpha\frac{\partial \phi_1}{\partial x_2} + (1 - \alpha)\frac{\partial \phi_2}{\partial x_2}}{\frac{\partial \phi_2}{\partial x_2}}. \tag{21}$$

The derivatives in Eqs. (20) and (21) are evaluated at $x_1(\alpha), x_2(\alpha)$, hence $r = r(\alpha), K = K(\alpha)$. They are functions of α as requested. $\qquad \square$

Note that zeros of the numerator and denominator in (20) can arise from simultaneous existence of both negative and positive externalities in which case $\alpha \frac{\partial \phi_1}{\partial x_1} + (1 - \alpha)\frac{\partial \phi_2}{\partial x_1}$ and/or $\alpha \frac{\partial \phi_1}{\partial x_2} + (1 - \alpha)\frac{\partial \phi_2}{\partial x_2}$ could be zero. Any situation like that would imply no existence of r for a certain α because of a break in domain of r or because $r = 0$, which violates the assumption on $r > 0$. For that to happen, α would have to be small if x_1 were the negative externality in the problem (i.e., the second player's payoff would be somehow preferred). Or, α would have to be large if x_2 were the negative externality in the problem (i.e., the first player's payoff would be somehow preferred). This suggests that, in the presence of positive and negative externalities, there must be no "extreme" preferences of the regulator for either player, for a Rosen equilibrium to be associated with a Pareto solution. We also notice that zeros can occur when $\frac{\partial \phi_1}{\partial x_1} = 0$ and $\frac{\partial \phi_2}{\partial x_2} = 0$. This would suggest stationary points of the players' payoffs, which should be a rather rare case because of the active constraint.

For the situations when both externalities are positive or both are negative we have the following corollary.

Corollary 1. *If all externalities are of the same sign i.e.,* $\frac{\partial \phi_i}{\partial x_{-i}} \cdot \frac{\partial \phi_{-i}}{\partial x_i} > 0$ *and neither player maximizes their payoff at an interior point i.e.,* $\frac{\partial \phi_i}{\partial x_i} \neq 0$, *then* $0 < r(\alpha) < \infty$.

If the players are coupled only through the constraint (i.e., no externalities), we have the following result.

Result 1. *If* $\phi_i(x_i, x_{-i}) = \phi_i(x_i)$ *(ϕ_i concave en x_i), then*

- Rosen's equilibria exist for any $\alpha \in [0, 1]$. This is so, because the pseudo-Hessian:

$$2 \begin{pmatrix} r\phi_1'' & 0 \\ 0 & \phi_2'' \end{pmatrix}$$

  has positive determinant and negative trace, hence is negative-definite. So, the game is DSC, which guarantees existence and uniqueness of equilibria.
- For all $\alpha \in (0, 1)$ such that a solution to (11) exists, there exists a bijection between $\alpha \in (0, 1)$ and $r \in (0, \infty)$ given by

$$r(\alpha) = \frac{\alpha}{1 - \alpha}.$$

This means that for this case, a set of Rosen equilibria corresponds to the set of constrained Pareto optima.

4.2. *Oligopoly*

We again consider the regulator problem (2) but, here, deal explicitly with n agents. As in the two-player game, a slack on a constraint would mean that there is no restriction for the regulator to enforce. Therefore, we consider the case of equality constraints. This allows us to reduce the number of active constraints to one.[1] So, $x = (x_i, x_{-i}) \in \overline{X}$ where $\overline{X}$ is the set of x, for which $h(x_i, x_{-i}) = 0$.

As before, functions $\phi_i(\cdot, \cdot)$, $i = 1, \ldots, n$ are differentiable payoff functions concave in the player's own decision variable.

The Lagrangian for the regulator's problem is:

$$L^P = \sum_{i=1}^{n-1} \alpha_i \phi_i(x) + \left(1 - \sum_{i=1}^{n-1} \alpha_i\right) \phi_n(x) + \mu h(x_1, \ldots, x_n). \tag{22}$$

The first-order conditions for a Pareto optimal solution are:

$$\sum_{i=1}^{n-1} \alpha_i \frac{\partial \phi_i(x)}{\partial x_i} + \left(1 - \sum_{i=1}^{n-1} \alpha_i\right) \frac{\partial \phi_i(x)}{\partial x_n} + \mu \frac{\partial h(x)}{\partial x_i} = 0, \quad i = 1, \ldots, n-1. \tag{23}$$

We assume that there exist $\alpha \in [0, 1]$, for which the above system can be solved.

For implementation, the regulator may seek $x_i(r)$, where $r = [r_1, r_2, \ldots, r_{n-1}, 1]$, that satisfy:

$$\left. \begin{array}{c} \max_{x_i} r_i \, \phi_i(x), \quad i = 1, \ldots, n-1 \\[2mm] \max_{x_n} \phi_n(x) \\[2mm] h(x) = 0. \end{array} \right\} . \tag{24}$$

The player Lagrangians are:

$$L_i^R = r_i \phi_i(x) + \lambda h(x), \quad i = 1, \ldots, n-1; \quad L_n^R = \phi_n(x) + \lambda h(x). \tag{25}$$

Vector $x(r)$ will be a Rosen's equilibrium (CCE) of game (24) if it satisfies the following first-order conditions:

$$\left. \begin{array}{c} \dfrac{\partial L_i^R}{\partial x_i} = r \dfrac{\partial \phi_i(x)}{\partial x_i} + \lambda \dfrac{\partial h(x)}{\partial x_i} = 0, \quad i = 1, \ldots, n-1 \\[4mm] \dfrac{\partial L_n^R}{\partial x_n} = \dfrac{\partial \phi_n(x)}{\partial x_n} + \lambda \dfrac{\partial h(x)}{\partial x_n} = 0. \end{array} \right\} . \tag{26}$$

We will assume that there exist $r_i \in (0, \infty)$, $i = 1, \ldots, n-1$, for which the above system can be solved.

Assuming as before that $\mu = K\lambda$, substituting in (23) and (26), yields

$$Kr_i \frac{\partial \phi_i(x)}{\partial x_i} = \sum_{i=1}^{n-1} \alpha_i \frac{\partial \phi_i(x)}{\partial x_i} + \left(1 - \sum_{i=1}^{n-1} \alpha_i\right) \frac{\partial \phi_n(x)}{\partial x_i}, \quad i = 1, \ldots, n-1 \tag{27}$$

$$K \frac{\partial \phi_n(x)}{\partial x_n} = \sum_{i=1}^{n-1} \alpha_i \frac{\partial \phi_i(x)}{\partial x_n} + \left(1 - \sum_{i=1}^{n-1} \alpha_i\right) \frac{\partial \phi_n(x)}{\partial x_n}. \tag{28}$$

---

[1]For example, suppose that we have two constraints that depend on three variables $h_1(x_1, x_2, x_3)$ and $h_2(x_1, x_2, x_3)$. Solving the second equation with respect to x_3, assuming a solution exists, and substituting x_3 for the obtained solution in the first equation, reduces the number of constraints to one.

Finally, we get for $i = 1, \ldots, n-1$

$$r_i(\alpha) = \frac{\frac{\partial \phi_n}{\partial x_n}}{\frac{\partial \phi_i}{\partial x_i}} \frac{\sum_{j=1}^{n-1} \alpha_j \frac{\partial \phi_j}{\partial x_i} + \left(1 - \sum_{j=1}^{n-1} \alpha_j\right) \frac{\partial \phi_n}{\partial x_i}}{\sum_{j=1}^{n-1} \alpha_j \frac{\partial \phi_j}{\partial x_n} + \left(1 - \sum_{j=1}^{n-1} \alpha_j\right) \frac{\partial \phi_n}{\partial x_n}} \tag{29}$$

and

$$K(\alpha) = \frac{\sum_{j=1}^{n-1} \alpha_j \frac{\partial \phi_j}{\partial x_n} + \left(1 - \sum_{j=1}^{n-1} \alpha_j\right) \frac{\partial \phi_n}{\partial x_n}}{\frac{\partial \phi_n}{\partial x_n}}. \tag{30}$$

Evidently, when dealing with many players, more "zeros" can occur in the numerator and denominator of (29) than in (20). Notwithstanding this complication, equivalents of Theorem 2 and Result 1 can be formulated rather easily while, interestingly, the wording of Corollary 1 will need not be modified.

5. Realization of Public Good Delivery

We will now analyze the seminal Rosen [1965]'s example about a duopoly with a positive and negative externalities, which we stylize as a public good delivery model. We will establish the values of α such that a solution to (11) exists with $\mu > 0$ and such that $0 < r(\alpha) < \infty$ is verified. We use this model to demonstrate nonexistence of the bijection $r(\alpha)$ even in a two-player context.

5.1. *A public good delivery model*

The interpretations and intuitions ascribed to the example from Rosen [1965] are ours.

Consider a rail network owned by a public firm and a private firm responsible for rolling stock and transportation.

Let x_1 be the tonnage of the goods transported through the network; let x_2 be the length of the tracks owned by the tracks' owner. The revenue of the transportation firm is proportional to the tonnage and to the tracks' length $\beta_1 x_1 x_2$.

In the absence of a discount price for super-large trains, perhaps due to an imperfect state of the tracks, a reasonable approximation of the cost function to the rolling stock firm may be $-\frac{\delta_1 x_1^2}{2}$: the more goods to transport, the more hardware needs to be maintained. Thus, the operator of the rolling stock has profit $\phi_1(x) = \beta_1 x_1 x_2 - \frac{\delta_1 x_1^2}{2}$ $(\beta_1 > 0, \delta_1 > 0)$.

The public firm operating the tracks is paid a fixed amount, which is normalized to zero. The costs of maintaining the tracks at level x_2 is $\beta_2 x_1 x_2 + \delta_2 x_2^2$ (where the first term is motivated by the destruction caused by tonnage x_1). Hence $\phi_2(x) = -\delta_2 x_2^2 - \beta_2 x_1 x_2$ $(\beta_2 > 0, \delta_2 > 0)$.

For social reasons, the government wants transportation activity (i.e., tracks and tonnage) $\gamma_1 x_1 + \gamma_2 x_2$ to be above level M. This can be written as $\gamma_1 x_1 + \gamma_2 x_2 - M \geq 0$.

Setting $\delta_1 = \delta_2 = \beta_1 = \beta_2 = \gamma_1 = \gamma_2 = M = 1$, the situation described above provides motivation for the following game model:

$$
\left.
\begin{array}{c}
\displaystyle\max_{x_1} \left(\phi_1(x) = -\frac{1}{2}x_1^2 + x_1 x_2 \right) \\[2mm]
\displaystyle\max_{x_2} (\phi_2(x) = -x_2^2 - x_1 x_2) \\[2mm]
g_1(x) = x_1 \geq 0 \\[2mm]
g_2(x) = x_2 \geq 0
\end{array}
\right\}. \tag{31}
$$

The solution (x_1, x_2) should allow for the coupling constraint:

$$
g_3(x) = x_1 + x_2 - 1 \geq 0. \tag{32}
$$

It is easy to see that the unconstrained equilibrium of game (31) is $\overline{x}_1 = 0, \overline{x}_2 = 0$. This solution clearly does not satisfy (32) and thus the regulator has a real problem to solve; if $\overline{x}_1 + \overline{x}_2 - 1 \geq 0$, then the regulator would not need to intervene.

In our economy, the regulator will prefer to implement a tax-assisted CCE rather than impose an efficient solution. In practical terms, the regulator has to first select an efficient solution from the many Pareto-optimal outcomes $\hat{x}(\alpha), \alpha \in [0, 1]$ and then look for how to implement it as CCE. To get the required CCE, the regulator would have to compute the KKT multipliers, that solve the coupled-constraint game (31)–(32), which are needed for the threat tax $T(\cdot)$ (in (10)).

5.2. *Regulator's solution*

We now know that the (unconstrained) Nash equilibrium is $x_1^* = 0, x_2^* = 0$ and hence the uncontrolled agents will not satisfy the constraint (32). So, the regulator must act to satisfy this constraint. We contend that he (or she) will select an efficient solution $\hat{x}_1(\alpha), \hat{x}_2(\alpha)$ and engineer a game for the agents to play, whose solution will coincide with the selected solution. Which efficient solution will the regulator select depends on his preferences represented by $\alpha \in [0, 1]$. However, as will be evident, not all preferred choices can be supported by an efficient solution and hence, by the engineered game.

Consider the regulator's problem of a satisfactory delivery of x_1 and x_2:

$$
\left.
\begin{array}{l}
\displaystyle\max P(x_1(\alpha), x_2(\alpha)) = \alpha \left(-\frac{1}{2}x_1^2 + x_1 x_2 \right) + (1 - \alpha)\left(-x_2^2 - x_1 x_2 \right) \\[2mm]
\text{s.t.} \quad g_3(x) = x_1 + x_2 - 1 \geq 0 \quad [x_1, x_2] \in \mathbf{R}_+^2.
\end{array}
\right\}. \tag{33}
$$

To use Theorem 2 we need no slack on g_3.

We note that if there were solutions $\hat{x}_1(\alpha), \hat{x}_2(\alpha)$ such that there was slack on g_3, then the public goods scarcity level M could be augmented and the problem re-scaled and the regulator's problem solution will satisfy the equality constraint

$$
h(x) = x_1 + x_2 - 1 = 0. \tag{34}
$$

We will now show that not all values of α generate a finite Pareto efficient solution. If the regulator had a very strong preference for the transportation firm ($i = 1$), α would be very close or equal to 1. It is easy from (33) to see that for $\alpha = 1$, the first agent's payoff grows to infinity in x_2. The longer the tracks, the larger the value of P. Hence, for α close to 1 the regulator's welfare P maximization problem has no solution even if, with a huge value of x_2, the constraint $g_3(\cdot)$ is always satisfied.

This is not the case when the values of α are small. For such values, P grows as either factor diminishes, but $g_3(\cdot)$ is not satisfied if both are small. Intuitively, there should exist a combination of x_1 and x_2 for which the constraint $g_3(\cdot)$ will be fulfilled with no slack (i.e., (34) will be satisfied).

We present two 3D snapshots of $P(x_1(\alpha), x_2(\alpha))$ in Fig. 1, for α equal to 0.9 and 0.5 to illustrate what happens to P for large and medium range α. The feasible region $X = \{x : x_1 + x_2 \geq 1\}$ is on the right-hand side of the coordinate system.

We notice that $P(x_1(\alpha), x_2(\alpha))$ appears concave[m] for all values of α. However, the location of the maximum changes.

The left panel of Fig. 1 depicts the Pareto programme for a large α of 0.9. We see that the function is unbounded in the feasibility region, for this value of α. This illustrates that if α takes on large values, there is no finite solution to the programme.

The right panel, which depicts the Pareto programme for $\alpha = 0.5$, shows that an unconstrained maximum is in the "scarcity" region. So, the constrained maximum for this α will be on the constraint.

We will now show that the cut-off value of α above which there is no solution to the regulator's problem is $\alpha_M = \frac{1}{2} + \frac{\sqrt{3}}{6} \approx 0.789$. For $\alpha > \alpha_M$, the maximum is unbounded; for $\alpha < x_M$, a maximum of the programme satisfies $h(\cdot)$.

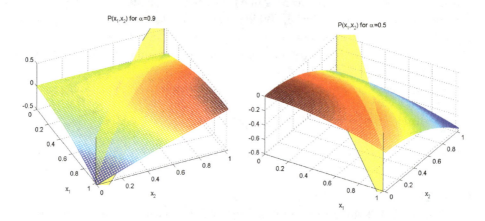

Fig. 1.　The $P(\cdot, \cdot)$ surface for two selected values of α (0.9 and 0.5).

[m]The Hessian $\begin{bmatrix} -a & 2a-1 \\ 2a-1 & 2a-2 \end{bmatrix}$ is (strictly) negative definite for $\alpha \in (0.211, 0.789)$.

To compute α_M let us examine how the programme behaves for large x_1 and x_2. Consider $x_1 + x_2 = M$ where $M \geq 1$. The maximizer of the programme

$$P_M(x_2(\alpha)) = \alpha\phi_1(M - x_2, x_2) + (1 - \alpha)\phi_2(M - x_2, x_2) \tag{35}$$

is $x_2^M = \frac{M}{3}\frac{3\alpha-1}{\alpha}$ and $\phi(x_2^M) = \frac{6\alpha^2-6\alpha+1}{6\alpha}M^2$. Given that $x_2^M \geq 0$ only for $\alpha \geq \frac{1}{3}$, $P_M(x_2^M) \to \infty$ if $M \to \infty$ when $\alpha \in (0.789, 1]$.

Consequently, the regulator's choices of α from this interval cannot be supported by a utilitarian maximum and hence cannot be implemented as CCE.

5.3. *Which CCE are available*

We will now establish for which $r > 0$ a coupled constraint equilibrium exists and is unique. For that purpose we compute the pseudo-Hessian for the game[n] (31) (also see (14)):

$$\mathcal{H} = \begin{bmatrix} -r & -\frac{1}{2} + \frac{1}{2}r \\ -\frac{1}{2} + \frac{1}{2}r & -2 \end{bmatrix}. \tag{36}$$

$\mathcal{H}$ is negative definite for

$$\frac{5}{2}r - \frac{1}{4} - \frac{1}{4}r^2 = -r^2 + 10r - 1 > 0 \tag{37}$$

i.e.,

$$5 - 2\sqrt{6} < r < 5 + 2\sqrt{6} \quad \text{or, approximately,} \quad 0.101 < r < 9.899. \tag{38}$$

So, we know that the CCE exists and is unique for r that satisfy (38).

We now compute the mapping $\alpha \rightsquigarrow r$ from (20)

$$r(\alpha) = \frac{1 - 6\alpha}{-2 + 3\alpha} \tag{39}$$

and plot it in Fig. 2. We see that $r > 0$ for $\alpha \in (\frac{1}{6}, \frac{2}{3})$.

Solving (39) for α when r equals the upper bound in (38) yields the largest $\bar{\alpha}$ for which a CCE exists and is unique; this is $\bar{\alpha} \approx 0.583$. Solving (39) for α when r equals the lower bound in (38) yields the smallest $\underline{\alpha}$, for which a CCE exists and is unique; this is $\underline{\alpha} \approx 0.191$. The interval $(\underline{\alpha}, \bar{\alpha})$ is included in $(\frac{1}{6}, \frac{2}{3})$, for which $r(\alpha) > 0$. Their intersection, which is

$$(\underline{\alpha}, \bar{\alpha}) \approx (0.191, 0.583) \tag{40}$$

defines the interval of α for which $r > 0$ and such that uniqueness of the corresponding equilibria is guaranteed. This interval is a subset of α, for which Pareto-efficient solutions exist that is $(0, 0.789)$. It is therefore the interval (40) for which the map (39) assigns r that generate a unique CCE.

---

[n] A sufficient condition for existence and uniqueness of CCE is $\mathcal{H} < 0$, see Sec. 3.

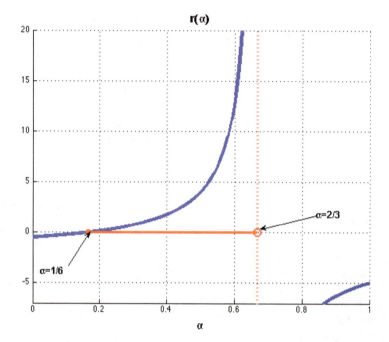

Fig. 2.   The map between α and r.

We can see that as the regulator attaches more weight to the first firm's payoff i.e., when α grows from $\underline{\alpha}$, to $\bar{\alpha}$ (i.e., from 0.191 to 0.583), the preferential treatment (measured by r) of this firm becomes increasingly strong. In particular, the marginal cost of the constraint's violation° can be five times smaller for this firm than for firm 2. This appears logical: the more income the regulator "wants" from firm 1, the smaller the marginal cost this firm should face.

We have also computed $\hat{\alpha} = \frac{\sqrt{6}}{6} \approx 0.4082 \in (\underline{\alpha}, \bar{\alpha})$ that *minimizes* the regulator's programme, which is a convex function of α. This suggests that a welfare-maximizing regulator will seek to implement an equilibrium *away* from $\hat{\alpha}$ and that a number close to $\bar{\alpha}$ will be a good candidate for that.

6. Concluding Remarks

We have proposed a mathematical framework for a novel approach to the solution of a politico-economic coordination problem, which is a realization of a constrained Pareto-efficient solution through a CCE *à la* Rosen. In other words, we have shown how to *engineer* a game whose equilibrium matches a Pareto-efficient solution. Theorem 2 formulates the necessary conditions when a correspondence between these two solution concepts exists for an oligopoly with externalities. The

---

°Measured by $\partial_+ T(\mu, r, \cdot)$; see (10).

conditions when the firms produce same-sign externalities or "nil" externalities, are formulated as Corollary 1 and Result 1, respectively.

We have provided a comprehensive illustration of how to use Theorem 2 to solve a problem of this kind. In particular, we have shown that if agents interact through positive and negative externalities, then the regulator's choices for his preferred solutions may exclude some extreme values of the marginal rate of substitution between the firms' payoffs (i.e., large and small α).

For the considered illustration, the array of unique equilibria that can support the regulator's choices is nonsymmetrical with respect to $\alpha = 0.5$ and excludes solutions in which the "public" firm's payoff would have contributed more than 58% toward the regulator's programme. On the other hand, heavy preferences of the "private" firm's payoff (i.e., small α) are also excluded. This suggests that the Pareto programmes supported by CCE are politically equilibrated and should be acceptable to the stockholders.

A caveat needs be made that the model considered in this chapter is deterministic and assumes symmetric information. Should any of these assumptions not be satisfied, the regulator who uses our approach will be prompting the agents to produce outputs with externalities which may be individually nonoptimal. If this was the case, the agents would start trading out the output-excess amounts until they would became individually optimal. However, allowing for "permit" trading surpasses the scope of this chapter.

Acknowledgments

The authors extend their thanks to Paul Calcott, Charles Figuières, Larry Karp, Vlado Petkov, Jack Robles, Wilbur Townsend and Georges Zaccour for comments. All errors are ours.

Appendix A. Rosen's Relative Weights in $\mathbf{R}^2_+$

Consider a game with payoffs $\phi_1(x), \phi_2(x)$ for which we know that the game has a unique equilibrium for a choice of r_1, r_2. The equilibrium first-order conditions are

$$\left.\begin{aligned} \frac{\partial \phi_1(x)}{\partial x_1} &= -\frac{\lambda(r_1, r_2)}{r_1} \cdot \frac{\partial h(x)}{\partial x_1} \\ \frac{\partial \phi_2(x)}{\partial x_2} &= -\frac{\lambda(r_1, r_2)}{r_2} \cdot \frac{\partial h(x)}{\partial x_2} \end{aligned}\right\}, \tag{A.1}$$

where $\lambda \geq 0$ is the shadow price of the common constraint of type (32).

Let us choose $r_1 = r, r \in (0, \infty)$ and $r_2 = 1$. The first-order conditions (A.1) become now

$$\left.\begin{aligned} \frac{\partial \phi_1(x)}{\partial x_1} &= -\frac{\lambda'(r, 1)}{r} \cdot \frac{\partial h(x)}{\partial x_1} \\ \frac{\partial \phi_2(x)}{\partial x_2} &= -\lambda'(r, 1) \cdot \frac{\partial h(x)}{\partial x_2} \end{aligned}\right\}, \tag{A.2}$$

where $\lambda' > 0$ is the KKT multiplier that corresponds to this choice or r.

We notice that conditions (A.1) are equivalent to (A.2) if

$$\left.\begin{aligned} \frac{\lambda(r_1, r_2)}{r_1} &= \frac{\lambda'(r, 1)}{r} \\ \frac{\lambda(r_1, r_2)}{r_2} &= \lambda'(r, 1) \end{aligned}\right\}. \qquad (A.3)$$

The above is true if and only if

$$r \equiv \frac{r_1}{r_2}. \qquad (A.4)$$

References

Arrow, K. J. and Debreu, G. [1954] Existence of an equilibrium for a competitive economy, *Econometrica* **22**(3), 265–290.

Aumann, R. J. [2008] Game engineering, in *Mathematical Programming and Game Theory for Decision Making*, eds. Neogy, S. K., Bapat, R. B., Das, A. K. & Parthasarathy, T. (World Scientific, Singapore), pp. 279–285.

Boucekkine, R., Krawczyk, J. B. and Vallée, T. [2010] Towards an understanding of trade-offs between regional wealth, tightness of a common environmental constraint and the sharing rule, *J. Econ. Dyn. Control* **34**(9), 1813–1835.

Contreras, J., Krawczyk, J. B. and Zuccollo, J. [2016] Economics of collective monitoring: A study of environmentally constrained electricity generators, *EURO XXV* **13**(3), 349–369.

Contreras, J., Krawczyk, J. B., Zuccollo, J. and García, J. [2013] Competition of thermal electricity generators with coupled transmission and emission constraints, *J. Energy Eng.* **139**(4), 239–252.

Dávila, J. and Eeckhout, J. [2008] Competitive bargaining equilibrium, *J. Econ. Theory* **139**(1), 269–294.

Drouet, L., Haurie, A., Moresino, F., Vial, J.-P., Vielle, M. and Viguier, L. [2008] An oracle based method to compute a coupled equilibrium in a model of international climate policy, *Comput. Manag. Sci.* **5**, 119–140.

Groves, T. and Ledyard, J. [1977] Optimal allocation of public goods: A solution to the "free rider" problem, *Econometrica* **45**(3), 783–810.

Harker, P. T. [1991] Generalized Nash games and quasivariational inequalities, *Eur. J. Oper. Res.* **4**, 81–94.

Haurie, A. [1994] Environmental coordination in dynamic oligopolistic markets, *Group Decis. Negot.* **4**, 46–67.

Haurie, A. and Krawczyk, J. B. [1997] Optimal charges on river effluent from lumped and distributed sources, *Environ. Model. Assess.* **2**, 177–189.

Haurie, A., Krawczyk, J. B. and Zaccour, G. [2012] *Games and Dynamic Games*, Business Series, Vol. 1 (World Scientific, Singapore).

Healy, P. J. and Mathevet, L. [2012] Designing stable mechanisms for economic environments, *Theor. Econ.* **7**(3), 609–661.

Kesselman, A., Leonardi, S. and Bonifaci, V. [2005] Game-theoretic analysis of internet switching with selfish users, *Proc. First Int. Workshop on Internet and Network Economics (WINE)*, Lectures Notes in Computer Science, Vol. 3828, pp. 236–245.

Krawczyk, J. B. [2005] Coupled constraint Nash equilibria in environmental games, *Resour. Energy Econ.* **27**, 157–181.

Krawczyk, J. B. [2007] Numerical solutions to coupled-constraint (or generalised) Nash equilibrium problems, *Comput. Manag. Sci.* **4**, 183–204.

Krawczyk, J. B. and Uryasev, S. [2000] Relaxation algorithms to find Nash equilibria with economic applications, *Environ. Model. Assess.* **5**, 63–73.

Nguyen, T. Vojnovic, M. [2011] Weighted proportional allocation, *Proc. ACM SIGMETRICS Joint Int. Conf. Measurement and Modeling of Computer Systems* (ACM), pp. 173–184.

Pang, J.-S. and Fukushima, M. [2005] Quasi-variational inequalities, generalized Nash equilibria and multi-leader-follower games, *Comput. Manag. Sci.* **1**, 21–56.

Rosen, J. B. [1965] Existence and uniqueness of equilibrium points for concave n-person games, *Econometrica* **33**(3), 520–534.

Tidball, M. and Zaccour, G. [2005] An environmental game with coupling constraints, *Environ. Model. Assess.* **10**, 153–158.

Chapter 14

Substitution, Complementarity, and Stability

Harborne W. Stuart Jr

Columbia Business School
New York, NY 10027, USA

hws7@gsb.columbia.edu

We provide necessary and sufficient conditions for a non-empty core in many-to-one assignment games. When players on the "many" side (buyers) are substitutes with respect to any given player on the other side (firms), we show that non-emptiness requires an additional condition that limits the competition among the buyers. When buyers are complements with respect to any given firm, a sufficient condition for non-emptiness is that buyers also be complements with respect to all of the firms, collectively. A necessary condition is that no firm can be guaranteed a profit when the core is non-empty.

Keywords: Cooperative game; matching; core.

1. Introduction

In this chapter, we provide necessary and sufficient conditions for a non-empty core in many-to-one assignment games. The many-to-one structure is used to model situations in which value is created by matching buyers — the "many" side — to individual firms — the "one" side. The core is used to model the competition between the players, so the question of non-emptiness of the core becomes a question of whether competition will lead to a stable outcome. We consider two scenarios. In the first, buyers are substitutes with respect to any given firm. In the second, buyers are complements with respect to any given firm. We show that non-emptiness of the core is not guaranteed in either scenario. With substitution, an additional condition limiting the competition among the buyers is required. With complementarity, we show that if buyers are complements with respect to all of the firms, collectively, then the core is non-empty. But this condition is not necessary. For a necessary condition, we show that no firm can be guaranteed a profit when the core is non-empty.

This chapter considers a many-to-one assignment game in the tradition of Kelso and Crawford [1982], but we take the abstract approach of Muto *et al.* [1988]. Generalizations of the assignment game of Shapley and Shubik [1971] typically derive the characteristic function from, say, the values of pair-wise matchings or from player preferences and costs. Examples include Kelso and Crawford [1982], Sotomayor [1992], Camiña [2006] and Ostrovsky [2008]. By contrast, in our model, we follow Muto *et al.* in taking the primitives to be the player set and the characteristic function. Our model generalizes the Muto *et al.* model by allowing buyers to be matched to competing firms, rather than to just one monopoly (or "big boss") firm.

The abstract approach of this chapter is motivated by an interest in exploring the implications of scale and network effects without specifying cost functions or quantifying network effects. For instance, the substitution scenario covers situations in which firms have decreasing returns-to-scale that are greater than any network effects, or in which firms have negative network effects greater than any scale effects. Similarly, the complementarity scenario covers situations in which positive network effects are greater than any scale effects, or in which increasing returns-to-scale are greater than any negative network effects (if they exist). For an example of the benefit of this abstraction, in the model of Kelso and Crawford [1982], decreasing returns-to-scale fail to guarantee stability. We show that this phenomenon is more general, and our model suggests a simple necessary and sufficient condition for stability with decreasing returns-to-scale.

Section 2 of the chapter presents the model and the results. In the subsection on complementarity, we highlight the difference between requiring that buyers be complements with respect to individual firms and requiring that buyers be complements with respect to all firms. In the latter case, the core is non-empty because the game will have a structure intimately related to a convex game (Shapley [1971]). In the former case, the core can be empty because when buyers are complements with respect to individual firms, they can become substitutes (competitors) with respect to all the firms collectively.

A brief discussion of the results is presented at the end.

2. Results

2.1. *Model*

To review, a TU game $(N; v)$ consists of a player set N and a mapping $v : 2^N \to \Re$ (the *characteristic function*). The TU game is *super-additive* if for any $S, T \subset N$ such that $S \cap T = \varnothing$, $v(S) + v(T) \le v(S \cup T)$. Let $x \in \Re^{|N|}$ denote an allocation where component x_i denotes player i's payoff. The *core* of a TU cooperative game $(N; v)$ is the set of allocations satisfying $\sum_{i \in N} x_i = v(N)$ and, for all $S \subseteq N$, $\sum_{i \in S} x_i \ge v(S)$. A TU cooperative game is *balanced* if for any collection of non-negative weights $\{\lambda_S\}_{S \subset N}$ satisfying $\sum_{S \subset N, S \ni j} \lambda_S = 1$ for all $j \in N$,

$$\sum_{S \subset N} \lambda_S v(S) \le v(N).$$

We will use the Bondareva [1963] and Shapley [1963] theorem: a TU cooperative game has a non-empty core if, and only if, it is balanced.

For notational ease, for any $S \subseteq N$, let $x(S) = \sum_{i \in S} x_i$. For expositional ease, we will describe our many-to-one matching game as a *firm-buyer* game. In the definition below, the set F is interpreted as the set of firms, and the set B is interpreted as the set of buyers.

Definition 1. A firm-buyer (FB) game is a super-additive TU game $(N; v)$ in which $N = F \cup B$, where $F \cap B = \varnothing$, $F \neq \varnothing$, and $B \neq \varnothing$, and in which the characteristic function is defined as follows: for all $S \subseteq N$, if $S \cap F = \varnothing$ or $S \cap B = \varnothing$, then $v(S) = 0$. Otherwise, let $\Im_S$ denote the collection of all partitions of $S \cap B$ with $|S \cap F|$ elements, and let $\{S_i\}_{i \in S \cap F}$ be a typical partition in $\Im_S$.[a] Then

$$v(S) = \max_{\{S_i\} \in \Im_S} \sum_{i \in S \cap F} v(\{i\} \cup S_i). \tag{1}$$

Let $\{B_i\}_{i \in F}$ denote a partition of B that maximizes $\sum_{i \in F} v(\{i\} \cup B_i)$. In other words,

$$v(N) = \sum_{i \in F} v(\{i\} \cup B_i). \tag{2}$$

Call any such partition of B an *efficient* partition.

In a FB game, value is created by matching buyers to individual firms. In words, Eq. (1) states that the value created by a coalition S is determined by optimally allocating the buyers in S to the firms in S.

The following lemma provides a necessary and sufficient condition for a non-empty core in a FB game. A corollary to this lemma will be used in the proof of Proposition 1 below. In both of these results, note that the core of a FB game is determined by coalitions formed by one firm and a set of buyers.

Lemma 1. *Consider a FB game. Fix an efficient partition $\{B_i\}_{i \in F}$ of B. The core is non-empty if, and only if, there exists a vector $x \in \Re_+^{|B|}$ such that for any $B_i \in \{B_i\}_{i \in F}$, and any $S \subseteq B$,*

$$v(\{i\} \cup B_i) - x(B_i) \geq v(\{i\} \cup S) - x(S). \tag{C1}$$

Further, the core contains the point in which each buyer $j \in B$ receives x_j, and each firm $i \in F$ receives $v(\{i\} \cup B_i) - x(B_i)$.

Proof. For sufficiency, we construct a core allocation, starting with a vector $x \in \Re_+^{|B|}$ satisfying condition (C1). Consider an efficient partition $\{B_i\}_{i \in F}$. For each

---

[a] To account for the possibility that a firm could have no buyers in a coalition, we allow an element of a partition to be empty.

$i \in F$, define $x_i = v(\{i\} \cup B_i) - x(B_i)$. Note that by taking S to be the empty set, condition (C1) implies that $x_i \geq v(\{i\}) = 0$. Then

$$x(N) = \sum_{i \in F} [v(\{i\} \cup B_i) - x(B_i)] + x(B)$$

$$= \sum_{i \in F} [v(\{i\} \cup B_i)] - x(B) + x(B)$$

$$= \sum_{i \in F} v(\{i\} \cup B_i) = v(N).$$

For any $S \subseteq N$ such that $S \cap F = \varnothing$ or $S \cap B = \varnothing$, $v(S) = 0$, and $x(S) \geq v(S)$ is immediate. So consider a $S \subseteq N$, such that $S \cap F \neq \varnothing$ and $S \cap B \neq \varnothing$. Let $\{S_i\}_{i \in S \cap F}$ be an optimal partition of $S \cap B$ as defined in Eq. (1). Then

$$x(S) = \sum_{i \in S \cap F} [v(\{i\} \cup B_i) - x(B_i)] + x(S \cap B)$$

$$= \sum_{i \in S \cap F} [v(\{i\} \cup B_i) - x(B_i) + x(S_i)]$$

$$\geq \sum_{i \in S \cap F} v(\{i\} \cup S_i) = v(S),$$

where the inequality follows from the condition of the lemma.

For necessity, consider a core allocation x. Note that (2) and the core conditions $x(\{i\} \cup B_i) \geq v(\{i\} \cup B_i), i \in F$, imply that $x(\{i\} \cup B_i) = v(\{i\} \cup B_i)$ for any B_i from any efficient partition. Thus,

$$v(\{i\} \cup B_i) - x(B_i) = x(\{i\} \cup B_i) - x(B_i) = x_i.$$

Using the core condition $x(\{i\} \cup S) \geq v(\{i\} \cup S)$, we have

$$x_i = x(\{i\} \cup S) - x(S)$$

$$\geq v(\{i\} \cup S) - x(S),$$

implying

$$v(\{i\} \cup B_i) - x(B_i) \geq v(\{i\} \cup S) - x(S). \qquad \square$$

Corollary 1. *Consider a FB game. Fix an efficient partition of B. The core is non-empty if, and only if, there exists a vector $x \in \Re_+^{|B|}$ that satisfies the following condition: for all $i \in F$, for any $T \subseteq B_i$, for any $S \subseteq B \backslash B_i$,*

$$x(S) - x(T) \geq v(\{i\} \cup S \cup B_i \backslash T) - v(\{i\} \cup B_i). \qquad (C2)$$

Proof. To show that condition C1 implies condition C2, consider a $T \subseteq B_i$ and a $S \subseteq B \backslash B_i$. Condition C1 gives

$$v(\{i\} \cup B_i) - x(B_i) \geq v(\{i\} \cup S \cup B_i \backslash T) - x(S \cup B_i \backslash T).$$

Rearranging gives condition C2:

$$x(S) - x(T) \geq v(\{i\} \cup S \cup B_i \backslash T) - v(\{i\} \cup B_i).$$

To show that condition C2 implies condition C1, consider a $S \subseteq B$. Let $T = B_i \cap S$, and let $R = S \backslash T$. Condition C2 gives

$$x(R) - x(B_i \backslash T) \geq v(\{i\} \cup R \cup T) - v(\{i\} \cup B_i).$$

Rearranging and subtracting $x(T)$ from both sides gives

$$v(\{i\} \cup B_i) - x(B_i) \geq v(\{i\} \cup R \cup T) - x(R \cup T)$$
$$= v(\{i\} \cup S) - x(S),$$

namely condition C1. □

The above results are not designed to provide intuition, but some interpretation is possible. In Lemma 1, the quantity $v(\{i\} \cup B_i) - x(B_i)$ can be viewed as firm i's profit. We can say that the core is non-empty if, and only if, there exist feasible firm profits such that no firm can do better with a different set of buyers. To interpret Corollary 1, think of the set T as a collection of firm i's customers. If a group of non-customers, described by the set S, contributes more value to firm i than the firm's customers, then this group of non-customers should capture more value than the firm's customers. The core is non-empty if, and only if, this is always the case.

2.2. Substitution

As described in the Introduction, the substitution condition is defined with respect to individual firms.

Condition 1. A FB game satisfies the *substitution (SUB)* condition if for any $i \in F$ and $S, T \subset B$,

$$v(\{i\} \cup S) + v(\{i\} \cup T) \geq v(\{i\} \cup S \cup T) + v(\{i\} \cup (S \cap T)).$$

Note that this condition implies that the marginal benefit to a firm of a group of buyers is decreasing in the presence of other buyers. For example, if $S \cap T = \varnothing$, rearranging the above equation yields:

$$v(\{i\} \cup S \cup T) - v(\{i\} \cup T) \leq v(\{i\} \cup S).$$

To show that the SUB condition does not ensure a non-empty core in a FB game, consider the following example. (Kelso and Crawford [1982, p. 1502] provide a similar example.)

Example 1. The set $F = \{a, b\}$; the set $B = \{1, 2, 3\}$; and the characteristic function is as follows:

$$v(\{a, 1\}) = v(\{a, 2\}) = 4; \quad v(\{a, 3\}) = 0,$$
$$v(\{a, 1, 3\}) = v(\{a, 2, 3\}) = v(\{a, 1, 2\}) = v(\{a, 1, 2, 3\}) = 4,$$

$$v(\{b,1\}) = v(\{b,2\}) = 4; \quad v(\{b,3\}) = 2,$$

$$v(\{b,1,2\}) = 7; \quad v(\{b,1,3\}) = v(\{b,2,3\}) = 4; \quad v(\{b,1,2,3\}) = 7,$$

$$v(N) = 8 \quad \text{e.g., } v(\{a,1\}) + v(\{b,2\}) \quad \text{or} \quad v(\{a,2\}) + v(\{b,1\}).$$

This example is consistent with the following story. Firm a has a capacity of one unit and zero cost of production. Firm b has a capacity of two units. To serve buyers 1 and 2, it can produce the first unit at zero cost and the second unit at a cost of 1. To serve only buyer 3 , firm b can produce one unit at a cost of 2. To serve buyer 3 and either buyer 1 or 2, firm b incurs a total cost of 4. Each buyer wants only one unit of product. Buyers 1 and 2 have a willingness-to-pay of 4 for either firm's product. Buyer 3 has a willingness-to-pay of 4 for firm b's product and 0 for firm a's product.

To demonstrate the emptiness of the core, consider the efficient partition in which buyer 1 is matched with firm a and buyer 2 is matched with firm b. Note that buyer 1's marginal contribution is 2:

$$v(N) - v(N\backslash\{1\}) = 8 - [v(\{a,2\}) + v(\{b,3\})]$$

$$= 2.$$

But note that

$$v(\{b,1,2\}) - v(\{b,2\}) = 3.$$

Buyer 1 could capture more value by joining a coalition with firm b and buyer 2. In this example, the emptiness of the core is due to the fact that a player's marginal contribution — e.g., 2 for buyer 1 — is less than its "outside option", namely $v(\{b,1,2\}) - v(\{b,2\})$.[b] This insight motivates the necessary and sufficient condition for non-emptiness in FB games satisfying the SUB condition.

Proposition 1. *Suppose a FB game satisfies the SUB condition. Consider an efficient partition $\{B_i\}_{i\in F}$. Then the game has a non-empty core if, and only if, for all $i \in F$, $j \in B_i$, and $k \in F\backslash\{i\}$,*

$$v(N) - v(N\backslash\{j\}) \geq v(\{j,k\} \cup B_k) - v(\{k\} \cup B_k).$$

To interpret the condition of Proposition 1, note that competition between the buyers is what limits a given buyer's marginal contribution, namely $v(N) - v(N\backslash\{j\})$. And a buyer j always has the threat of transacting with another firm, namely $v(\{j,k\} \cup B_k) - v(\{k\} \cup B_k)$, where $j \notin B_k$. So one can say that a stable outcome requires that buyers do not face too much competition with respect to their alternatives.

---

[b]The emptiness of the core can be seen directly by applying the Bondareva–Shapley theorem: $[v(\{b,1,2\}) + v(\{b,3\}) + v(\{a,1\}) + v(\{a,2,3\})]/2 > v(N)$. See, also, the discussion in Kelso and Crawford [1982, p. 1502].

It is important to note that in Example 1, the cost functions are well-behaved: each firm has increasing marginal costs. In fact, Example 1 is consistent with the conditions for non-emptiness in Kaneko [1976] with one important exception: one of the firms has to incur additional costs when serving one of the buyers. Kelso and Crawford [1982, Theorem 6] show that this exception is necessary for emptiness: with identical buyers, decreasing returns-to-scale would guarantee a non-empty core. But as the above example shows, the stability typically associated with increasing marginal costs can be lost when different buyers require different production costs.

The proof of Proposition 1 will use the following implication of the SUB condition. As an aside, this lemma shows that the SUB condition implies the substitution condition in Muto *et al.* [1988]. In that paper's model, there is only one firm. By taking $v(N) = v(\{i\} \cup B)$ and $v(N\backslash S) = v(\{i\} \cup B\backslash S)$, the condition of the lemma corresponds to the Muto *et al.* condition.[c]

Lemma 2. *In a FB game, the SUB condition implies that for any $i \in F$ and $S \subseteq B_i$.*

$$\sum_{j \in S}[v(N) - v(N\backslash\{j\})] \le v(\{i\} \cup B_i) - v(\{i\} \cup B_i\backslash S).$$

Proof. Consider a $S \subset B_i$ and a $j \in S$. By Eq. (1),

$$v(N\backslash\{j\}) \ge v(\{i\} \cup B_i\backslash\{j\}) + v(N\backslash(\{i\} \cup B_i)).$$

Adding $v(\{i\} \cup B_i)$ to both sides and rearranging, we have

$$v(N) - v(N\backslash\{j\}) \le v(\{i\} \cup B_i) - v(\{i\} \cup B_i\backslash\{j\}).$$

Then

$$\sum_{j \in S}[v(N) - v(N\backslash\{j\})] \le |S|v(\{i\} \cup B_i) - \sum_{j \in S} v(\{i\} \cup B_i\backslash\{j\})$$

$$\le |S|v(\{i\} \cup B_i) - [(|S| - 1)v(\{i\} \cup B_i) + v(\{i\} \cup B_i\backslash S)]$$

$$= v(\{i\} \cup B_i) - v(\{i\} \cup B_i\backslash S),$$

where the second inequality follows from repeated use of the SUB condition. $\square$

Proof of Proposition 1. For sufficiency, we construct a core allocation. For each $i \in F$, for each $j \in B_i$, let $x_j = v(N) - v(N\backslash\{j\})$ and let

$$x_i = v(\{i\} \cup B_i) - x(B_i).$$

From Lemma 2, $x(B_i) \le v(\{i\} \cup B_i)$, so $x_i \ge 0$ in this construction. It remains to show that x is a core allocation. Using Corollary 1, it suffices to show that for any

---

[c]In a game with just one firm, as in Muto *et al.*, the converse of Lemma 2 holds. With more than one firm, it can fail. See, for instance, Example 2 in the next section.

$T \subseteq B_i$, for any $S \subseteq B \backslash B_i$,

$$x(S) - x(T) \geq v(\{i\} \cup S \cup B_i \backslash T) - v(\{i\} \cup B_i).$$

Using the SUB condition,

$$v(\{i\} \cup B \backslash T) + v(\{i\} \cup S \cup B_i) \geq v(\{i\} \cup B) + v(\{i\} \cup S \cup B_i \backslash T).$$

Subtracting $v(\{i\} \cup B_i)$ from both sides and rearranging, we have

$$v(\{i\} \cup S \cup B_i) - v(\{i\} \cup B_i) - [v(\{i\} \cup B) - v(\{i\} \cup B \backslash T)]$$

$$\geq v(\{i\} \cup S \cup B_i \backslash T) - v(\{i\} \cup B_i).$$

It suffices to show than the left-hand side is greater than $x(S) - x(T)$. This is true if

$$x(T) \leq [v(\{i\} \cup B) - v(\{i\} \cup B \backslash T)]$$

and

$$x(S) \geq v(\{i\} \cup S \cup B_i) - v(\{i\} \cup B_i).$$

The first inequality is immediate from Lemma 2. For the second inequality, the condition of the proposition gives

$$x(S) = \sum_{j \in S} [v(N) - v(N \backslash \{j\})]$$

$$\geq \sum_{j \in S} [v(\{i, j\} \cup B_i) - v(\{i\} \cup B_i)]$$

$$\geq v(\{i\} \cup S \cup B_i) - v(\{i\} \cup B_i),$$

where the second inequality is due to repeated use of the SUB condition.

For necessity, if the core is non-empty, then $x_j \leq v(N) - v(N \backslash \{j\})$, $x(\{k\} \cup B_k) = v(\{k\} \cup B_k)$, and $x(\{j, k\} \cup B_k) \geq v(\{j, k\} \cup B_k)$. Using these three equations, we have

$$v(\{j, k\} \cup B_k) - v(\{k\} \cup B_k) \leq x(\{j, k\} \cup B_k) - x(\{k\} \cup B_k)$$

$$= x_j$$

$$\leq v(N) - v(N \backslash \{j\}). \qquad \square$$

2.3. *Complementarity*

Similar to the case of substitution, the complementarity condition is defined with respect to individual firms.

Condition 2. A FB game satisfies the *complementarity (CMP)* condition if for any $i \in F$ and $S, T \subset B$,

$$v(\{i\} \cup S) + v(\{i\} \cup T) \leq v(\{i\} \cup S \cup T) + v(\{i\} \cup (S \cap T)).$$

Note that this condition implies that the marginal benefit to a firm of a group of buyers is increasing in the presence of other buyers. For example, if $S \cap T = \varnothing$, rearranging the above equation yields:

$$v(\{i\} \cup S \cup T) - v(\{i\} \cup T) \geq v(\{i\} \cup S).$$

The CMP condition does not ensure a non-empty core in a FB game. Consider the following example.

Example 2. There are three firms, say a, b, and c, and three buyers, say 1, 2, and 3. Each firm has a fixed cost of ε and unlimited capacity at zero marginal cost of production. Each buyer has a willingness-to-pay for only one unit, and the following table lists the willingnesses-to-pay.

	w_1	w_2	w_3
Firm a	1	1	0
Firm b	1	0	1
Firm c	0	1	1

The following is a partial description of the characteristic function:

$$v(\{a, 1\}) = v(\{a, 2\}) = v(\{a, 1, 3\}) = v(\{a, 2, 3\}) = 1 - \varepsilon$$

$$v(\{a, 1, 2\}) = v(\{a, 1, 2, 3\}) = 2 - \varepsilon$$

$$v(N) - v(N \setminus \{j\}) = 1 - \varepsilon \quad \text{for } j \in \{1, 2, 3\}$$

$$v(N) - v(N \setminus \{i\}) = 0 \quad \text{for } i \in \{a, b, c\}$$

$$v(N) = 3 - 2\varepsilon \quad \text{e.g., } v(\{a, 1, 2\}) + v(\{b, 3\}).$$

In this game, the sum of the marginal contributions is less than $v(N)$, so the core must be empty. At first glance, it might seem that the complementarity of the buyers is the reason for the empty core. But complementarity *per se* does not necessarily lead to instability. In a convex game, all subsets of players are complements, and Shapley [1971] showed that these games always have a non-empty core. Furthermore, in any super-additive TU game, when a set of players necessary for value creation—like the buyers in a FB game — are complements when matched with all the other players, then the core will still be non-empty. We show this in Proposition 2 below. The result is based on the original Shapley [1971] result.

Proposition 2. *Consider a super-additive TU game $(N; v)$ satisfying the following property: there exists a subset $B \subset N$ such that for all $S \subseteq N$, if $S \cap B = \varnothing$, then $v(S) = 0$. In such a game, if for all $S, T \subseteq B$,*

$$v(S \cup (N \setminus B)) + v(T \cup (N \setminus B)) \leq v(S \cup T \cup (N \setminus B)) + v((S \cap T) \cup (N \setminus B)),$$

then the core is non-empty.

Proof. Arbitrarily label the buyers in set B from 1 to b, where $b = |B|$. Set

$$x_1 = v(N) - v(N \backslash \{1\})$$

$$x_2 = v(N \backslash \{1\}) - v(N \backslash \{1, 2\})$$

$$\vdots$$

$$x_b = v(N \backslash \{1, \ldots, b-1\}) - v(N \backslash B).$$

For all $i \in N \backslash B$, set

$$x_i = 0.$$

We will show that x is a core allocation. By construction and super-additivity, $x_i \geq 0$ for all $i \in N$. Because $v(N \backslash B) = 0$, $x(B) = v(N)$. For any $S \subset N$, if $S \cap B = \varnothing$, $x(S) \geq v(S)$ is immediate. So consider an $S \subset N$ such that $S \cap B \neq \varnothing$. Let $T = S \cap B$. Using the condition of the proposition, we know that for the highest-numbered buyer in T, say j,

$$x_j = v(\{j, \ldots, b\} \cup (N \backslash B)) - v(\{j+1, \ldots, b\} \cup (N \backslash B))$$

$$\geq v(\{j\} \cup (N \backslash B)) - v(N \backslash B).$$

For the next highest buyer in T, say j',

$$x_{j'} = v(\{j', \ldots, b\} \cup (N \backslash B)) - v(\{j'+1, \ldots, b\} \cup (N \backslash B))$$

$$\geq v(\{j, j'\} \cup (N \backslash B)) - v(\{j\} \cup (N \backslash B)).$$

Repeating for all buyers in T, and summing inequalities, we have

$$x(T) \geq v(T \cup (N \backslash B)) - v(N \backslash B)$$

$$= v(T \cup (N \backslash B))$$

$$\geq v(T \cup ((N \backslash B) \cap S))$$

$$= v(S),$$

where the second inequality is due to super-additivity. Because $x(S) = x(T)$, we have $x(S) \geq v(S)$, as required. $\square$

In the case of a FB game, the term $N \backslash B$ in Proposition 2 would be the set of firms, namely F. Returning to Example 2 and applying Proposition 2, the empty core shows that the buyers cannot be complements with respect to the group of firms, despite being complements with respect to any individual firm. It must be the case that there is a substitution effect between the buyers, despite the fact that each firm has unlimited capacity and zero marginal cost. We explore this effect after presenting the results using the CMP condition. The results use the following definition.

Definition 2. Call a collection of non-negative weights $\{\lambda_S\}_{S \subseteq B}$, *buyer-balanced* if for all $j \in B$,

$$\sum_{S \subseteq B, S \ni j} \lambda_S = 1.$$

A FB game satisfies *buyer-balancedness* (*BB*) if for any collection of buyer-balanced weights $\{\lambda_S\}_{S \subseteq B}$,

$$\sum_{S \subseteq B} \lambda_S v(F \cup S) \leq v(N).$$

With this definition, we have the following result.

Proposition 3. *Suppose a FB game satisfies the CMP condition. Then the game has a non-empty core if, and only if, it satisfies the BB condition.*

Proof. For necessity, we show that if the BB condition does not hold, the core must be empty. So suppose there exists a buyer-balanced collection of weights such that

$$\sum_{S \subseteq B} \lambda_S v(F \cup S) > v(N).$$

Using Eq. (1), note that for any $S \subseteq B$,

$$v(F \cup S) = \sum_{i \in F} v(\{i\} \cup S_i),$$

where S_i is the element of the partition of S matched with firm i. Thus, we can write

$$\sum_{S \subseteq B} \lambda_S v(F \cup S) = \sum_{S \subseteq B} \sum_{i \in F} \lambda_S v(\{i\} \cup S_i).$$

Consider the following application of the CMP condition:

$$\lambda_S v(\{i\} \cup S) + \lambda_T v(\{i\} \cup T) \leq \min\{\lambda_S, \lambda_T\}[v(\{i\} \cup S \cup T) + v(\{i\} \cup (S \cap T))]$$
$$+ (\lambda_T - \lambda_S) + v(\{i\} \cup T) + (\lambda_S - \lambda_T) + v(\{i\} \cup S).$$

Note that

$$S \cup T \supseteq T \supseteq S \cap T$$

and

$$S \cup T \supseteq S \supseteq S \cap T.$$

Additionally, note that the weight on a given buyer is the same in both the left-hand and right-hand side of the above inequality. Thus, repeated use of the CMP condition implies that for each firm $i \in F$, there exists a sequence $\{S_1^i, \ldots, S_{m_i}^i\}$ and weights $\lambda'(S_1^i), \ldots, \lambda'(S_{m_i}^i)$ such that

$$\sum_{S \subseteq B} \sum_{i \in F} \lambda_S v(\{i\} \cup S_i) \leq \sum_{i \in F} \sum_{S \in \{S_1^i, \ldots, S_{m_i}^i\}} \lambda'(S) v(\{i\} \cup S),$$

where for each $i \in F$,

$$S_1^i \supseteq \cdots \supseteq S_{m_i}^i$$

and for each $j \in B$,

$$\sum_{S \subseteq B, S \ni j} \lambda_S = \sum_{i \in F} \sum_{S \in \{S_1^i, \ldots, S_{m_i}^i\}, S \ni j} \lambda'(S).$$

Thus, the weights $\lambda'(S_1^i), \ldots, \lambda'(S_{m_i}^i)$, $i \in F$, are a buyer-balanced collection for the sequences $\{S_1^i, \ldots, S_{m_i}^i\}$, $i \in F$.

Because $S_1^i \supseteq \cdots \supseteq S_{m_i}^i$, a buyer $j \in S_{m_i}^i$ is in every one of these sets. Because we have a buyer-balanced collection, we therefore must have

$$\sum_{r=1}^{m_i} \lambda'(S_r^i) \leq 1.$$

For all $i \in F$, let

$$\lambda'(\{i\}) = 1 - \sum_{r=1}^{m_i} \lambda'(S_r^i).$$

Then

$$v(N) < \sum_{S \subseteq B} \lambda_S v(F \cup S)$$

$$\leq \sum_{i \in F} \sum_{S \in \{S_1^i, \ldots, S_{m_i}^i\}} \lambda'(S) v(\{i\} \cup S)$$

$$= \sum_{i \in F} \left(\lambda'(\{i\}) v(\{i\}) + \sum_{S \in \{S_1^i, \ldots, S_{m_i}^i\}} \lambda'(S) v(\{i\} \cup S) \right).$$

The last line above describes a balanced collection, so the core must be empty by the Bondareva–Shapley theorem.

For sufficiency, we show that an empty core implies that the BB condition does not hold. If the core is empty, there exists a balanced collection $\mathbf{S} \subset 2^N$ with weights $\lambda_S, S \in \mathbf{S}$, such that

$$\sum_{S \in \mathbf{S}} \lambda_S v(S) > v(N).$$

For all $T \subseteq B$, let

$$\mathbf{S}(T) = \{S \in \mathbf{S} : S \cap B = T\},$$

and let

$$\lambda'(T) = \sum_{S \in \mathbf{S}(T)} \lambda_S.$$

Using super-additivity,

$$v(N) < \sum_{S \in \mathbf{S}} \lambda_S v(S)$$

$$= \sum_{T \subseteq B} \sum_{S \in \mathbf{S}(T)} \lambda_S v(S)$$

$$\leq \sum_{T \subseteq B} \sum_{S \in \mathbf{S}(T)} \lambda_S v(S \cup F)$$

$$= \sum_{T \subseteq B} \lambda'(T) v(T \cup F).$$

For any $j \in B$,

$$\sum_{S \in \mathbf{S}, S \ni j} \lambda_S = 1 = \sum_{T \subseteq B, T \ni j} \sum_{S \in \mathbf{S}(T)} \lambda_S$$

$$= \sum_{T \subseteq B, T \ni j} \lambda'(T),$$

so $\{\lambda'(T)\}_{T \subseteq B}$ is a buyer-balanced collection, implying that buyer-balancedness does not hold. $\quad\square$

Loosely, Proposition 3 shows that with the CMP condition, using the Bondareva–Shapley result only requires balanced weights for coalitions of the form $F \cup S$, where $S \subseteq B$. For a more precise interpretation, we use the following corollary.

Corollary 2. *Consider a FB game $(N; v)$ satisfying the CMP condition. Let $N' = B$, and define the characteristic function v' as follows: for all $S \subseteq B$,*

$$v'(S) = v(S \cup F).$$

The game $(N'; v')$ has a non-empty core if, and only if, the game $(N; v)$ has a non-empty core.

Proof. The corollary follows from Proposition 3 and the fact that a buyer-balanced collection of weights for the game $(N; v)$ is a balanced collection of weights for the game $(N'; v')$. $\quad\square$

In contrast to the SUB condition, the CMP condition appears to produce a more limited result. The SUB condition almost guarantees a non-empty core — only one extra condition is required. But Corollary 2 shows that the CMP condition does not really resolve the question of whether or not the core is empty. Instead, the question of non-emptiness of the core can be answered by analyzing a smaller game. With the CMP condition, non-emptiness can be determined by applying the core conditions to a game with player set of size $|B|$, rather than of size $|F| + |B|$. Corollary 2 does imply the necessary result mentioned in the Introduction. When buyers are complements with respect to any firm, no firm can be guaranteed a profit if the core is non-empty.

Revisiting Example 2, we use Corollary 2 to isolate the substitution effect between the buyers. The set N' is $\{a, b, c\}$, and the characteristic function is as follows:

$$v'(S) = 1 - \varepsilon \quad \text{if } |S| = 1,$$

$$v'(S) = 2 - \varepsilon \quad \text{if } |S| = 2,$$

$$v'(N') = 3 - 2\varepsilon.$$

Each player has a payoff of $1 - \varepsilon$ on its own, but each player adds 1 by joining with another player. All three players are competing to be in a partnership of two, but there can only be one such partnership. (Mathematically, the substitution effect can be seen with two-player coalitions, e.g., $v(\{a, b\}) + v(\{b, c\}) = 4 - 2\varepsilon \geq v(\{a, b, c\}) + v(\{b\}) = 4 - 3\varepsilon$).

We close this section by noting that although the necessary and sufficient condition for the CMP scenario is not as definitive as the necessary and sufficient condition for the SUB scenario, definitive necessary conditions do exist. For example, the following corollary shows that the marginal contributions of a firm's buyers must (weakly) exceed the value created with the firm if the core is non-empty.

Corollary 3. *Consider a FB game satisfying the CMP condition. If the game has a non-empty core, then in any efficient partition of B,*

$$\sum_{j \in B_i} [v(N) - v(N \setminus \{j\})] \geq v(\{i\} \cup B_i)$$

for all $i \in F$.

3. Discussion

This chapter has considered two restrictions on the characteristic function in many-to-one assignment games. By restricting buyers to be substitutes with respect to a firm, we can model situations in which scale and network effects net out to a decreasing returns-to-scale effect. By restricting buyers to be complements to a firm, we can model situations in which scale and network effects net out to an increasing returns-to-scale effect. Many matching and assignment models assume a substitution condition, typically the gross substitutes condition of Kelso and Crawford [1982]. The gross substitutes condition implies decreasing returns-to-scale, and it is sufficient for a non-empty core. But it is not necessary. We identify a necessary and sufficient condition in the presence of decreasing returns-to-scale.

One might suspect that a complementarity assumption would also be assumed in matching and assignment models. Games in which all players are complements — convex games — have both non-empty cores and easily-computed extreme points.

Nonetheless, in games with "sides", complementarity is typically assumed between sides, but not within sides. Our results suggest one possible reason. Complementarity with respect to individual firms can create de-stabilizing competition effects in the presence of multiple firms.

Acknowledgments

The author thanks participants in the SING 13 conference for helpful comments. Financial support from Columbia Business School is gratefully acknowledged.

References

Bondareva, O. N. [1963] Certain applications of the methods of linear programming to the theory of cooperative games, *Problemy Kibernet.* **10**, 119–139.

Camiña, E. [2006] A generalized assignment game, *Math. Soc. Sci.* **52**, 152–161.

Kaneko, M. [1976] On the core and competitive equilibria of a market with indivisible goods, *Naval Res. Logist. Quart.* **23**, 321–337.

Kelso, A. S. and Crawford, V. P. [1982] Job matching, coalition formation, and gross substitutes, *Econometrica* **50**, 1483–1504.

Muto, S., Nakayama, M., Potters, J. and Tijs, S. [1988] On big boss games, *Econ. Stud. Quart.* **39**, 303–321.

Ostrovsky, M. [2008] Stability in Supply Chain Networks, *Amer. Econ. Rev.* **98**(3), 897–923.

Shapley, L. S. [1963] On balanced sets and cores, *Naval Res. Logist. Quart.* **14**, 453–460.

Shapley, L. S. [1971] Cores of Convex Games, *Int. J. Game Theory* **1**, 11–26.

Shapley, L. S. and Shubik, M. [1971] The assignment game I: The core, *Int. J. Game Theory* **1**, 111–130.

Sotomayor, M. [1992] The multiple partners game, in *Equilibrium and Dynamics: Essays in Honour of David Gale* (Palgrave Macmillan, UK), pp. 322–354.

PART II

COOPERATIVE GAMES AND AXIOMATIC VALUES

Chapter 15

On Analyzing Cost Allocation Problems: Cooperation Building Structures and Order Problem Representations

John Kleppe, Peter Borm, Ruud Hendrickx* and Hans Reijnierse

Department of Econometrics and Operations Research
School of Economics and Management, Tilburg University
P. O. Box 90153, 5000 LE Tilburg, The Netherlands

*R.L.P.Hendrickx@uvt.nl

To analyze cost allocation problems, this chapter identifies associated cooperation building structures, with joint cost functions, and corresponding efficient order problem representations, with individualized cost functions. This chapter presents an approach that, when applicable, offers a way not only to adequately model a cost allocation problem by means of a cooperative cost game, but also to construct a core element of such a game by means of a generalized Bird allocation. We apply the approach to both existing and new classes of cost allocation problems related to operational research problems: sequencing situations without initial ordering, maintenance problems, minimum cost spanning tree situations, permutation situations without initial allocation, public congestion network situations, traveling salesman problems, shared taxi problems and traveling repairman problems.

Keywords: Cost allocation problem; cooperation building structure; order problem representation; transferable utility game; generalized Bird allocation.

1. Introduction

The aim of this chapter is to investigate the issues of cooperation and allocation, with the notions of cooperation building structures and order problem representations (oprs) as a starting point. A (general) joint cost allocation problem is defined by a group of players, a set of alternatives and a joint cost function denoting a total cost for each alternative. Typically, the set of alternatives and associated joint costs stem from an operations research problem in which several players have a similar objective, e.g., to get their job processed on a machine, to get connected

---

*Corresponding author.

to a source or to be visited by a salesman. Well-known examples of *classes* of joint cost allocation problems are the class of traveling salesman problems (tsps) [Potters *et al.*, 1992] and the class of minimum cost spanning tree (mcst) situations [Claus and Kleitman, 1973; Bird, 1976]. We refer to Borm *et al.* [2001] for a comprehensive survey of this type of problems, usually summarized under the heading of operations research games.

A cost allocation problem gives rise to two main questions; which alternative should be realized and how should the costs of this alternative be divided? The literature in which these questions are addressed is extensive, but in most papers only one specific class of cost allocation problems is discussed. In this chapter we focus on the cost allocation question. However, our analysis is not limited to one specific class, as by using common underlying features we introduce an approach which can in principle be applied to *all* classes of cost allocation problems. We assume that there is a general consensus among the players that an alternative should be chosen with minimal total cost. Then transferable utility games are a tool designed to answer the cost allocation question. In general it is, however, not evident which transferable utility game best fits the cost allocation problem under consideration. Therefore, the first issue we address is finding an appropriate transferable utility game for a class of cost allocation problems. It is clear what the cost of the grand coalition in an appropriate game should be: simply the total cost of a cheapest alternative. Defining the cost of a coalition, however, is a modeling decision that is not always straightforward.

Following von Neumann and Morgenstern [1944], it is widely accepted that the cost of a coalition should somehow reflect the cost it can guarantee itself when its members would separate from the grand coalition and decide not to cooperate with players outside their coalition. This notion may, however, be rather ambiguous as it does not specify how the other players should (re)act to this defection. Should the coalition treat the outsiders as if they were not there? Should the coalition fear the worst-case scenario in which the outsiders act as unfavorably as possible? Or can the coalition expect that the complementary coalition just ignores them and simply minimizes its own costs? In this chapter we categorize classes of cost allocation problems on the basis of either a positive or a negative externality that players have on each other. We propose for each of the two categories a specific translation into an appropriate transferable utility game.

Typically, a class of cost allocation problems is characterized by a combinatorial or network problem, which in essence models the players' problem of creating a specific feasible structure (an alternative) with minimal total cost. A structure is *feasible* for a group of players if it meets all their objectives. In each minimum cost spanning tree situation, e.g., the objective of each player is to get connected to a single source, possibly via other players. This implies that for this class a feasible structure is a network containing a path from each player to the source, such as a spanning tree on the source and all players. It is this type of structure, with the corresponding cost function, that differentiates one class of cost allocation problems from another. The starting point of our approach is to explicitly specify this structure and to reformulate the cost allocation problem as a cooperation building structure. A *cooperation building structure* in particular specifies a set of (building) structures which is partially ordered in the sense that it is specified from which structures other structures can be realized. Moreover, each structure has its total (joint) costs and for each structure it is specified for which players this structure satisfies their individual (context-specific) objectives. Having a cooperation building structure as starting point, we consider the following approach consisting of several steps.

In the first step we represent each cooperation building structure of a particular class by an order problem representation. An *order problem representation* consists of three elements: the player set of the underlying cooperation building structure, the set of all possible orderings of the player set and an individualized cost function that describes for each player the marginal costs when (recursively) realizing a structure that is feasible for himself and all his predecessors, given the restriction that he can only *extend* the existing structure. Here "extending" means that the partial ordering given in the cooperation building structure has to be respected. Hence, by the use of an order problem representation we decompose the joint cost function of the cooperation building structure into individual parts.

The order problem representation forms the basis for the transferable utility game and the cost allocation defined further on in our approach. In order to continue with our approach, however, we need efficiency of the order problem representation. If for an optimal ordering (an ordering with minimal total cost) the sum of the individual costs exactly equals the total cost of a cheapest structure that is feasible for all players in the cooperation building structure, then the order problem representation is called *efficient*. If the order problem representation is not efficient for all underlying cooperation building structures within a certain class, then this class is not suited for our approach.

Now suppose that we have found an efficient order problem representation for each cooperation building structure within a certain class. The next step of our approach is to identify an appropriate transferable utility game for this class. For this, we differentiate between two types of order problem representations. In a *positive externality order problem representation* (*peopr*) the minimum cost for each (sub)group of players is obtained for an ordering in which they are ordered last, while in a *negative externality order problem representation* (*neopr*) the minimum cost for each group of players is obtained for an ordering in which they are ordered first. We argue that each positive externality order problem representation is appropriately modeled by the so-called *direct cost game*. Furthermore, we argue that each negative externality order problem representation is appropriately modeled by the dual of the direct cost game, called the *marginal cost game*.

The final step of our approach is the systematic determination of a reasonable cost allocation proposal for the underlying cost allocation problem that we started with. To this aim, we introduce for each order problem representation a generalized

Bird allocation. This allocation is inspired by Bird's tree allocation [Bird, 1976] for the class of minimum cost spanning tree situations, in the sense that each player contributes his individual marginal cost in an optimal ordering for the grand coalition.

We show that for each *negative* externality order problem representation a generalized Bird allocation is a core element of the *marginal* cost game. And for each *positive* externality order problem representation that satisfies predecessor order independence (*poi*; individual costs only depend on the set of predecessors and not on their exact ordering) each generalized Bird allocation is an element of the core of the *direct* cost game. Note that in both cases the transferable utility game under consideration is appropriate for the underlying cost allocation problem. Therefore, in these cases, on the basis of stability considerations, we consider the generalized Bird allocation to be a reasonable solution for the underlying cost allocation problem.

We apply our approach to several classes of cost allocation problems based on different types of operational research problems. We first discuss the class of sequencing situations without initial order, introduced by Klijn and Sánchez [2006], who propose two transferable utility games. Interestingly, by our procedure we derive an appropriate transferable utility game that is different from those two games, and we show that the core of our transferable utility game, which contains the generalized Bird allocation, is a subset of the core of the two games of Klijn and Sánchez [2006].

Second, we discuss the class of maintenance problems [Megiddo, 1978; Koster, 1999] in which a group of players has to maintain a (fixed) tree network. Koster [1999] defines the cost game in which the cost of a coalition is equal to the cost to maintain the smallest trunk in which all the players of the coalition are contained. Our approach supports this transferable utility game, and we derive that a generalized Bird allocation is an element of its core.

Third, we consider the class of minimum cost spanning tree situations. Claus and Kleitman [1973] and Bird [1976] implicitly assume that the edge between any two players can only be used if the two players cooperate with each other, and within this context Bird [1976] proposes the transferable utility game in which the cost for each coalition equals the cost of a minimum cost spanning tree on all players of the coalition plus the source. We, on the other hand, assume that all edges are publicly available. Nevertheless, our approach leads to the same transferable utility game. This enables us to provide an alternative proof by the use of order problem representations of the well-known fact that Bird's tree allocation is in the core of this game.

Fourth, we introduce and analyze the class of permutation situations without initial allocation. Permutation situations are introduced by Tijs *et al.* [1984], who assume that there is an initial allocation of machines to the players and that a coalition of players can interchange their machines between them. We assume on the contrary that there is no initial allocation of machinery. For this class of cost allocation problems our approach provides an appropriate transferable utility game for which a generalized Bird allocation is an element of its core.

Fifth, we consider the class of public congestion network situations, introduced by Kleppe *et al.* [2010]. We show that our approach supports the use of the marginal congestion cost game, as proposed by Kleppe *et al.* [2010], for public congestion network situations with *convex* cost functions. For the class of public congestion network situations with *concave* cost functions, however, we show that not every situation has an efficient order problem representation. Therefore, our approach cannot support the use of the direct congestion cost game, as proposed by Kleppe *et al.* [2010].

Sixth, we discuss the class of traveling salesman problems. For this class we come to the conclusion that even a reformulation as a cooperation building structure is not possible. A different type of modeling problem arises for the closely related (new) class of shared taxi problems (stps). Here we find an efficient order problem representation for all such problems. However, not all of them are either positive or negative externality order problem representations. This implies that we cannot categorize the order problem representations and cannot argue in favor of either the direct or the marginal cost game to model these kinds of problems. Finally, we discuss also the related class of traveling repairman problems (trps). Here, our approach gives an appropriate transferable utility game for which a generalized Bird allocation is an element of its core.

The structure of this chapter is as follows. In Sec. 2 we formally introduce cooperation building structures and order problem representations and describe our approach to find an appropriate transferable utility game, as well as a reasonable cost allocation. In the remainder of this chapter we apply this approach to several classes of cost allocation problems: sequencing situations without initial order (Sec. 3), maintenance problems (Sec. 4), minimum cost spanning tree situations (Sec. 5), permutation situations without initial allocation (Sec. 6), public congestion network situations (Sec. 7), and traveling salesman, shared taxi and traveling repairman problems (Sec. 8).

2. Cooperation Building Structures

A *cooperation building structure* is defined by $(N, (\mathcal{G}, \preceq), F, K)$, with N the finite set of players. $\mathcal{G}$ is a finite set of (building) *structures* with partial ordering $\preceq$. The function $F : \mathcal{G} \to 2^N$ indicates for each structure $G \in \mathcal{G}$ the group of players $F(G) \subseteq N$ for which the structure G is *feasible*, i.e., the context-specific notion of all players in $F(G)$ having their objective being met by structure G. We assume that there exists at least one structure that meets the objectives of all players in $N : F^{-1}(N) \neq \emptyset$. Furthermore, we assume that each player can extend a structure, as modeled by $\preceq$, to make it feasible to himself: for all $i \in N$, all $G \in \mathcal{G}$ with $i \notin F(G)$, there exists a $G' \in \mathcal{G}$ such that $G \preceq G'$ and $i \in F(G')$. Finally, the cost function $K : \mathcal{G} \to \mathbb{R}_+$ assigns to each structure $G \in \mathcal{G}$ a certain cost $K(G)$.

We assume K to satisfy the (monotonicity) condition that for all $G, G' \in \mathcal{G}$ with $G \preceq G'$ we have $K(G) \leq K(G')$.

Starting from a cost allocation problem, the ultimate aim is to allocate the costs of the cheapest structure that is feasible for N to the individual players in N in a reasonable way. Reformulating such a problem as a cooperation building structure is a tool that allows us to check if we can "adequately" individualize the joint cost function K. What we mean with "adequately" is explained below.

An *opr* is given by (N, Π, k), with N the finite player set and Π the set of orderings of the player set. Here an ordering is a bijection $\pi : \{1, \ldots, |N|\} \to N$. Finally, $k : \Pi \to \mathbb{R}^N$ is an *individualized cost function*, denoting for each ordering $\pi \in \Pi$ an individual cost $k^i(\pi)$ for each player $i \in N$.

A cooperation building structure $(N, (\mathcal{G}, \preceq), F, K)$ gives rise to an *opr* (N, Π, k) in the following way. Given an ordering $\pi \in \Pi$, we first define

$$k^{\pi(1)}(\pi) = \min_{G \in \mathcal{G} : \pi(1) \in F(G)} K(G),$$

and a structure in which this minimum is attained is denoted by $G^{\pi,1}$. Subsequently, we recursively define for $\ell \in \{2, \ldots, |N|\}$:

$$k^{\pi(\ell)}(\pi) = \min_{G \in \mathcal{G} : \pi(\ell) \in F(G), G \succeq G^{\pi(\ell-1)}} K(G) - K(G^{\pi, \ell-1}),$$

where (again) $G^{\pi,\ell}$ is a locus of the minimum, as the minimal marginal cost to create a feasible structure for player $\pi(\ell)$.

Note that in general, there might be multiple structures that minimize a player's individualized cost, so an *opr* is not unique. However, given a cooperation building structure, all minima are well defined, so a corresponding *opr* exists.

A corresponding *opr* is called efficient if an optimal ordering of N results in a feasible structure for N with minimal total cost. Formally, let (N, Π, k) be an *opr* corresponding to $(N, (\mathcal{G}, \preceq), F, K)$. Then (N, Π, k) is *efficient* if $\min_{\pi \in \Pi} \sum_{i \in N} k^i(\pi) = \min_{G \in \mathcal{G} : N = F(G)} K(G)$.

If we come to the conclusion that for each cooperation building structure within a certain class there exists an efficient *opr*, then the joint cost function K can be adequately individualized and we proceed to the next step of our approach. Otherwise we unfortunately have to conclude that the class of cooperation building structures under consideration is not suited for our approach.

As a result of the previous step, we may assume that we have a class of cooperation building structures, each of which possesses an *efficient opr*. In the second step we distinguish between two types of externalities and we define two games, each being appropriate for one type of externality.

Given the *opr* (N, Π, k) we denote by

$$\pi_S^* \in \operatorname*{argmin}_{\pi \in \Pi} \sum_{i \in S} k^i(\pi)$$

a cheapest or optimal ordering for coalition $S \subseteq N$.

By $\Pi_S \subseteq \Pi$ we denote the set of all orderings $\pi \in \Pi$ such that the players in $S \subseteq N$ are placed on the first $|S|$ position, i.e., $\pi^{-1}(i) < \pi^{-1}(j)$ for all $i \in S$, $j \in N \backslash S$. Note that $\Pi_N = \Pi$.

The *opr* (N, Π, k) is a *neopr* if for all $S \subseteq N$ there exists an optimal ordering π_S^* such that

$$\pi_S^* \in \Pi_S.$$

The *opr* (N, Π, k) is a *peopr* if for all $S \subseteq N$ there exists an optimal ordering π_S^* such that

$$\pi_S^* \in \Pi_{N \backslash S}.$$

Hence, in a *neopr* each group of players prefers to be ordered first, while for a *peopr* each group of players wants to be ordered last. An *opr* may not belong to any of the above two types, in which case our approach is not suited to suggest an appropriate TU-game for the underlying class of cooperation building structures.

We consider two types of TU-games.[a] For an efficient *opr* (N, Π, k) corresponding to a cooperation building structure we define the *direct cost game* (N, c_d) by

$$c_d(S) = \min_{\pi \in \Pi_S} \sum_{i \in S} k^i(\pi), \tag{2.1}$$

for all $S \subseteq N$. Hence, in calculating the cost of a coalition in the direct cost game it is assumed that this coalition is ordered first, i.e., before $N \backslash S$. The coalition S can optimize using the first $|S|$ positions of the ordering as slots for the players. Furthermore, we define the *marginal cost game* (N, c_m) as the dual of the direct cost game, i.e.,

$$c_m(S) = c_d(N) - c_d(N \backslash S), \tag{2.2}$$

for all $S \subseteq N$. The cost of coalition S in the marginal cost game reflects its marginal cost to the grand coalition.

Generally speaking, given a cooperation building structure the coalitional cost of an associated TU-game, $c(S)$, should be based on what this coalition, $S \subseteq N$, can *guarantee* itself when its members would separate from the grand coalition and decide not to cooperate with players outside their coalition.

Since in a *peopr* each group of players prefers to be ordered last, the cost of a coalition is minimal for an ordering in which all players *outside* the coalition are ordered first. However, the possible positive effect of the presence of other players should not be incorporated in the coalitional cost. Hence, it is reasonable to model a *peopr* by the direct cost game.

---

[a] A transferable utility (cost) game or TU-game is a pair (N, c), where N denotes the finite set of players and $c : 2^N \to \mathbb{R}$ is the characteristic function, assigning to every coalition $S \subseteq N$ of players a cost, $c(S)$, representing the total joint cost of this coalition of players when they cooperate. By convention, $c(\emptyset) = 0$.

We can motivate the use of the marginal cost game for a *neopr* as follows. If the players in S are not present (from the perspective of the costs of $N\backslash S$, equivalent with the players in S being on the last $|S|$ positions of an ordering) it is clear that $N\backslash S$ has a cost of $c_d(N\backslash S)$. The presence of the players in S (equivalent to the players in S not necessarily on the last $|S|$ positions) can by definition of a *neopr* only increase the cost of $N\backslash S$. Therefore, $c_d(N\backslash S)$ is the minimum amount coalition $N\backslash S$ ever has to pay in an ordering and is therefore a reasonable minimum contribution to $c(N)$. This implies that a reasonable maximum contribution for S to the total cost equals $c_m(S) = c_d(N) - c_d(N\backslash S)$.

Our approach to obtain appropriate coalitional costs for a *neopr* is well imbedded in the game theoretic literature; think, e.g., of bankruptcy situations [O'Neill, 1982; Aumann and Maschler, 1985; Thomson, 2003] in which an estate has to be divided among a number of players, each with his own claim on (part of) the estate. In order to determine a coalitional value in the corresponding bankruptcy game one first treats the players outside a coalition in an optimistic way (by giving them their full claim if possible), putting the value of a coalition equal to the remainder of the estate (or zero). Translating the idea of an efficient *opr* to the usual reward setting of bankruptcy situations, rather than the cost setting in the definition of *opr*, boils down to the following: for each ordering of the player set the players obtain their full claim one by one until there is no money left. Since clearly, each group of players prefers to be ordered first, this *opr* is a *neopr*. Consequently, our model suggests the marginal cost game to model bankruptcy situations. This game exactly boils down to the bankruptcy game used in Aumann and Maschler [1985].

Given a class of cooperation building structures for which we can find efficient order problem representations, the third and final step of our aproach introduces a general allocation rule for the costs of a cheapest structure that is feasible for all players. The quality or reasonability of this allocation rule is established using the core of the appropriate game as provided in the previous step.

We define a *generalized Bird allocation* $\beta(N, \Pi, k)$ associated with an *opr* (N, Π, k) such that each player contributes his individual cost according to an optimal ordering π_N^* for the grand coalition. Hence, such an allocation[b] is defined by

$$\beta^i(N, \Pi, k) = k^i(\pi_N^*), \tag{2.3}$$

for all $i \in N$. This type of allocation β is called a generalized Bird allocation, because it is a generalization of a Bird allocation for minimum cost spanning tree situations [Bird, 1976].

For analytical purposes we start by introducing two auxiliary TU-games. For an *opr* (N, Π, k) we define the *minimal cost game* (N, c_-) by

$$c_-(S) = \min_{\pi \in \Pi} \sum_{i \in S} k^i(\pi),$$

---

[b]Since π_N^* is not unique one should formally denote $\beta^i(N, \Pi, k; \pi_N^*)$, but following Bird [1976] we omit π_N^* in the expression.

for all $S \subseteq N$. The *maximal cost game* (N, c_+) is defined as the dual of the minimal cost game, i.e.,

$$c_+(S) = c_-(N) - c_-(N \backslash S),$$

for all $S \subseteq N$. Obviously, we obtain the following proposition.

Proposition 2.1. *Let* (N, Π, k) *be a neopr. Then the associated direct cost game* (N, c_d) *and minimal cost game* (N, c_-) *coincide. As a consequence, also the marginal cost game* (N, c_m) *and maximal cost game* (N, c_+) *coincide.*

The following theorem shows that the generalized Bird allocation of an order problem representation is an element of the core[c] of the corresponding maximal cost game.

Theorem 2.2. *Let* (N, Π, k) *be an opr with associated maximal cost game* (N, c_+). *Then* $\beta(N, \Pi, k) \in C(N, c_+)$.

Proof. By definition $\sum_{i \in N} \beta^i(N, \Pi, k) = \sum_{i \in N} k^i(\pi_N^*) = c_+(N)$. Furthermore, for all $S \subseteq N$,

$$c_+(S) = c_-(N) - c_-(N \backslash S)$$

$$= \sum_{i \in N} k^i(\pi_N^*) - \sum_{i \in N \backslash S} k^i(\pi_{N \backslash S}^*)$$

$$\geq \sum_{i \in N} k^i(\pi_N^*) - \sum_{i \in N \backslash S} k^i(\pi_N^*)$$

$$= \sum_{i \in S} k^i(\pi_N^*)$$

$$= \sum_{i \in S} \beta^i(N, \Pi, k).$$

Using Proposition 2.1, we immediately obtain the next result. □

Theorem 2.3. *Let* (N, Π, k) *be a neopr with associated marginal cost game* (N, c_m). *Then* $\beta(N, \Pi, k) \in C(N, c_m)$.

Hence, if for each cooperation building structure within a class one can define an efficient *neopr*, then β is a reasonable allocation for the underlying class of cost allocation problems, because β is an element of the core of the appropriate TU-game (N, c_m).

---

[c]The core $C(N, c)$ [Gillies, 1959] of a TU-game (N, c) consists of those cost allocations for which no coalition would be better off if it would separate itself and would pay its coalitional cost. It is given by $C(N, c) = \{x \in \mathbb{R}^N \mid x(N) = c(N), x(S) \leq c(S) \text{ for all } S \subseteq N\}$, with $x(S) = \sum_{i \in S} x^i$ for all $S \subseteq N$.

A similar result, placing a generalized Bird allocation in the core of the direct cost game associated with a *peopr*, is not immediate and requires an additional condition. This condition, which is called *predecessor order independence*, boils down to the idea that the *ordering* of the predecessors of each player should be irrelevant for his individual cost. Let (N, Π, k) be an *opr* and let

$$V^i(\pi) = \{ j \in N \mid \pi^{-1}(j) < \pi^{-1}(i) \}$$

denote the set of predecessors of player i given the ordering $\pi \in \Pi$. Then (N, Π, k) satisfies *poi* if, for all $i \in N$,

$$k^i(\pi) = k^i(\bar{\pi}),$$

for all $\pi, \bar{\pi} \in \Pi$, such that $V^i(\pi) = V^i(\bar{\pi})$.

Theorem 2.4. *Let (N, Π, k) be a peopr that satisfies poi and let (N, c_d) be the associated direct cost game. Then $\beta(N, \Pi, k) \in C(N, c_d)$.*

Proof. Note that $\sum_{i \in N} \beta^i(N, \Pi, k) = \sum_{i \in N} k^i(\pi_N^*) = c_d(N)$. Let $S \subseteq N$ and let $\hat{\pi}_S^* \in \Pi_S$ be such that $\sum_{i \in S} k^i(\hat{\pi}_S^*) = c_d(S)$. Recall that $\pi_{N \setminus S}^*$ is an optimal ordering for $N \setminus S$. Let $\pi_{N \setminus S}^*$ be such that the players in $N \setminus S$ are ordered in the last positions of $\pi_{N \setminus S}^*$. Since (N, Π, k) is a *peopr* such a $\pi_{N \setminus S}^*$ exists. Next define $\tilde{\pi}_N$ such that $\tilde{\pi}_N(t) = \hat{\pi}_S^*(t)$ for all $t \in \{1, \ldots, |S|\}$ and $\tilde{\pi}_N(t) = \pi_{N \setminus S}^*(t)$ for all $t \in \{|S| + 1, \ldots, n\}$. Then,

$$c_d(S) = \sum_{i \in S} k^i(\hat{\pi}_S^*)$$

$$\geq \sum_{i \in S} k^i(\hat{\pi}_S^*) + \sum_{i \in N \setminus S} k^i(\pi_{N \setminus S}^*) - \sum_{i \in N \setminus S} k^i(\pi_N^*)$$

$$= \sum_{i \in N} k^i(\tilde{\pi}_N) - \sum_{i \in N \setminus S} k^i(\pi_N^*)$$

$$\geq \sum_{i \in N} k^i(\pi_N^*) - \sum_{i \in N \setminus S} k^i(\pi_N^*)$$

$$= \sum_{i \in S} k^i(\pi_N^*)$$

$$= \sum_{i \in S} \beta^i(N, \Pi, k),$$

where the second equality follows from the fact that (N, Π, k) satisfies *poi*.   □

Hence, if for each cooperation building structure within a class we can define an efficient *peopr* and each such *peopr* satisfies *poi*, then β is a reasonable allocation for the underlying class of cost allocation problems, because β is an element of the core of the appropriate TU-game (N, c_d).

3. Sequencing Situations Without Initial Order

In this section we discuss the class of sequencing situations without initial order. The cost allocation problem associated with sequencing situations *with* initial order has been introduced by Curiel *et al.* [1989]. Note that an initial order gives each (group of) player(s) an *ex ante* position and cost. As a consequence, determining the cost for a subcoalition boils down to the same problem as for the grand coalition. In case there is no initial order the problem for a subcoalition is essentially different from the problem for the grand coalition, as the role of the outsiders is undetermined. Klijn and Sánchez [2006] initiated the analysis of sequencing situations without initial order.

In this section we reformulate a sequencing situation without initial order as a coalition building structure, derive an associated efficient *opr* and show that it is a *neopr*. Therefore, we suggest to model this class of cooperation building structures by the associated marginal cost game. Furthermore, we observe by the use of Theorem 2.3 that each associated generalized Bird allocation is an element of the core of this game. We also compare the marginal cost game with the two cost games proposed by Klijn and Sánchez [2006].

A *sequencing situation without initial order* is given by a triple[d] $Q = (N, p, \delta)$, with N being a finite set of players. Each player $i \in N$ owns one job that has to be processed on a single machine. The job of player i is with slight abuse of notation also denoted by i. The processing times of the jobs are given by $p = \{p^i\}_{i \in N}$ with $p^i > 0$ for all $i \in N$. Furthermore, with $\delta = \{\delta^i\}_{i \in N}$ each player has a cost function $c^i : [0, \infty) \to \mathbb{R}$ given by $c^i(t) = \delta^i t$ for $t \in [0, \infty)$, where $\delta^i > 0$. The expression $c^i(t)$ is interpreted as the cost incurred by agent i if his job is completed at time t. The optimal ordering of jobs for the grand coalition is obtained by putting them in a nonincreasing order of their urgency indices [Smith, 1956], which are defined as $u^i = \frac{\delta^i}{p^i}$ for all $i \in N$. Since it is assumed that only schedules are considered in which the jobs are processed without any breaks in between, the cost of player i is completely determined by an ordering of the player set $\pi \in \Pi$. Therefore, the cost function c^i is given by $c^i(\pi) = \delta^i \sum_{j \in (V^i(\pi) \cup \{i\})} p^j$ for all $i \in N$ and all $\pi \in \Pi$.

A sequencing situation without initial order $Q = (N, p, \delta)$ can be reformulated as a cooperation building structure $(N, (\mathcal{G}, \preceq), F, K)$ with $\mathcal{G} = \bigcup_{S \subseteq N : S \neq \emptyset} \Pi(S)$, where $\Pi(S)$ is the set of orderings of S.[e] Let $\pi, \pi' \in \mathcal{G}$. Assume $S \subseteq N$ and $T \subseteq N$ are such that $\pi \in \Pi(S)$ and $\pi' \in \Pi(T)$. Then $\pi \preceq \pi'$ if and only if $S \subseteq T$ and $(\pi')^{-1}(i) = \pi^{-1}(i)$ for all $i \in S$. For any $\pi \in \mathcal{G}$ with $\pi \in \Pi(S)$ we have $F(\pi) = S$ and $K(\pi) = \sum_{i \in S} (\delta^i \sum_{j \in V^i(\pi) \cup \{i\}} p^j)$.[f]

---

[d]Klijn and Sánchez [2006] use the notation (N, p, α).
[e]Note that this differs from Π_S, which is the set of orderings of N in which the players in S come first.
[f]It is readily checked that the additional assumptions $F^{-1}(N) \neq \emptyset$ and monotonicity of K are satisfied.

Note that $\preceq$ is modeled in such a way that in any *opr*, a player entering only has one choice to extend the existing structure, namely by leaving the order of the players already present undisturbed and joining at the end of the queue. Consequently, for a corresponding *opr* (N, Π, k) we have

$$k^i(\pi) = \delta^i \sum_{j \in V^i(\pi) \cup \{i\}} p^j,$$

for all $i \in N$ and all $\pi \in \Pi$. It is easily seen that such an *opr* is also efficient. The following proposition follows from the fact that $p^i > 0$ for all $i \in N$.

Proposition 3.1. *For a sequencing situation without initial order, a corresponding opr is a neopr.*

Since an *opr* is a *neopr*, we suggest to model the class of sequencing situations without initial order by the marginal cost game (N, c_m) as provided by (2.2). The next proposition follows by Theorem 2.3.

Proposition 3.2. *Let Q be a sequencing situation without initial order with corresponding opr (N, Π, k). Then $\beta(N, \Pi, k) \in C(N, c_m)$.*

Next we compare the marginal cost game to the cost games proposed by Klijn and Sánchez [2006] to model sequencing situations without initial order. In their tail game (N, c_{tail}) it is assumed that the players in $N \backslash S$ are ordered first and the players in S can only optimize their sequence in the tail of the ordering. This game is defined by

$$c_{\text{tail}}(S) = \min_{\pi \in \Pi_{N \backslash S}} \sum_{i \in S} k^i(\pi),$$

for all $S \subseteq N$. We obtain the following proposition.

Proposition 3.3. *Let Q be a sequencing situation without initial order with associated tail game (N, c_{tail}). Let (N, Π, k) be a corresponding opr with associated marginal cost game (N, c_m). Then $C(N, c_m) \subseteq C(N, c_{\text{tail}})$.*

Proof. By definition $c_m(N) = c_{\text{tail}}(N)$. Hence, it suffices to show that $c_m(S) \leq c_{\text{tail}}(S)$ for all $S \subseteq N$. Let $S \subseteq N$ and let $\hat{\pi}^*_{N \backslash S}$ be an ordering that is optimal for $N \backslash S$ such that $c_{\text{tail}}(S) = \sum_{i \in S} c^i(\hat{\pi}^*_{N \backslash S})$. Recall that π^*_N is an optimal ordering for N. Then

$$c_{\text{tail}}(S) = \sum_{i \in S} k^i(\hat{\pi}^*_{N \backslash S})$$

$$= \sum_{i \in N} k^i(\hat{\pi}^*_{N \backslash S}) - \sum_{i \in N \backslash S} k^i(\hat{\pi}^*_{N \backslash S})$$

$$\geq \sum_{i \in N} k^i(\pi_N^*) - \sum_{i \in N \setminus S} k^i(\hat{\pi}_{N \setminus S}^*)$$

$$= c_d(N) - c_-(N \setminus S)$$

$$= c_d(N) - c_d(N \setminus S)$$

$$= c_m(S).$$

Note that the equal sign between $c_d(N) - c_-(N \setminus S)$ and $c_d(N) - c_d(N \setminus S)$ follows from the fact that (N, Π, k) is a *neopr*. □

Klijn and Sánchez [2006] also introduce a pessimistic game (N, c_{pess}) from which they show that $C(N, c_{\text{tail}}) \subseteq C(N, c_{\text{pess}})$ for each sequencing situation without initial order. Therefore, the core of the marginal cost game is also a subset of the core of the pessimistic game. Moreover, each generalized Bird allocation is an element of the core of these three games.

4. Maintenance Problems

In this section we consider the class of maintenance problems [Megiddo, 1978; Koster, 1999]. In these problems the players have to maintain a tree network. This class of cost allocation problems contains the class of airport problems, introduced by Littlechild and Owen [1973]. We derive that there is an efficient *peopr* for each maintenance problem. Hence, on that basis we propose the direct cost game to model this class of cost allocation problems. It turns out that this game coincides with the game from literature, e.g., discussed in Koster [1999]. Furthermore, we obtain by Theorem 2.4 that a generalized Bird allocation is in the core of this game.

Formally, a *maintenance problem* is given by a triple $I = (N, T, \gamma)$, with N being the finite player set. The source is given by 0 and by S^0, $S \subseteq N$, we denote the set $S \cup \{0\}$. Here $T = (N^0, E)$ is a tree with the set of nodes N^0 and edge set E, such that the source 0 has only one adjacent edge. The objective of each player is that the unique path from his node to the source is maintained. The cost to maintain an edge is given by $\gamma : E \to \mathbb{R}_+$.

A maintenance problem can be reformulated as a cooperation building structure $(N, (\mathcal{G}, \preceq), F, K)$, with $\mathcal{G} = \bigcup_{S \subseteq N : S \neq \emptyset} \{(S^0, T(S^0))\}$, where $(S^0, T(S^0))$ is the smallest subtree of T spanning S^0. Moreover, $(S^0, T(S^0)) \preceq (U^0, T(U^0))$ if and only if $S^0 \subseteq U^0$, while $F(S^0, T(S^0))$ is the set of players connected to 0 by $T(S^0)$ and $K(S^0, T(S^0)) = \sum_{e \in T(S^0)} \gamma(e)$.

In a corresponding *opr* (N, Π, k), the individualized cost for a player represents the cost to maintain the edges with which he has to extend his predecessors' spanning tree to connect himself:

$$k^i(\pi) = \sum_{e \in T(V^i(\pi) \cup \{0,i\}) \setminus T(V^i(\pi) \cup \{0\})} \gamma(e),$$

for all $i \in N$ and $\pi \in \Pi$. Note that each ordering $\pi \in \Pi$ is optimal and that this opr (N, Π, k) is efficient. We obtain the following proposition by the nonnegativity of the cost function.

Proposition 4.1. *For a maintenance problem, a corresponding opr is a peopr.*

Since there an efficient *peopr*, we suggest to model the class of maintenance problems by the direct cost game (N, c_d) as provided by (2.1). This cost game coincides with the cost game in literature, e.g., discussed in Koster [1999]. Furthermore, since one can easily deduce that in this class an efficient *peopr* satisfies *poi*, we obtain by Theorem 2.4 the following proposition.

Proposition 4.2. *Let I be a maintenance problem with corresponding opr (N, Π, k). Then $\beta(N, \Pi, k) \in C(N, c_d)$.*

5. Minimum Cost Spanning Tree Situations

In this section we consider the class of *mcst* situations. Claus and Kleitman [1973] and Bird [1976] assume that the edge between any two players can only be used if the players cooperate with each other, and within this context Bird [1976] proposes a transferable utility game. In this section we assume instead that all edges are publicly available. Interestingly, however, our approach leads to the same transferable utility game. In particular, this enables us to provide an alternative proof of the fact that Bird's tree allocation is in the core of this game. We refer to Kleppe [2010] for an alternative proof of the fact that also the P-value [Branzei et al., 2004] is in the core of this game.

Formally, a *minimum cost spanning tree situation* is a triple $M = (N, 0, \gamma)$, with N being the finite player set and 0 the source. The objective of each player is to get connected to the source. By E_T we denote the set of all edges between pairs in $T \subseteq N^0$, i.e., (T, E_T) is the complete network on T. Further, $\gamma : E_{N^0} \to \mathbb{R}_+$ is a nonnegative cost function specifying the cost $\gamma(e)$ to construct an edge $e \in E_{N^0}$. Since in our approach each edge is publicly available, we assume without loss of generality that γ satisfies the triangle inequality, which means that $\gamma(j, \ell) \leq \gamma(j, i) + \gamma(i, \ell)$ for all $i, j, \ell \in N^0$. An edge is denoted by (j, ℓ), with $j, \ell \in N^0$.

Bird [1976] associates with each *mcst* situation $M = (N, 0, \gamma)$ the cooperative cost game (N, c_M), where $c_M(S)$, $S \subseteq N$, represents the minimum cost of a tree on S^0:

$$c_M(S) = \min\left\{ \sum_{e \in R} \gamma(e) \,\Big|\, R \subseteq E_{S^0} \text{ and } (S^0, R) \text{ is a tree} \right\},$$

for all $S \subseteq N$.

An *mcst* situation can be reformulated as a cooperation building structure $(N, (\mathcal{G}, \preceq), F, K)$, with $\mathcal{G} = \{(N^0, E) \mid E \subseteq E_{N^0}\}$, $(N^0, E) \preceq (N^0, E')$ if and only if $E \subseteq E'$, $F(N^0, E) = \{i \in N \mid E \text{ connects } i \text{ to } 0\}$, and $K(N^0, E) = \sum_{e \in E} \gamma(e)$.

In a corresponding *opr* (N, Π, k), the individualized cost k represents the minimum cost for a player to connect his node to the source (possibly via other nodes) given the tree constructed by his predecessors. To be precise,

$$k^i(\pi) = \min_{(i,j) \in E_{N^0}} \{\gamma(i,j) \mid j \in V^i(\pi) \cup \{0\}\}, \tag{5.1}$$

for all $i \in N$ and $\pi \in \Pi$.[g] Note that, contrary to the maintenance problems discussed in Sec. 4, a player entering actually has the problem to solve how to connect himself as cheaply as possible. There exists a π_N^* such that this procedure corresponds to *Prim's algorithm* [Prim, 1957] to obtain a minimum cost spanning tree. Therefore, the *opr* is efficient.

Let a minimum cost spanning tree for the grand coalition N be given by (N^0, R^*) and let e^i, for all $i \in N$, be the first edge on the unique path in (N^0, R^*) from player i to the source. Then *Bird's tree allocation*[h] $\dot\beta$ of M is obtained by assigning to each player $i \in N$ the cost of e^i, hence

$$\dot\beta^i(M) = \gamma(e^i), \tag{5.2}$$

for all $i \in N$.

Lemma 5.1. *Let (N, Π, k) be an opr of an mcst situation M. Then for every Bird allocation $\dot\beta(M)$ there exists a generalized Bird allocation $\beta(N, \Pi, k)$ such that $\dot\beta(M) = \beta(N, \Pi, k)$, and vice versa.*

Proof. Let (N^0, R^*) be a minimum cost spanning tree of M. According to Prim's algorithm (N^0, R^*) can be constructed starting from the source and adding the edges R^* one by one. The cost attributed to player $i \in N$ by the generalized Bird allocation is then, using (5.1), equal to the cost of connecting i as cheaply as possible to his predecessors. This connection is e^i by definition[i], so $\beta^i(N, \Pi, k) = \gamma(e^i) = \dot\beta^i(M)$.

For the other way around, it suffices to note that any optimal ordering by construction gives rise to a minimum cost spanning tree and that the same argument holds for why the Bird allocation equals the generalized Bird allocation. □

Since the individualized cost function k is only based upon the set of predecessors of each player, and not on their ordering, we obtain the following lemma.

Lemma 5.2. *Let M be an mcst situation. Then a corresponding opr (N, Π, k) satisfies poi.*

---

[g]Note that a player may have more than one cheapest option to connect himself to the source. Therefore, the tree constructed by this procedure need not be unique for a $\pi \in \Pi$. However, the choice of one player does not influence the options or costs of any other player. This implies that $k(\pi)$ is unique for each $\pi \in \Pi$.

[h]We denote this allocation by $\dot\beta$ instead of β to distinguish between a Bird allocation and a *generalized* Bird allocation defined in Sec. 2.

[i]Again, the edge may not be unique, but the corresponding costs are.

Let (N, Π, k) be an *opr* of $M = (N, 0, \gamma)$. The definition of k^i implies that the later player i is ordered, the more choice he has and therefore, the lower his individual cost. Hence, all players prefer to be ordered as late as possible. Since (N, Π, k) satisfies *poi* this implies that all groups of players want to be ordered as late as possible. Hence, (N, Π, k) is a *peopr*.

Proposition 5.3. *Let M be an mcst situation. Then a corresponding opr (N, Π, k) is a peopr.*

On the basis of Proposition 5.3, we suggest to model the class of *mcst* situations by the associated direct cost game (N, c_d) as provided by (2.1). This game coincides with the TU-game (N, c_M).

Proposition 5.4. *Let M be an mcst situation with associated cost game (N, c_M). Let (N, Π, k) be a corresponding opr with associated direct cost game (N, c_d). Then $(N, c_M) = (N, c_d)$.*

Proof. Let $S \subseteq N$. The coalitional cost for S according to (N, c_M) denotes the cost of the cheapest tree on S^0. Consider Prim's algorithm to obtain such a minimum cost spanning tree on S^0 and let $\pi \in \Pi_S$ be the ordering in which the players make their connection according to this algorithm. Then $c_M(S) = \sum_{i \in S} k^i(\pi) = c_d(S)$. $\qquad\square$

We are now ready to provide an alternative proof of the fact that each Bird's tree allocation is in the core of the game (N, c_M). The original result is due to Bird [1976].

Theorem 5.5. *Let M be an mcst situation. Then $\dot{\beta}(M) \in C(N, c_M)$.*

Proof. Let (N, Π, k) be a corresponding *opr* of *mcst* situation M, with associated direct cost game (N, c_d) and a generalized Bird allocation $\beta(N, \Pi, k)$ such that $\beta(N, \Pi, k) = \dot{\beta}(M)$ (Lemma 5.1). Since (N, Π, k) is a *peopr* (Proposition 5.3) that satisfies *poi* (Lemma 5.2) we obtain by Theorem 2.4 that $\dot{\beta}(M) \in C(N, c_d)$. Finally, by Proposition 5.4, $\dot{\beta}(M) \in C(N, c_M)$. $\qquad\square$

6. Permutation Situations Without Initial Allocation

In this section we introduce and analyze the class of permutation situations without initial allocation. Permutation situations are introduced by Tijs *et al.* [1984]. A *permutation situation without initial allocation* is given by a triple $P = (N, \Theta, \Gamma)$, with N begin the finite set of players, $\Theta = \{1, \ldots, |N|\}$ the finite set of machines and Γ an $N \times \Theta$ cost matrix. The element Γ_{ij} of this matrix denotes the cost for the use of machine $j \in \Theta$ by player $i \in N$. The idea is to allocate each machine to a different player with minimal total cost.

An optimal allocation for permutation situations without initial allocation P is denoted by $\alpha^*(P)$ with $(i, j) \in \alpha^*(P)$ an assignment of machine j to player i. One can find an optimal allocation by the Hungarian method [Kuhn, 1955]. In the paper by Tijs *et al.* [1984] it is assumed that there is an initial allocation of machines to players and that a coalition of players can interchange their machines between them. In the current framework, however, we assume there is no initial allocation.

A permutation situation without initial allocation P can be reformulated as a cooperation building structure $(N, (\mathcal{G}, \preceq), F, K)$, with $\mathcal{G} = \bigcup_{S \subseteq N : S \neq \emptyset} P(S)$, where $P(S) = \{P : S \to \Theta \mid P \text{ is injective}\}$. For $P, P' \in \mathcal{G}$ with $P \in P(S)$ and $P' \in P(T)$ we have $P \preceq P'$ if and only if $S \subseteq T$ and $P'(i) = P(i)$ for all $i \in S$. For $P \in P(S)$ we have $F(P) = S$ and $K(P) = \sum_{i \in S} \Gamma_{iP(i)}$.

In a corresponding *opr* (N, Π, k), the individualized cost function $k : \Pi \to \mathbb{R}^N$ is such that $k^i(\pi)$ denotes the cost of the cheapest possible machine player i can choose at his turn in π. Let $\Theta_t(\pi)$ denote the set of machines still available given that players $\pi(1), \ldots, \pi(t-1)$ have made their choice. Then

$$k^i(\pi) = \min_{j \in \Theta_{\pi^{-1}(i)}(\pi)} \Gamma_{ij},$$

for all $i \in N$. In Proposition 6.2 we show that this *opr* is efficient.

Lemma 6.1. *Let* $P = (N, \Theta, \Gamma)$ *be a permutation situation without initial allocation. In an optimal allocation of machines to players there is at least one player* $i \in N$ *that is assigned a machine* $j^* \in \Theta$ *such that* $\Gamma_{ij^*} \leq \Gamma_{ij}$ *for all* $j \in \Theta$.

Proof. Suppose that an optimal allocation is such that no such player exists. Then there exists at least one sequence of players such that a cheapest machine for each player is assigned to his follower in this sequence, with the final player of the sequence being assigned a cheapest machine of the first player. Let us then give each player the machine of his follower in the sequence. By this procedure the cost of the original allocation is decreased, which means that the original allocation was not optimal. This is a contradiction. $\qquad\square$

Proposition 6.2. *For a permutation situation without initial allocation, a corresponding opr is efficient.*

Proof. Let $P = (N, \Theta, \Gamma)$ be a permutation situation without initial allocation and let (N, Π, k) be a corresponding *opr*. We prove efficiency by providing an algorithm (Algorithm 1) that gives an optimal ordering π_N^* for the *opr* (N, Π, k).

Algorithm 1

Input: A permutation situation without initial allocation $P = (N, \Theta, \Gamma)$ *and an optimal assignment* $\alpha^*(P)$.

Output: An ordering π_N^* *for a corresponding opr* (N, Π, k).

1. Set $q = 1$ and $P_q = (N_q, \Theta_q, \Gamma_q) = (N, \Theta, \Gamma)$.
2. Let $\alpha^*(P_q)$ be an optimal assignment of P_q and let $t \in N_q$ be such that $(t, j^*) \in \alpha^*(P_q)$ with $j^* \in \Theta$ such that $\Gamma_{tj^*} \leq \Gamma_{tj}$ for all $j \in \Theta$.

3. Set $\pi(q) = t$.
4. Set $N_{q+1} = N_q \backslash \{t\}$, $\Theta_{q+1} = \Theta_q \backslash \{j^*\}$ and $\Gamma_{q+1} = N_{q+1} \times \Theta_{q+1}$.
5. If $q = n$ stop, otherwise set $q = q + 1$ and go to Step 2.

Algorithm 1 orders the players such that each player gets at his turn his cheapest machine in the reduced problem (the problem without his predecessors and the machines chosen by them). Note that this algorithm does not necessarily result in a unique ordering, as in step 2 there may be several optimal assignments $a^*(P_q)$ with several players t and several choices for j^* such that $(t, j^*) \in a^*(P_q)$. However, any choice results in an ordering and it follows by Lemma 6.1 that this ordering is optimal and leads to an optimal allocation. $\qquad\square$

Since the number of available machines decreases (and therefore the cost increases) with the position of a player in the ordering we have the following proposition.

Proposition 6.3. *Let P be a permutation situation without initial allocation. Then a corresponding efficient opr (N, Π, k) is a neopr.*

Since there is an efficient *neopr*, we suggest to model this class of cooperation building structures by the marginal cost game (N, c_m), which is determined by (2.2). The following proposition is a consequence of Proposition 6.3 and Theorem 2.3.

Proposition 6.4. *Let P be a permutation situation without initial allocation with a corresponding opr (N, Π, k). Then $\beta(N, \Pi, k) \in C(N, c_m)$.*

7. Public Congestion Network Situations

In this section we consider the class of public congestion network situations, as introduced by Kleppe *et al.* [2010]. In such a situation all players have to connect themselves to the root, while the cost of each arc depends on the number of its users. A *public congestion network situation*, or congestion network situation as we call it from here, is formally given by a triple $G = (N, 0, \gamma)$, where N is the finite set of players that has to be connected to the root 0. By A_S we denote the set of all arcs between pairs in $S \subseteq N^0$, i.e., (S, A_S) is the complete directed graph on S. For each arc $a \in A_{N^0}$ the function $\gamma_a : \{0, 1, \ldots, |N|\} \to \mathbb{R}_+$ associates each number of users of arc a with a corresponding cost. The function γ_a is nonnegative and (weakly) increasing. We assume that $\gamma_a(0) = 0$ for all $a \in A_{N^0}$. Elements of A_{N^0} are also denoted by (i, j), where $i, j \in N^0$. The arc (i, j) then denotes the connection between i and j in the direction from i to j. The cost function of an arc (i, j) is denoted by $\gamma_{i,j}$. A congestion network situation is called symmetric if $\gamma_{i,j} = \gamma_{j,i}$ for all $i, j \in N^0$.

In a congestion network situation each player chooses a path from his initial node to the root. A path between any two nodes i and j is denoted by $P(i, j)$ and is a sequence of arcs $((i_0, i_1), (i_1, i_2), \ldots, (i_{p-1}, i_p))$, such that $i_0 = i$, $i_p = j$ and

$i_r \neq i_s$ for all $r, s \in \{0, \ldots, p-1\}$, $r \neq s$. Furthermore, instead of $P(i, 0)$ we also write P^i.

A *network* is defined by an integer valued function $f : A_{N^0} \rightarrow \{0, 1, \ldots, |N|\}$, such that f assigns to each arc a number of users. The netdegree for a node $i \in N^0$ with respect to network f is defined by $\text{netdegree}^f(i) = \sum_{j \in N^0 \setminus \{i\}} f(i, j) - \sum_{j \in N^0 \setminus \{i\}} f(j, i)$. For a coalition $S \subseteq N$ the collection of all feasible networks connecting the members of S to the root is given by

$$F_S = \{f : A_{N^0} \rightarrow \{0, \ldots, |N|\} \mid \text{netdegree}^f(i) = 1 \text{ for all } i \in S,$$
$$\text{netdegree}^f(i) = 0 \text{ for all } i \in N \setminus S,$$
$$f(a) \in \{0, \ldots, |S|\} \text{ for all } a \in A_{N^0}\}.$$

An element of F_S is typically denoted by f_S. Note that in a feasible network for S each player of S is connected to the root by some path. However, as all arcs are publicly available these paths may consist of arcs between *any* two nodes in N^0. Each network f induces a directed graph (N^0, A_f), where A_f consists of all used arcs: $A_f = \{a \in A_{N^0} \mid f(a) > 0\}$. The cost of a network f is defined by

$$\gamma(f) = \sum_{a \in A_f} \gamma_a(f(a)).$$

A network for $S \subseteq N$ is called optimal if it is feasible with minimum total cost. Such an optimal network is denoted by $f_S^* \in F_S$.

A congestion network situation can be reformulated as a cooperation building structure $(N, (\mathcal{G}, \preceq), F, K)$ with $\mathcal{G} = \bigcup_{S \subseteq N : S \neq \emptyset} F_S$, where for $f, g \in \mathcal{G}$ with $f \in F_S$ and $g \in F_T$, we have $f \preceq g$ if and only if $S \subseteq T$ and $f \leq g$. For all $f_S \in F_S$, $F(f_S) = S$ and $K(f_S) = \sum_{a \in A_{f_S}} \gamma_a(f_S(a))$.

In a corresponding *opr* (N, Π, k), the individualized cost function k is such that, given the current network constructed by his predecessors, player i at his turn in π chooses the cheapest path to the root. Let $\mathcal{P}^i$ denote the set of all paths from player i to the root. Let f^t be a feasible network constructed by the players $\pi(1), \ldots, \pi(t-1)$ given $\pi \in \Pi$. Then

$$k^i(\pi) = \min_{P^i \in \mathcal{P}^i} \sum_{a \in P^i} \gamma_a\left(f^{\pi^{-1}(i)+1}(a)\right) - \gamma_a\left(f^{\pi^{-1}(i)}(a)\right),$$

for all $i \in N$.

7.1. *Convex congestion network situations*

A *convex congestion network situation* $G = (N, 0, \gamma)$ is a congestion network situation in which all γ_a are convex. A cost function γ_a, $a \in A_{N^0}$, is convex if $\gamma_a(r+1) - \gamma_a(r) \geq \gamma_a(r) - \gamma_a(r-1)$ for all $r \in \{1, \ldots, |N|-1\}$.

In Kleppe *et al.* [2010] each convex congestion network situation is associated with direct and marginal cost games. To distinguish between these two games and the games defined in Sec. 2, the direct cost game of Kleppe *et al.* [2010] is called the *direct congestion cost game* and we denote it by (N, c_{dG}), with

$$c_{dG}(S) = \min_{f_S \in F_S} \gamma(f_S),$$

for all $S \subseteq N$. Similarly, the marginal cost game of Kleppe *et al.* [2010] is called the *marginal congestion cost game*, denoted by (N, c_{mG}), with

$$c_{mG}(S) = c_{dG}(N) - c_{dG}(N \backslash S),$$

for all $S \subseteq N$.

Example 7.1. An example of a symmetric convex congestion network situation $G = (N, 0, \gamma)$ is given in Fig. 1. In this situation there are three players, which are denoted by 1, 2 and 3, and the root, which is denoted by 0. The numbers on the arcs represent the total usage costs for each number of users. All coalitional costs for the direct and marginal congestion cost games associated with this congestion network situation are given in the next table.

S	$\{1\}$	$\{2\}$	$\{3\}$	$\{1,2\}$	$\{1,3\}$	$\{2,3\}$	N
$c_{dG}(S)$	1	3	2	7	6	8	14
$c_{mG}(S)$	6	8	7	12	11	13	14

The coalitional costs for the direct congestion cost game are determined by the cost of an optimal network for S in the absence of $N \backslash S$. The optimal network for $S = \{2,3\}$, e.g., is given by f_S^* with $f_S^*(2,3) = 1$, $f_S^*(3,1) = f_S^*(1,0) = 2$ with $\gamma(f_S^*) = 8$. By duality we obtain the marginal congestion cost game.

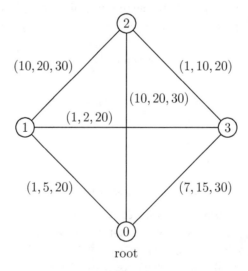

Fig. 1.   A symmetric convex congestion network situation.

Let $G = (N, 0, \gamma)$ be a convex congestion network situation and let $S \subseteq N$ with feasible network f_S. Let $D_{f_S} = \{D^i_{f_S}\}_{i \in S}$ be a *decomposition* of f_S into $|S|$ parts such that $D^i_{f_S}$ corresponds to a specific path P^i, $i \in S$, from player i to the root. Given a network such a decomposition need not be unique, in the sense that we cannot distinguish which arcs are used by which players.

Example 7.2. Consider the convex congestion network situation of Fig. 1, with the optimal network f^*_N given by $f^*_N(2, 3) = f^*_N(3, 1) = f^*_N(3, 0) = 1$ and $f^*_N(1, 0) = 2$. This network does not have a unique decomposition. $D_{f^*_N}$ is a decomposition of f^*_N, where $D^1_{f^*_N}$ corresponds to the path $((1, 0))$, $D^2_{f^*_N}$ corresponds to $((2, 3), (3, 0))$ and $D^3_{f^*_N}$ corresponds to $((3, 1), (1, 0))$. Another decomposition is $\hat{D}_{f^*_N}$, where $\hat{D}^1_{f^*_N}$ corresponds to the path $((1, 0))$, $\hat{D}^2_{f^*_N}$ corresponds to $((2, 3), (3, 1), (1, 0))$ and $\hat{D}^3_{f^*_N}$ corresponds to $((3, 0))$. Note that $D_{f^*_N}$ and $\hat{D}_{f^*_N}$ are the only possible decompositions.

Proposition 7.1. *Let G be a convex congestion network situation with associated marginal congestion cost game (N, c_{mG}). Let (N, Π, k) be a corresponding opr with associated marginal cost game (N, c_m). Then $(N, c_{mG}) = (N, c_m)$.*

Proof. Clearly, it suffices to show that $c_{dG}(S) = c_d(S)$ for all $S \subseteq N$. To prove the equivalence of the direct congestion cost game and the direct cost game we provide an algorithm (Algorithm 2) that produces for all $S \subseteq N$ an optimal ordering π^*_S for *opr* (N, Π, k). It then follows that this ordering π^*_S results in an optimal network for S in the absence of $N \backslash S$, f^*_S.

For a network f with decomposition $\{D^i_f\}_{i \in S}$, write $f \backslash D^i_f$ for the network obtained from f by taking out D^i_f.

Algorithm 2

Input: A convex congestion network situation $G = (N, 0, \gamma)$, a coalition $S \subseteq N$ and an optimal network f^*_S.

Output: An optimal ordering π^*_S for the opr (N, Π, k) for coalition $S \subseteq N$.

1. Set $q = 1$ and set network $\bar{f}^q = f^*_S$.
2. Let $D_{\bar{f}^q}$ be a path decomposition of $\bar{f}^q$.
3. Let player $i \in N$ be such that $D^i_{\bar{f}^q}$ corresponds to the shortest path from i to 0, given the network $\bar{f}^q \backslash D^i_{\bar{f}^q}$.
4. Set $\pi^*_S(|S| + 1 - q) = i$.
5. Set $\bar{f}^{q+1} = \bar{f}^q \backslash D^i_{\bar{f}^q}$.
6. If $q = |S|$ stop, otherwise set $q = q + 1$ and return to step 2.

Note that in step 2 there may be more than one path decomposition. Any choice here, however, results in an optimal ordering. It is furthermore important to realize that in step 3 there always exists such a player.

We conclude that for $S \subseteq N$ with optimal network f^*_S the outcome π^*_S of Algorithm 2 results in an optimal network for S, i.e., $\sum_{i \in S} k^i(\pi^*_S) = \gamma(f^*_S)$. □

Proposition 7.1 leads to the following corollary.

Corollary 7.2. *For a convex congestion network situation, a corresponding opr is efficient.*

Example 7.3. Reconsider the convex congestion network situation G of Fig. 1. The next table gives for each ordering $\pi \in \Pi$ the corresponding individualized cost allocation $k(\pi)$ for a corresponding *opr* (N, Π, k).

π	$k^1(\pi)$	$k^2(\pi)$	$k^3(\pi)$
$(1,2,3)$	1	6	7
$(1,3,2)$	1	8	5
$(2,1,3)$	4	3	7
$(2,3,1)$	8	3	5
$(3,1,2)$	4	8	2
$(3,2,1)$	8	6	2

Consider the ordering $\pi = (2,1,3)$. The shortest path from player 2 to the root, given that no others formed a network, is the path $((2,3),(3,1),(1,0))$ with a cost of 3. Then player 1 takes the path $((1,0))$ with a cost of 4. Finally, player 3 takes the path $((3,0))$ with a cost of 7.

All coalitional costs for the direct and marginal cost games associated with the corresponding *opr* (N, Π, k) are given in the next table.

S	$\{1\}$	$\{2\}$	$\{3\}$	$\{1,2\}$	$\{1,3\}$	$\{2,3\}$	N
$c_d(S)$	1	3	2	7	6	8	14
$c_m(S)$	6	8	7	12	11	13	14

Note that they exactly coincide with the costs of the direct congestion cost game (N, c_{dG}) and the marginal congestion cost game (N, c_{mG}), respectively.

The following proposition follows from the convexity of the cost function γ.

Proposition 7.3. *For a convex congestion network situation, a corresponding opr is a neopr.*

Since there is an efficient *neopr*, we suggest to model the class of convex congestion network situations by the associated marginal cost game (N, c_m) as given by (2.2), which coincides by Proposition 7.1 with the marginal congestion cost game (N, c_{mG}), used in Kleppe *et al.* [2010]. Moreover, a generalized Bird allocation will provide a core element of this game.

Proposition 7.4. *Let G be a convex congestion network situation with corresponding opr (N, Π, k). Then $\beta(N, \Pi, k) \in C(N, c_{mG})$.*

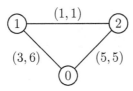

Fig. 2.   A concave congestion network situation.

7.2. *Concave congestion network situations*

In this subsection we consider *concave* congestion network situations. A concave congestion network situation $G = (N, 0, \gamma)$ is a congestion network situation in which all cost functions γ_a are concave. A cost function γ_a, $a \in A_{N^0}$, is concave if $\gamma_a(r+1) - \gamma_a(r) \le \gamma_a(r) - \gamma_a(r-1)$ for all $r \in \{1, \ldots, |N| - 1\}$.

For a concave congestion network situation the direct congestion cost game is denoted by (N, c_{dG}). Kleppe *et al.* [2010] propose to use the direct congestion cost game to model concave congestion network situations. In this subsection we show that our procedure does not support (nor reject) this choice, since in this case a corresponding *opr* (N, Π, k) need not be efficient. This is seen in the following example.

Example 7.4. Consider the symmetric concave congestion network situation $G = (N, 0, \gamma)$ of Fig. 2.

The optimal network f_N^* is given by $f_N^*(1,2) = 1$ and $f_N^*(2,0) = 2$, with a cost of 6. However, for a corresponding *opr* (N, Π, k), both orderings $\pi^1 = (1,2)$ and $\pi^2 = (2,1)$ result in network f_N given by $f_N(2,1) = 1$ and $f_N(1,0) = 2$, with a cost of 7.

When a corresponding *opr* (N, Π, k) is not efficient, the underlying concave congestion network situation G and the *opr* do not describe the same cost allocation problem. Consequently, our approach cannot be used to find an appropriate TU-game for the class of concave congestion network situations and, in particular it cannot be used to support the use of the direct congestion cost game.

8. Traveling Salesman and Related Problems

In this section we consider three closely related classes of cost allocation problems; next to the well-known traveling salesman problems, we study shared taxi problems and traveling repairman problems.

8.1. *Traveling salesman problems*

In this subsection we discuss the class of traveling salesman problems [Potters *et al.*, 1992]. In a traveling salesman problem, the objective of each player is to be visited by a single salesman. In a weighted network with root 0 and other nodes corresponding to the players in N, a salesman tour for $S \subseteq N$ is a circuit starting in 0, visiting all players in S exactly once and returning to 0. The associated cost is the travel time of the salesman: the sum of the (nonnegative) weights on all the edges in the circuit.

The main issue when modeling a *tsp* as a cooperation building structure is how to model the structures $\mathcal{G}$ and, crucially, the extendability relation $\preceq$. If we model $\mathcal{G}$ to consist of all salesman tours over all $S \subseteq N$, then it is unclear what a cost monotonic extension of such a structure boils down to. When a new player $i \in N$ joins, one can of course just append the loop $0 \rightarrow i \rightarrow 0$ to the existing tour to a obtain a feasible structure, but clearly, this is not going to be efficient.

One might get the idea that the link from the last player in a tour back home, which leads to the fact that the costs of $|N| + 1$ edges are to be shared by $|N|$ players, is the reason our approach is not suited for the class of *tsps*. To study the effect of this extra edge we introduce a closely related class of cost allocation problems in the next subsection.

8.2. *Shared taxi problems*

In this subsection we introduce and analyze the class of *shared taxi problems*. The idea is as follows. A group of players is at a particular location and shares a single taxi in order to reach their individual destinations. The cost of the taxi only depends on the distance from the starting point to the final individual destination. The group of players wants to minimize the total cost.

Formally, a *shared taxi problem* is given by a triple $H = (N, 0, \gamma)$, where N is the finite set of players that shares the taxi, which starts at node 0. The function $\gamma : E_{N^0} \rightarrow \mathbb{R}_+$ is a nonnegative cost function, which can be viewed as the taxi cost from one node to another. We consider publicly available networks, which implies that the taxi is able to use the edge between any two nodes of the network. As a result, we can impose without loss of generality that the function γ satisfies the triangle inequality.

The cheapest way to drop off all the players is with a trip, where a trip on N is a path from node 0 to a node $i \in N$ such that all players in N are visited *exactly once*. The ordering of the players in this trip is given by $\pi \in \Pi$. It follows that the total cost for a group of players $S \subseteq N$ to use the taxi equals $\gamma(0, \pi(1)) + \gamma(\pi(1), \pi(2)) + \cdots + \gamma(\pi(|S| - 1), \pi(|S|))$, with $\pi \in \Pi_S$.

Note that the class of *stps* is closely related to the class of *tsps*, in the sense that the only difference with the latter is that for an *stp* the link from the last player back to node 0 is excluded from the structure (and the costs).

We reformulate an *stp* $(N, 0, \gamma)$ as a cooperation building structure $(N, (\mathcal{G}, \preceq), F, K)$, where $\mathcal{G} = \bigcup_{S \subseteq N : S \neq \emptyset} \Pi(S)$ contains for all S all the trips $(0, \pi(1), \ldots, \pi(|S|))$ on S. For $\pi, \pi' \in \mathcal{G}$ with $\pi \in \Pi(S), \pi' \in \Pi(T)$ we have $\pi \preceq \pi'$ if and only if $S \subseteq T$ and $(\pi')^{-1}(i) = \pi^{-1}(i)$ for all $i \in S$. For all $\pi \in \Pi(S)$, $F(\pi) = S$ and $K(\pi) = \gamma(0, \pi(1)) + \gamma(\pi(1), \pi(2)) + \cdots + \gamma(\pi(|S| - 1), \pi(|S|))$.

For a corresponding *opr* (N, Π, k) the individualized cost function $k : \Pi \to \mathbb{R}^N$ is given by

$$k^i(\pi) = \gamma(\pi(\pi^{-1}(i) - 1), i),$$

for all $i \in N$ and $\pi \in \Pi$. Hence, the individualized cost of a player denotes the cost for the taxi to travel from the node of his direct predecessor to his own node.

It is easily seen that such an *opr* is efficient. Hence, the change from a *tsp* to an *stp* by not including the edge back to node 0 in the structure is an important difference for our approach. The following example, however, shows that there exists an *stp* H for which the corresponding *opr* is neither a *neopr* nor a *peopr*.

Example 8.1. Let Fig. 3 represent the *stp* $H = (N, 0, \gamma)$. The numbers on the edges denote the cost for the taxi to travel from one node to another. The individual costs for a given ordering according to a corresponding *opr* (N, Π, k) are given in the next table.

π	$k^1(\pi)$	$k^2(\pi)$	$k^3(\pi)$
$(1,2,3)$	1	2	3
$(1,3,2)$	1	3	4
$(2,1,3)$	2	2	4
$(2,3,1)$	4	2	3
$(3,1,2)$	4	2	5
$(3,2,1)$	2	3	5

Each player pays for the edge directly leading to him. Therefore, $\pi = (1, 2, 3)$ leads to the individualized costs $k^1(\pi) = 1$, $k^2(\pi) = 2$ and $k^3(\pi) = 3$. Since an optimal ordering for player 3 is either $(1, 2, 3)$ or $(2, 3, 1)$ it is not beneficial for all

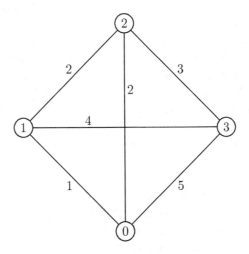

Fig. 3.   A shared taxi problem.

coalitions to be ordered first. Hence, the *opr* (N, Π, k) is no *neopr*. Further, as an optimal ordering for player 1 is either $(1, 2, 3)$ or $(1, 3, 2)$ it is also not beneficial for all coalitions to be ordered last. Therefore, the *opr* (N, Π, k) is no *peopr* either.

Example 8.1 demonstrates that a corresponding *opr* is neither a *neopr* nor a *peopr*, which implies that we cannot argue in favor of either the direct cost game or the marginal cost game for the class of *stps*.

We can also conclude that the fact that our approach cannot be applied to *tsps* is not (only) due to the link from the last player back home. Part of the reason also lays in the fact that the presence of other players can have either a positive or a negative effect on the cost of a player, as illustrated in the *stp* setting in which an efficient *opr* does exist.

8.3. *Traveling repairman problems*

In this subsection we consider situations in which several players need to be visited by a single repairman. These players, as well as the repairman, are not located at the same place and therefore, the repairman has to decide on a specific way to visit all the players. The cost for each player depends on the time he has to wait for the arrival of the repairman. As a result we obtain the problem of finding a route that minimizes the total waiting time of the players. This operations research problem is known as a traveling repairman problem [Afrati *et al.*, 1986]. A natural example of the above situation is to think of the players as factories with broken machinery that needs to be repaired. In this case the costs reflect opportunity costs of production.

Note that a traveling repairman problem can be seen as a special type of sequencing situation in which the processing time of a player depends on his predecessor, e.g., due to changeover costs. The difference between the class of traveling repairman problems on the one hand and the classes of traveling salesman and shared taxi problems on the other is that for the latter two classes the goal is to minimize the travel time of the salesman or taxi, while for the class of traveling repairman problems the players' waiting time is to be minimized. So in a sense, the model already contains individualized costs.

Formally, a *traveling repairman problem* is given by a triple $T = (N, 0, \gamma)$, where N is the finite set of players (nodes) that has to be visited by the repairman. He starts at node 0, called *home*. The function $\gamma : E_{N^0} \to \mathbb{R}_+$ is a nonnegative cost function, which can be viewed as the travel time of the repairman from one node to another.

In a *trp* the single repairman has to visit all players. We consider publicly available networks, which means that the repairman is able to use the edge between any two nodes of the network. As a result, we can impose without loss of generality that γ satisfies the triangle inequality. Hence, a *trp* can be reformulated as a cooperation building structure $(N, (\mathcal{G}, \preceq), F, K)$, where $(\mathcal{G}, \preceq)$ and F are exactly the same as in the shared taxi problems of Sec. 8.2 and for $\pi \in \Pi(S)$,

$K(\pi) = \gamma(0, \pi(1)) + [\gamma(0, \pi(1)) + \gamma(\pi(1), \pi(2))] + \cdots + [\gamma(0, \pi(1)) + \gamma(\pi(1), \pi(2)) + \cdots + \gamma(\pi(|S| - 1), \pi(|S|))].$

A corresponding *opr* (N, Π, k) satisfies

$$k^i(\pi) = \gamma(0, \pi(1)) + \gamma(\pi(1), \pi(2)) + \cdots + \gamma(\pi(\pi^{-1}(i) - 1), i),$$

for all $i \in N$. It follows immediately that such an *opr* is also efficient. From the fact that γ satisfies the triangle inequality we obtain the next proposition.

Proposition 8.1. *For a trp, a corresponding opr is a neopr.*

Since a corresponding *opr* is an efficient *neopr*, we suggest to model the class of *trps* by the marginal cost game (N, c_m) as given by (2.2). Therefore, we also have the following proposition.

Proposition 8.2. *Let T be a trp with corresponding opr (N, Π, k). Then $\beta(N, \Pi, k) \in C(N, c_m)$.*

References

Afrati, F., Cosmadakis, S., Papadimitriou, C., Papageorgiou, G. and Papakostantinou, N. [1986] The complexity of the traveling repairman problem, *RAIRO Theor. Inf. Appl.* **20**, 79–87.

Aumann, R. and Maschler, M. [1985] Game theoretic analysis of a bankruptcy problem from the Talmud, *J. Econ. Theory* **36**, 195–213.

Bird, C. [1976] On cost allocation for a spanning tree: A game theoretic approach, *Networks* **6**, 335–350.

Borm, P., Hamers, H. and Hendrickx, R. [2001] Operations research games: A survey, *Top* **9**, 139–216.

Branzei, R., Moretti, S., Norde, H. and Tijs, S. [2004] The *P*-value for cost sharing in minimum cost spanning tree situations, *Theory Decis.* **56**, 47–61.

Claus, A. and Kleitman, D. [1973] Cost allocation for a spanning tree, *Networks* **3**, 289–304.

Curiel, I., Pederzoli, G. and Tijs, S. [1989] Sequencing games, *Eur. J. Oper. Res.* **40**, 344–351.

Gillies, D. [1959] Solutions to general non-zero-sum games, in *Contributions to the Theory of Games IV*, eds. Tucker, A. and Luce, R. (Princeton University Press, princeton), pp. 47–85.

Kleppe, J. [2010] *Modeling Interactive Behavior, and Solution Concepts*, Ph.D. thesis, Tilburg University, Tilburg, The Netherlands.

Kleppe, J., Quant, M. and Reijnierse, H. [2010] Public congestion network situations and related games, *Networks* **55**, 368–378.

Klijn, F. and Sánchez, E. [2006] Sequencing games without initial order, *Math. Methods Oper. Res.* **63**, 53–62.

Koster, M. [1999] *Cost Sharing in Production Situations and Network Exploitation*, Ph.D. thesis, Tilburg University, Tilburg, The Netherlands.

Kuhn, H. [1955] The Hungarian method for the assignment problem, *Nav. Res. Logist. Q.* **2**, 83–97.

Littlechild, S. and Owen, G. [1973] A simple expression for the Shapley value in a special case, *Manage. Sci.* **20**, 370–372.

Megiddo, N. [1978] Computational complexity of the game theory approach to cost allocation for a tree, *Math. Oper. Res.* **3**, 189–196.

von Neumann, J. and Morgenstern, O. [1944] *Theory of Games and Economic Behavior* (Princeton University Press, Princeton).

O'Neill, B. [1982] A problem of rights arbitration from the Talmud, *Math. Soc. Sci.* **2**, 345–371.

Potters, J., Curiel, I. and Tijs, S. [1992] Traveling salesman games, *Math. Program.* **53**, 199–211.

Prim, R. [1957] Shortest connection networks and some generalizations, *Bell Syst. Tech. J.* **36**, 1389–1401.

Smith, W. [1956] Various optimizers for single-stage production, *Nav. Res. Logist. Q.* **3**, 59–66.

Thomson, W. [2003] Axiomatic and game theoretic analysis of bankruptcy and taxation problems: A survey, *Math. Soc. Sci.* **45**, 249–297.

Tijs, S., Parthasarathy, T. Potters, J. and Rajendra Prasad, V. [1984] Permutation games: Another class of totally balanced games, *OR Spektrum* **6**, 119–123.

Chapter 16

Coalition Formation with Externalities: The Case of the Northeast Atlantic Mackerel Fishery in a Pre- and Post-Brexit Context

Evangelos Toumasatos

SNF — Centre for Applied Research
Norwegian School of Economics, N-5045 Bergen, Norway

Department of Business and Management Science
Norwegian School of Economics, N-5045 Bergen, Norway

evangelos.toumasatos@snf.no

Stein Ivar Steinshamn

Department of Business and Management Science
Norwegian School of Economics, N-5045 Bergen, Norway

stein.steinshamn@nhh.no

The partition function approach is applied to study coalition formation in the Northeast Atlantic mackerel fishery in the presence of externalities. Atlantic mackerel is mainly exploited by the European Union (EU), the United Kingdom (UK), Norway, the Faroe Islands and Iceland. Two games are considered. First, a four-player game where the UK is still a member of the EU. Second, a five-player game where the UK is no longer a member of the union. Each game is modeled in two stages. In the first stage, players form coalitions following a predefined set of rules. In the second stage, given the coalition structure that has been formed, each coalition chooses the economic strategy that maximizes its own net present value of the fishery, given the behavior of the other coalitions. The game is solved using backward induction to obtain the set of Nash equilibria coalition structures in pure strategies, if any. We find that the current management regime is among the stable coalition structures in all eight scenarios of the four-player game but in only one case of the five-player game. In addition, stability in the five-player game is sensitive to the growth function applied and the magnitude of the stock elasticity parameter.

Keywords: Mackerel dispute; straddling fish stock; Brexit; game theory; externalities; coalition formation; coalition structure stability.

1. Introduction

The 1982 United Nations Convention on the Law of the Sea (UNCLOS) recognized a 200 nautical mile Exclusive Economic Zone (EEZ) stretching from the baseline of a coastal state [United Nations, 1982]. The establishment of the EEZ has fundamentally changed the management of world marine-captured fisheries by recognizing property rights. Thus, allowing coastal states to manage their stocks for their own benefit. However, such regime has inadequately addressed issues arising from internationally shared fishery resources,[a] e.g., unregulated fishing, over-capitalization, excessive fleet size, etc. [United Nations, 1995; Munro, 2008]. Therefore, if the harvesting activities of one coastal state have a significant negative effect on the harvesting opportunities of the other coastal state(s), a coordinated plan for sustainable management from all parties is required.

This need for cooperation has led to the adoption of the 1995 United Nations Fish Stocks Agreement (UNFSA), which supplements and strengthens the 1982 UNCLOS by addressing the problems related to the conservation and management of internationally shared fishery resources [United Nations, 1995]. According to UNFSA, the exploitation of a shared fish stock within its spatial distribution should be coordinated by a coalition of all interest parties through a UN sanctioned Regional Fisheries Management Organization (RFMO), e.g., the Northeast Atlantic Fisheries Commission (NEAFC). Membership into an RFMO is open both to nations in the region, i.e., coastal states, and distant nations with interest in the fisheries concerned, as long as they agree to abide by the RFMO's conservation and management measures.

Although UNFSA has established robust international principles and standards for the conservation and management of shared fish stocks [Balton and Koehler, 2006], the fact that RFMOs lack the necessary coercive enforcement power, either to exclude non-members from harvesting or to set the terms of entry for new members, has caused doubts over the long-term viability of such regional management mechanisms [McKelvey et al., 2002]. These two inter-related problems, namely, the "interloper problem" [Bjørndal and Munro, 2003] and the "new member problem" [Kaitala and Munro, 1993], merge when a nation with no past interest in a particular shared fishery starts exploiting the resource. In this case, the interests of the traditional fishing nations (incumbents) and the new entrant(s) are strongly opposed. On the one hand, incumbents face the prospect of having to give up a share of their quotas to the new entrant(s) in order to join their coalition and exploit the resource sustainably; whereas on the other hand, it might be more profitable for the new entrant(s) not to join and therefore harvest without having to abide by the coalition's conservation measures.

The aforementioned situation gives rise to the free-rider problem due to stock externalities, i.e., the effect of this period's harvest on next period's stock level

---

[a]See FAO [2011] and Gulland [1980] for a categorization of shared fish stocks.

[Bjørndal, 1987]. Stock externalities, which occur when the cost of fishing changes as the population of fish is altered, are negative externalities [Smith, 1969; Agnello and Donnelley, 1976]. That is, a nation's harvesting activities lead to less fishing opportunities for another nation and therefore increase the other's nation fishing cost. As nations start cooperating, the externality is internalized and thus the external cost is reduced. The externality disappears if all nations cooperate together. Because the reduction of the negative externality leads to higher benefits for all nations, not only the ones cooperating. This can be referred to a positive externality.

The intuition is as follows. Assume that a cooperative agreement, which aims to preserve a fish stock by limiting the number of catches and thus increasing its population, is signed by a group of nations. A nation who is not part of such agreement can still enjoy the positive effects that the agreement has on the fish stock level without having to reduce its fishing activities. Therefore, a free-rider (noncooperating nation or coalition of nations[b]) can enjoy a lower cost of fishing without having to mitigate its fishing strategy. Because of the free-rider problem, cooperative agreements among all interest parties in a fishery have not always been possible to achieve.

The importance of externalities emanating from coalition formation where the economic performance of a coalition including singletons,[c] is affected by the structure of other distinct coalitions has been studied both within game theoretic and fisheries literature. Bloch [1996], Yi [1997] and Ray and Vohra [1999], among others, have established the theoretical framework to analyze coalition formation in the presence of externalities, also referred as endogenous coalition formation, using the partition function approach introduced by Thrall and Lucas [1963]. The advantage of those models to the ones using the traditional characteristic function approach is that they consider all possible coalition structures and compute coalition values for every one of them, instead of fixating on some. Thus, stability of different coalition structures, i.e., partial cooperation, can be tested and externalities across coalitions can be captured.

Within the fisheries literature, Pintassilgo [2003] and Pham Do and Folmer [2003] have introduced the partition function approach to fishery games. Pintassilgo [2003] applied this method to the Northern Atlantic bluefin tuna. Pham Do and Folmer [2003] studied the feasibility of coalitions smaller than the grand coalition. Kronbak and Lindroos [2007] applied different sharing rules to study the stability of a cooperative agreement for the Baltic cod in the presence of externalities. They stated that even though the benefit from cooperation is high enough for a cooperative agreement to be reached, its stability is very sensitive to the sharing

---

[b]It is possible, although not usual, that a shared fishery is managed by more than one cooperative agreements, where the signatories of one agreement differ from the signatories of the other agreement. An example presented in Munro [2003] consists of 14 independent Pacific Island nations, which were coalesced into two sub-coalitions. If this is the case, then a coalition of nations can free-ride on another coalition.

[c]A coalition consisting of one member.

rule applied due to free-riding effects. For more comprehensive reviews on coalition games and fisheries, as well as game theory and fisheries, see Kaitala and Lindroos [2007], Lindroos *et al.* [2007], Bailey *et al.* [2010] and Hannesson [2011].

In this chapter, we implement the partition function approach to study coalition formation in the Northeast Atlantic mackerel fishery. Atlantic mackerel is a highly migratory and straddling stock making extensive annual migrations in the Northeast Atlantic. The stock consists of three spawning components, namely, the southern, the western and the North Sea component, which mix together during its annual migration pattern. As a result, the exploitation of mackerel in different areas cannot be separated. Thus, all three spawning components are evaluated as one stock by the International Council for the Exploration of the Sea (ICES) since 1995 [ICES CM, 1996].

Because of the wide geographic range that mackerel is distributed, it is exploited by several nations both in their EEZs and the high seas. Traditionally, mackerel has been cooperatively exploited by the European Union[d] (EU), Norway and the Faroe Islands, with the latter taking only a small proportion of the overall catch until 2010 (2%[e] on average). Also, the NEAFC, of which the three nations are members, allocates a share of the mackerel quota to Russia (7% on average), which can fish mackerel in the high seas. In the last decade, however, mackerel has extended its distribution and migration pattern starting to appear in the Icelandic and Greenlandic economic zones. Although the causes of such northward expansion are not fully understood, increased sea surface temperatures in the northeast Atlantic [Pavlov *et al.*, 2013] and high population size of the mackerel stock [Hannesson, 2012] are mostly referred in the literature.

Due to mackerel's distributional shifting, Iceland, which in the past had requested and been denied to be recognized as a coastal state for the management of mackerel, has begun fishing mackerel at increasingly large quantities in 2008 (approximately 18% of the total catch). In 2009, the Faroese, having observed the quantities that Iceland was harvesting, withdrew from the cooperative agreement with the EU and Norway on the grounds that their quota was very low. A bilateral agreement between the EU and Norway was not reached until 2010. Since then, and despite many rounds of consultations, no consensus agreement by all four nations has been reached. However, in 2014, the Faroe Islands together with Norway and the EU signed a five-year arrangement, which is still in

---

[d]It is assumed that the EU acts as a nation in this context due to the fact that all of its members abide by the CFP. The CFP gives the EU exclusive competence when it comes to negotiating and signing fisheries agreements with non-EU nations. Therefore, EU member states are no longer able to negotiate fisheries agreements by themselves. This is a common assumption when analyzing fishery games that include the EU as a player, see Kennedy [2003], Hannesson [2012], Ellefsen [2013] and Jensen *et al.* [2015].

[e]Unless otherwise stated, all computations in this chapter are based on ICES [2016a] advice report 9.3.39, tables 9.3.39.12 and 9.3.39.14.

place, determining the total allowable catch (TAC) and the relative share for each participant.

In the past, several authors have closely examined the so-called mackerel dispute among the EU, Norway, Iceland and the Faroe Islands. Ellefsen [2013] applied the partition function approach to study the effects of Iceland's entry into the fishery. He considered two games, a three-player game among the EU, Norway and the Faroe Islands, and a four-player game where he included Iceland. His results indicated that the grand coalition is potentially stable, i.e., it is stable for some but not all sharing rules, in the three-player but not in the four-player game. Hannesson [2012, 2013] studied the outcome of cooperation assuming different migratory scenarios of the mackerel stock. He found out that if the migrations are stock-dependent, then minor players, like Iceland and the Faroe Islands, are in a weak position to bargain. The opposite is true if the migrations are purely random or fixed. Jensen *et al.* [2015] tried to empirically explain the outcome of the mackerel crisis after Iceland's entry into the fishery. They considered two strategies for all nations, namely, cooperation and noncooperation. They concluded that noncooperation is a dominant strategy for each player.

The purpose of this chapter is to investigate how the UK's decision to withdraw from the EU is likely to affect the current management regime in the mackerel fishery. The UK, which has been a member of the EU since 1973, voted on 26 June 2016 to leave the Union. Nine months later, on 29 March 2017, the British government officially initiated Brexit by invoking Article 50 of the EU's Lisbon Treaty. This will lead to the conclusion of an international agreement between the two parties by the 29th of March 2019 unless the European Council extends this period. Such agreement will define the terms of the UK's disengagement from the European legal system, internal market and other policies, including the Common Fisheries Policy (CFP) [Sobrino Heredia, 2017]. Being a member state of the EU, the UK has not been directly involved in the negotiations for the mackerel quota but represented by the EU, which allocates fishing opportunities to member states based on the principle of relative stability, i.e., a fixed percentage of the quota based on historical catch levels. Thus, after Brexit is concluded, the UK will have to negotiate on its behalf with the remaining coastal states regarding its share of the mackerel quota, which will most likely be based on the principle of zonal attachment, i.e., each party's share of the quota should be proportional to the catchable stock found in its EEZ [Churchill and Owen, 2010].

In what follows, we focus on two games: (i) a four-player game where the UK is still part of the EU and (ii) a five-player game where the UK is allowed to make its own decisions. The remaining players/nations considered are Norway, Iceland and the Faroe Islands. Both games are analyzed using the partition function approach. That is, we investigate how players are likely to organize themselves in coalitions, which result in the formation of a coalition structure. The objective of a coalition is to maximize its own net present value (NPV) of the fishery, given the behavior of the other coalitions in the coalition structure. The optimal strategies and payoffs of the games are derived as pure Nash equilibria between coalitions in a coalition structure. Finally, stability of a coalition structure is tested and the set of the Nash equilibria coalition structures is obtained.

The chapter is structured as follows. In Secs. 2 and 3, we lay out the bioeconomic and game theoretic models employed. The empirical model specification is presented in Sec. 4. In Sec. 5, we report the solution of both games, evaluate the stability of the coalition structures and discuss the results. Finally, Sec. 6 summarizes our main findings and concludes.

2. Bioeconomic Model

The bioeconomic model we expand on is a deterministic stock–recruitment model introduced by Clark [1973].[f] The model is in discrete time between seasons but continuous within them. Also, it is linear in the control variable, i.e., harvest.

The spawning stock biomass (SSB) of a fishery at the beginning of a period t, for $t = 0, 1, 2, \ldots, \infty$, is referred to as the recruitment R_t. The harvested biomass in a period t is denoted by H_t and must be between zero and the recruitment, $0 \leq H_t \leq R_t$. The SSB at the end of a period is the difference between the recruitment and the harvest and is called the escapement S_t, $S_t = R_t - H_t$. The SSB at the beginning of the next period R_{t+1} is a function of the SSB at the end of the current period S_t, $R_{t+1} = F(S_t)$. The schema below illustrates the stock dynamics between time periods:

$$R_t \rightarrow H_t \rightarrow S_t \rightarrow R_{t+1} = F(S_t) \ldots.$$

The function $F(S)$, which is usually referred to as the stock–recruitment relationship, is assumed to be continuous, increasing, concave and differentiable in $[0, K]$ with $F(0) = 0$ and $F(K) = K$, where $K > 0$ is the carrying capacity of the fishery.

Note that only harvest mortality occurs during a period t. Natural mortality is accounted for within the stock–recruitment relationship, which can be viewed as the net recruitment function or the "natural" production function [Clark and Munro, 1975].

2.1. *Cooperative management*

Suppose now that a shared fishery, like the Northeast Atlantic mackerel, is cooperatively managed by a coalition whose members are all the relevant coastal states, also referred to as grand coalition. The goal of the grand coalition is to maximize the NPV of the fishery over an infinite horizon subject to the biological constraint.

---

[f]Important contributors towards the development of stock–recruitment models have also been Reed [1974] and Jaquette [1974] who analyzed the stochastic stock–recruitment models in discrete time.

The maximization problem can be expressed as follows:

$$\underset{S_t}{\text{maximize}} \sum_{t=0}^{\infty} \gamma^t \Pi(R_t, S_t)$$

$$\text{subject to } R_{t+1} = F(S_t)$$

$$0 \leq S_t \leq R_t,$$

where $\Pi(R_t, S_t)$ is the joint profit from the fishery for each period, which is defined as the difference between gross revenue and total cost. Two assumptions are made when specifying the net revenue function. First, the demand curve is assumed to be infinitely elastic, i.e., each harvested unit of fish can be sold at a fixed price p. Hence, the gross revenue from the fishery is expressed as $\text{TR}(R_t, S_t) = p(R_t - S_t)$. Second, the unit cost of harvest is assumed to be density-dependent, i.e., it increases as the size of the stock decreases. Thus, for a given stock size x, the unit cost of harvest is equal to $c(x)$, which is a continuous and decreasing function. Consequently, the total cost of harvest within one period is defined as $\text{TC}(R_t, S_t) = \int_{S_t}^{R_t} c(x)dx$. To sum up, the joint profit in period t can be written as

$$\Pi(R_t, S_t) = p(R_t - S_t) - \int_{S_t}^{R_t} c(x)dx.$$

Clark [1973] showed that if the profit function is specified as above, then the optimal harvest strategy that maximizes the NPV of the fishery is given by a "bang–bang" strategy with equilibrium escapement S^*

$$H_t = \begin{cases} R_0 - S^*, & t = 0, \\ F(S^*) - S^*, & t \geq 1, \end{cases}$$

i.e., for the initial period, the stock should be depleted to the equilibrium escapement level and then harvest the difference between optimal recruitment and escapement. The optimal escapement level is independent of t and must satisfy the so-called "golden rule":

$$\pi(S^*) = \gamma F'(S^*)\pi[F(S^*)], \tag{1}$$

where $\pi(x)$ is the marginal profit defined as $\pi(x) = p - c(x)$. The interpretation of the "golden rule" is straightforward, a cooperatively managed fishery is exploited until the marginal profit of harvesting the last unit of the stock is equivalent to the marginal profit of letting that unit grow and be harvested in the next period.

2.2. *Noncooperative management*

Although cooperative management is the desired outcome from the perspective of stock conservation, it is often the case that shared fisheries are noncooperatively managed. In this subsection, we generalize the above model in order to allow for noncooperative behavior among nations. First, we describe how the mackerel stock is exploited in the presence of two or more distinct coalitions. Then, we specify coalition's i maximization problem and derive the noncooperative "golden rule".

If the mackerel fishery is noncooperatively managed, then a number of coalitions[g] interacting with each other must exist. Each coalition acts on its own, aiming to maximize its own NPV of the fishery, which is potentially detrimental to other coalitions. Coalitions are assumed to harvest mackerel in the EEZs of their members. Furthermore, we ignore mackerel exploitation on international waters for the following reasons. First, the size of the high seas territory where mackerel potentially exists is relatively small and remote, compared to the rest of its habitat. Second, mackerel is mainly exploited on the high seas by Russia, which receives a small proportion of the total quota and is not directly involved in the management of the stock.

Let θ_l be the share of the mackerel stock that only appears in the EEZ of nation l for a whole year. The share of the mackerel stock that coalition i enjoys is simply the sum of its members' shares, i.e., $\theta_i = \sum_{l \in i} \theta_l$. For example, if EU and NO form a coalition, then $\theta_{(EU,NO)} = \theta_{EU} + \theta_{NO}$. Parameter θ is assumed to be stationary, i.e., constant through all time periods. For details on the specification of the share parameter, see Sec. 4.

Although each coalition exploits mackerel in its own zone, the stock–recruitment relationship specified in the beginning of this section still holds for the aggregated population level, i.e., $R_{t+1} = F(S_t)$. Let m be the number of coalitions that noncooperatively manage the mackerel fishery. The share parameter θ_i, where $i = 1, 2, \ldots, m$, enables us to work out the share of recruitment R_{it} for each coalition in a time period, i.e., $R_{it} = \theta_i R_t$. After mackerel harvesting activities, H_{it} are performed by all coalitions, the escapement from the zone of each coalition is $S_{it} = R_{it} - H_{it}$. The total recruitment for the next time period is determined by the total escapement of the current period through the stock–recruitment relationship on the aggregated escapement level S_t, where $S_t = \sum_{i=1}^{m} S_{it}$. The schema below illustrates such process when three coalitions exist, $m = 3$.

$$R_t \begin{cases} R_{1t} = \theta_1 R_t \longrightarrow H_{1t} \longrightarrow S_{1t} \\ R_{2t} = \theta_2 R_t \longrightarrow H_{2t} \longrightarrow S_{2t} \\ R_{3t} = \theta_3 R_t \longrightarrow H_{3t} \longrightarrow S_{3t} \end{cases} S_t = \sum_{i=1}^{3} S_{it} \longrightarrow R_{t+1} = F(S_t) \ldots.$$

Based on the above setting, a coalition i maximizes its own NPV of the fishery subject to its recruitment share R_i, the escapement strategies of the other coalitions S_j and the stock–recruitment relationship. Such maximization problem can

---

[g]The term coalition is typically used to refer to situations where two or more entities, e.g., companies, political parties, nations, etc., cooperate together to achieve a goal. However, within the game theory literature, the term is used as follows: given a set of players, any subset of the given set can be a coalition. Thus, according to game theorists, an individual player acting on its behalf can be a coalition. Coalitions consisting of only one player are usually referred to as singletons.

be expressed as follows:

$$\underset{S_{it}}{\text{maximize}} \sum_{t=0}^{\infty} \gamma^t \Pi_i(R_{it}, S_{it})$$

$$\text{subject to } R_{it} = \theta_i R_t,$$

$$S_t = S_{it} + \sum_{j=1}^{m-1} S_{jt} \quad i \neq j, \tag{2}$$

$$R_{t+1} = F(S_t),$$

$$0 \leq S_{it} \leq R_{it}.$$

Here, $\Pi_i(R_{it}, S_{it})$ is the profit for coalition i for each period and is specified as in the cooperative case, i.e.,

$$\Pi_i(R_{it}, S_{it}) = p(R_{it} - S_{it}) - \int_{S_{it}}^{R_{it}} c_i(x)dx.$$

The optimal harvest strategy that maximizes the NPV for coalition i is given by a target escapement strategy with equilibrium escapement S_i^*:

$$H_{it} = \begin{cases} R_{i0} - S_i^* = \theta_i R_0 - S_i^*, & t = 0 \\ R_i - S_i^* = \theta_i F\left(S_i^* + \sum_{j=1}^{m-1} S_j\right) - S_i^*, & t \geq 1, \end{cases}$$

i.e., for the first period, the initial recruitment of coalition i should be depleted to its equilibrium escapement level, and then harvest the difference between its recruitment share and its optimal escapement. The recruitment share of coalition i is determined by its share and the stock–recruitment relationship, which depends on the optimal escapement of coalition i and the escapement strategies of other coalitions j. The optimal escapement level is independent of t and must satisfy the following "golden-rule" (see Appendix A.1 for the proof):

$$\pi_i(S_i^*) = \gamma \theta_i F'(S) \pi_i[\theta_i(F(S)], \tag{3}$$

where $\pi_i(x)$ is the marginal profit for coalition i defined as $\pi_i(x) = p - c_i(x)$ and S is the aggregated escapement defined as $S = S_i^* + \sum_{j=1}^{m-1} S_j$.

It is evident from the noncooperative golden rule (3) that the optimal escapement strategy S_i^* of coalition i depends on the escapement strategies of other coalitions j. Therefore, in order for coalition i to be able to determine its optimal escapement strategy S_i^*, it has to have some information regarding the escapement strategies of the remaining coalitions j.

Suppose that coalition i makes an educated guess about the escapement strategies of all the remaining coalitions j based on the information it possesses. Coalition i is now able to compute its optimal escapement strategy S_i^* by substituting its educated guess in (3). If all coalitions act in the same manner, i.e., they make an educated guess for the strategies of their counterparts, substitute in (3), and compute their escapement strategies, then all educated guesses that have been made will probably differ from the escapement strategies that have been computed. Suppose now that some sort of updating based on the newly computed escapement strategies takes place and updates the information set of the coalitions allowing them to adjust their escapement strategies on the new information. Then, all coalitions will have to recompute their escapement strategies based on the new information. This process will keep repeating until no coalition can further gain by adjusting its escapement strategy, then the Nash equilibrium is reached.

Since this chapter's intention is to compute the Nash equilibrium escapement strategies for the coalitions formed and not to derive the optimal escapement paths for these coalitions, there is no need to make any further specification upon the information coalitions have and how this information is updated. The Nash equilibrium escapement strategies can be obtained by solving a system of equations as will be shown in the next section.

Finally, the noncooperative "golden-rule" is a generalization of the cooperative one. To see this, assume that all nations cooperate and the grand coalition is formed. The stock share of the grand coalition is equal to one, $\theta_i = 1$, and since no other coalition exists, the aggregated escapement is equivalent to the optimal escapement of the grand coalition, $S = S_i^*$. Thus, the two rules are equivalent under full cooperation.

3. Game Theoretic Model

A coalition game with externalities is modeled in two stages. In the first stage, players, i.e., nations, form coalitions following a predefined set of rules. For our fishery game, we adopt the simultaneous-move "Open Membership" game described in Yi and Shin [1995]. According to this rule, players can freely form coalitions as long as no player is excluded from joining a coalition. This type of coalition game is in line with how membership is established within an RFMO according to Article 8(3) of the UNFSA. Also, it is the *de facto* framework used so far to analyze coalition games in fisheries.

Let $N = \{1, 2, \ldots, n\}$ be the set of players. A coalition C is a subset of N, i.e., $C \subseteq N$, with 2^n being the number of coalitions that can be formed, including the empty set. The coalition(s) formed in the first stage lead to a coalition structure CS $= \{C_1, C_2, \ldots, C_m\}$, where $1 \leq m \leq n$. A coalition structure has at least one coalition, i.e., full cooperation, and at most n coalitions, i.e., full noncooperation. The formal definition of a coalition structure as provided in Yi [1997] states that a coalition structure is a partition of the players N into disjoint, nonempty and exhaustive coalitions, i.e., $C_i \cap C_j = \varnothing$ for all $i, j = 1, 2, \ldots, m$ and $i \neq j$, and $\bigcup_{i=1}^{m} C_i = N$. This means that within a coalition structure, each player belongs only to one coalition and some players may be alone in their coalitions.

Given the coalition structure that has been formed in the first stage, in the second stage, each coalition chooses the economic strategy that maximizes its own

NPV of the fishery given the behavior of the other coalitions. If the grand coalition is formed, then the total NPV of the fishery is maximized. The economic strategies in the second stage game, as well as the respective payoffs, are pure strategy Nash equilibria.[h] Given the optimal strategies in the second stage of the game, the Nash equilibria coalition structures in pure strategies are the ones that satisfy the stability criteria.

The game is solved using backward induction to obtain the set of stable coalition structures, if any. First, we fix all coalition structures. Then, we compute optimal strategies and payoffs for all coalitions in every coalition structure. Finally, we check which coalition structures satisfy the stability criteria.

3.1. *Second stage of coalition formation*

Let $K = \{CS_1, CS_2, \ldots, CS_\kappa\}$ be the set of coalition structures and κ the number of coalition structures that can be formed.[i] From the κ coalition structures, the $\kappa - 1$ consist of two or more coalitions, which noncooperatively manage the fishing resource. The κth coalition structure contains only one coalition, the grand coalition, that cooperatively manages the stock.

For a given coalition structure $CS_k = \{C_1, C_2, \ldots, C_m\}$, where $k = 1, 2, \ldots, \kappa$, we denote the payoff of coalition C_i, where $i = 1, 2, \ldots, m$, as $v_i(S_i, S)$. The coalitional payoff depends on the escapement strategy of the coalition, S_i, and the overall escapement strategy profile of the coalition structure, $S = S_i + \sum_{j=1}^{m-1} S_j$.[j] Also, the set of feasible escapement strategies for any coalition i is between zero, i.e., harvest everything, and its recruitment, i.e., harvest nothing, $S_i \in [0, R_i]$.

The equilibrium escapement strategies S_i^* for all coalitions C_i in a coalition structure CS_k are derived as a Nash equilibrium between coalition C_i and coalitions C_j where $j = 1, 2, \ldots, m-1$, $i \neq j$ and $C_i \cup C_j = CS_k$, and must satisfy the following m inequalities:

$$v_i\left(S_i^*, S_i^* + \sum_{j=1}^{m-1} S_j^*\right) \geq v_i\left(S_i, S_i + \sum_{j=1}^{m-1} S_j^*\right),$$

$$\forall C_i \in CS_k; \quad S_i, S_i^* \in [0, R_i]; \quad S_j^* \in [0, R_j]; \quad i, j = 1, 2, \ldots, m; \quad i \neq j,$$

i.e., for every coalition C_i, the optimal escapement strategy S_i^* must maximize the coalitional payoff, given the optimal escapement strategies of the other coalitions S_j^*. In other words, the equilibrium escapement strategy profile of a coalition structure requires that no coalition can get better off by deviating from its escapement

---

[h]No mixed strategies are considered when solving this game.
[i]The number of coalition structures κ depends on the number of players and is referred to as the Bell number within combinatorial mathematics.
[j]Games where a player's or a coalition's payoff depends only upon its own strategy (S_i in our setting), and a linear aggregate of the full strategy profile (S in our setting) are also called aggregate games, see Martimort and Stole [2012] for additional details and applications.

strategy, i.e., optimal escapement strategies are the best responses. If the grand coalition is formed, the above decision rule reduces to a single inequality:

$$v(S^*) \geq v(S), \quad S, S^* \in [0, R],$$

i.e., the optimal escapement level must maximize the grand coalition's payoff.

In order to determine the equilibrium escapement strategy profile of a coalition structure CS_k, the maximization problem (2) as specified in Sec. 2.2 must be repeatedly solved for every coalition C_i within a coalition structure CS_k until no coalition can further increase its NPV by adjusting its escapement strategy, given the escapement strategies of the other coalitions. However, as described in the same subsection, such maximization problem boils down to a single expression, the "golden-rule", specified in (3). Therefore, in order to determine the equilibrium escapement strategy profile of a coalition structure, we solve the following system of m equations:

$$\pi_i(S_i) = \gamma \theta_i F'(S) \pi_i[\theta_i(F(S))], \quad \forall C_i \in \mathrm{CS}_k; \quad i = 1, 2, \ldots, m, \quad \text{where}$$

$$S = \sum_{i=1}^{m} S_i, \quad i = 1, 2, \ldots, m. \tag{4}$$

These equations refer to the "golden-rules" that coalitions within a coalition structure apply in order to determine their escapement strategies. The overall escapement, S, is a linear aggregate of the full strategy profile and captures how coalitions interact with each other through their escapement strategies. Note that in the case of the grand coalition, the above system of equations consists of only one equation, which is equivalent to the cooperative "golden-rule" (1).

It should be obvious by now that the equilibrium escapement strategies depend on the coalition structure that is formed and on the parameters of the model. The coalitions formed are assumed to be asymmetric. They are differentiated by parameter θ_i, the share of mackerel stock that occurs in the EEZ(s) of a coalition, and their marginal cost of harvest, $c_i(x)$. Some coalitions may have equivalent shares, if their members are of the same type, see Sec. 4 for additional details. These asymmetries ensure that escapement strategies across coalitions are different and depend upon the form of the coalition structure. Thus, a unique payoff, which depends on the coalition structure, can be computed for every coalition in a coalition structure.

The coalitional payoff, which is equivalent to the NPV of the fishery over an infinite time horizon and depends on the escapement strategy profile of the coalition structure formed, can be written as follows:

$$v_i(S_i^*, S^*) = \sum_{t=0}^{\infty} \gamma^t \Pi_i(R_{it}, S_{it}) = \Pi_i(\theta_i R_0, S_i^*) + \frac{\gamma}{1-\gamma} \Pi_i[\theta_i F(S^*), S_i^*], \tag{5}$$

where R_0 is the initial recruitment and $S^* = S_i^* + \sum_{j=1}^{m-1} S_j^*$ is the optimal escapement strategy profile of a coalition structure. While specifying the coalitional payoff, it is important to remember that two things are assumed. First, the initial recruitment is high enough to allow for the prescribed harvest strategy in the first period, i.e., $S_i^* \leq \theta_i R_0$, $\forall C_i \in CS_k$. If this is not the case, the stock should not be harvested but allowed to grow until recruitment exceeds escapement. For our mackerel case, the initial recruitment is high enough to sustain all escapement strategies as feasible. Second, the fishing fleet capacity required to implement such harvest strategies (initial depletion and steady state harvest) exists. If the necessary capacity does not exist, the following situations arise: (i) there exists sufficient capacity to harvest the steady state quantity but not to deplete the stock to the steady state in one period and (ii) no sufficient capacity exists to harvest the steady state quantity.[k,l] If case (i) occurs, then the initial depletion of the stock to the steady state escapement level would take a couple of periods depending on the capacity of the current fishing fleet. On the other hand, if case (ii) occurs, we will never reach the "true" steady state prescribed by the optimal escapement strategy. In the long run, however, a nation would increase its fishing fleet capacity to meet the optimal escapement strategy, either by investing in more fishing vessels or by shifting vessels that operate in less profitable stocks. Since mackerel is one of the most valuable stocks in the Northeast Atlantic region and, in order not to complicate things by endogenously determining the fishing fleet capacity, we assume that the necessary capacity for implementing the prescribed strategies exists for all nations.

3.2. *First stage of coalition formation*

Our analysis is in line with the internal and external stability concepts of d'Aspremont *et al.* [1983] and what is defined as potential internal stability by Eyckmans and Finus [2004]. These concepts have been used to test a coalition's stability in both characteristic and partition function games.[m]

We start by introducing the notion of an embedded coalition, which is extensively used throughout this subsection. An embedded coalition is a pair (C_i, CS_k) consisting of a coalition and a coalition structure which contains that coalition, $C_i \in CS_k$. Let $V(C_i, CS_k)$ denote the payoff of an embedded coalition[n] and $V_x(C_i, CS_k)$ denote the payoff received by subcoalition x of the embedded coalition (C_i, CS_k), $x \subset C_i$. The subscript x may refer to an individual player (see internal

---

[k]For a formal analysis of these two cases, see Clark [1972].

[l]If a capacity constraint is to be included, then instead of harvesting $\max(R - S, 0)$, our sequence of harvest strategies should satisfy the following: $\max[\min(R - S, \text{Cap}), 0]$, i.e., if $S < R$, then harvest their difference if it is below the fishing fleet capacity Cap or harvest the capacity, otherwise do not harvest and let the stock grow.

[m]See, among others, Pintassilgo *et al.* [2010] and Liu *et al.* [2016] for applications of these concepts on fishery games in partition function form.

[n]Note that the payoff of an embedded coalition is equivalent to the coalitional payoff specified in Sec. 3.2, given that the coalition structure in which the coalitional payoff refers to is the same, i.e., $V(C_i, CS_k) \equiv v_i(S_i^*, S^*)$ if the coalition structure that v_i refers to is equivalent to CS_k.

stability condition below) or a coalition of players (see external stability condition below). The following relationship holds: $\sum_{x \in C_i} V_x(C_i, \mathrm{CS}_k) = V(C_i, \mathrm{CS}_k)$.

An embedded coalition (C_i, CS_k) is internal stable if none of its members l, $l \in C_i$, has incentives to leave and form a singleton coalition C^l, where $C^l = \{l\}$. Such condition can be written as follows:

$$V_l(C_i, \mathrm{CS}_k) \geq V(C^l, \mathrm{CS}_k^l), \quad \forall l \in C_i, \tag{6}$$

where $\mathrm{CS}_k^l = \{(\mathrm{CS}_k \backslash C_i), (C_i \backslash l), (C^l)\}$ stands for a coalition structure formed from the original coalition structure CS_k in which coalition C_i is split into two coalitions: $(C_i \backslash l)$ and (C^l). In other words, given an embedded coalition (C_i, CS_k), the payoff a member l receives as a member of coalition C_i must be higher or equal to the payoff that l can receive if it leaves the coalition in order to form a singleton coalition. If this is true for all the members, then the embedded coalition (C_i, CS_k) is internal stable. Note that the remaining form of the coalition structure is assumed to be unaffected by l's deviation, i.e., the remaining members of the said coalition do not leave after l leaves and the remaining coalitions in the coalition structure, if any, do not merge or split. This assumption is equivalent to the ceteris paribus assumption. By definition, all embedded coalitions which are singletons are always internal stable.

In an open membership game, where membership into a coalition is free for all players, a second condition ensuring that outsiders do not have incentives to join a coalition is needed. Such condition is referred to as external stability. An embedded coalition (C_i, CS_k) is external stable if no other embedded coalition (C_j, CS_k), singleton or not, in the coalition structure CS_k has incentives to join coalition (C_i, CS_k). Such condition can be written as follows:

$$V(C_j, \mathrm{CS}_k) \geq V_j(C_j^i, \mathrm{CS}_k^j), \quad \forall C_j \in \mathrm{CS}_k; \; C_j \neq C_i, \tag{7}$$

where $C_j^i = C_j \cup C_i$ stands for a coalition formed if coalitions C_i and C_j merge, and $\mathrm{CS}_k^j = \{(\mathrm{CS}_k \backslash (C_j, C_i)), (C_j^i)\}$ stands for a coalition structure formed from the original coalition structure CS_k in which coalitions C_i and C_j are merged into one coalition: (C_j^i). That is to say, given a coalition structure CS_k, the payoff an embedded coalition (C_j, CS_k) receives must be higher or equal to the payoff C_j can receive if it joins coalition C_i and forms a larger coalition. If this is true for all coalitions other than C_i within coalition structure CS_k, then the embedded coalition (C_i, CS_k) is external stable. Again, the remaining form of the coalition structure is assumed to be unaffected by the mergence. By definition, the grand coalition is always external stable.

So far, our analysis has been within the context of d'Aspremont *et al.* [1983] applied for embedded coalitions. Testing stability within this context requires the division of the coalitional payoff among coalition members. For instance, it is impossible to test for internal stability without the knowledge of the individual payoff a coalition member receives (left-hand side of (6)). Likewise, external stability requires information regarding the payoff the merging coalition will receive after the merger takes place (right-hand side of (7)). Hence, a sharing rule is needed in order to split the coalitional payoff. Consequently, the stability of a coalition is going to depend upon such sharing rule.

The existing literature on sharing rules that can be applied to partition function games is not so extensive compared to the one for characteristic function games.[o] Specifying a sharing rule for games in partition form is not an easy undertaking because of the complexity of the partition function. A common issue is that for a given coalition, the coalitional payoff is not unique since the same coalition can belong to more than one coalition structures.[p] Some authors have proposed different weighted rules in order to determine a unique coalitional payoff.[q] However, these approaches do not provide a unique solution unless the weight parameters are fully specified.

In order to avoid these issues and since the main objective of this chapter is to determine the set of stable coalition structures and not to distribute the gains of cooperation among cooperating nations, we adopt Eyckmans and Finus [2004] concept of potential internal stability. An embedded coalition (C_i, CS_k) is potentially internal stable if the sum of the free-riding payoffs of its members l, $l \in C_i$, does not exceed its coalitional payoff, i.e.,

$$V(C_i, CS_k) \geq \sum_{l \in C_i} V(C^l, CS_k^l), \tag{8}$$

where $C^l = \{l\}$ is a singleton coalition and $CS_k^l = \{(CS_k \backslash C_i), (C_i \backslash l), (C^l)\}$ stands for a coalition structure formed from the original coalition structure CS_k in which coalition C_i is split into two coalitions: $(C_i \backslash l)$ and (C^l). $V(C^l, CS_k^l)$ is the free-riding payoff that a coalition member l can receive if it leaves coalition C_i and forms the singleton coalition C^l, ceteris paribus. By definition, a singleton embedded coalition is always potential internal stable.

A clear advantage of condition (8) over (6) is that it can test for internal stability in the absence of a sharing rule. If an embedded coalition is potentially internal stable, then there exist some allocation schemes which can ensure internal stability. On the other hand, if potential internal stability does not hold, then no sharing rule can make an embedded coalition internal stable [Pintassilgo *et al.*, 2010].

Clearly, potential internal stability is a necessary condition for internal stability. By the same token, a necessary condition for external stability is needed in order to be able to determine stability in the absence of a sharing rule. An embedded coalition (C_i, CS_k) is potentially external stable if for all other embedded coalitions

---

[o] The coalitional payoff of a game in characteristic form is independent of the coalition structure.
[p] To see this point, consider a four-player game and the following two coalition structures: $CS_1 = \{12, 3, 4\}$ and $CS_2 = \{12, 34\}$. In both coalition structures, players 1 and 2 form a coalition. Players 3 and 4 act as singletons in CS_1 and also form a coalition in CS_2. The payoff of coalition (12) depends on the coalition structure that it belongs, and the coalition structure that contains coalition (12) is not unique.
[q] See Macho-Stadler *et al.* [2007], Pham Do and Norde [2007] and De Clippel and Serrano [2008] for examples.

(C_j, CS_k) the following inequality holds:

$$V(C_j, \mathrm{CS}_k) \geq V(C_j^i, \mathrm{CS}_k^j) - \sum_{l \in C_i} V(C^l, \mathrm{CS}_k^{jl}), \quad \forall C_j \in \mathrm{CS}_k; \ C_j \neq C_i, \qquad (9)$$

where $C_j^i = C_j \cup C_i$ stands for a coalition formed if coalitions C_i and C_j merge, and $\mathrm{CS}_k^j = \{(\mathrm{CS}_k \backslash (C_j, C_i)), (C_j^i)\}$ stands for a coalition structure formed from the original coalition structure CS_k in which coalitions C_i and C_j are merged into one coalition: (C_j^i). In addition, $C^l = \{l\}$ is a singleton coalition and $\mathrm{CS}_k^{jl} = \{(\mathrm{CS}_k^j \backslash C_j^i), (C_j^i \backslash l), (C^l)\}$ stands for a coalition structure formed from coalition structure CS_k^j in which coalition C_j^i is split into two coalitions: $(C_j^i \backslash l)$ and (C^l). $V(C_j^i, \mathrm{CS}_k^j)$ is the payoff coalition C_j^i receives after the merger occurs, ceteris paribus (hereinafter, the joint payoff). And, $V(C^l, \mathrm{CS}_k^{jl})$ is the free-riding payoff that a member l of coalition C_i receives if it leaves coalition C_j^i, ceteris paribus. Thus, given a coalition structure CS_k, an embedded coalition (C_i, CS_k) is potentially external stable if and only if the payoff of all other embedded coalitions C_j in CS_k is greater than the joint payoff minus the sum of the free-riding payoffs of coalition's C_i members. In other words, in order for coalition C_j not to be willing to merge with coalition C_i, its potential share of the joint payoff must be lower than its current payoff. The potential share of the joint payoff that coalition C_j is entitled to is the remainder of the joint payoff after all members of coalition C_i have received their free-riding payoffs. By definition, the grand coalition is always potentially external stable.

Having defined the necessary conditions for an embedded coalition to be internal and external stable, in the absence of a sharing rule, we can now proceed in defining the necessary conditions for a coalition structure to be stable. As in the case of a coalition, stability of a coalition structure in an open membership game requires that the coalition structure is both internal and external stable.

Before we start analyzing the two conditions, let us take a step back and visualize what internal and external stability of a coalition structure is. Figure 1 depicts the coalition structures for a four-player game. The nodes represent coalition structures. The arcs represent mergers of two coalitions when followed upward and split of a coalition into two subcoalitions when followed downward. In a four-player game, there exist four levels in total. A coalition structure level is a subset of the coalition structure set that consists of coalition structures with equal number of coalitions. In our example, the third level subset is composed of coalition structures that have only two coalitions. A stable coalition structure should not move upwards or downwards in the graph but remain in its position. This occurs if all embedded coalitions in a coalition structure do not have incentives to merge or split.

The split part is the easiest to test as it merely requires all embedded coalitions of a coalition structure to be internal stable. If this is true, then the coalition structure cannot be downgraded, i.e., move downwards in the graph. Using the

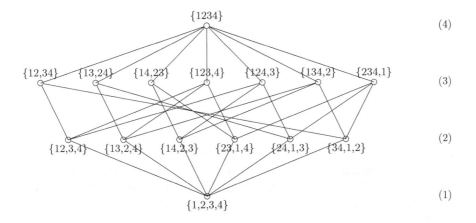

$$\{1234\} \tag{4}$$

$$\{12,34\} \quad \{13,24\} \quad \{14,23\} \quad \{123,4\} \quad \{124,3\} \quad \{134,2\} \quad \{234,1\} \tag{3}$$

$$\{12,3,4\} \quad \{13,2,4\} \quad \{14,2,3\} \quad \{23,1,4\} \quad \{24,1,3\} \quad \{34,1,2\} \tag{2}$$

$$\{1,2,3,4\} \tag{1}$$

Fig. 1. Coalition structure graph for a four-player game.

notion of potential internal stability, such condition can be written as follows:

$$V(C_i, \mathrm{CS}_k) \geq \sum_{l \in C_i} V(C^l, \mathrm{CS}_k^l), \quad \forall C_i \in \mathrm{CS}_k. \tag{10}$$

Therefore, if all embedded coalitions of a coalition structure are potentially internal stable, then the coalition structure is potentially internal stable, which is a necessary condition for internal stability to hold.

On the other hand, the merge part of our argument is not so straightforward to test. This is because it is not equivalent as saying that all embedded coalitions of a coalition structure should be external stable. If we say so, then some externally stable coalition structures will fail to pass the test and considered as externally unstable. To see this point, suppose that external stability of a coalition structure requires all of its embedded coalitions to be external stable. Consider the following coalition structure: $\mathrm{CS}_{11} = \{123, 4\}$. According to the aforementioned definition, CS_{11} is external stable if coalitions (123) and (4) are external stable. That is to say that coalition (123) does not want to merge with (4) and coalition (4) does not want to merge with (123). This sounds like a valid definition for a coalition structure to be external stable, and, as a matter of fact, it is. If all embedded coalitions of a coalition structure are external stable, then the coalition structure cannot be upgraded, i.e., move upwards in the graph.

Suppose now that one of the two embedded coalitions of CS_{11} is not external stable. Is this assumption going to upgrade CS_{11} permanently and therefore making it "truly" external unstable? Let coalition (123) be the only external stable coalition. In other words, (4) does not want to merge with (123) but (123) wants to merge with (4). Since not all embedded coalitions are external stable, by definition, coalition structure CS_{11} is not external stable. Therefore, upgrade into coalition structure $\mathrm{CS}_{15} = \{1234\}$ occurs. But, we know that only coalition (123) is better off under the new coalition structure since by assumption, it is the only coalition that wants to merge. Thus, coalition (4) deviates and coalition structure $CS_{11} = \{123, 4\}$ forms again.

The question now becomes: is it possible, given a pair of embedded coalitions, that only one has incentives to join the other? The short answer is yes. Typically, games with positive externalities are superadditive, i.e., $V(C_i \cup C_j, CS_k) \geq V(C_i, CS_k^i) + V(C_j, CS_k^j)$, where $CS_k^i = CS_k^j = \{(CS_k \backslash (C_i \cup C_j)), ((C_i \cup C_j) \backslash C_i)\}$. Superadditivity means that a merger between two embedded coalitions generates a payoff at least equal to the sum of the individual payoffs. The superadditivity property may or may not hold across the entire game but it holds for at least some embedded coalitions, at least it does in the game analyzed in this chapter.

Back to our question. Suppose that the superadditive property holds between the embedded coalitions of CS_{11} and CS_{15}, i.e., $V(1234, \{1234\}) \geq V(123, \{123, 4\}) + V(4, \{123, 4\})$. If this is true, then coalition (123) is better off under the mergence (strict inequality) or indifferent (equality). This is because the individual payoff of coalition (4) under CS_{11} is also its free-riding payoff. That is, after the mergence occurs, if coalition (4) deviates, it cannot receive a payoff greater than the payoff it already receives. Therefore, after mergence, coalition (123) receives at least its individual payoff. However, after mergence, coalition (4) may not necessarily receive its individual payoff. This is because, coalition (123) must receive a payoff which is at least as high as the sum of the free-riding payoffs of its members, i.e., $V_{123}(1234, \{1234\}) \geq \sum_{l \in (123)} V(l, \{(1234 \backslash l), (l)\})$. Therefore, the potential payoff that coalition (4) can receive cannot exceed the difference between the joint payoff and the sum of the free-riding payoffs of coalition (123), i.e., $V_4(1234, \{1234\}) \leq V(1234, \{1234\}) - \sum_{l \in (123)} V(l, \{(1234 \backslash l), (l)\})$. If $V_4(1234, \{1234\})$ is greater than $V(4, \{123, 4\})$, then coalition (4) has incentives to merge, otherwise it does not. It should be clear by now, that given a pair of coalitions, (C_1, C_2), the fact that C_1 wants to merge with C_2 does not imply that C_2 also wants to merge with C_1. In order for C_2 to be willing to merge, its payoff under the mergence should be greater than its individual payoff and this depends on the magnitude of the free-riding payoffs of C_1 members.

Even if the entire game is superadditive, i.e., at least some coalitions want to merge, the free-riding effects of these coalitions may be so strong that they make it impossible for mergence to occur. And, it is because of strong free-riding effects that superadditive games with externalities cannot necessarily sustain the grand coalition as a stable outcome.

So far, we have argued that requiring all embedded coalitions of a coalition structure to be external stable does not necessarily provide us with the set of all external stable coalition structures. So, is there a rule that when applied can give us the set of all external stable coalition structures? The answer is yes. Such condition requires that, given a coalition structure CS_k, all possible embedded coalitions pairs $[(C_i, CS_k), (C_j, CS_k)]$, $\forall C_i, C_j \in CS_k$ and $C_i \neq C_j$, are not willing to merge. An embedded coalition pair is not willing to merge if at least one of its embedded coalitions do not want to merge. Such conditions can be written as follows:

(A) $V(C_i, \mathrm{CS}_k) \geq V(C_i^j, \mathrm{CS}_k^i) - \sum_{l \in C_j} V(C^l, \mathrm{CS}_k^{il}), \quad C_i \neq C_j; \ C_i, C_j \in \mathrm{CS}_k,$

$$(11)$$

(B) $V(C_j, \mathrm{CS}_k) \geq V(C_j^i, \mathrm{CS}_k^j) - \sum_{l \in C_i} V(C^l, \mathrm{CS}_k^{jl}), \quad C_j \neq C_i; \ C_j, C_i \in \mathrm{CS}_k.$

$$(12)$$

Condition A (B) is equivalent to the potential external stability condition (9) but only with respect to coalition C_i (C_j). That is, if A is true, then C_i does not want to merge with C_j, i.e., C_j is potentially external stable with respect to C_i. Similarly, if B is true, then C_j does not want to merge with C_i, i.e., C_i is potentially external stable with respect to C_j. If one of the two conditions holds, i.e., A $\lor$ B, then the pair $[(C_i, \mathrm{CS}_k), (C_j, \mathrm{CS}_k)]$ will not merge and therefore is considered as external stable. If this is true for all possible pairs within a coalition structure, i.e.,

$$\mathrm{A} \lor \mathrm{B}, \quad \forall C_i, C_j \in \mathrm{CS}_k; \quad C_i \neq C_j, \tag{13}$$

then the coalition structure is potentially external stable, which is a necessary condition for external stability to hold. A coalition structure is stable if it is both internal and external stable, i.e., stability of a coalition structure requires conditions (10) and (13) to hold simultaneously. An illustration of the stability concepts applied in this chapter is provided through a small numerical example in Appendix A.2.

4. Empirical Model

Before proceeding with the specification of functional forms and parameters, we first identify the different coalition structures in the four- and five-player games. The four-player game consists of the following nations: the EU, Norway, the Faroe Islands and Iceland. The total number of coalitions and coalition structures that are likely to occur in a four-player game are 15 and are depicted in Tables 1 and 2. The five-player game consists of the following nations: the EU, the UK, Norway, the Faroe Islands and Iceland. The total number of coalitions and coalition structures that are likely to occur in this game are 31 and 52 and are shown in Tables 3 and 4.

The singleton coalition of EU in the four-player game is treated to be equivalent to the coalition of EU and UK in the five-player game. As a consequence, all of the

Table 1. List of all coalitions for the four-player game.

No.	Coalition	No.	Coalition	No.	Coalition
1	(EU)	6	(EU,FO)	11	(EU,NO,FO)
2	(NO)	7	(EU,IS)	12	(EU,NO,IS)
3	(FO)	8	(NO,FO)	13	(EU,FO,IS)
4	(IS)	9	(NO,IS)	14	(NO,FO,IS)
5	(EU,NO)	10	(FO,IS)	15	(EU,NO,FO,IS)

Table 2.   List of all possible coalition structures for the four-player game.

No.	Coalition structure	No.	Coalition structure	No.	Coalition structure
1	(EU),(NO),(FO),(IS)	6	(NO,IS),(EU),(FO)	11	(EU,NO,FO),(IS)
2	(EU,NO),(FO),(IS)	7	(FO,IS),(EU),(NO)	12	(EU,NO,IS),(FO)
3	(EU,FO),(NO),(IS)	8	(EU,NO),(FO,IS)	13	(EU,FO,IS),(NO)
4	(EU,IS),(NO),(FO)	9	(EU,FO),(NO,IS)	14	(NO,FO,IS),(EU)
5	(NO,FO),(EU),(IS)	10	(EU,IS),(NO,FO)	15	(EU,NO,FO,IS)

Table 3.   List of all coalitions for the five-player game.

No.	Coalition	No.	Coalition	No.	Coalition	No.	Coalition
1	(EU)	9	(EU,IS)	17	(EU,UK,FO)	25	(NO,FO,IS)
2	(UK)	10	(UK,NO)	18	(EU,UK,IS)	26	(EU,UK,NO,FO)
3	(NO)	11	(UK,FO)	19	(EU,NO,FO)	27	(EU,UK,NO,IS)
4	(FO)	12	(UK,IS)	20	(EU,NO,IS)	28	(EU,UK,FO,IS)
5	(IS)	13	(NO,IS)	21	(EU,FO,IS)	29	(EU,NO,FO,IS)
6	(EU,UK)	14	(NO,FO)	22	(UK,NO,FO)	30	(UK,NO,FO,IS)
7	(EU,NO)	15	(FO,IS)	23	(UK,NO,IS)	31	(EU,UK,NO,FO,IS)
8	(EU,FO)	16	(EU,UK,NO)	24	(UK,FO,IS)		

coalition structures that are likely to occur in the four-player game are also likely to reoccur in the five-player game. For example, CS_1 in the four-player game is equivalent to CS_2 in the five-player game, etc. However, the set of stable coalition structures is not necessarily equivalent between the two games. This is due to the fact that in the five-player game, we allow for the UK to make its own decisions and these decisions may not necessarily be aligned to the ones EU and UK as cooperators may implement. For the remaining of the chapter and unless explicitly stated all figures related to EU refer to the five-player game and do not take UK into consideration. Table 5 provides a concrete list of all the symbols used in this chapter.

4.1.   *Stock–recruitment relationship*

In order to capture the relationship between a period's escapement S_t and next period's recruitment R_{t+1}, a function $F(S)$ is needed, where $R_{t+1} = F(S_t)$. One functional form, introduced by Ricker [1954] is $F(S) = aSe^{-bS}$. This function has the property of overcompensation, i.e., it reaches a peak and then descends asymptotically towards $R = 0$, $\lim_{S \to \infty} F(S) = 0$. Another functional form, proposed by Beverton and Holt [1957] is $F(S) = \frac{aS}{b+S}$. This one does not decrease but instead increases asymptotically towards $R = a$, $\lim_{S \to \infty} F(S) = a$. Both functions are well known among the models that have been developed to fit stock–recruitment curves to data sets.[r] We estimate and make use of both when running our model. By doing so, we are able to test how sensitive the set of stable coalition structures is to the biological constraint of our model.

---

[r]See Iles [1994] for a review.

Table 4. List of all possible coalition structures for the five-player game.

No.	Coalition structure	No.	Coalition structure	No.	Coalition structure	No.	Coalition structure
1	(EU),(UK),(NO),(FO),(IS)	14	(EU,FO),(UK,NO),(IS)	27	(EU,UK,NO),(FO),(IS)	40	(EU,NO,FO),(UK,IS)
2	(EU,UK),(NO),(FO),(IS)	15	(EU,UK),(NO,IS),(FO)	28	(EU,UK,FO),(NO),(IS)	41	(EU,NO,IS),(UK,FO)
3	(EU,NO),(UK),(FO),(IS)	16	(EU,NO),(UK,IS),(FO)	29	(EU,UK,IS),(NO),(FO)	42	(EU,FO,IS),(UK,NO)
4	(EU,FO),(UK),(NO),(IS)	17	(EU,IS),(UK,NO),(FO)	30	(EU,NO,FO),(UK),(IS)	43	(UK,NO,FO),(EU,IS)
5	(EU,IS),(UK),(NO),(FO)	18	(EU,UK),(FO,IS),(NO)	31	(EU,NO,IS),(UK),(FO)	44	(UK,NO,IS),(EU,FO)
6	(UK,NO),(EU),(FO),(IS)	19	(EU,FO),(UK,IS),(NO)	32	(EU,FO,IS),(UK),(NO)	45	(UK,FO,IS),(EU,NO)
7	(UK,FO),(EU),(NO),(IS)	20	(EU,IS),(UK,FO),(NO)	33	(UK,NO,FO),(EU),(IS)	46	(NO,FO,IS),(EU,UK)
8	(UK,IS),(EU),(NO),(FO)	21	(EU,NO),(FO,IS),(UK)	34	(UK,NO,IS),(EU),(FO)	47	(EU,UK,NO,FO),(IS)
9	(NO,IS),(EU),(UK),(FO)	22	(EU,FO),(NO,IS),(UK)	35	(UK,FO,IS),(EU),(NO)	48	(EU,UK,NO,IS),(FO)
10	(NO,FO),(EU),(UK),(IS)	23	(EU,IS),(NO,FO),(UK)	36	(NO,FO,IS),(EU),(UK)	49	(EU,UK,FO,IS),(NO)
11	(FO,IS),(EU),(UK),(NO)	24	(UK,NO),(FO,IS),(EU)	37	(EU,UK,NO),(FO,IS)	50	(EU,NO,FO,IS),(UK)
12	(EU,UK),(NO,FO),(IS)	25	(UK,FO),(NO,IS),(EU)	38	(EU,UK,FO),(NO,IS)	51	(UK,NO,FO,IS),(EU)
13	(EU,NO),(UK,FO),(IS)	26	(UK,IS),(NO,FO),(EU)	39	(EU,UK,IS),(NO,FO)	52	(EU,UK,NO,FO,IS)

Table 5.   List of symbols and abbreviations.

Symbol	Description	Value	Unit
Sets			
N	Players		
K	Coalition structures		
Subscripts			
n	Number of players	$4, 5$	
m	Number of coalitions in a CS	$1, 2, \ldots, n$	
κ	Number of CSs	$15, 52$	
t	Time index	$0, 1, 2, \ldots, \infty$	
l	Player index	$1, 2, \ldots, n$	
i, j	Coalition index	$1, 2, \ldots, m$	
k	CS index	$1, 2, \ldots, \kappa$	
Variables			
S_i	Escapement of coalition i in a CS		Thousand tonnes
S	Total escapement		Thousand tonnes
R	Total recruitment		Thousand tonnes
H	Total harvest		Thousand tonnes
V_i	NPV of coalition i in a CS (embedded coalition)[a]		Million NOK
V_{CS}	Total NPV of a CS[b]		Million NOK
Parameters			
p	Price	10	NOK/kg
r	Discount rate	5%	
θ_l	Share of mackerel stock in player's l EEZ	cf. Table 7	
a	Stock–recruitment parameter	cf. Table 6	
b	Stock–recruitment parameter	cf. Table 6	
c_i	Cost parameter of coalition i	cf. Tables 9 and 10	
β	Stock elasticity parameter	$1.0, 0.6, 0.3$	
$\bar{R}$	Base year recruitment	$4{,}887$	Thousand tonnes
$\bar{H}_l$	Base year harvest of player l	cf. Table 8	
ψ	Cost–revenue ratio	0.78	
Abbreviations			
CS	Coalition structure		
EU	European Union		
UK	United Kingdom		
NO	Norway		
FO	Faroe Islands		
IS	Iceland		
NPV	Net present value		

[a] V_i is equivalent to $V(C_i, \mathrm{CS}_k)$ and should not be confused with $V_x(C_i, \mathrm{CS}_k)$. We make use of compact notation in order to convenience ourselves in the presentation of the results.
[b] $V_{\mathrm{CS}} = \sum_{i \in \mathrm{CS}_k} V(C_i, \mathrm{CS}_k)$.

Both functions are nonlinear, thus before proceeding with the regressions, we linearize them. The Ricker stock–recruitment relationship becomes

$$R_t = aS_{t-1}e^{-bS_{t-1}} \Leftrightarrow \ln(R_t)$$

$$= \ln(a) + \ln(S_{t-1}) - bS_{t-1} \Leftrightarrow \ln\left(\frac{R_t}{S_{t-1}}\right) = \ln(a) - bS_{t-1}. \qquad (14)$$

Table 6. Results from fitting recruitment and escapement data on the Ricker and Beverton–Holt functions.

Functional form	Parameters		Adjusted R^2
	a	b	
Ricker	1.6784	9.73×10^{-5}	0.35
	(0.000)	(0.000)	
Beverton–Holt	10,977	5,965	0.88
	(0.000)	(0.000)	

Note: p-values of the transformed regression in parentheses.

Similarly, the Beverton–Holt function becomes

$$R_t = \frac{aS_{t-1}}{b + S_{t-1}} \Leftrightarrow \frac{1}{R_t} = \frac{1}{a} + \frac{b}{a}\frac{1}{S_{t-1}}. \tag{15}$$

We fit Eqs. (14) and (15) using Ordinary Least Squares on recruitment and escapement data. The data used are obtained from ICES [2016a] advice report 9.3.39 Table 9.3.39.14. In particular, the following columns covering the period between 1980 and 2015 are used: (i) SSB (Spawning time) and (ii) Landings. According to ICES [2014], the estimate of the SSB at spawning time in the year in which the TAC applies, taking into account of the expected catch (Annex 9.3.17.1 Management plan harvest control rule). In the beginning of Sec. 2, we define the recruitment of a fishery as the unexploited SSB at the beginning of a period. If we identify that the

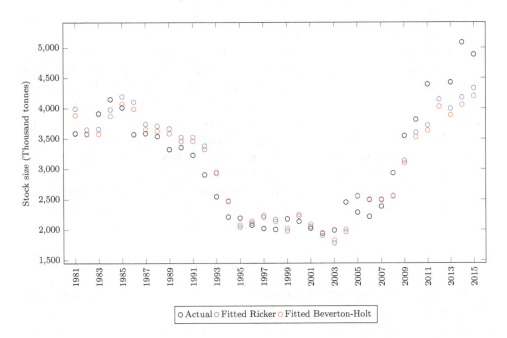

o Actual o Fitted Ricker o Fitted Beverton-Holt

Fig. 2.   Actual and fitted development of the mackerel stock 1981–2015.

beginning of a period occurs when spawning takes place, then the terms recruitment and SSB are equivalent. Moreover, landings refer to the mackerel biomass landed in all ports in the Northeast Atlantic area in a respective year, which is equivalent to the total harvested biomass. Therefore, the difference between SSB and landings represents the escapement of the stock in a particular period/year.

The parameters a and b in Eqs. (14) and (15) are estimated after the time lag as well as transformation for variables R and S have been taken into account. The results of the regression are shown in Table 6. Figure 2 shows the actual development of the mackerel stock and the fitted curves for both stock–recruitment functions on the escapement data. Both functions can trace the actual mackerel stock reasonably well.

4.2. *Share of mackerel stock*

As we have already mentioned in Sec. 2.2, θ_l denotes the share of the mackerel stock that only appears in the EEZ of nation l during the whole year. We believe that the share parameters consist of two dimensions, namely, time and space. Time refers to the percentage of months in a year that mackerel appears in the EEZ of a nation. And space refers to the percentage of the mackerel stock that appears in the EEZ of a nation. Multiplication of the two percentages for nation l yields parameter θ_l.

For the dimension of time, we base our analysis on the annual migration pattern of the mackerel stock and the time it spends on the respective EEZs of the nations concerned in this chapter. The migration pattern of mackerel is divided into two elements, namely, a pre-spawning migration and a post-spawning one [ICES, 2016b]. From late summer to autumn, the pre-spawning migration starts from the feeding grounds in the North and Nordic seas. This migration phase includes shorter or longer halts in deep waters along the edge of the continental shelf where mackerel shoals overwinter until they reach the spawning grounds south down the west coast of Scotland and Ireland, and along the shelf break waters between Spain and Portugal. The stock is targeted by Norwegian, British and European vessels when it overwinters (fourth quarter) and by European and British vessels afterwards (first quarter). After spawning occurs, the post-spawning migration towards the feeding grounds begins. No significant catches occur during this migration, which takes place in spring (second quarter). During summer, the stock is more spread as it feeds in Northern waters. At this time, Norwegian, Icelandic and Faroese vessels are active (third quarter).

According to the mackerel migration pattern, we conclude that the stock occurs 50% of the time in the Norwegian EEZ (third and fourth quarters), 50% of the time in the European and British EEZs (fourth and first quarters), and 25% of the time in the Icelandic and Faroese EEZs (third quarter).

For the spatial distribution, unfortunately, no data exist that measure the amount of mackerel that appears in a specific geographical area within the

Table 7. Shares of mackerel stock in player's l EEZ.

	EU[a]	UK	NO	FO	IS
Mackerel share in %, θ_l	25.0	25.0	25.0	12.5	12.5

Note: See Table 5 for abbreviations.
[a]Mackerel share for EU refers to the five-player game, which does not include UK. Mackerel share for EU in the four-player game is equivalent to the sum of EU and UK mackerel shares, i.e., 50%.

Northeast Atlantic. Therefore, we make the simplifying assumption that approximately half of the stock appears in the EEZ of a nation during mackerel's annual migration pattern. That is, the space percentage that appears in the EEZ of a nation is constant and equal to 50% for all nations. Table 7 shows the share of the mackerel stock that appears in the EEZ of the nations we consider in this chapter, calculated as the product of the two dimensions analyzed here. As already mentioned in Sec. 2.2, the share of the mackerel stock of coalition i is computed as the sum of the individual shares of its members.

4.3. Unit cost of harvest

As we discuss in Sec. 2, the coalitional unit cost of harvest $c_i(x)$ is a continuous and decreasing function with respect to stock size and the total cost within one period is specified as $\text{TC}_i(R_{it}, S_{it}) = \int_{S_{it}}^{R_{it}} c_i(x)dx$. Total costs can be also expressed to be proportionate with fishing effort E_i, that is $\text{TC}_i(E_i) = c_i E_i$, where c_i is a cost parameter. Furthermore, we define the harvest production function of a coalition to be $H_i = E_i x^\beta$, where β is the stock elasticity and is assumed to be the same for all coalitions. Solving the harvest production function with respect to fishing effort and substituting in the total cost function yields $\text{TC}_i(H_i, x) = c_i H_i x^{-\beta}$. Dividing with harvest, the unit cost of harvest can be expressed as $c_i(x) = c_i x^{-\beta}$. Substituting for the unit cost of harvest in the initial total cost expression and solving the integral provides us with an analytic expression for the total cost of harvest of coalition i. Note that for values of $\beta = 1$ and $\beta \in (0,1)$, the integral yields different solutions.[s] Thus,

$$TC_i(R_{it}, S_{it}) = \begin{cases} c_i \ln\left(\dfrac{R_{it}}{S_{it}}\right), & \beta = 1, \\[2ex] c_i \dfrac{1}{1-\beta}(R_{it}^{1-\beta} - S_{it}^{1-\beta}), & 0 < \beta < 1. \end{cases} \tag{16}$$

[s]For $\beta = 0$, total cost becomes proportional to harvest and the unit cost of harvest is no longer stock-dependent. Constant stock density ($\beta = 0$) implies that the equilibrium escapement strategy profile of a coalition structure as specified in Sec. 3.2 (system of equations (4)) cannot be obtained. This is because marginal profit at the beginning and the end of a harvesting period is no longer different and the noncooperative golden rule becomes $1 = \gamma \theta_i F'(S)$.

Due to lack of uniformly reported cost data across the nations considered in this chapter as well as the short-length of some of these series, the cost parameters cannot be estimated through statistical procedures. Instead, the cost coefficients c_i for all coalitions are calibrated at the level which ensures that for base year harvest, $\bar{H}_i = \sum_{l \in C_i} \bar{H}_l$, and base year recruitment, $\bar{R}_i = \theta_i \bar{R}$, total cost is the estimated base year proportion of total revenue ψ, i.e.,

$$
c_i = \begin{cases} \psi p \bar{H}_i \ln\left(\dfrac{\bar{R}_i}{\bar{R}_i - \bar{H}_i}\right)^{-1}, & \beta = 1, \\[2mm] \psi p \bar{H}_i (1-\beta)[\bar{R}_i^{\,1-\beta} - (\bar{R}_i - \bar{H}_i)^{1-\beta}]^{-1}, & 0 < \beta < 1. \end{cases} \tag{17}
$$

The cost–revenue ratio ψ is equal to 0.78 and is assumed to be equal for all nations. Its computation is based on operating expenses and operating revenues of licensed Norwegian purse seiners for the year 2015 obtained from the report: profitability survey on the Norwegian fishing fleet, Table G 20 [Norwegian Directorate of Fisheries, 2015].

Table 8.   Base year (2015) harvest for EU, UK, Norway, Faroe Islands and Iceland. Units: Thousand tonnes.

	EU[a]	UK	NO	FO	IS
Base year harvest, $\bar{H}_l$	269.929	247.986	242.231	108.412	169.333

Note: See Table 5 for abbreviations.
[a]Base year harvest for EU refers to the five-player game, which does not include UK. Base year harvest for EU in the four-player game is equivalent to the sum of EU and UK base year harvests, i.e., 517.915 thousand tonnes.

Table 9.   Cost parameters for coalitions i in the four-player game for different stock elasticity levels.

Coalition, C_i	Cost parameter, c_i			
	$\beta = 1$	$\beta = 0.6$	$\beta = 0.3$	$\beta = 0.1$
(EU)	17,032.48	788.13	78.59	16.90
(NO)	8,587.07	522.52	63.99	15.78
(FO)	4,346.93	347.26	52.16	14.74
(IS)	4,086.31	334.84	51.24	14.65
(EU,NO)	25,619.69	1,006.85	88.83	17.60
(EU,FO)	21,379.94	903.27	84.14	17.28
(EU,IS)	21,120.84	896.80	83.84	17.26
(NO,FO)	12,934.16	668.07	72.35	16.44
(NO,IS)	12,675.85	660.17	71.93	16.40
(FO,IS)	8,436.17	517.08	63.66	15.75
(EU,NO,FO)	29,967.05	1,106.10	93.10	17.88
(EU,NO,IS)	29,708.49	1,100.46	92.87	17.86
(EU,FO,IS)	25,468.83	1,003.35	88.68	17.59
(NO,FO,IS)	17,023.72	787.89	78.58	16.90
(EU,NO,FO,IS)	34,056.20	1,194.40	96.75	18.11

Note: See Table 5 for abbreviations.

Base year harvest for all nations, $\bar{H}_l$, and base year recruitment for the entire mackerel fishery, $\bar{R}$, are obtained from ICES [2016a] advice report 9.3.39. Recruitment for year 2015 is provided from Table 9.3.39.14 of the report and is equivalent to 4,887 thousand tonnes for the entire mackerel fishery. Individual harvest levels for year 2015 are provided from Table 9.3.39.12 of the ICES report and are depicted in Table 8. Base year harvest for coalition i, $\bar{H}_i$, is defined as the sum of the base year quantities of its members l, i.e., $\bar{H}_i = \sum_{l \in C_i} \bar{H}_l$. Base year recruitment for coalition i, $\bar{R}_i$, is defined as the product of the coalition's share of mackerel stock θ_i and overall base year recruitment, i.e., $\bar{R}_i = \theta_i \bar{R}$.

The price p is equivalent to $10\,\mathrm{NOK/kg}$. The stock elasticity β is not estimated empirically and is therefore varied when running our model in order to capture a

Table 10.   Cost parameters for coalitions i in the five-player game for different stock elasticity levels.

Coalition, C_i	Cost parameter, c_i			
	$\beta = 1$	$\beta = 0.6$	$\beta = 0.3$	$\beta = 0.1$
(EU)	8,469.55	518.29	63.73	15.76
(UK)	8,562.75	521.65	63.94	15.77
(NO)	8,587.07	522.52	63.99	15.78
(FO)	4,346.93	347.26	52.16	14.74
(IS)	4,086.31	334.84	51.24	14.65
(EU,UK)	17,032.48	788.13	78.59	16.90
(EU,NO)	17,056.91	788.80	78.62	16.90
(EU,FO)	12,817.18	664.50	72.16	16.42
(EU,IS)	12,557.15	656.52	71.73	16.39
(UK,NO)	17,149.83	791.33	78.75	16.91
(UK,FO)	12,909.92	667.33	72.32	16.43
(UK,IS)	12,651.26	659.41	71.89	16.40
(NO,IS)	12,934.16	668.07	72.35	16.44
(NO,FO)	12,675.85	660.17	71.93	16.40
(FO,IS)	8,436.17	517.08	63.66	15.75
(EU,UK,NO)	25,619.69	1,006.85	88.83	17.60
(EU,UK,FO)	21,379.94	903.27	84.14	17.28
(EU,UK,IS)	21,120.84	896.80	83.84	17.26
(EU,NO,FO)	21,404.30	903.88	84.16	17.29
(EU,NO,IS)	21,145.41	897.41	83.87	17.27
(EU,FO,IS)	16,905.74	784.67	78.42	16.88
(UK,NO,FO)	21,497.00	906.19	84.27	17.29
(UK,NO,IS)	21,238.92	899.75	83.97	17.27
(UK,FO,IS)	16,999.26	787.22	78.55	16.89
(NO,FO,IS)	17,023.72	787.89	78.58	16.90
(EU,UK,NO,FO)	29,967.05	1,106.10	93.10	17.88
(EU,UK,NO,IS)	29,708.49	1,100.46	92.87	17.86
(EU,UK,FO,IS)	25,468.83	1,003.35	88.68	17.59
(EU,NO,FO,IS)	25,493.32	1,003.92	88.70	17.59
(UK,NO,FO,IS)	25,586.54	1,006.08	88.79	17.60
(EU,UK,NO,FO,IS)	34,056.20	1,194.40	96.75	18.11

Note: See Table 5 for abbreviations.

range of possibilities. We set β equal to 1, 0.6, 0.3 and 0.1.[t] Tables 9 and 10 show the cost parameters for all coalitions in both the four- and five- player games for all realizations of the stock elasticity.

5. Numerical Results and Discussion

Having defined all parameters and functional forms, the solution process of the game is as follows. First, optimal escapement strategies for all coalitions in a coalition structure are computed by solving the system of equations presented in (4). The sum of the optimal escapement strategies determines the optimal recruitment through Ricker (Eq. (14)) or the Beverton–Holt (Eq. (15)) stock–recruitment function. Then, recruitment and harvest levels for all coalitions in a coalition structure are calculated following the framework described in the beginning of Sec. 2.2. The coalitional payoff of all coalitions in a coalition structure is determined through Eq. (5). This process is repeated for all coalition structures in both games. Finally, internal and external stability of a coalition structure is tested using conditions (10) and (13). Both games are solved eight times in total, two times for each stock–recruitment function (Ricker and Beverton–Holt) and four times for all the different variations of the stock elasticity parameter. Due to limited reporting space, all result tables are placed in a supplementary report, which can be obtained from the authors by request.

Before proceeding with the discussion of stable coalition structures, we point out three facts regarding the overall results of these games. First and foremost, our results indicate that positive externalities occur in the mackerel fishery since when coalitions merge to form a larger coalition, outside coalitions not affected by the merger are better off. According to Yi [1997], this result is the defining feature of coalition games with positive externalities. The members of merging coalitions increase the stock level and hence reduce their cost of fishing in order to internalize the positive externality which affects them. Noncooperating coalitions benefit from the merger by free-riding on the merging coalitions' stock increase.

Second, because of this internalization, aggregate escapement and recruitment increase as the degree of cooperation between coalition structures increases. Figures 3, 4, 8 and 9 show the escapement and recruitment development across coalition structures in the four- and five-player games for both stock–recruitment functions and all realizations of the stock elasticity. Escapement and recruitment levels are almost the same for both stock–recruitment functions. For stock elasticities equal to 0.3, 0.6 and 1.0, the Ricker function gives slightly higher levels of escapement and recruitment. The opposite is true when stock elasticity is equal to 0.1 for most coalition structures. Furthermore, the lower the stock elasticity, the higher the depletion

---

[t]For models which empirically estimate the stock elasticity, see Nøstbakken [2006] and Ekerhovd and Steinshamn [2016].

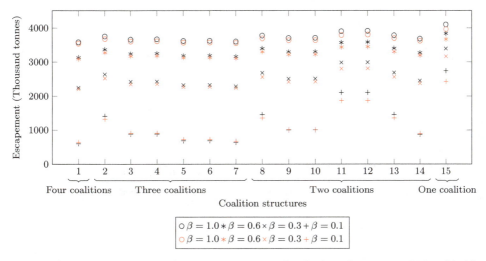

Fig. 3.   Aggregate escapement of a coalition structure for the four-player game; Ricker (black) and Beverton–Holt (red) functions; and different realizations of stock elasticity, β.

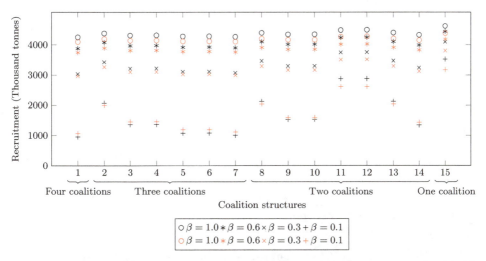

Fig. 4.   Aggregate recruitment of a coalition structure for the four-player game; Ricker (black) and Beverton–Holt (red) functions; and different realizations of stock elasticity, β.

of the stock and thus its growth. This effect is mitigated as the number of coalitions within a coalition structure decreases.

Harvest, which is defined as the difference between recruitment and escapement, is depicted in Figs. 5 and 10. It is not clear whether it increases or not as we move to more cooperative behaviors. For stock elasticities equal to 0.6 and 1.0, it decreases and for stock elasticities equal to 0.1 and 0.3, it increases. This is due to the fact that in stock–recruitment models, as escapement increases, harvest increases from zero to a maximum, i.e., the maximum sustainable yield (MSY) point, and afterwards

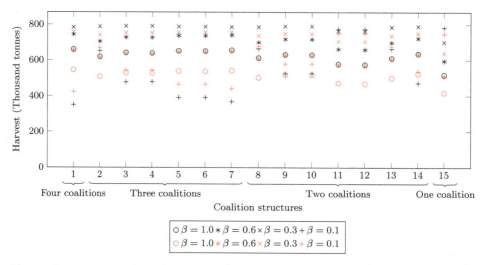

Fig. 5.   Aggregate harvest of a coalition structure for the four-player game; Ricker (black) and Beverton–Holt (red) functions; and different realizations of stock elasticity, β.

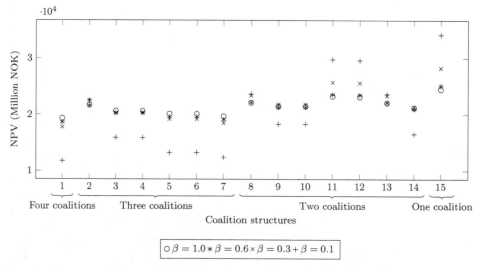

Fig. 6.   Aggregate NPV of a coalition structure for the four-player game; Ricker function; and different realizations of stock elasticity, β.

decreases back to zero, i.e., the carrying capacity point. The MSY points in our model occur at approximately 2,482 and 2,162 thousand tonnes for the Ricker and the Beverton–Holt functions, respectively. Thus, all escapement levels before (after) these points lead to an increased (decreased) growth rate and therefore harvest, which explains the change in harvest.

Third, the aggregated value of a coalition structure increases as the number of coalitions within decreases. Figures 6, 7, 11 and 12 show this increase for both

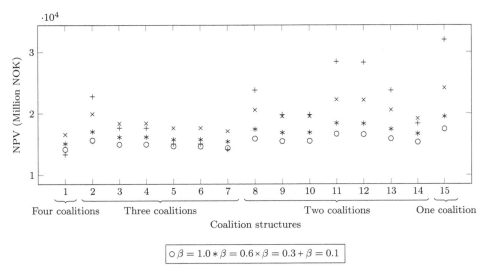

Fig. 7.   Aggregate NPV of a coalition structure for the four-player game; Beverton–Holt function; and different realizations of stock elasticity, β.

stock–recruitment functions and all realizations of the stock elasticity for both four- and five-player games. The fact that cooperative behaviors generate more value indicates that incentives for cooperation among nations exist. However, these incentives must exceed the free-riding benefits in order for cooperation to succeed.

In the four-player game, the grand coalition structure is not a stable outcome in all eight cases, that is, the sum of the free-riding payoffs of the players exceeds the payoff of the grand coalition, thus, making it impossible for any sharing rule to stabilize it. Table 11 shows the set of stable coalition structures in the four-player game for all eight cases. The set of stable coalition structures, which is the same for all cases but the Beverton–Holt with $\beta=1$, consists of all coalition structures that consist of two coalitions, where one of them is a singleton. In addition, the coalition structure representing the current management regime, i.e., $\mathrm{CS}_{11} = \{(\mathrm{EU}, \mathrm{NO}, \mathrm{FO}), (\mathrm{IS})\}$, is among the stable ones. Recall, that by stability, we mean that in the presence of some but not all sharing rules, the coalitions within a coalition structure do not have incentives to merge or split.

In the five-player game, again, the grand coalition structure cannot be sustained as an optimal outcome. The set of stable coalition structures is depicted in Table 12 for all eight cases. For both stock–recruitment functions and for stock elasticity levels equal to 0.6 and 0.3, all coalition structures consisting of two coalitions, where none of them is a singleton, are stable, namely, CS_{37} to CS_{46}.

For the two extreme stock elasticities, the set of stable coalition structures differs across the stock–recruitment functions as well as between the middle elasticities. For $\beta = 1$, $\mathrm{CS}_{37} = \{(\mathrm{EU}, \mathrm{UK}, \mathrm{NO}), (\mathrm{FO}, \mathrm{IS})\}$ is no longer stable for both stock–recruitment functions, but $\mathrm{CS}_{27} = \{(\mathrm{EU}, \mathrm{UK}, \mathrm{NO}), (\mathrm{FO}), (\mathrm{IS})\}$ becomes stable.

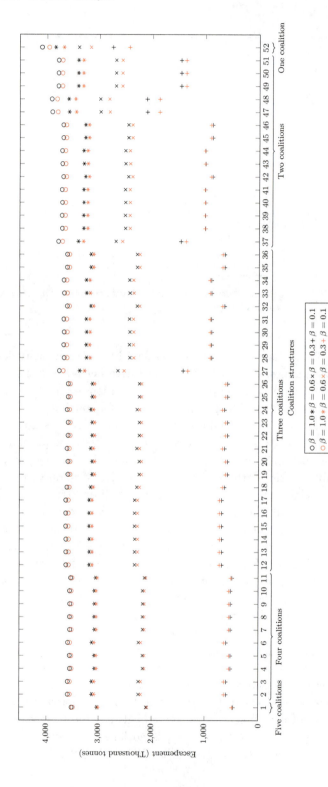

Fig. 8.   Aggregate escapement of a coalition structure for the five-player game; Ricker (black) and Beverton–Holt (red) functions; and different realizations of stock elasticity, β.

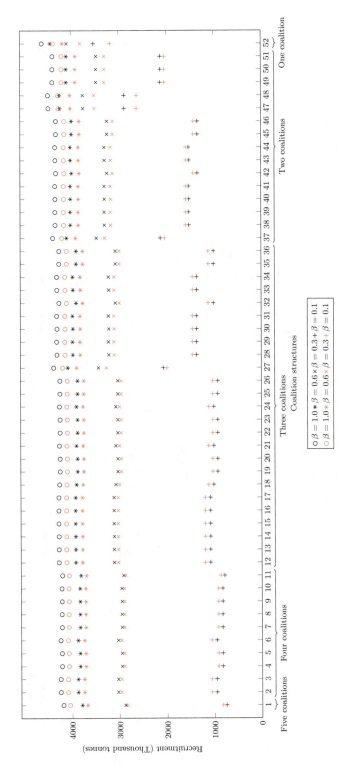

Fig. 9. Aggregate recruitment of a coalition structure for the five-player game; Ricker (black) and Beverton–Holt (red) functions; and different realizations of stock elasticity, β.

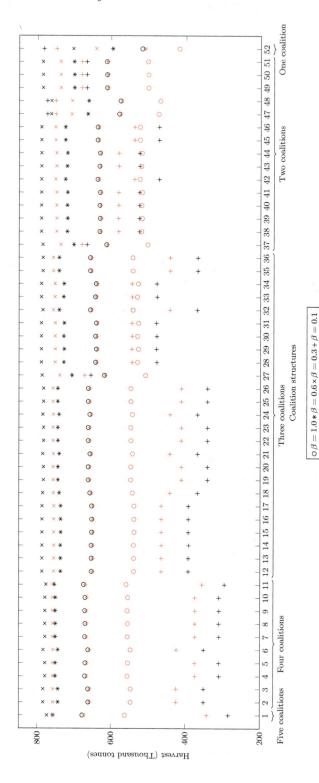

Fig. 10.   Aggregate harvest of a coalition structure for the five-player game; Ricker (black) and Beverton–Holt (red) functions; and different realizations of stock elasticity, β.

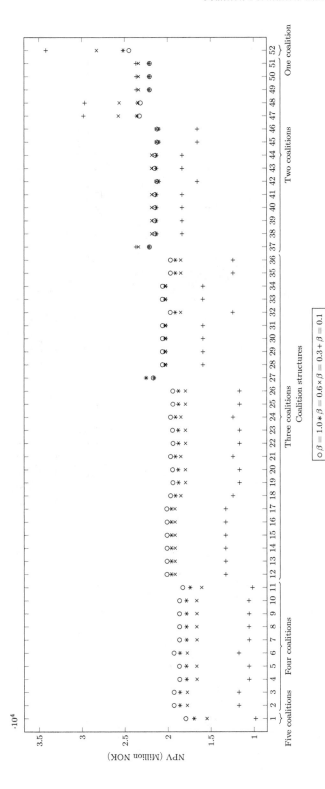

Fig. 11.   Aggregate NPV of a coalition structure for the five-player game; Ricker function; and different realizations of stock elasticity, β.

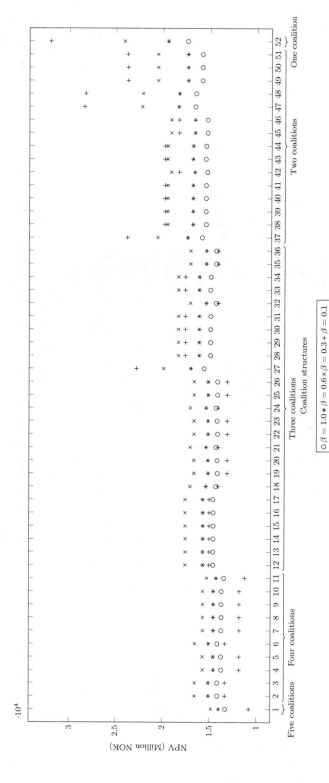

Fig. 12.   Aggregate NPV of a coalition structure for the five-player game; Beverton–Holt function; and different realizations of stock elasticity, β.

Table 11. Nash equilibria coalition structures for the four-player game for the Ricker and Beverton–Holt stock–recruitment relationships and different realizations of stock elasticity.

Ricker				Beverton–Holt			
$\beta = 1$	$\beta = 0.6$	$\beta = 0.3$	$\beta = 0.1$	$\beta = 1$	$\beta = 0.6$	$\beta = 0.3$	$\beta = 0.1$
11	11	11	11	11	11	11	11
12	12	12	12	12	12	12	12
13	13	13	13	13	13	13	13
14	14	14	14		14	14	14

Note: See Table 2 for which coalition structures the indices refer to.

Table 12. Nash equilibria coalition structures for the five-player game for the Ricker and Beverton–Holt stock–recruitment relationships and different realizations of stock elasticity.

Ricker				Beverton–Holt			
$\beta = 1$	$\beta = 0.6$	$\beta = 0.3$	$\beta = 0.1$	$\beta = 1$	$\beta = 0.6$	$\beta = 0.3$	$\beta = 0.1$
27	37	37	37	27	37	37	37
38	38	38	38	38	38	38	38
39	39	39	39	39	39	39	39
40	40	40	40	40	40	40	40
41	41	41	41	41	41	41	41
42	42	42	42	43	42	42	42
43	43	43	43	44	43	43	43
44	44	44	44		44	44	44
45	45	45	45		45	45	45
46	46	46	46		46	46	46
			47				49
			48				50
			49				51
			50				
			51				

Note: See Table 4 for which coalition structures the indices refer to.

Thus, according to our results, if the mackerel fishery is uniformly distributed, then Iceland and the Faroe Islands do not have incentives to cooperate with each other any more, given that the remaining nations cooperate. In addition to CS$_{37}$, coalition structures 42, 45 and 46 are no longer stable when $\beta = 1$ for the Beverton–Holt case. These coalition structures consist of two coalitions where in one coalition, a major player (EU, UK or NO) cooperate with the two minors (FO and IS) and in the other, the remaining major players cooperate together.

For $\beta = 0.1$, coalitions structures 47 to 51 also become stable for the Ricker case, but only coalition structures 49, 50 and 51 for the Beverton–Holt case. These coalition structures consist of two coalitions, where one of them is a singleton.

Compared to the four-player game, where the set of stable coalition structures remains the same in almost all the cases, in the five-player game, stability of some coalition structures is sensitive to the stock–recruitment function and the stock elasticity parameter. Interesting enough, the stable coalition structures in the four-player game are no longer stable for most of the cases in the five-player game. Recall that the singleton coalition of (EU) in the four-player game is equivalent to the coalition of (EU, UK) in the five-player game. The five-player game coalition structures, which are equivalent to the stables ones in the four-player game are $CS_{11} \equiv CS_{47}$, $CS_{12} \equiv CS_{48}$, $CS_{13} \equiv CS_{49}$ and $CS_{14} \equiv CS_{46}$.

The current management regime, i.e., CS_{11} and CS_{47} in the four- and five-player games, respectively, is stable only in one case in the five-player game (Ricker; $\beta = 0.1$), in contrast to the four-player game, where it is stable in all eight cases. This is also true for CS_{12} or $CS_{48} = \{(EU, UK, NO, IS), (FO)\}$. Coalition structure 13, i.e., $CS_{49} = \{(EU, UK, FO, IS), (NO)\}$ in the five-player game, occurs only when $\beta = 0.1$ irrespective of the stock–recruitment function. The only four-player game coalition structure that remains stable in all but one (Beverton–Holt; $\beta = 1$) of the five-player game cases is CS_{14}, i.e., $CS_{46} = \{(NO, FO, IS), (EU, UK)\}$.

On the other hand, some stable coalition structures in the five-player game are not stable in the four-player game, namely, CS_{27}, CS_{37}, CS_{38} and CS_{39}. The common property of these coalition structures is that the EU and the UK belong to the same coalition. This change in stability between the two games is due to the relative magnitude of the free-riding payoff of the EU in the four-player game and the sum of the free-riding payoffs of the EU and the UK together in the five-player game. In general, the smaller the free-riding payoff, the higher the chance that the external stability condition will not be satisfied, i.e., a coalition will have incentives to merge with another coalition. In the four-player game, the free-riding payoff of the EU is low enough to make it profitable for other coalitions to want to merge with the coalition that it belongs, thus the external stability condition does not hold and therefore, the respective coalition structures in the four-player game, i.e., CS_2, CS_8, CS_9 and CS_{10}, are not stable. In the five-player game, however, the free-riding payoffs of the EU and the UK together are high enough that is no longer profitable for other coalitions to merge with the coalition that they belong, and therefore making these coalition structures stable.

Having determined the set of stable coalition structures, we now ask ourselves how likely they are to form in reality. From the four stable coalition structures of the four-player game, we know that only CS_{12} has been formed in the mackerel fishery. So, from the stable coalition structures of the five-player game, which ones are likely to occur in reality? Or, to put it another way, which ones are unlikely to occur? In what follows, we discuss which coalitions we believe are likely or not to occur post-Brexit based on our intuition of the relations between all five parties.

First, is the cooperation among the EU, the UK, Norway and the Faroe Islands as defined by the current five-year management plan, likely to continue after the conclusion of the Brexit's negotiations? The agreement itself will cease to exist since the UK will no longer be represented by the EU and therefore must sign its own agreements. In general, after Brexit, the UK will have to negotiate fisheries agreements with other coastal states as well as with the EU. Regarding, straddling and highly migratory fish stocks, such as mackerel, international law requires that all interest parties must cooperate, directly or through RFMOs, that is, the NEAFC in case of Atlantic mackerel. Thus, one possibility is that post-Brexit relationships in the mackerel fishery will be similar or close to the existing ones. Of course, the relative TAC shares of the EU, the UK and the other parties may change depending on the outcome of the negotiations.

After the conclusion of Brexit, the UK will have sovereign control over the resources in its EEZ, and therefore, the principle of equal access[u] will cease to apply in British waters and access will now be determined by the criteria set out in UNCLOS. In other words, access in British fisheries will no longer be regulated by European law but by international law. At the moment, EU vessels harvesting in UK's EEZ catch more fish inside British waters than UK vessels catch in the Union's EEZ. Particularly, in 2015, EU vessels caught 683,000 tonnes, i.e., 484 million GBP in revenue, in UK waters, whereas UK vessels caught 111,000 tonnes, i.e., 114 million GBP in revenue, in European waters [Brexit White Paper, 2017].[v] Regarding mackerel, the vast majority of catches taken by the EU occurs within the UK EEZ [Doering *et al.*, 2017]. According to a recent study by Le Gallic *et al.* [2017], if the UK prohibits the EU fleet from accessing fishing stocks within its EEZ, it will cause great loss of revenues for these vessels. Even if the EU redistributes quotas inside its EEZ, it is unlikely that it will compensate for the loss of such important fishing grounds.

From the above, it seems like the UK has all the bargaining power when it comes to negotiating a post-Brexit agreement with the EU. However, this is not true. The UK depends primarily on the EU market for its fishery exports. For the period 2001–2016, 68% on average of the total value generated by fishery exports came from the EU, i.e., 1204 million EUR. As far as mackerel is concerned, since 2010, on average, more than 60% of UK's annual mackerel exports go to the EU market, generating on average 70 million EUR.[w] Thus, the EU, which is an important trading partner of the UK when it comes to fishery products, might introduce trade barriers, if its access to British waters is limited or denied.

Furthermore, is it possible that cooperation between the EU and Norway will fall apart post-Brexit? Europe and Norway have a long tradition of positive relations, not only in fisheries but across many sectors, and it is doubtful that Norway will act unilaterally, especially if EU and UK agree to cooperate after UK's withdrawal. The bilateral agreement between the EU and Norway covering the North Sea and the

---

[u]Fishing vessels registered in the EU fishing fleet register have equal access to all European waters and resources that are managed under the CFP.

[v]Provisional Statistics — UK Fleet Landings from other EU Member States waters: 2015, Marine Management Organization, February 2017. These figures do not include fish caught by third country vessels, for example, Norway, in UK waters, or fish caught by UK fisherman in third country waters.

[w]The data one obtained from the European Market Observatory for Fisheries and Aquaculture Products (EUMOFA).

Atlantic is the Union's most important international fisheries agreement in terms of both the exchange of fishing opportunities and joint fisheries management measures [Doering *et al.*, 2017].[x] Although this agreement is not related to the management of the mackerel stock,[y] a possible conflict between the EU and Norway regarding the management of mackerel could undermine it. In addition, access of Norwegian fishery products to EU's internal market may also be undermined. As far as Norway is concerned, Brexit is going to make fishery resources in EU waters less attractive, given that when it comes to quota exchanges, stocks in UK waters are more important for Norway than those in EU waters [Sobrino Heredia, 2017]. Still, the fact that the EU is a very important trading partner for Norway gives both players more or less equal bargaining power when it comes to negotiating their post-Brexit relationship. The value of Norwegian mackerel exports to the EU excluding the UK was on average 475 million NOK for the period 2007–2016, whereas to the UK for the same period, was valued at 62 million NOK on average.[z,aa] Thus, making the EU a more significant trading partner for Norway regarding mackerel.

Finally, how are Iceland and the Faroe Islands likely to behave post-Brexit? There has been no indication so far that Iceland is willing to cooperate with the remaining states to jointly determine the mackerel quota. Given its history of disputes, it is highly unlikely that it will cooperate unless it is allowed to maintain its quota or offered something else in exchange for reducing it. However, Iceland may be interested to strengthen its relations with an independent UK and perhaps willing to compromise in the prospect of a future agreement with the UK. As far as the Faroe Islands are concerned, they will most probably keep cooperating with the EU and Norway, given that their post-Brexit quota is close to the current one. Like Iceland, they may also be interested to strengthen their relations with the UK. In general, the UK will have to work closely with Norway, Iceland and the Faroe Islands in order to ensure access in one another's waters.

6. Conclusion

In this chapter, we analyze how cooperation is likely to occur in the Northeast Atlantic mackerel fishery after the Brexit negotiations are concluded. To do so, we have considered two games: a four-player game, which treats the EU and the UK as one coalition acting together, and a five-player game where the UK is a distinct player acting on its behalf. For our bioeconomic model of the mackerel fishery,

---

[x]The agreement was first enforced on 16 June 1981 for a 10-year period, after that has been tacitly renewed for successive six-year periods. The last renewal tool place in 2015.
[y]The stocks that this agreement refers to are cod, plaice and haddock.
[z]The data are obtained from Statistics Norway, Table: 09283: Exports of fish, by country/trade region/continent.
[aa]Value of Norwegian fish, crustaceous animals and mollusc exported to the EU excluding the UK was on average 34,637 million NOK. That is, 90% higher compared to the respective exports in the UK, which were amounted to 3073 million NOK on average.

we assume a density-dependent stock–recruitment relationship. Both games are solved multiple times for different stock–recruitment functions and levels of the stock elasticity.

We find that positive stock externalities are indeed present in both games since outsiders are better off when a merger between coalitions occurs. The members of a coalition are able to reduce their fishing cost by internalizing the positive externality, thus increasing the stock level. This allows outsiders to free-ride on them by benefiting from the increase in the stock. As expected, escapement and recruitment as well as the aggregated value a coalition structure generates increase as the number of coalitions within a coalition structure decreases. That is, cooperation leads to higher profits as well as higher stock preservation. However, in order for cooperation to be achieved, the free-riding payoffs of the cooperating nations must not exceed their aggregate coalitional payoff.

In both games, the grand coalition cannot be sustained as an optimal outcome for all scenarios evaluated. The current management regime, however, is found to be a stable outcome in all eight cases of the four-player game, but only in one case of the five-player game. This is also true for all the remaining, but one, stable coalition structures of the four-player game. In addition, some nonstable coalition structures in the four-player game become stable in the five-player game. This occurs because the free-riding payoff of the EU in the four-player game is less than the sum of the free-riding payoffs of the EU and the UK in the five-player game, and therefore making the external stability condition for those coalition structures to only hold in the five-player game. Moreover, in the four-player game, the set of stable coalition structures remains the same in almost all cases, whereas in the five-player game, stability depends on the stock–recruitment function as well as the magnitude of the stock elasticity.

As far as the future of the mackerel fishery is concerned, we believe that the EU and Norway will keep cooperating post-Brexit. In the event that the UK restricts access to the EU's fleet within its waters, then perhaps, Norway will have to give a percentage of its quota to the EU in order to maintain access to the European market. In case of a "hard" Brexit, i.e., no compromises between the EU and the UK during the negotiations, the UK will most likely set its mackerel quota unilaterally. It goes without saying that if this happens, then the pressure on the mackerel stock will increase even more, especially if Iceland continues not to cooperate. However, both the EU and Norway could respond harshly by introducing trade sanctions, as they have already done to Icelandic and Faroese catches in 2013. If a "soft" Brexit occurs, then relationships in the mackerel fishery may be similar or close to existing ones, but the relative shares of the TAC may change depending on the outcome of the negotiations.

A natural extension of the current research is to consider issue-linkage, i.e., link the current cooperative arrangement (exploitation of the mackerel fishery) with a second one, for example, cooperation over the mackerel trade. Then, perhaps, the set of stable coalition structures could indicate that cooperation in the mackerel fishery in the presence of externalities is strengthened due to the threat of sanctions in the mackerel trade.

Acknowledgments

The authors are grateful to Leif K. Sandal, Sturla F. Kvamsdal and others for valuable comments that led to the improvement of the current work. Financial support from the Norwegian Research Council through the MESSAGE Project (Grant No. 255530/E40) is gratefully acknowledged.

Appendix A

A.1. *Proof of noncooperative "golden-rule"*

The logic of the proof is similar to the one presented by Clark [2010, p. 91]. The profit of coalition i in period t is

$$\Pi_i(R_{it}, S_{it}) = p(R_{it} - S_{it}) - \int_{S_{it}}^{R_{it}} c_i(x)dx = \int_{S_{it}}^{R_{it}} [p - c_i(x)]dx,$$

where $\pi_i(x) = p - c_i(x)$ is the marginal profit of coalition i. Let $\phi_i(x)$ be the antiderivative of $\pi_i(x)$, then we can express the profit of coalition i as

$$\Pi_i(R_{it}, S_{it}) = \phi_i(R_{it}) - \phi_i(S_{it}).$$

Therefore, the NPV of coalition i becomes

$$V_i = \sum_{t=0}^{\infty} \gamma^t [\phi_i(R_{it}) - \phi_i(S_{it})].$$

Substituting for the recruitment share of coalition i, $R_{it} = \theta_i R_t$, and for the stock–recruitment relationship, $R_t = F(S_{t-1})$ for $t \geq 1$, the first term of the NPV expression yields

$$\sum_{t=0}^{\infty} \gamma^t \phi_i(R_{it}) = \phi_i(R_{i0}) + \sum_{t=1}^{\infty} \gamma^t \phi_i[\theta_i F(S_{t-1})]$$

$$= \phi_i(R_{i0}) + \sum_{t=0}^{\infty} \gamma^{t+1} \phi_i[\theta_i F(S_t)].$$

Finally, substituting the above term in the NPV of coalition i, we obtain

$$V_i = \phi_i(R_{i0}) + \sum_{t=0}^{\infty} \gamma^{t+1} \phi_i[\theta_i F(S_t)] - \sum_{t=0}^{\infty} \gamma^t \phi_i(S_{it})$$

$$= \phi_i(R_{i0}) + \sum_{t=0}^{\infty} \gamma^t [\gamma \phi_i[\theta_i F(S_t)] - \phi_i(S_{it})].$$

Now, coalition i is enabled to set out the optimal escapement strategy, given the escapement strategies of the other coalitions, namely, coalition i to choose the escapement level S_{it} for each time period $t = 0, 1, 2, \ldots, \infty$ by solving the following maximization problem:

$$\underset{S_{it}}{\text{maximize}} \ \gamma\phi_i[\theta_i F(S_t)] - \phi_i(S_{it})$$

$$\text{subject to} \ S_t = S_{it} + \sum_{j=1}^{m-1} S_{jt}, \quad i \neq j.$$

Substituting for S_t in the objective function and taking the first-order condition we get

$$\left[\gamma\phi_i \left[\theta_i F \left(S_{it} + \sum_{j=1}^{m-1} S_{jt} \right) \right] - \phi_i(S_{it}) \right]'$$

$$= \gamma\phi_i' \left[\theta_i F \left(S_{it} + \sum_{j=1}^{m-1} S_{jt} \right) \right] \theta_i \frac{dF(S_{it} + \sum_{j=1}^{m-1} S_{jt})}{dS_{it}} - \phi_i'(S_{it})$$

$$= \gamma\pi_i \left[\theta_i F \left(S_{it} + \sum_{j=1}^{m-1} S_{jt} \right) \right] \theta_i \frac{dF(S_{it} + \sum_{j=1}^{m-1} S_{jt})}{dS_{it}} - \pi_i(S_{it}) = 0. \quad \text{(A.1)}$$

It can be shown that the derivative of the stock–recruitment function $F(S)$ with respect to coalition's i escapement, S_i is equivalent to the derivative of $F(S)$ with respect to the aggregate escapement S. The proof makes use of the chain rule and the fact that the derivative of the aggregate escapement with respect to coalition's i escapement is one, i.e.,

$$\frac{dS}{dS_i} = \frac{d(S_i + \sum_{j=1}^{m-1} S_j)}{dS_i} = 1.$$

Thus,

$$\frac{dF(S_i + \sum_{j=1}^{m-1} S_j)}{dS_i} = \frac{dF(S_i + \sum_{j=1}^{m-1} S_j)}{d(S_i + \sum_{j=1}^{m-1} S_j)} \frac{d(S_i + \sum_{j=1}^{m-1} S_j)}{dS_i} = \frac{dF(S)}{dS} = F'(S).$$

Let $S_{it} = S_i^*$, solve (1), then we can rewrite it as follows:

$$\pi_i(S_i^*) = \gamma\theta_i F'(S)\pi_i[\theta_i F(S)],$$

where $S = S_i^* + \sum_{j=1}^{m-1} S_j$ is the aggregate escapement and it depends on the optimal escapement strategy of coalition i and the escapement strategies of the other coalitions j.

A.2. *Illustration of coalition structure stability concepts*

Consider a three-player coalition formation game of the class studied in this chapter. Let $N = \{a, b, c\}$ be the set of players. Table A.1 depicts the payoffs of all embedded coalitions in this game. The property of superadditivity holds for the entire game,

Table A.1.   Embedded coalition payoffs.

CS_k	$V(C_1, CS_k)$	$V(C_2, CS_k)$	$V(C_3, CS_k)$
$\{a, b, c\}$	2	4	1
$\{ab, c\}$	7	2	
$\{ac, b\}$	4	5	
$\{bc, a\}$	6	4	
$\{abc\}$	10		

i.e., the joint payoff of two embedded coalitions belonging in the same coalition structure is at least as high as their individual payoffs.

Suppose we want to test if coalition structure $\{ab, c\}$ is stable. According to Sec. 3.2, a coalition structure is stable if all of its embedded coalitions are potentially internal and external stable. The tested coalition structure consists of two coalitions: (ab) and (c).

Let us test for potential internal stability first. Coalition (c) is a singleton and therefore is always internal stable. In order for coalition (ab) to be potentially internal stable, the payoff of (ab) given coalition structure $\{ab, c\}$ must be greater or equal to the free-riding payoffs of its members, ceteris paribus. The free-riding payoffs are determined as follows. Consider player a first, if player a leaves coalition (ab), then the new coalition structure, ceteris paribus, is $\{a, b, c\}$. Similarly, if player b leaves coalition (ab), then the new coalition structure, ceteris paribus, is $\{a, b, c\}$. Note that the new coalition structures are the same in both deviations; this is not always the case as we will see in the next case. Having determined the new coalition structures, we can now compare the payoffs and test if coalition (ab) is potentially internal stable:

$$V(ab, \{ab, c\}) \geq V(a, \{a, b, c\}) + V(b, \{a, b, c\}) \Rightarrow 7 \geq 2 + 4 = 6.$$

Since the above inequality holds, we can conclude that coalition (ab) is potentially internal stable. Seeing that both coalitions (ab) and (c) are potentially internal stable, we can conclude that coalition structure $\{ab, c\}$ is potentially internal stable. We move on to test for potential internal stability.

Coalition structure $\{ab, c\}$ consists of only one pair of embedded coalitions, i.e., $[(ab, \{ab, c\}), (c, \{ab, c\})]$. In order for $\{ab, c\}$ to be external stable, at least one of the two embedded coalitions should not have incentives to merge. Let us start with (ab), if (ab) merges with (c), then the new coalition structure, ceteris paribus, will be $\{abc\}$ but player c must receive at least her free-riding payoff which occurs if she deviates from the new coalition (abc). If player c leaves (abc), the new coalition structure, ceteris paribus, will be $\{ab, c\}$. Thus, the potential external stability condition for coalition (ab) with respect to coalition (c) requires the following:

$$V(ab, \{ab, c\}) \geq V(abc, \{abc\}) - V(c, \{ab, c\}) \Rightarrow 7 \geq 10 - 2 = 8.$$

Since the above inequality does not hold, we can conclude that coalition (ab) does have incentives to merge with coalition (c) and therefore, (c) is not potentially external stable with respect to (ab). However, coalition structure $\{ab, c\}$ may still be external stable as long as coalition (c) is better off without the mergence. If (c) merges with (ab), then the new coalition structure, ceteris paribus, will be $\{abc\}$ but players a and b must receive at least their free-riding payoffs. If player a leaves (abc), the new coalition structure, ceteris paribus, will be $\{bc, a\}$. Similarly, if player b leaves (abc), the new coalition structure, ceteris paribus, will be $\{ac, b\}$. Thus, the potential external stability condition for coalition (c) with respect to coalition (ab) requires the following:

$$V(c, \{ab, c\}) \geq V(abc, \{abc\}) - V(a, \{bc, a\}) - V(b, \{ac, b\}) \Rightarrow 2 \geq 10 - 4 - 5 = 1.$$

Since the above inequality holds, coalition (c) does not have incentives to merge with coalition (ab) and therefore, (ab) is potentially external stable with respect to coalition (c). Since $[(ab, \{ab, c\}), (c, \{ab, c\})]$ is the only embedded coalition pair of coalition structure $\{ab, c\}$ and $(c, \{ab, c\})$ is not willing to merge, we can conclude that coalition structure $\{ab, c\}$ is potentially external stable. Because coalition structure $\{ab, c\}$ is both potentially internal and external stable, we can conclude that $\{ab, c\}$ is a stable coalition structure.

Following the same procedure, it can be showed that coalition structures $\{ac, b\}$ and $\{bc, a\}$ are also stable. The singleton coalition structure $\{a, b, c\}$ is not potentially external stable since all the players have incentives to form a coalition with at least one more player. The grand coalition structure $\{abc\}$ is not potentially internal stable since the sum of the free-riding payoffs of its members exceeds the payoff of the grand coalition, i.e.,

$$V(abc, \{abc\}) \geq V(a, \{bc, a\}) + V(b, \{ac, b\}) + V(c, \{ab, c\}) \Rightarrow 10 \geq 4 + 5 + 2 = 11.$$

This also verifies the fact that superadditive games with externalities cannot necessarily sustain the grand coalition as a stable outcome.

References

Agnello, R. J. and Donnelley, L. P. [1976] Externalities and property rights in the fisheries, *Land Econ.* **52**, 518–529.

Bailey, M., Sumaila, U. R. and Lindroos, M. [2010] Application of game theory to fisheries over three decades, *Fish. Res.* **102**, 1–8.

Balton, D. A. and Koehler, H. R. [2006] Reviewing the United Nations fish stocks treaty, *Sustain. Dev. Law Policy* **7**, 5–9.

Beverton, R. J. H. and Holt, S. J. [1957] *On the Dynamics of Exploited Fish Populations*, Fisheries Investigations Series, Vol. 2 (Ministry of Agriculture, London), p. 19.

Bjørndal, T. [1987] Production economics and optimal stock size in a North Atlantic fishery, *Scand. J. Econ.* **89**, 145–164.

Bjørndal, T. and Munro, G. R. [2003] The management of high seas fisheries, in *The International Yearbook of Environmental and Resource Economics 2003/2004*, eds. Folmer, H. and Tietenberg, T. (Elgar, Cheltenham, UK), pp. 1–35.

Bloch, F. [1996] Sequential formation of coalitions in games with externalities and fixed payoff division, *Games Econ. Behav.* **14**, 90–123.

Brexit White Paper [2017] The United Kingdom's exit from and new partnership with the European Union, Department for Exiting the European Union and The Rt Hon David Davis MP, February 2017, Available at: https://www.gov.uk/government/publications/the- united- kingdoms- exit- from- and- new- partnership- with-the-euro pean-union-white-paper.

Churchill, R. R. and Owen, D. [2010] *The EU Common Fisheries Policy: Law and Practice* (Oxford University Press, Oxford), p. 640.

Clark, C. W. [1972] The dynamics of commercially exploited natural animal populations, *Math. Biosci.* **13**, 149–164.

Clark, C. W. [1973] Profit maximization and the extinction of animal species, *J. Political Econ.* **81**, 950–961.

Clark, C. W. [2010] *Mathematical Bioeconomics: The Mathematics of Conservation* (Wiley, New Jersey).

Clark, C. W. and Munro, G. R. [1975] The economics of fishing and modern capital theory: A simplified approach, *J. Environ. Econ. Manage.* **2**, 92–106.

d'Aspremont, C., Jacquemin, A., Gabszewicz, J. J. and Weymark, J. A. [1983] On the stability of collusive price leadership, *Can. J. Econ.* **16**, 17–25.

De Clippel, G. and Serrano, R. [2008] Marginal contributions and externalities in the value, *Econometrica* **76**, 1413–1436.

Doering, R., Kempf, A., Belschner, T., Berkenhagen, J., Bernreuther, M., Hentsch, S., Kraus, G., Raetz, H.-J., Rohlf, N., Simons, S., Stransky, C. and Ulleweit, J. [2017] Research for PECH committee — Brexit consequences for the common fisheries policy-resources and fisheries: A case study, European Parliament, Policy Department for Structural and Cohesion Policies, Brussels.

Ekerhovd, N. A. and Steinshamn, S. I. [2016] Economic benefits of multi-species management: The pelagic fisheries in the Northeast Atlantic, *Mar. Resour. Econ.* **31**, 193–210.

Ellefsen, H. [2013] The stability of fishing agreements with entry: The northeast atlantic mackerel, *Strat. Behav. Environ.* **3**, 67–95.

Eyckmans, J. and Finus, M. [2004] An almost ideal sharing scheme for coalition games with externalities, FEEM Working Paper No. 155.04, Available at SSRN: https://ssrn.com/abstract=643641.

FAO [2011] Article 7-Fisheries Management, in Code of conduct for responsible fisheries, Special edition, Rome.

Gulland, J. A. [1980] Some problems of the management of shared stocks, FAO Fisheries Technical Paper No. 206, Rome.

Hannesson, R. [2011] Game theory and fisheries, *Annu. Rev. Resour. Econ.* **3**, 181–202.

Hannesson, R. [2012] Sharing the Northeast Atlantic mackerel, *ICES J. Mar. Sci.* **70**, 256–269.

Hannesson, R. [2013] Sharing a migrating fish stock, *Mar. Resour. Econ.* **28**, 1–17.

ICES [2014] Mackerel in the Northeast Atlantic (combined Southern, Western, and North Sea spawning components), ICES Advice 2014, Book 9, Section 3.17b.

ICES [2016a] Mackerel (Scomber scombrus) in subareas 1–7 and 14, and in divisions 8.a–e and 9.a (Northeast Atlantic), ICES Advice 2016, Book 9, Section 3.39.

ICES [2016b] Stock Annex: Mackerel (Scomber scombrus) in subareas 1–7 and 14 and divisions 8.a–e, 9.a (the Northeast Atlantic and adjacent waters), ICES, Available at: http://www.ices.dk/sites/pub/Publication%20Reports/Stock%20Annexes/2016/mac-nea_SA.pdf.

ICES CM [1996] Report of the working group on the assessment of mackerel, horse mackerel, sardine and anchovy, Copenhagen, 10–19 October 1995. ICES Doc. C.M. 1996/Assess:7.

Iles, T. C. [1994] A review of stock–recruitment relationships with reference to flatfish populations, *Neth. J. Sea Res.* **32**, 399–420.

Jaquette, D. L. [1974] A discrete time population control model with setup cost, *Oper. Res.* **22**, 298–303.

Jensen, F., Frost, H., Thøgersen, T., Andersen, P. and Andersen, J. L. [2015] Game theory and fish wars: The case of the Northeast Atlantic mackerel fishery, *Fish. Res.* **172**, 7–16.

Kaitala, V. and Munro, G. R. [1993] The management of high seas fisheries, *Mar. Resour. Econ.* **8**, 313–329.

Kaitala, V. and Lindroos, M. [2007] Game theoretic applications to fisheries, in *Handbook of Operations Research in Natural Resources*, Chapter 11, eds. Weintraub, A., Romero, C., Bjørndal, T. and Epstein, R. (Springer), pp. 200–215.

Kennedy, J. [2003] Scope for efficient multinational exploitation of Nort-East Atlantic mackerel, *Mar. Resour. Econ.* **18**, 55–60.

Kronbak, L. G. and Lindroos, M. [2007] Sharing rules and stability in coalition games with externalities, *Mar. Resour. Econ.* **22**, 137–154.

Le Gallic, B., Mardle, S. and Metz, S. [2017] Research for PECH committee — Common fisheries policy and brexit — Trade and economic related issues, European Parliament, Policy Department for Structural and Cohesion Policies, Brussels.

Lindroos, M., Kaitala, V. and Kronbak, L. G. [2007] Coalition games in fisheries Economics, in *Advances in Fisheries Economics: Festschrift in Honour of Professor Gordon Munro*, Chapter 11, eds. Bjørndal, T., Gordon, D., Arnason, R. and Sumaila, U. R. (Blackwell), pp. 184–195.

Liu, X., Lindroos, M. and Sandal, L. [2016] Sharing a fish stock when distribution and harvest costs are density dependent, *Environ. Resour. Econ.* **63**, 665–686.

Macho-Stadler, I., Perez-Castrillo, D. and Wettstein, D. [2007] Sharing the surplus: An extension of the Shapley value for environments with externalities, *J. Econ. Theory* **135**, 339–356.

Martimort, D. and Stole, L. [2012] Representing equilibrium aggregates in aggregate games with applications to common agency, *Games Econ. Behav.* **76**, 753–772.

McKelvey, R. W., Sandal, L. K. and Steinshamn, S. I. [2002] Fish wars on the high seas: A straddling stock competitive model, *Int. Game Theor. Rev.* **4**, 53–69.

Munro, G. R. [2003] On the management of shared fish stocks, Papers presented at the Norway-FAO expert consultation on the management of shared fish stocks, FAO Fisheries Report, No. 695, pp. 2–29.

Munro, G. R. [2008] Game theory and the development of resource management policy: The case of international fisheries, *Environ. Dev. Econ.* **14**, 7–27.

Norwegian Directorate of Fisheries [2015] Profitability survey on the Norwegian fishing fleet, Norwegian Directorate of Fisheries, Bergen, Norway, Available at: http://www.fiskeridir.no/Yrkesfiske/Statistikk-yrkesfiske/Statistiske-publikasjoner/Loenns omhetsundersoekelse-for-fiskefartoey.

Nøstbakken, L. [2006] Cost structure and capacity in Norwegian pelagic fisheries, *Appl. Econ.* **38**, 1877–1887.

Pavlov, A. K., Tverberg, V., Ivanov, B. V., Nilsen, F., Falk-Petersen, S. and Granskog, M. A. [2013] Warming of Atlantic water in two west Spitsbergen fjords over the last century (1912–2009), *Polar Res.* **32**, 11206.

Pham Do, K. and Folmer, H. [2003] International fisheries agreements: The feasibility and impacts of partial cooperation (Center Discussion Paper; Vol. 2003–2052). Microeconomics, Tilburg.

Pham Do, K. and Norde, H. [2007] The Shapley value for partition function form games, *Int. Game Theor. Rev.* **9**, 353–360.

Pintassilgo, P. [2003] A coalition approach to the management of high seas fisheries in the presence of externalities, *Nat. Resour. Model.* **16**, 175–197.

Pintassilgo, P., Finus, M. and Lindroos, M. [2010] Stability and success of regional fisheries management organisations, *Environ. Resour. Econ.* **46**, 377–402.

Ray, D. and Vohra, R. [1999] A theory of endogenous coalition structures, *Games Econ. Behav.* **26**, 286–336.

Reed, W. J. [1974] A stochastic model for the economic management of a renewable resource animal, *Math. Biosci.* **22**, 313–337.

Ricker, W. E. [1954] Stock and recruitment, *J. Fish. Res. Board Canada* **11**, 559–623.

Sobrino Heredia, J. M. [2017] Research for PECH committee — Common fisheries policy and BREXIT — Legal framework for governance, European Parliament, Policy Department for Structural and Cohesion Policies, Brussels.

Smith, V. L. [1969] On models of commercial fishing, *J. Political Econ.* **77**, 181–198.

Thrall, R. M. and Lucas, W. F. [1963] N–person games in partition function form, *Nav. Res. Logist.* **10**, 281–298.

United Nations [1982] United Nations convention on the law of the sea, UN Doc. A/Conf.62/122.

United Nations [1995] United Nations conference on straddling fish stocks and highly migratory fish stocks. Agreement for the implementation of the provisions of the United Nations convention on the law of the sea of 10 December 1982 relating to the conservation and management of straddling fish stocks and highly migratory fish stocks, UN Doc. A/Conf./164/37.

Yi, S. S. [1997] Stable coalition structures with externalities, *Games Econ. Behav.* **20**, 201–237.

Yi, S. S. and Shin, H. [1995] Endogenous formation of coalitions in oligopoly, Dartmouth College Department of Economics WP No. 95–2.

Chapter 17

Stable Marketing Cooperation in a Differential Game for an Oligopoly

Mario A. García-Meza*

Facultad de Economía, Contaduría y Administración
Universidad Juárez del Estado de Durango
Fanny Anitua y Priv. de la Loza s/n
Durango, Durango 34000, México

mario.agm@ujed.mx

Ekaterina Viktorovna Gromova

Faculty of Applied Mathematics and Control Processes
St. Petersburg State University, Universitetskii Prospekt 35
Petergof, Saint Petersburg 198504, Russia

e.v.gromova@spbu.ru

José Daniel López-Barrientos

Facultad de Ciencias Actuariales
Universidad Anáhuac México, Campus Norte
Av. Universidad Anáhuac, 46, Col. Lomas Anáhuac
Huixquilucan, Edo. de México 52786, México

daniel.lopez@anahuac.mx

In this chapter, we develop a dynamic model of an oligopoly playing an advertising game of goodwill accumulation with random terminal time. The goal is to find a cooperative solution that is time-consistent, considering a dynamic accumulation of goodwill with depreciation for a finite number of firms.

Keywords: Differential games; stability of cooperation; random terminal time; advertising cooperation.

1. Introduction

Advertising is an essential feature of economic activity. Most of the businesses in an economy are constantly using some kind of advertising tools as a mean to reach

---

*Corresponding author.

their consumers and compete within themselves in the market. A way we can think they do this is by making an investment on their reputation as a brand, which results in higher sales volumes. In this view, we can think of a firm's *goodwill* as an asset that the firms can provide themselves in order to increase their profitability. As any asset, goodwill can be subject to some kind of depreciation, because of competition within the firms or just a natural tendency of clients to forget about the brand. The insight of a model of advertising with these characteristics was first developed by Nerlove and Arrow [1962] in their seminal paper. Their work is the first modern development of a model focused on advertising with the use of a stock of goodwill as a state variable. There, the authors provide the optimal price policy to be used under such circumstances.

A natural extension of this model is the competition of an oligopoly with advertising policy as a mean to increase competitive profits. That kind of competition can be found in the model by Fershtman [1984], along with a steady state for the game theoretical model.

In an advertising competition, we can consider that the firms are looking forward to use the policy that optimizes their profits, given that their competitors have the same problem with similar incentives. That policy is a function of the state dynamics of their goodwill. If we follow a model similar to that of Nerlove and Arrow [1962], then its dynamics depends on the advertising level and a given rate of goodwill depreciation.

A full account of capital accumulation games can be found in Dockner *et al.* [2005] and applications of the scheme to marketing games are surveyed in Jørgensen and Zaccour [2004]. This chapter is an instance of such games and an extension of Jørgensen and Gromova [2016], which considers a goodwill accumulation game on a 3-player oligopoly with cooperative advertising. This represents a different focus in the objectives from previous research and shows optimal cooperative strategies between firms and its sustainability in time.

In this work, we generalize the results to an n-player oligopoly game that contemplates the case where goodwill is subject to depreciation. This might be a convenient abstraction to account for the customers forgetting about the brand, which forces the firms to keep advertising in order to maintain relevance in the market.

The model presented in this chapter also introduces uncertainty about the period of time in which the game will develop. This offers flexibility for analysis in industries where the product has a life cycle. Also, it allows to consider the risk that involves making a joint advertising campaign: if one of the producers of the coalition does not comply with the expected value of the product, the whole coalition's reputation is at stake. This might cause an abrupt end of the game that is accounted for.

Finally, the main result of this chapter is the construction of an Imputation Distribution Procedure (IDP) for this type of game. This scheme allows to allocate the profits from collaboration between the players such that they do not have any incentives to deviate from the coalition.

Consider then, the situation in which a group of firms in a particular industry want to create a joint advertising campaign. An example of this situation can be found in franchises or in a union of producers of a homogeneous product, like dairy milk. It is not uncommon to find commodity producers creating such unions to protect themselves against economic risks. In this case, the individual firms might be able to create a brand on their own, but still have incentives to try to increase the goodwill-dependent profit of the whole industry. For such a cartel, two questions are in order: what is the advertising level rule that optimizes the coalition profit? and, if the profit is greater in a coalition for the players, how are the spoils of joint effort shared between them?

The reputation of the firms will depend positively on the advertising level of the coalition as long as the customers receive a product with a steady perceived value. It is therefore important that the firms in the coalition create a homogeneous product. Consider the case in which there is an inherent risk that the firms may fail to deliver such value. That risk is given exogenously by our analysis, but we can think that it can come for a variety of externalities such as managerial mistakes, underfunding, a failure in the supply chain or bankruptcy. Under these circumstances, there is a tradeoff in making a joint campaign: the firms will share the risk of delivering an undervalued product. Such an occurrence will result in an instantaneous termination of the advertising campaign.

In this chapter, we make such model in which the firms of an industry with homogeneous firms create a joint campaign to improve their industrywise profits through their goodwill. We assume that firms make a promise of a certain quality to their clients. If the clients do not meet their expectations, they will stop the consumption of the product and the game is finished for the firm (see Ariely and Norton [2009] about consuming expectancies). Thus, a joint campaign implies sharing the risk of failure in delivering the promised value to customers.

In the present chapter, we find optimal rules for advertising expenditures of firms in cooperative and noncooperative games and find the equilibrium path of the game and the stable profit allocation of the coalition. The risk of failure of the firms and its resulting termination of the campaign is represented by a random terminal time of the game.

The chapter is organized as follows. In Sec. 2, we state the model of the game and the payoffs of the players. The setup of the problem includes a random terminal time, which presents a computation challenge that is addressed in Sec. 3. Section 4 shows the equilibrium for a noncooperative game, that is, for a game where no agreement was achieved before the start of the game. If an agreement is achieved, we turn to a fully cooperative game, whose results are shown in Sec. 5. The earnings of a cooperative agreement are then shared among the firms. We use the IDP scheme to arrive at a time-consistent cooperation agreement between the agents in Sec. 6. The concluding remarks of the work are shown in Sec. 7.

2. The Model

Consider an oligopolistic market of n firms belonging to the set $N = \{1, \ldots, n\}$, competing for sales volume of its own particular brand of a homogeneous product or service. The firms can increase their sales and hence their profits by using advertising.

Let $a_i(t) \geq 0$ be the rate of advertising efforts of firm i. Let $G_i(t)$ be the stock of goodwill of firm i, which summarizes the effect of the accumulation of past and present marketing efforts in the demand of the product. Naturally, it is to be expected that this effect is increasing in the advertising volume and wears off in proportion to the current level of the stock itself at a constant depreciation rate (cf. Nerlove and Arrow [1962]).

Suppose that the advertising goodwill of firm i evolves according to the differential equation

$$\dot{G}_i = \kappa a_i - \delta G_i; \quad G_i(0) = g_0^i > 0, \quad i \in N, \tag{1}$$

where $\delta > 0$ is the common depreciation rate and κ is the effect of advertising on the accumulation of goodwill. A big value of κ can be found in a market that is very receptive to advertising. Since the firms in this game are symmetrical in this sense, all the values of the effects of advertising on goodwill are the same. Furthermore, the measure of the goodwill is such that each dollar spent on advertising provides one unit of goodwill. Therefore, without loss of generality, we can set $\kappa = 1$.

Let $G(t) \triangleq (G_1(t), \ldots, G_n(t))$ be the joint goodwill of the firms in the market. For every player $i \in N$, the goal is to maximize her profit in the period of time $[t_0, T_i]$, where without loss of generality, $t_0 = 0$ and T_i is a random variable with distribution function $F(t)$. Therefore, the expected payoff for player $i \in N$ from time t, and until the (random) moment of her departure from the system is described by

$$J_i(a_i) = \mathbb{E} \left(\int_t^{T_i} (\pi s_i(\tau) - C(a_i)) \mathrm{d}\tau \right),$$

where π represents the marginal profit of the market, $C(a_i)$ is the cost of advertising efforts, and $s_i(\tau)$ stands for the sales volume of the ith firm at time τ. Therefore, the payoff from time t onwards can be stated as

$$J_i(a_i) = \int_0^\infty \int_0^t (\pi s_i(\tau) - C(a_i)) \mathrm{d}\tau \, \mathrm{d}F_i(t). \tag{2}$$

For the sake of concreteness and simplicity, the cost of advertising efforts will be assumed to be of the form $C(a_i) = \frac{c}{2} a_i^2$. Now, the sales volume of firm i can be defined as a function of the joint goodwill from all the firms. This means that advertising efforts affect positively the whole market as well as a firm's own sales. We follow Reynolds [1991] and define the sales profits before advertising expenditures by means of a function that depends on the joint goodwill in the market $G(\tau)$. That is,

$$s_i(t) := \left[\theta - \sum_{h=1}^{n} G_h(t)\right] G_i(t).$$

This function is a typical example of an oligopolistic profit function, where θ is the demand intercept (net of marginal cost of production). The interpretation of this kind of function is that, since sales are influenced positively by goodwill, the output of products by the firms can flood the market when the goodwill of the industry is high, thus, reducing prices and diminishing profits for all the players. This can be easily seen in

$$\frac{\partial s_i(t)}{\partial G_i} = \theta - G_i(t) - \sum_{j=1}^{n} G_j(t).$$

A necessary condition for this expression to be positive is given by

$$\sum_{j=1}^{n} G_j(t) + G_i(t) < \theta. \tag{3}$$

It is easy to verify also that $\frac{\partial^2 s_i(t)}{\partial G_i^2} < 0$. The interpretation of this lies in the idea that there is a threshold where the sales volume affects the profitability of the product, that is, the goodwill can present positive, but decreasing returns if (3) does not hold. That is, if $\sum_{j\neq i} G_j(t) + 2G_i(t) < \theta$. The goodwill of rival companies affect negatively the firm's own sales, as can be verified by the inequalities $\frac{\partial s_i}{\partial G_j(t)} < 0$, $\frac{\partial^2 s_i}{\partial G_i(t)G_j(t)} < 0$ for any $j \neq i$. This implies that the marketing efforts made by firms are both in the form of *generic* and *brand* advertising (cf. Bass *et al.* [2005]).

Therefore, the payoff of agent i is given by

$$\int_0^\infty \int_0^t \left(\pi \left[\theta - \sum_{h=1}^{n} G_h\right] G_i - \frac{c}{2} a_i^2\right) d\tau \, dF_i(t). \tag{4}$$

For both cooperative and noncooperative scenarios, the game is supposed to be autonomous, which means that advertising strategies and value functions do not depend on time.

3. Random Terminal Time

The game is played in a random time horizon. The class of differential games with random terminal times was first considered by Petrosjan and Murzov [1966] and further developments can be found in Petrosjan and Shevkoplyas [2000].

For every player $i \in N$, let us define the terminal time for the campaign with the random variable T_i with known probability distribution function. This characteristic can be caused by externalities or by the nature of the product (or service) itself, for example, when production can have failures, which might affect the real or perceived value of the product (cf. Shevkoplyas [2014]). This situation can vary from changes in prices to complete lack of stock, leading the firm to stop in the campaign in order to avoid a case of *cognitive dissonance*, i.e., inconsistencies between the expected value of the product and the real value of it (cf. Festinger [1962]).

Since the analysis at hand concerns the cooperation of firms in the advertising campaign, it is natural to think that the failure to deliver from one of the firms in the coalition can adversely affect the goodwill of the industry in general, and so the campaign should stop as soon as the first firm stops. Therefore, the terminal time of the game is defined by $T = \min\{T_i : i = 1, \ldots, n\}$. The formulation of a cooperative differential game with asymmetric random terminal times has been considered previously in Kostyunin *et al.* [2014] and extended in Gromova *et al.* [2016].

Let $\{T_i : i = 1, \ldots, n\}$ be the set of terminal times for all players in N, defined by independent random variables, with corresponding probability distribution functions $\{F_i(t) : i = 1, \ldots, n\}$. We start our analysis by imposing a condition on these functions. Such condition will enable us to write the performance index of each player's (4) in terms of a density function f that is common to all players.

Assumption 1. The cumulative distribution function of the random variable T_i is such that

(a) it is absolutely continuous with respect to Lebesgue's measure, and $F_i'(t) = f_i(t)$;
(b) it belongs to the so-called *exponential family of distributions*. That is,

$$F_i(t) = \int_0^t \lambda_i(\tau) e^{-\int_0^t \lambda_i(s)ds} d\tau,$$

for some *hazard function* $\lambda_i(\cdot)$,

for all $i = 1, 2, \ldots, n$.

Now, we introduce the cumulative distribution function for the random terminal time of the game and the hazard function for this random variable. To do all these, we use Theorem 6 in Chap. V of Mood *et al.* [2004].

Proposition 1. *Define the random variable* $T := \min\{T_i : i = 1, \ldots, n\}$. *Under Assumption 1, the probability distribution function* $F(t)$ *of* T *has the form*

$$F(t) = 1 - \prod_{i=1}^{n}(1 - F_i(t)).$$

Corollary 1. *For the random variable* $T = \min\{T_i : i = 1, \ldots, n\}$, *the hazard function* $\rho(t)$ *can be calculated by*

$$\rho(t) = \sum_{i=1}^{n} \lambda_i(t). \tag{5}$$

In the sequence, we will consider that the hazard rate of the termination time T_i is constant, that is $\lambda_i(t) = \lambda_i > 0$ for $i = 1, \ldots, n$ and $t > 0$. It is well known that this yields T_i that has an exponential distribution and $f_i(t) = \lambda_i e^{-\lambda_i t}$. Then, Proposition 1 gives us that, for $T = \min\{T_i : i = 1, \ldots, n\}$,

$$1 - F(t) = e^{-\sum_{i=1}^{n} \lambda_i t} \tag{6}$$

and by Corollary 1, $\rho(t) = \sum_{i=1}^{n} \lambda_i$ for all $t > 0$. It has been shown in Shevkoplyas [2014] that a game in the form described in (4) can be reduced to a form that can be solved by dynamic programming methods. Thus, our problem turns into

$$J_i(a_i) = \int_0^\infty (1 - F(t)) \left(\pi \left[\theta - \sum_{h=1}^{n} G_h(t) \right] G_i(t) - \frac{c}{2} a_i^2 \right) dt. \tag{7}$$

Plug (6) and (5) into (7) to get the payoff function for the ith player:

$$J_i(a_i) = \int_0^\infty e^{-\rho t} \left(\pi \left[\theta - \sum_{h=1}^{n} G_h(t) \right] G_i(t) - \frac{c}{2} a_i^2 \right) dt. \tag{8}$$

That is, all players are discounting their profit according to the hazard function of the game, which in turn depends on the probability of having to stop the campaign to maintain a good reputation for the industry.

4. Noncooperative Game

In this section, we obtain a Nash equilibrium for the game in a competitive scenario. This can happen if the firms cannot reach an agreement before the game starts, so they decide to play with competitive strategies in the interval of time $t \in [0, \infty)$. From now on, we will omit the time argument except when it is necessary to remove ambiguity. Let $V_i(G)$ be a continuously differentiable value function of firm i. To find a Nash equilibrium, it is necessary to solve the Hamilton–Jacobi–Bellman (HJB) equation

$$\rho V_i(G) = \max_{a_i \geq 0} \left\{ \pi \left[\theta - \sum_{h=1}^{n} G_h \right] G_i - \frac{c}{2} a_i^2 + \sum_{i=1}^{n} \frac{\partial V_i}{\partial G_i} (a_i - \delta G_i) \right\}. \tag{9}$$

A maximization of the right-hand side yields the closed-loop (feedback) advertising rates

$$a_i(G_i) = \begin{cases} \dfrac{1}{c} \dfrac{\partial V_i}{\partial G_i} & \text{if } \displaystyle\sum_{h=1}^{n} \frac{\partial V_i}{\partial G_i} > 0, \\[4mm] 0 & \text{if } \displaystyle\sum_{h=1}^{n} \frac{\partial V_i}{\partial G_i} \leq 0. \end{cases} \tag{10}$$

Inserting these advertising rates in the HJB equation (9) and solving for the value function, we get

$$V_i(G) = \frac{\pi}{\rho}\left[\theta - \sum_{h=1}^{n} G_h\right]G_i + \frac{1}{2\rho c}\left(\frac{\partial V_i}{\partial G_i}\right)^2 - \frac{\delta}{\rho}\frac{\partial V_i}{\partial G_i}G_i \quad \text{if } a_i > 0,$$

$$V_i(G) = \frac{\pi}{\rho}\left[\theta - \sum_{h=1}^{n} G_h\right]G_i - \frac{\delta}{\rho}\frac{\partial V_i}{\partial G_i}G_i \quad \text{if } a_i = 0.$$

(11)

Since $\frac{1}{2\rho c}\left(\frac{\partial V_i}{\partial G_i}\right)^2 > 0$, it is easy to verify that the value function is larger when the advertising rate is positive, thus, for all players, the strategy (10) represents a payoff-dominant solution. This means that they will always prefer to choose a strategy with positive advertising rates.

Note that (11) is a nonhomogeneous Clairaut differential system (see Evans [1997], Chap. 3.1). Therefore, we use the ansatz:

$$V_i(G) = \alpha + \gamma_A G_i + \frac{\epsilon_A}{2}G_i^2 + (\eta_A G_i + \gamma_B)\sum_{j\neq i} G_j$$

$$+ \frac{\epsilon_B}{2}\sum_{j\neq i} G_j^2 + \sum_{j\neq i}\sum_{k=j}^{n} \eta_B G_j G_k,$$

(12)

where $\alpha, \gamma_A, \epsilon_A, \eta_A, \gamma_B, \epsilon_B, \eta_B$ are constants to be determined and ϵ_A is negative.

From this follows that for every firm, the partial derivative of the value function is given by

$$\frac{\partial V_i}{\partial G_i} = \gamma_A + \epsilon_A G_i + \sum_{j\neq i} \eta_A G_j.$$

(13)

Inserting (12) and (13) into (11) leads us to the values of the parameters

$$\alpha = \frac{\gamma_A^2}{2\rho c}, \quad c\rho\gamma_A = \theta c\pi + \gamma_A\epsilon_A - c\delta\gamma_A,$$

$$c\rho\epsilon_A = \epsilon_A^2 - 2\pi c - 2c\delta\epsilon_A, \quad c\rho\eta_A = \epsilon_A\eta_A - c\pi - \delta c\eta_A,$$

$$c\rho\gamma_B = \eta_A\gamma_A, \quad c\rho\epsilon_B = c\rho\eta_B = \eta_A^2.$$

This system of equations admits a feasible solution, namely,

$$\gamma_A = \frac{\theta}{2\mu}\left(\sqrt{c^2(\rho+2\delta)^2 + 8\pi c} - \rho c\right) > 0,$$

$$\gamma_B = -\frac{1}{2\mu^2}\left(\frac{\theta}{\rho}(4\pi + c(2\delta^2 + 2\delta\rho + \rho^2)) + \frac{1}{\pi^2}\sqrt{c^2(\rho+2\delta)^2 + 8\pi c}\right) < 0,$$

$$\epsilon_A = \frac{1}{2}\left(2\delta c + \rho c - \sqrt{8\pi c + c^2(2\delta+\rho)^2}\right) < 0,$$

$$\epsilon_B = \eta_B = \frac{1}{2\mu^2}\left(\frac{1}{\rho}(4\pi + c(2\delta^2 + 2\delta\rho + \rho^2)) + \frac{1}{\pi^2}\sqrt{c^2(\rho+2\delta)^2 + 8\pi c}\right) > 0,$$

$$\eta_A = \frac{1}{2\mu}\left(c\rho - \sqrt{c^2(\rho+2\delta)^2 + 8\pi c}\right) < 0,$$

where $\mu \triangleq \frac{1}{\pi}(c\delta(\rho + \delta) + 2\pi)$. This solution is feasible in the sense that if the firm has a profit margin of $\pi = 0$, then the value function will be equal to zero as well. This can be easily verified and implies that the firm will not advertise under these circumstances, given there is no revenue flow from it. After some algebraic manipulations, we can see that $\gamma_A = -\theta\eta_A$ and $\epsilon_A = \mu\eta_A + \delta c$. This fact, along with (13), yields

$$\frac{\partial V_i}{\partial G_i} = \eta_A \left(\mu G_i + \sum_{j \neq i} G_j - \theta \right) + \delta c G_i. \tag{14}$$

Let $\mu G_i + \sum_{j \neq i} G_j < \theta$, where $\mu \geq 2$, depending on the level of the depreciation. If the level of depreciation is positive, then $\mu > 2$ and the inequality becomes more restrictive than the one shown in (3), i.e., a smaller level of goodwill G_i is needed for the inequality to apply (see Remark 1 for more details on the meaning of the amount μ). Since η_A is negative, (14) is positive. The optimal advertising rate depends on the derivative of the value function. Therefore, inserting (14) in (10), we find that

$$a_i(G_i) = \varphi_i(G_i) = \frac{\eta_A}{c} \left(\mu G_i + \sum_{j \neq i} G_j - \theta \right) + \delta G_i, \tag{15}$$

where φ_i denotes the optimal advertising rate for firm i. We claim that the n-tuple $\varphi = (\varphi_1, \ldots, \varphi_n)$ constitutes a Nash equilibrium for the noncooperative game. To verify this claim, we refer to Theorems 2 and 3 of Gromova and López-Barrientos [2016], and their subsequent proofs, including the statement that V_i constitutes the optimal value of the game for each player (see also Theorem 2.1 in Jørgensen and Zaccour [2004]). To see this, note that the statement of the problem fulfills the conditions in those results to be seen rather as n separate control problems.

Here, we can see that the advertising level includes a term proportional to the goodwill in order to keep up with depreciation. Inserting the optimal level of advertising in (15) into the goodwill dynamics (1), we can obtain the equilibrium path of the goodwill

$$\dot{G}_i = \frac{\eta_A}{c} \left(\mu G_i + \sum_{j \neq i} G_j - \theta \right). \tag{16}$$

We solve the system of differential equations for all firms, obtaining a symmetrical goodwill path

$$G_i(t) = \exp\left\{ \frac{\eta_A}{c}(\mu + n - 1)t \right\} \left(g_0 - \frac{\theta}{\mu + n - 1} \right) + \frac{\theta}{\mu + n - 1}. \tag{17}$$

Since $\eta_A < 0$, the goodwill of the firms converges to $\frac{\theta}{\mu+n-1}$ as $t \to \infty$.

Figure 1 shows the goodwill dynamics for a player when the hazard rate ranges from 0 to 2. Note that the goodwill converges to $\frac{\theta}{\mu+n-1}$, but this number is slightly

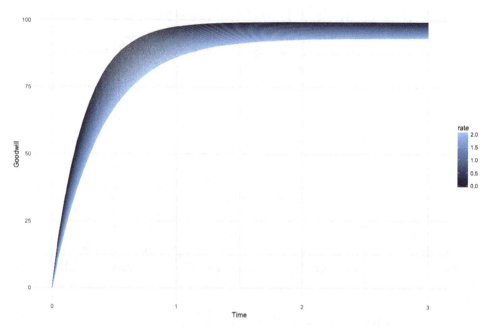

Fig. 1.   Noncooperative goodwill dynamics for a player given $g_0 = 0$ for $n = 3$, $\theta = 400$, $\pi = 1$, $c = 1/2$, $\delta = 1/4$, and $\rho \in [0, 2]$.

different for each hazard rate. A higher rate accounts for a lower goodwill in the long run. Recall, by (5), that ρ is the sum of the hazard rates of the players. Note that a higher hazard rate means a smaller parameter η_A, which, as we can note by (15), causes a lower rate of advertising. In Fig. 2, we can see the goodwill dynamics for a player when the number of players ranges from 3 to 40.

Remark 1. In the special case that the depreciation of advertising is zero ($\delta = 0$), as analyzed in Jørgensen and Gromova [2016], a level of depreciation equal to zero implies that $\mu = 2$ and, therefore, for n players, the optimal path for the players is given by

$$\dot{G}_i = \frac{\eta_A}{c}\left(2G_i + \sum_{j\neq i} G_j - \theta\right),$$

which, for $n = 3$, has the same solution as (13) in Jørgensen and Gromova [2016]. That is,

$$G_i(t) = \exp\left\{\frac{4\eta_A}{c}t\right\}\left(g_0 - \frac{\theta}{4}\right) + \frac{\theta}{4}.$$

Here, the parameter $\eta_A = 1/4(c\rho - \sqrt{c^2\rho^2 + 8\pi c})$. Thus, our model is consistent with that in Jørgensen and Gromova [2016] and extends it in concordance with the assumption of the presence of a nonzero depreciation rate δ.

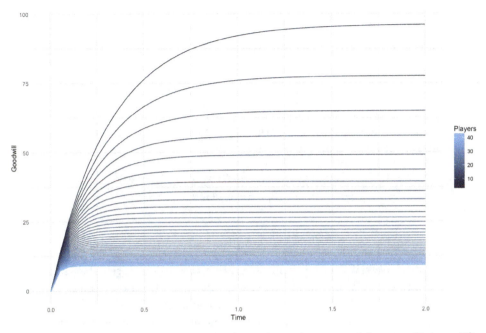

Fig. 2. Noncooperative goodwill dynamics for a player given $g_0 = 0$ for $n = \{3, 4, \ldots, 40\}$, $\theta = 400$, $\pi = 1$, $c = 1/2$, and $\delta = 1/4$.

5. Cooperative Game

In a cooperative game, the players arrive at an agreement at the beginning of the time horizon. They negotiate the amount of expenditure in advertising in order to maximize the industry's profit. The negotiations concern as well a rule to share the joint profit achieved by such grand coalition. The point of the collaboration is to maximize a joint payoff criterion and to avoid giving incentives to the firms to leave the coalition, for if any of them left the group, we would end up with the competitive game we analyzed in Sec. 4. We define the joint performance index of the coalition by the following rule of correspondence:

$$J(a_1, \ldots, a_n) = \int_0^\infty e^{-\rho t} \sum_{i=1}^n \left(\pi \left[\theta - \sum_{h=1}^n G_h \right] G_i - \frac{c}{2} a_i^2 \right) dt. \tag{18}$$

In this case, the firms agree to maximize the sum of their individual profits over time. As with the noncooperative game, this problem can be solved using dynamic programming techniques. The discounted payoff (18) is still subject to the same state dynamics in (1). Let $V(G)$ be a continuously differentiable function that solves the HJB equation:

$$\rho V(G) = \max_{a_1, \ldots, a_n \geq 0} \left\{ \pi \sum_{i=1}^n \left[\theta - \sum_{h=1}^n G_h \right] G_i - \frac{c}{2} \sum_{i=1}^n a_i^2 + \sum_{i=1}^n \frac{\partial V}{\partial G_i} (a_i - \delta G_i) \right\}.$$

The maximization of the right-hand side yields advertising rates

$$a_i(G) = \begin{cases} \dfrac{1}{c}\dfrac{\partial V(G)}{\partial G_i} & \text{if } \dfrac{\partial V(G)}{\partial G_i} > 0, \\[3mm] 0 & \text{if } \dfrac{\partial V(G)}{\partial G_i} \leq 0. \end{cases} \tag{19}$$

By inserting these optimal advertising rates into the right-hand side of the HJB equation, we obtain the following value functions:

$$V(G) = \frac{\pi}{\rho}\sum_{i=1}^{n}\left[\theta - \sum_{h=1}^{n} G_h\right]G_i + \frac{1}{2\rho c}\sum_{i=1}^{n}\left(\frac{\partial V}{\partial G_i}\right)^2 - \frac{\delta}{\rho}\sum_{i=1}^{n}\frac{\partial V}{\partial G_i}G_i \quad \text{if } a_i > 0,$$

$$V(G) = \frac{\pi}{\rho}\sum_{i=1}^{n}\left[\theta - \sum_{h=1}^{n} G_h\right]G_i - \frac{\delta}{\rho}\sum_{i=1}^{n}\frac{\partial V(G)}{\partial G_i}G_i \quad \text{if } a_i = 0. \tag{20}$$

Just as we observed in the noncooperative case, we can see that a positive advertising rate is a dominant strategy, since the value function is larger in every case. This can be easily seen if we take into account that $\frac{1}{2\rho c}\sum_{i=1}^{n}(\frac{\partial V}{\partial G_i})^2 > 0$. As with (11), we observe that (20) is a nonhomogeneous Clairaut differential system, so we guess that the value function that solves the HJB problem for positive advertising rates has the form

$$V(G) = \alpha + \gamma\sum_{i=1}^{n} G_i + \frac{\eta}{2}\left(\sum_{i=1}^{n} G_i\right)^2 = \alpha + \gamma Y + \frac{\eta}{2}Y^2, \tag{21}$$

where $Y \triangleq \sum_{i=1}^{n} G_i$ and α, γ, η are parameters to be determined. For every player $i \in N$, we have that

$$\frac{\partial V}{\partial G_i} = \gamma + \eta\sum_{j=1}^{n} G_i = \gamma + \eta Y. \tag{22}$$

Therefore, equating (21) with (20) and using (22), we arrive at the values

$$\alpha = \frac{n}{2\rho c}\gamma^2,$$

$$\rho\gamma = \pi\theta + \frac{n\gamma\eta}{c} - n\delta\gamma, \tag{23a}$$

$$\rho\eta = \frac{n\eta^2}{c} - 2\pi - 2n\delta\eta,$$

which have as unique feasible solutions

$$\eta = \frac{\rho c - \sqrt{(2nc\delta + c\rho)^2 + 8nc\pi}}{2n} + \delta c < 0, \tag{23b}$$

$$\gamma = \frac{-\theta(\rho c - \sqrt{(2nc\delta + c\rho)^2 + 8nc\pi})}{2n\mu} > 0. \tag{23c}$$

Note that $\gamma = -\frac{\theta}{\mu}(\eta - \delta c)$. We obtain the optimal advertising rate for each firm i by substituting (22) into (19):

$$a_i(G) = \frac{\gamma + \eta Y}{c} = \gamma/c - \frac{\gamma\mu}{c\theta}Y + \delta Y.$$

Substituting this value in the restriction given by (1) and adding the goodwills of the firms, we get the coalition goodwill dynamics

$$\dot{Y}(t) = \sum_{i=1}^{n} \dot{G}_i(t) = \frac{\gamma n}{c} - \frac{\gamma\mu n}{c\theta}Y, \quad Y_0 = \sum_{i=1}^{n} g_0^i.$$

Solving the differential equation, we get the coalition goodwill

$$Y(t) = \frac{\theta}{\mu} + \exp\left\{\left(\frac{\eta n}{c} - \delta\right)t\right\}\left(Y_0 - \frac{\theta}{\mu}\right). \tag{24}$$

Note that if $\delta = 0$, the cooperative goodwill for a game with $n = 3$ turns into

$$Y(t) = \frac{\theta}{2} + \exp\left\{\frac{3\eta}{c}t\right\}\left(Y_0 - \frac{\theta}{2}\right),$$

which is consistent with the results shown in Jørgensen and Gromova [2016].

Just as in the noncooperative case, the same effect in the goodwill can be observed for different hazard rates of the players.

Since the dynamics of the goodwill for all the firms are modeled by (1), the goodwill of the firms under cooperation is $Y(t)/n$. Note that this means that the

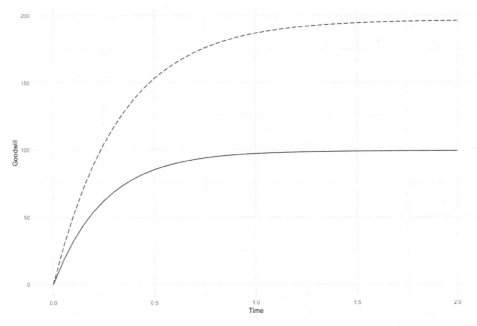

Fig. 3.   Goodwill dynamics for player $i \in N$ given $g_0 = Y_0 = 0$ in cooperative (dashed line) and noncooperative (thick line) scenarios for $n = 3, \rho = 0.05, \theta = 400, \pi = 1, c = 1/2, \delta = 1/4$.

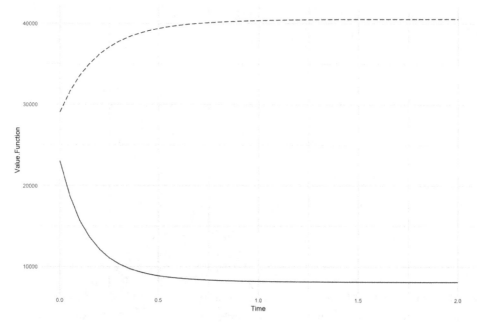

Fig. 4.   Value function for player $i \in N$ given $g_0 = 0$ in competitive (thick line) and noncooperative (dashed line) scenarios for $n = 3, \rho = 0.05, \theta = 400, \pi = 1, c = 1/2, \delta = 1/4$.

firms will have an individually lower goodwill under cooperation than they would if
they were competing among them noncooperatively. Figure 3 shows a comparison
between individual goodwill and joint goodwill for the industry as a whole.

Figure 4 shows the value function for the noncooperative game and it's cooperative counterpart in a dashed line. Note that the value function for a cooperative
game is nondecreasing, while the noncooperative value function behaves oppositely,
for it is nonincreasing; both values, remain stable after they reach stability.

Note that this behavior in the value functions mean that, as long as the initial
value function for the coalition is larger than the one of any individual, the value of
a joint advertising campaign will always be larger than a noncooperative one. This
means that the coalition will remain stable, provided that its members will share
the marginal profits from coalition in a way that staying in the coalition remains
more profitable throughout the complete period of the game.

6. Stability of Cooperation

In Sec. 5, we dealt with the acquisition of profits for the grand coalition. The
objective of this section is to arrive at a distribution of the profits between players
such that they have no incentives to deviate from the coalition. Since the structures that model the features of all players are given by (1) and (2), we use the
egalitarian share for all players and then use the IDP proposed by Petrosjan (cf.
Petrosjan [1977], Petrosjan and Zaccour [2003] and Petrosjan and Danilov [1985]).

This procedure has been shown to arrive at a distribution of the profits such that the players have no incentive to deviate at any moment in time.

Let (G, t) describes the state of the game, and $\Gamma(G, t)$ be the subgame starting at time $t \in [0, \infty)$. Denote by G^N the trajectory of goodwill accumulation by the grand coalition. Suppose that at some moment within the time-horizon, the players want to renegotiate the agreement for the game $\Gamma(G^N, t)$. If player i finds that she can obtain a profit from leaving the coalition, that is, that her individual payoff for playing noncooperatively is larger than the cooperative one, then it is to her best interest to leave. The same will be true for all other players in the coalition, which means that the game will be noncooperative from that time onwards.

Let $\phi(V, G^N, t) = (\phi_1(V, G^N, t), \ldots, \phi_n(V, G^N, t)) \equiv (\phi_1(t), \ldots, \phi_n(t))$ be the egalitarian share of the profits from the game played by the grand coalition. That is, the value function of the players in the grand coalition with their share of goodwill,

$$\phi_i(t) = V(G^N(t))/n, \tag{25}$$

$$\phi_i(t) = \frac{\alpha}{n} + \frac{\gamma}{n}Y(t) + \frac{\eta}{2n}Y(t)^2. \tag{26}$$

Since the cooperative goodwill in (24) is unique and all players are using an equal goodwill path, then $G_i^N = Y(t)/n$ is the cooperative goodwill of firm i. Therefore, (26) turns into

$$\phi_i(t) = \frac{\alpha}{n} + \gamma G_i^N + \frac{\eta}{2}n(G_i^N)^2.$$

Using (24), we get

$$\phi_i(t) = \frac{\alpha}{n} - \frac{\eta}{2n}\frac{\theta^2}{\mu^2} + \frac{\eta}{n}\left(\frac{1}{2}\exp\left\{\left(\frac{\eta n}{c} - \delta\right)t\right\}\left(Y_0 - \frac{\theta}{\mu}\right)\right)^2 - \frac{\delta c}{n}Y(t). \tag{27}$$

Denote by $\beta_i(t)$ the profit to be allocated to player $i \in N$ at an instant of time and $\beta(t) = (\beta_1(t), \ldots, \beta_n(t))$ is the vector of profit allocations, that is, the vector of *shares of the players*. We say that this vector is an IDP if

$$\phi_i(0) = \int_0^\infty e^{-\rho t}\beta_i(t)dt, \quad i = 1, \ldots, n. \tag{28}$$

In other words, the share of the players in the grand coalition (since $t = 0$) is built upon the compounded allocated profits.

Remark 2. Note that the symbol $\phi_i(0)$ stands for $\phi_i(V, G, 0)$. This implies that (28) is well defined for every possible trajectory of G (even when (3) does not hold!).

Recall that the discount rate is built upon the risk of failure of a system that makes the firms to stop the advertising campaign. For an IDP to be consistent, the following condition (cf. Petrosjan and Shevkoplyas [2000]) has to hold at $t \in [0, \infty)$:

$$\phi_i(V, G, t) = \int_0^t e^{-\rho\tau}\beta_i(\tau)d\tau + e^{-\rho t}\phi_i(V, G^N, t),$$

which can also be stated in terms of β_i as

$$\beta_i(t) = \rho\phi_i(V, G^N, t) - \frac{d}{dt}\phi_i(V, G^N, t).$$

The interpretation of this statement is that, at any moment in time, the firms can reevaluate the sum of the profit up to the moment and the discounted future profit of staying in the coalition. This evaluation must satisfy the so-called *subgame consistency* condition:

$$\phi_i(t) \geq V_i(t), \quad i \in N.$$

Our developments ensure that this condition is met. For details on the proof, see Petrosjan [1977] and further in Petrosjan and Danilov [1985] and Petrosjan and Zaccour [2003]. Using (25) we get as a time-consistent IDP the following:

$$\beta_i(t) = \rho\frac{V(G^N(t))}{n} - \frac{d}{dt}\frac{V(G^N(t))}{n}.$$

Recall from (21) that $V(G^N(t)) = \alpha + \gamma Y + \frac{\eta}{2}Y^2$. Therefore, using the parameters obtained in (23a) and (23b), we obtain

$$\beta_i(t) = \frac{\rho\alpha}{n} - \frac{\eta\rho}{2n}\frac{\theta^2}{\mu^2} + \frac{\eta}{n}\left(\exp\left\{\left(\frac{n\eta}{c} - \delta\right)t\right\}\left(Y_0 - \frac{\theta}{\mu}\right)\right)^2\left(\frac{\rho}{4} - \left(\frac{\eta n}{c} - \delta\right)\right)$$

$$+ \frac{\delta c}{n}\left[\left(\frac{n\eta}{c} - \delta\right)\exp\left\{\left(\frac{n\eta}{c}\delta\right)t\right\}\left(Y_0 - \frac{\theta}{\mu}\right) - \rho Y(t)\right].$$

This is the value of the game's IDP. Note that in the long run, it will stabilize in $\frac{\rho\alpha}{n} - \frac{\eta\rho}{2n}\frac{\theta^2}{\mu^2} - \frac{\delta\rho c}{n}\frac{\theta}{\mu}$. It can be verified with the use of the value of η that $\frac{\eta\rho}{2n}\frac{\theta^2}{\mu^2} > \frac{\delta\rho c}{n}\frac{\theta}{\mu}$ and, therefore, the IDP is always positive.

7. Concluding Remarks

In this chapter, we presented a differential game for a competition with advertising goodwill accumulation for n players with random terminal time goodwill dynamics given by (1) and performance index (2). The goodwill system (1) behaves similarly to that of physical capital, where depreciation plays a role in it, and therefore influencing the optimal quantity of advertising.

The use of random terminal time allows us to make a more realistic model of an advertising campaign focused on goodwill, where the campaign can be forced to stop by sudden changes in the promised value by any of the firms in the coalition. Since the products are nondifferentiated, the failure in the delivery of expected value by any of the firms affects the whole industry. It is worth to mention that, in spite of the fact that the structures of the dynamics and the payoff are symmetrical, the firms are not supposed to share the same random time of termination. Instead, we have worked with the first time that a company is forced to leave the market.

The risk of failure can be reflected in a discounting rate for the profit functional. As shown in Fig. 1, in the long run, a higher risk yields a lower goodwill for the players and thus to the industry.

We identified the cooperative optimal advertising and goodwill trajectory, therefore arriving at a performance index for the firms acting cooperatively in a marketing campaign. We use the egalitarian solution of profit distribution and arrive at the computation of a time-consistent IDP.

Acknowledgments

The first author was supported by scholarship from Mexico Council of Science and Technology, CONACYT. The formulation of the cooperative differential game with asymmetric random terminal times and the construction of the time-consistent solution has been done by Ekaterina Gromova and supported by grant 17-11-01079 from the Russian Science Foundation.

References

Ariely, D. and Norton, M. I. [2009] Conceptual consumption, *Annu. Rev. Psychol.* **60**, 475–499.

Bass, F. M., Krishnamoorthy, A., Prasad, A. and Sethi, S. P. [2005] Generic and brand advertising strategies in a dynamic duopoly, *Mark. Sci.* **24**(4), 556–568.

Dockner, E., Jørgensen, S., Van Long, N. and Sorger, G. [2005] *Differential Games in Economics and Management Science* (Cambridge University Press, Cambridge).

Evans, L. C. [1997] *Partial Differential Equations* (American Mathematical Society, Providence, RI).

Fershtman, C. [1984] Goodwill and market shares in oligopoly, *Economica* **51**(203), 271–281.

Festinger, L. [1962] *A Theory of Cognitive Dissonance* (Stanford University Press, CA, Stanford).

Gromova, E. V. and López-Barrientos, J. D. [2016] A differential game model for the extraction of nonrenewable resources with random initial times: The cooperative and competitive cases, *Int. Game Theor. Rev.* **18**(2), 1640004.

Gromova, E. V., Tur, A. V. and Balandina, L. I. [2016] A game-theoretic model of pollution control with asymmetric time horizons, *Contrib. Game Theor. Manage.* **9**(1), 170–179.

Jørgensen, S. and Gromova, E. V. [2016] Sustaining cooperation in a differential game of advertising goodwill accumulation, *Eur. J. Oper. Res.* **254**(1), 294–303.

Jørgensen, S. and Zaccour, G. [2004] *Differential Games in Marketing* (Springer Science + Business Media, New York).

Kostyunin, S., Palestini, A. and Shevkoplyas, E. [2014] On a nonrenewable resource extraction game played by asymmetric firms, *Journal of Optimization Theory and Applications* **163**(2), 660–673.

Mood, A., Graybill, F. and Boes, D. [2004] *Introduction to the Theory of Probability and Statistics* (McGraw-Hill).

Nerlove, M. and Arrow, K. J. [1962] Optimal advertising policy under dynamic conditions, *Economica* **29**(14), 129–142.

Petrosjan, L. A. [1977] Stable solutions of differential games with many participants., *Viestnik of Leningrad University* **19**, 46–52.

Petrosjan, L. A. and Danilov, N. N. [1985] *Cooperative Differential Games and Their Applications* (Izd. Tomsk Univesity, Tomsk).

Petrosyan, L. A. and Murzov, N. V. [1966] Game-theoretic problems of mechanics, *Litovsk. Math. Sb.* **6**, 423–433.

Petrosjan, L. A. and Shevkoplyas, E. V. [2000] Cooperative differential games with stochastic time, *Vestn. St. Petersburg Univ. Math.* **33**(4), 14–18.

Petrosjan, L. A. and Zaccour, G. [2003] Time-consistent Shapley vallue allocation of pollution cost reduction, *J. Econ. Dyn. Control* **27**(3), 381–398.

Reynolds, S. S. [1991] Dynamic oligopoly with capacity adjustment costs, *J. Econ. Dyn. Control* **15**(3), 491–514.

Shevkoplyas, E. V. [2014] Optimal solutions in differential games with random duration, *J. Math. Sci.* **199**(6), 715–722.

Chapter 18

A Cooperative Dynamic Environmental Game of Subgame Consistent Clean Technology Development

David W. K. Yeung

Center of Game Theory, St. Petersburg State University
St. Petersburg 198904, Russia

SRS Consortium for Advanced Study in Cooperative Dynamic Games
Hong Kong Shue Yan University, Hong Kong, P. R. China

dwkyeung@hksyu.edu

Leon A. Petrosyan

Faculty of Applied Mathematics-Control Processes
St. Petersburg State University
St. Petersburg 198904, Russia

Spbuoasis7@petrlink.ru

Cooperative adoption and development of clean technology play a key role to effectively solving the continual worsening industrial pollution problem. For cooperation over time to be credible, a subgame consistency solution which requires the agreed-upon optimality principle to remain in effect throughout the collaboration duration has to hold. In this chapter, we present a cooperative dynamic game of collaborative environmental management with clean technology development. A subgame consistent cooperative scheme is derived. It is the first time that cooperative dynamic environmental games with clean technology development are analyzed. Given that there exist discrete choices of production techniques and switching to clean technology brings about cost savings and improved effectiveness, the group optimal solution cannot be obtained with standard differentiable optimization techniques. To overcome this problem the joint optimal solutions for all the possible patterns of production techniques are computed and the pattern with the highest joint payoff is then selected. The analysis widens the scope of study in collaborative environmental management.

Keywords: Cooperative dynamic games; subgame consistency; clean technology development; environmental management.

1. Introduction

Due to the global nature of environmental effects and trade, unilateral response on the part of one nation is often ineffective. Cooperation in environmental management holds out the best promise of effective action (see Dockner and Long [1993],

Dutta and Radner [2006], Jørgensen and Zaccour [2001], Fredj *et al.* [2004], Breton *et al.* [2005, 2006], Petrosyan and Zaccour [2003], Rubio and Ulph [2007], Yeung [2007, 2008], Yeung and Petrosyan [2008], and Li [2014]). Collaborative pollution control schemes with policy instruments like carbon reduction or emission permits under the conventional production technique often imply limitations on output growth and hence become offer less attractiveness for participants. Joint effort in pollution reduction through internalization of the spill-over environmental costs and coordinated adoption of environmentally clean techniques provide a promising starting point (see Yeung [2014]). However, existing clean techniques are in general more expensive therefore cooperative schemes involving pollution reduction and adoption of clean technology may not be able to meet the needs of industrial growth. Collaborative development of environmentally clean technique into efficient and affordable means of production plays a key role to effectively solving the continual worsening global industrial pollution problem and meeting the industrial growth needs.

In addition, for cooperation over time to be sustainable, subgame consistency which requires the agreed-upon optimality principle to remain in effect throughout the collaboration duration has to hold. Cooperative dynamic games that have identified subgame consistent solutions can be found in Jørgensen and Zaccour [2001], Petrosyan and Zaccour [2003], Yeung [2007], and Yeung and Petrosyan [2004, 2006, 2008, 2010]. To illustrate the notion of subgame consistent we present the following numerical example. Consider a 3-nation dynamic game with three stages of actions and a terminal stage with terminal payoffs contingent upon the terminal state. At stage 1 with initial state x_1^0 the noncooperative payoffs of nations 1–3 (which includes the sum of the gains in stages 1–3 plus the terminal payoff) are respectively

$$V^1(1, x_1^0) = 267.35, \quad V^2(1, x_1^0) = 301.5, \quad \text{and} \quad V^3(1, x_1^0) = 101.$$

The nations agree to cooperate and maximize their joint payoff. The maximized joint payoffs at stages 1–3 along the cooperative trajectory $\{x_1^0, x_2^*, x_3^*\}$ are

$$W(1, x_1^0) = 836.5, \quad W(2, x_2^*) = 613, \quad W(3, x_3^*) = 404.98, \quad \text{and} \quad W(4, x_4^*) = 204.11.$$

The terminal payoffs of the nations at stage 4 generated by cooperative actions are:

$$V^1(4, x_4^*) = 43.53, \quad V^2(4, x_4^*) = 88.16, \quad V^3(4, x_4^*) = 72.42.$$

The noncooperative payoffs of nations 1–3 in stages 2 and 3 along the cooperative trajectory are:

$$V^1(2, x_2^*) = 188.25, \quad V^2(2, x_2^*) = 226.74, \quad V^3(2, x_2^*) = 91.02,$$
$$V^1(3, x_3^*) = 115.66, \quad V^2(3, x_3^*) = 158.6, \quad V^3(3, x_3^*) = 86.02.$$

At stage 1, the nations agree to maximize their joint payoff and share the cooperative payoff proportional to their noncooperative payoffs, that is the cooperative payoff that nation i is entitled to obtain is

$$\xi^i(1, x_1^0) = \frac{V_1^i(1, x_1^0)}{\sum_{j=1}^n V_1^j(1, x_1^0)} W(1, x_1^0), \quad \text{for } i \in \{1, 2, 3\}.$$

In stage $t \in \{2, 3\}$, the state variables become x_2^* and x_3^* and according to the agreed-upon sharing mechanism

$$\xi^i(t, x_t^*) = \frac{V_t^i(t, x_t^*)}{\sum_{j=1}^n V_t^j(t, x_t^*)} W(t, x_t^*), \quad \text{for } i \in \{1, 2, 3\} \text{ and } t \in \{2, 3\}. \quad (1.1)$$

Subgame consistency implies that in each of the stages $t \in \{1, 2, 3\}$ the cooperative payoffs of the nations have to be proportional to their noncooperative payoffs, that means condition (1.1) has to hold in all stages $t \in \{1, 2, 3\}$. Using (1.1) a subgame consistent payoff distribution can be calculated in the following table:

t	Nation 1's Coop. payoff $\xi^1(t, x_t^*)$	Nation 2's Coop. payoff $\xi^2(t, x_t^*)$	Nation 3's Coop. payoff $\xi^3(t, x_t^*)$
1	334	376.3	126.2
2	228.05	274.69	110.26
3	130.01	178.28	96.69

One can readily observed at each of stage $t \in \{1, 2, 3\}$ the payoff distribution agreed-upon initially holds. In a subgame consistent solution the agreed-upon optimality is maintained in every stage/subgame so negotiation at any subsequent stage would yield the same optimality principle. Hence no nation has any incentive to deviate from the plan. Subgame consistency is indeed crucial to a stable and sustainable cooperative scheme.

In this chapter, we present a cooperative dynamic environmental game which involves joint pollution abatements, coordinated tax levies internalizing spill-over damage costs, and cooperative development of environmentally clean technology. A subgame consistent cooperative solution entailing group optimality and individual rationality which includes subgame consistent abatement efforts, levies and investments in clean technology development together with an imputation distribution procedure which ensures subgame consistency is derived. Given that there exist discrete choices of production techniques, switching to clean technology is gradual over time, and cooperative development of clean technology brings about cost savings and improved effectiveness of research effort, the group optimal solution cannot be obtained with differentiable optimization techniques applied to the joint payoff function under cooperation. To overcome this nondifferentiable optimization problem we compute the joint optimal solutions for all the possible patterns of production techniques adopted and select the pattern with the highest joint payoff.

The analysis is the first dynamic environmental game with clean technology development and it widens the scope of study in global environmental management. The organization of the chapter is as follows. Section 2 presents a dynamic game model with production technique choices. Noncooperative outcomes are characterized in Sec. 3. Cooperative arrangements, group optimal actions, solution state trajectories, and time-consistent solution are examined in Sec. 4. An imputation distribution mechanism bringing about the proposed time-consistent solution is derived and scrutinized in Sec. 5. Discussion and remarks are given in Sec. 6.

2. Game Formulation

In this section, we present a T stages and n nations dynamic environmental game model with clean technology development. There are two types of production techniques available to each nation's industrial sector: conventional technique and environmentally clean technique. In the initial stage industrial sectors pay more for using clean technique. The amount of pollutants emitted by the clean technique is less than that emitted by conventional technique and so are the local environmental impacts to the nation itself and its adjacent nations. The average cost of producing a unit of output with conventional technique in nation j is c^j while that of producing a unit of output with clean technique is $\hat{c}_1^j$ in the initial stage 1.

2.1. *Clean technology development*

Many of the existing environment preserving technologies are not fully developed and cost efficient. Further development of these techniques would be reflected in the lower unit cost of production with clean technology. Let ϑ_t^i denote the level of clean technology of nation i at stage t. The dynamics of the level of environment preserving technology of nation i is governed by the difference equation

$$\vartheta_{t+1}^i = \vartheta_t^i + z^i w_t^i, \quad \vartheta_1^i = \vartheta_1^{i(0)}, \quad \text{for } i \in N, \tag{2.1}$$

where w_t^i is nation i's effort (inputs) in developing clean technology and $z^i w_t^i$ is the advancement in clean technology brought about by w_t^i.

The cost of nation i's effort w_t^i is $\chi^i(w_t^i)^2$. If the state of clean technology is ϑ_t^i, nation i's cost of production with the clean technique at stage t is

$$\hat{c}_t^i = \underline{c}^i - \sigma_t^i \vartheta_t^i, \quad \text{for } i \in N, \tag{2.2}$$

where $\underline{c}^i > c_t^i$ and σ_t^i is a nonnegative parameter.

There is an upper ceiling of the level of clean technology denoted by $\bar{\vartheta}^i$. When the highest level of clean technology has been reached, investment in clean technology development is ineffective. In addition, even when the ceiling of clean technology has been reached, the cost of using the technology $\hat{c}_t^i = \underline{c}^i - \sigma_t^i \bar{\vartheta}^i$ is positive for all $i \in N$. In stage t, we let the set of nations using conventional technique be denoted by S_t^1 and the set of nations using clean technique by S_t^2. The industrial sectors can switch their production techniques in any stage. For notational convenience, we denote $\vartheta_t = (\vartheta_t^1, \vartheta_t^2, \ldots, \vartheta_t^n)$.

2.2. *Impacts and accumulation dynamics of pollutants*

Industrial production emits pollutants into the environment and the amount of pollution created by different nations' outputs may be different. For an output of q_t^i produced by nation i using conventional technique, there will be a local short term environmental impact (cost) of $\varepsilon_i^i q_t^i$ on nation i itself and a local impact of $\varepsilon_j^i q_t^i$ on its neighbor nation j. For an output of $\hat{q}_t^i$ produced by nation i using clean technique, there will be a local short term environmental impact (cost) of $\hat{\varepsilon}_i^i \hat{q}_t^i$ on nation i itself and a local impact of $\hat{\varepsilon}_j^i \hat{q}_t^i$ on its neighbor nation j. Let $\underline{K_t^i}$ denote the set of nations that are affected by the output of nation i. The local impact using conventional technique is higher than that using clean technique. Let $\bar{K}_t^i$ denote the subset of nations whose outputs produce local environmental impacts to nation i. Nation i will receive local environmental impacts from its adjacent nations measured as $\varepsilon_i^j q_t^j$ for $j \in \bar{K}_t^i \cap S_t^1$, and $\hat{\varepsilon}_i^\ell \hat{q}_t^\ell$ for $\ell \in \bar{K}_t^i \cap S_t^2$. A linear approximation is adopted to measure the environmental impacts of outputs. Moreover, industrial output creates long-term environmental impacts by building up existing pollution stocks like greenhouse gas, CFC, carbon, and atmospheric particulates. Each nation adopts its own pollution abatement policy to reduce pollutants in the environment. At the initial stage 1, the level of pollution is $x_1 = x^0$. The dynamics of pollution accumulation is governed by the difference equation:

$$x_{t+1} = x_t + \sum_{i_t \in S_t^1} a^{i_t} q_t^{i_t} + \sum_{\ell_t \in S_t^2} \hat{a}^{\ell_t} \hat{q}_t^{\ell_t} - \sum_{j=1}^n b_j u_t^j (x_t)^{1/2} - \delta x_t, \quad x_1 = x_{1(0)}, \quad (2.3)$$

where a^{i_t} is the amount of pollution created by a unit of nation i_t's output using conventional technique, $\hat{a}^{\ell_t}$ is the amount of pollution created by a unit of nation ℓ_t's output using clean technique, u_t^j is the pollution abatement effort of nation j at stage t, $b_j u_t^j (x_t)^{1/2}$ is the amount of pollution removed by u_t^j units of abatement effort from nation j, and δ is the natural rate of decay of the pollutants.

There is a physical limit on pollution abatement. In particular, the total level of pollution abatement cannot exceed the existing (net of natural degradation) pollution stock, that is $\sum_{j=1}^n b_j u_t^j (x_t)^{1/2} \le (1 - \delta) x_t$. The damage (cost) of x_t amount of pollution to nation is $h^j x_t$. The cost of u_t^j units of abatement effort is $c_i^a (u_t^j)^2$. The convex abatement cost reflects increasing marginal cost.

2.3. *Industry equilibria and nations' objectives*

The value of nation i's industrial output is $[\alpha_t^i q_t^i - \beta_t^i (q_t^i)^2]$. Let v_t^i denote the tax rate imposed by nation i on industrial output produced by conventional technique in stage t, and $\hat{v}_t^i$ denote the tax rate imposed on output produced by clean technique.

The profit of the industrial sector of nation $i_t \in S_t^1$ and that of the industrial sector of nation $\ell_t \in S_t^1$ in stage t can be expressed respectively as

$$\pi_t^{i_t} = [\alpha_t^{i_t} q_t^{i_t} - \beta_t^{i_t} (q_t^{i_t})^2] - c^{i_t} q_t^{i_t} - v_t^{i_t} q_t^{i_t}, \quad \text{for } i_t \in S_t^1, \tag{2.4}$$

and

$$\hat{\pi}_t^{\ell_t} = [\alpha_t^{\ell_t} \hat{q}_t^{\ell_t} - \beta_t^{\ell_t} (\hat{q}_t^{\ell_t})^2] - \hat{c}_t^{\ell_t} \hat{q}_t^{\ell_t} - \hat{v}_t^{\ell_t} \hat{q}_t^{\ell_t}, \quad \text{for } \ell_t \in S_t^2. \tag{2.5}$$

For nontrivial positive output possibilities the parametric conditions $\alpha_t^i > c^i + \varepsilon_i^i$ for all $i \in N$ and $t \in \kappa$ are imposed. In each stage t the industrial sector of nation $i_t \in S_t^1$ seeks to maximize (2.4) and the industrial sector of nation $\ell_t \in S_t^1$ seeks to maximize (2.5). The first-order condition for a market equilibrium in stage t yields

$$\alpha_t^{i_t} - 2\beta_t^{i_t} q_t^{i_t} = c^{i_t} + v_t^{i_t}, \quad \text{for } i_t \in S_t^1; \quad \text{and}$$
$$\alpha_t^{\ell_t} - 2\beta_t^{\ell_t} \hat{q}_t^{\ell_t} = \hat{c}_t^{\ell_t} + \hat{v}_t^{\ell_t}, \quad \text{for } \ell_t \in S_t^2. \tag{2.6}$$

Condition (2.6) shows that the industrial sectors will produce up to a point where marginal revenue (the left-hand side of the equations) equals the unit cost of production plus tax of a unit of output produced (the right-hand-side of the equations). In particular, nation i's industrial sector will choose to use clean technique if $c^i + v_t^i > \hat{c}^i + \hat{v}_t^i$, otherwise it would choose to use conventional technique.

The nations have to promote business interests and at the same time bear the costs brought about by pollution. In particular, each nation maximizes the net gains in the industrial sector minus the sum of expenditures on pollution abatement, clean technology development cost and local and global damages from pollution. The payoff of nation $i_t \in S_t^1$ at stage t can be expressed as:

$$[\alpha_t^{i_t} q_t^{i_t} - \beta_t^{i_t} (q_t^{i_t})^2] - c^{i_t} q_t^{i_t} - \varepsilon_{i_t}^{i_t} q_t^{i_t} - \sum_{j \in \bar{K}_t^{i_t} \cap S_t^1} \varepsilon_{i_t}^j q_t^j - \sum_{\zeta \in \bar{K}_t^{i_t} \cap S_t^2} \hat{\varepsilon}_{i_t}^\zeta \hat{q}_t^\zeta$$
$$- \chi^{i_t} (w_t^{i_t})^2 - c_{i_t}^a (u_t^{i_t})^2 - h_t^{i_t} x_t; \tag{2.7}$$

and the payoff of nation $\ell_t \in S_t^2$ at stage t can be expressed as:

$$[\alpha_t^{\ell_t} \hat{q}_t^{\ell_t} - \beta_t^{\ell_t} (\hat{q}_t^{\ell_t})^2] - \hat{c}_t^{\ell_t} \hat{q}_t^{\ell_t} - \hat{\varepsilon}_{\ell_t}^{\ell_t} \hat{q}_t^{\ell_t} - \sum_{j \in \bar{K}_t^{\ell_t} \cap S_t^1} \varepsilon_{\ell_t}^j q_t^j - \sum_{\zeta \in \bar{K}_t^{\ell_t} \cap S_t^2} \hat{\varepsilon}_{\ell_t}^\zeta \hat{q}_t^\zeta$$
$$- \chi^{\ell_t} (w_t^{\ell_t})^2 - c_{\ell_t}^a (u_t^{\ell_t})^2 - h_t^{\ell_t} x_t. \tag{2.8}$$

The nations' planning horizon is from stage 1 to stage T. It is possible that T may be very large. The discount rate is r. A terminal appraisal of pollution damage is $g^i(\bar{x}^i - x_{T+1})$ and a technology premium $g^{(P)i} \vartheta_{T+1}^i$ will be credited to nation i at stage $T+1$, where $g^i \geq 0$ and $g^{(P)i} \geq 0$. Each one of the n nations seeks to maximize the sum of the discounted payoffs over the T stages plus the terminal appraisal. In particular, nation i would seek to maximize the objective

$$\sum_{t=1}^{T} \left[[\alpha_t^i \bar{q}_t^i - \beta_t^i (\bar{q}_t^i)^2] - \bar{c}^i \bar{q}_t^i - \bar{\varepsilon}_i^i \bar{q}_t^i - \sum_{j \in \bar{K}_t^i \cap S_t^1} \varepsilon_i^j q_t^j \right. $$

$$\left. - \sum_{\zeta \in \bar{K}_t^i \cap S_t^2} \varepsilon_i^\zeta \hat{q}_t^\zeta - \chi^i (w_t^i)^2 - c_i^a (u_t^i)^2 - h_t^i x_t \right] \left(\frac{1}{1+r} \right)^{t-1}$$

$$+ [g^i(\bar{x}^i - x_{T+1}) + g^{(P)i} \vartheta_{T+1}^i] \left(\frac{1}{1+r} \right)^T, \quad i \in N, \tag{2.9}$$

where $\bar{q}_t^i = q_t^i$, $\bar{c}_t^i = c^i$ and $\bar{\varepsilon}_i^i = \varepsilon_i^i$ if industrial sector i uses conventional technique; and $\bar{q}_t^i = \hat{q}_t^i$, $\bar{c}_t^i = \hat{c}_t^i = (\underline{c}^i + \mu_t^i \vartheta_t^i)$ and $\bar{\varepsilon}_i^i = \hat{\varepsilon}_i^i$ if industrial sector i uses environment-preserving technique.

The problem of maximizing objectives (2.9) subject to pollution dynamics (2.7) and clean technology development dynamics (2.6) is a dynamic game between these n nations. Note that each nation would use taxes to achieve the game optimal output determined by the solution of the game.

Remark 2.1. Note that the damage from the pollution stock is taken as linear approximation of the size of the pollution stock. The cost of pollution abatement and the cost of clean technology research effort are quadratic reflecting increasing marginal cost. The amount of pollution reduced by pollution abatement is a factor of the product of pollution abatement effort and the square-root of the pollution stock. In particular, if the stock is zero there is no reduction possible and the reduction is related to the square-root of the stock. The objective functions of the nations are quadratic in the controls and linear in the state while the state dynamics is not linear in the state and the nations' controls. Thus the game is not a linear quadratic dynamic game. However, the game equilibrium Hamilton–Jacobi–Bellman equations are similar to the type generated by a linear quadratic dynamic game.

3. Noncooperative Outcomes

In this section, we discuss the solution to the dynamic environmental game of clean technology development involving the maximization of different nations' objectives (2.9) subject to dynamics (2.1) and (2.3). In particular, each nation has to determine its preferred levels of output, clean technology development investment and abatement efforts to maximize its objective in (2.9). Under a noncooperative framework, a feedback Nash equilibrium solution can be characterized as follows.

Theorem 3.1. *A set of policies* $\{q_t^{it*} = \phi_t^{it}(x, \vartheta), \hat{q}_t^{it*} = \hat{\phi}_t^{it}(x, \vartheta), u_t^{it*} = v_t^{it}(x, \vartheta),$ $u_t^{it*} = \hat{v}_t^{it}(x, \vartheta), w_t^{it*} = \omega_t^{it}(x, \vartheta), \hat{w}_t^{it*} = \hat{\omega}_t^{it}(x, \vartheta),$ *for* $t \in \kappa$ *and* $i_t \in S_t^1$ *and* $\hat{i}_t \in S_t^2\}$ *provides a feedback Nash equilibrium solution to the game (2.1), (2.3) and*

(2.9) *if there exist functions* $V^{i_t}(t, x, \vartheta)$ *and* $\hat{V}^{\hat{i}_t}(t, x, \vartheta)$, *for* $t \in \kappa$ *and* $i_t \in S_t^1$ *and* $\hat{i}_t \in S_t^2$, *such that the following recursive relations are satisfied:*

$$
\begin{aligned}
V^{i_t}(t, x, \vartheta) = \max_{q_t^{i_t}, u_t^{i_t}, w_t^{i_t}} & \left\{ \left[[\alpha_t^{i_t} q_t^{i_t} - \beta_t^{i_t}(q_t^{i_t})^2] - c^{i_t} q_t^{i_t} - \varepsilon_{i_t}^{i_t} q_t^{i_t} \right. \right. \\
& - \sum_{j \in \bar{K}_t^{i_t} \cap S_t^1} \varepsilon_{i_t}^j \phi_t^j(x, \vartheta) - \sum_{\varsigma \in \bar{K}_i^{i_t} \cap S_t^2} \hat{\varepsilon}_i^\varsigma \hat{\phi}_t^\varsigma(x, \vartheta) - \chi^{i_t}(w_t^{i_t})^2 \\
& \left. - c_{i_t}^a (u_t^{i_t})^2 - h_t^{i_t} x \right] \left(\frac{1}{1+r} \right)^{t-1} + \bar{V}^{i_t} \left[t+1, x + \sum_{\substack{j \in S_t^1 \\ j \neq i_t}} a^j \phi_t^j(x, \vartheta) \right. \\
& + \sum_{\varsigma \in S_t^2} \hat{a}^\varsigma \hat{\phi}_t^\varsigma(x, \vartheta) + a^{i_t} q_t^{i_t} - \sum_{\substack{j \in S_t^1 \\ j \neq i_t}} b_j v_t^j(x, \vartheta) x^{1/2} \\
& - \sum_{j \in S_t^2} b_j \hat{v}_t^j(x, \vartheta) x^{1/2} - b_{i_t} u_t^{i_t} x^{1/2} - \delta x, \vartheta^{i_t} \\
& + z^{i_t} w_t^{i_t}, \underline{\vartheta^{j \in S_t^1 \backslash i_t} + z^{j \in S_t^1 \backslash i_t} w_t^{j \in S_t^1 \backslash i_t}(x, \vartheta),} \\
& \left. \left. \underline{\vartheta^{\ell \in S_t^2} + \hat{z}^{\ell \in S_t^2} \hat{\omega}_t^{\ell \in S_t^2}(x, \vartheta)} \right] \right\}, \quad \text{for } t \in \kappa,
\end{aligned}
$$

$$
V^{i_t}(T+1, x, \vartheta) = [g^{i_t}(\bar{x}^{i_t} - x) + g^{(P)i_t} \vartheta^{i_t}] \left(\frac{1}{1+r} \right)^T, \quad \text{for } i_t \in S_t^1; \quad \text{and}
$$

$$
\begin{aligned}
\hat{V}^{\hat{i}_t}(t, x, \vartheta) = \max_{\hat{q}_t^{\hat{i}_t}, u_t^{\hat{i}_t}, w_t^{\hat{i}_t}} & \left\{ \left[[\alpha_t^{\hat{i}_t} \hat{q}_t^{\hat{i}_t} - \beta_t^{\hat{i}_t}(\hat{q}_t^{\hat{i}_t})^2] - (\underline{c}^{\hat{i}_t} - \sigma_t^{\hat{i}_t} \vartheta^i) \hat{q}_t^{\hat{i}_t} - \hat{\varepsilon}_{i_t}^{\hat{i}_t} \hat{q}_t^{\hat{i}_t} \right. \right. \\
& - \sum_{j \in \bar{K}_t^{\hat{i}_t} \cap S_t^1} \varepsilon_{i_t}^j \phi_t^j(x, \vartheta) - \sum_{\varsigma \in \bar{K}_t^{\hat{i}_t} \cap S_t^2} \hat{\varepsilon}_{\hat{i}_t}^\varsigma \hat{\phi}_t^\varsigma(x, \vartheta) - \chi^{\hat{i}_t}(w_t^{\hat{i}_t})^2 \\
& \left. - c_{\hat{i}_t}^a (u_t^{\hat{i}_t})^2 - h_t^{\hat{i}_t} x \right] \left(\frac{1}{1+r} \right)^{t-1} + \bar{V}^{\hat{i}_t} \left[t+1, x + \sum_{j \in S_t^1} a^j \phi_t^j(x, \vartheta) \right. \\
& + \sum_{\substack{\varsigma \in S_t^2 \\ \varsigma \neq \hat{i}_t}} \hat{a}^\varsigma \hat{\phi}_t^\varsigma(x, \vartheta) + \hat{a}^{\hat{i}_t} \hat{q}_t^{\hat{i}_t} - \sum_{j \in S_t^1} b_j v_t^j(x, \vartheta) x^{1/2}
\end{aligned}
$$

$$- \sum_{\substack{j \in S_t^2 \\ j \neq \hat{i}_t}} b_j \hat{v}_t^j(x, \vartheta) x^{1/2} - b_{\hat{i}_t} u_t^{\hat{i}_t} x^{1/2} - \delta x, \vartheta^{\hat{i}_t}$$

$$+ \hat{z}^{\hat{i}_t} \hat{w}_t^{\hat{i}_t}, \underline{\vartheta^{j \in S_t^1} + z^{j \in S_t^1} \omega_t^{j \in S_t^1}}(x, \vartheta),$$

$$\left. \left. \left. \underline{\vartheta^{\ell \in S_t^2 \setminus \hat{i}_t} + \hat{z}^{\ell \in S_t^2 \setminus \hat{i}_t} \hat{\omega}_t^{\ell \in S_t^2 \setminus \hat{i}_t}}(x, \vartheta) \right] \right] \right\}, \quad for \ t \in \kappa,$$

$$\hat{V}^{\hat{i}_t}(T + 1, x, \vartheta) = [g^{\hat{i}_t}(\bar{x}^{\hat{i}_t} - x) + g^{(P)\hat{i}_t} \vartheta^{\hat{i}_t}] \left(\frac{1}{1+r} \right)^T, \quad for \ \hat{i}_t \in S_t^2;$$

$$(3.1)$$

$$c^{i_t} + \varepsilon_{i_t}^{i_t} - a^{i_t} \bar{V}_{x_{t+1}}^{i_t}(t + 1, x_{t+1}, \vartheta_{t+1})(1 + r)^{t-1}$$

$$< (\underline{c}^{i_t} - \sigma_t^{i_t} \vartheta_t) + \hat{\varepsilon}_{i_t}^{i_t}$$

$$- \hat{a}^{i_t} \bar{V}_{x_{t+1}}^{i_t}(t + 1, x_{t+1}^{i_t \to S_t^2}, \vartheta_{t+1}^{i_t \to S_t^2})(1 + r)^{t-1}, \quad for \ i_t \in S_t^1 \quad and \quad (3.2)$$

$$c^{\hat{i}_t} + \varepsilon_{\hat{i}_t}^{\hat{i}_t} - a^{\hat{i}_t} \bar{V}_{x_{t+1}}^{\hat{i}_t}(t + 1, x_{t+1}^{\hat{i}_t \to S_t^1}, \vartheta_{t+1}^{\hat{i}_t \to S_t^1})(1 + r)^{t-1}$$

$$\geq (\underline{c}^{\hat{i}_t} - \sigma_t^{\hat{i}_t} \vartheta_t) + \varepsilon_{\hat{i}_t}^{\hat{i}_t} - \hat{a}^{\hat{i}_t} \bar{V}_{x_{t+1}}^{\hat{i}_t}(t + 1, x_{t+1}, \vartheta_{t+1})(1 + r)^{t-1}, \quad if \ \hat{i}_t \in S_t^2,$$

where $\bar{V}^i[t + 1, x_{t+1}, \vartheta_{t+1}, y_{t+1}] = V^i[t + 1, x_{t+1}, \vartheta_{t+1}, y_{t+1}]$ *if* i *uses conventional technology in stage* $t + 1$ *and* $\bar{V}^i[t + 1, x_{t+1}, \vartheta_{t+1}, y_{t+1}] = \hat{V}^i[t + 1, x_{t+1}, \vartheta_{t+1}, y_{t+1}]$ *if* i *uses environment preserving technology in stage* $t + 1$; $\vartheta^{j \in S_t^1 \setminus i_t} + z^{j \in S_t^1 \setminus i_t} \omega_t^{j \in S_t^1 \setminus i_t}(x, \vartheta)$ *is the vector containing* $\vartheta^j + z^j \omega_t^j(x, \vartheta)$ *for* $j \in S_t^1$ *and* $j \neq i_t$, *and* $\vartheta^{\ell \in S_t^2} + \hat{z}^{\ell \in S_t^2} \hat{\omega}_t^{\ell \in S_t^2}(x, \vartheta)$ *is the vector containing* $\vartheta^\ell + \hat{z}^\ell \hat{\omega}_t^\ell(x, \vartheta)$ *for* $\ell \in S_t^2$; $x_{t+1}^{i_t \to S_t^2}$ *and* $\vartheta_{t+1}^{i_t \to S_t^2}$ *are the state variables in stage* $t + 1$ *if nation* $i_t \in S_t^1$ *adopts environmental preserving technology in stage* t; *and* $x_{t+1}^{\hat{i}_t \to S_t^1}$ *and* $\vartheta_{t+1}^{\hat{i}_t \to S_t^1}$ *are the state variables in stage* $t + 1$ *if nation* $\hat{i}_t \in S_t^2$ *adopts conventional technology in stage* t.

Proof. See Appendix A.   □

Performing the indicated maximization in (3.1) we obtain:

$$\alpha_t^{i_t} - 2\beta_t^{i_t} q_t^{i_t} = c^{i_t} + \varepsilon_{i_t}^{i_t} - a^{i_t} \bar{V}_{x_{t+1}}^{i_t}(t + 1, x_{t+1}, \vartheta_{t+1})(1 + r)^{t-1}, \quad for \ i_t \in S_t^1;$$

$$(3.3)$$

and

$$\alpha_t^{\hat{i}_t} - 2\beta_t^{\hat{i}_t} q_t^{\hat{i}_t} = (\underline{c}^{\hat{i}_t} - \sigma_t^{\hat{i}_t} \vartheta_t) + \hat{\varepsilon}_{\hat{i}_t}^{\hat{i}_t}$$

$$- \hat{a}^{\hat{i}_t} \bar{V}_{x_{t+1}}^{\hat{i}_t}(t + 1, x_{t+1}, \vartheta_{t+1})(1 + r)^{t-1}, \quad for \ \hat{i}_t \in S_t^2. \quad (3.4)$$

In view of (2.4) and (2.5), the left-hand side of (3.3) and that of (3.4) reflect the marginal revenues to the industrial sectors. To motivate the industrial sectors to produce outputs as given in (3.3) nation i_t has to impose a tax v_t^{it} equaling $\varepsilon_{i_t}^{it} - a^{it}\bar{V}_{x_{t+1}}^{it}(t+1, x_{t+1}, \vartheta_{t+1})(1+r)^{t-1}$ on a unit of output produced with conventional technique. Similarly, nation $\hat{i}_t$ has to impose a tax $\hat{v}_t^{it}$ equaling $\varepsilon_{\hat{i}_t}^{it} - \hat{a}^{it}\bar{V}_{x_{t+1}}^{it}(t+1, x_{t+1}, \vartheta_{t+1})(1+r)^{t-1}$ on a unit of output produced with clean technique to arrive at (3.4). The term $\varepsilon_{i_t}^{it} - a^{it}\bar{V}_{x_{t+1}}^{it}(t+1, x_{t+1})(1+r)^{t-1}$ reflects the marginal social cost to nation i_t brought about by a unit of output produced with conventional technique. The term $\hat{\varepsilon}_{\hat{i}_t}^{it} - \hat{a}^{it}\bar{V}_{x_{t+1}}^{it}(t+1, x_{t+1})(1+r)^{t-1}$ reflects the marginal social cost to nation $\hat{i}_t$ brought about by a unit of output produced with clean technique. Though conventional technique emits higher level of pollutants, nations have no incentive to switch to clean technique if the sum of marginal cost of producing the output and the nation's social cost resulted from using conventional technique is lower than that resulted from using clean technique.

Using Theorem 3.1 the game equilibrium strategies can be obtained as:

$$\phi_t^{it}(x, \vartheta) = [\alpha_t^{it} - c^{it} - \varepsilon_{i_t}^{it} + a^{it}\bar{V}_{x_{t+1}}^{it}(t+1, x_{t+1}, \vartheta_{t+1})(1+r)^{t-1}]/2\beta_t^{it},$$

$$v_t^{it}(x, \vartheta) = -\frac{b_{i_t}}{2c_{i_t}^a}\bar{V}_{x_{t+1}}^{it}(t+1, x_{t+1}, \vartheta_{t+1})(1+r)^{t-1}x^{1/2},$$

$$\omega_t^{it}(x, \vartheta) = \frac{z^{it}}{2\chi^{it}}\bar{V}_{\vartheta_{t+1}^{it}}^{it}(t+1, x_{t+1}, \vartheta_{t+1})(1+r)^{t-1}, \quad \text{for } i_t \in S_t^1; \quad \text{and}$$

$$\hat{\phi}_t^{it}(x, \vartheta) = \alpha_t^{\hat{i}t} - (\underline{c}^{\hat{i}t} - \sigma_t^{\hat{i}t}\vartheta^{\hat{i}t}) \tag{3.5}$$

$$- \hat{\varepsilon}_{\hat{i}_t}^{it} + \hat{a}^{it}\bar{V}_{x_{t+1}}^{it}(t+1, x_{t+1}, \vartheta_{t+1})(1+r)^{t-1}]/2\hat{\beta}_t^{it},$$

$$\hat{v}_t^{it}(x, \vartheta) = -\frac{b_{\hat{i}_t}}{2c_{\hat{i}_t}^a}\bar{V}_{x_{t+1}}^{\hat{i}t}(t+1, x_{t+1}, \vartheta_{t+1})(1+r)^{t-1}x^{1/2},$$

$$\hat{\omega}_t^{it}(x, \vartheta) = \frac{\hat{z}^{it}}{2\hat{\chi}^{it}}\bar{V}_{\vartheta_{t+1}^{it}}^{\hat{i}t}(t+1, x_{t+1}, \vartheta_{t+1})(1+r)^{t-1}, \quad \text{for } \hat{i}_t \in S_t^2.$$

The value functions of the game (2.1), (2.3) and (2.9) satisfying Theorem 3.1 can be obtained as the following.

Proposition 3.1.

$$V^{it}(t, x, \vartheta) = \left(A_t^{i_i}x + H_t^{i_i}\vartheta^{it} + J_t^{i_i}(\vartheta^{it})^2 + \sum_{\substack{j=1 \\ j\neq i_t}}^n K_t^{it(j)}\vartheta^j + C_t^{it}\right)\left(\frac{1}{1+r}\right)^{t-1},$$

for $i_t \in S_t^1$,

$$\hat{V}^{\hat{i}_t}(t,x,\vartheta) = \left(\hat{A}_t^{\hat{i}_i} x + \hat{H}_t^{\hat{i}_i} \vartheta^{\hat{i}_t} + \hat{J}_t^{\hat{i}_i} (\vartheta^{\hat{i}_t})^2 + \sum_{\substack{j=1 \\ j \neq \hat{i}_t}}^{n} \hat{K}_t^{\hat{i}_t(j)} \vartheta^j + \hat{C}_t^{\hat{i}_t} \right) \left(\frac{1}{1+r} \right)^{t-1},$$

$$\text{for } \hat{i}_t \in S_t^2, \quad \text{for } t \in \kappa;$$

(3.6)

with $A_t^{i_t}, H_t^{i_t}, J_t^{i_t}, C_t^{i_t}, K_t^{i_t(j)}, \hat{A}_t^{\hat{i}_t}, \hat{H}_t^{\hat{i}_t}, \hat{J}_t^{\hat{i}_t}, \hat{K}_t^{\hat{i}_t(j)}$ and $\hat{C}_t^{\hat{i}_t}$ being constants involving the model parameters; for $t \in \kappa$.

Proof. See Appendix B. $\qquad\qquad\qquad\qquad\qquad\qquad\qquad\qquad\qquad$ □

4. Cooperative Clean Technology Development and Pollution Control

Now consider the case when all the nations want to collaborate and tackle the pollution problem together. Cooperation suggests the possibility of socially optimal and group efficient solutions to decision problems involving interactive strategic actions.

4.1. *Gains in cooperative clean technique development*

Research and development leading to the advancement of technology which include scientific knowledge, technical know-how and information technology have public goods property. In the case of cooperative development in clean technology, two positive externalities appear. First, savings in developmental costs from less duplication, sharing of knowledge and economy of scales arise. In particular the costs of the nation's clean technology development efforts at stage t under cooperation are:

$$\chi^{(S_t^1)i_t}(w_t^{i_t})^2 \quad \text{for nation } i_t \in S_t^1, \quad \text{and}$$

$$\hat{\chi}^{(S_t^2)\hat{i}_t}(w_t^{\hat{i}_t})^2 \quad \text{for nation } \hat{i}_t \in S_t^2.$$

In particular, $\hat{\chi}^{(S_t^2)j}(w_t^j)^2 \leq \chi^j(w_t^j)^2$ and $\chi^{(S_t^1)j}(w_t^j)^2 \leq \chi^j(w_t^j)^2$.

Second, under cooperative development, the effectiveness of cooperative research effort is higher than the effectiveness of research effort under individual pursuit. The evolution of clean technology indicator under cooperation becomes

$$\vartheta_{t+1}^{i_t} = \vartheta_t^{i_t} + z_t^{(S_t^1)i_t} w_t^{i_t}, \quad \vartheta_1^{i_t} = \vartheta_{1(0)}^{i_t}, \quad \text{for } i_t \in S_t^1, \quad \text{and}$$

$$\vartheta_{t+1}^{\hat{i}_t} = \vartheta_t^{\hat{i}_t} + \hat{z}_t^{(S_t^2)\hat{i}_t} w_t^{\hat{i}_t}, \quad \vartheta_1^{\hat{i}_t} = \vartheta_{1(0)}^{\hat{i}_t}, \quad \text{for } \hat{i}_t \in S_t^2,$$

(4.1)

where $z_t^{(S_t^1)i_t} \geq z^{i_t}$ and $\hat{z}_t^{(S_t^2)\hat{i}_t} > z^{\hat{i}_t}$.

4.2. *Cooperative optimization*

An important feature of collaborative scheme is group optimality. In particular, group optimality ensures that all potential gains from cooperation are captured.

Consider the case when all the nations agree to act cooperatively so that the joint payoff will be maximized. To ensure group optimality, the nations would seek to maximize their joint payoff under cooperation. Since two technique choices are available they have to determine which nations would use which type of techniques over the T stages. Let M^γ be a $\kappa \times n$ matrix reflecting the pattern of technique choices by the n nations over the T stages. In particular, according to pattern M^γ, the set of nations that use conventional technique is $S_t^{M^\gamma [1]}$ and the set of nations that use clean technique is $S_t^{M^\gamma [2]}$ in stage $t \in \kappa$. To select the controls which would maximize joint payoff under pattern M^γ the nations have to solve the following optimal control problem which maximizes:

$$
\sum_{t=1}^{T} \left\{ \sum_{i_t \in S_t^{M^\gamma [1]}} \left[[\alpha_t^{it} q_t^{it} - \beta_t^{it}(q_t^{it})^2] - c^{it} q_t^{it} - \varepsilon_{i_t}^{it} q_t^{it} - \sum_{j \in \bar{K}_t^{it} \cap S_t^{M^\gamma [1]}} \varepsilon_{i_t}^{j} q_t^{j} \right. \right.
$$

$$
\left. - \sum_{\zeta \in \bar{K}_t^{it} \cap S_t^{M^\gamma [2]}} \hat{\varepsilon}_{i_t}^{\zeta} \hat{q}_t^{\zeta} - \chi^{(S_t^{M^\gamma [1]})i_t}(w_t^{it})^2 - c_{i_t}^{a}(u_t^{it})^2 - h_t^{it} x_t \right] \left(\frac{1}{1+r} \right)^{t-1}
$$

$$
+ \sum_{\hat{i}_t \in S_t^{M^\gamma [2]}} \left[[\hat{\alpha}_t^{it} \hat{q}_t^{it} - \hat{\beta}_t^{it}(\hat{q}_t^{it})^2] - (\underline{c}^{it} - \sigma_t^{it} \vartheta_t^{it}) \hat{q}_t^{it} - \hat{\varepsilon}_{i_t}^{it} \hat{q}_t^{it} - \sum_{j \in \bar{K}_t^{it} \cap S_t^{M^\gamma [1]}} \varepsilon_{i_t}^{j} q_t^{j} \right.
$$

$$
\left. - \sum_{\zeta \in \bar{K}_t^{it} \cap S_t^{M^\gamma [2]}} \hat{\varepsilon}_{i_t}^{\zeta} \hat{q}_t^{\zeta} - \hat{\chi}^{(S_t^{M^\gamma [2]})i_t}(\hat{w}_t^{it})^2 - h_t^{it} x_t \right] \left(\frac{1}{1+r} \right)^{t-1} \right\}
$$

$$
+ \sum_{i=1}^{n} [g^i(\bar{x}^i - x_{T+1}) + g^{(P)i} \vartheta_{T+1}^i] \left(\frac{1}{1+r} \right)^{T} \tag{4.2}
$$

subject to the dynamics

$$
x_{t+1} = x_t + \sum_{\ell_t \in S_t^{M^\gamma [1]}} a^{\ell_t} q_t^{\ell_t} + \sum_{\hat{\ell}_t \in S_t^{M^\gamma [2]}} \hat{a}^{\ell_t} \hat{q}_t^{\ell_t} - \sum_{\ell_t \in S_t^{M^\gamma [1]}} b_{\ell_t} u_t^{\ell_t}(x_t)^{1/2}
$$

$$
- \sum_{\hat{\ell}_t \in S_t^{M^\gamma [1]}} b_{\hat{\ell}_t} \hat{u}_t^{\ell_t}(x_t)^{1/2} - \delta x_t, \quad x_1 = x_{1(0)},
$$

$$
\vartheta_{t+1}^{it} = \vartheta_t^{it} + z_t^{(S_t^{M^\gamma [1]})i_t} w_t^{it}, \quad \vartheta_1^{it} = \vartheta_{1(0)}^{it}, \quad \text{for } i_t \in S_t^{M^\gamma [1]}, \quad \text{and}
$$
$$
\hat{\vartheta}_{t+1}^{it} = \hat{\vartheta}_t^{it} + \hat{z}_t^{(S_t^{M^\gamma [2]})i_t} \hat{w}_t^{it}, \quad \hat{\vartheta}_1^{it} = \hat{\vartheta}_{1(0)}^{it}, \quad \text{for } \hat{i}_t \in S_t^{M^\gamma [1]}. \tag{4.3}
$$

The solution to the optimal control problems (4.2)–(4.3) can be characterized as follows.

Theorem 4.1. *A set of strategies* $\{q_t^{\ell_t *} = \psi_t^{(M^\gamma)\ell_t}(x,\vartheta),\ \hat{q}_t^{\hat{\ell}_t *} = \hat{\psi}_t^{(M^\gamma)\hat{\ell}_t}(x,\vartheta),$
$u_t^{\ell_t *} = \mu_t^{(M^\gamma)\ell_t}(x,\vartheta),\ \hat{u}_t^{\hat{\ell}_t *} = \hat{\mu}_t^{(M^\gamma)\hat{\ell}_t}(x,\vartheta),\ w_t^{\ell_t *} = \varpi_t^{(M^\gamma)\ell_t}(x,\vartheta),\ \hat{u}_t^{\hat{\ell}_t *} =$
$\hat{\varpi}_t^{(M^\gamma)\hat{\ell}_t}(x,\vartheta)$ *for* $t \in \kappa$ *and* $\ell_t \in S_t^{M^\gamma[1]}$ *and* $\hat{\ell}_t \in S_t^{M^\gamma[2]}\}$ *constitutes an optimal
solution to the control problems* (4.2)–(4.3) *if there exist functions* $W^{M^\gamma}(t,x,\vartheta)$,
for $t \in \kappa$, *such that the following recursive relations are satisfied:*

$$W^{M^\gamma}(T+1,x,\vartheta) = \sum_{i=1}^{n}[g^i(\bar{x}^i - x) + g^{(P)i}\vartheta^i]\left(\frac{1}{1+r}\right)^T,$$

$$W^{M^\gamma}(t,x,\vartheta) = \max_{u_t^{\ell_t},q_t^{\ell_t},w_t^{\ell_t},\ell_t \in S_t^{M^\gamma[1]};\hat{u}_t^{\hat{\ell}_t},\hat{q}_t^{\hat{\ell}_t},\hat{w}_t^{\hat{\ell}_t},\hat{\ell}_t \in S_t^{M^\lambda[2]}} \left\{ \sum_{i_t \in S_t^{M^\gamma[1]}} \left[[\alpha_t^{i_t} q_t^{i_t} - \beta_t^{i_t}(q_t^{i_t})^2] \right.\right.$$

$$- c^{i_t} q_t^{i_t} - \varepsilon_{i_t}^{i_t} q_t^{i_t} - \sum_{j \in \bar{K}_t^{i_t} \cap S_t^{M^\gamma[1]}} \varepsilon_{i_t}^{j} q_t^{j} - \sum_{\zeta \in \bar{K}_t^{i_t} \cap S_t^{M^\gamma[2]}} \hat{\varepsilon}_{i_t}^{\zeta} \hat{q}_t^{\zeta}$$

$$\left. - \chi^{(S_t^{M^\gamma[1]})i_t}(w_t^{i_t})^2 - c_{i_t}^a(u_t^{i_t})^2 - h_t^{i_t} x_t \right] \left(\frac{1}{1+r}\right)^{t-1}$$

$$+ \sum_{\hat{i}_t \in S_t^{M^\gamma[2]}} \left[[\alpha_t^{\hat{i}_t}\hat{q}_t^{\hat{i}_t} - \beta_t^{\hat{i}_t}(\hat{q}_t^{\hat{i}_t})^2] - (\underline{c}^{\hat{i}_t} - \sigma_t^{\hat{i}_t}\vartheta^{\hat{i}_t})\hat{q}_t^{\hat{i}_t} - \hat{\varepsilon}_{\hat{i}_t}^{\hat{i}_t}\hat{q}_t^{\hat{i}_t} \right.$$

$$- \sum_{j \in \bar{K}_t^{\hat{i}_t} \cap S_t^{M^\gamma[1]}} \varepsilon_{\hat{i}_t}^{j} q_t^{j} - \sum_{\zeta \in \bar{K}_t^{\hat{i}_t} \cap S_t^{M^\gamma[2]}} \hat{\varepsilon}_{\hat{i}_t}^{\zeta} \hat{q}_t^{\zeta} - \hat{\chi}^{(S_t^{M^\gamma[2]})\hat{i}_t}(\hat{w}_t^{\hat{i}_t})^2$$

$$\left. - c_{\hat{i}_t}^a(\hat{u}_t^{\hat{i}_t})^2 - h_t^{\hat{i}_t} x \right]\left(\frac{1}{1+r}\right)^{t-1} + W^{M^\gamma}\left[t+1, x + \sum_{\ell_t \in S_t^{M^\gamma[1]}} a^{\ell_t} q_t^{\ell_t} \right.$$

$$+ \sum_{\hat{\ell}_t \in S_t^{M^\gamma[2]}} \hat{a}^{\hat{\ell}_t}\hat{q}_t^{\hat{\ell}_t} - \sum_{\ell_t \in S_t^{M^\gamma[1]}} b_{\ell_t} u_t^{\ell_t}(x)^{1/2} - \sum_{\hat{\ell}_t \in S_t^{M^\gamma[2]}} b_{\hat{\ell}_t} \hat{u}_t^{\hat{\ell}_t}(x)^{1/2}$$

$$- \delta x, \underline{\vartheta_t^{i_t \in S_t^{M^\gamma[1]}} + z_t^{(S_t^{M^\gamma[1]})i_t \in S_t^{M^\gamma[1]}} w_t^{i_t \in S_t^{M^\gamma[1]}}},$$

$$\left.\left. \underline{\vartheta_t^{\hat{i}_t \in S_t^{M^\gamma[2]}} + \hat{z}_t^{(S_t^{M^\gamma[2]})\hat{i}_t \in S_t^{M^\gamma[2]}} w_t^{\hat{i}_t \in S_t^{M^\gamma[2]}}} \right] \right\}, \quad \text{for } t \in \kappa, \quad (4.4)$$

where $\underline{\vartheta_t^{i_t \in S_t^{M^\gamma[1]}} + z_t^{(S_t^{M^\gamma[1]})i_t \in S_t^{M^\gamma[1]}} w_t^{i_t \in S_t^{M^\gamma[1]}}}$ *is the vector containing* $\vartheta_t^{i_t} +$
$z_t^{(S_t^{M^\gamma[1]})i_t} w_t^{i_t}$, *for* $i_t \in S_t^{M^\gamma[1]}$; *and* $\underline{\vartheta_t^{\hat{i}_t \in S_t^{M^\gamma[2]}} + \hat{z}_t^{(S_t^{M^\gamma[2]})\hat{i}_t \in S_t^{M^\gamma[2]}} w_t^{\hat{i}_t \in S_t^{M^\gamma[2]}}}$ *is the
vector containing* $\vartheta_t^{\hat{i}_t} + z_t^{(S_t^{M^\gamma[2]})\hat{i}_t} w_t^{\hat{i}_t}$, *for* $\hat{i}_t \in S_t^{M^\gamma[2]}$.

Proof. The results in (4.4) satisfy the standard optimality conditions in discrete-time dynamic programming. □

Performing the indicated maximization in (4.4) yields the optimal controls under cooperation as:

$$\psi_t^{(M^\gamma)\ell_t}(x,\vartheta) = \left[\alpha_t^{\ell_t} - c^{\ell_t} - \varepsilon_{\ell_t}^{\ell_t} - \sum_{j\in \underline{K}_t^{\ell_t}} \varepsilon_j^{\ell_t}\right.$$

$$\left. + a^{\ell_t} W_{x_{t+1}}^{M^\gamma}(t+1, x_{t+1}, \vartheta_{t+1})(1+r)^{t-1}\right]/2\beta_t^{\ell_t}, \quad \text{for } \ell_t \in S_t^{M^\gamma[1]}, \quad \text{and}$$

$$\hat{\psi}_t^{(M^\gamma)\hat{\ell}_t}(x,\vartheta) = \left[\alpha_t^{\hat{\ell}_t} - (\underline{c}^{\hat{\ell}_t} - \sigma_t^{\hat{\ell}_t}\vartheta^{\hat{\ell}_t}) - \varepsilon_{\hat{\ell}_t}^{\hat{\ell}_t} - \sum_{j\in \underline{K}_t^{\hat{\ell}_t}} \hat{\varepsilon}_j^{\hat{\ell}_t}\right.$$

$$\left. + \hat{a}^{\hat{\ell}_t} W_{x_{t+1}}^{M^\gamma}(t+1, x_{t+1}, \vartheta_{t+1})(1+r)^{t-1}\right]/2\beta_t^{\hat{\ell}_t}, \quad \text{for } \hat{\ell}_t \in S_t^{M^\gamma[2]};$$

$$\mu_t^{(M^\gamma)\ell_t}(x,\vartheta) = -\frac{b_{\ell_t}}{2c_{\ell_t}^a} W_{x_{t+1}}^{M^\gamma}(t+1, x_{t+1}, \vartheta_{t+1})(1+r)^{t-1}x^{1/2}, \quad \text{for } \ell_t \in S_t^{M^\gamma[1]}, \quad \text{and}$$

$$\hat{\mu}_t^{(M^\gamma)\hat{\ell}_t}(x,\vartheta) = -\frac{b_{\hat{\ell}_t}}{2c_{\hat{\ell}_t}^a} W_{x_{t+1}}^{M^\gamma}(t+1, x_{t+1}, \vartheta_{t+1})(1+r)^{t-1}x^{1/2}, \quad \text{for } \hat{\ell}_t \in S_t^{M^\gamma[2]};$$

$$\varpi_t^{(M^\gamma)\ell_t}(x,\vartheta) = \frac{z_t^{(S_t^{M^\gamma[1]})\ell_t}}{2\chi^{(S_t^{M^\gamma[1]})\ell_t}} W_{\vartheta_{t+1}^{\ell_t}}^{M^\gamma}(t+1, x_{t+1}, \vartheta_{t+1})(1+r)^{t-1}, \quad \text{for } \ell_t \in S_t^{M^\gamma[1]}, \quad \text{and}$$

$$\hat{\varpi}_t^{(M^\gamma)\hat{\ell}_t}(x,\vartheta) = \frac{\hat{z}_t^{(S_t^{M^\gamma[2]})\hat{\ell}_t}}{2\hat{\chi}^{(S_t^{M^\gamma[2]})\hat{\ell}_t}} W_{\vartheta_{t+1}^{\hat{\ell}_t}}^{M^\gamma}(t+1, x_{t+1}, \vartheta_{t+1})(1+r)^{t-1}, \quad \text{for } \hat{\ell}_t \in S_t^{M^\gamma[2]}. \quad (4.5)$$

Proposition 4.1. *System (4.4) admits a solution*

$$W^{M^\gamma}(t,x,\vartheta) = \left(A_t^{M^\gamma}x + \sum_{i=1}^n H_t^{(M^\gamma)i}\vartheta^i + \sum_{i=1}^n J_t^{(M^\gamma)i}(\vartheta^i)^2 + C_t^{M^\gamma}\right)\left(\frac{1}{1+r}\right)^{t-1},$$

$$t \in \kappa, \qquad (4.6)$$

$$W^{M^\gamma}(T+1,x,\vartheta) = \sum_{i=1}^n [g^i(\bar{x}^i - x_{T+1}) + g^{(P)i}\vartheta_{T+1}^i]\left(\frac{1}{1+r}\right)^T, \quad \text{for } t \in \kappa.$$

where $A_t^{M^\gamma}$, $H_t^{(M^\gamma)i}$, $J_t^{(M^\gamma)i}$ *and* $C_t^{M^\gamma}$ *are constants involving the model parameters.*

Proof. Similar to the proof of Proposition 3.1, we first substitute the optimal controls in (4.5) into (4.4). Then by invoking Proposition 4.1 the system in (4.4) yields an equation with the left-hand side being

$$A_t^{M^\gamma} x + \sum_{i=1}^{n} H_t^{(M^\gamma)i} \vartheta^i + \sum_{i=1}^{n} J_t^{(M^\gamma)i} (\vartheta^i)^2 + C_t^{M^\gamma}$$

and the right-hand side being an expression linear in x, ϑ^1, $\vartheta^2, \ldots, \vartheta^n$, $(\vartheta^1)^2$, $(\vartheta^2)^2, \ldots, (\vartheta^n)^2$. Hence $A_t^{M^\gamma}$, $H_t^{(M^\gamma)i}$, $J_t^{(M^\gamma)i}$ and $C_t^{M^\gamma}$ are constants involving the model parameters.   □

The technology pattern M^γ which yields the highest joint payoff $W^{M^\gamma}(t, x, \vartheta)$ will be adopted in the cooperative scheme. Let us denote the technique pattern that yields the highest joint payoff by M^*. Using Proposition 4.1 and (4.5) the optimal cooperative strategies under technology pattern M^* can be obtained in closed-form. The optimal technology pattern M^* yields plan for phasing out conventional technology and the introduction of clean technology at different stages for different nations. To induce the industrial sector to produce the socially optimal levels of output with the desired technique, we invoke the industry equilibrium condition (2.6) and the socially optimal output condition in (4.5) to obtain the optimal tax rates

$$v_t^{i_t} = \varepsilon_{i_t}^{i_t} + \sum_{j \in \underline{K}_t^{i_t}} \varepsilon_j^{i_t} - a^{i_t} A_{t+1}^{M^*} (1+r)^{-1} \quad \text{on output using conventional technique}$$

for $i_t \in S_t^{M^*[1]}$, and

$$\hat{v}_t^{\hat{i}_t} = \hat{\varepsilon}_{\hat{i}_t}^{\hat{i}_t} + \sum_{j \in \underline{\hat{K}}_t^{\hat{i}_t}} \hat{\varepsilon}_j^{\hat{i}_t} - \hat{a}^{\hat{i}_t} A_{t+1}^{M^*} (1+r)^{-1} \quad \text{on output using clean technique}$$

(4.7)

for $\hat{i}_t \in S_t^{M^*[2]}$.

Substituting the optimal control strategies from (4.5) into (4.2) yields the dynamics of pollution accumulation under cooperation as:

$$x_{t+1} = \sum_{\ell_t \in S_t^{M^*[1]}} \left[\alpha_t^{\ell_t} - c^{\ell_t} - \varepsilon_{\ell_t}^{\ell_t} - \sum_{j \in \underline{K}_t^{\ell_t}} \varepsilon_j^{\ell_t} + a^{\ell_t} A_{t+1}^{M^*} (1+r)^{-1} \right] / 2\beta_t^{\ell_t}$$

$$\times \sum_{\hat{\ell}_t \in S_t^{M^*[2]}} \left[\alpha_t^{\hat{\ell}_t} - (\underline{c}^{\hat{\ell}_t} - \sigma_t^{\hat{\ell}_t} \vartheta_t^{\hat{\ell}_t}) - \hat{\varepsilon}_{\hat{\ell}_t}^{\hat{\ell}_t} - \sum_{j \in \underline{\hat{K}}_t^{\hat{\ell}_t}} \hat{\varepsilon}_j^{\hat{\ell}_t} + \hat{a}^{\hat{\ell}_t} A_{t+1}^{M^*} (1+r)^{-1} \right] / 2\beta_t^{\hat{\ell}_t}$$

$$+ \left[1 + \sum_{j=1}^{n} \frac{(b_j)^2}{2c_j^a} A_{t+1}^{M^*} (1+r)^{-1} - \delta \right] x_t, \quad x_1 = x_{1(0)}. \quad (4.8)$$

Using Proposition 4.1 and (4.5) the optimal clean technology input can be expressed as:

$$\varpi_t^{(M^*)\ell_t}(x,\vartheta) = \frac{z_t^{(S_t^{M^*[1]})\ell_t}}{2\chi^{(S_t^{M^*[1]})\ell_t}}[H_{t+1}^{(M^*)\ell_t} z_t^{(S_t^{M^*[1]})\ell_t} + 2J_{t+1}^{(M^*)\ell_t} z_t^{(S_t^{M^*[1]})\ell_t}\vartheta^{\ell_t}](1+r)^{-1}$$

$$\div \left[1 - \frac{z_t^{(S_t^{M^*[1]})\ell_t}}{\chi^{(S_t^{M^*[1]})\ell_t}} J_{t+1}^{(M^*)\ell_t} z_t^{(S_t^{M^*[1]})\ell_t}(1+r)^{-1}\right], \quad \text{for } \ell_t \in S_t^{M^*[1]};$$

and

$$\hat{\varpi}_t^{(M^*)\hat{\ell}_t}(x,\vartheta) = \frac{\hat{z}_t^{(S_t^{M^*[2]})\hat{\ell}_t}}{2\hat{\chi}^{(S_t^{M^*[2]})\hat{\ell}_t}}[H_{t+1}^{(M^*)\hat{\ell}_t} \hat{z}_t^{(S_t^{M^*[2]})\hat{\ell}_t} + 2J_{t+1}^{(M^*)\hat{\ell}_t} \hat{z}_t^{(S_t^{M^*[2]})\hat{\ell}_t}\vartheta^{\hat{\ell}_t}](1+r)^{-1}$$

$$\div \left[1 - \frac{\hat{z}_t^{(S_t^{M^*[2]})\hat{\ell}_t}}{\hat{\chi}^{(S_t^{M^*[2]})\hat{\ell}_t}} J_{t+1}^{(M^*)\hat{\ell}_t} \hat{z}_t^{(S_t^{M^*[2]})\hat{\ell}_t}(1+r)^{-1}\right], \quad \text{for } \hat{\ell}_t \in S_t^{M^*[2]}.$$

$$(4.9)$$

Substituting the optimal control strategy (4.9) into (4.3) yields the dynamics of clean technology development under cooperation

$$\vartheta_{t+1}^{it} = \vartheta_t^{it} + z_t^{(S_t^{M^*[1]})it}\varpi_t^{(M^*)\ell_t}(x,\vartheta), \quad \vartheta_1^{it} = \vartheta_{1(0)}^{it}, \quad \text{for } i_t \in S_t^{M^*[1]}, \quad (4.10)$$

$$\vartheta_{t+1}^{\hat{i}t} = \vartheta_t^{\hat{i}t} + \hat{z}_t^{(S_t^{M^*[2]})\hat{i}t}\hat{\varpi}_t^{(M^*)\hat{\ell}_t}(x,\vartheta), \quad \vartheta_1^{\hat{i}t} = \vartheta_{1(0)}^{\hat{i}t}, \quad \text{for } \hat{i}_t \in S_t^{M^*[2]}. \quad (4.11)$$

The cooperative state dynamics (4.8) and (4.10)–(4.11) is a system of $n+1$ first-order linear difference equations which can be solved by standard techniques. We use $\{x_k^*, \vartheta_k^*\}_{k=t}^{T+1}$ to denote the solution of the system (4.8) and (4.10)–(4.11).

4.3. *Subgame consistent collaboration*

A crucial factor for dynamic cooperation to be sustainable is that the solution has to be subgame consistent in the sense that the specific optimality principle chosen at the outset must remain in effect at any instant of time throughout the game along the optimal state trajectory. Hence none of the participating nations would have an incentive to depart from the collaborative scheme. Cooperation will cease if any of the nations refuses to act accordingly at any time within the game horizon. Since nations are asymmetric and the number of nations may be large, a reasonable optimality principle for gain distribution is to share the gain from cooperation proportional to the nations' relative sizes of noncooperative payoffs. Such sharing principle fulfills individual rationality. Note that this optimality principle satisfies individual rationality because each nation's payoff under cooperation is higher than its payoff under noncooperation.

Let $\xi^\ell(t, x_t^*, \vartheta_t^*)$ denote nation ℓ's imputation (payoff under cooperation) covering the stages t to T under the agreed-upon optimality principle along the cooperative trajectories $\{x_k^*, \vartheta_K^*\}_{k=t}^{T+1}$. To achieve subgame consistency the agreed-upon optimality principle must be maintained at every stage of collaboration. Hence the solution imputation scheme has to satisfy the following.

Condition 4.1.

$$\xi^\ell(t, x_t^*, \vartheta_t^*) = \frac{\bar{V}^\ell(t, x_t^*, \vartheta_t^*)}{\sum_{i \in S_t^1} V^i(t, x_t^*, \vartheta_t^*) + \sum_{j \in S_t^2} \hat{V}^j(t, x_t^*, \vartheta_t^*)} W^{M^*}(t, x_t^*, \vartheta_t^*),$$

$$\text{for all } \ell \in N \text{ and all } t \in \kappa, \tag{4.12}$$

where $\bar{V}^\ell(t, x_t^*, \vartheta_t^*) = V^\ell(t, x_t^*, \vartheta_t^*)$ if $\ell \in S_t^1$ and $\bar{V}^\ell(t, x_t^*, \vartheta_t^*) = \hat{V}^\ell(t, x_t^*, \vartheta_t^*)$ if $\ell \in S_t^2$.

Crucial to the derivation of a time-consistent solution is the formulation of an imputation distribution mechanism that would lead to the realization of Condition 4.1. This will be done in the next section.

5. Imputation Distribution Mechanism

To design an imputation distribution scheme over time so that the agreed-upon imputation in Condition 4.1 can be realized, we apply the techniques developed in Yeung and Petrosyan [2010]. In formulating an imputation distribution procedure we let $B_t^\ell(x_t^*, \vartheta_t^*)$ denote the payment that nation ℓ will receive at stage t under the cooperative agreement given the states are x_t^* and ϑ_t^* at stage $t \in \kappa$. The payment scheme involving $B_t^\ell(x_t^*, \vartheta_t^*)$ constitutes an imputation distribution procedure in the sense that along the optimal state trajectories $\{x_k^*, \vartheta_K^*\}_{k=t}^{T+1}$ the imputation to nation ℓ over the stages from t to T can be expressed as:

$$\xi^\ell(t, x_t^*, \vartheta_t^*) = \sum_{\zeta=t}^{T} B_\zeta^\ell(x_\zeta^*, \vartheta_\zeta^*) \left(\frac{1}{1+r}\right)^{\zeta-1} + [g^\ell(\bar{x}^\ell - x_{T+1}) + g^{(P)\ell}\vartheta_{T+1}^\ell] \left(\frac{1}{1+r}\right)^{T},$$

$$\text{for } \ell \in N \text{ and } t \in \kappa. \tag{5.1}$$

Theorem 5.1. *A payment*

$$B_t^\ell(x_t^*, \vartheta_t^*) = (1+r)^{t-1}[\xi^\ell(t, x_t^*, \vartheta_t^*) - \xi^\ell(t+1, x_{t+1}^*, \vartheta_{t+1}^*)], \quad \text{for } \ell \in N$$

given to nation $\ell \in N$ *at stage* $t \in \{1, 2, \dots, T-1\}$, *and a payment*

$$B_T^\ell(x_T^*, \vartheta_T^*) = (1+r)^{T-1}\Bigg[\xi^\ell(T, x_T^*, \vartheta_T^*) - [g^\ell(\bar{x}^\ell - x_{T+1}^*)$$

$$+ g^{(P)\ell}\vartheta_{T+1}^\ell] \left(\frac{1}{1+r}\right)^{T}\Bigg], \tag{5.2}$$

given to nation $\ell \in N$ *at stage* T *would lead to the realization of the imputation* $\{\xi^\ell(t, x_t^*, \vartheta_t^*), \text{ for } t \in \kappa \text{ and } \ell \in N\}$.

Proof. Making use of (5.1), one can arrive at:

$$\xi^\ell(t, x_t^*, \vartheta_t^*) = \sum_{\zeta=t}^{h-1} B_\zeta^\ell(x_\zeta^*, \vartheta_\zeta^*) \left(\frac{1}{1+r}\right)^{\zeta-1} + \xi^\ell(h, x_h^*, \vartheta_h^*), \qquad (5.3)$$

for $\ell \in N$ and $t \in \kappa$ and $h \in \{t+1, t+2, \ldots, T\}$.

From (5.3) one can obtain

$$B_t^\ell(x_t^*, \vartheta_t^*) \left(\frac{1}{1+r}\right)^{t-1} = \xi^\ell(t, x_t^*, \vartheta_t^*) - \xi^\ell(t+1, x_{t+1}^*, \vartheta_{t+1}^*),$$

for $\ell \in N$ and $t \in \kappa$.

Note that $B_t^\ell(x_t^*, \vartheta_t^*)(\frac{1}{1+r})^{t-1}$ is the present value (as from initial stage 1) of a payment $B_t^\ell(x_t^*, \vartheta_t^*)$ that will be given nation ℓ at stage t. Hence if a payment as specified in (5.2) is given to nation ℓ at stage $t \in \kappa$, the imputation $\{\xi^\ell(t, x_t^*, \vartheta_t^*),$ for $t \in \kappa$ and $\ell \in N\}$ can be realized by showing that

$$\sum_{\zeta=t}^{T} B_\zeta^\ell(x_\zeta^*, \vartheta_\zeta^*) \left(\frac{1}{1+r}\right)^{\zeta-1} + g^\ell(\bar{x}^\ell - x_{T+1}) \left(\frac{1}{1+r}\right)^T$$

$$= \sum_{\zeta=t}^{T} [\xi^\ell(\zeta, x_\zeta^*, \vartheta_\zeta^*) - \xi^\ell(\zeta+1, x_{\zeta+1}^*, \vartheta_{\zeta+1}^*)] = \xi^\ell(t, x_t^*, \vartheta_t^*), \qquad (5.4)$$

for $\ell \in N$ and $t \in \kappa$, given that $\xi^\ell(T+1, x_{T+1}^*, \vartheta_{T+1}^*) = g^\ell(\bar{x}^\ell - x_{T+1}^*)(\frac{1}{1+r})^T$. $\quad\square$

Invoking Theorem 5.1 and Condition 4.1 the payment (in present value terms) to nation ℓ in stage $t \in \kappa$ can be obtained as:

$$B_t^\ell(x_t^*, \vartheta_t^*) \left(\frac{1}{1+r}\right)^{t-1}$$

$$= \xi^\ell(t, x_t^*, \vartheta_t^*) - \xi^\ell(t+1, x_{t+1}^*, \vartheta_{t+1}^*)$$

$$= \frac{\bar{V}^\ell(t, x_t^*, \vartheta_t^*)}{\sum_{i \in S_t^1} V^i(t, x_t^*, \vartheta_t^*) + \sum_{j \in S_t^2} \hat{V}^j(t, x_t^*, \vartheta_t^*)} W^{M^*}(t, x_t^*, \vartheta_t^*)$$

$$- \frac{\bar{V}^\ell(t+1, x_{t+1}^*, \vartheta_{t+1}^*)}{\sum_{i \in S_{t+1}^1} V^i(t+1, x_{t+1}^*, \vartheta_{t+1}^*) + \sum_{j \in S_{t+1}^2} \hat{V}^j(t+1, x_{t+1}^*, \vartheta_{t+1}^*)}$$

$$W^{M^*}(t+1, x_{t+1}^*, \vartheta_{t+1}^*), \quad \text{for } \ell \in N, t \in \kappa, \qquad (5.5)$$

where $\bar{V}^\ell(t, x_t^*, \vartheta_t^*) = V^\ell(t, x_t^*, \vartheta_t^*)$ if $\ell \in S_t^1$ and $\bar{V}^\ell(t, x_t^*, \vartheta_t^*) = \hat{V}^\ell(t, x_t^*, \vartheta_t^*)$ if $\ell \in S_t^2$.

Formula (5.5) indeed provides an imputation distribution procedure leading to the satisfaction of Condition 4.1 and hence a time-consistent solution will be obtained. Under cooperation, nations would use the optimal cooperative strategies (4.5). Substituting the states x_t^* and ϑ_t^* and these cooperative strategies into the nations' payoff at stage t in (2.7) and (2.8) with reference to the chosen technique pattern M^*, one can obtain the payoffs that these nations receive in stage t. We use $\zeta_t^i(x_t^*, \vartheta_t^*)$ to denote the payment at stage t that nation ℓ receives when nations are using the optimal cooperative strategies.

According to Theorem 5.1, the payoff that nation ℓ should receive under the agreed-upon optimality principle is $B_t^\ell(x_t^*, \vartheta_t^*)$. Hence a transfer payment

$$\chi_t^\ell(x_t^*, \vartheta_t^*) = B_t^\ell(x_t^*, \vartheta_t^*) - \zeta_t^i(x_t^*, \vartheta_t^*)$$

has to be given to nation ℓ in stage t, for $\ell \in N$ and $t \in \kappa$.

Though cooperation in environmental control holds out the best promise of effective action, limited success has been observed. Conventional multinational joint initiatives like the 2015 Paris Agreement on Climate Change and Kyoto Protocol can hardly be expected to offer a long-term solution because first, the plans are limited only to emissions reduction which is unlikely be able to offer an effective mean to halt the accelerating trend of environmental deterioration, and second, there is no guarantee that participants will always be better off within the entire duration of the agreement (as some states in the US had already refused immediate adoption of the Paris Agreement). Adoption and development of clean technology play a key role to effectively solving the continual worsening global pollution problem amid many nations' growing economies.

The above analysis presents a subgame consistent cooperative solution of a cooperative environmental scheme which entails (i) group optimality that brings forth the maximized joint payoff of the participating nations, (ii) a cooperative payoff higher than the noncooperative payoff for each participant, (iii) sustainable cooperation as the agreed-upon optimality principle will be in effect throughout the cooperative duration, (iv) efficient development of clean technology and the capturing of positive externalities, (v) pollution abatement internalizing pollution externalities, (vi) the determination of an optimal pattern of conventional and clean technologies over the different stages of cooperation, and (vii) a technically viable cooperative scheme for nations which seek a higher payoff. Three notable phenomena concerning cooperative environmental management are developed in this the analysis. First, cooperative development and adoption of clean technology on top of pollution abatement provide an effective mean to control pollution in the growing global economy. Second, the optimal technology pattern yields a detailed plan for phasing out conventional technology and the introduction of clean technology at different stages for different nations. Third, a subgame consistent payoff distribution scheme would guarantee the maintenance of the agreed-upon optimality principle and no nations would have incentive to deviate from the cooperation scheme.

6. Conclusions

The chapter is the first analysis of cooperative dynamic environmental games with clean technology development technology switching. A subgame consistent solution is derived. In particular, it resolves the problem of making technology switching for individual nations. The derived closed-form solutions of the nation's nonco-operative payoff allow the derivation of noncooperative payoffs along the optimal trajectories. The analysis also resolves the problem of identifying the Pareto opti-mal cooperation strategies amid different research costs and different advancements in clean technology brought about by research effort under different patterns of nations adopting conventional technology and clean technology. Despite the rather complicated mathematical derivations due to the complexity of a model the paper yields analytically tractable results. The analysis represents the first application of cooperative environmental control with a subgame consistent solution entailing technology choice and technology development. It lays a foundation for develop-ing an effective measure to tackle the catastrophe-bound environmental problem. Further developments in theory and policy along this line are expected.

Appendix A. Proof of Theorem 3.1

If nation $i_t \in S_t^1$ adopts conventional technique and nation $\hat{i}_t \in S_t^2$ adopts clean technique, the results in (3.1) satisfy the optimality conditions in dynamic program-ming and the Nash equilibrium. Hence a feedback Nash equilibrium is characterized. See Basar and Olsder [1999] and Yeung and Petrosyan [2012]. The inequalities in (3.2) yield the conditions justifying why nation $i_t \in S_t^1$ adopts conventional tech-nique and nation $\hat{i}_t \in S_t^2$ adopts clean technique. To prove this we perform the indicated maximization in (3.1) and obtain:

$$\alpha_t^{i_t} - 2\beta_t^{i_t} q_t^{i_t} = c^{i_t} + \varepsilon_{i_t}^{i_t} - a^{i_t}\bar{V}_{x_{t+1}}^{i_t}(t+1, x_{t+1}, \vartheta_{t+1})(1+r)^{t-1},$$

$$\text{for } i_t \in S_t^1; \tag{A.1}$$

and

$$\alpha_t^{\hat{i}_t} - 2\beta_t^{\hat{i}_t} q_t^{\hat{i}_t} = (\underline{c}^{\hat{i}_t} - \sigma_t^{\hat{i}_t}\vartheta_t) + \hat{\varepsilon}_{\hat{i}_t}^{\hat{i}_t} - \hat{a}^{\hat{i}_t}\bar{V}_{x_{t+1}}^{\hat{i}_t}(t+1, x_{t+1}, \vartheta_{t+1})(1+r)^{t-1},$$

$$\text{for } \hat{i}_t \in S_t^2. \tag{A.2}$$

In view of (2.4) and (2.5), the left-hand side of (A.1) and that of (A.2) reflect the marginal revenues to the industrial sectors. To motivate the industrial sectors to produce outputs as given in (A.1) nation i_t has to impose a tax $v_t^{i_t}$ equaling $\varepsilon_{i_t}^{i_t} - a^{i_t}\bar{V}_{x_{t+1}}^{i_t}(t+1, x_{t+1}, \vartheta_{t+1})(1+r)^{t-1}$ on a unit of output produced with conventional technique. Similarly, nation $\hat{i}_t$ has to impose a tax $\hat{v}_t^{\hat{i}_t}$ equaling $\hat{\varepsilon}_{\hat{i}_t}^{\hat{i}_t} - \hat{a}^{\hat{i}_t}\bar{V}_{x_{t+1}}^{\hat{i}_t}(t+1, x_{t+1}, \vartheta_{t+1})(1+r)^{t-1}$ on a unit of output produced with clean technique to arrive at (A.2). At stage t the unit cost plus unit tax to the industrial sector of nation $i_t \in S_t^1$ for using conventional technique is

$$c^{i_t} + \varepsilon_i^{i_t} - a^{i_t}\bar{V}_{x_{t+1}}^i(t+1, x_{t+1}, \vartheta_{t+1})(1+r)^{t-1}, \tag{A.3}$$

and the unit cost plus unit tax to the industrial sector of nation $\hat{i}_t \in S_t^2$ for using clean technique is

$$(\underline{c}^{\hat{i}_t} - \sigma_t^{\hat{i}_t}\vartheta^{\hat{i}_t}) + \hat{\varepsilon}_{\hat{i}_t}^{\hat{i}_t} - \hat{a}^{\hat{i}_t}\bar{V}_{x_{t+1}}^{\hat{i}_t}(t+1, x_{t+1}, \vartheta_{t+1})(1+r)^{t-1}. \tag{A.4}$$

The industrial sector would adopt the technique which costs (production cost plus tax) less. Therefore for nation $i_t \in S_t^1$ to use the conventional technology in stage t it implies

$$c^{it} + \varepsilon_{i_t}^{it} - a^{it} \bar{V}_{x_{t+1}}^{it}(t+1, x_{t+1}, \vartheta_{t+1})(1+r)^{t-1}$$

$$< (\underline{c}^{it} - \sigma_t^{it} \vartheta_t) + \hat{\varepsilon}_{i_t}^{it} - \hat{a}^{it} \bar{V}_{x_{t+1}}^{it}(t+1, x_{t+1}^{it \to S_t^2}, \vartheta_{t+1}^{it \to S_t^2})(1+r)^{t-1}, \quad \text{(A.5)}$$

where $x_{t+1}^{it \to S_t^2}$ and $\vartheta_{t+1}^{it \to S_t^2}$ are the state variables in stage $t+1$ if nation $i_t \in S_t^1$ adopts environmental preserving technology in stage t.

Similarly, for nation $\hat{i}_t \in S_t^2$ to use the conventional technology in stage t it implies

$$\hat{c}^{it} + \hat{\varepsilon}_{\hat{i}_t}^{it} - \hat{a}^{it} \bar{V}_{x_{t+1}}^{\hat{i}t}(t+1, x_{t+1}^{\hat{i}t \to S_t^1}, \vartheta_{t+1}^{\hat{i}t \to S_t^1})(1+r)^{t-1}$$

$$\geq (\hat{\underline{c}}^{it} - \sigma_t^{\hat{i}t} \vartheta_t) + \varepsilon_{\hat{i}_t}^{\hat{i}t} - \hat{a}^{\hat{i}t} \bar{V}_{x_{t+1}}^{\hat{i}t}(t+1, x_{t+1}, \vartheta_{t+1})(1+r)^{t-1}, \quad \text{(A.6)}$$

where $x_{t+1}^{\hat{i}t \to S_t^1}$ and $\vartheta_{t+1}^{\hat{i}t \to S_t^1}$ are the state variables in stage $t+1$ if nation $\hat{i}_t \in S_t^2$ adopts conventional technology in stage t.

Hence Theorem 3.1 follows. $\qquad\qquad\qquad\qquad\qquad\qquad\qquad\qquad\qquad\qquad$ □

Appendix B. Proof of Proposition 3.1

Invoking Proposition 3.1 the game equilibrium strategies in (3.5) can be expressed as:

$$\phi_t^{it}(x, \vartheta) = [\alpha_t^{it} - c^{it} - \varepsilon_{i_t}^{it} + a^{it} \bar{A}_{t+1}^{it}(1+r)^{-1}]/2\beta_t^{it},$$

$$v_t^{it}(x, \vartheta) = -\frac{b_{i_t}}{2c_{i_t}^a} \bar{A}_{t+1}^{it}(1+r)^{-1} x^{1/2},$$

$$w_t^{it}(x, \vartheta) = \frac{z^{it}}{2\chi^{it}}[\bar{H}_{t+1}^{it} z^{it} + 2\bar{J}_{t+1}^{it} z^{it} \vartheta^{it}](1+r)^{-1} \div \left[1 - \frac{z^{it}}{\chi^{it}} \bar{J}_{t+1}^{it} z^{it}(1+r)^{-1}\right],$$

$$\text{for } i_t \in S_t^1;$$

$$\hat{\phi}_t^{it}(x, \vartheta) = [\hat{\alpha}_t^{it} - (\hat{\underline{c}}^{it} - \sigma_t^{it} \vartheta^{it}) - \hat{\varepsilon}_{\hat{i}_t}^{it} + \hat{a}^{it} \bar{A}_{t+1}^{it}(1+r)^{-1}]/2\hat{\beta}_t^{it},$$

$$\hat{v}_t^{it}(x, \vartheta) = -\frac{\hat{b}_{i_t}}{2c_{\hat{i}_t}^a} \bar{A}_{t+1}^{\hat{i}t}(1+r)^{-1} x^{1/2},$$

$$\hat{w}_t^{it}(x, \vartheta) = \frac{z^{\hat{i}t}}{2\chi^{\hat{i}t}}[\bar{H}_{t+1}^{\hat{i}t} z^{\hat{i}t} + 2\bar{J}_{t+1}^{\hat{i}t} z^{\hat{i}t} \vartheta^{\hat{i}t}](1+r)^{-1} \div \left[1 - \frac{z^{\hat{i}t}}{\chi^{\hat{i}t}} \bar{J}_{t+1}^{\hat{i}t} z^{\hat{i}t}(1+r)^{-1}\right],$$

$$\text{for } \hat{i}_t \in S_t^2. \qquad\qquad\qquad\qquad\qquad\qquad\qquad\qquad\qquad\qquad\qquad \text{(B.1)}$$

Using Proposition 3.1 and substituting the game equilibrium strategies in (B.1) into (3.1) yield a system of equations such that:

(i) For $i_t \in S_t^1$, the left-hand side of the equation is $(A_t^{ii}x + H_t^{ii}\vartheta^{it} + J_t^{ii}(\vartheta^{it})^2 + \sum_{\substack{j=1 \\ j \neq i_t}}^n K_t^{it(j)}\vartheta^j + C_t^{it})$ and the right-hand side is an expression linear in x, $(\vartheta^{it})^2$ and $\vartheta^1, \vartheta^2, \ldots, \vartheta^n$;

(ii) For $\hat{i}_t \in S_t^2$, the left-hand side of the equation is $(\hat{A}_t^{\hat{i}i}x + \hat{H}_t^{\hat{i}i}\vartheta^{\hat{i}t} + \hat{J}_t^{\hat{i}i}(\vartheta^{\hat{i}t})^2 + \sum_{\substack{j=1 \\ j \neq \hat{i}_t}}^n \hat{K}_t^{\hat{i}t(j)}\vartheta^j + \hat{C}_t^{\hat{i}t})$ and the right-hand side is an expression linear in x, $(\vartheta^{\hat{i}t})^2$ and $\vartheta^1, \vartheta^2, \ldots, \vartheta^n$; and

(iii) $(A_{T+1}^i x + H_{T+1}^i \vartheta^i + J_{T+1}^i(\vartheta^i)^2 + \sum_{\substack{j=1 \\ j \neq i_t}}^n K_{T+1}^{i(j)}\vartheta^j + C_{T+1}^i) = [g^i(\bar{x}^i - x_{T+1}) + g^{(P)i}\vartheta_{T+1}^i]$, for $i \in N$.

Starting backwards from $A_{T+1}^i = g^i$, $H_{T+1}^i = g^{(P)i}$, $C_{T+1}^i = g^i\bar{x}^i$, $J_{T+1}^i = 0$ and $K_{T+1}^{i(j)} = 0$, one can derive A_t^{it}, H_t^{it}, J_t^{it}, C_t^{it}, $K_t^{it(j)}$, $\hat{A}_t^{\hat{i}t}$, $\hat{H}_t^{\hat{i}t}$, $\hat{J}_t^{\hat{i}t}$, $\hat{K}_t^{\hat{i}t(j)}$ and $\hat{C}_t^{\hat{i}t}$ for $t = T$ according to the condition (3.2) that determines the choice of technology in stage t:

$$c^{it} + \varepsilon_{i_t}^{it} - a^{it}\bar{A}_{t+1}^{it}(1+r)^{-1} < (\underline{c}^{it} - \sigma_t^{it}\vartheta_t) + \hat{\varepsilon}_{i_t}^{it}$$
$$- \hat{a}^{it}\bar{A}_{t+1}^{it}(1+r)^{-1}, \quad \text{for } i_t \in S_t^1, \quad \text{and}$$

$$\hat{c}^{it} + \varepsilon_{\hat{i}_t}^{\hat{i}t} - a^{\hat{i}t}\bar{A}_{t+1}^{\hat{i}t}(1+r)^{-1} \geq (\underline{c}^{\hat{i}t} - \sigma_t^{\hat{i}t}\vartheta_t) + \varepsilon_{\hat{i}_t}^{\hat{i}t} \tag{B.2}$$
$$- \hat{a}^{\hat{i}t}\bar{A}_{x_{t+1}}^{\hat{i}t}(1+r)^{-1}, \quad \text{if } \hat{i}_t \in S_t^2.$$

Repeating the process from stage $T-1$ to stage 1 yields all the coefficients for A_t^{it}, H_t^{it}, J_t^{it}, C_t^{it}, $K_t^{it(j)}$, $\hat{A}_t^{\hat{i}t}$, $\hat{H}_t^{\hat{i}t}$, $\hat{J}_t^{\hat{i}t}$, $\hat{K}_t^{\hat{i}t(j)}$ and $\hat{C}_t^{\hat{i}t}$.

Hence Proposition 3.1. follows. $\square$

References

Basar, T. and Olsder, G. J. [1999] *Dynamic Noncooperative Game Theory*, 2nd edition (SIAM, Philadelphia).

Breton, M., Zaccour, G. and Zahaf, M. [2005] A differential game of joint implementation of environmental projects, *Automatica* **41**, 1737–1749.

Breton, M., Zaccour, G. and Zahaf, M. [2006] A game-theoretic formulation of joint implementation of environmental projects, *Eur. J. Oper. Res.* **168**, 221–239.

Dockner, E. J. and Long, N. V. [1993] International pollution control: Cooperative versus noncooperative strategies, *J. Environ. Econ. Manage.* **25**, 13–29.

Dutta, P.-K. and Radner, R. [2006] Population growth and technological change in a global warming Model, *Econ. Theory* **29**(2), 251–270.

Fredj, K., Martín-Herrán, G. and Zaccour, G. [2004] Slowing deforestation pace through subsidies: A differential game, *Automatica* **40**, 301–309.

Jørgensen, S. and Zaccour, G. [2001] Time consistent side payments in a dynamic game of downstream pollution, *J. Econ. Dyn. Control* **25**, 1973–1987.

Li, S. [2014] A differential game of transboundary industrial pollution with emission permits trading, *J. Optim. Theory Appl.* **163**, 42–659.

Petrosyan, L. and Zaccour, G. [2003] Time-consistent Shapley value allocation of pollution cost reduction, *J. Econ. Dyn. Control* **27**, 381–398.

Rubio, S. and Ulph, A. [2007] An infinite-horizon model of dynamic membership of international environmental agreements, *J. Environ. Econ. Manage.* **54**(3), 296–310.

Yeung, D. W. K. [2007] Dynamically consistent cooperative solution in a differential game of transboundary industrial pollution, *J. Optim. Theory Appl.* **134**(1), 143–160.

Yeung, D. W. K. [2008] Dynamically consistent solution for a pollution management game in collaborative abatement with uncertain future payoffs, in D. W. K. Yeung and L. A. Petrosyan (Guest Eds.) *Special Issue on Frontiers in Game Theory: In Honour of John F. Nash, Int. Game Theory Rev.* **10**(4), 517–538.

Yeung, D. W. K. [2014] Dynamically consistent collaborative environmental management with production technique choices, *Ann. Oper. Res.* **220**(1), 181–204.

Yeung, D. W. K. and Petrosyan, L. A. [2004] Subgame consistent cooperative solutions in stochastic differential games, *J. Optim. Theory Appl.* **120**, 651–666.

Yeung, D. W. K. and Petrosyan, L. A. [2006] *Cooperative Stochastic Differential Games* (Springer-Verlag, New York).

Yeung, D. W. K. and Petrosyan, L. A. [2008] A cooperative stochastic differential game of transboundary industrial pollution, *Automatica*, **44**, 1532–1544.

Yeung, D. W. K. and Petrosyan, L. A. [2010] Subgame consistent solutions for cooperative stochastic dynamic games, *J. Optim. Theory Appl.* **145**(3), 579–596.

Yeung, D. W. K. and Petrosyan, L. A. [2012] *Subgame Consistent Economic Optimization: An Advanced Cooperative Dynamic Game Analysis* (Birkhäuser, Boston).

Chapter 19

Partnership's Profit Sharing: Linear and Nonlinear Contracts

Yigal Gerchak* and Eugene Khmelnitsky[†]

Department of Industrial Engineering
Tel Aviv University, Tel-Aviv 69978, Israel
*yigal@post.tau.ac.il

[†]*xmel@tau.ac.il*

Suppose that one party proposes to another a contract for sharing an uncertain profit which maximizes the former's expected utility, with respect to its beliefs, subject to a constraint on the latter's expected utility, with respect to the latter's beliefs. It turns out that the optimal contract, which we find, can be nonmonotone, as well as nonlinear, in the realized profit. To avoid the implausible lack of monotonicity, we formulate and solve a model constrained to have monotone increasing profits for both partners. If beliefs are identical, the (unconstrained) contract is shown to be monotone, and under certain conditions, linear. That might explain one famous contract from the history of jazz. If the other party can be assumed risk neutral, the linear contract reduces to the former receiving a constant amount, and the latter the residual net profit, as in the case of another famous contract from the history of jazz. Since in the type of partnerships, we have in mind the partners are always motivated to exert high effort due to other factors like reputation, our setting has no moral hazard or adverse selection, and the partnerships do not involve a large initial investment.

Keywords: Contracts; partnerships; risk-aversion; jazz history; optimal control.

1. Introduction

A partnership, sometimes referred to as a "syndicate" [Wilson, 1968], is an arrangement in which two or more individuals cooperate to advance their mutual interests and share the profits and liabilities of a business venture. These type of partnerships are especially common in human-capital-intensive professional services [Levin and Tadelis, 2005]. Various arrangements are possible: all partners might share

---

*Corresponding author.

liabilities and profits equally, or some partners may have limited liabilities (e.g., Levin and Tadelis [2005]. In a broad sense, a partnership is any cooperative endeavor undertaken by multiple parties. These parties can be governments, nonprofits, businesses, individuals, or a combination, and the goals of the partnership can vary widely. Borch [1962], to whom Wilson [1968] attributes a pioneering contribution, considered such issues in the context of reinsurance markets. There may or may not be a written agreement governing the partnership, but it is generally a good idea to lay out specific terms at the outset, so that disagreement can be settled according to predetermined rules. In some jurisdictions such an agreement is legally required.

Partnerships are widespread in common law jurisdictions such as the United States, Britain and the Commonwealth, especially in the professional services industry. While partnerships with informal or formal contracts have existed for centuries, the academic investigation of the economics of partnerships and contracts essentially started in the late 1960s [Wilson, 1968]; see also Brousseau and Glachan [2002]. While the partnerships we are interested in here have no "principal" as such, one party usually suggests to the other(s) a certain type of arrangement. If accepted, that becomes the contract between them. Future income is typically uncertain. From an economic perspective, one can view the initiating partner as seeking to maximize its expected utility from the partnership, while providing the other partner(s) with some minimum level of expected utility. See, for example, Christensen and Feltham [2005] and Kadan and Swinkels [2013]. In more general Principal–Agent settings, contracts should provide for the "right" incentives for the agent(s). Our settings, on the other hand, is a type of partnership where the partners can be assumed to always exert high effort, since the individual reputations and income outside the partnership depends on the success of this partnership. Thus, our setting is simple in that it has no "action" or incentive aspects. Also, the partners in our models are not the stereotypical "one partner provides "sweat equity" while the other is (only) an investor".

Our intent is to "explain" these contracts in terms of the partners', possibly different, risk attitudes and their, possibly different, beliefs about the amounts of future profits. To this end, we initially formulate the problem of determining an optimal contract, which aims at maximizing the expected utility of the leading partner subject to a given expected utility of the other one. The case where both parties are risk-neutral is treated separately. Later, for the sake of greater generality, we ask what type of preferences and beliefs, and relationships among them, give rise to, general or special, linear contracts. Note that we do not necessarily assume identical HARA utility functions, which in the literature constitute sufficient conditions for optimal contracts being linear (e.g., Christensen and Feltham [2003]).

We pay particular attention to conditions for the optimality of linear contracts, and even of a constant wage to initiating party, as was the case in the famous contracts, respectively, between Duke Ellington and Irwin Mills and between Louis Armstrong and Joe Glaser [Ward and Burns, 2000, p. 214], the details of which are given in the Appendix.

2. Model and Optimality Conditions

Let $u_1(\cdot)$ and $u_2(\cdot)$ be the utility functions of partners 1 and 2, respectively. Partner 1 is assumed to know the utility of partner 2. The VNM-utility functions are assumed to be increasing, continuous, concave and almost everywhere differentiable, $u_i'(\cdot) > 0$, $u_i''(\cdot) \leq 0$, and DARA i.e., $-u''/u'$ decreasing [Pratt, 1964], $i = 1, 2$. However, the functions are not assumed to be necessarily identical, nor are they assumed to be necessarily of HARA type (e.g., Christensen and Feltham [2003]).

Let $\varphi_1(x)$ and $\varphi_2(x)$ be the pdf's of the profit, X, which represent the beliefs of partners 1 and 2, respectively; the two functions are defined over a finite common support, $\varphi_i(x) > 0$ $i = 1, 2$ $\forall x \in [x_0, x_1]$. Previous literature assumed identical beliefs. Partner 1 is assumed to know the other's beliefs. The corresponding cdf's are $F_1(x)$ and $F_2(x)$. What we are looking for are, $c_1(x)$ and $c_2(x)$, the shares of the partners of realized profit x. We denote by B the expected utility guaranteed to partner 2.

The problem of finding optimal functions $c_1(x)$ and $c_2(x)$ can be stated as follows:

$$\max_{c_1(\cdot),\ c_2(\cdot)} \int_{x_0}^{x_1} u_1(c_1(x))\varphi_1(x)dx \tag{1}$$

s.t.

$$c_1(x) + c_2(x) = x, \tag{2}$$

$$E[u_2(c_2(x))] = \int_{x_0}^{x_1} u_2(c_2(x))\varphi_2(x)dx = B.$$

By changing B one obtains the Pareto frontier of the profit divisions.

The last constraint can be rewritten as a two-point, boundary value differential equation,

$$\frac{dz(x)}{dx} = u_2(c_2(x))\varphi_2(x), \quad z(x_0) = 0, \quad z(x_1) = B, \tag{3}$$

so that the problems (1)–(3) has a canonical form of optimal control, (e.g., Hartl *et al.* [1995]).

The optimality conditions for problems (1)–(3) are derived by means of the maximum principle, which states that if the control functions $c_1(x)$ and $c_2(x)$ and state variable $z(x)$ solve the problem, then there exists a costate variable, $\psi(x)$, that satisfies the following conditions (e.g., Hartl *et al.* [1995]):

— the optimal control maximizes the Hamiltonian function, H, at each x,

$$\{c_1(x), c_2(x)\} = \arg\max H, \tag{4}$$

where

$$H(x) = u_1(c_1(x))\varphi_1(x) + \psi(x)u_2(c_2(x))\varphi_2(x), \tag{5}$$

— costate equation,

$$\frac{d}{dx}\psi(x) = -\frac{\partial H}{\partial z} = 0, \quad \text{i.e., } \psi(x) = \psi \text{ is a constant.} \tag{6}$$

Note that the constant ψ is not given in advance. Its value depends on the parameters and functions that constitute problems (1)–(3). Therefore, ψ and $c_i(x)$ are not independent of each other, and as a result, the maximization of (5) implies Pareto optimality at an aggregate level of profit, rather than Pareto optimality for every realization of x.

The combination of (2) and (4)–(6) gives

$$u_1'(c_1(x)) = \psi u_2'(x - c_1(x))\frac{\varphi_2(x)}{\varphi_1(x)}. \tag{7}$$

From (7) it follows that the value of ψ is positive, which guarantees the Legendre–Clebsch condition (sufficient condition for local optimality) to hold

$$\frac{\partial^2 H}{\partial c_1^2} = u_1''(c_1(x))\varphi_1(x) + \psi u_2''(x - c_1(x))\varphi_2(x) < 0.$$

The analysis of (7) leads to the optimal division of the profit between the two parties, as shown in Sec. 3. In Sec. 5, we "reverse" the initial question, by asking what type of utility functions and probability distributions, and relations among them, result in linear contracts being optimal. The answer to the original question is also provided indirectly.

3. Analysis of the Model

This section develops a method for determining the optimal contract for general beliefs and general strictly concave utility functions of the two partners. The case of linear utility functions (risk neutral partners) is considered in Sec. 7.

By differentiating (7) w.r.t x, we obtain

$$u_1''(c_1(x))c_1'(x) = \psi u_2''(x - c_1(x))(1 - c_1'(x))\frac{\varphi_2(x)}{\varphi_1(x)}$$

$$+ \psi u_2'(x - c_1(x))\frac{\varphi_2'(x)\varphi_1(x) - \varphi_1'(x)\varphi_2(x)}{\varphi_1^2(x)}.$$

Solving for $c_1'(x)$ results in

$$c_1'(x) = \psi\frac{u_2''(x - c_1(x))\varphi_1(x)\varphi_2(x) + u_2'(x - c_1(x))(\varphi_2'(x)\varphi_1(x) - \varphi_1'(x)\varphi_2(x))}{u_1''(c_1(x))\varphi_1^2(x) + \psi u_2''(x - c_1(x))\varphi_1(x)\varphi_2(x)}. \tag{8}$$

Example 1 below solves (8) in a closed form and generalizes the exponential beliefs function case presented in Wilson [1968].

Example 1. The beliefs of the partners are distributed exponentially, $\varphi_i(x) = \lambda_i e^{-\lambda_i x}$, $x \geq 0$, and the utility functions of the partners are also exponential,

$$u_i(c) = U_i \cdot (1 - e^{-\mu_i c}), \quad i = 1, 2.$$

The differential equation (8) reduces to

$$c_1'(x) = \frac{\psi U_2 \mu_2 \lambda_2 (\mu_2 - \lambda_1 + \lambda_2)}{U_1 \mu_1^2 \lambda_1 e^{-(\mu_1 + \mu_2) c_1(x)} e^{(\mu_2 - \lambda_1 + \lambda_2) x} + \psi U_2 \mu_2^2 \lambda_2}. \tag{9}$$

The solution of (9) is obtained analytically

$$c_1(x) = \frac{\mu_2 - \lambda_1 + \lambda_2}{\mu_1 + \mu_2} x + \frac{1}{\mu_1 + \mu_2} \ln \frac{U_1 \mu_1 \lambda_1}{\psi U_2 \mu_2 \lambda_2}. \tag{10}$$

The value of ψ is obtained from $E_2[u_2(c_2(X))] = B$,

$$\psi = \frac{U_1 \mu_1 \lambda_1}{U_2 \mu_2 \lambda_2} \left(\frac{U_2 \lambda_2 (\mu_1 + \mu_2)}{(U_2 - B)(\mu_1 \mu_2 + \lambda_1 \mu_2 + \lambda_2 \mu_1)} \right)^{1 + \frac{\mu_1}{\mu_2}}$$

and after substituting in (10), we get

$$c_1(x) = \frac{\mu_2 - \lambda_1 + \lambda_2}{\mu_1 + \mu_2} x + \frac{1}{\mu_2} \ln \left(\frac{U_2 - B}{U_2} \cdot \frac{\mu_1 \mu_2 + \lambda_1 \mu_2 + \lambda_2 \mu_1}{\lambda_2 (\mu_1 + \mu_2)} \right). \tag{11}$$

Note that if $\lambda_1 = \lambda_2 + \mu_2$, then

$$c_1(x) = \frac{1}{\mu_2} \ln \left(\frac{U_2 - B}{U_2} \cdot \frac{\lambda_2 + \mu_2}{\lambda_2} \right),$$

that is a fixed wage contract where partner 1's share is constant, is optimal. For $\lambda_2 - \mu_1 < \lambda_1 < \lambda_2 + \mu_2$ the shares of the two partners grow linearly with x (see Fig. 1). In the complementary regions of the (λ_1, λ_2) plane, one of the shares decreases linearly.

The numerical procedure presented below is useful if no closed form solution to (8) is available. The procedure determines the optimal contract $c_1(x)$ by calculating ψ and boundary condition $c_1(0)$, and by integrating (8) over the profit domain.

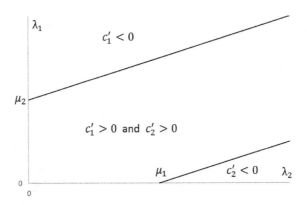

Fig. 1.   The regions of positive and negative growth of partners' shares.

Numerical solution method

Step 1. Set a positive ψ.

Step 2. Solve, analytically or numerically, the equation, $u_1'(y) = \psi u_2'(-y)\frac{\varphi_2(0)}{\varphi_1(0)}$ and obtain y. We assume here, without loss of generality, that $0 \in [x_0, x_1]$.

Step 3. Integrate (8) with the boundary condition $c_1(0) = y$ and obtain $c_1(x)$ for $x \in [x_0, x_1]$.

Step 4. Set $c_2(x) = x - c_1(x)$, and calculate $E_2[u_2(c_2(X))]$.

Step 5. If $E_2[u_2(c_2(X))]$ is sufficiently close to B, stop. Otherwise, adjust ψ (decrease ψ if $E_2[u_2(c_2(X))]$ is greater than B, and increase ψ if $E_2[u_2(c_2(X))]$ is smaller than B). Go to Step 2.

Contrary to Example 1, the next example considers a scenario where $\varphi_i(x)$ and $u_i(c)$ belong to different families of functions, so that no analytical solution of (8) is available. The result obtained by the above procedure shows strongly nonlinear and nonmonotone behavior of partners' shares.

Example 2. The beliefs of the partners are distributed normally, $\varphi_i(x) \sim N(\mu_i, \sigma_i)$, and utility functions of the partners are logarithmic, $u_i(c) = U_i \cdot \ln(1 + v_i c)$, $i = 1, 2$. Figure 2 presents optimal partners' shares calculated numerically for $B = 1, U_1 = 1$, $U_2 = 4$, $v_1 = 1, v_2 = 0.2, \mu_1 = 2, \mu_2 = 1, \sigma_1 = 2, \sigma_2 = 1$. For the utility functions to be well defined, we verify the condition $1 + v_i c_i(x) > 0$, while carrying out the steps of the numerical method.

The possibility of $c_i(x)$ behavior other than monotone increasing, demonstrated in Examples 1 and 2 is surprising. Intuitively, this can be explained as follows. At an interval of the profit domain where partner 1 has more optimistic beliefs than partner 2, partner 2 agrees to give up his share to some extent. At another interval with opposite beliefs (optimistic for partner 2 and pessimistic for partner 1) partner 1 agrees to give up some of his share. The path, $c_i(x)$, over the entire domain may

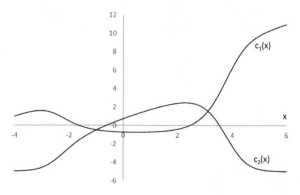

Fig. 2.   Partners' shares in the scenario of Example 2.

have, in such a case, a nonmonotone increasing shape. In Example 1, for instance, where the beliefs are distributed exponentially, high values of λ_1 correspond to small partner 1's expectation of the overall profit. Therefore, partner 1 agrees to give up his share for high profit values that are unlikely according to him. As a result (see (11)), $c_1(x)$ takes on high values for small x and decreases as the profit grows.

In the next section, we impose a constraint that forces $c_i(x)$ to be monotone increasing.

4. Constrained Model

This section imposes additional constraint on a contract, which requires that the shares of the two partners do not decline when the profit grows. The formulation of problems (1)–(3) modifies so that $c_1(x)$ becomes a new state variable, while its derivative, denoted by $w(x)$, controls the two shares. That is

$$c'_1(x) = w(x), \quad 0 \le w(x) \le 1. \tag{12}$$

The maximum principle also changes and states that, if $w(x)$ is the optimal control, then there exist two costate variables, $\psi(x)$ and $\xi(x)$ that satisfy the following conditions:

— the optimal control maximizes the Hamiltonian function, H, at each x,

$$w(x) = \arg \max H, \tag{13}$$

where

$$H(x) = u_1(c_1(x))\varphi_1(x) + \psi(x)u_2(c_2(x))\varphi_2(x) + \xi(x)w(x), \tag{14}$$

— new costate equation in addition to (6) is

$$\xi'(x) = -\frac{\partial H}{\partial c_1} = -u'_1(c_1(x))\varphi_1(x)$$

$$+ \psi u'_2(x - c_1(x))\varphi_2(x), \quad \xi(x_0) = \xi(x_1) = 0. \tag{15}$$

The maximization of the Hamiltonian (14) relates $w(x)$ to $\xi(x)$ as follows:

$$w(x) = \begin{cases} 1, & \text{if } \xi(x) > 0, \\ 0, & \text{if } \xi(x) < 0, \\ \in [0,1], & \text{if } \xi(x) = 0. \end{cases} \tag{16}$$

Equation (16) indicates that the optimal contract, $c_1(x)$, consists of a sequence of arcs, each from one of the three categories:

— entire marginal profit is allocated to partner 1, $c'_1(x) = 1$ (type 1 arc);
— entire marginal profit is allocated to partner 2, $c'_1(x) = 0$ (type 2 arc);
— marginal profit is divided between the two partners, $0 < c'_1(x) < 1$ (singular arc).

The last category is termed singular, since the division of the profit over the arc cannot be explicitly determined from the Hamiltonian maximization, and additional analysis is required. Let a singular arc occur at an interval of x. Then, at that interval, $\xi(x) = 0$ and $\xi'(x) = 0$. Now, from (15), we obtain

$$u_1'(c_1(x))\varphi_1(x) = \psi u_2'(x - c_1(x))\varphi_2(x)$$

the condition that has already been studied and resolved in the unconstrained model.

Having determined the value of $w(x)$ on each arc category, we identify sub-intervals of x, where the singular arcs can potentially occur, and connect the sub-intervals with the type 1 and type 2 arcs by solving the system of state–costate differential equations (8) and (15). The details of the method can be found in Khmelnitsky [2002].

Example 3. The beliefs of the partners are distributed piece-wise constant

$$\varphi_1(x) = \begin{cases} 1/6, & \text{if } 0 \le x \le 3 \text{ and } 6 \le x \le 9, \\ 0, & \text{otherwise,} \end{cases} \qquad \varphi_2(x) = \begin{cases} 1/3, & \text{if } 3 \le x \le 6, \\ 0, & \text{otherwise} \end{cases}$$

and utility functions of the partners are exponential, $u_1(c) = 3(1 - e^{-0.1c})$, $u_2(c) = 5(1 - e^{-0.05c})$. The expected utility of partner 2 is $B = 0.6$.

The solution method identifies the optimal contract consisting of a type 2 arc followed by a type 1 arc with no singular arcs. The arcs switch at $x = 4.68$. The value of ψ is 1.0168. The state–costate dynamics are shown in Fig. 3. We observe that in this case, partner 1 has a guaranteed income when profit is low, but it shares in the profit when it is high. Partner 2 shares in the profit when it is low (at the same rate that partner 1 shares when profit is high), but has a fixed income when profit is high (higher than partner 1 fixed income when profit is low).

The next three sections deal with three important particular cases of problem (1)–(3) and (12). In the first one an inverse analysis of the model is undertaken in

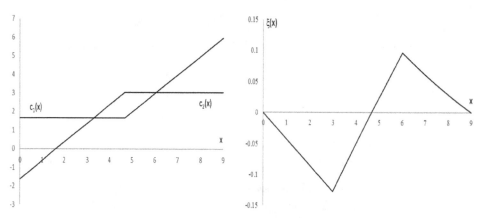

Fig. 3.   Partners' shares $c_1(x)$ and $c_2(x)$, and the costate variable $\xi(x)$ in the scenario of Example 3.

order to fully specify the scenarios that yield a linear contract, that is the contract where the shares of the two partners grow linearly with profit. The second case assumes that the two partners have identical beliefs, and since this case allows more insights to be gained, presents additional characterization of optimal contract types. The third case analyzes an optimal sharing policy for risk neutral partners.

5. Partners' Shares Grow Linearly

Suppose that, as in the partnership of Ellington and Mills, the shares of the partners are linear in x, $c_1(x) = p \cdot x + K$, $c_2(x) = (1 - p) \cdot x - K$, $0 < p < 1$. In that story $K = 0$. The following proposition establishes the necessary and sufficient conditions for such contract to be optimal. The proposition assumes that the belief functions satisfy the likelihood-ratio order of probability distributions (e.g., Shaked and Shanthikumar [1994]),

$$\frac{\varphi_1'(x)}{\varphi_1(x)} < \frac{\varphi_2'(x)}{\varphi_2(x)} \quad \forall x. \tag{17}$$

Proposition 1. *For any $u_1(c)$, and $\phi_1(x)$ and $\phi_2(x)$ ordered with respect to (17), the partners' shares are linear, i.e., $c_1(x) = px + K$, $c_2(x) = (1-p)x - K$, $0 < p < 1$, if and only if there exist constants, $0 < p < 1$, $A > 0$ and K, such that*

$$u_2(c) = B + A \left(\int_{x_0}^{x_1} F_2(x) \frac{\varphi_1(x)}{\varphi_2(x)} u_1'(px + K) dx - \int_{\frac{c+K}{1-p}}^{x_1} \frac{\varphi_1(x)}{\varphi_2(x)} u_1'(px + K) dx \right). \tag{18}$$

Proof. Necessity. Let $c_1(x) = px + K$ for some p and K. Then, by changing variable $c = (1 - p)x - K$, optimality condition (7) results in

$$u_2'(c) = \frac{u_1'\left(p\frac{c+K}{1-p} + K\right)}{\psi} \frac{\varphi_1\left(\frac{c+K}{1-p}\right)}{\varphi_2\left(\frac{c+K}{1-p}\right)}. \tag{19}$$

The solution of (19) is

$$u_2(c) = M - \frac{1-p}{\psi} \int_{\frac{c+K}{1-p}}^{x_1} \frac{\varphi_1(z)}{\varphi_2(z)} u_1'(pz + K) dz, \tag{20}$$

where M is a constant, which is determined from the given expected utility of partner 2

$$B = E[u_2(c_2(X))] = M - \frac{1-p}{\psi} \int_{x_0}^{x_1} \varphi_2(x) \int_x^{x_1} \frac{\varphi_1(z)}{\varphi_2(z)} u_1'(pz + K) dz \, dx. \tag{21}$$

By changing the order of integration in (21) and denoting $A = \frac{1-p}{\psi}$, we obtain

$$B = M - A \int_{x_0}^{x_1} F_2(z) \frac{\varphi_1(z)}{\varphi_2(z)} u_1'(pz + K) dz.$$

That is

$$M = B + A \int_{x_0}^{x_1} F_2(z) \frac{\varphi_1(z)}{\varphi_2(z)} u_1'(pz + K) \, dz. \qquad (22)$$

By combining (20) and (22), we obtain (18) stated in the proposition. The likelihood-ratio ordered beliefs (17) guarantee the negativity of the second derivative of $u_2(c)$.

Sufficiency. Suppose there exist constants K, $0 < p < 1$ and $A > 0$ such that

$$u_2(c) = B + A \left(\int_{x_0}^{x_1} F_2(x) \frac{\varphi_1(x)}{\varphi_2(x)} u_1'(px + K) dx - \int_{\frac{c+K}{1-p}}^{x_1} \frac{\varphi_1(x)}{\varphi_2(x)} u_1'(px + K) dx \right).$$

Then,

$$u_2'(c) = A \frac{u_1'\left(p\frac{c+K}{1-p} + K\right) \varphi_1\left(\frac{c+K}{1-p}\right)}{1 - p} \frac{1}{\varphi_2\left(\frac{c+K}{1-p}\right)}. \qquad (23)$$

By substituting (23) into (7), one obtains

$$u_1'(c_1(x)) = \psi u_2'(x - c_1(x)) \frac{\varphi_2(x)}{\varphi_1(x)} = \psi \frac{A}{1-p} u_1'(px + K) \frac{u_2'(x - c_1(x))}{u_2'((1-p)x - K)}. \qquad (24)$$

Since there exists $\psi = \frac{1-p}{A}$ such that $c_1(x) = px + K \ \forall x$ satisfies optimality condition (24), such partner 1's share is optimal. $\qquad \square$

Proposition 1 presents a parametrized family (with three parameters p, K and A) of utility functions. In case the utility function of partner 2 belongs to that family, the optimal contract allocates a linear share, $px + K$, to partner 1 and the rest of the profit to partner 2. We note that Proposition 1 holds for only very specific cases, where the partners' utilities and beliefs are strongly related.

Statements similar to that of Proposition 1, which determine the conditions for the optimal policy to be a constant wage of one of the partners, are given in Propositions 2 and 3.

Proposition 2. *For any $u_1(c)$, partner 1 share is constant, i.e., $c_1(x) = K$, where K satisfies the equation*

$$E_2[u_2(X - K)] = B,$$

if and only if

$$\frac{F_1(x)}{F_2(x)} \geq \frac{E_2[u_2'(X - K) \mid X \leq x]}{E_2[u_2'(X - K)]}, \quad \forall x \in (x_0, x_1].$$

Proof. First note that since Hamiltonian (14) is concave in $c_1(x)$, the conditions (13)–(16) are necessary and sufficient for an optimal control [Kamien and Schwartz, 1971]).

Let $c_1(x) = K$ and $c_2(x) = x - K$ for all x. Then, the value of K is determined from the given expected utility of partner 2, as stated in the proposition. The optimal solution consists of a single arc of type 2, which implies that the costate variable, $\xi(x)$, is nonpositive along the entire path, $\xi(x) \leq 0$ for all x(see (16)). By integrating costate equation (15), we obtain

$$\xi(x) = -u_1'(K)F_1(x) + \psi \int_{x_0}^{x} u_2'(y - K)\varphi_2(y)dy \leq 0. \tag{25}$$

The value of ψ is found from the costate boundary condition, $\xi(x_1) = 0$, as

$$\psi = \frac{u_1'(K)}{E_2[u_2'(X - K)]}.$$

Now, the proposition is obtained by substituting ψ in inequality (25) and by using the conditional expectation formula,

$$\int_{x_0}^{x} u_2'(y - K)\varphi_2(y)dy = F_2(x)E_2[u_2'(X - K) \,|\, X \leq x]. \qquad \square$$

Proposition 3. *For any $u_2(c)$, partner 2 share is constant, i.e., $c_2(x) = -K$, where K satisfies the equation*

$$u_2(-K) = B,$$

if and only if

$$\frac{F_1(x)}{F_2(x)} \leq \frac{E_1[u_1'(X + K)]}{E_1[u_1'(X + K) \,|\, X \leq x]}, \qquad \forall\, x \in (x_0, x_1].$$

Proof. Similar to the proof of Proposition 2. $\qquad \square$

Note that if it is known in advance that the contract is linear, or one restricts oneself to such contract, the individual rationality constraint (3) may be sufficient to characterize the parameters of the contract (without explicitly considering the objective function). As that constraint becomes

$$\int_{x_0}^{x_1} u_2((1 - p)x - K)\varphi_2(x)dx = B.$$

Then,

(i) if $p = 0$, need K such that $\int_{x_0}^{x_1} u_2(x - K)\varphi_2(x)dx = B$,

(ii) if $p = 1$, need $K = -u_2^{-1}(B)$,

(iii) if $K = 0$, need p that satisfies $\int_{x_0}^{x_1} u_2((1 - p)x)\varphi_2(x)dx = B$.

6. Identical Beliefs of the Partners

If the partners have identical beliefs, $\varphi_1(x) = \varphi_2(x) = \varphi(x) \,\forall\, x$, then, the optimal contract is obtained from (8), which in this case reduces to

$$c_1'(x) = \frac{\psi u_2''(x - c_1(x))}{u_1''(c_1(x)) + \psi u_2''(x - c_1(x))}.$$

Contrary to the general case considered in Sec. 3, the differential equation for $c_1(x)$ guarantees that the shares of the two partners are nondecreasing in x. The probability distribution of $\varphi(x)$ impacts the costate multiplier ψ and the partners' shares $c_i(x)$ through the expected utilities, as shown in Propositions 4 and 5 below. The propositions specify Propositions 1 and 2 for the case of identical beliefs.

Proposition 4. *If the partners have identical beliefs, a linear contract is optimal if and only if*

$$u_2(c_2(x)) - E[u_2(c_2(X))] = A(u_1(c_1(x)) - E[u_1(c_1(X))]), \quad \forall x, \qquad (26)$$

where A is an arbitrary constant.

Proof. (26) is obtained from (18) by integration by parts. $\qquad\qquad\square$

Proposition 4 generalizes a result in the literature [Christensen and Feltham, 2003], where the optimality of a linear contract was proven under the condition of identical beliefs and utility functions belonging to the HARA class and having identical risk cautiousness defined as

$$\frac{u'(c)u'''(c)}{[u''(c)]^2} - 1. \qquad (27)$$

This is proven in the following.

Corollary 1. *If the two partners have identical beliefs and identical risk cautiousness, and $u_1(c)$ and $u_2(c)$ belong to the HARA class, then the optimal contract is linear.*

Proof. In order to verify the optimality of the linear contract, $c_1(x) = px + K$, we seek constants, p, K and A that satisfy (26). Let

$$A = \frac{E[u_2(c_2(X))]}{E[u_1(c_1(X))]},$$

then from (26), we obtain

$$u_2(c_2(x)) = Au_1(c_1(x)). \qquad (28)$$

A HARA utility function has the form

$$u(c) = \frac{1-\gamma}{\gamma}\left(\frac{ac}{1-\gamma} + b\right)^{\gamma} \qquad (29)$$

for some $a > 0$, γ and b. By differentiating (29) and substituting in (27), one obtains that identical risk cautiousness of the partners implies equal γ. By substituting (29) in (28), we have

$$\frac{1-\gamma}{\gamma}\left(\frac{a_2c_2(x)}{1-\gamma} + b_2\right)^{\gamma} = A\frac{1-\gamma}{\gamma}\left(\frac{a_1c_1(x)}{1-\gamma} + b_1\right)^{\gamma}. \qquad (30)$$

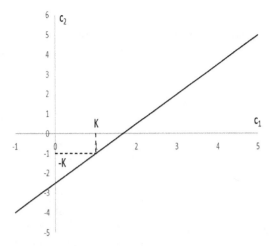

Fig. 4. $K = 1$, $p = 0.4$.

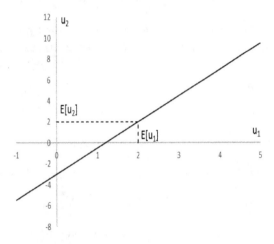

Fig. 5. $A = 2.5$, $E[u_1(c_1(x))] = E[u_2(c_2(x))] = 2$.

Expression (30) is rewritten as

$$\frac{a_2(x - c_1(x))}{1 - \gamma} + b_2 = A^{1/\gamma}\left(\frac{a_1 c_1(x)}{1 - \gamma} + b_1\right)$$

and is solved by the linear contract

$$c_1(x) = \frac{a_2}{a_2 + a_1 A^{1/\gamma}}x + (1 - \gamma)\frac{b_2 - b_1 A^{1/\gamma}}{a_2 + a_1 A^{1/\gamma}}. \qquad \square$$

Figures 4 and 5 illustrate (26) and show the linearity of partner 2's share and utility as functions of partner 1's share and utility, respectively, when profit x implicitly runs over the domain, $x \in [x_0, x_1]$. The next example demonstrates the applicability of Proposition 4.

Example 4. Let $u_1(c) = a_1 - b_1 e^{-\mu_1 c}$ and $u_2(c) = a_2 - b_2 e^{-\mu_2 c}$, $a_2 > B$.

In order to verify the optimality of the linear contract, $c_1(x) = px + K$, we seek constants, p, K and A that satisfy (26). Namely,

$$a_2 - b_2 e^{-\mu_2((1-p)x-K)} - B = A(a_1 - b_1 e^{-\mu_1(px+K)} - a_1 + b_1 e^{-\mu_1 K} E[e^{-\mu_1 pX}]).$$

$$(31)$$

For $p = \frac{\mu_2}{\mu_1 + \mu_2}$, (31) is rewritten as

$$a_2 - b_2 e^{\mu_2 K} e^{-\frac{\mu_1\mu_2}{\mu_1+\mu_2}x} - B = A b_1 e^{-\mu_1 K}\left(-e^{-\frac{\mu_1\mu_2}{\mu_1+\mu_2}x} + E[e^{-\frac{\mu_1\mu_2}{\mu_1+\mu_2}X}]\right).$$

The last equality is true for

$$A = \frac{b_2}{b_1}\left(\frac{a_2 - B}{b_2 E\left[e^{-\frac{\mu_1\mu_2}{\mu_1+\mu_2}X}\right]}\right)^{1+\frac{\mu_1}{\mu_2}} \quad \text{and} \quad K = \frac{1}{\mu_2}\ln\frac{a_2 - B}{b_2 E\left[e^{-\frac{\mu_1\mu_2}{\mu_1+\mu_2}X}\right]}. \quad (32)$$

Note that the value of K is independent of a_1 and b_1, and that $K = 0$ if

$$E\left[e^{-\frac{\mu_1\mu_2}{\mu_1+\mu_2}X}\right] = \frac{a_2 - B}{b_2}.$$

Proposition 5. *If the partners have identical beliefs, then for arbitrary $u_1(c)$ and $\varphi(x)$, partner 1's share is constant if and only if there exist constants, K and $A > 0$, such that if*

$$u_2(c) - E[u_2(X - K)] = A(c - E[X - K]), \quad (33)$$

then, $c_1(x) = K$, $c_2(x) = x - K$.

Proof. In the particular case of identical beliefs, Equation (18) reduces to

$$u_2(c) = B + A\int_{x_0}^{x_1} F_2(x)dx - A\int_{c+K}^{x_1} dx = B + A(c + K - E[X]).$$

Equivalently, $u_2(c) - B = A(c - E[X - K])$. □

Propositions 2 and 5 outline the scenarios where the fixed wage contract is optimal. The scenarios are divided into two groups. The first group, which fits a wide range of applications, requires risk neutrality of partner 2 and identical beliefs of the partners. The case of Armstrong and Glaser probably belongs to this group of scenarios, since Glaser can be considered as being risk neutral. The second group covers only very specific cases, each requiring that the three functions — partner 2's utility and partners' beliefs — be related and satisfy the conditions of Proposition 2. Example 5 presents a scenario from the first group.

Example 5. Partner 2 is risk neutral, i.e., $u_2(c) = U \cdot c$, where U is a constant, and the beliefs of the partners are identical.

Since the given $u_2(c)$ belongs to the parameterized family defined in (18) with the parameters $A = U$, $p = 0$ and $K = E[X] - B/U$, from Proposition 5 it follows that regardless of partner 1's utility function, the optimal contract allocates a fixed wage, K, to partner 1 and the rest of the profit to partner 2. Figure 6 illustrates

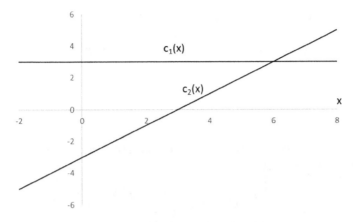

Fig. 6.   The optimal $c_1(x)$ and $c_2(x)$.

the solution for $-2 \le x \le 8$, $E[X] = 4$, $A = B = 1$. Note that $c_2(-2) = -5$ and $c_2(8) = 5$ and this is the range of partner 2's profits. The profit of partner 1 is constant, $c_1(x) = K = 3 \; \forall \, x$.

7. Risk Neutral Partners

Consider a scenario where the two partners are risk neutral, $u_1(c) = U_1 \cdot c$ and $u_2(c) = U_2 \cdot c$. The partners have different beliefs, so that the equality $F_2(x) = F_1(x)$ holds only at a finite number of points x. The unconstrained problems (1)–(3) are not well-posed in such a scenario. The optimality conditions either do not hold (for different beliefs), or hold for any partner shares (for identical beliefs). The constrained problem provides an optimal solution, which consists of arcs of types 1 and 2 only. No singular arcs occur, since from the solution of the costate dynamic equations it follows that:

$$\psi = U_1/U_2 \quad \text{and} \quad \xi(x) = U_1(F_2(x) - F_1(x)) \tag{34}$$

and the singular arc condition, $\xi(x) = 0$, does not hold. With respect to (16), the number of arcs equals the number of times the difference $F_2(x) - F_1(x)$ changes sign (plus one). The following case exemplifies a closed-form solution of the problem where the two partners are risk neutral.

Example 6. Let the beliefs of the partners be distributed normally with equal mean, μ, but different standard deviations, σ_1 and σ_2. For σ_1 smaller than σ_2, $F_1(x) < F_2(x)$ for $x < \mu$ and $F_1(x) > F_2(x)$ for $x > \mu$. From (16) and (34) it follows that the optimal $c_1(x)$ consists of a sequence of two arcs, $2 \rightarrow 1$, with the switching point located at the common mean (since $F_1(x)$ and $F_2(x)$ are equal at μ), i.e.,

$$c_1(x) = \begin{cases} x + C - \mu, & \text{if } x < \mu, \\ C, & \text{if } x \ge \mu, \end{cases} \qquad c_2(x) = \begin{cases} \mu - C, & \text{if } x < \mu, \\ x - C, & \text{if } x \ge \mu, \end{cases} \tag{35}$$

where the constant C is obtained from the given expected utility of partner 2,

$$\int_{-\infty}^{\mu} U_2(\mu - C)\varphi_2(x)dx + \int_{\mu}^{\infty} U_2(x - C)\varphi_2(x)dx = B$$

as $C = \mu - \frac{B}{U_2} + \frac{\sigma_2}{\sqrt{2\pi}}$. Now, the expected utility of partner 1 is

$$\int_{-\infty}^{\mu} U_1(x + C - \mu)\varphi_1(x)dx + \int_{\mu}^{\infty} U_1 C \varphi_1(x)dx = U_1\left(\mu - \frac{B}{U_2} + \frac{\sigma_2 - \sigma_1}{\sqrt{2\pi}}\right).$$

The opposite case, $\sigma_1 > \sigma_2$, is considered similarly, and the expected utility of partner 1 for both cases is,

$$U_1\left(\mu - \frac{B}{U_2} + \frac{|\sigma_1 - \sigma_2|}{\sqrt{2\pi}}\right).$$

Not so intuitive, the expected utility of partner 1 grows linearly with the absolute difference between the standard deviations no matter whose belief uncertainty is higher.

8. Conclusions

This chapter states the problem of determining the shares of two partners and solves it by optimal control. The optimality conditions allow gaining insights into the properties of the optimal solution and construct a numerical method for calculating the optimal shares, as functions of the profit. After analyzing the general case, motivated by famous profit sharing arrangements from the history of jazz, we prove necessary and sufficient conditions for the optimal contract to be of a "fixed wage" type, or linear. The major conclusions are

(1) In a general scenario where the partners' beliefs and utility functions are unrelated, the optimal shares can exhibit nonlinear and even nonmonotone-increasing behavior. A numerical method suggested approximates the partners' optimal shares.
(2) The numerical method is generalized to a constrained problem where the shares are forced to be nondecreasing.
(3) The scenarios where a linear contract is optimal are limited to only very specific cases, where the partners' utilities and beliefs are correlated; for example, identical beliefs and utilities belonging to the same family of HARA functions with equal risk cautiousness.
(4) The scenarios where a fixed wage contract is optimal require either risk neutrality of partner 2 and identical beliefs of the partners, or very specific conditions imposed on the partner's characteristics.

Our assumption that partner 1 knows partner 2's beliefs is strong. Though note that most of the literature assumes identical, common-knowledge, beliefs. A more nuanced Bayesian approach to mutual beliefs should be used in future research. Another generalization would be to make the duration of the contract into a decision variable. Some researchers [Sappington, 1983] favor adding "limited liability" constraints that guarantee certain realized income (utility) to partner 2. Perhaps in the future the effect of such constraints could be explained within our partnership setting.

An alternative approach to profit sharing is bargaining. For example, one can use the Nash Bargaining solution, which maximizes the product of expected utilities (e.g., Friedman [1986]). The optimality condition obtained will be similar to our (7). The exact relations among the results of bargaining concepts and the approach presented need to be explored further.

Appendix: The Stories

Louis Armstrong (1901–1971), the great jazz trumpeter and singer, was extremely popular and successful until the early 1930's. But in 1935, upon returning home from an extensive tour in Britain, during the great depression, he "sailed into a world of trouble" [Ward and Burns, 2000, p. 214]. Besides numerous personal problems and resulting legal ones, "when he got back to Chicago he couldn't seem to find steady work... Even Louis Armstrong had finally hit hard times". [Ward and Burns, 2000, p. 214]. He then turned to Joe Glaser, who used to be a manager in the mid 20's of a mob related cafe and dance hall where Armstrong also played. The Repeal of prohibition, however, created a situation, where "... minor hoodlooms like Glaser found themselves hard pressed to make a living" [Ward and Burns, 2000, p. 214].

Armstrong supposedly proposed to Glaser the following arrangement: "You get me the jobs, you collect the money. You pay me one thousand dollars every week free and clear" (worth about $17,650 in 2016). "You pay off the band, the travel and hotel expenses, my income tax, and you take everything that's left" [Ward and Burns, 2000, p. 214]. While the two men are thought never to have had a written contract (!), Glaser indeed became Armstrong's manager, a very successful arrangement that lasted more than three decades. It is not clear until when the above arrangement was kept, but it is likely that it was used in the early years of the partnership.

Another famous earlier contract from the history of jazz was struck in 1926 between the future great band-leader, composer and pianist Edward Kennedy "Duke" Ellington (then young and largely unknown) and Irwin Mills, already a successful music publisher, to become Ellington manager (it lasted until 1939). The arrangement apparently assigned 45–55% of the profits to Mills.[a] This partnership was also quite successful.

[a]Sources differ on that figure [Ward and Burns, 2000, Wikipedia (2)], and a written contract is not available. The agreement apparently also "allowed" Mills to add his name as co-composer of some Ellington's greatest hits (according to some then band members, undeservedly) that may also come with a share of royalties.

References

Borch, K. [1962] Equilibrium in a reinsurance market, *Econometrica* **30**(3), 424–444.

Brousseau, E. and Glachan, J.-M. [2002] *The Economics of Contracts: Theory and Applications* (Cambridge University Press).

Christensen, P. O. and Feltham, G. A. [2003] *Economics of Accounting*, Vol. I: *Information in Markets* (Springer).

Christensen, P. O. and Feltham, G. A. [2005] *Economics of Accounting*, Vol. II: *Performance Evaluation* (Springer).

Friedman, J. W. [1986] *Game Theory with Applications to Economics* (Oxford University Press).

Hartl, R. F., Sethi, S. P. and Vickson, R. G. [1995] A survey of the maximum principles for optimal control problems with state constraints, *SIAM Review* **37**, 181–218.

Kadan, O. and Swinkels, J. M. [2013] On the moral hazard problem without the first-order approach, *J. Econ. Theory* **148**, 2313–2343.

Kamien, M. I. and Schwartz, N. L. [1971] Sufficient conditions in optimal control theory, *J. Econ. Theory* **3**, 207–214.

Khmelnitsky, E. [2002] A combinatorial, graph-based solution method for a class of continuous-time optimal control problems, *Math. Oper. Res.* **27**, 312–325.

Levin, J. and Tadelis, S. [2005] Profit sharing and the role of professional partnerships, *Quart. J. Econ.* **120**(1), 131–171.

Pratt, J. W. [1964] Risk aversion in the small and in the large, *Econometrica* **32**, 122–136.

Sappington, D. [1983] Limited liability contracts between principal and agent, *J. Econ. Theory* **29**, 1–21.

Shaked, M. and Shanthikumar, J. G. [1994] *Stochastic Orders and Their Applications* (Academic Press).

Ward, G. C. and Burns, K. [2000] *Jazz: A History of America's Music*, Knopf.

Wilson, R. [1968] The theory of syndicates, *Econometrica* **36**(1), 119–132.

Chapter 20

Stackelberg Oligopoly TU-Games: Characterization and Nonemptiness of the Core

Dongshuang Hou

Department of Applied Mathematics, Northwestern Polytechnical University
710072, Xi'An, P. R. China

dshhou@126.com

Aymeric Lardon

Université Côte d'Azur, GREDEG
CNRS, Valbonne, France

aymeric.lardon@unice.fr

T. S. H. Driessen

Faculty of Electrical Engineering, Mathematics and Computer Science
Department of Applied Mathematics, University of Twente, P. O. Box 217
7500 AE Enschede, The Netherlands

t.s.h.driessen@ewi.utwente.nl

In this chapter, we consider the dynamic setting of Stackelberg oligopoly TU-games in γ-characteristic function form. Any deviating coalition produces an output at a first period as a leader and then, outsiders simultaneously and independently play a quantity at a second period as followers. We assume that the inverse demand function is linear and that firms operate at constant but possibly distinct marginal costs. First, we show that the core of any Stackelberg oligopoly TU-game always coincides with the set of imputations. Second, we provide a necessary and sufficient condition, depending on the heterogeneity of firms' marginal costs, under which the core is nonempty.

Keywords: Stackelberg oligopoly; TU-game; set of imputations; core.

1. Introduction

Usually, oligopoly situations are modeled by means of noncooperative games. Every profit-maximizing firm pursues Nash strategies and the resulting outcome is not Pareto optimal. Yet, it is known that firms are better off by forming cartels and that Pareto efficiency is achieved when all the firms merge together. A problem faced by the members of a cartel is the stability of the agreement and noncooperative game theory predicts that the cartel members always have an incentive to deviate from the agreed-upon output decision.

However, in some oligopoly situations firms do not always behave noncooperatively and if sufficient communication is feasible it may be possible for firms to sign agreements. A question is then whether it is possible for firms to agree all together and coordinate their decision to achieve Pareto efficiency. For that, we consider a fully cooperative approach for oligopoly situations. Under this approach, firms are allowed to sign binding agreements in order to form cartels called coalitions. Under such an assumption cooperative games called oligopoly TU(Transferable Utility)-games can be defined and the existence of stable collusive behaviors is then related to the nonemptiness of the core. Aumann [1959] proposes two approaches in order to define cooperative games: according to the max–min approach, every cartel computes the total profit which it can guarantee itself regardless of what outsiders do; the min–max approach consists in computing the minimal profit for which outsiders can prevent the cartel members from getting more. These two-steps approaches lead to consider the α and β-characteristic functions, respectively. The set of Cournot oligopoly TU-games in α and β-characteristic function forms has been studied by Zhao [1999a, 1999b], Norde *et al.* [2002] and Driessen and Meinhardt [2005] among others. However, these approaches can be questioned in various oligopoly situations since outsiders probably cause substantial damages upon themselves by minimizing the profit of the cartel. This is why Chander and Tulkens [1997] propose an alternative approach where external firms choose their strategy individually as a best reply to the cartel action in a simultaneous game. This leads to consider the γ-characteristic function which is more appropriate in the context of oligopoly industries.[a] The set of Cournot oligopoly TU-games in γ-characteristic function form has been analyzed by Lardon [2012].

In many oligopolies, the formation of a cartel can only be inferred from the observation of quantity changes in the industry. In this case, a cartel takes place secretly and is revealed to firms outside only when the new quantities have been played. Therefore, it makes sense to assume that every cartel possesses a first-mover advantage by acting as a leader at a first period while firms outside play their individual best reply strategies as followers at a second period. For general TU-games, Marini and Currarini [2003] associate a two-stage structure with the γ-characteristic function. Under standard assumptions on the payoff functions,[b]

---

[a] Another interesting approach in oligopolistic markets is due to Lekeas and Stamatopoulos [2014]. We refer to Bloch and van den Nouweland [2014] for a detailed presentation of expectations rules in general partition function games.

[b] Every player's payoff function is assumed twice differentiable and strictly concave on its own strategy set.

they prove that if the players and externalities are symmetric[c] and the game has strategic complementarities then the equal division solution belongs to the core. They also study a quantity competition setting where strategies are substitutes for which they show that the same mechanism underlying their core result determines the nonemptiness of the core.[d] In this chapter, we consider the dynamic structure associated with the γ-characteristic function for oligopoly situations. The introduction of this temporal sequence leads to consider a specific set of cooperative oligopoly games, i.e., the set of Stackelberg oligopoly TU-games in γ-characteristic function form. Thus, contrary to Cournot oligopoly TU-games in γ-characteristic function form in which all the firms simultaneously choose their strategies, every deviating coalition produces an output at a first period as a leader and then, outsiders simultaneously and independently play a quantity at a second period as followers. We assume that the inverse demand function is linear and firms operate at constant but possibly distinct marginal costs. Hence the payoff (profit) functions are not necessarily identical. First, in Stackelberg oligopoly TU-games, we show that the core always coincides with the set of imputations. The reason is that the first-mover advantage gives too much power to singletons so that the worth of every deviating coalition is less than or equal to the sum of its members' individual worths except for the grand coalition. Second, we provide a necessary and sufficient condition under which the core is nonempty. We prove that this condition holds if and only if firms' marginal costs are not too heterogeneous. When the number of firms becomes large, even weak heterogeneity of firms' marginal costs ensures the nonemptiness of the core. Surprisingly, in case the inverse demand function is strictly concave, we provide an example in which the opposite result holds, i.e., when the heterogeneity of firms' marginal costs increases, the core becomes larger.

The remainder of this chapter is structured as follows. In Sec. 2, we recall some basic definitions in cooperative game theory and show that Stackelberg oligopoly TU-games in γ-characteristic function form are well-defined. In Sec. 3, for Stackelberg oligopoly TU-games, we prove that the core is always equal to the set of imputations and we provide a necessary and sufficient condition under which the core is nonempty. Section 4 gives some concluding remarks.

2. The Model

2.1. *TU-games*

Before introducing Stackelberg oligopoly TU-games in γ-characteristic function form [Chander and Tulkens, 1997], we recall some basic definitions in cooperative game theory. Given a set of players N, we call a subset $S \in 2^N \backslash \{\emptyset\}$, a **coalition**.

---

[c]The players are symmetric if they have identical payoff functions and strategy sets. Externalities are symmetric if they are either positive or negative.
[d]Lehrer and Scarsini [2013] have also dealt with an alternative core solution concept, called the intertemporal core, for dynamic cooperative games.

The **size** $s = |S|$ of coalition S is the number of players in S. A **TU-game** (N, v) is a **set function** $v : 2^N \to \mathbb{R}$ with the convention that $v(\emptyset) = 0$, which assigns a number $v(S) \in \mathbb{R}$ to every coalition $S \in 2^N \backslash \{\emptyset\}$. The number $v(S)$ is the worth of coalition S and represents the maximal amount of monetary benefits due to the mutual cooperation among the members of the coalition. We denote by G the **set of TU-games** where (N, v) is a representative element of G.

In the framework of the division problem of the benefits $v(N)$ of the grand coalition N among the players, any allocation scheme of the form $x = (x_i)_{i \in N} \in \mathbb{R}^n$ is supposed to meet, besides the efficiency principle $\sum_{i \in N} x_i = v(N)$, the so-called individual rationality condition in that each player is allocated at least the individual worth, i.e., $x_i \geq v(\{i\})$ for all $i \in N$. For any TU-game $(N, v) \in G$, these two conditions lead to consider the **set of imputations** defined as

$$I(N, v) = \left\{ x \in \mathbb{R}^n : \sum_{i \in N} x_i = v(N) \text{ and } \forall i \in N, x_i \geq v(\{i\}) \right\}.$$

The best known set-valued solution concept called **core** requires the group rationality condition in that the aggregate allocation to the members of any coalition is at least its coalitional worth

$$C(N, v) = \left\{ x \in I(N, v) : \forall S \in 2^N \backslash \{\emptyset, N\}, \sum_{i \in S} x_i \geq v(S) \right\}.$$

Given any imputation in the core, the grand coalition N could form and distribute its worth $v(N)$ as payoffs to its members in such a way that no coalition can contest this allocation scheme by breaking off from the grand coalition.

2.2. *Strategic Stackelberg oligopoly games*

We introduce strategic Stackelberg oligopoly games which will permit to define the γ-characteristic function in the framework of Stackelberg oligopoly TU-games. A **Stackelberg oligopoly situation** is a quadruplet $(L, F, (C_i)_{i \in N}, p)$ defined as

(1) the disjoint finite **sets of leaders and followers** L and F respectively where $L \cup F = \{1, 2, \ldots, n\}$ is the **set of firms** denoted by N;
(2) for every $i \in N$, an **individual cost function** $C_i : \mathbb{R}_+ \to \mathbb{R}_+$;
(3) an **inverse demand function** $p : \mathbb{R}_+ \to \mathbb{R}$ which assigns to any aggregate quantity $X \in \mathbb{R}_+$ the unit price $p(X)$.

Throughout this chapter, we assume that

(a) firms have no production capacity constraint;
(b) firms operate at constant but possibly distinct marginal costs

$$\forall i \in N, \quad \exists c_i \in \mathbb{R}_+ : C_i(y_i) = c_i y_i,$$

where c_i is firm i's marginal cost, and $y_i \in \mathbb{R}_+$ is the quantity produced by firm i;

(c) firms face the linear inverse demand function

$$p(X) = a - X,$$

where $X \in \mathbb{R}_+$ is the **total production** of the industry and $a \in \mathbb{R}_+$ is the intercept of the inverse demand function p such that $a \geq 2n \times \max\{c_i : i \in N\}$.

Given assumptions (a), (b) and (c), a Stackelberg oligopoly situation is summarized by the 4-tuple $(L, F, (c_i)_{i \in N}, a)$. Without loss of generality, we assume that the firms are ranked according to their marginal costs, i.e., $c_1 \leq \cdots \leq c_n$. For notational convenience, for any coalition $S \in 2^N \setminus \{\emptyset\}$, we denote the **minimal coalitional cost** by $\underline{c}_S = \min\{c_i : i \in S\}$ and by $i_S \in S$ the firm in S with the smallest index that operates at marginal cost $\underline{c}_S$.

The **strategic Stackelberg oligopoly game** associated with the Stackelberg oligopoly situation $(L, F, (c_i)_{i \in N}, a)$ is a quadruplet $\Gamma_{\text{so}} = (L, F, (X_i, \pi_i)_{i \in N})$ defined as

(1) the disjoint finite **sets of leaders and followers** L and F, respectively, where $N = L \cup F$ is the **set of firms**;
(2) for every $k \in N$, an **individual strategy set** X_k such that

   — for every leader $i \in L$, $X_i = \mathbb{R}_+$ where $x_i \in X_i$ represents the quantity produced by leader i. We denote by $X_L = \prod_{i \in L} X_i$ the set of strategy profiles of the leaders where $x_L = (x_i)_{i \in L}$ is a representative element of X_L;
   — for every follower $j \in F$, X_j is the set of mappings $x_j : X_L \to \mathbb{R}_+$ where $x_j(x_L)$ represents the quantity produced by follower j given leaders' strategy profile $x_L \in X_L$. We denote by $X_F = \prod_{j \in F} X_j$ the set of strategy profiles of the followers where $x_F = (x_j)_{j \in F}$ is a representative element of X_F;
(3) for every $k \in N$, an **individual profit function** $\pi_k : X_L \times X_F \to \mathbb{R}_+$ such that

   — for every $i \in L$, $\pi_i : X_L \times X_F \to \mathbb{R}_+$ is defined as

$$\pi_i(x_L, x_F(x_L)) = p(X)x_i - c_i x_i;$$

   — for every $j \in F$, $\pi_j : X_L \times X_F \to \mathbb{R}_+$ is defined as

$$\pi_j(x_L, x_F(x_L)) = p(X)x_j(x_L) - c_j x_j(x_L),$$

where $X = \sum_{i \in L} x_i + \sum_{j \in F} x_j(x_L)$ is the **total production**.

Given a strategic Stackelberg oligopoly game $\Gamma_{\text{so}} = (L, F, (X_i, \pi_i)_{i \in N})$, every leader $i \in L$ produces an output $x_i \in X_i$ at a first period while every follower $j \in F$ plays a quantity $x_j(x_L) \in X_j$ at a second period given leaders' strategy profile $x_L \in X_L$.[e] We denote by $\mathcal{G}_{\text{so}}$ the **set of strategic Stackelberg oligopoly games**.

---

[e]In the framework of cooperative Stackelberg oligopoly games, leaders are assumed to be able to correlate their strategies at the first period according to a partial agreement equilibrium concept.

In case there is a single leader and multiple followers, Sherali *et al.* [1983] prove the existence and uniqueness of the Nash equilibrium in strategic Stackelberg oligopoly games under standard assumptions on the inverse demand function and the individual cost functions, i.e., the inverse demand function is twice differentiable, strictly decreasing and satisfies

$$\forall X \in \mathbb{R}_+, \quad \frac{dp}{dX}(X) + X\frac{d^2p}{dX^2}(X) \le 0,$$

and the individual cost functions are twice differentiable and convex. In particular, they show that the convexity of followers' reaction functions with respect to leader's output is crucial for the uniqueness of the Nash equilibrium. Assumptions (a), (b) and (c) ensure that Sherali *et al.*'s result [1983] holds on $\mathcal{G}_{so}$ so that any strategic Stackelberg oligopoly game $\Gamma_{so} = (L, F, (X_i, \pi_i)_{i \in N}) \in \mathcal{G}_{so}$ such that $|L| = 1$ admits a unique Nash equilibrium.

2.3. *Towards Stackelberg oligopoly TU-games*

Now, we want to define Stackelberg oligopoly TU-games in γ-characteristic function form. In a dynamic oligopoly "à la Stackelberg", this assumption implies that the coalition members produce an output at a first period, thus anticipating outsiders' reaction who simultaneously and independently play a quantity at a second period.

Partial agreement equilibrium is a solution concept which generalizes Nash equilibrium and permits to formalize the possibility for some firms to cooperate facing other firms acting individually. For any coalition $S \in 2^N \setminus \{\emptyset\}$, where $S = L$ is the set of leaders and $N \setminus S = F$ is the set of followers, the **coalition profit function** $\pi_S : X_S \times X_{N \setminus S} \to \mathbb{R}$ is defined as

$$\pi_S(x_S, x_{N \setminus S}(x_S)) = \sum_{i \in S} \pi_i(x_S, x_{N \setminus S}(x_S)).$$

Moreover, **followers' individual best reply strategies** $\tilde{x}_{N \setminus S} : X_S \to X_{N \setminus S}$ are defined as

$$\forall j \in N \setminus S, \quad \forall x_S \in X_S, \quad \tilde{x}_j(x_S) \in \arg \max_{x_j(x_S) \in X_j} \pi_j(x_S, \tilde{x}_{N \setminus (S \cup \{j\})}(x_S), x_j(x_S)).$$

For any coalition $S \in 2^N \setminus \{\emptyset\}$ and the induced strategic Stackelberg oligopoly game $\Gamma_{so} = (S, N \setminus S, (X_i, \pi_i)_{i \in N}) \in \mathcal{G}_{so}$, a strategy profile $(x_S^*, \tilde{x}_{N \setminus S}(x_S^*)) \in X_S \times X_{N \setminus S}$ is a **partial agreement equilibrium** under S if

$$\forall x_S \in X_S, \quad \pi_S(x_S^*, \tilde{x}_{N \setminus S}(x_S^*)) \ge \pi_S(x_S, \tilde{x}_{N \setminus S}(x_S)),$$

and

$$\forall j \in N \setminus S, \quad \forall x_j \in X_j, \quad \pi_j(x_S^*, \tilde{x}_{N \setminus S}(x_S^*)) \ge \pi_j(x_S^*, \tilde{x}_{N \setminus (S \cup \{j\})}(x_S^*), x_j(x_S^*)).$$

For any coalition $S \in 2^N \setminus \{\emptyset\}$ and the induced strategic Stackelberg oligopoly game $\Gamma_{so} = (S, N \setminus S, (X_i, \pi_i)_{i \in N}) \in \mathcal{G}_{so}$, the associated **Stackelberg oligopoly**

TU-game in γ-characteristic function form, denoted by (N, v_γ), is defined as

$$v_\gamma(S) = \pi_S(x_S^*, \tilde{x}_{N\setminus S}(x_S^*)),$$

where $(x_S^*, \tilde{x}_{N\setminus S}(x_S^*)) \in X_S \times X_{N\setminus S}$ is a partial agreement equilibrium under S. We denote by $G_{so}^\gamma \subseteq G$ the **set of Stackelberg oligopoly TU-games in γ-characteristic function form**.

In the framework of Stackelberg oligopoly TU-games, we argue that the γ-characteristic function is well-defined. Given any strategic Stackelberg oligopoly game $\Gamma_{so} = (S, N\setminus S, (X_i, \pi_i)_{i\in N}) \in \mathcal{G}_{so}$, by assumptions (a), (b) and (c), and by the definition of the partial agreement equilibrium, any deviating coalition $S \in 2^N\setminus\{\emptyset\}$ can be represented by firm $i_S \in S$ (the firm in S with the smallest marginal cost) acting as a single leader while the other firms in coalition S play a zero output. It follows from Sherali *et al.*'s [1983] result that the induced strategic Stackelberg oligopoly game $\Gamma'_{so} = (\{i_S\}, N\setminus S, (X_i, \pi_i)_{i\in\{i_S\}\cup N\setminus S}) \in \mathcal{G}_{so}$ has a unique Nash equilibrium. Hence the strategic Stackelberg oligopoly game Γ_{so} admits at least one partial agreement equilibrium under S. Indeed, in case there are at least two firms operating at the minimal marginal cost $\underline{c}_S$, the most efficient firms in coalition S can coordinate their output decision and reallocate the Nash equilibrium output of firm i_S among themselves. Hence there can exist several partial agreement equilibria under S derived from the unique Nash equilibrium in Γ'_{so} and which support the unique worth $v_\gamma(S)$. The following proposition goes further by expressing the worth of any deviating coalition.

Proposition 2.1. *For any coalition $S \in 2^N\setminus\{\emptyset\}$ and the induced strategic Stackelberg oligopoly game $\Gamma_{so} = (S, N\setminus S, (X_i, \pi_i)_{i\in N}) \in \mathcal{G}_{so}$, it holds that*

$$v_\gamma(S) = \frac{1}{4(n-s+1)}\left(a + \sum_{j\in N\setminus S} c_j - \underline{c}_S(n-s+1)\right)^2.$$

Proof. Take any coalition $S \in 2^N\setminus\{\emptyset\}$ and consider the induced strategic Stackelberg oligopoly game $\Gamma_{so} = (S, N\setminus S, (X_i, \pi_i)_{i\in N}) \in \mathcal{G}_{so}$. In order to compute the worth $v_\gamma(S)$ of coalition S, we have to successively solve the maximization problems derived from the definition of the partial agreement equilibrium.

First, we consider the profit maximization program of any follower $j \in N\setminus S$ at the second period. The first-order conditions imply that any follower's individual best reply strategy $\tilde{x}_j(x_S) \in X_j$ is given by

$$\forall j \in N\setminus S, \quad \forall x_S \in X_S, \quad \tilde{x}_j(x_S) = \frac{1}{(n-s+1)}\left(a - \sum_{i\in S} x_i + \sum_{k\in N\setminus S} c_k\right) - c_j.$$

$$(1)$$

Second, given $\tilde{x}_{N\setminus S}(x_S) \in X_{N\setminus S}$ we consider the profit maximization program of coalition S at the first period. Since the firms have no capacity constraint, it follows that the above profit maximization program of coalition S is equivalent to the profit maximization program of firm $i_S \in S$ given that the other members in S play a zero output. The first-order condition implies that the unique maximizer $x_{i_S}^* \in X_{i_S}$ is given by

$$x_{i_S}^* = \frac{1}{2}\left(a + \sum_{j \in N \setminus S} c_j - \underline{c}_S(n - s + 1)\right). \tag{2}$$

By (1) and (2), for any $j \in N \setminus S$, it holds that

$$\tilde{x}_j(x_S^*) = \tilde{x}_j(x_{i_S}^*, 0_{S \setminus \{i_S\}})$$

$$= \frac{1}{2(n - s + 1)}\left(a + \sum_{k \in N \setminus S} c_k + \underline{c}_S(n - s + 1)\right) - c_j. \tag{3}$$

By (2) and (3), we deduce that

$$v_\gamma(S) = \pi_S(x_S^*, \tilde{x}_{N \setminus S}(x_S^*))$$

$$= \pi_{i_S}((x_{i_S}^*, 0_{S \setminus \{i_S\}}), \tilde{x}_{N \setminus S}(x_{i_S}^*, 0_{S \setminus \{i_S\}}))$$

$$= \frac{1}{4(n - s + 1)}\left(a + \sum_{j \in N \setminus S} c_j - \underline{c}_S(n - s + 1)\right)^2,$$

which completes the proof. $\qquad\square$

Thus, the worth of any deviating coalition is increasing with respect to outsiders' marginal costs and decreasing with respect to the smallest marginal cost among its members. Note that condition $a \geq 2n \times \max\{c_i : i \in N\}$ (assumption (c)) ensures that equilibrium outputs in (2) and (3) are positive.

3. Core Results

In this section, based on the description of the γ-characteristic function in a Stackelberg oligopoly TU-game, we characterize the core by proving that it always coincides with the set of imputations. In addition, we provide a necessary and sufficient condition under which the core is nonempty. Finally, we prove that this condition holds if and only if firms' marginal costs are not too heterogeneous. This last result extends Marini and Currarini's [2003] core nonemptiness result for oligopoly situations.

3.1. *Characterization of the core*

We characterize the core of any Stackelberg oligopoly TU-game in γ-characteristic function form by showing that it always coincides with the set of imputations. To this end, we will use some additional notations. Given a set of marginal costs $\{c_i\}_{i\in N}$ and any coalition $S \in 2^N\backslash\{\emptyset\}$ let $\alpha(S) = \sum_{j\in S\backslash\{is\}}(c_S - c_j)^2$ and denote by

$$A_1(S) = \frac{1}{2}\sum_{j\in S\backslash\{is\}}\sum_{k\in S\backslash\{is\}}(c_j - c_k)^2; \quad B_1(S) = (s-1)(\alpha(S) - A_1(S));$$

$$C_1(S) = -(s-1)(s\alpha(S) + A_1(S)); \quad D_1(S) = -(s-1)(\alpha(S) + A_1(S)).$$

We define the functions $f_1 : \mathbb{N} \times 2^N\backslash\{\emptyset\} \to \mathbb{R}$ and $f_2 : \mathbb{N} \times 2^N\backslash\{\emptyset\} \to \mathbb{R}$ as

$$f_1(n, S) = 3A_1(S)n^2 + (3A_1(S) + 2B_1(S))n + A_1(S) + B_1(S) + C_1(S);$$

$$f_2(n, S) = A_1(S)n^3 + B_1(S)n^2 + C_1(S)n + D_1(S).$$

We first establish the following lemma.

Lemma 3.1. *Let $\{c_i\}_{i\in N}$ be a set of marginal costs. Then, for any $n \geq 3$ and any coalition $S \in 2^N\backslash\{\emptyset\}$ such that $s \in \{2, \ldots, n-1\}$ it holds that (i) $f_1(n, S) \geq 0$, and (ii) $f_2(n, S) \geq 0$.*

Proof. First, we show point (i). For any $n \geq 3$ and any coalition $S \in 2^N\backslash\{\emptyset\}$ such that $s = n - 1$ it holds that

$$f_1(n, S) = (n^2 - 4)\alpha(S) + (n^2 + 5n + 5)A_1(S)$$

$$\geq 0. \tag{4}$$

Then, we show that for any $n \geq 3$ and any coalition $S \in 2^N\backslash\{\emptyset\}$ such that $s \in \{2, \ldots, n-1\}$, $f_1(n, S) \geq 0$. We proceed by a double induction on the number of firms $n \geq 3$ and the size $s \in \{2, \ldots, n-1\}$ of coalition S, respectively.

Initialization: Assume that $n = 3$ and take any coalition $S \in 2^N\backslash\{\emptyset\}$ such that $s = 2$. By (4), it holds that $f_1(3, S) \geq 0$.

Induction hypothesis: Assume that for any $n \leq k$ and for any coalition $S \in 2^N\backslash\{\emptyset\}$ such that $s \in \{2, \ldots, n-1\}$, $f_1(n, S) \geq 0$.

Induction step: We want to show that for $n = k + 1$ and for any coalition $S \in 2^N\backslash\{\emptyset\}$ such that $s \in \{2, \ldots, k\}$, $f_1(k + 1, S) \geq 0$. It follows from (4) that for any coalition $S \in 2^N\backslash\{\emptyset\}$ such that $s = k$, $f_1(k + 1, S) \geq 0$. It remains to show that for any coalition $S \in 2^N\backslash\{\emptyset\}$ such that $s \in \{2, \ldots, k-1\}$, $f_1(k + 1, S) \geq 0$. Take any coalition $S \in 2^N\backslash\{\emptyset\}$ such that $s \in \{2, \ldots, k-1\}$. Then it follows from the definition of f_1 and the induction hypothesis that

$$f_1(k + 1, S) = f_1(k, S) + 6A_1(S)k + 6A_1(S) + 2B_1(S)$$

$$= f_1(k, S) + A_1(S)(6k - 2s + 8) + 2(s-1)\alpha(S)$$

$$\geq 0,$$

which concludes the proof of point (i).

Then, we show point (ii). For any $n \geq 3$ and any coalition $S \in 2^N \setminus \{\emptyset\}$ such that $s = n - 1$, it holds that

$$f_2(n, S) = (n^2 - 3n + 2)\alpha(S) + (n^2 + n + 2)A_1(S)$$

$$\geq 0. \tag{5}$$

Then, we show that for any $n \geq 3$ and any coalition $S \in 2^N \setminus \{\emptyset\}$ such that $s \in \{2, \ldots, n - 1\}$, $f_2(n, S) \geq 0$. We proceed by a double induction on the number of firms $n \geq 3$ and the size $s \in \{2, \ldots, n - 1\}$ of coalition S, respectively.

Initialization: Assume that $n = 3$ and take any coalition $S \in 2^N \setminus \{\emptyset\}$ such that $s = 2$. By (5), it holds that $f_2(3, S) \geq 0$.

Induction hypothesis: Assume that for any $n \leq k$ and for any coalition $S \in 2^N \setminus \{\emptyset\}$ such that $s \in \{2, \ldots, n - 1\}$, $f_2(n, S) \geq 0$.

Induction step: We want to show that for $n = k + 1$ and for any coalition $S \in 2^N \setminus \{\emptyset\}$ such that $s \in \{2, \ldots, k\}$, $f_2(k + 1, S) \geq 0$. It follows from (5) that for any coalition $S \in 2^N \setminus \{\emptyset\}$ such that $s = k$, $f_2(k + 1, S) \geq 0$. It remains to show that for any coalition $S \in 2^N \setminus \{\emptyset\}$ such that $s \in \{2, \ldots, k - 1\}$, $f_2(k + 1, S) \geq 0$. Take any coalition $S \in 2^N \setminus \{\emptyset\}$ such that $s \in \{2, \ldots, k - 1\}$. Then it follows from the definitions of f_1 and f_2, the induction hypothesis and point (i) of Lemma 3.1 that

$$f_2(k + 1, S) = f_2(k, S) + f_1(k, S)$$

$$\geq 0,$$

which concludes the proof of point (ii).   □

The following notations will be useful for the sequel. Given a set of marginal costs $\{c_i\}_{i \in N}$ and any coalition $S \in 2^N \setminus \{\emptyset\}$, we denote by

$$A_2(S) = \frac{(n - s)(s - 1)}{4n(n - s + 1)} \text{ (note that for any } s \in \{2, \ldots, n - 1\}, A_2(S) > 0);$$

$$B_2(S) = \frac{1}{2n} \sum_{i \in S} \left(\sum_{j \in N \setminus \{i\}} c_j - nc_i \right) - \frac{1}{2(n - s + 1)} \left(\sum_{j \in N \setminus S} c_j - \underline{c}_S(n - s + 1) \right);$$

$$C_2(S) = \frac{1}{4n} \sum_{i \in S} \left(\sum_{j \in N \setminus \{i\}} c_j - nc_i \right)^2 - \frac{1}{4(n - s + 1)} \left(\sum_{j \in N \setminus S} c_j - \underline{c}_S(n - s + 1) \right)^2.$$

These quantities will be used in the proof of the following proposition.

Proposition 3.2. *Let $(N, v_\gamma) \in G^\gamma_{so}$ be a Stackelberg oligopoly TU-game. Then, it holds that*

$$C(N, v_\gamma) = I(N, v_\gamma).$$

Proof. First, assume that $n = 2$. By the definitions of the core and the set of imputations, it holds that $C(N, v_\gamma) = I(N, v_\gamma)$.

Then, assume that $n \geq 3$. The core is equal to the set of imputations if and only if

$$\forall S \in 2^N \backslash \{\emptyset\} : s \in \{2, \ldots, n-1\}, \quad v_\gamma(S) \leq \sum_{i \in S} v_\gamma(\{i\}).$$

In order to prove the above condition, take any coalition $S \in 2^N \backslash \{\emptyset\}$ such that $s \in \{2, \ldots, n-1\}$. By Proposition 2.1, we deduce that

$$\sum_{i \in S} v_\gamma(\{i\}) - v_\gamma(S) = A_2(S)a^2 + B_2(S)a + C_2(S).$$

Now, we define the mapping $P_S : \mathbb{R} \to \mathbb{R}$ as

$$P_S(y) = A_2(S)y^2 + B_2(S)y + C_2(S),$$

so that $P_S(a) = \sum_{i \in S} v_\gamma(\{i\}) - v_\gamma(S)$. We want to show that for any $y \in \mathbb{R}$, $P_S(y) \geq 0$. It follows from $A_2(S) > 0$ that the minimum of P_S is obtained at point $y^* \in \mathbb{R}$ such that

$$y^* = -\frac{B_2(S)}{2A_2(S)}.$$

The minimum of P_S is then equal to

$$P_S(y^*) = \frac{1}{4n(n-s)(s-1)} f_2(n, S),$$

where f_2 is defined as in Lemma 3.1. Hence, it follows from point (ii) of Lemma 3.1 that $P_S(y^*) \geq 0$, which implies that for any $y \in \mathbb{R}$, $P_S(y) \geq 0$. In particular, we conclude that $P_S(a) \geq 0$, and so $\sum_{i \in S} v_\gamma(\{i\}) - v_\gamma(S) \geq 0$. $\qquad\square$

The proof of Proposition 3.2 brings to light that the first-mover advantage gives too much power to singletons so that the worth of any deviating coalition is less than or equal to the sum of its members' individual worths except for the grand coalition.

3.2. Nonemptiness of the core

Now, we provide a necessary and sufficient condition for the nonemptiness of the core of any Stackelberg oligopoly TU-game in γ-characteristic function form.

Theorem 3.3. *Let* $(N, v_\gamma) \in G_{so}^\gamma$ *be a Stackelberg oligopoly TU-game. Then,* $C(N, v_\gamma) \neq \emptyset$ *if and only if*

$$2a(\bar{c}_N - \underline{c}_N) \geq \frac{(n+1)^2}{n} \sum_{j \in N} c_j^2 - \frac{(n+2)}{n} \left(\sum_{j \in N} c_j\right)^2 - \underline{c}_N^2, \tag{6}$$

where $\bar{c}_N = \sum_{i \in N} c_i/n$ *is the average cost of the grand coalition.*

Proof. It follows from Proposition 3.2 that the core is nonempty if and only if $\sum_{i \in N} v_\gamma(\{i\}) \leq v_\gamma(N)$. By Proposition 2.1, it holds that

$$\sum_{i \in N} v_\gamma(\{i\}) = \frac{1}{4n} \sum_{i \in N} \left(a + \sum_{j \in N \setminus \{i\}} c_j - nc_i \right)^2$$

$$= \frac{1}{4n} \sum_{i \in N} \left(a + \sum_{j \in N \setminus \{i\}} c_j - nc_i - \underline{c}_N + \underline{c}_N \right)^2$$

$$= \frac{1}{4n} \sum_{i \in N} \left[(a - \underline{c}_N)^2 + 2(a - \underline{c}_N) \left(\sum_{j \in N \setminus \{i\}} c_j - nc_i + \underline{c}_N \right) \right.$$

$$\left. + \left(\sum_{j \in N \setminus \{i\}} c_j - nc_i + \underline{c}_N \right)^2 \right]$$

$$= v_\gamma(N) + \frac{1}{4n} \left[2a \left(n\underline{c}_N - \sum_{i \in N} c_i \right) + \sum_{i \in N} \left(\sum_{j \in N \setminus \{i\}} c_j - nc_i \right)^2 - n\underline{c}_N^2 \right]$$

$$= v_\gamma(N) + \frac{1}{4n} \left[2a \left(n\underline{c}_N - n\bar{c}_N \right) + (n+1)^2 \sum_{j \in N} c_j^2 \right.$$

$$\left. - (n+2) \left(\sum_{j \in N} c_j \right)^2 - n\underline{c}_N^2 \right],$$

which permits to conclude that $\sum_{i \in N} v_\gamma(\{i\}) \leq v_\gamma(N)$ if and only if inequality (6) holds. □

Inequality (6) has the following interpretation: it stipulates that the core is nonempty if and only if the difference between the average cost $\bar{c}_N$ and the minimum cost $\underline{c}_N$ in the industry is greater than or equal to an expression which depends positively on the sum of the square of the marginal costs and negatively on the square of the sum of the marginal costs, the square of the minimum cost and the intercept of the demand. In case all the firms have the same marginal cost, the both sides of inequality (6) are equal to zero which implies that $\sum_{i \in N} v_\gamma(\{i\}) = v_\gamma(N)$, and so $(v_\gamma(\{i\}))_{i \in N} = (v_\gamma(N)/n)_{i \in N}$ is the unique core element.

Furthermore, note that a deviating coalition in a dynamic oligopoly "à la Stackelberg" could always obtain its equilibrium profit in a static oligopoly "à la Cournot" by choosing Cournot quantities to which every follower would reply with its own Cournot quantity.[f] As a consequence, on the basis of the same inverse demand function and individual cost functions, the worth of any deviating coalition in Stackelberg competition is greater than or equal to its worth in Cournot competition. Hence, we conclude that the core of a Stackelberg oligopoly TU-game is always included in the core of a Cournot oligopoly TU-game as illustrated in the following example.

Example 3.4. Consider the Stackelberg and Cournot oligopoly TU-games denoted by (N, v_γ^{St}) and (N, v_γ^{Co}), respectively, where $N = \{1, 2\}$ and the inverse demand function is given by $p(X) = 10 - X$. Firm i's marginal cost are given by $c_1 = 0$ and $c_2 = \delta$ where $0 \leq \delta \leq 2.5$ (assumption (c)). The worth of any coalition $S \in 2^N \setminus \{\emptyset\}$ is given in the following table:

S	$\{1\}$	$\{2\}$	$\{1, 2\}$
$v_\gamma^{St}(S)$	$\dfrac{(10 + \delta)^2}{8}$	$\dfrac{(10 - 2\delta)^2}{8}$	25
$v_\gamma^{Co}(S)$	$\dfrac{(10 + \delta)^2}{9}$	$\dfrac{(10 - 2\delta)^2}{9}$	25

Thus, it holds that $\emptyset \neq C(N, v_\gamma^{St}) \subset C(N, v_\gamma^{Co})$ for any $\delta \leq 2.5$.

The following proposition gives a more relevant expression of inequality (6). When the difference between any two successive marginal costs is constant, it provides an upper bound on the heterogeneity of firms' marginal costs under which inequality (6) holds.

Proposition 3.5. *Let* $(N, v_\gamma) \in G_{so}^\gamma$ *be a Stackelberg oligopoly TU-game such that*

$$\exists \delta \in \mathbb{R}_+ : \forall i \in \{1, \ldots, n - 1\}, c_{i+1} = c_i + \delta. \tag{7}$$

Then, inequality (6) *holds if and only if*

$$\delta \leq \delta^*(n) = \frac{12(a - \underline{c}_N)}{n^3 + 3n^2 + 6n - 2}. \tag{8}$$

Proof. It follows from (7) that inequality (6) can be expressed as

$$\sum_{i \in N} \left(\delta \frac{(n^2 - 2ni + n)}{2} - c_i \right)^2 - n\underline{c}_N^2 \leq an(n - 1)\delta.$$

[f]Note that outsiders' reaction functions are the same in Stackelberg and Cournot oligopolies.

By noting that

$$\sum_{i\in N} c_i^2 - n\underline{c}_N^2 = \delta\underline{c}_N n(n-1) + \delta^2\frac{n(n-1)(2n-1)}{6},$$

we deduce that the above inequality is equivalent to

$$\delta\left[\sum_{i\in N}\left(\frac{n^2-2ni+n}{2}\right)^2 + \frac{n(n-1)(2n-1)}{6}\right]$$

$$\le (a-\underline{c}_N)n(n-1) + \sum_{i\in N}(n^2-2ni+n)c_i. \tag{9}$$

It remains to compute the two sums in inequality (9). First, it holds that

$$\sum_{i\in N}\left(\frac{n^2-2ni+n}{2}\right)^2 = \frac{n^3(n+1)^2}{4} - n^2(n+1)\sum_{i\in N}i + n^2\sum_{i\in N}i^2$$

$$= \frac{1}{12}(2n^4(n+1) - n^3(n+1)^2). \tag{10}$$

Then, it holds that

$$\sum_{i\in N}c_i = n\underline{c}_N + \delta\sum_{i=1}^{n-1}i$$

$$= n\left(\underline{c}_N + \delta\frac{(n-1)}{2}\right),$$

and

$$\sum_{i\in N}ic_i = \underline{c}_N\sum_{i=1}^{n}i + \delta\sum_{i=1}^{n}i(i-1)$$

$$= \underline{c}_N\frac{n(n+1)}{2} + \delta\frac{n(n+1)(2n-2)}{6}.$$

Hence, we deduce that

$$\sum_{i\in N}(n^2-2ni+n)c_i = (n^2+n)\sum_{i\in N}c_i - 2n\sum_{i\in N}ic_i$$

$$= \frac{1}{6}(\delta n^2(n+1)(1-n)). \tag{11}$$

By (10) and (11), we conclude that inequality (9) is equivalent to (8). □

By noting that

$$\frac{d\delta^*}{dn}(n) = -\frac{36(a-\underline{c}_N)(n^2+2n+2)}{(n^3+3n^2+6n-2)^2}$$

$$< 0,$$

and

$$\frac{d^2\delta^*}{dn^2}(n) = \frac{72(a - \underline{c}_N)(2n^4 + 8n^3 + 15n^2 + 20n + 14)}{(n^3 + 3n^2 + 6n - 2)^3}$$

$$> 0,$$

we deduce that the bound $\delta^*(n)$ is strictly decreasing and strictly convex with respect to the number of firms n. Moreover, when n tends to infinity its limit is equal to 0. So, the more the number of firms is, the less the heterogeneity of firms' marginal costs must be in order to ensure the nonemptiness of the core. This result extends Marini and Currarini's core allocation result (2003) and shows that their result crucially depends on the symmetric players assumption.

We have proved that when the heterogeneity of firms' marginal costs increases the core becomes smaller. Surprisingly, in case the inverse demand function is strictly concave, the following example shows that the opposite result may hold, i.e., when the heterogeneity of firms' marginal costs increases the core becomes larger. This result illustrates the difficulty to extend the analysis to general inverse demand function.

Example 3.6. Consider the three Stackelberg oligopoly TU-games $(N, v_\gamma^1) \in G_{so}^\gamma$, $(N, v_\gamma^2) \in G_{so}^\gamma$ and $(N, v_\gamma^3) \in G_{so}^\gamma$ where $N = \{1, 2\}$ and the inverse demand function is given by $p(X) = 10 - X^2$. Firm i's marginal cost in the TU-game (N, v_γ^k), $k \in \{1, 2, 3\}$, is denoted by c_i^k. We assume that $c_1^k = 2$ for any $k \in \{1, 2, 3\}$, $c_2^1 = 2$, $c_2^2 = 4$ and $c_2^3 = 5$. The worth of any coalition $S \in 2^N \setminus \{\emptyset\}$ is given in the following table:

S	$\{1\}$	$\{2\}$	$\{1, 2\}$
$v_\gamma^1(S)$	4.70	4.70	8.71
$v_\gamma^2(S)$	6.19	2.02	8.71
$v_\gamma^3(S)$	7.05	1.05	8.71

Thus, it holds that $\emptyset = C(N, v_\gamma^1) \subset C(N, v_\gamma^2) \subset C(N, v_\gamma^3) \neq \emptyset$, and so when the heterogeneity of firms' marginal costs increases the core becomes larger.

4. Concluding Remarks

In this chapter, we have considered Stackelberg oligopoly TU-games in γ-characteristic function form [Chander and Tulkens, 1997]. Based on the description of the γ-characteristic function in any Stackelberg oligopoly TU-game, we have first proved that the core always coincides with the set of imputations. Then, we have provided a necessary and sufficient condition under which the core is nonempty. Finally, we have proved that this condition depends on the heterogeneity of firms' marginal costs, i.e., the core is nonempty if and only if firms' marginal costs are not too heterogeneous. Instead of quantity competition, we can associate a two-stage structure with the γ-characteristic function in a price competition setting.

Marini and Currarini's [2003] core nonemptiness result applies to this setting and they provide examples in which the core of the sequential Bertrand oligopoly TU-games is included in the core of the classical Bertrand oligopoly TU-games. A question concerns the effect of the heterogeneity of firms' marginal costs on the nonemptiness of the core. This is left for future work.

Acknowledgments

The first author is financially supported by National Science Foundation of China (NSFC) through Grant No. 71601156, 71671140, 71571143 as well as Shaanxi Province Science and Technology Research and Development Program (No. 2016JQ7008).

References

Aumann, R. [1959] Acceptable points in general cooperative n-person games, in *Contributions to the Theory of Games*, Vol. 4, Annals of Mathematics Studies, Vol. 40, eds. Tucker, A. W. and Luce, R. D. (Princeton University Press, Princeton).

Bloch, F. and van den Nouweland, A. [2014] Expectation formation rules and the core of partition function games, *Games Econ. Behav.* **88**, 538–553.

Chander, P. and Tulkens, H. [1997] The core of an economy with multilateral environmental externalities, *Int. J. Game Theory* **26**, 379–401.

Driessen, T. S. and Meinhardt, H. I. [2005] Convexity of oligopoly games without transferable technologies, *Math. Soc. Sci.* **50**, 102–126.

Lardon, A. [2012] The γ-core of cournot oligopoly games with capacity constraints, *Theory Decis.* **72**, 387–411.

Lehrer, E. and Scarsini, M. [2013] On the core of dynamic cooperative games, *Dyn. Games Appl.* **3**(3), 359–373.

Lekeas, P. V. and Stamatopoulos, G. [2014] Cooperative oligopoly games with boundedly rational firms, *Ann. Oper. Res.* **223**(1), 255–272.

Marini, M. A. and Currarini, S. [2003] A sequential approach to the characteristic function and the core in games with externalities, in *Advances in Economic Design* (Springer, Berlin, Heidelberg), pp. 233–251.

Norde, H., Pham Do, K. H. and Tijs, S. [2002] Oligopoly games with and without transferable technologies, *Math. Soc. Sci.* **43**, 187–207.

Sherali, H. D., Soyster, A. L. and Murphy, F. H. [1983] Stackelberg-Nash-Cournot equilibria: Characterizations and computations, *Oper. Res.* **31**(2), 253–276.

Zhao, J. [1999a] A necessary and sufficient condition for the convexity in oligopoly games, *Math. Soc. Sci.* **37**, 189–204.

Zhao, J. [1999b] A β-core existence result and its application to oligopoly markets, *Games Econ. Behav.* **27**, 153–168.

Chapter 21

Strong Strategic Support of Cooperation in Multistage Games

Leon Petrosyan

St Petersburg State University
Universitetskaya emb., 7/9, St. Petersburg
199034, Russia

l.petrosyan@spbu.ru

The problem of cooperation in repeated and multistage games is considered. The strong equilibrium (equilibrium stable against deviations of coalitions) with payoffs which can be attained under cooperation is constructed for a wide class of such games. The new solution concept based on solutions of stage games is introduced and in some cases this solution is a subset of the core defined for repeated and multistage games in a classical way. It is also proved that this newly introduced solution concept is strongly time consistent. The strong time consistency of the solution is a very important property since in case it does not take place players in reality in some time instant in subgame on cooperative trajectory may switch from the previously selected optimal solution to any other optimal solution in the subgame and as result realize the solution which will not be optimal in the whole game.

Keywords: Strong equilibrium; strongly time consistency; cooperative game; core.

1. Introduction

It is a very exceptional event when in static games maximal joint payoff of players is attained in some Nash equilibrium [Nash, 1951; Rubinstein, 1994]. And since the set of all Nash equilibrium contains the set of strong equilibrium (any strong equilibrium is Nash equilibrium, but converse is not true) it is much less possible that it is attained in a strong equilibrium (equilibrium stable against the deviation of coalitions). The definition of strong Equilibrium points was first introduced by Aumann [1959] for infinite supergame (infinitely repeated stage game) with payoffs defined as limits of mean payoffs, when the number of plays (stages) tends to infinity. The situation is different in repeated, multistage, stochastic and differential games. For these classes of games in many cases such Nash equilibrium can be constructed and in this case we can say that the cooperation is strategically supported. If in these games one can construct a strong equilibrium (equilibrium stable against the deviation of coalitions) with such payoffs of players that their sum takes maximal value we can say that the cooperation is strongly strategically supported [Petrosyan *et al.*, 2017]. In repeated, multistage, stochastic and differential games the corresponding Nash equilibrium is constructed using the ideas coming from famous Folk theorems. Folk theorems are well known in game theory [Aumann, 1959; Fudenberg and Maskin, 1986; Maschler *et al.*, 2013; Myerson, 1986]. By using the so-called "punishment" strategies they show the possibility to attain in equilibrium preferable outcomes. These outcomes are stable against deviations of single players. Since cooperation gives in general better payoffs for players than individually rational behavior it is also important to make the cooperation stable not only against the deviations of single players but also against the deviations of coalitions.

In this chapter, we try to construct a mechanism based on newly introduced analogue of characteristic function which guarantees under some natural conditions the coalition-proofness for multistage and as a result also for repeated games. The basic difficulty in constructing the corresponding strong equilibrium consists in the fact that it is not easy to apply "punishment" for the coalitions deviating from cooperation. In the simple case when only one shot deviations and only of one coalition are allowed the result is published in Petrosyan *et al.* [2017]. But the approach which was presented in Petrosyan *et al.* [2017] cannot be extended for the general case when different members of deviating coalitions deviate in different stage games and in different time instants, because immediately after the first deviation the non-deviating players cannot identify the deviating coalition and thus cannot realize the "punishment" against the coalition which decides to end up with cooperation. To solve this problem, we introduce new "punishment" strategies for players: if they (for instance player i) see the first time (on the next step after deviation) the deviation of one or more players, the non-deviating players start to use their optimal strategies in a zero sum game, playing against the coalition of all players (except himself). In practice, this means that each non-deviating player has to start the play against all others immediately after the first deviation by someone takes place. We derive the conditions under which this way of behavior guarantees that each coalition deviating from cooperation will lose. Based on this approach, we construct the analogue of characteristic function for each stage game and introduce the new characteristic function for the whole multistage game. Using this newly defined characteristic function we define the analogue of the core and prove its strongly time consistency [Petrosyan, 1993]. Time consistency is subgame perfection but only for subgames along the cooperative trajectory. To clarify the problem it is important to mention that if we try to define time-consistency for Nash equilibrium then it is clear that every Nash equilibrium is time-consistent but of course not subgame-perfect. It is important that we define the notion of time-consistency for cooperative multistage TU games. We require that the optimal solution (it can be the Shapley Value, core, nucleous and others) in the whole multistage game truncated to the subgame with initial conditions on the cooperative trajectory (trajectory maximizing the sum of players payoffs) remains optimal in this subgame.

2. Repeated Games

2.1. *Game description*

Consider an infinitely repeated game G with finite stage game Γ played on each stage. Introduce the following notations

$$\Gamma = \langle N; U_1, \ldots, U_i, \ldots, U_n; K_1, \ldots, K_i, \ldots, K_n \rangle,$$

here N is the set of players, U_i the set of strategies of player i, $K_i(u_1, \ldots, u_n)$, $u_i \in U_i$ the payoff function of player i, ($i \in N$). If on the stage m ($1 \leq m \leq \infty$) the n-tuple of strategies $u^m = (u_1^m, \ldots, u_i^m, \ldots, u_n^m)$ is used, payoffs in G are defined as:

$$H_i(u_1(\cdot), \ldots, u_i(\cdot), \ldots, u_n(\cdot)) = \sum_{m=1}^{\infty} \delta^{m-1} K_i(u_1^m, \ldots, u_i^m, \ldots, u_n^m)$$

$$= \sum_{m=1}^{\infty} \delta^{m-1} K_i(u^m) = H_i(u(\cdot)), \quad i \in N, \tag{1}$$

where $u_1(\cdot) = (u_1^1, \ldots, u_1^m, \ldots), \ldots, u_i(\cdot) = (u_i^1, \ldots, u_i^m, \ldots), \ldots, u_n(\cdot) = (u_n^1, \ldots, u_n^m, \ldots)$.

We suppose in addition that at stage m when selecting strategies u_i^m in stage game Γ player i knows the choices of other players on previous stages and remember their choices, thus in fact strategies are functions of histories $h^m = (u_1^1, \ldots, u_1^{m-1}; \ldots; u_i^1, \ldots, u_i^{m-1}; \ldots; u_n^1, \ldots, u_n^{m-1})$. Thus for correctness we have to write $u_i^m(h^m)$ but to simplify notations we shall only keep it in mind, which means that we shall denote the choice of player $i \in N$ in Γ by u_i^m having in mind that his choice is the function of histories (not only from current position).

Denote by $\bar{u}(\cdot) = (\bar{u}_1(\cdot), \ldots, \bar{u}_i(\cdot), \ldots, \bar{u}_n(\cdot))$ the n-tuple of strategies for which

$$\sum_{i \in N} H_i(\bar{u}) = \max_{u(\cdot)} \sum_{i \in N} H_i(u). \tag{2}$$

Since the game Γ is finite such n-tuple of strategies always exists. One can select the strategies in G as $\bar{u}_i(\cdot) = (\bar{u}_i^1, \ldots, \bar{u}_i^m, \ldots,) \ i \in N$ such that

$$\sum_{i \in N} K_i(\bar{u}_1, \ldots, \bar{u}_i, \ldots, \bar{u}_n) = \max_{u_1, \ldots, u_i, \ldots, u_n} \sum_{i \in N} K_i(u_1, \ldots, u_i, \ldots, u_n). \tag{3}$$

Then the repeated game G is realized in which $\bar{u}_i^m = \bar{u}_i$ for all $m = 1, \ldots, n$. From (1)–(3), we get

$$\sum_{i \in N} H_i(\bar{u}) = \sum_{i \in N} \left(\sum_{m=1}^{\infty} \delta^{m-1} K_i(\bar{u}_1^m, \ldots, \bar{u}_n^m) \right)$$

$$= \sum_{i \in N} \left(\sum_{m=1}^{\infty} \delta^{m-1} K_i(\bar{u}_1, \ldots, \bar{u}_n) \right) = \frac{1}{1-\delta} \sum_{i \in N} K_i(\bar{u}_1, \ldots, \bar{u}_n). \tag{4}$$

Strategies $(\bar{u}_1, \ldots \bar{u}_n)$ we will call "cooperative" strategies in Γ. Introduce the characteristic function $V(S)$, $S \subset N$ in Γ in the classical sense [Neumann and Morgenstern, 1951], i.e., as value of zero-sum game between coalition S as first player and coalition N/S as second with the payoff of coalition S equal to the sum of payoffs of its members, by definition

$$V(N) = \sum_{i \in N} K_i(\bar{u}_1, \ldots, \bar{u}_n). \tag{5}$$

It is easy to see that the characteristic function $\bar{V}(S)$, $S \subset N$ in repeated game G will be

$$\bar{V}(S) = \frac{1}{1-\delta} V(S), \quad S \subset N. \tag{6}$$

Definition 1. The n-tuple of strategies $(\hat{u}_1, \ldots, \hat{u}_i, \ldots, \hat{u}_n) = \hat{u}$ is called Strong Equilibrium if for all $S \subset N$ and all $u_S = \{u_i, i \in S\}$ the following inequalities hold

$$\sum_{i \in S} K_i(\hat{u}) \geq \sum_{i \in S} K_i(\hat{u} \| u_S), \tag{7}$$

where $(\hat{u} \| u_S)$ is an n-tuple of strategies in which players from $N \backslash S$ use strategies $\hat{u}_i$ and players from coalition S arbitrary strategies.

When the discount factor goes to one we obtain a strong equilibrium a la Aumann.

2.2. Associated zero-sum games

Consider a family of zero-sum games $\Gamma_{N \backslash i, i}$ based on game Γ between coalition $N \backslash i$ as first player and coalition consisting from a single player $\{i\}$ as second. The payoff of coalition $N \backslash i$ is equal to the sum of payoffs of players from this coalition.

Let $(\bar{\mu}_{N \backslash i}, \bar{\mu}_i)$ be the saddle point in mixed strategies in this game $\Gamma_{N \backslash i, i}$. Consider the n-tuple of strategies

$$\bar{\mu} = (\bar{\mu}_1, \ldots, \bar{\mu}_i, \ldots, \bar{\mu}_n)$$

and define

$$\bar{W}(S) = \max_{\mu_S} \sum_{i \in S} E_i(\mu_S; \bar{\mu}_{N \backslash S}),$$

where $\mu_S = \{\mu_i, i \in S\}$, $\bar{\mu}_{N \backslash S} = \{\bar{\mu}_i, i \in N \backslash S\}$ and $E_i(\mu_S; \bar{\mu}_{N \backslash S})$ is the mathematical expectation of the payoff of player i when strategies $(\mu_S; \bar{\mu}_{N \backslash S})$ are played.

$\bar{W}(S)$ is the maximal payoff that the coalition S can guarantee if other players use strategies $\bar{\mu}_i$, $i \in N \backslash S$ which represent optimal mixed strategies in the games $\Gamma_{N \backslash i, i}$.

The following lemma holds.

Lemma 1.

$$\bar{W}(S) \geq V(S), \quad S \subset N, \quad \bar{W}(N) = V(N).$$

Proof. For all $\mu(S)$ we have

$$\sum_{i \in S} E_i(\mu_S, \bar{\mu}_{N \backslash S}) \geq \min_{\mu_{N \backslash S}} \sum_{i \in S} E_i(\mu_S, \mu_{N \backslash S}).$$

Since the above inequality is true for all μ_S, taking maximum from both sides of this inequality we get

$$\bar{W}(S) = \max_{\mu_S} \sum_{i \in S} E_i(\mu_S, \bar{\mu}_{N \backslash S}) \geq \max_{\mu_S} \min_{\mu_{N \backslash S}} \sum_{i \in S} E_i(\mu_S, \mu_{N \backslash S}) = V(S).$$

The equality $\bar{W}(N) = V(N)$ is trivial. The lemma is proved. $\qquad \square$

Suppose that there exists a solution to the following inequalities

$$\sum_{i \in S} \alpha_i > \bar{W}(S), \quad S \subset N, \quad S \neq N, \quad \sum_{i \in N} \alpha_i = \bar{W}(N) = V(N). \tag{8}$$

Consider the modification G^α of game G. The difference between these two games consists only in payoffs in stage game Γ which are realized at each stage of game G. If the cooperative strategies $\bar{u} = (\bar{u}_1, \ldots, \bar{u}_n)$ are used the payoffs in stage games of game G are equal to $\alpha = (\alpha_1, \ldots, \alpha_n)$, where α satisfies (8). For other strategy combinations in stage games of game G and game G^α the payoffs coincide.

The following theorem holds [Petrosyan *et al.*, 2018]. For the completeness of the exposition we give below the proof of Theorem 1.

Theorem 1. *Suppose condition* (8) *is satisfied, then there exist a* $\delta \in (0, 1)$ *such that in the game* G^α *there exists a strong equilibrium with payoffs* $\alpha_i \frac{1}{1-\delta}$. *These payoffs coincide with payoffs in the game* G^α *when the cooperative strategies are played.*

Proof. Denote $\alpha(S) = \sum_{i \in S} \alpha_i$ and by $|S|$ the number of players in coalition S. If players from coalition S use cooperative strategies and as payoffs under cooperation vector $\alpha = (\alpha_1, \ldots, \alpha_n)$ that satisfies (8) is used the payoff of coalition S in G^α will be equal to $\frac{\alpha(S)}{1-\delta}$.

Suppose that players from S deviate from cooperation and the last deviating player deviates at stage m. Since the coalition S consists of a finite number of members such m always exists. Denote by M the maximal payoff that the player from

S can get by deviating from cooperation. Then it is clear that after the deviation the maximal possible payoff of the deviating coalition S cannot exceed

$$\sum_{l=0}^{m} \delta^l |S| M + \sum_{l=m+1}^{\infty} \delta^l \bar{W}(S) = |S| M \sum_{l=0}^{m} \delta^l + \delta^{m+1} \sum_{l=0}^{\infty} \delta^l \bar{W}(S)$$

$$= |S| M \sum_{l=0}^{m} \delta^l + \delta^{m+1} \frac{\bar{W}(S)}{1 - \delta}. \tag{9}$$

Since starting from stage $m + 1$ onwards, the players outside S are getting the information about the deviation of the first deviating player they can use strategies $\bar{\mu}_i$ optimal in the associated stage game $\Gamma_{N \backslash i, i}$. As result, starting from this stage, the maximal payoff that the coalition S can gurantee herself in future stage games coincides with the value $\bar{W}(S)$ (the payoff of coalition S using her best reply strategies against the strategies $\bar{\mu}_i$ of players $i \in N \backslash S$). At the same time, by cooperation, the coalition gets

$$\frac{\alpha(S)}{1 - \delta} = \left(\sum_{l=0}^{m} \delta^l \right) \alpha(S) + \delta^{m+1} \frac{\alpha(S)}{1 - \delta}. \tag{10}$$

We show that there exists a $\delta' \in (0, 1)$ such that for all $\delta \in [\delta', 1)$ we have

$$\left(\sum_{l=0}^{m} \delta^l \right) \alpha(S) + \delta^{m+1} \frac{\alpha(S)}{1 - \delta} > |S| M \sum_{l=0}^{m} \delta^l + \delta^{m+1} \frac{\bar{W}(S)}{1 - \delta}$$

or

$$\sum_{l=0}^{m} \delta^l (\alpha(S) - |S| M) + \frac{\delta^{m+1}}{1 - \delta} (\alpha(S) - \bar{W}(S)) > 0.$$

The last inequality can be rewritten in the form

$$\frac{\delta^{m+1}}{1 - \delta} (\alpha(S) - \bar{W}(S)) > \sum_{l=0}^{m} \delta^l (|S| M - \alpha(S)). \tag{11}$$

It is clear that for $\delta \in (0, 1)$ the following inequality holds

$$\sum_{l=0}^{m} \delta^l (|S| M - \alpha(S)) < (m + 1)(|S| M - \alpha(S)).$$

Then if we could find such $\delta' \in (0, 1)$, $\delta \in [\delta', 1)$ that

$$\frac{\delta^{m+1}}{1 - \delta} (\alpha(S) - \bar{W}(S)) > (m + 1)(|S| M - \alpha(S)),$$

then for such δ we will trivially have

$$\frac{\delta^{m+1}}{1 - \delta} > (m + 1) \frac{|S| M - \alpha(S)}{\alpha(S) - \bar{W}(S)}.$$

Denote

$$(m + 1) \frac{|S| M - \alpha(S)}{\alpha(S) - \bar{W}(S)} = A,$$

Evidently $A \geq 0$ and

$$\frac{\delta^{m+1}}{1 - \delta} > 0$$

There always exists a $\delta' \in (0, 1)$ such that for all $\delta \in [\delta', 1)$ the inequality

$$\frac{\delta^{m+1}}{1 - \delta} > A$$

holds, and also condition (11). The theorem is proved. $\qquad\qquad\square$

The above inequality holds when δ is sufficiently close to 1 and the theorem holds. But also the Folk theorem will hold since any strong equilibrium is also Nash equilibrium (unfortunately the constructed equilibrium will not be subgame-perfect).

It follows from Lemma 1 that the vector $\alpha = (\alpha_1, \ldots, \alpha_n)$ is an imputation in a cooperative version of stage game Γ and belongs to the core of the stage game. Based on the cooperative version of stage games one can easily construct the cooperative version of the game G^{α}. This can be done by introducing the characteristic function in G^{α} by formulas

$$\tilde{V}(S) = \frac{V(S)}{1 - \delta}.$$

It can be easily seen that vectors of the form $\tilde{\alpha} = \alpha \frac{1}{1-\delta}$ are imputations in G^{α}, and if α belongs to the core of the stage game, imputation $\tilde{\alpha}$ belongs to the core of game G^{α}. And it follows from Theorem 1 that each imputation from the core of G^{α} which additionally satisfies conditions

$$\sum_{i \in S} \alpha_i > \bar{W}(S) \tag{12}$$

can be strategically supported by strong equilibrium. Denote the set of all imputations satisfying (12) by $\bar{C}$. This set as it follows from Lemma 1 is a subset of the core in G and can be called quasicore. It can be proved that the quasicore $\bar{C}$ is strongly time consistent. We shall prove this in the next section for general multistage games which include repeated games as special case.

3. Multistage Games

3.1. *Game description*

Suppose that an infinite graph-tree $\bar{G} = (Z, T)$, $T_z \subset Z$, $|T_z| \leq l$ is given. In each vertex $z \in Z$ of the tree a finite simultaneous n-person game $\Gamma(z)$ is defined. For simplicity suppose also that the number of possible different games $F(z)$ for all $z \in Z$ is finite and the player set N is the same in all stage games. At the first stage of game G in the root z_1 of the tree $\bar{G}$ the game $\Gamma(z_1)$ is played

$$\Gamma(z_1) = \langle N; U_1^{z_1}, \ldots, U_i^{z_1}, \ldots, U_n^{z_1}; K_1^{z_1}, \ldots, K_i^{z_1}, \ldots, K_n^{z_1} \rangle. \tag{13}$$

Suppose that in the game $\Gamma(z_1)$ players choose strategies $(u_1(z_1), \ldots, u_n(z_1))$, $u_i(z_1) \in U_i^{z_1}$, $i \in N$, then the game passes to the vertex $z_2 = T(z_1; u_1(z_1), \ldots, u_n(z_1))$ and in vertex z_2 the simultaneous game $\Gamma(z_2)$ is played. In general case at stage $m-1$, in vertex z_{m-1} the simultaneous game $\Gamma(z_{m-1})$ is played. Suppose that in $\Gamma(z_{m-1})$ players choose strategies $(u_1(z_{m-1}), \ldots, u_n(z_{m-1}))$ then the game passes to the vertex

$$z_m = T(z_{m-1}; u_1(z_{m-1}), \ldots, u_n(z_{m-1}))$$

and in this vertex the simultaneous game

$$\Gamma(z_m) = \Gamma(T(z_{m-1}; u_1(z_{m-1}), \ldots, u_n(z_{m-1})))$$

is played and so on. As result we get an infinite sequence of simultaneous stage games $\Gamma(z_1), \ldots, \Gamma(z_m), \ldots$.

As in the previous section, define the strategy $u_i(\cdot)$ of player i in G which in every vertex of the tree $\bar{G}$ determines the choice of $u_i(z)$ as function of game history. We suppose that there exists the n-tuple of strategies $\hat{u}(\cdot) = (\hat{u}_1(\cdot), \ldots, \hat{u}_n(\cdot))$ that maximizes the sum of players' payoffs in G and we shall call this n-tuple "cooperative strategies", and the corresponding sequence of stage games (or sequence of realized vertexes) "cooperative trajectory".

The cooperative behavior generates the sequence of simultaneous stage games

$$\Gamma(\hat{z}_m) = \Gamma(T(\hat{z}_{m-1}; \hat{u}_1(\hat{z}_{m-1}), \ldots, \hat{u}_n(\hat{z}_{m-1}))),$$

where $\hat{u}_1(\hat{z}_{m-1}), \ldots, \hat{u}_n(\hat{z}_{m-1})$ $\hat{u}_i(\hat{z}) \in U_i^z$, $i = 1, \ldots, n$ are cooperative strategies of players and $\hat{z}_1, \ldots, \hat{z}_m, \ldots$ is the corresponding trajectory. Thus in every simultaneous game on cooperative trajectory joint players payoff is equal to

$$L(\hat{z}_m) = \sum_{i \in N} K_i^{\hat{z}_m}(\hat{u}_1(\hat{z}_m), \ldots \hat{u}_n(\hat{z}_m)).$$

Let

$$L = \min_m L(\hat{z}_m).$$

Minimum is attained since the number of different simultaneous games Γ in G is finite.

3.2. Associated zero-sum games

Consider two types of associated zero-sum games $\Gamma_{S,N\setminus S}(z)$, $\Gamma_{N\setminus i,i}(z)$ based on stage games and two types of associated zero-sum games $G_{S,N\setminus S}(z)$, $G_{N\setminus i,i}(z)$ based on subgames of multistage game $G(z)$. The game $\Gamma_{S,N\setminus S}$ ($G_{S,N\setminus S}$) is a zero-sum game played by player S as first player and player $N\setminus S$ as second with payoff of player S equal to the sum of payoffs of players in S. The game $\Gamma_{N\setminus i,i}$ ($G_{N\setminus i,i}$) is a zero-sum game played by player $N\setminus\{i\}$ as first player and player $\{i\}$ as second with payoff of player $N\setminus\{i\}$ equal to the sum of payoffs of players in $N\setminus\{i\}$ ($\Gamma_{N\setminus i,i}$ ($G_{N\setminus i,i}$) coincides with $\Gamma_{S,N\setminus S}$ ($G_{S,N\setminus S}$) when $S = N\setminus\{i\}$).

Let $(\bar{\mu}_{N\setminus i}(z), \bar{\mu}_i(z))$ be the saddle point in $\Gamma_{N\setminus i,i}(z)$ in mixed strategies. Consider the n-tuple of strategies $\bar{\mu}(z) = (\bar{\mu}_1(z), \ldots, \bar{\mu}_n(z))$ and introduce the function

$$\bar{W}(z, S) = \max_{\mu_S(z)} \sum_{i \in S} E_i(\mu_S(z), \ \bar{\mu}_{N\setminus S}(z)), \tag{14}$$

where $\mu_S(z) = \{\mu_i(z), i \in S\}$, $\bar{\mu}_{N\setminus S}(z) = \{\bar{\mu}_i(z), i \in N\setminus S\}$ and $E_i(\mu_S(z); \bar{\mu}_{N\setminus S}(z))$ is the mathematical expectation of the payoff of player i when the strategies $(\mu_S, \bar{\mu}_{N\setminus S})$ are played. The function $\bar{W}(z, S)$ is defined for each stage game $\Gamma(z)$. Introduce the new function

$$\bar{\bar{W}}(S) = \max_{z \in G} \bar{W}(z, S), \quad S \in N, \quad S \neq N. \tag{15}$$

Since the number of different stage games is finite the maximum in (15) is attained for all S. Call the function $\bar{\bar{W}}(S)$ generalized characteristic function (GCF). In general $\bar{\bar{W}}(S)$ is not superadditive, but is monotone. From the definition it follows that $\bar{\bar{W}}(S) \geq \bar{W}(z, S)$ for all S, and suppose that $\bar{\bar{W}}(z, N) \geq \bar{\bar{W}}(S)$ and $L \geq \bar{\bar{W}}(S)$ for all $z \in Z$, $S \subset N$. Denote by D the set of vectors $\alpha = (\alpha_1, \ldots, \alpha_n)$ such that

$$\sum_{i \in S} \alpha_i \geq \bar{\bar{W}}(S), \quad S \subset N, \quad S \neq N$$

$$\sum_{i \in N} \alpha_i = L. \tag{16}$$

Suppose that $D \neq \emptyset$ and there exists a vector $\alpha = (\alpha_1, \ldots, \alpha_n)$ such that

$$\sum_{i \in S} \alpha_i > \bar{\bar{W}}(S), \quad S \subset N, \quad S \neq N,$$

$$\sum_{i \in N} \alpha_i = L. \tag{17}$$

In games $\Gamma(\hat{z}_k)$ along cooperative trajectory define vectors $\hat{\alpha}^{\hat{z}_k} = (\hat{\alpha}_1^{\hat{z}_k}, \ldots, \hat{\alpha}_n^{\hat{z}_k})$, where

$$\hat{\alpha}_i^{\hat{z}_k} = \alpha_i + \frac{L(\hat{z}_k) - \sum_{i \in N} \alpha_i}{n},$$

where $\alpha = (\alpha_1, \ldots, \alpha_n)$ satisfies (17).

Similar to Sec. 2, we construct the modification G^α of game G. Games G^α and G differ only by payoffs in stage games $\Gamma(\hat{z}_k)$ defined along cooperative trajectory. If in G the cooperative strategies $(\hat{u}_1(\cdot), \ldots, \hat{u}_n(\cdot))$ (that means that in stage games $\Gamma(\hat{z}_k)$ strategies $(\hat{u}_1(\hat{z}_k), \ldots, \hat{u}_n(\hat{z}_k))$ are used), the payoffs when using this strategies are defined as

$$\hat{\alpha}_i^{\hat{z}_k} = \alpha_i + \frac{L(\hat{z}_k) - \sum_{i \in N} \alpha_i}{n}, \quad k = 1, \ldots, l, \ldots, i \in N. \tag{18}$$

In all other cases, the payoffs in $\Gamma(\hat{z}_k)$ coincide with those of stage game of game G. The theorem similar to Theorem 1 from Sec. 2 holds.

Theorem 2. *Suppose condition* (17) *is satisfied, then there exists a* $\delta \in (0,1)$ *such that in the game* G^α *there exists a strong Nash equilibrium with payoffs of players equal to*

$$\frac{\alpha_i}{1-\delta} + \sum_{l=0}^{\infty} \delta^l \frac{L(\hat{z}_{l+1}) - \sum_{i \in N} \alpha_i}{n}.$$

These payoffs coincide with payoffs of players in G^α *under cooperation.*

One can find the proof of Theorem 2 in Petrosyan *et al.* [2018].

We introduce the characteristic function for every subgame $G(z)$ of multistage game G in the classical way, as value of the zero-sum game $G_{S,N \backslash S}(z)$ between coalition S and $N \backslash S$. Denote this characteristic function by $\tilde{V}(z, S)$.

Consider now the cooperative trajectory (path) in G, $\hat{z} = (\hat{z}_1, \ldots, \hat{z}_m, \ldots)$ and subgames $G(\hat{z}_m)$, $m = 1, \ldots, l, \ldots$ along this trajectory (path). In each subgame, define the core $C(\hat{z}_m)$ as the set of imputations $\alpha^m = (\alpha_1^m, \ldots, \alpha_n^m)$ such that

$$\sum_{i \in S} \alpha_i^m \geq \tilde{V}(\hat{z}_m, S), \quad \sum_{i \in N} \alpha_i^m = \tilde{V}(\hat{z}_m, N) \tag{19}$$

and the set of imputations $M(\hat{z}_m)$ such that $\alpha^m \in M(\hat{z}_m)$ can be represented in the form

$$\alpha^m = \sum_{k=m}^{\infty} \delta^{k-m} \alpha^{\hat{z}^k},$$

where

$$\hat{\alpha}_i^{\hat{z}_k} = \alpha_i + \frac{L(\hat{z}_k) - \sum_{i \in N} \alpha_i}{n}, \quad k = m, m+1, \ldots$$

and $\alpha \in D$ (see (18)).

For all $\alpha^m \in M(\hat{z}_m)$ we have

$$\sum_{i \in S} \alpha_i^m \geq \bar{\bar{W}}(S) \frac{1}{1-\delta}, \quad \sum_{i \in N} \alpha_i^m = V(\hat{z}_m, N). \tag{20}$$

The set $M(\hat{z}_m)$ is analogue of the core if under characteristic function we understand the function $W'(\hat{z}_m, S) = \bar{\bar{W}}(S)\frac{1}{1-\delta}$ ($S \subset N$, $S \neq N$), $W'(\hat{z}_m, N) = V(\hat{z_m}, N)$. We call it quasicore.

It is well known that the core can be void and it is clear that the set $M(z)$ can be also void.

Remind the notion of strongly time-consistency of the solution.

Definition 2. The solution $M(z)$ (given set of imputations) in $G(z)$ is strongly time consistent if for any given cooperative trajectory

$$\hat{z}_1, \ldots, \hat{z}_m, \ldots (\hat{z}_1 = z)$$

and any imputation $\bar{\alpha} \in M(\hat{z}_1)$ there exist a sequence of vectors

$$\beta_1, \ldots, \beta_m, \ldots$$

such that the following condition holds

$$M(\hat{z}_1) \supset \sum_{l=1}^{m-1} \beta_l \delta^{l-1} \oplus \delta^m M(\hat{z}_m), \quad \bar{\alpha} = \sum_{l=1}^{\infty} \beta_l \delta^{l-1},$$

where $a \oplus cB$, $a \in R^n$, $c \in R^1$, $B \subset R^n$ is the set of vectors of the form

$$a \oplus cB = \{a + cb, b \in B\}.$$

The definition has sense when the sets $M(\hat{z}_m)$ are not void.

The sequence of vectors $\beta_1, \ldots, \beta_m, \ldots$ is called imputation distribution proce-dure (IDP).

Theorem 3. *The quasicore $M(\hat{z}_1)$ is strongly time consistent.*

Proof. For every $\bar{\alpha} \in M(\hat{z}_1)$ we can construct IDP as $\beta_l = \hat{\alpha}^{\hat{z}_l}$ (see (18)). By the definition of the set $M(\hat{z}_1)$ every imputation from this set can be represented in the form $\bar{\alpha} = (\bar{\alpha}_1, \ldots, \bar{\alpha}_i, \ldots, \bar{\alpha}_n)$, where

$$\bar{\alpha}_i = \sum_{m=1}^{\infty} \delta^{m-1} \hat{\alpha}_i^{\hat{z}_m} = \sum_{m=1}^{\infty} \delta^{m-1} \left(\alpha_i + \frac{L(\hat{z}_m) - \sum_{i \in N} \alpha_i}{n} \right),$$

where $\alpha = (\alpha_1, \ldots, \alpha_n)$ belongs to the set D defined for the stage games $\Gamma(z)$, $z \in Z$. And similarly every imputation from $M(\hat{z}_m)$ has the form

$$\sum_{k=m}^{\infty} \delta^{k-m} \hat{\alpha}'^{\hat{z}_k} \in M(\hat{z}_m),$$

where $\hat{\alpha}'^{\hat{z}_k}$ satisfies (18).

By putting $\beta_l = \hat{\alpha}^{\hat{z}_l}$ for $l = 1, \ldots, m-1$ and $\beta_l = \hat{\alpha}'^{\hat{z}_l}$ for $l = m, \ldots$ we have

$$\sum_{l=1}^{m-1} \beta_l \delta^{l-1} \oplus \delta^l M(\hat{z}_m) \subset M(\hat{z}_1)$$

which completes the proof. $\qquad \square$

From Theorem 3, it follows that there exists a strong equilibrium in G with payoffs $\bar{\alpha} = (\bar{\alpha}_1, \ldots, \bar{\alpha}_n)$, which coincides with payoffs under cooperation if these payoffs are elements of quasicore. Thus in cases when in subgames along the coop-erative trajectory the quasicore is not void it can be strategically supported by strong equilibrium and at the same time this quasicore is a strongly time consistent solution in the dynamic game under consideration.

Acknowledgment

This research was supported by the Russian Science Foundation (Grant No. 17-11-01079).

References

Aumann, R.-J. [1959] Acceptable points in general cooperative n-person games, in *Contributions to the Theory of Games IV*, Annals of Mathematics Studies, R. D. Luce and A. W. Tucker (eds.), Vol. 40, Princeton University Press, Princeton, pp. 287–324.

Fudenberg, D. and Maskin, E. [1986] The folk theorem in repeated games with discounting or with incomplete information, *Econometrica* **54**(3), 533–554.

Maschler, M., Solan, E. and Zamir, S. [2013] Game Theory (Cambridge University Press).

Myerson, R.-B. [1986] Multistage games with communication, *Econometrica* **54**, 323–358.

Nash, J. [1951] Non-cooperative games, *Ann. Math.* **54**(2), 286–295.

Neumann, J. and Morgenstern, O. [1947] *Theory of Games and Economic Behavior* (Princeton University Press, Princeton).

Petrosyan, L. [1993] Strongly time-consistent differential optimality principles, *Vestnik St. Petersburg Univ. Math.* **26**(4), 40–46 (in Russian).

Petrosyan, L., Chistyakov, S. and Pankratova, Ya. [2017] Existence of strong equilibrium in repeated and multistage games, in *Constructive Nonsmooth Analysis and Related Topics Dedicated to the Memory of V.F. Demyanov*, CNSA 2017, Institute of Electrical and Electronics Engineers Inc., 7974003, pp. 255-257, DOI: 0.1109/CNSA.2017.7974003.

Petrosjan, L. A. and Pankratova, Y. B. [2018] Construction of strong equilibria in a class of infinite nonzero-sum games, *Trudy Inst. Mat. Mekh. UrO RAN.* **24**(1), 165–174 (in Russian).

Rubinstein, A. [1994] Equilibrium in Supergames, in *Essays in Game Theory*, Nimrod Migiddo (ed.), Springer-Verlag, pp. 17–27.

Chapter 22

A Solution Concept Related to "Bounded Rationality" for Some Two-Echelon Models

Joaquin Sanchez-Soriano* and Natividad Llorca[†]

Center of Operations Research (CIO)
Miguel Hernandez University of Elche
Avda. Universidad s/n, Elche, E-03202, Spain
*joaquin@umh.es

†nllorca@umh.es

Two-echelon models describe situations in which there are two differentiated groups of agents. Some examples of these models can be found in supply chain problems, transportation problems or two-sided markets. In this chapter, we deal with two-sided transportation problems which can be used to describe a wide variety of logistic and market problems. We approach the problem from the perspective of cooperative games and study some solution concepts closely related to the game theoretical concept of core, but rather than focus specifically on the core of a transportation game, we introduce and study a new solution concept, a core catcher, which can be motivated by a kind of bounded rationality which can arise in these cooperative contexts.

Keywords: Cooperative games; two-echelon models; transportation games; core; core catcher; bounded rationality.

1. Introduction

Two-echelon models describe situations in which there are two differentiated groups of agents, who interact among themselves in order to optimize their objective functions. These models can be found in relevant logistic situations such as supply chain problems, transportation problems and in socio-economic problems such as two-sided markets, among others. When agents in the same group collaborate, then it is said that horizontal cooperation takes place and when agents in different groups collaborate, then it is referred to as vertical cooperation.

---

*Corresponding author.

Both horizontal and vertical collaborations imply the need to reach agreements. These agreements will include, among other issues, how the profit obtained from collaboration is distributed between the collaborating agents. Cooperative game theory provides solutions to obtain convincing allocations of this profit. Nevertheless, a major drawback of this approach is related to the computational complexity of the proposed solutions. Therefore, this might lead to limited ability of agents in the process of choosing an alternative, because in real-life problems agents may not spend an unbounded amount of resources to evaluate all the possibilities for finding an optimal outcome [Simon, 1972; Deng and Fang, 2008]. Consequently, an approach to the concept of (bounded) rationality in situations in which calculation and analysis themselves are costly and/or limited would be worthy of consideration [Aumann, 1997].

In a procedure to select an allocation of the profit obtained from cooperation, a solution which meets certain criterion of stability should be reached. One criterion could be, for example, belong to the core of the game. In this sense, this concept would play the role of a necessary condition for an allocation proposal to be accepted, but not a sufficient condition. This means that any allocation which is finally agreed meets the criterion, but not any allocation satisfying the criterion must necessarily be accepted.

In this chapter, we focus on cooperation in two-sided transportation problems which can be used to model a wide variety of logistic and market problems. In these problems, allocations in the core can be determined by using duality, but very often they are at the boundary of the core which might make them unacceptable for some agents. Moreover, the set of allocations in the core determined by duality might consist of only one point and then if it is rejected no other alternatives using this procedure could be proposed. On the other hand, if an agent proposes an allocation of the profit, it is not easy, in general, to reject this if it does not belong to the core of the game, due to the computational cost necessary to check it. For these reasons, we examine the following questions:

(1) Can optimal solutions of the two-sided transportation problem help to achieve an allocation of the profit obtained from collaboration?
(2) What is the relationship between the set of allocations obtained from the optimal solutions of the transportation problem and the core of the transportation game?
(3) How can the set of all allocations based on the optimal solutions be used as a criterion for accepting or rejecting an allocation proposal?
(4) Is the set of allocations based on optimal solutions related to bounded rationality?

To answer these questions, we first introduce 2-games associated with optimal solutions of transportation problems and study their cores. We then construct a catcher for the core of the transportation game by considering the intersection of all 2-games associated with optimal solutions of the underlying transportation problem and study its properties. We relate this concept to bounded rationality when calculation and analysis themselves are costly and/or limited. For this reason and because of its interesting properties, we propose this core catcher as an alternative to the core of the game as a necessary condition for an allocation to be accepted.

2. Literature Review

In game theory literature, we can find numerous papers devoted to the aforementioned problems. Cachon and Netessine [2004], Leng and Parlar [2005] and Nagarajan and Sosic [2008] have reviewed the literature describing supply chain and game theory. Some recent papers, studying different aspects of cooperation in supply chain are Arthanari *et al.* [2015], Zhang *et al.* [2017], Zhou *et al.* [2018] and Johari and Hosseini-Motlagh [2018], among others.

Assignment and transportation problems can also be interpreted as two-echelon models. Since Shapley and Shubik [1972], where cooperative assignment games associated with assignment problems were introduced, different generalizations have been developed. Three of these extensions are quasi-assignment games [Auriol and Marchi, 2002], infinite assignment games [Llorca *et al.*, 2004] and assignment games with externalities [Eriksson *et al.*, 2011; Gudmundsson and Habis, 2017]. Quasi-assignment games are assignment games in which some inequalities are converted into equalities. Assignment games with externalities are assignment games in which negative externalities come from ill will among players. See Izquierdo *et al.* [2012] for a survey on assignment games and related markets. On the other hand, (cooperative) transportation games, which are studied in Sanchez-Soriano *et al.* [2001, 2002] can be seen as extensions of assignment games. Moreover, transportation problems are also an important issue in logistics and the cooperation among the agents involved is worth analysis (see, for example, Mason *et al.* [2007], Cruijssen *et al.* [2007]). This work falls within the scope of two-echelon models such as those described in Operations Research for transportation problems.

Two-sided markets can be also interpreted as two-echelon models. Likewise, transportation and assignment problems can be seen as two-sided market situations and conversely. Thompson [1980] studied an extension of the market model described in Shapley and Shubik [1972]. Moreover, other two-sided market situations are studied in Fragnelli *et al.* [2007] and Sanchez-Soriano and Fragnelli [2010].

A classic issue in cooperative game theory is how to distribute the profit generated by the cooperating players. One way to do this is to use allocations in the core [Gillies, 1953] of the game. Transportation games have nonempty core. However, in order to fully determine the core of a transportation game one has, on the one hand, to evaluate an important amount of constraints and, on the other hand, to solve the same number of transportation problems for computing the characteristic function. This means that in a negotiation procedure of succesive proposals of the allocation of the total revenues it may be extremely difficult for the players to check whether a proposed allocation is stable in the sense of the core, in order to accept it or not.

A rational player behaves in such a way that a greater payoff is preferred to a smaller one, i.e., among all possible choices he will select the one which gives him the maximum payoff. A first problem related to this issue is that, on many occasions, the number of available options is huge, which implies that it is impossible to use an algorithm for finding the optimal payoff. To overcome this problem, bounded rationality has been introduced, mainly in the context of non-cooperative games. The concept of "bounded rationality" was coined by Simon in his pioneering works cf. Simon [1957, 1972]. The paper by Rubinstein [1990] and his book [Rubinstein, 1998] have become cornerstones in this field. For cooperative games, bounded rationality is closely related to the computational complexity either to check whether an allocation satisfies certain conditions or to calculate itself [Deng and Papadimitriou, 1994; Papadimitriou and Yannakakis, 1994; Deng and Fang, 2008]. In this sense, Simon [1972] pointed out that real-life agents may not spend an unbounded amount of resources to evaluate all the possibilities for an optimal outcome. Following these ideas, we will introduce a new solution concept in this chapter, rather than focus on the core of a transportation game. This concept is a core catcher from a theoretical point of view and can be interpreted as based on a kind of bounded rationality in the cooperative setting.

For assignment games with externalities, Eriksson *et al.* [2011] introduce a new notion of stable outcomes based on a certain assumption of bounded rationality, which is related to the so-called naive players. When there are bids from one side to the other in order to deviate from a certain outcome (a matching and a distribution of the profit), a naive player behaves by accepting any bid from the other side that increases his payoff, believing that among agents on the same side only he and the agent affected by his bid will have their payoffs affected, and expecting that the agent on the same side affected by his bid will accept any bid from the other side that increases his payoff. In this chapter, we consider transportation games which implies that we do not have a one-to-one matching but a many-to-many matching. Additionally, every matching may have different (discrete) levels from one to a certain integer number which depends on the structure of the problem. Moreover, we consider a solution which is built from the primal optimal solutions of the underlying transportation problem and we then relate it to bounded rationality by considering a procedure of successive proposals of allocation of the total revenue obtained from cooperation.

The rest of the chapter is organized as follows. In the next section, we introduce the 2-games [van de Nouweland *et al.*, 1996] associated with transportation problems. Using these games, we obtain a core catcher for transportation games in Sec. 4. Section 5 presents a numerical study and then an approach to bounded rationality in the cooperative framework by means of this core catcher. Section 6 concludes.

3. On 2-Games Arising from Two-Sided Transportation Situations

A two-sided transportation problem describes a situation in which demands at several points for a certain product need to be covered by supplies from other locations. The transportation of one unit from a supply point to a demand point generates a profit. The goal of the cooperating agents is to maximize the total profit from transport. More formally, let P be the set of supply points (origins) and Q the demand points (destinations).[a] The supply of the product at point $i \in P$ equals s_i units and the demand at point $j \in Q$ is d_j units. Both s_i and d_j are (positive) integer numbers for all $i \in P$ and $j \in Q$, as we assume that the goods are indivisible. We do not consider those suppliers and demanders who own zero units because they do not contribute to improve the profit. The revenue of sending one unit from supply point i to demand point j is r_{ij}, a non-negative real number. All profits are gathered in the matrix $R = [r_{ij}]_{i \in P, j \in Q}$. Hence, a transportation problem can be described by the tuple (P, Q, R, s, d) where $s = (s_i)_{i \in P}$ and $d = (d_j)_{j \in Q}$ are the vectors containing respectively the supplies and demands of the goods. A transportation plan for a transportation problem is a matrix $\mu = [\mu_{ij}]$, where μ_{ij} is the number of units sent from supply point i to destination point j.

We are interested in a related cooperative game with transferable utility (TU) for each transportation problem. A TU-game is a pair (N, v), where $N = \{1, 2, \ldots, n\}$ is the set of players and $v : 2^N \to \mathbb{R}$ is the characteristic function, satisfying $v(\emptyset) = 0$. For each $S \subset N$, $v(S)$ is interpreted as the gain that the players in S can obtain by themselves. We often identify a TU-game (N, v) with its characteristic function v, and denote by G^N the set of all TU-games with set of players N.

Given a transportation problem, the corresponding transportation game (N, v) is a cooperative TU-game with player set $N = P \cup Q$. Each agent owns only one node and, moreover, each node (with all its supply or demand) is owned by only one player. Let $S \subset N$, $S \neq \emptyset$, be a coalition of players and define $P_S = P \cap S$ and $Q_S = Q \cap S$. If $S = P_S$ then there are no demand agents present in S and therefore the supply agents in S cannot get rid of their goods. In this case the worth $v(S)$ of coalition S equals zero. Similarly, if $S = Q_S$ then the demand agents in S cannot receive any unit of the goods and $v(S) = 0$. This means that when there is exclusively horizontal cooperation the worth of the cooperation is zero. Otherwise, the worth $v(S)$ depends upon the possible transportation plans. Therefore, for cooperation to be profitable there must be vertical cooperation, regardless of horizontal cooperation. A transportation plan $\mu(S)$ for coalition S is a transportation plan for the transportation problem $(P_S, Q_S, [r_{ij}]_{i \in P_S, j \in Q_S}, (s_i)_{i \in P_S}, (d_j)_{j \in Q_S})$. In this

---

[a]In a supply chain we can distinguish different agents such as suppliers, manufacturers or producers, distributors or warehousers, retailers and customers. The flow of transport goes from the suppliers to the customers. In two-sided transportation problems only two of these types of agents are considered.

case,

$$v(S) = \max\left\{ \sum_{(i,j)\in P_S \times Q_S} r_{ij}\mu_{ij} \,\middle|\, \mu(S) \text{ is a transportation plan for } S \right\} \quad (1)$$

is the worth of coalition S.

One of the main issues in cooperative game theory is how to divide the total profit derived from cooperation. One way to share this profit among the players in N is to do so according to an element in the core. The *core* of a game $v \in G^N$ is the set of payoff vectors

$$C(v) = \left\{ x \in \mathbb{R}^N \,\middle|\, \sum_{i\in S} x_i \geq v(S), \ \forall S \subset N, \ \sum_{i\in N} x_i = v(N) \right\}. \quad (2)$$

Given a set of players $N = \{1, 2, \ldots, n\}$, Θ^N denotes the set of permutations on N, i.e.,

$$\Theta^N = \{\theta : N \to N \,|\, \sigma \text{ is bijective}\}. \quad (3)$$

Let $v \in G^N$ and $\theta \in \Theta^N$, the marginal worth vector $x^\theta(v) = (x_1^\theta(v), \ldots, x_n^\theta(v)) \in \mathbb{R}^n$ with respect to the order θ is given by $x_i^\theta(v) = v(P_i^\theta \cup \{i\}) - v(P_i^\theta) \ \forall i \in N$, where P_i^θ represents the set of predecessors of i in θ, i.e., $P_i^\theta = \{j \in N \,|\, \theta(j) < \theta(i)\}$. These vectors are efficient and satisfy at least $|N|$ constrains which determine the core. Furthermore, they are out of the core or an extreme point of it. In general, a marginal worth vector does not belong to the core of the game. In Weber [1978] it is proved that $C(v)$ is always included in the convex hull of all marginal worth vectors. Nevertheless, if the game is convex, i.e., $v(S) + v(T) \leq v(S \cap T) + v(S \cup T) \ \forall S, T \subset N$, these vectors are in the core cf. Shapley [1971]. Therefore, for convex games $C(v)$ coincides with the convex hull of all marginal vectors.

A special class of TU-games named the k-games was introduced by van de Nouweland *et al.* [1996] to analyze a specific problem arising from telecommunications.

Definition 1. A cooperative TU-game (N, v) is called a k-game if $v(S) = \sum_{T \subset S : |T| = k} v(T)$, for each $S \subset N$.

Non-negative k-games exhibit interesting properties such as that they are convex and for $k = 2$ the Nucleolus and the Shapley value coincide. We are interested in building a game associated with a transportation problem which has the structure of a 2-game.

Definition 2. Let (P, Q, R, s, d) be a transportation problem and μ an optimal transportation plan for it. We define the transportation game associated with μ as the TU-game $(P \cup Q, v_\mu)$ with $v_\mu(S) = \sum_{(i,j)\in P_s \times Q_s} r_{ij}\mu_{ij}$, for all $S \subset N = P \cup Q$.

Obviously $(P \cup Q, v_\mu)$ is a 2-game such that

$$v_\mu(P \cup Q) = \sum_{(i,j) \in P \times Q} r_{ij} \mu_{ij}$$

$$= \max \left\{ \sum_{(i,j) \in P \times Q} r_{ij} \mu'_{ij} \,\middle|\, \mu'(N) \text{ is a transportation plan for } N \right\}$$

$$= v(N). \tag{4}$$

This transportation game can be interpreted as the game that the agents face after deciding which optimal transportation plan μ is to be carried out. Thus, in a certain way, this game can be considered as an *a posteriori* transportation game, whereas the transportation game defined before would be the *a priori* transportation game.

Definition 3. Let (P, Q, R, s, d) be a transportation problem and μ an optimal transportation plan. A distribution matrix associated with μ is defined by

$$D_\mu(z) = [(z_{ij,o}, z_{ij,d})]_{i=1,2,\ldots,n}^{j=1,2,\ldots,m}, \tag{5}$$

where $z_{ij,o} + z_{ij,d} = r_{ij} \mu_{ij}$, $z_{ij,o} \geq 0$, $z_{ij,d} \geq 0$, $\forall (i,j) \in P \times Q$, with $n = |P|$ and $m = |Q|$.

This distribution matrix associated with μ means that each pair origin-destination shares the profit related to the matching connecting both agents; i.e., with this system the only relevant transfers are those between agents matched by the optimal transportation plan μ. The next result shows these distribution matrices always provide allocations in the core of the corresponding game, which is a relevant property in cooperative games.

Theorem 1. *Let (P, Q, R, s, d) be a transportation problem and μ an optimal transportation plan. Let $(P \cup Q, v_\mu)$ be the corresponding 2-game and $D_\mu(z)$ a distribution matrix associated with μ. Then,*

$$(x; y) = \left(\sum_{j \in Q} z_{1j,o}, \ldots, \sum_{j \in Q} z_{nj,o}; \sum_{i \in P} z_{i1,d}, \ldots, \sum_{i \in P} z_{im,d} \right) \in C(v_\mu). \tag{6}$$

Proof. Since μ is an optimal transportation plan and $D_\mu(z)$ a distribution matrix associated with μ, we have

$$v_\mu(P \cup Q) = \sum_{(i,j) \in P \times Q} r_{ij} \mu_{ij} = \sum_{(i,j) \in P \times Q} (z_{ij,o} + z_{ij,d}) \tag{7}$$

$$= \sum_{i \in P} \sum_{j \in Q} z_{ij,o} + \sum_{j \in Q} \sum_{i \in P} z_{ij,d}. \tag{8}$$

Thus $(x; y)$ is an imputation. On the other hand, it holds

$$x(P_S) + y(Q_S) = \sum_{i \in P_S} \sum_{j \in Q} z_{ij,o} + \sum_{j \in Q_S} \sum_{i \in P} z_{ij,d} \tag{9}$$

$$= \sum_{i \in P_S} \left\{ \sum_{j \in Q_S} z_{ij,o} + \sum_{j \in Q \backslash Q_S} z_{ij,o} \right\}$$

$$+ \sum_{j \in Q_S} \left\{ \sum_{i \in P_S} z_{ij,d} + \sum_{i \in P \backslash P_S} z_{ij,d} \right\} \tag{10}$$

$$= \sum_{i \in P_S} \sum_{j \in Q_S} z_{ij,o} + \sum_{j \in Q_S} \sum_{i \in P_S} z_{ij,d} + \sum_{i \in P_S} \sum_{j \in Q \backslash Q_S} z_{ij,o}$$

$$+ \sum_{j \in Q_S} \sum_{i \in P \backslash P_S} z_{ij,d} \tag{11}$$

$$= \sum_{i \in P_S} \sum_{j \in Q_S} (z_{ij,o} + z_{ij,d}) + \sum_{i \in P_S} \sum_{j \in Q \backslash Q_S} z_{ij,o} + \sum_{j \in Q_S} \sum_{i \in P \backslash P_S} z_{ij,d} \tag{12}$$

$$= \sum_{i \in P_S} \sum_{j \in Q_S} r_{ij} \mu_{ij} + \sum_{i \in P_S} \sum_{j \in Q \backslash Q_S} z_{ij,o} + \sum_{j \in Q_S} \sum_{i \in P \backslash P_S} z_{ij,d}$$

$$\geq v_\mu(S), \quad \forall S \subset N. \tag{13}$$

Therefore $(x; y)$ is in the core of v_μ. $\square$

An interesting issue is whether each imputation in $C(v_\mu)$ can be written as a distribution matrix associated with μ; i.e., if $z \in \mathbb{R}_+^{2(n \times m)}$ exists such that each $(x; y) \in C(v_\mu)$ can be decomposed in the following way:

$$x_i = \sum_{j=1}^m z_{ij,o}, \quad i = 1, \ldots, n, \tag{14}$$

$$y_j = \sum_{i=1}^n z_{ij,d}, \quad j = 1, \ldots, m, \tag{15}$$

$$r_{ij} \mu_{ij} = z_{ij,o} + z_{ij,d}, \quad \forall (i, j) \in P \times Q. \tag{16}$$

The positive answer to this question is stated in the next theorem.

Theorem 2. *Let* (P, Q, R, s, d) *be a transportation problem,* μ *an optimal transportation plan, and* v_μ *the characteristic function of the corresponding 2-game. Then, for all* $(x; y) \in C(v_\mu)$, *there is a distribution matrix associated with* μ, $D_\mu(z)$, such that

$$(x; y) = \left(\sum_{j \in Q} z_{1j,o}, \dots, \sum_{j \in Q} z_{nj,o}; \sum_{i \in P} z_{i1,d}, \dots, \sum_{i \in P} z_{im,d} \right). \tag{17}$$

Proof. The result is equivalent to show that the system (14)–(16) has at least a non-negative solution for each $(x; y) \in C(v_\mu)$. On the other hand, it is known that v_μ is a convex game. Then, $C(v_\mu)$ is the convex hull of all marginal vectors and it suffices to prove that the system (14)–(16) has a non-negative solution for each marginal vector. In this case, a marginal vector has the following structure:

$$x_i^\theta = \sum_{j \in Q,\, \theta(j) < \theta(i)} r_{ij} \mu_{ij}, \quad i = 1, \dots, n, \tag{18}$$

$$y_j^\theta = \sum_{i \in P, \theta(i) < \theta(j)} r_{ij} \mu_{ij}, \quad j = 1, \dots, m, \tag{19}$$

where $\theta \in \Theta^{n+m}$, and Θ^{n+m} is the set of all permutations of $n + m$ elements. Note that, for these vectors, the system (14)–(16) has the non-negative solution

$$z_{ij,o} = \begin{cases} r_{ij} \mu_{ij} & \text{if } \theta(i) > \theta(j), \\ 0 & \text{if } \theta(i) < \theta(j), \end{cases} \tag{20}$$

$$z_{ij,d} = \begin{cases} r_{ij} \mu_{ij} & \text{if } \theta(i) < \theta(j), \\ 0 & \text{if } \theta(i) > \theta(j). \end{cases} \tag{21}$$

Taking into account that given x^* and y^* solutions for $Ax = b$ and $Ay = c$, respectively, $z^* = \lambda x^* + (1 - \lambda)y^*$ is a solution for $Az = \lambda b + (1 - \lambda)c$, $\forall 0 \leq \lambda \leq 1$, and the above-mentioned result by Shapley [1971] on the convex hull of all marginal worth vectors, we obtain the desired conclusion. □

Theorems 1 and 2 show how to obtain all allocations in the core of the game associated with an optimal transportation plan μ, besides their structure. The core can contain many different allocations, for this reason it is interesting to know how large the core of a transportation game associated with an optimal transportation plan is. It is well-known [Shapley, 1971] that the number of extreme points of the core in convex games is at most $l!$, where l represents the number of players. However, in the class of our 2-games we can find a better upper bound.

Proposition 1. Let (P, Q, R, s, d) be a transportation problem, μ an optimal transportation plan and $(P \cup Q, v_\mu)$ the corresponding 2-game. The number of extreme points of $C(v_\mu)$ is at most 2^a, where a represents the number of matchings in μ.

Proof. Let $\Lambda_\mu = \{(i_1, j_1), \dots, (i_a, j_a)\}$ be the set of all matchings in the optimal transportation plan μ. Consider the map $\Delta_\mu : \Theta^{n+m} \to \{0, 1\} \times \overset{a}{\cdots} \times \{0, 1\}$ such that for each $\theta \in \Theta^{n+m}$

$$\Delta_\mu^k(\theta) = \begin{cases} 1, & \text{if } \theta(i_k) > \theta(j_k), \\ 0, & \text{if } \theta(i_k) < \theta(j_k), \end{cases} \qquad k = 1, \ldots, a. \tag{22}$$

It is easy to check that the marginal worth vectors of v_μ are the elements $\{(u_1^\theta, \ldots, u_n^\theta; v_1^\theta, \ldots, v_m^\theta) \mid \theta \in \Theta^{n+m}\}$, where

$$u_i^\theta = \sum_{k:i_k=i} \Delta_\mu^k(\theta) r_{ij} \mu_{ij}, \quad \forall i \in P, \tag{23}$$

$$v_j^\theta = \sum_{k:j_k=j} (1 - \Delta_\mu^k(\theta)) r_{ij} \mu_{ij}, \quad \forall j \in Q. \tag{24}$$

From the definition of Δ_μ, it is clear that there are at most 2^a different marginal vectors. Since v_μ is a 2-game (and, therefore, convex), its core is the convex hull of such vectors. □

As a consequence, $C(v_\mu)$ can be determined by means of an easy procedure and, in general, with far less calculations than the core, which is a relevant property for cooperative games [Deng and Fang, 2008].

4. A Catcher of the Core

In this section, we obtain a new solution concept, a core catcher, of a transportation game by means of the 2-games associated with the optimal transportation plans for the corresponding transportation problem. Since a transportation problem (P, Q, R, s, d) may have more than one optimal transportation plan, it would seem to be of interest to consider a set that takes into account all of them. Making use of the 2-games associated with the optimal transportation plans, we can define the following set:

$$CC(v) = \bigcap_{\mu \in \Im} C(v_\mu), \tag{25}$$

where v is the associated transportation game, v_μ is the transportation game associated with the optimal transportation plan μ and $\Im$ is the set of all optimal transportation plans. $CC(v)$ contains all the allocations that are stable regardless of the optimal transportation plan considered, which seems to be a good property.

The next theorem states that the set $CC(v)$ contains the core of the transportation game.

Theorem 3. *Let (P, Q, R, s, d) be a transportation problem, $\Im$ its set of optimal transportation plans, and (N, v) the related transportation game. Then, $C(v) \subset CC(v) = \bigcap_{\mu \in \Im} C(v_\mu)$.*

Proof. It is straightforward that $v(S) \geq v_\mu(S) \ \forall S \subset N, \forall \mu \in \Im$. Furthermore, $v(P \cup Q) = v_\mu(P \cup Q) \ \forall \mu \in \Im$. Therefore, if $(x; y) \in C(v)$, this imputation is also in each $C(v_\mu)$. □

In view of Theorem 3, we can introduce the following definition.

Definition 4. Let (P, Q, R, s, d) be a transportation problem and $\mathfrak{I}$ its set of optimal transportation plans. We call the set $CC(v) = \bigcap_{\mu \in \mathfrak{I}} C(v_\mu)$ the core catcher.

Likewise, several results can be derived from the previous theorem. Firstly, each imputation in $C(v)$ can be obtained by means of an optimal transportation plan μ (see, Sanchez-Soriano [2003, 2006] for other solutions for transportation games based on optimal solutions of the primal problem and their relationships with the core of the game), where transfers occur only between the supply and demand agents who are connected by this solution. As a result, the optimal solution does not affect the final distribution in a quantitative way, but in a qualitative manner by showing who the players involved in the utility (maybe money) transfers are.

Secondly, the game $(P \cup Q, v_\mu)$ can be read as an "*a posteriori*" transportation game associated with the transportation problem, i.e., when the optimal transportation plan μ has been implemented by the players. However, (N, v) can be seen as an "*a priori*" transportation game related to the transportation problem, i.e., before the optimal transportation plan is fixed.

Thirdly, we should point out that if $\mu = \lambda\mu^1 + (1 - \lambda)\mu^2$, $\lambda \in [0, 1]$, then $C(v_{\mu^1}) \cap C(v_{\mu^2}) \subset C(v_\mu)$. Because if we consider $x \in C(v_{\mu^1}) \cap C(v_{\mu^2})$, then

$$x(S) \geq v_{\mu^1}(S), \quad \forall S \subset N, \tag{26}$$

$$x(S) \geq v_{\mu^2}(S), \quad \forall S \subset N. \tag{27}$$

Thus

$$x(S) \geq \lambda v_{\mu^1}(S) + (1 - \lambda)v_{\mu^2}(S) = \lambda \sum_{(i,j) \in P_S \times Q_S} r_{ij}\mu_{ij}^1 + (1 - \lambda) \sum_{(i,j) \in P_S \times Q_S} r_{ij}\mu_{ij}^2$$

$$= v_\mu(S), \quad \forall S \subset N. \tag{28}$$

This property implies that in Theorem 3 the relevant optimal solutions are those which are basic (extreme points of the polytope defined by the set of constraints of the transportation problem), because the remainder is a linear combination of these. Moreover, the opposite is not true as the next example shows.

Example 1. Let us consider the 3-players transportation game defined by

$$(P = \{1, 2\}, Q = \{3\}, R = (1, 1), s = (2, 2), q = 2). \tag{29}$$

It is easy to check that $C(v) = \{(0, 0; 2)\}$. The corresponding transportation problem has three optimal transportation plans:

$$\mu^1 = \begin{bmatrix} 2 \\ 0 \end{bmatrix}, \quad \mu^2 = \begin{bmatrix} 1 \\ 1 \end{bmatrix}, \quad \mu^3 = \begin{bmatrix} 0 \\ 2 \end{bmatrix}, \tag{30}$$

where $\mu^2 = \frac{1}{2}\mu^1 + \frac{1}{2}\mu^3$. However, $C(v_{\mu^1}) \cap C(v_{\mu^3}) = C(v)$ and $C(v_{\mu^2}) = \{(x_1, x_2; 2 - x_1 - x_2) \mid 0 \leq x_1 \leq 1 \text{ and } 0 \leq x_2 \leq 1\}$.

Finally, this core catcher can be rewritten in terms of the maximum of a finite set of games. Given the games (N, v_1) and $(N, v_2) \in G^N$, the game with characteristic function $v_M = \max\{v_1, v_2\}$ is defined in the following way

$$v_M(S) = \max\{v_1(S), v_2(S)\}, \quad \forall S \subset N. \tag{31}$$

Proposition 2. *Let* (P, Q, R, s, d) *be a transportation problem and* $\Im$ *its set of optimal transportation plans. Then,* $C(\max_{\mu \in \Im}\{v_\mu\}) = \bigcap_{\mu \in \Im} C(v_\mu)$.

Proof. To show that $C(v_M) \subset \bigcap_{\mu \in \Im} C(v_\mu)$, where $v_M = \max_{\mu \in \Im}\{v_\mu\}$, let $x \in C(v_M)$ then

$$x(S) \geq v_M(S) \geq v_\mu(S), \quad \forall S \subset N \quad \text{and} \quad \forall \mu \in \Im. \tag{32}$$

Therefore, $x \in C(v_\mu) \forall \mu \in \Im$.

On the other hand, to see that $\bigcap_{\mu \in \Im} C(v_\mu) \subset C(v_M)$ consider $x \in \bigcap_{\mu \in \Im} C(v_\mu)$, then $x(S) \geq v_\mu(S), \forall \mu \in \Im$ and $\forall S \subset N$. Thus $x(S) \geq \max_{\mu \in \Im}\{v_\mu(S)\} = v_M(S)$, $\forall S \subset N$. $\qquad\square$

Corollary 1. *Let* (P, Q, R, s, d) *be a transportation problem and* (N, v) *the related transportation game. Then,* $C(v) \subset C(\max_{\mu \in \Im}\{v_\mu\})$.

We have only taken into account, in order to bound or approach the core, the optimal solutions of the transportation problem so far. Nevertheless, if we consider all feasible transportation plans, then it is possible to identify the core of the corresponding transportation game through the maximum of certain games. To do so, we set out to study the decomposition of the transportation games as a maximum of a finite set of games and the 2-games associated with the feasible transportation plans. The decomposition of cooperative games in terms of a maximum of convex games is analyzed in Llerena and Rafels [2006].

Theorem 4. *Let* (P, Q, R, s, d) *be a transportation problem and* (N, v) *the related transportation game. Then,* $v = \max_{\mu \in \mathbb{F}}\{v_\mu\}$, *where* $\mathbb{F}$ *is the set of feasible transportation plans for the corresponding transportation problem.*

Proof. By the definition of the transportation game, $v(S)$ is the maximal profit that players in S can obtain by themselves. Thus, $v(S) \geq \max_{\mu \in \mathbb{F}}\{v_\mu(S)\}, \forall S \subset N$. On the other hand, if we consider the optimal transporation plan, μ'_S, for the transportation problem corresponding to the agents in S, and we add a feasible (or optimal) transportation plan, $\mu'_{N \setminus S}$, for the problem restricted to $N \setminus S$, then by joining μ'_S and $\mu'_{N \setminus S}$, we obtain a feasible transportation plan, μ', for the whole problem which satisfies $v_{\mu'}(S) = v(S)$. As a result, $\max_{\mu \in \mathbb{F}}\{v_\mu(S)\} \geq v(S)$ $\forall S \subset N$. $\qquad\square$

Corollary 2. *Let* (P, Q, R, s, d) *be a transportation problem and* (N, v) *the related transportation game. Then,* $C(v) = C(\max_{\mu \in \mathbb{F}}\{v_\mu\})$.

The last result is not true if we change the maximum operator for the intersection operator. In fact, we only have to recall that the value for the grand coalition in the case of feasible plans can be smaller than in those which are optimal plans and, therefore, the intersection can be empty. On the other hand, we should point out that all possible transportation plans are necessary to determine completely the core of a transportation game, while only the optimal transportation plans are used to fully determine the core catcher.

To finalize our analysis of the core catcher, we examine several examples which show certain aspects of this new solution concept in the framework of transportation games.

Example 2. Let us consider the transportation game with

$$(P = \{1\}, Q = \{2, 3\}, R = (2, 3), s = 4, d = (3, 2)). \tag{33}$$

In Fig. 1, the imputation set, the core, the core catcher and the core element obtained by duality are depicted. The pale gray area corresponds to the additional imputations of $C(v_\mu)$ in relation to $C(v)$ (in this case, the optimal transportation plan is unique). Note that the core provides a symmetric treatment of players 2 and 3, however, in the original situation they are not symmetric. This happens because the characteristic function of the transportation game simplifies the situation considering only the worth for each coalition. Thus, when dealing with this, we are losing certain qualitative information of the original situation. In this way, $\bigcap_{\mu \in \mathfrak{I}} C(v_\mu)$ can be seen as a solution concept in this context which takes into account a different information about the game.

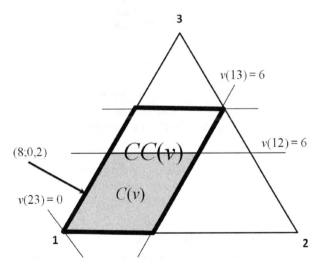

Fig. 1. The imputation set, the core, the core catcher and the only core element determined by duality for the transportation game in Example 2.

Example 3. Let us consider the following transportation game with four players:

$$(P = \{1\}, Q = \{2, 3, 4\}, R = (1, 2, 1), s = 3, d = (1, 2, 1)). \qquad (34)$$

In this case, $C(v) = \bigcap_{\mu \in \Im} C(v_\mu)$, with

$$C(v) = \bigcap_{\mu \in \Im} C(v_\mu) = \{(x; y) \in \mathbb{R}^4 \mid y_2 = y_4 = 0, x_1 \geq 2, x_1 + y_3 = 5\}. \qquad (35)$$

If we take $d' = (2, 2, 2)$, then the core of the game

$$C(v') = \{(x; y) \in \mathbb{R}^4 \mid y_2 = y_4 = 0, x_1 \geq 3, x_1 + y_3 = 5\}. \qquad (36)$$

However, $\bigcap_{\mu \in \Im} C(v_\mu)$ is the same as in the original situation.

To sum up, we have shown that the core is more sensitive to small changes in the problem than its catcher. This is a very interesting feature of $\bigcap_{\mu \in \Im} C(v_\mu)$ as a solution concept for transportation games.

5. Bounded Rationality and Numerical Study

In the framework of non-cooperative games, the concept of bounded rationality has been introduced to deal with different ideas of relaxing the classical assumption of perfect rationality of the players. However, in cooperative game theory this aspect of human behavior has not been faced, at least, with the same profusion in the literature. In this section, a definition of bounded rationality based on computational complexity is introduced following the ideas pointed out in Simon [1972], Deng and Papadimitriou [1994], Papadimitriou and Yannakakis [1994] and Deng and Fang [2008], among others. We consider the following definition of bounded rationality related to the limited time to analyze or make a proposal.

Definition 5. A procedure for cooperative games is said to be κ-Υ-τ-rational if the following holds

$$f(n)\tau \leq \Upsilon, \quad \forall n \leq \kappa, \qquad (37)$$

where n is the number of players, $f(n)$ is the highest number of basic operations needed to run the procedure, κ is a nonnegative integer number, τ is the processing or computation capacity and Υ is the maximum time available to obtain an answer.

Note that we consider that bounded rationality not only depends on the game itself, but also on other circumstances related to the particular situation we are analyzing such as the time available or the real technical capacity of the tools used by players. Therefore full rationality implies the assumption of no bounds for the time and/or infinite technical capacity. In the definition the worst case, in the running of the procedure, has been considered, so we can use the Big-Oh approach of the computational complexity theory [Arora and Barak, 2009] to analyze the bounded rationality of different procedures. Once we have established a definition about what we mean by bounded rationality in a procedure, we can define when a solution for cooperative games satisfies bounded rationality.

Definition 6. A solution for cooperative games is said to be κ-Υ-τ-rational if the best procedure known to compute it is κ-Υ-τ-rational.

This definition of bounded rationality gives a measure of the possibilities of a solution to be effectively applied and follows the ideas in Deng and Papadimitriou [1994] and Deng and Fang [2008] about considering computational complexity as an extra factor in considering rationality and fairness of a solution concept. In this sense, a solution with a higher κ and lower Υ and τ will be preferable to other solution with a lower κ and higher Υ and τ.

Let us now consider a cooperative game whose players agree to accept a distribution only if any particular coalition cannot be better off, but this does not mean that an allocation satisfying that condition cannot be rejected. In such a situation, the players' rationality can be established via the core of the game, which we know in transportation games is always non-empty. Assume that the procedure to reach an agreement on a distribution of the profit is through proposals which are accepted or rejected by the players, but if they do not belong to the core of the game, then they are rejected with certainty. This is a clear and, theoretically, easy process. Nevertheless, it may be impossible to check if a given distribution belongs to the core in practice. That is, maybe the players are not able to reject a proposal which is not in the core, because they cannot know even the characteristic function of the game. On the contrary, they can reject an imputation which does not satisfy a weaker condition, for instance, it does not belong to the core of a certain 2-game, v_μ, in the case of transportation models. Therefore, we can consider that the players will reject an imputation if it is not in $C(v_\mu)$, for a certain optimal transportation plan μ. Assume that every player uses this methodology to accept or reject, but each one can choose a different optimal transportation plan. In this situation, the imputation which will be implemented belongs to the intersection of all $C(v_\mu)$. Thus $CC(v) = \bigcap_{\mu \in \Im} C(v_\mu)$ can be proposed as a solution concept for some two-echelon models as those described by two-sided transportation problems, because it seems suitable in such environments. As pointed out in the Introduction, for instance, assignment models are a particular class of transportation situations. Thus, the ideas developed in this part can be applied to a great variety of two-echelon models.

In order to illustrate the above-mentioned approach to bounded rationality in this setting, let us consider a 10×10 transportation problem, which has been randomly generated limiting the per unit profit to the integers from 0 to 9, and supplies and demands from 1 to 9.

$$P = \{1, 2, 3, 4, 5, 6, 7, 8, 9, 0\}, \tag{38}$$

$$Q = \{1', 2', 3', 4', 5', 6', 7', 8', 9', 0'\}, \tag{39}$$

$$R = \begin{bmatrix} 7 & 6 & 3 & 7 & 0 & 2 & 8 & 8 & 7 & 6 \\ 0 & 4 & 5 & 5 & 5 & 9 & 6 & 6 & 5 & 0 \\ 1 & 6 & 9 & 2 & 4 & 4 & 4 & 1 & 0 & 6 \\ 6 & 6 & 0 & 8 & 5 & 6 & 5 & 8 & 4 & 9 \\ 2 & 2 & 5 & 5 & 8 & 6 & 5 & 2 & 9 & 3 \\ 0 & 0 & 7 & 2 & 2 & 1 & 7 & 1 & 1 & 7 \\ 2 & 9 & 2 & 7 & 5 & 0 & 5 & 9 & 5 & 2 \\ 5 & 6 & 0 & 6 & 0 & 0 & 9 & 8 & 5 & 7 \\ 7 & 2 & 6 & 3 & 4 & 8 & 3 & 9 & 9 & 0 \\ 5 & 2 & 0 & 6 & 6 & 5 & 8 & 0 & 0 & 1 \end{bmatrix}, \tag{40}$$

$$s = (4, 9, 5, 5, 3, 6, 2, 8, 9, 8), \tag{41}$$

$$d = (6, 1, 9, 2, 1, 3, 7, 8, 6, 9). \tag{42}$$

An optimal transportation plan — alternate solutions exist — for this problem is given in (43), where only the strictly positive transports are indicated

$$\mu = \begin{bmatrix} 4 & - & - & - & - & - & - & - & - & - \\ - & - & - & - & - & 3 & - & - & - & - \\ - & - & 5 & - & - & - & - & - & - & - \\ - & - & - & - & - & - & - & - & - & 5 \\ - & - & - & - & 1 & - & - & - & 2 & - \\ - & - & 4 & - & - & - & - & - & - & 2 \\ - & 1 & - & - & - & - & - & 1 & - & - \\ - & - & - & - & - & 2 & 4 & - & 2 & \\ 2 & - & - & - & - & - & 3 & 4 & - & \\ - & - & - & 2 & - & - & 5 & - & - & - \end{bmatrix}. \tag{43}$$

From Theorem 3, we know that $C(v)$ is contained in $C(v_\mu)$. On the other hand, we need to solve $2^{20} - 2^{11} + 1 = 1046529$ transportation problems to determine completely the characteristic function. Furthermore, if we want to know whether a given imputation $(x; y)$ belongs or not to the core of the game, then we have to check (in the worst case) the same number of constraints. On the contrary, if we want to confirm whether the same imputation is in $C(v_\mu)$ or not, then according to Theorem 2 it is enough to solve the system (44) and (45). It has 20 constraints associated with the players, and 18 more which correspond to matchings in the optimal transportation plan μ. Likewise, many of the following equations can straightforwardly be solved by substitution.

$$
\begin{array}{ll}
(1)\ x_1 = a_{11}, & (1')\ y_1 = b_{11} + b_{91}, \\
(2)\ x_2 = a_{26}, & (2')\ y_2 = b_{72}, \\
(3)\ x_3 = a_{33}, & (3')\ y_3 = b_{33} + b_{63}, \\
(4)\ x_4 = a_{40}, & (4')\ y_4 = b_{04}, \\
(5)\ x_5 = a_{55} + a_{59}, & (5')\ y_5 = b_{55}, \\
(6)\ x_6 = a_{63} + a_{60}, & (6')\ y_6 = b_{26}, \\
(7)\ x_7 = a_{72} + a_{78}, & (7')\ y_7 = b_{87} + b_{07}, \\
(8)\ x_8 = a_{87} + a_{88} + a_{80}, & (8')\ y_8 = b_{78} + b_{88} + b_{98}, \\
(9)\ x_9 = a_{91} + a_{98} + a_{99}, & (9')\ y_9 = b_{59} + b_{99}, \\
(0)\ x_0 = a_{04} + a_{07}, & (0')\ y_0 = b_{40} + b_{60} + b_{80},
\end{array}
\tag{44}
$$

$$
\begin{array}{ll}
(11)\ 28 = a_{11} + b_{11}, & (78)\ 9 = a_{78} + b_{78}, \\
(26)\ 27 = a_{26} + b_{26}, & (87)\ 18 = a_{87} + b_{87}, \\
(33)\ 45 = a_{33} + b_{33}, & (88)\ 32 = a_{88} + b_{88}, \\
(40)\ 45 = a_{40} + b_{40}, & (80)\ 14 = a_{80} + b_{80}, \\
(55)\ 8 = a_{55} + b_{55}, & (91)\ 14 = a_{91} + b_{91}, \\
(59)\ 18 = a_{59} + b_{59}, & (98)\ 27 = a_{98} + b_{98}, \\
(63)\ 28 = a_{63} + b_{63}, & (99)\ 36 = a_{99} + b_{99}, \\
(60)\ 14 = a_{60} + b_{60}, & (04)\ 12 = a_{04} + b_{04}, \\
(72)\ 9 = a_{72} + b_{72}, & (07)\ 40 = a_{07} + b_{07}.
\end{array}
\tag{45}
$$

In view of the previous facts, it is clear that to use $C(v_\mu)$ is easier than $C(v)$. Thus, it is not difficult to conclude that, on many occasions, the players are unable to reject an imputation which is not in the core, because they cannot compute the characteristic function of the game. However, it is possible that they can reject an imputation which is not in the core of v_μ, since it is highly manoeuvrable and easy to calculate. Moreover, it is reasonable to use it in the context of bounded rationality in two-sided transportation models. Nevertheless, one can object to this argument since with the great breakthrough in the field of computers nowadays, it is possible that in the near future there will be no difference and, then, the core would be preferred.

Assume that we have two algorithms to check whether an imputation belongs to the core of the game and to $C(v_\mu)$. Further, suppose that the computer can perform each basic task in $\tau = 10^{-6}$, independently of the size of the chains involved. In this situation, we will have the following processing times for both algorithms, in the case of transportation problems with sizes 10×10, 15×15 and 20×20:

10×10 (20 players)		
Approx. time for...	$C(v)$	$C(v_\mu)$
Determine charact. function	2h 20'	—
Check it is in	2h 20'	0.13"
End of process	**4h 40'**	**0.13"**
15×15 (30 players)		
Approx. time for...	$C(v)$	$C(v_\mu)$
Determine charact. function	99 days 12h	—
Check it is in	99 days 12h	0.43"
End of process	**199 days**	**0.43"**
20×20 (40 players)		
Approx. time for...	$C(v)$	$C(v_\mu)$
Determine charact. function	279 years	—
Check it is in	279 years	1.01"
End of process	**558 years**	**1.01"**

The calculation of the above times assumes that the algorithm has to check all the constraints (the worst case). On the other hand, the requirements of memory in both algorithms should be mentioned. Bearing in mind that the need for storing certain information is equal to the length of the chain which contains this information in binary code, we can derive the requirements of memory. The length of the chains needed for the storage of the characteristic function of a transportation game, for sizes as 10×10, 15×15 and 20×20 are 37675044, 1.0736763×10^9 and 1.0995095×10^{12}, respectively, which represent 4.7, 6173.6 and 7696566.7 Megabytes. Nevertheless, the requirements for the algorithm of $C(v_\mu)$ are only related to those needed to describe the matrix associated with the optimal transportation plan, which in the case of a 20×20 problem are 25744 bytes in the worst scenario.

To be precise, we next analyze the computational complexity of both procedures in order to determine their kind of bounded rationality according to Definition 6. In both procedures, transportation problems must be solved, the computational complexity for obtaining the optimal solution is proportional to $m \log(m)(mn + n \log n)$, where $m \geq n$ [Kleinschmidt and Schannath, 1995]. On the other hand, the determination of the characteristic function involves the solution of $2^{n+m} - 2^n - 2^m + 1$ transportation problems, this implies that the computational complexity for determining the characteristic function of the game is at least exponential. The checking step in both procedures is as follows. In the case of the core, the worst case involves $2^{n+m} - 2^n - 2^m + 1$ basic operations. For the core of a 2-game associated with an optimal transportation plan, the checking step consists of solving a sparce linear system of which computational complexity is $O(m^2)$, where $m \geq n$ [Tinney and Hart, 1972]. The next proposition shows what level of bounded rationality the core catcher of transportation games has.

Proposition 3. *The checking step of the procedure using the core catcher of trans-portation games is better than* $\mathrm{int}\left(\frac{1}{2}\sqrt[3]{\frac{\Upsilon}{\tau}}\right)$*-$\Upsilon$-$\tau$-rational.*

Proof. First, we note that the number of floating operations in the Gauss method to solve a (dense) $n \times n$ linear system is given by

$$\sum_{i=1}^{n-1}\{2(n+1-(i-1))(n-i)\} + \sum_{i=1}^{n}(3n-3i+1) = \frac{2}{3}n^3 + \frac{5}{2}n^2 - \frac{13}{6}n, \qquad (46)$$

where the first sum corresponds to the triangulation phase, and the second sum corresponds to the substitution-and-solution phase. In our case, the linear system to be solved has $2(n+m-1)$ variables — if we use a basic optimal solution — and $2(n+m)-1$ constraints. Since we know that the system is solvable, then we can consider a $2(n+m-1) \times 2(n+m-1)$ linear system, because one of the linear equations must be redundant.

Now we have the following,

$$\frac{16}{3}(n+m-1)^3 + 10(n+m-1)^2 - \frac{13}{3}(n+m-1) \leq \frac{\Upsilon}{\tau}. \qquad (47)$$

Solving Eq. (47) by using Wolfram Alpha LLC. [2018], we obtain that the only relevant case for our purpose is when $\frac{\Upsilon}{\tau} > 11.0508$. In this case, taking $a = \frac{\Upsilon}{\tau}$, we have that

$$n+m-1 \leq \frac{\sqrt[3]{2\sqrt{3}\sqrt{15552a^2 - 165240a - 73177} + 432a - 2295}}{8\sqrt[3]{3^2}}$$

$$+ \frac{127}{8\sqrt[3]{3}\sqrt[3]{2\sqrt{3}\sqrt{15552a^2 - 165240a - 73177} + 432a - 2295}} - \frac{5}{8}$$

$$= \Gamma\left(\frac{\Upsilon}{\tau}\right). \qquad (48)$$

We can derive that

$$\mathrm{int}\left(\frac{1}{2}\sqrt[3]{\frac{\Upsilon}{\tau}}\right) \leq \mathrm{int}\left(\Gamma\left(\frac{\Upsilon}{\tau}\right)\right). \qquad (49)$$

Therefore, the result holds. □

In view of this analysis, the core catcher is a more suitable criterion than the core for rejecting allocation proposals regarding computational bounded rationality. Of course, the core of the game is a better criterion than the core catcher with regard to other properties. However, in real-life situations the computational bounded rationality matters.

6. Conclusions

Horizontal and vertical cooperation is an interesting issue in logistic and market problems because of its possible economic benefits, for example saving costs or increasing total profits. In order to reach a collaboration agreement, agents involved must settle for an allocation of the profits. How to determine and select an allocation are interesting questions. In this chapter, these questions have been studied for two-sided transportation problems because they can be used to describe a variety of logistic and market problems.

First, 2-games associated with the optimal solutions of the two-sided transportation game have been introduced, and the structure of their cores has been determined. All these cores contain the core of the transportation game. It is interesting to note that these allocations are based on the primal optimal solution of the underlying transportation problem which is different from the usual approach in operations research games in which dual optimal solutions are used.

A core catcher for the core of the transportation game is defined by the intersection of the cores of the 2-games associated with the optimal solutions of the problem. The core catcher keeps more information about the problem and is less sentitive to small changes in the problem than the core of the game. Therefore, it seems a suitable solution concept for this kind of problem.

Regarding the selection of an allocation, a reasonable criterion is that the selected allocation of the profit belongs to the core of the game, which is individually and coalitionally rational. In this sense, the core would play the role of a necessary condition for an allocation to be accepted, this role would be analogous to that played by the Nash equilibrium in non-cooperative games. However, the core can be empty, and checking whether an allocation is in the core can be computationally very costly -even impossible-, which encourages us to think of other weaker concepts of rationality. Thus, following the ideas in Deng and Papadimitriou [1994], Aumann [1997], and Deng and Fang [2008], we have introduced a new concept of (bounded) rationality related to the computational cost of a solution. To the best of our knowledge this (bounded) rationality definition is new in the literature. In the particular case of two-sided transportation problems, we have shown that the core catcher has a reasonable behavior with regard to this concept of bounded rationality. Therefore, in our opinion, the core catcher is a reasonable candidate to replace the core as a necessary condition in a procedure of choice of an allocation of the profit when cooperation holds.

Acknowledgments

The authors are thankful for the helpful comments and suggestions to improve the contents of this work. Financial support from the Ministerio de Economia y Competitividad (MINECO) of Spain and FEDER funds Under Project MTM2014-54199-P and from Fundacion Seneca de la Region de Murcia through Grant 19320/PI/14 is gratefully acknowledged.

References

Arora, S. and Barak, B. [2009] *Computational Complexity: A Modern Approach* (Cambridge University Press, Cambridge).

Arthanari, T., Carfi, D. and Musolino, F. [2015] Game theoretic modeling of horizontal supply chain coopetition among growers, *Int. Game Theory Rev.* **17**, Article ID:1540013.

Aumann, R. [1997] Rationality and bounded rationality, *Games Econ. Behav.* **21**, 2–14.

Auriol, I. and Marchi, E. [2002] Quasi-assignment cooperative games, *Int. Game Theory Rev.* **4**, 173–182.

Cachon, G. and Netessine, S. [2004] Game theory in supply chain analysis in *Handbook of Quantitative Supply Chain Analysis: Modeling in the Ebusiness Era, eds.* Simchi-Levi, D., Wu, S. D. and Shen, Z.-J. (Kluwer Academic Publishers), pp. 13–66.

Cruijssen, F., Dullaert, W. and Fleuren, H. [2007] Horizontal cooperation in transport and logistics: A literature review, *Transp. J.* **46**, 22–39.

Deng, X. and Fang, Q. [2008] Algorithmic cooperative game theory, in *Pareto Optimality, Game Theory and Equilibria*, Springer Optimization and its Applications, Vol. 17, eds. Chinchuluun, A., Pardalos, P. M., Migdalas, A. and Pitsoulis, L. (Springer), pp. 159–186.

Deng, X. and Papadimitriou, C. [1994] On the complexity of cooperative game solution concepts, *Math. Operat. Res.* **19**, 257–266.

Eriksson, K., Jansson, F. and Vetander, T. [2011] The assignment game with negative externalities and bounded rationality, *Int. Game Theory Rev.* **13**, 443–459.

Fragnelli, V., Llorca, N. and Tijs, S. [2007] Balancedness of infinite permutation games and related classes of games, *Int. Game Theory Rev.* **9**, 425–435.

Gillies, D. B. [1953] Some theorems on n-person games, PhD thesis, Princeton University, Princeton.

Gudmundsson, J. and Habis, H. [2017] Assignment games with externalities revisited, *Econo. Theory Bull.* **5**, 247–257.

Izquierdo, J. M., Nuñez, M. and Rafels, C. [2012] A survey on assignment games and related markets, *Boletin de Estadistica e Investigacion Operativa* **28**, 220–246. (In Spanish).

Johari, M. and Hosseini-Motlagh, S. [2018] Coordination of cooperative promotion efforts with competing retailers in a manufacturer-retailer supply chain, *Uncertain Supply Chain Manag.* **6**, 25–48.

Kleinschmidt, P. and Schannath, H. [1995] A strongly polynomial algorithm for the transportation problem, *Math. Program.* **68**, 1–13.

Leng, M. and Parlar, M. [2005] Game theoretic applications in supply chain management: A review, *INFOR* **43**, 187–220.

Llerena, F. and Rafels, C. [2006] The vector lattice structure of the n-person TU games, *Games Econ. Behav.* **54**, 373–379.

Llorca, N., Sanchez-Soriano, J., Tijs, S. and Timmer, J. [2004] The core and related solution concepts for infinite assignment games, *TOP* **12**, 331–350.

Mason, R., Lalwani, L. and Boughton, R. [2007] Combining vertical and horizontal collaboration for transport optimisation, *Supply Chain Manag. An Int. J.* **12**, 187–199.

Nagarajan, M. and Sosic, G. [2008] Game-theoretic analysis of cooperation among supply chain agents: Review and extensions, *Europ. J. Operat. Res.* **187**, 719–745.

Papadimitriou, C. H. and Yannakakis, M. [1994] On complexity as bounded rationality, in *Proc. of the 26th ACM Symp. on the Theory of Computing*, Montreal, Quebec, Canada, pp. 726–733.

Rubinstein, A. [1990] New directions in economic theory: Bounded rationality, *Rev. Española de Economia* **7**, 3–15.

Rubinstein, A. [1998] *Modeling Bounded Rationality* (MIT Press, Cambridge, Massachusetts).

Sanchez-Soriano, J. [2003] The pairwise egalitarian solution, *Europ. J. Operat. Res.* **150**, 220–231.

Sanchez-Soriano, J. [2006] Pairwise solutions and the core of transportation situations, *Europ. J. Operat. Res.* **175**, 101–110.

Sanchez-Soriano, J. and Fragnelli, V. [2010] Two-sided market situations with existing contracts, *Social Choice Welf.* **34**, 295–313.

Sanchez-Soriano, J., Llorca, N., Tijs, S. and Timmer, J. [2002] On the core of semi-infinite transportation games with infinite divisible goods, *Europ. J. Operat. Res.* **109**, 41–60.

Sanchez-Soriano, J., Lopez, M. A. and Garcia-Jurado, I. [2001] On the core of transportation games, *Math. Soc. Sci.* **41**, 215–225.

Shapley, L. S. [1971] Cores of convex games, *Int. J. Game Theory* **1**, 11–26.

Shapley, L. S. and Shubik, S. [1972] The assignment game I: The core, *Int. J. Game Theory* **1**, 111–130.

Simon, H. A. [1957] A behavioral model of rational choice, in *Models of Man, Social and Rational: Mathematical Essays on Rational Human Behavior in a Social Setting* (John Wiley and Sons, NY, USA).

Simon, H. A. [1972] Theories of bounded rationality, in *Decision and Organization*, eds. McGuire, C. B. and Radner, R. (North Holland, Amsterdam), pp. 161–176.

Thompson, G. L. [1980] Computing the core of a market game, in *Extremal Methods and Systems Analysis*. Lecture Notes in Economics and Mathematical Systems, Vol. 174, eds. Fiacco, A. V. and Kortanek, K. O. (Springer, Berlin), pp. 312–334.

Tinney, W. and Hart, C. [1972] Power flow solutions by Newton's method, *IEEE Trans. Power Apparatus Syst.* **86**, 1449–1456.

van den Nouweland, A., Borm, P., van Golstein Brouwers, W., Groot Bruinderink, R. and Tijs, S. [1996] A game theoretic approach to problems in telecommunication, *Manag. Sci.* **42**, 294–303.

Weber, R. J. [1978] Probabilistics Values for Games. Cowles Foundation Discussion Paper 471R, Yale University.

Wolfram Alpha LLC. [2018] Wolfram|Alpha. http://www.wolframalpha.com/input/?i= solve+(16%2F3)*x%5E3+%2B+10*x%5E2+-(13%2F3)*x%3C%3Da (access March 28, 2018).

Zhang, P., He, Y. and Shi, C. [2017] Transshipment and coordination in a two-echelon supply chain, *RAIRO-Operat. Res.* **51**, 729–747.

Zhou, Y. W., Li, J. and Zhong, Y. [2018] Cooperative advertising and ordering policies in a two-echelon supply chain with risk-averse agents, *Omega* **75**, 97–117.

Chapter 23

Intrinsic Comparative Statics of a Nash Bargaining Solution

Michael R. Caputo

Department of Economics, University of Central Florida
P. O. Box 161400, Orlando, FL 32816-1400, USA

mcaputo@ucf.edu

A generalization of the class of bargaining problems examined by Engwerda and Douven [(2008) On the sensitivity matrix of the Nash bargaining solution, *Int. J. Game Theory* **37**, 265–279] is studied. The generalized class consists of nonconvex bargaining problems in which the feasible set satisfies the requirement that the set of weak Pareto-optimal solutions can be described by a smooth function. The intrinsic comparative statics of the aforesaid class are derived and shown to be characterized by a symmetric and positive semidefinite matrix, and an upper bound to the rank of the matrix is established. A corollary to this basic result is that a Nash bargaining solution is intrinsically a locally nondecreasing function of its own disagreement point. Other heretofore unknown results are similarly deduced from the basic result.

Keywords: Nash bargaining solution; disagreement point; comparative statics.

1. Introduction

The formal study of bargaining theory began with the seminal work of Nash [1950]. The subject matter of this chapter, on the other hand, was initiated by Thomson [1987] and Chun and Thomson [1988] during their investigation of the disagreement point monotonicity of various bargaining solutions. Thomson [1987] introduced two versions of monotonicity of some relevance to the present work. To quote Thomson [2009, p. 19], "The disagreement point represents the agents' fall-back positions. Thus, it is natural to require that if an agent's disagreement utility increases, he should not lose. This is the property of *disagreement point monotonicity*. A related property, *strong disagreement point monotonicity* says that such a change should benefit none of the other agents." Note that the aforesaid definitions of monotonicity are also referred to as *d-monotonicity* and *strong d-monotonicity*, respectively.

Building on these definitions, Engwerda and Douven [2008, Definition 1] offered local, i.e., differential, versions of the above two monotonicity definitions. Briefly, a Nash bargaining solution is said to be *locally strongly d-monotonic* if the partial derivative of the ith component of a Nash bargaining solution vector with respect to the ith disagreement point is greater than or equal to zero and that with respect to the jth disagreement point is less than or equal to zero, for every agent i and for all $j \neq i$, whereas the said solution is said to be *locally d-monotonic* if it meets just the first condition. Engwerda and Douven [2008, Theorem 1] used the classical approach to comparative statics, that is, the implicit function theorem, to calculate the Jacobian matrix of partial derivatives of a Nash bargaining solution with respect to the disagreement point vector. Corollary 1 of Engwerda and Douven [2008] gives a sufficient condition for a Nash bargaining solution to be locally d-monotonic.[a] Their Theorem 2 is a more potent result, as it gives a necessary and sufficient condition for a Nash bargaining solution to be locally strongly d-monotonic.

The present work complements and extends that by Thomson [1987] and Engwerda and Douven [2008] by considering a generalization of the latter's bargaining problem and deriving basic results. The generalization contemplated here assumes that the set of weak Pareto-optimal solutions is described by a smooth function, i.e., it drops the strictly decreasing and strongly concave assumptions. The generalization thus results in a nonconvex bargaining problem. Moreover, instead of determining necessary and sufficient conditions, or sufficient conditions, for a differentiable Nash bargaining solution to satisfy the above two versions of d-monotonicity, as did Engwerda and Douven [2008], the extension investigated here is basic, to wit, the determination of the *intrinsic comparative statics* of a locally differentiable Nash bargaining solution. Such comparative statics are defined as those implied by (i) the assumption that a locally differentiable solution to the posed optimization problem exists, and (ii) the basic assumptions inherent to the theory underlying the problem. As a result, intrinsic comparative statics rely only on the necessary conditions of an optimization problem. They therefore avoid the use of sufficient optimality conditions, *ad hoc* assumptions placed on the objective and constraint functions that transcend those required for a differential characterization of its comparative statics, and assumptions which are not inherent to the theory underlying the problem.

Consider, say, the price-taking and profit-maximizing model of a firm, as typified by Silberberg [1990, Chap. 4]. The intrinsic comparative statics of the model are summarized by the statement that the Jacobian matrix comprised of the partial derivatives of the factor demand functions and the negative of the output supply

---

[a] Engwerda and Douven [2008] call their result d-monotonic, but seeing as it is based on the Jacobian matrix of partial derivatives of the Nash bargaining solution with respect to the disagreement points — a differential characterization — the term locally d-monotonic is perhaps more apt in view of their Definition 1.

function with respect to the factor and output prices is symmetric and negative semidefinite. This comparative statics result is intrinsic to the theory, seeing as it follows solely from the assumptions that a locally differentiable solution to the profit maximization problem exists and that the firm is a price taker, the latter being a basic assumption of the theory. The central result derived here is of the same fundamental nature. Moreover, though basic, its derivation is surprisingly compact, as it relies on recent developments in differential methods of comparative statics due to Partovi and Caputo [2006].

In particular, it is shown that the intrinsic comparative statics of a Nash bargaining solution are characterized by a symmetric and positive semidefinite matrix, and an upper bound to the rank of the matrix is provided. It follows from this basic result that a Nash bargaining solution is intrinsically locally d-monotonic. Furthermore, both of these basic results hold independently of the slope and curvature of the Pareto frontier, i.e., they are independent of monotonicity and curvature assumptions that might be placed on the function that describes the Pareto frontier. In other words, the fact that a Nash bargaining solution is locally d-monotonic depends only on the assumption that the function describing the Pareto frontier is $C^{(2)}$. Finally, heretofore unknown symmetry properties of a Nash bargaining solution are derived and elucidated.

2. Preliminaries

To begin, an *n-person bargaining problem* is a pair $(S, \mathbf{d})$, where $S \subset \mathbb{R}^n$ is called the *feasible set*, $\mathbb{R}^n$ the *utility space*, and $\mathbf{d} \overset{\text{def}}{=} (d_1, d_2, \ldots, d_n)$ the *disagreement point* — note that vectors are henceforth denoted in boldface type. Two classes of bargaining problems are often contemplated in the economics literature, namely: (i) $\bar{\Sigma}^n$, where the feasible set $S \subset \mathbb{R}^n$ is convex and compact, such that there exists an $\mathbf{x} \in S$ with $\mathbf{x} > \mathbf{d}$, vector inequality notation implied, and (ii) Σ^n, a subclass of $\bar{\Sigma}^n$ often referred to as the class of comprehensive bargaining problems, obtained by considering just those elements in $S \subset \mathbb{R}^n$ satisfying the additional property that whenever $\mathbf{x} \in S$ and $\mathbf{d} < \bar{\mathbf{x}} \leq \mathbf{x}$, then $\bar{\mathbf{x}} \in S$.

Following Engwerda and Douven [2008, p. 267], a generalization of a subclass of bargaining problems that they denoted by Σ_P^n is under investigation here. In the subclass Σ_P^n, the feasible set is assumed to satisfy the additional requirement that the set P of weak Pareto-optimal solutions can be described by a strictly decreasing and strongly concave $C^{(2)}$ function $\varphi(\cdot)$. Formally,

$$\Sigma_P^n \overset{\text{def}}{=} \{\mathbf{x} \in \Sigma^n \mid \mathbf{x} > \mathbf{d} \text{ and } x_n \leq \varphi(\mathbf{x}_-)\},$$

where $\mathbf{x}_- \overset{\text{def}}{=} (x_1, x_2, \ldots, x_{n-1})$, $\partial\varphi(\mathbf{x}_-)/\partial x_i < 0$, $i = 1, 2, \ldots, n-1$, and for all $\mathbf{h} \in \mathbb{R}^{n-1}$ and $\mathbf{h} \neq \mathbf{0}_{n-1}$, $\sum_{i=1}^{n-1}\sum_{j=1}^{n-1}[\partial^2\phi(\mathbf{x}_-)/\partial x_i\partial x_j]h_i h_j < 0$, the latter being equivalent to the strong concavity of $\varphi(\cdot)$. This class of bargaining problems occurs with some regularity in applied economics, as documented by Engwerda and Douven [2008, p. 267]. The generalization of the subclass Σ_P^n considered here (i) represents the set of weak Pareto-optimal solutions in the more general form $g(\mathbf{x}) \geq 0$, and (ii) does not assume that $g(\cdot)$ is strictly decreasing and strongly concave. The representation of the set of weak Pareto-optimal solutions in the form $g(\mathbf{x}) \geq 0$ also has the added advantage of keeping the formulation of the bargaining problem symmetric, thereby making for a crisp derivation of the basic results.

The bargaining problem of interest is given by the constrained maximization problem

$$\max_{x_1, x_2, \ldots, x_n} \left\{ f(\mathbf{x}; \mathbf{d}) \overset{\text{def}}{=} \prod_{j=1}^{n} [x_j - d_j] \text{ s.t. } g(\mathbf{x}) \geq 0 \right\}. \tag{1}$$

Let $f_{x_i}(\mathbf{x}; \mathbf{d}) \overset{\text{def}}{=} \partial f(\mathbf{x}; \mathbf{d})/\partial x_i$ and $f_{d_\alpha}(\mathbf{x}; \mathbf{d}) \overset{\text{def}}{=} \partial f(\mathbf{x}; \mathbf{d})/\partial d_\alpha$, and similarly for $g(\mathbf{x})$, so that partial differentiation with respect to a decision variable or parameter is signified by a variable or parameter appearing as a subscript on a function. Furthermore, the ensuing assumptions are maintained throughout and explained subsequently:

(A1) The function $g(\cdot) : \mathbb{R}^n \to \mathbb{R}$ is $C^{(2)}$ on its domain.

(A2) There exists an optimal solution to the bargaining problem defined by Eq. (1) for each value of $\mathbf{d}$ in some open set D, denoted by $\mathbf{x} = \mathbf{x}^N(\mathbf{d})$, where $x_j^N(\mathbf{d}) > d_j$, $j = 1, 2, \ldots, n$, with corresponding value of the Lagrange multiplier $\lambda = \lambda^N(\mathbf{d})$.

(A3) The function $\mathbf{x}^N(\cdot) \in C^{(1)}$ for all $\mathbf{d} \in D$.

The above are natural assumptions to make given the focus on a differential characterization of the intrinsic comparative statics of bargaining problem (1). Supposition (A1) is analogous to that made by Engwerda and Douven [2008, p. 268]. In contrast to Engwerda and Douven [2008, p. 268], however, no assumption regarding the monotonicity or concavity of the function $g(\cdot)$ is made, which is the generalization contemplated here, as remarked above and demonstrated below. Moreover, only necessary conditions of optimality are employed in what ensures, whereas Engwerda and Douven [2008, p. 268] assumed that the second-order sufficient condition of the bargaining problem held at a Nash bargaining solution when they assumed that the Hessian matrix of the objective function was invertible at the said point. Invertibility is important in their approach, as they used the implicit function theorem to characterize the comparative statics of a Nash bargaining solution, and furthermore, required that certain matrices were invertible in their analysis. Such assumptions are dispensed with here as the implicit function theorem is eschewed by employing the differential approach of Partovi and Caputo [2006] to derive the intrinsic comparative statics of the bargaining problem.

The lack of other assumptions in the present work necessitates assumptions (A2) and (A3), as does the aforementioned focus on obtaining a differential characterization of the intrinsic comparative statics of problem (1). That is to say, as the analysis henceforth relies only on necessary conditions, suppositions (A2) and (A3) are essential. In sum, the approach taken here avoids the introduction of assumptions such as the aforesaid monotonicity and concavity properties that transcend those needed for the discovery of intrinsic results.

The minimal assumptions employed here imply that the bargaining problem defined by Eq. (1) is not a convex optimization problem. Even so, as long as assumptions (A1)–(A3) hold, the results to follow apply to such a class of nonconvex bargaining problems, thereby representing a nontrivial extension of existing qualitative results for bargaining problems. But as pointed out by Kaneko [1980] and Xu and Yoshihara [2006], the lack of convexity in problem (1) means that assumption (A3) may not hold because $\mathbf{x}^N(\cdot)$ may be a correspondence and thus be only upper hemi-continuous. In such cases, other assumptions must be introduced to ensure the local differentiability of $\mathbf{x}^N(\cdot)$. One such assumption is that a Nash bargaining solution be locally isolated, which it is if the bargaining problem were convex. Another is that the usual second-order sufficient condition holds, seeing as the implicit function theorem could be invoked in this case to ensure the local differentiability of $\mathbf{x}^N(\cdot)$. Such stipulations are worthwhile to keep in mind in what follows given the lack of assumptions in the present setting.

The Lagrangian for problem (1) is defined as

$$L(\mathbf{x}, \lambda; \mathbf{d}) \overset{\text{def}}{=} \prod_{j=1}^{n} [x_j - d_j] + \lambda g(\mathbf{x}). \tag{2}$$

By assumptions (A1) and (A2), $\mathbf{x} = \mathbf{x}^N(\mathbf{d})$ satisfies the usual Karush–Kuhn–Tucker first-order necessary conditions for each $\mathbf{d} \in D$, to wit,

$$L_{x_i}(\mathbf{x}, \lambda; \mathbf{d}) = \prod_{j=1, j \neq i}^{n} [x_j - d_j] + \lambda g_{x_i}(\mathbf{x}) = 0, \quad i = 1, 2, \ldots, n, \tag{3}$$

$$L_\lambda(\mathbf{x}, \lambda; \mathbf{d}) = g(\mathbf{x}) \geq 0, \quad \lambda \geq 0, \quad \lambda g(\mathbf{x}) = 0. \tag{4}$$

By assumption (A2), $x_j^N(\mathbf{d}) > d_j$, $j = 1, 2, \ldots, n$, and so by way of Eq. (3), $\lambda^N(\mathbf{d}) g_{x_i}(\mathbf{x}^N(\mathbf{d})) < 0$, $i = 1, 2, \ldots, n$. But this deduction and Eq. (4) imply that $\lambda^N(\mathbf{d}) > 0$, $g(\mathbf{x}^N(\mathbf{d})) \equiv 0$, and $g_{x_i}(\mathbf{x}^N(\mathbf{d})) < 0$ for all $\mathbf{d} \in D$ and $i = 1, 2, \ldots, n$. In other words, in an open neighborhood of a Nash bargaining solution, the function that describes the Pareto frontier is necessarily strictly decreasing and hence the classical nondegenerate constraint qualification is satisfied, the Lagrange multiplier associated with the Pareto frontier is positive, and a Nash bargaining solution necessarily lies on the boundary of the Pareto frontier. These results will be referred to later and are sufficiently important to record in the following lemma.

Lemma 1. *Given assumptions* (A1)–(A3), *a Nash bargaining solution of problem* (1) *satisfies* $\lambda^N(\mathbf{d}) > 0, g(\mathbf{x}^N(\mathbf{d})) \equiv 0$, *and* $g_{x_i}(\mathbf{x}^N(\mathbf{d})) < 0$, *for all* $\mathbf{d} \in D$ *and* $i = 1, 2, \ldots, n$.

Note that Lemma 1 shows that the monotonicity assumption employed by Engwerda and Douven [2008, p. 268] may be dispensed with for the purposes of comparative statics.

Now observe that the standard second-order necessary condition of problem (1), namely, at a solution of problem (1), the Hessian matrix of $L(\cdot)$ with respect to $\mathbf{x}$ is negative semidefinite in directions orthogonal to the gradient vector $g_{\mathbf{x}}(\mathbf{x})$, does not imply that $g(\cdot)$ is strongly concave. As a result, strong concavity of $g(\cdot)$ is not intrinsic to bargaining problem (1). This conclusion, however, does not imply that the assumption of strong concavity is not useful. Indeed, Engwerda and Douven [2008] convincingly demonstrate the important role that it plays in their qualitative results. In sum, the set of assumptions employed here does indeed result in a generalization of the class of bargaining problems studied by Engwerda and Douven [2008].

With the preliminary matters addressed, attention is now turned toward the main result of the note — the derivation of the intrinsic qualitative properties of a Nash bargaining solution of the bargaining problem defined by Eq. (1).

3. Intrinsic Qualitative Properties

Under the foregoing stipulations and Lemma 1, the differential comparative statics methodology of Partovi and Caputo [2006, 2007] can be applied to problem (1). Moreover, as the constraint in problem (1) does not depend on the vector $\mathbf{d}$, Theorem 1 of Partovi and Caputo [2006, 2007] shows that partial derivatives are sufficient to characterize its intrinsic comparative statics and that the $n \times n$ matrix $\mathbf{\Omega}(\mathbf{d}) = [\mathbf{\Omega}_{\alpha\beta}(\mathbf{d})]$, $\alpha, \beta = 1, 2, \ldots, n$, the typical element of which is given by

$$\mathbf{\Omega}_{\alpha\beta}(\mathbf{d}) \overset{\text{def}}{=} \sum_{i=1}^{n} L_{x_i d_\alpha}(\mathbf{x}^N(\mathbf{d}), \lambda^N(\mathbf{d}); \mathbf{d}) \frac{\partial x_i^N(\mathbf{d})}{\partial d_\beta}, \tag{5}$$

is symmetric and positive semidefinite. Furthermore, Theorem 1 of Partovi and Caputo [2006, 2007] is intrinsic to all differentiable optimization problems, as it relies only on the $C^{(2)}$ nature of the objective and constraint functions and the assumption that a locally differentiable solution to the problem exists. Accordingly, the intrinsic comparative statics of the bargaining problem defined by Eq. (1) are found by implementing the formula in Eq. (5). In addition, note that by Theorem 4 of Partovi and Caputo [2006], the matrix $\mathbf{\Omega}(\mathbf{d})$ has a maximum rank of $n-1$.

To begin the derivation, note that the definition of $L(\cdot)$ in Eq. (1) implies that

$$L_{x_i d_\alpha}(\mathbf{x}, \lambda; \mathbf{d}) = -\prod_{j=1, j \neq i, j \neq \alpha}^{n} [x_j - d_j][1 - \delta_{i\alpha}], \quad i, \alpha = 1, 2, \ldots, n, \tag{6}$$

where $\delta_{i\alpha}$ is Kronecker's delta. Upon evaluating Eq. (6) at $\mathbf{x} = \mathbf{x}^N(\mathbf{d})$ and using Eq. (5), the typical element of the $n \times n$ matrix $\mathbf{\Omega}(\mathbf{d}) = [\mathbf{\Omega}_{\alpha\beta}(\mathbf{d})]$ takes the form

$$\mathbf{\Omega}_{\alpha\beta}(\mathbf{d}) = -\sum_{i=1}^{n} \prod_{j=1, j \neq i, j \neq \alpha}^{n} [x_j^N(\mathbf{d}) - d_j][1 - \delta_{i\alpha}] \frac{\partial x_i^N(\mathbf{d})}{\partial d_\beta}, \quad \alpha, \beta = 1, 2, \ldots, n, \tag{7}$$

or equivalently,

$$
\Omega_{\alpha\beta}(\mathbf{d}) = - \sum_{i=1,i\neq\alpha}^{n} \prod_{j=1,j\neq i,j\neq\alpha}^{n} [x_j^N(\mathbf{d}) - d_j] \frac{\partial x_i^N(\mathbf{d})}{\partial d_\beta}, \quad \alpha,\beta = 1,2,\ldots,n. \quad (8)
$$

In order to put Eq. (8) into a form suited for the purposes at hand, one may proceed as follows.

First observe that because $g(\mathbf{x}^N(\mathbf{d})) \equiv 0$ for all $\mathbf{d} \in D$ by Lemma 1, it follows that

$$
\sum_{k=1}^{n} g_{x_k}(\mathbf{x}^N(\mathbf{d})) \frac{\partial x_k^N(\mathbf{d})}{\partial d_\beta} \equiv 0, \quad \beta = 1,2,\ldots,n. \quad (9)
$$

Next, multiply the identity form of the first-order necessary condition in Eq. (3) by $\partial x_i^N(\mathbf{d})/\partial d_\beta$ and sum the resulting expression over i to arrive at

$$
\sum_{i=1}^{n} \prod_{j=1,j\neq i}^{n} [x_j^N(\mathbf{d}) - d_j] \frac{\partial x_i^N(\mathbf{d})}{\partial d_\beta}
$$
$$
+ \lambda^N(\mathbf{d}) \sum_{i=1}^{n} g_{x_i}(\mathbf{x}^N(\mathbf{d})) \frac{\partial x_i^N(\mathbf{d})}{\partial d_\beta} \equiv 0, \quad \beta = 1,2,\ldots,n. \quad (10)
$$

By Eq. (9), the latter summation in Eq. (10) vanishes identically, hence Eq. (10) reduces to

$$
\sum_{i=1}^{n} \prod_{j=1,j\neq i}^{n} [x_j^N(\mathbf{d}) - d_j] \frac{\partial x_i^N(\mathbf{d})}{\partial d_\beta} \equiv 0, \quad \beta = 1,2,\ldots,n. \quad (11)
$$

Now observe that Eq. (11) can be equivalently written as

$$
\sum_{i=1,i\neq\alpha}^{n} \prod_{j=1,j\neq i}^{n} [x_j^N(\mathbf{d}) - d_j] \frac{\partial x_i^N(\mathbf{d})}{\partial d_\beta}
$$
$$
+ \prod_{j=1,j\neq\alpha}^{n} [x_j^N(\mathbf{d}) - d_j] \frac{\partial x_\alpha^N(\mathbf{d})}{\partial d_\beta} \equiv 0, \quad \alpha,\beta = 1,2,\ldots,n. \quad (12)
$$

But seeing as $x_\alpha^N(\mathbf{d}) > d_\alpha$, $\alpha = 1,2,\ldots,n$, by assumption (A2), it follows that Eq. (12) is equivalent to

$$
- \sum_{i=1,i\neq\alpha}^{n} \prod_{j=1,j\neq i}^{n-1} \frac{[x_j^N(\mathbf{d}) - d_j]}{[x_\alpha^N(\mathbf{d}) - d_\alpha]} \frac{\partial x_i^N(\mathbf{d})}{\partial d_\beta}
$$
$$
\equiv \prod_{j=1,j\neq\alpha}^{n} \frac{[x_j^N(\mathbf{d}) - d_j]}{[x_\alpha^N(\mathbf{d}) - d_\alpha]} \frac{\partial x_\alpha^N(\mathbf{d})}{\partial d_\beta}, \quad \alpha,\beta = 1,2,\ldots,n. \quad (13)
$$

Furthermore, note that the expression on the left-hand side of Eq. (13) can be written as

$$
- \sum_{i=1,i\neq\alpha}^{n} \prod_{j=1,j\neq i}^{n} \frac{[x_j^N(\mathbf{d}) - d_j]}{[x_\alpha^N(\mathbf{d}) - d_\alpha]} \frac{\partial x_i^N(\mathbf{d})}{\partial d_\beta}
$$

$$
\equiv - \sum_{i=1,i\neq\alpha}^{n} \prod_{j=1,j\neq i,j\neq\alpha}^{n} [x_j^N(\mathbf{d}) - d_j] \frac{\partial x_i^N(\mathbf{d})}{\partial d_\beta}, \quad \alpha, \beta = 1, 2, \ldots, n. \quad (14)
$$

It then follows from Eqs. (13) and (14) and transitivity that Eq. (8) can be expressed as

$$
\Omega_{\alpha\beta}(\mathbf{d}) = \prod_{j=1,j\neq\alpha}^{n} \frac{[x_j^N(\mathbf{d}) - d_j]}{[x_\alpha^N(\mathbf{d}) - d_\alpha]} \frac{\partial x_\alpha^N(\mathbf{d})}{\partial d_\beta}, \quad \alpha, \beta = 1, 2, \ldots, n. \quad (15)
$$

In summary, the following basic comparative statics result has been established.

Theorem 1. *Under assumptions* (A1)–(A3), *the intrinsic comparative statics of a Nash bargaining solution* $\mathbf{x} = \mathbf{x}^N(\mathbf{d})$ *of problem* (1) *are given by the* $n \times n$ *symmetric and positive semidefinite matrix* $\Omega(\mathbf{d}) = [\Omega_{\alpha\beta}(\mathbf{d})], \alpha, \beta = 1, 2, \ldots, n,$ *where*

$$
\Omega(\mathbf{d}) = \left[\Omega_{\alpha\beta}(\mathbf{d}) \right]_{\alpha,\beta=1,2,\ldots,n} = \left[\prod_{j=1,j\neq\alpha}^{n} \frac{[x_j^N(\mathbf{d}) - d_j]}{[x_\alpha^N(\mathbf{d}) - d_\alpha]} \frac{\partial x_\alpha^N(\mathbf{d})}{\partial d_\beta} \right]_{\alpha,\beta=1,2,\ldots,n}. \quad (16)
$$

Moreover, $\mathrm{rank}[\Omega(\mathbf{d})] \leq n - 1.$

Theorem 1 gives the intrinsic comparative statics of a Nash bargaining solution of bargaining problem (1) and is the central result of this work. It is heretofore unknown and thus worthy of several extended comments. First, because the matrix $\Omega(\mathbf{d})$ is positive semidefinite, its diagonal elements are nonnegative, hence

$$
\prod_{j=1,j\neq\alpha}^{n} \frac{[x_j^N(\mathbf{d}) - d_j]}{[x_\alpha^N(\mathbf{d}) - d_\alpha]} \frac{\partial x_\alpha^N(\mathbf{d})}{\partial d_\alpha} \geq 0, \quad \alpha = 1, 2, \ldots, n. \quad (17)
$$

Seeing as $x_j^N(\mathbf{d}) > d_j, j = 1, 2, \ldots, n,$ by assumption (A2), the following corollary is immediate.

Corollary 1. *Under assumptions* (A1)–(A3), *a Nash bargaining solution* $\mathbf{x} = \mathbf{x}^N(\mathbf{d})$ *of problem* (1) *satisfies* $\partial x_\alpha^N(\mathbf{d})/\partial d_\alpha \geq 0, \alpha = 1, 2, \ldots, n.$

Corollary 1 asserts that an increase in an agent's own disagreement point does not lower the agent's attainable utility, i.e., an increase in an agent's own disagreement point cannot make the agent worse off. This is a local result, as it holds only for "small" changes in the disagreement point. It is natural, therefore, to call the comparative statics result in Corollary 1 local *d*-monotonicity. This property is intrinsic to the bargaining problem defined by Eq. (1) in view of the fact that it is predicated only on the basic assumptions (A1)–(A3). That is to say, Corollary 1 is independent of assumptions that transcend those required for a differential representation of the basic comparative statics of problem (1), as well as assumptions that are not an inherent part of the underlying economic theory. In effect, Corollary 1 is the differential analogue to Theorem 1 of Thomson [1987].

Corollary 1 shows that the strong concavity assumption of Engwerda and Douven [2008, Corollary 1] is not required for establishing the local d-monotonicity of a Nash bargaining solution. In other words, Corollary 1 demonstrates that the said assumption can be dispensed with, as local d-monotonicity is an intrinsic property of a Nash bargaining solution.

Now consider the heretofore unknown reciprocity properties of a Nash bargaining solution implied by the symmetry of $\boldsymbol{\Omega}(\mathbf{d})$. In general, Theorem 1 shows that

$$\prod_{j=1,j\neq\alpha}^{n} \frac{[x_j^N(\mathbf{d}) - d_j]}{[x_\alpha^N(\mathbf{d}) - d_\alpha]} \frac{\partial x_\alpha^N(\mathbf{d})}{\partial d_\beta} \equiv \prod_{j=1,j\neq\beta}^{n} \frac{[x_j^N(\mathbf{d}) - d_j]}{[x_\beta^N(\mathbf{d}) - d_\beta]} \frac{\partial x_\beta^N(\mathbf{d})}{\partial d_\alpha},$$

$$\alpha, \beta = 1, 2, \ldots, n, \ \alpha \neq \beta. \tag{18}$$

Unlike the prototypical profit-maximizing and cost-minimizing models of a firm, the reciprocity conditions do not take the form of the equality of cross-partial derivatives. Nonetheless, seeing as $x_j^N(\mathbf{d}) > d_j$, $j = 1, 2, \ldots, n$, by assumption (A2), the ensuing corollary is also immediate.

Corollary 2. *Under assumptions* (A1)–(A3), *a Nash bargaining solution* $\mathbf{x} = \mathbf{x}^N(\mathbf{d})$ *of problem* (1) *satisfies* $\text{sign}[\partial x_\alpha^N(\mathbf{d})/\partial d_\beta] = \text{sign}[\partial x_\beta^N(\mathbf{d})/\partial d_\alpha], \alpha, \beta = 1, 2, \ldots, n, \alpha \neq \beta.$

Corollary 2 demonstrates that the symmetry of $\boldsymbol{\Omega}(\mathbf{d})$ implies that all symmetric pairs of cross-partial derivatives of a Nash bargaining solution have the same sign. This property, however, does not imply that the off-diagonal terms of $\boldsymbol{\Omega}(\mathbf{d})$ are either all positive or all negative, nor that they are equal in magnitude. For example, it is possible that $\partial x_1^N(\mathbf{d})/\partial d_2 > 0$ and $\partial x_2^N(\mathbf{d})/\partial d_1 > 0$, while $\partial x_1^N(\mathbf{d})/\partial d_3 < 0$ and $\partial x_3^N(\mathbf{d})/\partial d_1 < 0$. It therefore follows from Corollary 2 that a Nash bargaining solution is not intrinsically locally strongly d-monotonic. Note that these symmetry results are also heretofore unknown differential properties of a Nash bargaining solution, as neither Thomson [1987] nor Engwerda and Douven [2008] investigated such matters.

In concluding the section, it is useful to briefly consider the case of two agents, i.e., $n = 2$. First recall that $g(\mathbf{x}^N(\mathbf{d})) \equiv 0$ for all $\mathbf{d} \in D$ by Lemma 1, from which it follows that

$$\frac{\partial x_2^N(\mathbf{d})}{\partial d_1} \equiv -\frac{g_1(\mathbf{x}^N(\mathbf{d}))}{g_2(\mathbf{x}^N(\mathbf{d}))} \frac{\partial x_1^N(\mathbf{d})}{\partial d_1} \leq 0, \tag{19}$$

$$\frac{\partial x_1^N(\mathbf{d})}{\partial d_2} \equiv -\frac{g_2(\mathbf{x}^N(\mathbf{d}))}{g_1(\mathbf{x}^N(\mathbf{d}))} \frac{\partial x_2^N(\mathbf{d})}{\partial d_2} \leq 0, \tag{20}$$

using Corollary 1. In view of the fact that $\Omega_{12}(\mathbf{d}) \equiv \Omega_{21}(\mathbf{d})$, it follows from Eq. (18) that

$$\frac{[x_2^N(\mathbf{d}) - d_2]}{[x_1^N(\mathbf{d}) - d_1]} \frac{\partial x_1^N(\mathbf{d})}{\partial d_2} = \frac{[x_1^N(\mathbf{d}) - d_1]}{[x_2^N(\mathbf{d}) - d_2]} \frac{\partial x_2^N(\mathbf{d})}{\partial d_1}. \tag{21}$$

Finally, note that the first-order necessary conditions given by Eq. (3) imply that

$$\frac{[x_2^N(\mathbf{d}) - d_2]}{[x_1^N(\mathbf{d}) - d_1]} = \frac{g_1(\mathbf{x}^N(\mathbf{d}))}{g_2(\mathbf{x}^N(\mathbf{d}))}. \tag{22}$$

The ensuing basic result then follows from Eqs. (19)–(22).

Corollary 3. *Under assumptions* (A1)–(A3) *and* $n = 2$, *a Nash bargaining solution of problem* (1) *satisfies* $\partial x_1^N(\mathbf{d})/\partial d_1 \equiv \partial x_2^N(\mathbf{d})/\partial d_2 \geq 0$, $\partial x_2^N(\mathbf{d})/\partial d_1 \equiv [\frac{g_1(\mathbf{x}^N(\mathbf{d}))}{g_2(\mathbf{x}^N(\mathbf{d}))}]^2 \partial x_1^N(\mathbf{d})/\partial d_2 \leq 0$, *and* $[\partial x_1^N(\mathbf{d})/\partial d_1][\partial x_2^N(\mathbf{d})/\partial d_2] - [\partial x_1^N(\mathbf{d})/\partial d_2][\partial x_2^N(\mathbf{d})/\partial d_1] \equiv 0$.

Corollary 3 asserts that when there are two agents, a fundamental property of a Nash bargaining solution is that it is locally strongly d-monotonic, a result that has been noted by Engwerda and Douven [2008, p. 271], for they wrote that "In the two-player case, the Nash solution is always strongly d-monotonic." But Corollary 3 also asserts three previously unknown properties that are intrinsic to a Nash bargaining solution in the two-agent setting. The first is that the own disagreement point effects are quantitatively identical for each agent. The second is that the cross disagreement point effects obey a rather simple reciprocity condition that depends only on the slope of the Pareto frontier. And the third is that the product of the own disagreement point effects is identical to the product of the cross disagreement point effects. It also follows from Corollary 3 that $\partial x_2^N(\mathbf{d})/\partial d_1 \equiv \partial x_1^N(\mathbf{d})/\partial d_2$ if and only if $g_1(\mathbf{x}^N(\mathbf{d})) \equiv g_2(\mathbf{x}^N(\mathbf{d}))$ or $x_2^N(\mathbf{d}) - d_2 \equiv x_1^N(\mathbf{d}) - d_1$.

4. Summary and Conclusion

The intrinsic comparative statics of a Nash bargaining solution have been success-fully derived using recent advances in the theory of differential comparative statics due to Partovi and Caputo [2006]. They take the form of a symmetric and positive semidefinite matrix whose maximum rank cannot exceed the number of agents less one. From this basic result it was shown that local d-monotonicity is an intrinsic property of a Nash bargaining solution and that symmetric pairs of cross disagree-ment point comparative statics have the same sign. On the other hand, it was shown that a Nash bargaining solution is not intrinsically locally strongly d-monotonic, in general. These results complement and extend the work of Thomson [1987] and Engwerda and Douven [2008].

In the case of two agents, stronger results were obtained, viz., the Nash bargain-ing solution is intrinsically locally strongly d-monotonic. Moreover, in the two-agent setting, it was shown that own disagreement point effects are quantitatively iden-tical for each agent and that the cross disagreement point effects obey a simple reciprocity condition that depends only on the slope of the Pareto frontier. A natural extension of the present work would investigate the same basic matters for other types of bargaining solutions.

References

Chun, Y. and Thomson, W. [1988] Monotonicity properties of bargaining solutions when applied to economics, *Math. Social Sci.* **15**, 11–27.

Engwerda, J. C. and Douven, R. C. [2008] On the sensitivity matrix of the Nash bargaining solution, *Int. J. Game Theory* **37**, 265–279.

Kaneko, M. [1980] An extension of the Nash bargaining problem and the Nash social welfare function, *Theory Decis.* **12**, 135–148.

Nash, J. F. [1950] "The bargaining problem, *Econometrica* **18**, 155–162.

Partovi, M. H. and Caputo, M. R. [2006] A complete theory of comparative statics for differentiable optimization problems, *Metroeconomica* **57**, 31–67.

Partovi, M. H. and Caputo, M. R. [2007] Erratum: A complete theory of comparative statics for differentiable optimization problems, *Metroeconomica* **58**, 360.

Silberberg, E. [1990] *The Structure of Economics: A Mathematical Analysis*, 2nd edition (McGraw-Hill, New York).

Thomson, W. [1987] Monotonicity of bargaining solutions with respect to the disagreement point, *J. Econ. Theory* **42**, 50–58.

Thomson, W. [2009] Bargaining and the theory of cooperative games: John Nash and beyond, Rochester Center for Economic Research, Working Paper 554.

Xu, Y. and Yoshihara, N. [2006] Alternative characterizations of three bargaining solutions for nonconvex problems, *Games Econ. Behav.* **57**, 86–92.

Chapter 24

Optimal Fair Division for Measures with Piecewise Linear Density Functions

Jerzy Legut

Department of Mathematics
Wrocław University of Technology
Wrocław, Poland

jerzy.legut@pwr.wroc.pl

A nonlinear programming method is used for finding an optimal fair division of the unit interval $[0, 1)$ among n players. Preferences of players are described by nonatomic probability measures $\mu_1, \ldots, \mu_n$ with piecewise linear (PWL) density functions. The presented algorithm can be applied for obtaining "almost" optimal fair divisions for measures with arbitrary density functions approximable by PWL functions. The number of cuts needed for obtaining such divisions is given.

Keywords: Fair division; cake cutting; optimal partitioning of a measurable space.

1. Introduction

Suppose we are given a cake to be divided among n players. Let measurable space $\{[0, 1), \mathcal{B}\}$ represent the cake and nonatomic probability measures $\mu_1, \ldots, \mu_n$ defined on the σ-algebra $\mathcal{B}$ of Borel measurable sets describe individual preferences of each player. The measures $\mu_1, \ldots, \mu_n$ are used by the players to evaluate the size of sets $A \in \mathcal{B}$. Denote by $I = \{1, 2, \ldots, n\}$ the set of numbered players. By an ordered partition $P = \{A_i\}_{i=1}^n$ of the cake among the players $i \in I$ is meant a collection of $\mathcal{B}$-measurable disjoint subsets $A_1, \ldots, A_n$ of $[0, 1)$ whose union is equal to $[0, 1)$. Let $\mathbb{P}$ stand for the set of all measurable partitions $P = \{A_i\}_{i=1}^n$ of $[0, 1)$.

The problem of fair division of the cake is the task to divide $[0, 1)$ among the players $i \in I$ in a way that would be "fair" according to some fairness notion accepted by all players. In classic fair division problem we are interested in giving the ith person a set $A_i \in \mathcal{B}$ such that $\mu_i(A_i) \geq 1/n$ for $i \in I$. A simple and well-known method for realizing a fair division (of a cake) for two players is "for one to cut, the other to choose". Steinhaus in 1944 asked whether the fair procedure could be found for dividing a cake among n participants for $n > 2$. He found a solution for $n = 3$ and Banach and Knaster (cf. Knaster [1946]) showed that the solution for $n = 2$ could be extended to arbitrary n. An interesting challenge in fair division theory is to find the best partition $P = \{A_i\}_{i=1}^n \in \mathbb{P}$ satisfying $\mu_i(A_i) = \mu_j(A_j)$ for all $i, j \in I$.

Definition 1. A partition $P^* = \{A_i^*\}_{i=1}^n \in \mathbb{P}$ is said to be an *optimal fair division* if

$$v := \min_{i \in I}[\mu_i(A_i^*)] = \sup_{P \in \mathbb{P}} \min_{i \in I}[\mu_i(A_i)]. \tag{1}$$

The number v will be called as an *optimal value* of the fair division problem. We will also call the optimal fair division interchangeably as the *optimal partition* of a measurable space $\{[0, 1), \mathcal{B}\}$. The existence of optimal partitions follows from the theorem of Dvoretzky *et al.* [1951].

Theorem 1. *If $\mu_1, \mu_2, \ldots, \mu_n$ are nonatomic countably additive finite measures defined on the measurable space $\{\mathcal{X}, \mathcal{B}\}$ then the range $\boldsymbol{\mu}(\mathbb{P})$ of the mapping $\boldsymbol{\mu}$: $\mathbb{P} \to \mathbb{R}^n$ defined by*

$$\boldsymbol{\mu}(P) = (\mu_1(A_1), \ldots, \mu_n(A_n)), \quad P = \{A_i\}_{i=1}^n \in \mathbb{P}$$

is convex and compact in $\mathbb{R}^n$.

Unfortunately finding optimal partitions for given measures in general case is not easy. In the literature of the fair division field there are known only few results concerning algorithms for finding optimal partitions. Legut and Wilczyński [2012] showed how to obtain optimal partitions for two players. In turn Dall'Aglio *et al.* [2015] presented an algorithm for finding an optimal partition in case of measures with simple density functions. There are many results for the estimation of the optimal value v. The first estimation of the number was given by Elton *et al.* [1986] and further by Legut [1988]. An interesting algorithm for finding the bounds for the optimal value was found by Dall'Aglio and Di Luca [2015].

A general form of the optimal partitions could be helpful in some cases for finding constructive methods of optimal partitioning of a measurable space. Let $S = \{s = (s_1, \ldots, s_n) \in \mathbb{R}^n, s_i \geq 0, i \in I, \sum_{i=1}^n s_i = 1\}$ be the $(n-1)$-dimensional simplex. We assume that all nonatomic measures $\mu_1, \mu_2, \ldots, \mu_n$ are absolutely continuous with respect to the same measure ν (e.g., $\nu = \sum_{i=1}^n \mu_i$). Denote by $f_i = d\mu_i/d\nu$ the Radon–Nikodym derivatives, i.e.,

$$\mu_i(A) = \int_A f_i \, d\nu, \quad \text{for } A \in \mathcal{B} \text{ and } i \in I.$$

For $p = (p_1, \ldots, p_n) \in S$ and $i \in I$ define the following measurable sets:

$$B_i(p) = \bigcap_{j=1, j \neq i}^{n} \{x \in [0, 1) : p_i f_i(x) > p_j f_j(x)\},$$

$$C_i(p) = \bigcap_{j=1}^{n} \{x \in [0, 1) : p_i f_i(x) \geq p_j f_j(x)\}.$$

Legut and Wilczyński [1988] using a minmax theorem of Sion (cf. Aubin [1980]) proved the following.

Theorem 2. *There exists a point $p^* \in S$ and a corresponding optimal partition $P^* = \{A_i^*\}_{i=1}^n$ satisfying*

(i) $B_i(p^*) \subset A_i^* \subset C_i(p^*)$,
(ii) $\mu_1(A_1^*) = \mu_2(A_2^*) = \cdots = \mu_n(A_n^*)$

Moreover, any partition $P^* = \{A_i^*\}_{i=1}^n$ which satisfies (i) *and* (ii) *is optimal.*

Presented here in less general form Theorem 2 describes the basic structure of the optimal partitions.

2. The Main Result

In this section, we show how to obtain an optimal partition for measures with piecewise linear (PWL) density functions. Suppose we are given n PWL functions defined on the unit interval $[0, 1)$

$$f_i(x) = \sum_{j=1}^{m} (c_{ij} x + d_{ij}) I_{[a_j, a_{j+1})}(x), \quad \int_0^1 f_i(x) \, dx = 1, \quad i \in I,$$

where $\{[a_j, a_{j+1})\}_{j=1}^m$ is a partition of the interval $[0, 1)$ such that

$$[0, 1) = \bigcup_{j=1}^{m} [a_j, a_{j+1}), \quad a_1 = 0, \ a_{m+1} = 1, \ a_{j+1} > a_j, \ j = 1, \ldots, m. \tag{2}$$

By $I_A(x)$ we denote here the indicator of the set $A \in \mathcal{B}$.

We assume that

$$c_{ij} x + d_{ij} \geq 0 \quad \text{for all } x \in [a_j, a_{j+1}), \ i \in I, \ j = 1, \ldots, m.$$

From now on we consider nonatomic probability measures $\mu_1, \ldots, \mu_n$ given by

$$\mu_i(A) = \int_A f_i dx, \quad \text{for } A \in \mathcal{B}, \ i \in I. \tag{3}$$

Throughout this chapter and without loss of generality we consider only left side closed and right side open intervals unless they are otherwise defined. Consider partitions of each interval $[a_j, a_{j+1})$, $j = 1, \ldots, m$ into n subintervals by cuts in points $b_k^{(j)}$, $k = 1, \ldots, n-1, j = 1, \ldots, m$ such that

$$[a_j, a_{j+1}) = \bigcup_{k=1}^{n} [b_{k-1}^{(j)}, b_k^{(j)}),$$

where $b_0^{(j)} = a_j$, $b_n^{(j)} = a_{j+1}$, $b_{k+1}^{(j)} \geq b_k^{(j)}$, $k = 1, \ldots, n-1, j = 1, \ldots, m$.

If $b_{k-1}^{(j)} = b_k^{(j)}$ for some $k = 1, \ldots, n$ we put $[b_{k-1}^{(j)}, b_k^{(j)}) = \emptyset$.

For simplicity we will also denote $B_{kj} := [b_{k-1}^{(j)}, b_k^{(j)})$, $k = 1, \ldots, n, j = 1, \ldots, m$.

Now we construct an assignment of each subintervals B_{kj} to each player $i \in I$.

Let p_j, q_j, $j = 1, \ldots, m$ be integers satisfying $0 \leq p_j \leq q_j \leq n$ and

$$\#\{i : i \in I, c_{ij} < 0\} = p_j,$$

$$\#\{i : i \in I, c_{ij} = 0\} = q_j - p_j,$$

$$\#\{i : i \in I, c_{ij} > 0\} = n - q_j,$$

where by $\#A$ we denote the number of elements of a finite set A.

For each interval $[a_j, a_{j+1})$, $j = 1, \ldots, m$, we define permutations $\sigma_j : I \to I, j = 1, \ldots, m$ satisfying the following conditions:

(1) If $p_j > 0$ we define $\sigma_j(k) \in \{i : i \in I, c_{ij} < 0\}$ for $k = 1, \ldots, p_j$ such that

$$\frac{d_{\sigma_j(k)j}}{c_{\sigma_j(k)j}} \geq \frac{d_{\sigma_j(k+1)j}}{c_{\sigma_j(k+1)j}}, \quad k = 1, \ldots, p_j - 1. \tag{4}$$

(2) If $q_j - p_j > 0$ we define $\sigma_j(k) \in \{i : i \in I, c_{ij} = 0\}$ for $k = p_j + 1, \ldots, q_j$ such that

$$\sigma_j(k) \leq \sigma_j(k+1), \quad k = p_j + 1, \ldots, q_j - 1. \tag{5}$$

(3) If $n - q_j > 0$ we define $\sigma_j(k) \in \{i : i \in I, c_{ij} > 0\}$ for $k = q_j + 1, \ldots, n$ such that

$$\frac{d_{\sigma_j(k)j}}{c_{\sigma_j(k)j}} \geq \frac{d_{\sigma_j(k+1)j}}{c_{\sigma_j(k+1)j}}, \quad k = q_j + 1, \ldots, n - 1. \tag{6}$$

Permutations σ_j, $j = 1, \ldots, m$ define one-to-one assignment of the subintervals $B_{ij} \subset [a_j, a_{j+1})$, $i \in I, j = 1, \ldots, m$ such that player $i \in I$ receives subinterval $B_{\sigma_j^{-1}(i)j}$. Finally we obtain a partition $\{B_i\}_{i=1}^n$ of the unit interval defined by

$$B_i = \bigcup_{j=1}^{m} B_{\sigma_j^{-1}(i)j}, \quad i \in I.$$

The following theorem presents an algorithm for obtaining an optimal fair division.

Theorem 3. *Let the collection of numbers* $z^*, \{c_k^{(j)}\}, k = 1, \ldots, n-1, j = 1, \ldots, m$ *be a solution of the following nonlinear programming* (NLP) *problem*

$$\max z \tag{7}$$

subject to quadratic constraints

$$z = \sum_{j=1}^{m} \mu_i(B_{\sigma_j^{-1}(i)j}) = \sum_{j=1}^{m} \int_{B_{\sigma_j^{-1}(i)j}} f_i dx, \quad i = 1, \ldots, n,$$

with respect to variables $z, \{b_k^{(j)}\}$ $k = 1, \ldots, n-1, j = 1, \ldots, m$ *satisfying the following inequalities*

$$0 = a_1 \le b_1^{(1)} \le \cdots \le b_{n-1}^{(1)} \le a_2,$$

$$a_2 \le b_1^{(2)} \le \cdots \le b_{n-1}^{(2)} \le a_3,$$

$$\vdots$$

$$a_m \le b_1^{(m)} \le \cdots \le b_{n-1}^{(m)} \le a_{m+1} = 1.$$

Then, the partition $\{C_i\}_{i=1}^n$ *of the unit interval* $[0, 1)$ *defined by*

$$C_i = \bigcup_{j=1}^{m} C_{\sigma_j^{-1}(i)j}, \quad i \in I, \tag{8}$$

where

$$C_{\sigma_j^{-1}(i)j} = [c_{\sigma_j^{-1}(i)-1}^{(j)}, c_{\sigma_j^{-1}(i)}^{(j)}), \tag{9}$$

and $c_0^{(j)} = a_j$, $c_n^{(j)} = a_{j+1}$, $j = 1, \ldots, m$ *is an optimal fair division and* z^* *is the optimal value.*

To prove Theorem 3 we need first to show three lemmas.

Lemma 1. *Assume we are given two non-negative linear functions* $g_i(x) = c_i x + d_i$, $i = 1, 2$ *defined on an interval* $[a, b), 0 \le a < b \le 1$. *Assume that*

$$c_1 c_2 > 0, \quad \text{and} \quad \frac{d_1}{c_1} < \frac{d_2}{c_2}.$$

Then, for any number $s \in (a, b)$ *there exists a number* $t \in (a, b)$ *such that*

$$\int_a^s g_1(x)\, dx < \int_t^b g_1(x)\, dx, \quad \int_s^b g_2(x)\, dx < \int_a^t g_2(x)\, dx. \tag{10}$$

Moreover if $\dfrac{d_1}{c_1} = \dfrac{d_2}{c_2}$ *then for any* $s \in (a, b)$ *we can find* $t \in (a, b)$ *such that*

$$\int_a^s g_1(x)\, dx = \int_t^b g_1(x)\, dx, \quad \int_s^b g_2(x)\, dx = \int_a^t g_2(x)\, dx. \tag{11}$$

Proof. Assume first that $c_i < 0$, $i = 1, 2$. Consider the following equations with respect to the variables $s, t \in (a, b)$

$$\int_a^s g_i(x)\, dx = \int_t^b g_i(x)\, dx, \quad i = 1, 2. \tag{12}$$

After easy calculations we obtain the circle equations

$$\left(s + \frac{d_i}{c_i}\right)^2 + \left(t + \frac{d_i}{c_i}\right)^2 = \left(a + \frac{d_i}{c_i}\right)^2 + \left(b + \frac{d_i}{c_i}\right)^2, \quad i = 1, 2, \tag{13}$$

with respect to the variables $s, t \in (a, b)$. Let $h_i : [a, b] \to [a, b], i = 1, 2$ denote functions obtained from Eq. (13) by expressing variable t in terms of variable s. The explicit formulas of these functions are not important for our considerations. It is easy to see that two circles (13) intersect in two points with coordinates $(a, b) = (a, h_i(a))$ and $(b, a) = (b, h_i(b))$, $i = 1, 2$. It follows from our assumptions that $c_2 b + d_2 \geq 0$ and $d_i \geq 0$, $i = 1, 2$. Since

$$b \leq -\frac{d_2}{c_2} < -\frac{d_1}{c_1},$$

the radius of the first circle is greater than the radius of the second one. It means that the arc of the first circle connecting points (a, b) and (b, a) lies above the arc of the second one. Therefore, both functions h_i, $i = 1, 2$, are convex and

$$h_1(s) > h_2(s), \quad \text{for all } s \in (a, b). \tag{14}$$

Let $s \in (a, b)$ and $t_i = h_i(s)$, $i = 1, 2$. Thus $t_2 < t_1$ and

$$\int_a^s g_i(x)\, dx = \int_{t_i}^b g_i(x)\, dx, \quad i = 1, 2. \tag{15}$$

Hence, for any $t \in (t_2, t_1)$ the following hold

$$\int_a^s g_1(x)\, dx = \int_{t_1}^b g_1(x)\, dx < \int_t^b g_1(x)\, dx$$

and the first inequality of (10) is proved. It follows from (15) for $i = 2$ that

$$\int_s^b g_2(x)\, dx = \int_a^b g_2(x)\, dx - \int_a^s g_2(x)\, dx$$

$$= \int_a^b g_2(x)\, dx - \int_{t_2}^b g_2(x)\, dx = \int_a^{t_2} g_2(x)\, dx.$$

Then we obtain the second inequality of (10)

$$\int_s^b g_2(x)\, dx < \int_a^t g_2(x)\, dx,$$

where $t \in (t_2, t_1)$.

In case of $\frac{d_1}{c_1} = \frac{d_2}{c_2}$ we have $h_1(s) = h_2(s)$ for all $s \in [a, b]$ and then $t_2 = t_1$. The inequalities (11) follow immediately from (15).

In case of $c_i > 0$, $i = 1, 2$ the analogous arguments prove the inequalities (10) and (11). The only difference is that instead of convex we define concave functions $h_i : [a, b] \to [a, b], i = 1, 2$ for which the inequality (14) holds. $\square$

Lemma 2. *Let ν_i, $i = 1, 2$ be measures defined on the measurable subsets of an interval $[a, b), 0 \le a < b \le 1$ as follows*

$$\nu_i(A) = \int_A g_i dx, \quad \text{for } A \subset [a, b), \ i = 1, 2,$$

where $g_i(x) = c_i x + d_i$, $i = 1, 2$ are non-negative linear functions defined on the interval $[a, b]$. Assume that one of the following cases is met

(1) $c_1 c_2 > 0$, $i = 1, 2$ *and* $\frac{d_1}{c_1} < \frac{d_2}{c_2}$,
(2) $c_1 > 0$, $c_2 < 0$,
(3) $c_1 > 0$, $c_2 = 0$,
(4) $c_1 = 0$, $c_2 < 0$.

Then, for any $s \in (a, b)$ there exists $t \in (a, b)$ such that

$$\nu_1([a, s)) < \nu_1([t, b)), \quad \nu_2([s, b)) < \nu_2([a, t)). \tag{16}$$

Moreover if $\frac{d_1}{c_1} = \frac{d_2}{c_2}$ or $c_1 = 0$, $c_2 = 0$ the following equalities are satisfied

$$\nu_1([a, s)) = \nu_1([t, b)), \quad \nu_2([s, b)) = \nu_2([a, t)). \tag{17}$$

Proof. For the first case the inequality (16) follows immediately from Lemma 1. It is not difficult to check the inequalities (16) for the cases 2, 3 and 4. The equalities (17) can be easily concluded from Lemma 1. $\square$

Lemma 3. *Suppose that for fixed j, $j = 1, \ldots, m$ we are given a partition of the interval $[a_j, a_{j+1})$ into n measurable sets A_{ij}. We assume that each set A_{ij} is a finite union of some disjoint subintervals belonging to $[a_j, a_{j+1})$ and*

$$[a_j, a_{j+1}) = \bigcup_{i \in I} A_{ij}.$$

Then, there exists a partition $\{[b_{k-1}^{(j)}, b_k^{(j)})\}_{k=1}^n$, with $b_0^{(j)} = a_j$, $b_n^{(j)} = a_{j+1}$ of the interval $[a_j, a_{j+1})$ and a permutation $\sigma_j : I \to I$, $j = 1, \ldots, m$ satisfying three conditions (4)–(6) and the following inequalities

$$\mu_i\left([b_{\sigma_j^{-1}(i)-1}^{(j)}, b_{\sigma_j^{-1}(i)}^{(j)})\right) \ge \mu_i(A_{ij}), \quad i \in I,$$

where μ_i, $i \in I$ denote measures defined by (3).

Proof. Suppose that an interval $[a, s) \subset A_{i_1 j}$ and $[s, b) \subset A_{i_2 j}$ for players $i_1, i_2 \in I$, where $a_j \le a < s < b \le a_{j+1}$. Denote

$$d_1 = d_{i_1 j}, \ d_2 = d_{i_2 j}, \ c_1 = c_{i_1 j}, \ c_2 = c_{i_2 j}, \tag{18}$$

and

$$\nu_1 = \mu_{i_1}, \quad \nu_2 = \mu_{i_2}.$$

Now we apply Lemma 2 for measures ν_i, $i = 1, 2$ and numbers (18). For each of the cases 1–4 from Lemma 2 applied for numbers (18) we can find number $t \in (a, b)$ such that

$$\nu_1([a, s)) < \nu_1([t, b)), \quad \nu_2([s, b)) < \nu_2([a, t)). \tag{19}$$

In case of

$$c_1 = 0, \quad c_2 = 0 \quad \text{for } i_1 > i_2 \tag{20}$$

we put $t = a + b - s$ and the equalities (17) hold. Now we define sets $A'_{i_1 j}$, $A'_{i_2 j}$ by the following modification of the sets $A_{i_1 j}, A_{i_2 j}$:

$$A'_{i_1 j} = (A_{i_1 j} \setminus [a, s)) \cup [t, b), \quad A'_{i_2 j} = (A_{i_1 j} \setminus [s, b)) \cup [a, t).$$

It follows from the inequalities (19) and (17) for the case (20) that after the modification the evaluations of sets $A'_{i_1 j}, A'_{i_2 j}$ by the players i_1, i_2 are not decreased

$$\mu_{i_1}(A'_{i_1 j}) \geq \nu_{i_1}(A_{i_1 j}), \quad \mu_{i_2}(A'_{i_2 j}) \geq \nu_{i_2}(A_{i_1 j}).$$

Now for any two players $i_1, i_2 \in I$ we modify sets $A_{i_1 j}, A_{i_2 j}$ for all neighboring intervals for which the cases 1–4 from Lemma 2 and the case (20) is met. In case of $\frac{d_{i_1 j}}{c_{i_1 j}} = \frac{d_{i_2 j}}{c_{i_2 j}}$ the modifications are not unique, then we modify sets A_{ij}, $i \in I$ in such a way to obtain sets B_{ij}, $i \in I$ which are unions of neighboring intervals and can be denoted by

$$B_{ij} = [b^{(j)}_{\sigma_j^{-1}(i)-1}, \quad b^{(j)}_{\sigma_j^{-1}(i)}), \quad i \in I, \ j = 1, \ldots, m,$$

where $\sigma_j : I \to I$, $j = 1, \ldots, m$ is the permutation satisfying (4)–(6) with $b^{(j)}_0 = a_j$, $b^{(j)}_n = a_{j+1}$. Because in each step of modifications the measures of modified sets do not decrease, we have

$$\mu_i([b^{(j)}_{\sigma_j^{-1}(i)-1}, b^{(j)}_{\sigma_j^{-1}(i)})) \geq \mu_i(A_{ij}), \quad i \in I,$$

and the proof of Lemma 3 is completed. □

Now we can prove Theorem 3.

Proof of Theorem 3. Let a partition $\{C_i\}_{i=1}^n$ be defined by (8) and (9). Suppose $\{A_i\}_{i=1}^n$ is an optimal fair division satisfying

$$\min_{i \in I}[\mu_i(A_i)] > \min_{i \in I}[\mu_i(C_i)]. \tag{21}$$

Note that, for any $p = (p_1, \ldots, p_n) \in S$ and PWL density functions f_i, $i \in I$, sets $\{x \in [0, 1) : p_i f_i(x) > p_j f_j(x)\}$ and $\{x \in [0, 1) : p_i f_i(x) \geq p_j f_j(x)\}$ are unions of finite number of intervals. It follows from Theorem 2 that each A_i also must be a union of finite number of intervals. Let $A_{ij} = A_i \cap [a_j, a_{j+1})$, $i \in I, j = 1, \ldots, m$. We conclude from Lemma 3 applied to sets A_{ij} that there exists a partition

$\{[b_{k-1}^{(j)}, b_k^{(j)})\}_{k=1}^n$, with $b_0^{(j)} = a_j$, $b_n^{(j)} = a_{j+1}$ of the interval $[a_j, a_{j+1})$ and a permutation $\sigma_j : I \to I$, $j = 1, \ldots, m$ satisfying three conditions (4)–(6) and the following inequalities

$$\mu_i([b_{\sigma_j^{-1}(i)-1}^{(j)}, b_{\sigma_j^{-1}(i)}^{(j)})) \geq \mu_i(A_{ij}), \quad i \in I.$$

Hence from (21) we have

$$\sum_{j=1}^m \mu_i(B_{\sigma_j^{-1}(i)j}) \geq \mu_i(A_i) > \mu_i(C_i), \quad i \in I, \tag{22}$$

where $B_{\sigma_j^{-1}(i)j} = [b_{\sigma_j^{-1}(i)-1}^{(j)}, b_{\sigma_j^{-1}(i)}^{(j)})$. The inequalities (22) contradict assumption (21), since the collection of numbers $\{c_k^{(j)}\}$, $k = 1, \ldots, n-1$, $j = 1, \ldots, m$ is a solution of the NLP problem (7). This contradiction proves that the partition $\{C_i\}_{i=1}^n$ is an optimal fair division. It follows immediately from Definition 1 that the number z^* is the optimal value. $\qquad\square$

Now we present an example illustrating the algorithm described in the proof of Theorem 3.

Example 1. Consider problem of fair division for three players. Assume that each player $i = 1, 2, 3$ estimates the measurable subsets of the unit interval $[0, 1)$ using measures μ_i, $i = 1, 2, 3$ defined respectively by density functions

$$f_i(x) = \sum_{j=1}^3 (c_{ij}x + d_{ij})I_{[a_j, a_{j+1})}(x), \quad i = 1, 2, 3,$$

where the numbers c_{ij} and d_{ij} are elements of the matrices

$$[c_{ij}] = \begin{bmatrix} -2 & 1 & -1 \\ -1 & 1 & 1 \\ -\frac{1}{2} & 0 & -2 \end{bmatrix}, \quad [d_{ij}] = \begin{bmatrix} 2 & 0 & \frac{5}{4} \\ \frac{5}{4} & \frac{1}{4} & \frac{1}{4} \\ \frac{1}{2} & \frac{7}{4} & \frac{13}{4} \end{bmatrix},$$

and $a_1 = 0$, $a_2 = \frac{1}{2}$, $a_3 = \frac{3}{4}$, $a_4 = 1$. Now we construct an optimal fair division. For $c_{ij} \neq 0$ define new matrix $[e_{ij}]$ with elements $e_{ij} = \frac{d_{ij}}{c_{ij}}$ (the element e_{32} is not defined)

$$[e_{ij}] = \begin{bmatrix} -1 & 0 & -\frac{5}{4} \\ -\frac{5}{4} & \frac{1}{4} & \frac{1}{4} \\ -1 & & -\frac{13}{8} \end{bmatrix}.$$

Divide the players into three groups depending on the sign of c_{ij} separately on each interval:

$$[0, \tfrac{1}{2}) : \{1, 2, 3\}, \{\emptyset\}, \{\emptyset\},$$
$$[\tfrac{1}{2}, \tfrac{3}{4}) : \{\emptyset\}, \{3\}, \{1, 2\},$$
$$[\tfrac{3}{4}, 1) : \{1, 3\}, \{\emptyset\}, \{2\},$$

where in the first group we have $c_{ij} < 0$, in the second $c_{ij} = 0$ and in the third one $c_{ij} > 0$. Analyzing the columns of the matrix $[e_{ij}]$ we define permutations $\sigma_j : \{1, 2, 3\} \to \{1, 2, 3\}, j = 1, 2, 3$ satisfying conditions (4)–(6). Ranking players according to the ratio e_{ij} in each of the three groups we obtain

$$[0, \tfrac{1}{2}) : \frac{d_{11}}{c_{11}} = \frac{d_{31}}{c_{31}} > \frac{d_{21}}{c_{21}} \text{ and } \sigma_1(1) = 1, \sigma_1(2) = 3, \sigma_1(3) = 2,$$
$$[\tfrac{1}{2}, \tfrac{3}{4}) : \frac{d_{22}}{c_{22}} > \frac{d_{12}}{c_{12}} \text{ and } \sigma_2(1) = 3, \sigma_2(2) = 2, \sigma_2(3) = 1,$$
$$[\tfrac{3}{4}, 1) : \frac{d_{13}}{c_{13}} > \frac{d_{33}}{c_{33}} \text{ and } \sigma_3(1) = 1, \sigma_3(2) = 3, \sigma_3(3) = 2.$$

(Because of the equality $\frac{d_{11}}{c_{11}} = \frac{d_{31}}{c_{31}}$ we can also alternatively define $\sigma_1(1) = 2, \sigma_1(2) = 1, \sigma_1(3) = 2$). Permutations $\sigma_j, j = 1, 2, 3$ establish the assignment of subintervals $\{[b_{k-1}^{(j)}, b_k^{(j)})\}_{k=1}^{3}, (b_0^{(j)} = a_j, b_n^{(j)} = a_{j+1})$ to each player $i = 1, 2, 3$ such that

player $i = 1$ receives set $B_1 = [0, b_1^{(1)}) \cup [b_2^{(2)}, \tfrac{3}{4}) \cup [\tfrac{3}{4}, b_1^{(3)})$,

player $i = 2$ receives set $B_2 = [b_2^{(1)}, \tfrac{1}{2}) \cup [b_1^{(2)}, b_2^{(2)}) \cup [b_2^{(3)}, 1)$ and

player $i = 3$ receives set $B_3 = [b_1^{(1)}, b_2^{(1)}) \cup [\tfrac{1}{2}, b_1^{(2)}) \cup [b_1^{(3)}, b_2^{(3)})$.

Formulate the NLP problem (7) as follows

$$\max z$$

subject to constraints

$$z = \int_0^{b_1^{(1)}} f_1(x)dx + \int_{b_2^{(2)}}^{3/4} f_1(x)dx + \int_{3/4}^{b_1^{(3)}} f_1(x)dx$$

$$= -\frac{3}{8} + 2b_1^{(1)} - [b_1^{(1)}]^2 - \frac{[b_2^{(2)}]^2}{2} + \frac{5b_1^{(3)}}{4} - \frac{[b_1^{(3)}]^2}{2},$$

$$z = \int_{b_2^{(1)}}^{1/2} f_2(x)dx + \int_{b_1^{(2)}}^{b_2^{(2)}} f_2(x)dx + \int_{b_2^{(3)}}^{1} f_2(x)dx$$

$$= \frac{5}{4} - \frac{5b_2^{(1)}}{4} + \frac{[b_2^{(1)}]^2}{2} - \frac{b_1^{(2)}}{4} - \frac{[b_1^{(2)}]^2}{2} + \frac{b_2^{(2)}}{4} + \frac{[b_2^{(2)}]^2}{2} - \frac{b_2^{(3)}}{4} - \frac{[b_2^{(3)}]^2}{2},$$

$$z = \int_{b_1^{(1)}}^{b_2^{(1)}} f_3(x)dx + \int_{1/2}^{b_1^{(2)}} f_3(x)dx + \int_{b_1^{(3)}}^{b_2^{(3)}} f_3(x)dx$$

$$= -\frac{7}{8} - \frac{b_1^{(1)}}{2} + \frac{[b_1^{(1)}]^2}{4} + \frac{b_2^{(1)}}{2} - \frac{[b_2^{(1)}]^2}{4}$$

$$+ \frac{7b_1^{(2)}}{4} - \frac{13b_1^{(3)}}{4} + [b_1^{(3)}]^2 + \frac{13b_2^{(3)}}{4} - [b_2^{(3)}]^2,$$

with respect to the variables z, $\{b_k^{(j)}\}$ $k = 1, 2, j = 1, 2, 3$ and satisfying the inequalities

$$0 \le b_1^{(1)} \le b_2^{(1)} \le \frac{1}{2} \le b_1^{(2)} \le b_2^{(2)} \le \frac{3}{4} \le b_1^{(3)} \le b_2^{(3)} \le 1.$$

Using some software we get the following solution

$$z = 0.465276, \quad b_1^{(1)} = b_2^{(1)} = 0.26852,$$

$$b_1^{(2)} = b_2^{(2)} = b_1^{(3)} = 0.75 \quad \text{and} \quad b_2^{(3)} = 0.766019.$$

Hence the optimal value of the fair division problem $v = 0.465276$ and the optimal partition $\{B_1, B_2, B_3\}$ is given by $B_1 = [0, b_1^{(1)})$, $B_2 = [b_2^{(1)}, \frac{1}{2}) \cup [b_2^{(3)}, 1)$, $B_3 = [\frac{1}{2}, b_2^{(3)})$.

3. Final Remarks

(1) The algorithm of obtaining an optimal fair division described in Theorem 3 can be used for finding "almost" optimal partitions for measures with arbitrary density functions approximable by PWL functions.

(2) It follows from Theorem 3 that for obtaining an optimal division for measures with PWL density functions defined on m intervals we need at most $mn - 1$ cuts of the unit interval $[0, 1)$.

(3) The presented method can be easily generalized for finding α-optimal partitions defined by Legut and Wilczyński [1988].

Acknowledgment

The author is grateful for the comments and suggestions to improve this work.

References

Aubin, J. P. [1980] *Mathematical Methods of Game and Economic Theory* (North-Holland Publishing Company, Amsterdan).

Dall'Aglio, M. and Di Luca [2015] Bounds for α-optimal partitioning of a measurable space based on several efficient partitions, *J. Math. Anal. Appl.* **425**, 854–863.

Dall'Aglio, M., Legut, J. and Wilczyński, M. [2015] On finding optimal partitions of a measurable space, *Math. Appl.* **43**(2), 157–172.

Elton, J., Hill, T. and Kertz, R. [1986] Optimal partitioning inequalities for non-atomic probability measures, *Trans. Am. Math. Soc.* **296**, 703–725.

Knaster, B. [1946] Sur le probleme du partage pragmatique de H. Steinhaus, *Ann. Soc. Polon. Math.* **19**, 228–230.

Dvoretzky, A., Wald, A. and Wolfowitz, J. [1951] Relations among certain ranges of vector measures, *Pacific J. Math.* **1**, 59–74.

Legut, J. [1988] Inequalities for α-optimal partitioning of a measurable space, *Proc. Am. Math. Soc.* **104**, 1249–1251.

Legut, J. and Wilczyński, M. [2012] How to obtain a range of nonatomic vector measure in $\mathbb{R}^2$, *J. Math. Anal. Appl.* **394**, 102–111.

Legut, J. and Wilczyński, M. [1988] Optimal partitioning of a measurable space, *Proc. Am. Math. Soc.* **104**, 262–264.

Chapter 25

An Extension of the Solidarity Value for Environments with Externalities

Julio Rodríguez-Segura

Facultad de Economa, UASLP
Consejo Estatal de Población (COESPO)
San Luis Potosí, México

julio_98_12@hotmail.com

Joss Sánchez-Pérez[*]

Facultad de Economía, UASLP
Av. Pintores s/n, Col. B. del Estado 78213
San Luis Potosí, México

joss.sanchez@uaslp.mx

In this chapter, we propose an axiomatic extension for the Solidarity value of Nowak and Radzik [1994] A solidarity value for n-person transferable utility games, *Int. J. Game Theor.* **23**, 43–48] to the class of games with externalities. This value is characterized as the unique function that satisfies linearity, symmetry, efficiency and average nullity. In this context, we discuss a key subject of how to extend the concept of average marginal contribution to settings where externalities are present.

Keywords: Solidarity value; games with externalities; TU games; marginal contribution.

1. Introduction

The problem of how to fairly divide a surplus obtained through cooperation is one of the most fundamental issues studied in coalitional game theory and it is relevant to a wide range of economic and social situations. These issues are often difficult to resolve especially in environments with externalities, where the benefits of a group depends not only on its members, but also on the arrangement of agents outside the group. This is the general problem to which this chapter contributes. In this line, such problem was effectively modeled in Lucas and Thrall [1963] by the

---

[*]Corresponding author.

concept of games in partition function form: A partition function assigns a value to each pair consisting of a coalition and a coalition structure which includes that coalition. The advantage of this model is that it takes into account both internal factors (coalition itself) and external factors (coalition structure) that may affect cooperative outcomes and allows to go deeper into cooperation problems. Thus, it is closer to real-life situations but more complex to analyze.

To study solutions for the problem of fair distribution of the surplus generated by a group of people who are willing to cooperate with one another, one can take a normative approach (called axiomatic solutions or values). Given the coalitions and their sets of feasible payoffs as primitives, the question tackled is the identification of final payoffs awarded to each agent.

In the absence of externalities,[a] Shapley [1953] obtained a remarkable uniqueness result. It is characterized by two standard axioms and the *marginality* axiom [see Young, 1985], that is, if the marginal contributions of a player in two games are the same, then his value should be the same. Alternatively, it can be expressed by the *nullity* axiom, that is, if all the marginal contributions of a player in a game are zero, then the player should obtain a zero payoff.

Alternatively, Nowak and Radzik [1994] developed another coalitional value (called the Solidarity value) for TU games, where the rule followed to share the benefits among the players is less competitive[b] than the rule used in the Shapley value and it reflects some social behavior of players in coalitions. Such a value is based on the assumption that if a coalition S forms, then the players who contribute to S more than the average marginal contribution of a member to S support in some sense their weaker partners in S.

Values for games with externalities can be found in Myerson [1977], Bolger [1989], and more recently, Macho-Stadler *et al.* [2007], De Clippel and Serrano [2008], McQuillin [2009], Hu and Yang [2010], and Grabisch and Funaki [2012]. All of them are in some way extensions of the Shapley value for games with externalities.[c]

On the other hand, it is not easy to find literature related with extensions of the Solidarity value for environments with externalities. Hernández-Lamoneda *et al.* [2009] propose one extension and, as far as we know, there are no other proposals in the literature. These authors follow an axiomatic approach and characterize a value using an standard translation of the original axioms to games with externalities.[d]

In this chapter, we offer a solution concept that is conceptually an extension (for games with externalities) of the Solidarity value for TU games. Briefly, what we do is to study how the original axiomatization can be adapted to settings when externalities are considered. In this context, we prove that such extension is the unique

---

[a]Modelled by games in characteristic function form (TU games).
[b]It is very easy to find real-life examples where the groups formed seek to protect their weaker members by giving them a share of the gains obtained by the group.
[c]A complete characterization of a family of extensions of the Shapley value is presented in Sánchez-Pérez [2015].
[d]The precise definitions will be provided in Sec. 2.

solution that satisfies the axioms of linearity, symmetry, efficiency and average nullity. This value establishes a fair compromise between the coalitional effects and externality effects and therefore, well reflects the role of coalition structures in determining players' final payoffs whereas the value defined by Hernández-Lamoneda *et al.* [2009] ignores the externality side.

Considering an extension of the Solidarity value would allow us to enrich the analysis from a theoretical point of view and from a practical perspective. At the theoretical level it would be possible to formulate (considering the influence of coalition structures) new useful axioms to characterize solutions under the principle of average marginal contribution. In a practical sense, such an extension responds to situations in which, in addition to externalities, social and collective interests predominate.

Some examples of scenarios where it would be appropriate to apply a Solidarity value (for environments where externalities are present) are: Humanitarian aid to territories affected by a natural disaster, where the formation of blocks exogenous to them could damage support for the victims; Situations in which the distribution of some social benefit or resource indispensable for the quality of life tends to favor certain alliances for political convenience; Family saving funds where the way in which investors are organized outside of an alliance affects the profits of the same and where the savers assume a position of solidarity by not expecting that some of their relatives lose their patrimony.

Applying some extension of the Shapley value to situations similar to those mentioned above may not be the best alternative since agents receive payoffs according to their productivity and so, some of them may be excluded from any benefit or vital resource. Whereas, the solution proposed in this chapter is even more robust as it considers all possible ways in which external agents can organize themselves and assumes a social behavior of the agents.

The chapter is organized as follows. We first recall the main basic features of games in partition function form in the next section. A brief description of the axioms used in the characterization is provided in Sec. 3. The Solidarity value for games with externalities is characterized in an axiomatic way in Sec. 4. Finally, a brief comparison between our extension and related literature is provided in Sec. 5.

2. Framework and Conventions

Let $N = \{1, 2, \ldots, n\}$ be a fixed nonempty finite set, and let the members of N be interpreted as players in some game situation. Given N, PT denotes the set of partitions of N, so

$$\{S_1, S_2, \ldots, S_m\} \in \text{PT} \quad \text{iff} \bigcup_{i=1}^{m} S_i = N, \quad S_j \cap S_k = \varnothing \,\forall j \neq k.$$

By convention, $\{\varnothing\} \in Q$ for every $Q \in \text{PT}$. For $S \subseteq N$, $P(S)$ denotes que set of partitions that contains coalition S; i.e., $P(S) = \{Q \in \text{PT} \,|\, S \in Q\}$.

Also, let $\text{ECL} = \{(S, Q) \,|\, S \in Q \in \text{PT}\}$ be the set of *embedded coalitions*, that is the set of coalitions together with specifications as to how the other players are aligned. The embedded coalition (S, Q) is called nontrivial if $S \neq \varnothing$.

Definition 1. A game with externalities is a mapping

$$w : \text{ECL} \to \mathbb{R}$$

with the property that $w(\varnothing, Q) = 0$ for every $Q \in \text{PT}$. The set of games with externalities with player set N is denoted by G, i.e.,

$$G = G^{(n)} = \{w : \text{ECL} \to \mathbb{R} \,|\, w(\varnothing, Q) = 0 \,\forall Q \in \text{PT}\}.$$

The value $w(S, Q)$ represents the payoff of coalition S, given the coalition structure Q forms. In this kind of games, the worth of some coalition depends not only on what the players of such coalition can jointly obtain, but also on the way the other players are organized. We assume that, in any game situation, the universal coalition N (embedded in $\{N\}$) will actually form, so that the players will have $w(N, \{N\})$ to divide among themselves. But we also anticipate that the actual allocation of this worth will depend on all the other potential worths $w(S, Q)$, as they influence the relative bargaining strengths of the players.

For $Q \in \text{PT}$, $S \in Q$ and $i, k \in N$, we define $Q_{-S} = Q \backslash \{S\}$, $S_{-k} = S \backslash \{k\}$, $S_{+k} = S \cup \{k\}$, and Q^i denotes the member of Q to which i belongs. Additionally, we will denote the cardinality of a set by its corresponding lower-case letter, for instance $n = |N|$, $s = |S|$, $q = |Q|$, and so on.

Given $w_1, w_2 \in G$ and $c \in \mathbb{R}$, we define the sum $w_1 + w_2$ and the product cw_1, in G, in the usual form, i.e.,

$$(w_1 + w_2)(S, Q) = w_1(S, Q) + w_2(S, Q) \quad \text{and} \quad (cw_1)(S, Q) = cw_1(S, Q)$$

respectively. It is easy to verify that G is a vector space with these operations.

A *solution* is a function $\varphi : G \to \mathbb{R}^n$. If φ is a solution and $w \in G$, then we can interpret $\varphi_i(w)$ as the utility payoff whose player i should expect from the game w.

Now, the group of permutations of N, $S_n = \{\theta : N \to N \,|\, \theta \text{ is bijective}\}$, acts on 2^N and on ECL in the natural way; i.e., for $\theta \in S_n$:

$$\theta(S) = \{\theta(i) \,|\, i \in S\},$$

$$\theta(S_1, \{S_1, S_2, \ldots, S_l\}) = (\theta(S_1), \{\theta(S_1), \theta(S_2), \ldots, \theta(S_l)\}).$$

And also, S_n acts on the space of payoff vectors, $\mathbb{R}^n$:

$$\theta(x_1, x_2, \ldots, x_n) = (x_{\theta^{-1}(1)}, x_{\theta^{-1}(2)}, \ldots, x_{\theta^{-1}(n)}).$$

3. The Axioms

In the cooperative game theory framework, reasonable requirements to impose on a solution are the usual linearity, symmetry, and efficiency axioms. Next, we define them.

Axiom 1 (Linearity). *A solution φ is linear if $\varphi(w_1 + w_2) = \varphi(w_1) + \varphi(w_2)$ and $\varphi(cw_1) = c\varphi(w_1)$, for all $w_1, w_2 \in G$ and $c \in \mathbb{R}$.*

The axiom of linearity means that when a group of players shares the benefits (or costs) stemming from two different issues, how much each player obtains does not depend on whether they consider the two issues together or one by one. Hence, the agenda does not affect the final outcome. Also, the sharing does not depend on the unit used to measure the benefits.

Axiom 2 (Symmetry). *A solution φ is said to be symmetric if and only if $\varphi(\theta \cdot w) = \theta \cdot \varphi(w)$ for every $\theta \in S_n$ and $w \in G$, where the game $\theta \cdot w$ is defined as*

$$(\theta \cdot w)(S, Q) = w[\theta^{-1}(S, Q)].$$

Symmetry means that player's payoffs do not depend on their names. The payoff of a player is only derived from his influence on the worth of the coalitions.

Axiom 3 (Efficiency). *A solution φ is efficient if $\sum_{i \in N} \varphi_i(w) = w(N, \{N\})$ for all $w \in G$.*

We assume that the grand coalition forms and we leave issues of coalition formation out of this chapter. Efficiency then simply means that the value must be feasible and exhaust all the benefits from cooperation, given that everyone cooperates.

In order to present the last axiom, define, for any $(S, Q) \in ECL$ such that $S \neq \varnothing$ and any $w \in G$:

$$\overline{A}^w(S, Q) = \frac{1}{s} \sum_{j \in S} \frac{1}{|P(S_{-j})|} \sum_{Q' \in P(S_{-j})} [w(S, Q) - w(S_{-j}, Q')].$$

We can interpret $\overline{A}^w(S, Q)$ as an average marginal contribution[e] of a member of S, given the coalition structure Q. In the computation of $w(S, Q) - w(S_{-j}, Q')$, notice that Q' is a partition of N that reflects the dynamics of coalition formation (among players in $N \backslash S_{-j}$) once player j has left the main coalition S. Even when players in $N \backslash S_{-j}$ form a coalitional structure completely different from the original one (Q), the essence of the idea of marginal contribution remains. The reason relies from the fact that other agents' behavior could be a consequence of the departure of j from S; thus, this possibility must be taken into account.

Also note that if S contains only one player, then $\overline{A}^w(S, Q) = w(S, Q)$.

Definition 2. $i \in N$ is called a null player in the game w if $\overline{A}^w(S, Q) = 0$ for every $(S, Q) \in ECL$ such that $i \in S$.

Now, we give the following version of a nullity axiom for games with externalities:

Axiom 4 (Average nullity). *If $i \in N$ is a null player in $w \in G$, then $\varphi_i(w) = 0$.*

---

[e] Just like its counterpart for TU games [c.f. Nowak and Radzik, 1994].

Nowak and Radzik [1994] proved that these four axioms characterize a unique value (called the Solidarity value) in the class of games with no externalities.[f] If v denotes a game with no externalities (where $v : 2^N \to \mathbb{R}$ is a function that gives the worth of each coalition, independently of the partition structure), then the Solidarity value Sol is defined as:

$$\text{Sol}_i(v) = \sum_{S \ni i} \frac{(n-s)!(s-1)!}{n!} A^v(S), \tag{1}$$

where in turn, $A^v(S) = \frac{1}{s} \sum_{j \in S} [v(S) - v(S_{-j})]$.

4. Characterization

In this section, we show that there is a unique value satisfying the axioms of linearity, symmetry, efficiency, and average nullity.

Lemma 1. *The collection of games* $\{e_{(T,K)} \,|\, (T,K) \in ECL\}$ *defined as*

$$e_{(T,K)}(S,Q) = \begin{cases} 1 & \text{if } (T,K) = (S,Q), \\[2mm] \left[\binom{s}{t} |P(T)| \right]^{-1} & \text{if } T \subsetneq S, \\[2mm] 0 & \text{otherwise} \end{cases} \tag{2}$$

constitutes a basis of G.

Proof. The number of elements in the collection equals the dimension of the space. It remains to show that they are linearly independent, that is,

$$\sum_{(T,K) \in ECL} \gamma_{(T,K)} e_{(T,K)} = 0_G \tag{3}$$

implies that $\gamma_{(T,K)} = 0$ for each $(T,K) \in ECL$. Notice that 0_G denotes the zero game, i.e., $0_G(S,Q) = 0$ for every $(S,Q) \in ECL$.

Suppose on the contrary that there exists a collection $\{\gamma_{(T,K)}\}_{(T,K)\in ECL}$ of real numbers satisfying (3) and such that $\gamma_{(T,K)} \neq 0$ for $(T,K) \in ECL$. Let (T',K') be an embedded coalition such that:

(a) $\gamma_{(T',K')} \neq 0$;
(b) $\gamma_{(T,K)} = 0$ for any $(T,K) \in ECL$ such that $T \subsetneq T'$.

---

[f] A game is with no externalities if and only if the worth that the players in a coalition S can jointly obtain if this coalition formed is independent of the way the other players are organized. This means that in a game with no externalities, the characteristic function satisfies $w(S,Q) = w(S,Q')$ for any two partitions $Q, Q' \in PT$ and any coalition S which belongs both to Q and Q'. Hence, the worth of a coalition S can be written without reference to the organization of the remaining players, $w(S) := w(S,Q)$ for all $Q \ni S$, $Q \in PT$.

Now, since (according to (2)) $e_{(T,K)}(T', K') = 0$ if T is not contained in T' and the property (b) of (T', K'), then

$$\left[\sum_{(T,K)\in ECL} \gamma_{(T,K)}e_{(T,K)}\right](T', K') = \left[\gamma_{(T',K')}e_{(T',K')}\right](T', K')$$

$$= \gamma_{(T',K')}.$$

Finally, from Eq. (3) it follows that $\gamma_{(T',K')} = 0$, a contradiction. $\square$

The aim of this section is to generalize the Solidarity value to the class of games in partition function form. Next, we prove the following

Lemma 2. *Every $i \in N\backslash T$ is a null player in the game $e_{(T,K)}$.*

Proof. Let $(T, K) \in ECL$ and $i \in N\backslash T$. We will show that

$$\overline{A}^{e_{(T,K)}}(S, Q) = e_{(T,K)}(S, Q) - \frac{1}{s}\sum_{j\in S}\frac{1}{|P(S_{-j})|}\sum_{Q'\in P(S_{-j})} e_{(T,K)}(S_{-j}, Q') = 0$$

for every $(S, Q) \in ECL$ such that $i \in S$. The proof is carried out in three cases:

(i) The conclusion is quite obvious if $T \not\subseteq S$. Indeed, since $T \not\subseteq S_{-i}$ for all $S \ni i$, then $e_{(T,K)}(S, Q) = e_{(T,K)}(S_{-i}, Q') = 0$ for every $Q' \in P(S_{-i})$. Therefore $\overline{A}^{e_{(T,K)}}(S, Q) = 0$ for all $(S, Q) \in ECL$ such that $i \in S$.

(ii) If $T = S_{-i}$, then $e_{(T,K)}(S, Q) = \left[\binom{s}{s-1}|P(S_{-i})|\right]^{-1} = \frac{1}{s|P(S_{-i})|}$. Also, notice that $e_{(T,K)}(S_{-j}, Q') = 0$ whenever $j \neq i$ or $Q' \neq K$. Therefore,

$$\overline{A}^{e_{(T,K)}}(S, Q) = \frac{1}{s|P(S_{-i})|} - \frac{1}{s}\sum_{\substack{j\in S \\ j=i}}\frac{1}{|P(S_{-j})|}\sum_{\substack{Q'\in P(S_{-j}) \\ Q'=K}} e_{(T,K)}(S_{-j}, Q')$$

$$= 0.$$

(iii) Finally, if $T \subset S_{-i}$, then $e_{(T,K)}(S, Q) = \left[\binom{s}{t}|P(T)|\right]^{-1} = \frac{t!(s-t-1)!}{s!|P(T)|}$. Thus,

$$\overline{A}^{e_{(T,K)}}(S, Q) = \frac{t!(s-t-1)!}{s!|P(T)|} - \frac{1}{s}\sum_{j\in S}\frac{1}{|P(S_{-j})|}\sum_{Q'\in P(S_{-j})} e_{(T,K)}(S_{-j}, Q')$$

$$= \frac{t!(s-t-1)!}{s!|P(T)|} - \frac{1}{s}\sum_{j\in S\backslash T}\frac{1}{|P(S_{-j})|}\sum_{Q'\in P(S_{-j})} \left[\binom{s-1}{t}|P(T)|\right]^{-1}$$

$$= \frac{t!(s-t-1)!}{s!|P(T)|} - \frac{1}{s}\sum_{j\in S\backslash T}\frac{1}{|P(S_{-j})|} \left[\binom{s-1}{t}|P(T)|\right]^{-1}|P(S_{-j})|$$

$$= \frac{t!(s-t-1)!}{s!|P(T)|} - \frac{1}{s}(s-t)\left[\binom{s-1}{t}|P(T)|\right]^{-1} = 0.$$

And so, i is a null player in the game $e_{(T,K)}$.   □

We can now state the main result of this section.

Theorem 1. *The solution $\varphi : G \to \mathbb{R}^n$ given by*

$$\varphi_i(w) = \sum_{\substack{(S,Q)\in\text{ECL} \\ i\in S}} \frac{(n-s)!(s-1)!}{n! \cdot |P(S)|} \cdot \overline{A}^w(S,Q) \tag{4}$$

for each $i \in N$ and each $w \in G$; is the unique solution satisfying linearity, symmetry, efficiency, and average nullity axioms.

Proof. For the existence part, it is straightforward to check that the solution given by (4) satisfies linearity, symmetry, and average nullity axioms. We show the efficiency.

From the proof of Lemma 2, we have that $\overline{A}^{e_{(T,K)}}(S,Q) = 0$ for every $(T,K) \in \text{ECL}\setminus\{(S,Q)\}$ and then, $\overline{A}^{e_{(T,K)}}(T,K) = 1$. In this way, for any $i \in T$ it holds that $\varphi_i(e_{(T,K)}) = \frac{(n-t)!(t-1)!}{n! \cdot |P(T)|}$. If $w \in G$, then there exist real numbers $\{\delta_{(T,K)}\}_{(T,K)\in\text{ECL}}$ such that $w = \sum_{(T,K)\in\text{ECL}} \delta_{(T,K)}e_{(T,K)}$. Thus, by linearity of φ, $\varphi(w) = \sum_{(T,K)\in\text{ECL}} \delta_{(T,K)}\varphi(e_{(T,K)})$ and

$$\sum_{i\in N}\varphi_i(w) = \sum_{(T,K)\in\text{ECL}} \delta_{(T,K)} \sum_{i\in N}\varphi_i\left(e_{(T,K)}\right)$$

$$= \sum_{(T,K)\in\text{ECL}} \delta_{(T,K)} \sum_{i\in T} \frac{(n-t)!(t-1)!}{n! \cdot |P(T)|} = \sum_{(T,K)\in\text{ECL}} \delta_{(T,K)} \frac{(n-t)!(t-1)!}{n! \cdot |P(T)|}t$$

$$= \sum_{(T,K)\in\text{ECL}} \delta_{(T,K)} \frac{(n-t)!t!}{n! \cdot |P(T)|} = \sum_{(T,K)\in\text{ECL}} \delta_{(T,K)}e_{(T,K)}(N,\{N\})$$

$$= w(N,\{N\}).$$

Now, we prove uniqueness. Let ψ be a solution satisfying the four axioms. For any $w \in G$, $w = \sum_{(T,K)\in\text{ECL}} \delta_{(T,K)}e_{(T,K)}$. Then, by linearity, $\psi(w) = \sum_{(T,K)\in\text{ECL}} \delta_{(T,K)}\psi(e_{(T,K)})$. We will show that $\psi(w)$ is determined for all $w \in G$ and by the previous discussion, it is therefore sufficient to show that $\psi(e_{(T,K)})$ is determined for all $(T,K) \in \text{ECL}$. In this way, ψ is unique for each $e_{(T,K)}$ and so for w.

Notice that,

(a) $e_{(T,K)}(N,\{N\}) = \left[\binom{n}{t}|P(T)|\right]^{-1}$.

(b) From Lemma 2, player i is a null player in $e_{(T,K)}$ whenever $i \notin T$.

(c) If $j,l \in T$ and $\theta \in S_n$ is such that $\theta(j) = l$ and $\theta(T) = T$, then $\theta \cdot e_{(T,K)} = e_{(T,K)}$ and by symmetry: $\psi_j(e_{(T,K)}) = \psi_j(\theta \cdot e_{(T,K)}) = \psi_{\theta(j)}(e_{(T,K)}) = \psi_l(e_{(T,K)})$.

Hence, by the efficiency axiom:

$$\psi_i(e_{(T,K)}) = \begin{cases} \frac{1}{t}\left[\binom{n}{t}|P(T)|\right]^{-1} & \text{if } i \in T, \\ 0 & \text{if } i \notin T. \end{cases}$$

And so,

$$\psi_i(w) = \sum_{\substack{(T,K)\in \text{ECL} \\ T \ni i}} \frac{\delta_{(T,K)}}{t}\left[\binom{n}{t}|P(T)|\right]^{-1}, \quad \forall i \in N. \qquad \square$$

Remark 1. According to the value given by (4), a productivity principle is taken into account, as the players' marginal contributions are used in the calculation. Moreover, it also exhibits a redistribution effect, as it not only takes into account the player's own marginal contribution to the coalition that he belongs to, but also the marginal contributions of the remaining players in the coalition.

Remark 2. The solution we have characterized in the previous theorem is an extension of the Solidarity value. First, notice that the subset of G formed by the games w such that $w(S, Q) = w(S) \forall (S, Q) \in \text{ECL}$, can be identified with the set of TU games. So, taking into account formula (4), it coincides with the Solidarity value for these games. Formally:

Corollary 1. *Let $w \in G$. If for every $S \subseteq N$ it holds that $w(S, Q) = w(S, Q')$ $\forall Q, Q' \in P(S)$, then φ given by (4) equals the Solidarity value for TU games.*

Remark 3. As the reader might guess, the proposed solution for games with externalities (4) can be obtained as the Solidarity value (given by (1)) of an expected game in characteristic function form.

The idea behind the definition of the expected game is the following: In an environment with externalities, the worth of a group of players is influenced by the way the outside players are organized. So, what should then be the worth "assigned" to that group of players? An obvious candidate is to take an average of the different worths of this group for all the possible organizations of the other players. Repeating this process for all groups leads to an "average" game with no externalities.

Formally, with each game $w \in G$, we associate a TU game v_w on N defined for every coalition $S \subseteq N$ by

$$v_w(S) = \frac{1}{|P(S)|} \sum_{Q \in P(S)} w(S, Q).$$

Then, the solution given by (4) is computed as

$$\varphi(w) = \text{Sol}(v_w).$$

5. A Comment on Related Literature

Hernández-Lamoneda *et al.* [2009] prove that there is exactly one solution Φ that satisfies the axioms of linearity, symmetry, efficiency, and an alternative version of the nullity axiom. Moreover, it is given by

$$\Phi_i(w) = \sum_{\substack{(S,Q)\in\text{ECL}\\ i\in S}} \frac{(n-s)!(s-1)!}{n!} \cdot C^w(S,Q)$$

for each $i \in N$ and each $w \in G$, where $C^w(S,Q) = \frac{1}{s}\sum_{j\in S}\left[w(S,Q) - w(S_{-j}, \{S_{-j}, \{j\}\} \cup Q_{-S})\right]$. The amount $C^w(S,Q)$ measures an average marginal contribution of a member of S (given the coalition structure Q), with an specific restriction of the organization of players after player j leaves coalition S. In this respect, player j simply leaves S and then he/she acts alone, whereas the other coalitions remain unchanged.

As usual, here the version of nullity establishes that $\Phi_i(w) = 0$ whenever $C^w(S,Q) = 0$ for every $(S,Q) \in \text{ECL}$ such that $i \in S$.

Although the value Φ represents an extension of the Solidarity value for games with externalities, it ignores the externality side since its computation depends on amounts that consider only particular types of partitions.[g] In this way, some information present in the partition function has to be discarded as a consequence of the combination of those axioms.

On the other hand, the value given by (4) tries to avoid such problem. The concept of average nullity might be more suitable, since this version includes a broader range of information, not only takes into account a given coalitional configuration that measures the marginal contribution of an agent, but also takes into account all possible structures that coalitions may form once a given agent leaves the block. Ignoring these possibilities leads to not look completely the big picture and information that might be relevant when defining criteria for allocation rules.

Now, we provide an example that demonstrates the advantage of our extension value over Φ.

Example 1. Three neighboring villages ($N = \{i, j, k\}$), live near a dam. Due to an unfortunate natural disaster, the dam collapses affecting the villages closest to it (j and k), while village i does not suffer any damage. The economic cost of rebuilding villages j and k is 1 million dollars each. If these villages decide to cooperate and share resources to address the problem, the cost decreases to 3/4 of this amount.

On the other hand, if village i takes a solidarity posture and helps any of its neighbors (either j or k), then the (joint) cost of rebuilding reduces to 1/2. In addition, the alliance between i and either j or k, indirectly benefits the community that is outside the coalition (now with a cost of 4/5), as the recovery of its neighbor will help to obtain useful resources for its own recovery.

---

[g]That is, for the computation of $C^w(S,Q)$ (and consequently for Φ), a player j leaves coalition S and stays alone; the possibilities of j joining any other nonempty coalition in Q are discarded.

Finally, if i decides to support both villages j and k, their potential to help decrease (by the fact that it is more difficult to serve two communities that focus on one). For this reason the grand coalition will incur a cost of $7/12$ of the initial amount.

The above situation can be modeled with the following game:

Partition			$w(S, Q)$		
$\{i\}$	$\{j\}$	$\{k\}$	0	1	1
$\{i\}$	$\{j,k\}$		0	3/4	
$\{j\}$	$\{i,k\}$		4/5	1/2	
$\{k\}$	$\{i,j\}$		4/5	1/2	
$\{i,j,k\}$			7/12		

It is not difficult to verify that player i is a null player in the sense of Hernández-Lamoneda *et al.* [2009] and so, despite the solidarity behavior of this village, the solution does not reflect such behavior ($\Phi_i(w) = 0$). Thus,

$$\Phi(w) = (0, 7/24, 7/24)$$

whereas our extension value provides

$$\varphi(w) = (1/60, 17/60, 17/60)$$

which seems to be a more reasonable solution for this game w than the one of Hernández-Lamoneda *et al.* One can say that if some kind of solidarity of player i with players j and k is assumed, then such a fact should be reflected by the game w itself as well as by the final allocation. Please notice that we do not want to say that our value is the only right solution concept even for this particular example. We would rather like to point out that it could be used to take into account different ways of defining marginal contributions.

Acknowledgment

J. Sánchez-Pérez acknowledges support from CONACYT research grant 283644. J. Rodríguez-Segura acknowledges support from CONACYT grant 519971.

References

Bolger, E. M. [1989] A set of axioms for a value for partition function games, *Int. J. Game Theor.* **18**(1), 37–44.

De Clippel, G. and Serrano, R. [2008] Marginal contributions and externalities in the value, *Econometrica* **6**, 1413–1436.

Grabisch, M. and Funaki, Y. [2012] A coalition formation value for games with externalities, *Eur. J. Oper. Res.* **221**(1), 175–185.

Hernández-Lamoneda, L. Sánchez-Pérez, J. and Sánchez-Sánchez, F. [2009] The class of efficient linear symmetric values for games in partition function form, *Int. Game Theor. Rev.* **11**(3), 369–382.

Hu, C.-C. and Yang, Y.-Y. [2010] An axiomatic characterization of a value for games in partition function form, *SERIEs* **1**(4), 475–487.

Lucas, W. F. and Thrall, R. M. [1963] *n*-person games in partition function form, *Nav. Res. Logis. Q.* **10**, 281–298.

Macho-Stadler, I., Pérez-Castrillo, D. and Wettstein, D. [2007] Sharing the surplus: An extension of the Shapley value for environments with externalities, *J. Econ. Theor.* **135**, 339–356.

McQuillin, B. [2009] The extended and generalized Shapley value: Simultaneous consideration of coalitional externalities and coalitional structure, *J. Econ. Theor.* **144**(2), 696–721.

Myerson, R. B. [1977] Values of games in partition function form, *Int. J. Game Theor.* **6**(1), 23–31.

Nowak, A. and Radzik, T. [1994] A solidarity value for *n*-person transferable utility games, *Int. J. Game Theor.* **23**, 43–48.

Sánchez-Pérez, J. [2015] A note on a class of solutions for games with externalities generalizing the Shapley value, *Int. Game Theor. Rev.* **17**(3), 1–12.

Shapley, L. [1953] A value for *n*-person games, in Kuhn & Tucker (eds.), *Contribution to the Theory of Games*, Annals of Mathematics Studies, Princeton University Press, Vol. 2, pp. 307–317.

Young, H. P. [1985] Monotonic solutions of cooperative games, *Int. J. Game Theor.* **14**, 65–72.

Chapter 26

On Harsanyi Dividends and Asymmetric Values

Pierre Dehez

CORE, University of Louvain Voie du Roman Pays 34
1348 Louvain-la-Neuve, Belgium

pierre.dehez@uclouvain.be

The concept of dividend in transferable utility games was introduced by Harsanyi [1959], offering a unifying framework for studying various valuation concepts, from the Shapley value to the different notions of values introduced by Weber. Using the decomposition of the characteristic function used by Shapley to prove uniqueness of his value, the idea of Harsanyi was to associate to each coalition a dividend to be distributed among its members to define an allocation. Many authors have contributed to that question. We offer a synthesis of their work, with a particular attention to restrictions on dividend distributions, starting with the seminal contributions of Vasil'ev, Hammer, Peled and Sorensen and Derks, Haller and Peters, until the recent papers of van den Brink, van der Laan and Vasil'ev.

Keywords: Harsanyi dividends; Weber set; weighted Shapley values; core.

1. Introduction

We begin with a chronological enumeration of the concepts that will be formally defined and interrelated in this chapter.

The notion of game with transferable utility, defined by a set of players and a characteristic function, was introduced by von Neumann and Morgenstern [1944]. The characteristic function associates to each subset (coalition) of players a real number measuring what that coalition can do at best in terms of some commodity-money.

The concept of *value* of a transferable utility game was introduced by Shapley in 1953. His initial idea was to define what a player may reasonably expect from playing a game. However, by requiring players' evaluations to be consistent in order to achieve efficiency (an exact distribution of the social output) and symmetry (equal treatment of equals), the Shapley value is more of a normative tool.

The concept of *dividend* was introduced by Harsanyi [1959]. His idea is to associate to each coalition a dividend (positive or negative) that can be distributed among its members to define an allocation of the social surplus. The dividends are identified with the coefficients of the decomposition of the characteristic function used by Shapley to prove uniqueness of his value. The set of allocations that results from all possible distributions of the dividends is an object that has been studied in the 1970s independently by Vasil'ev in papers published in Russian, and by Hammer, Peled and Sorensen in a paper published in a Belgian operations research journal. While the latter used the name *"selectope"*, here we shall retain the term *"Harsanyi set"*. Derks *et al.* [2000] popularized that concept in a paper published in *International Journal of Game Theory*. At that time, they were not aware of the contributions of Vasil'ev. These became known with the publication by Vasil'ev and van der Laan [2002] of a paper containing all the results known by that time. Since then, a number of papers have been published, in particular by Derks *et al.* [2006, 2010].

Shapley [1971] has characterized geometrically the core of a convex game using the concept of marginal contribution vector that associates allocations to players' orderings: for a given ordering, each player receives his marginal contribution, following the ordering. He shows that the core of a convex game is the nonempty and bounded polyhedral convex set whose vertices are precisely these marginal contribution vectors.

Weber [1988] has introduced the notion of *probabilistic values* that allocates to each player his expected marginal contribution computed with respect to a probability distribution independent of the game's data. *Quasi-values* are probabilistic values obtained by considering probability distributions ensuring efficiency and the Shapley value is the unique efficient and symmetric quasi-value. Weber also defines the concept of *random order value* as the expected marginal contribution vector, computed with respect to a given probability distribution over players' orderings. He shows that random order values are quasi-value, the Shapley value being the random order value corresponding to the uniform probability distribution. Moreover, he shows that the core is a subset of the set of random order values, equivalently defined as the convex hull of the marginal contribution vectors. This set is known as the *Weber set* and, following Shapley's characterization of the core, the two sets coincide on the class of convex games, and only for these games as shown by Ichiishi [1981].

In his doctoral thesis, Shapley [1953a] did also consider the possibility for symmetric players to be treated differently. The *asymmetric* version of the value is obtained by introducing *exogenous* weights in order to cover asymmetries that are not included in the underlying game. The weighted Shapley value has been axiomatized later, in particular by Kalai and Samet [1987] without explicit reference to weights, by Hart and Mas-Colell [1989] using a generalized potential function and by Dehez [2011] in a cost sharing context along the lines suggested by Shapley [1981]. The value associated to positive weights is obtained as a weighted division of dividends. As a consequence, weighted values are Harsanyi payoffs. Alternatively, it can be defined as the expected marginal contribution vector corresponding to a probability distribution over players' orderings derived from weights. The *Shapley set* is the set of all weighted values obtained by considering all possible weight systems, including zero weights. Monderer *et al.* [1992] show the Shapley set contains the core, an inclusion *"somewhat surprising in light of the difference in concept behind these solutions"* to quote the authors. Weighted values being random order values, the Shapley set is contained in the Weber set and the three solution sets coincide when applied to convex games.

The Harsanyi set turns out to be the largest solution set. It includes the Weber set. Hammer *et al.* [1977] and Vasil'ev [1981] show that the Harsanyi set and the core coincide if and only if they apply to almost positive games — games whose dividends of multiplayer coalitions are nonnegative. Consequently, positive games being convex, the four solution sets — core, Shapley, Weber and Harsanyi sets — coincide when applied to almost positive games.

All these solutions have been characterized axiomatically. They can also be characterized starting with the Harsanyi set and imposing restrictions on dividend distributions. A natural restriction is *monotonicity*. It requires that the share of a player in the dividend of a coalition does not increase if the coalition is enlarged. Even if such restrictions reduce considerably the set of possible dividend distributions, it is not sufficient to generate a particular solution set. Billot and Thisse [2005] claim that the set of Harsanyi payoff vectors resulting from monotonic dividend distributions coincides with the core if (and only if) the game is convex. We show that this is actually true only for 3-player games! Assuming a monotonic dividend distribution, individual rationality obtains in 3-player superadditive games and in 4-player convex games. Beyond four players, we show that there is no hope. Under a stronger monotonicity condition, Vasil'ev [1988] shows that Harsanyi payoff vectors are random order values. Under the assumption that the distributions of dividends within coalitions are compatible with Bayesian updating, Derks *et al.* [2000] show that the resulting Harsanyi subset coincides with the Shapley set.

The chapter is organized as follows. Section 2 introduces transferable utility games. Solution concepts are then defined and interrelated in Sec. 3, with a particular attention to weighted Shapley values and the case where some players are assigned a zero weight. In Sec. 4, we review the axiomatic characterizations of the Shapley, Weber and Harsanyi sets. We then look at the characterization of the Weber and Shapley sets by way of restrictions on dividend distributions. We finally consider the subsets of Harsanyi payoffs that result from graph structures on the set of players. Section 5 offers concluding remarks and Appendix A gathers intermediary results.

2. Transferable Utility Games

2.1. *Characteristic functions*

Cooperative games cover situations in which a group of individuals cooperate on a common project with the objective of maximizing the resulting collective gain. It is assumed that *utility is transferable* through some commodity-money, allowing for transfers (side-payments) between players. A cooperative game with transferable utility is defined by a player set N and a *characteristic function* v that associates to each *coalition* $S \subset N$ a real number $v(S)$ that represents its (potential) worth defined as the gain that it can realize without the participation of the others. In particular, $v(i)$ is what player i could obtain alone and the value of the game $v(N)$ is the maximum amount that the "grand coalition" is able to generate. By convention, we set $v(\emptyset) = 0$.

Notation. Set inclusion is denoted by $\subset$ and *strict* inclusion by $\emptyset$. For a given subset S, $S\backslash i$ denotes the subset obtained by *subtracting* i from S. Upper-case letters are used to denote sets and the corresponding lower-case letters to denote their sizes: $s = |S|, t = |T|, \ldots$. Coalitions $\{i, j, k, \ldots\}$ are sometimes written as $ijk \ldots$. For any given set T, we denote by $\mathcal{C}(T) = \{S \subset T \mid S \neq \emptyset\}$ the collection of nonempty subsets of T and by $\mathcal{C}_i(T) = \{S \subset T \mid i \in S\}$ the collection of subsets of T containing player i. Given a vector $x \in \mathbf{R}^n$ and a subset $S \subset N$, it will be convenient to write $x(S) = \sum_{i \in S} x_i, x_S = (x_i \mid i \in S)$ and $x = (x_S, x_{N\backslash S})$ with the convention $x(\emptyset) = 0$. Vectors are compared following the sequence $x \geq y, x > y$ and $x \gg y$. In some instances, the summation sign Σ will be used without reference to a set when there is no ambiguity. For a given finite set A, we denote by $\Delta(A)$ the set of all probability distributions on A.

Two games (N, v') and (N, v'') on a common player set are said to be *strategically equivalent* if there exist $a > 0$ and $b \in \mathbf{R}^n$ such that $v''(S) = a \, v'(S) + b(S)$. This defines an *equivalence relation*. In particular, a game (N, v) and its 0-normalization (N, v_0) defined by

$$v_0(S) = v(S) - \sum_{i \in S} v(i)$$

are strategically equivalent. A game can be restricted to a subset of its player set. Formally, given a game (N, v) and a subset $R \subset N$, the restriction to R is the game (R, v_R) defined by $v_R(S) = v(S)$ for all $S \subset R$. Given two games (N, v') and (N, v'') on a common set of players, the game *sum* (N, v) is simply defined by $v(S) = v(S') + v(S'')$ for all $S \subset N$.

2.2. *Superadditivity and monotonicity*

A game (N, v) is *superadditive* if getting together is beneficial, or at least harmless:

$$v(S) + v(T) \leq v(S \cup T) \quad \text{for all } S \text{ and } T \text{ such that } S \cap T = \emptyset.$$

A game (N, v) is *monotonic* if $S \subset T \Rightarrow v(S) \leq v(T)$. It implies that the largest surplus is generated by the grand coalition. There is no direct relation between superadditivity and monotonicity except for the following lemma.[a]

Lemma 1. *Consider a game (N, v) such that $v(i) \geq 0$ for all $i \in N$. Superadditivity then implies that the characteristic function v is monotonic and positive valued.*

A game (N, v) is *0-monotonic* if its 0-normalization (N, v_0) is a monotonic game. As a consequence of Lemma 1, the 0-normalization of a superadditive game is a monotonic and positive-valued game and thereby, superadditive games are 0-monotonic.

0-monotonicity implies that the inequalities $\sum_{i \in S} v(i) \leq v(S)$ hold for all $S \subset N$. A coalition S is said to be *essential* if the inequality is strict. If equality holds, S is said to be inessential. Obviously, for 0-monotonic games, if a coalition is inessential, so are all its subcoalitions. A game is said to be essential if the grand coalition is essential and a game whose coalitions are all inessential is an *additive* game.

Superadditivity is a natural assumption that is satisfied in most economic and social situations. It ensures that allocating *exactly* the value of a game among the players if efficient: no partition of players can form and generate a total gain larger than the value of the game. This is not ensured by 0-monotonicity for games with more than three players.

2.3. *Harsanyi dividends*

We denote by $G(N)$ the set of all set functions on the finite set N. It is a *vector space* that is formally equivalent to $\mathbf{R}^{2^n-1}$. Shapley [1953b] shows that the collection of *unanimity games* (N, u_T) defined for all $T \in \mathcal{C}(N)$ by $u_T(S) \in \{0, 1\}$ and $u_T(S) = 1$ if and only if $T \subset S$ forms a basis of the vector space $G(N)$: for any given set function v on N, there exists a unique $(2^n - 1)$-dimensional vector $\alpha(N, v) = (\alpha_T(N, v) \,|\, T \in \mathcal{C}(N))$ such that

$$v(S) = \sum_{T \in \mathcal{C}(N)} \alpha_T(N, v)\, u_T(S) = \sum_{T \in \mathcal{C}(S)} \alpha_T(N, v). \tag{1}$$

Following Harsanyi [1959, 1963], α_T is interpreted as the *dividend* accruing to coalition T. By (1), $v(N)$ is the sum of the dividends of all coalitions. Hence, an allocation of $v(N)$ can be obtained by distributing the dividends of every coalition among its members.[b] Dividends can be defined recursively, starting with $\alpha_{\varnothing} = 0$, as follows:

$$\alpha_T = v(T) - \sum_{S \varnothing T} \alpha_S \quad \text{for all } T \subset N \tag{2}$$

---

[a]Proofs of lemmas are in Appendix A.
[b]To keep notation simple, the dependence of dividends on the game will sometimes be omitted.

i.e.,

$$\alpha_{\{i\}} = v(i),$$

$$\alpha_{\{ij\}} = v(ij) - v(i) - v(j),$$

$$\alpha_{\{ijk\}} = v(ijk) - v(ij) - v(ik) - v(jk) + v(i) + v(j) + v(k), \dots.$$

Alternatively, the collection of α_T's is the *unique* solution of the linear system (1):

$$\alpha_T(N, v) = \sum_{S \in \mathcal{C}(T)} (-1)^{t-s} v(S) \quad (T \subset N, T \neq \emptyset). \tag{3}$$

Additivity of the dividends follows from (2):

$$\alpha_T(N, v' + v'') = \alpha_T(N, v') + \alpha_T(N, v'') \quad \text{for all } T \subset N.$$

Two games (N, v') and (N, v'') are *disjoint* if no dividend are simultaneously different from zero: $\alpha_T(N, v') \cdot \alpha_T(N, v'')$ for all $T \subset N$.

Remark 1. The dividends associated to the unanimity games (N, u_S) are given by:

$$\alpha_T(N, u_S) = \begin{cases} 1 & \text{if } T = S, \\ 0 & \text{otherwise.} \end{cases}$$

Remark 2. The dividends associated to an additive game are all zero except for single players. The dividends of 0-monotonic games associated to inessential *multi-player* coalitions are equal to zero. Furthermore, the dividends associated to the 0-normalization (N, v_0) of an arbitrary game (N, v) are unchanged except for singletons:

$$\alpha_{\{i\}}(N, v_0) = 0 \quad \text{for all } i \in N,$$

$$\alpha_T(N, v_0) = \alpha_T(N, v) \quad \text{for all } T \subset N, \ t \geq 2.$$

2.4. *Positive games*

Dividends can be negative or positive. A game is (totally) *positive* if its dividends are all nonnegative.[c] The term *almost positive* is used for games whose dividends of *multi-player* coalitions are nonnegative. Equivalently, a game is almost positive if its 0-normalization is positive. Obviously, 2-player 0-monotonic games are almost positive and unanimity games are positive.

Remark 3. Positive games are monotonic. This can easily be seen from (1).

2.5. *Marginal contributions*

The *marginal contribution* of player i to coalition S is defined by $v(S) - v(S \backslash i)$. It is the value added of player i to coalition S and it is obviously zero for all coalitions

---

[c]Totally positive games were introduced and systematically studied by Vasil'ev [1975, 1981].

of which he is not a member. For 0-monotonic games, marginal contributions are bounded below by individual worth. Indeed, for all $S \subset N$ such that $i \in S$, we have

$$v(S) - \sum_{j \in S} v(j) \geq v(S \backslash i) - \sum_{\substack{j \in S \\ j \neq i}} v(j) \Rightarrow v(S) - v(S \backslash i) \geq v(i). \tag{4}$$

Two players i and j are *symmetric* in a game (N, v) if they contribute equally to all coalitions to which they belong: $v(S) - v(S \backslash i) = v(S) - v(S \backslash j)$ for all $S \subset N$ such that $i, j \in S$. A player i is *null* in a game (N, v) if he never contributes: $v(S) - v(S \backslash i) = 0$ for all $S \subset N$.

Player i is *necessary* for player j in a game (N, v) if the marginal contributions of j are zero in all coalition not containing i: $v(S) = v(S \backslash j)$ for all $S \not\ni i$. In particular, $v(j) = 0$. Using (2), van den Brink *et al.* [2014] prove the following Lemma.

Lemma 2. *If player i is necessary for player j in a game (N, v), then $\alpha_T(N, v) = 0$ for all $T \subset N \backslash i$ such that $j \in T$.*

Remark 4. A player is null *if and only if* the dividends associated to coalitions containing that player are all equal to zero. This is an immediate consequence of (2). See also Lemma 2.

Given a player set N, we denote by Π_N the set of all players' orderings. The *marginal contribution vector* $\mu^\pi(N, v)$ associated to the players' ordering $\pi = (i_1, \ldots, i_n) \in \Pi_N$ is the vector of dimension n defined by:

$$\mu^\pi_{i_1}(N, v) = v(i_1) - v(\emptyset) = v(i_1),$$

$$\mu^\pi_{i_k}(N, v) = v(i_1, \ldots, i_k) - v(i_1, \ldots, i_{k-1}) \quad (k = 2, \ldots, n)$$

i.e.,

$$\mu^\pi_i(N, v) = v(\pi^i) - v(\pi^i \backslash i) \quad (i = 1, \ldots, n). \tag{5}$$

Here, π^i denotes the set of players preceding i in π, i included. There are $n!$ marginal contribution vectors, not necessarily all distinct.

Looking at strategically equivalent games, if $v''(S) = a \, v'(S) + b(S)$ for some $a > 0$ and $b \in \mathbf{R}^n$, the following identities prevail:

$$\mu^\pi_i(N, v'') = a \, \mu^\pi_i(N, v') + b_i \quad (i = 1, \ldots, n). \tag{6}$$

2.6. *Convex games*

A game (N, v) is *convex* (or *supermodular*) if $v(S) + v(T) \leq v(S \cup T) + v(S \cap T)$ for all S and $T \subset N$. Obviously, convexity implies superadditivity and a game that is strategically equivalent to a convex game is itself convex. It is easily verified that unanimity games are convex. As a consequence, positive games are convex as positive linear combinations of convex games, and almost positive games are convex as well by strategic equivalence.

Shapley [1971] shows that a game is convex *if and only if* players' marginal contributions do not decrease with coalition size:

$$i \in S \subset T \Rightarrow v(S) - v(S \setminus i) \leq v(T) - v(T \setminus i).$$

Hence convexity means *increasing returns to size* and marginal contributions are *maximal* at the grand coalition N.

For any given player set N, the set of superadditive games, the set of monotonic games, the set of 0-monotonic games and the set of convex games are *convex cones*, denoted $\mathrm{SG}(N)$, $\mathrm{MG}(N)$, $\mathrm{MG}_0(N)$ and $\mathrm{CG}(N)$, respectively. They are indeed closed under addition and positive scalar multiplication.[d] The set of positive games is a convex cone as well. It is denoted by $G^+(N)$. Following Remark 3, we have the following sequences of inclusions:

$$G^+(N) \subset \mathrm{CG}(N) \subset \mathrm{SG}(N) \subset \mathrm{MG}_0(N) \quad \text{and} \quad G^+(N) \subset \mathrm{MG}(N).$$

An arbitrary game (N, v) can be decomposed in a difference between two positive (and thereby convex) games. Indeed (1) can be written as

$$v(S) = \sum_{T \in \mathcal{C}(N)} \alpha_T(N, v)\, u_T(S) = v^+(S) - v^-(S) \quad \text{for all } S \subset N,$$

where

$$v^+(S) = \sum_{T:\, \alpha_T > 0} \alpha_T(N, v)\, u_T(S) \quad \text{and} \quad v^-(S) = \sum_{T:\, \alpha_T < 0} -\alpha_T(N, v)\, u_T(S). \quad (7)$$

The dividends associated to these two games are given by

$$\alpha_T(N, v^+) = \mathrm{Max}(0, \alpha_T(N, v)),$$

$$\alpha_T(N, v^-) = -\mathrm{Min}(0, \alpha_T(N, v)).$$

Convex games form an interesting class of games because solution concepts tend to agree when applied to convex games.[e] Moreover, many interesting economic situations can be modeled as convex games, like production games with increasing returns, bankruptcy games [Aumann and Maschler, 1985] and airport games [Littlechild and Owen, 1973]. Positive games are convex and monotonic games with interesting properties and applications, like river games [Ambec and Sprumont, 2002], queuing games [Maniquet, 2003] and liability games [Dehez and Ferey, 2013].

3. Values and Solution Sets

3.1. *Basic properties*

Given a game (N, v), the problem is to allocate $v(N)$ between the n players. Given a player set N, a *value* is a mapping that associates a payoff vector $\varphi(N, v) \in \mathbf{R}^n$ to any game (N, v).

---

[d]A nonempty set X is a convex cone if and only if $x, y \in X$ and $\alpha \geq 0 \Rightarrow \alpha x + y \in X$.
[e]See Maschler et al. (1972).

A *solution set* is a mapping Φ that associates a *subset* $\Phi(N, v)$ of payoff vectors to any game (N, v). Basic properties that a solution set should ideally possess are the following:

- *Nonemptiness:* $\Phi(N, v) \neq \emptyset$ for all game (N, v).
- *Efficiency:* $x \in \Phi(N, v) \Rightarrow x(N) = v(N)$.
- *Individual rationality:* $x \in \Phi(N, v) \Rightarrow x_i \geq v(i)$ for all $i \in N$.
- *Covariance:* $x \in \Phi(N, v) \Rightarrow ax + b \in \Phi(N, a\, v + b)$ for all $a > 0$ and $b \in \mathbf{R}^n$.
- *Convexity:* $\Phi(N, v)$ is a convex set.

Nonemptiness implies restrictions on the class of games on which the solution applies. As such, efficiency is an accounting identity. It does not necessarily imply full efficiency except for superadditive games. Indeed, there may exist a partition $(S_1, \ldots, S_k)$ of the grand coalition such that $\sum v(S_h) > v(N)$. Individually rationality is a minimal requirement to be imposed on allocations: no player will ever accept to take part in a collective project if his remuneration falls short of what he could secure by himself. A solution is covariant if, once it has been applied to a game, it can be extended to all strategically equivalent games. Convexity is a natural requirement in a world where utility is transferable. Corresponding properties apply to values. Ideally, a value should be covariant and define an efficient and individually rational allocation.

3.2. *Imputations*

Imputations are efficient and individually rational allocations. This defines the *imputation set*:

$$I(N, v) = \{x \in \mathbf{R}^n \mid x(N) = v(N), \ x(i) \geq v(i) \text{ for all } i \in N\}.$$

The class of games (N, v) satisfying the inequality $\sum v(i) \leq v(N)$ is the largest class of games on which the imputation set is a well-defined solution. It includes 0-monotonic games. If the game is essential, $I(N, v)$ is a regular simplex of dimension $n - 1$. If instead the game is inessential, $I(N, v)$ is reduced to the singleton $(v(1), \ldots, v(n))$. The imputation set is the largest solution set satisfying the above requirements.

Remark 5. For 0-monotonic games, marginal contributions as defined by (5) are imputations. Indeed, for a given players' ordering, adding the $n!$ vectors results in $v(N)$ and, according to (4), marginal contributions are bounded below by individual worth's. Efficiency and individual rationality then follow.

3.3. *Stable allocations: The core*

The *core* is the set of imputations that no coalition can improve upon:

$$C(N, v) = \{x \in \mathbf{R}^n \mid x(N) = v(N), \ x(S) \geq v(S) \text{ for all } S \subset N\}. \tag{8}$$

The core extends rationality from individuals to coalitions: given an allocation, a coalition that receives less than what it could secure for itself is in a position to object. In this sense, core allocations are "stable".[f]

The core is a *polytope* i.e., a bounded polyhedral convex set. It is indeed bounded and results from the intersection of finitely many closed half spaces. It is a subset of the imputation set and it may be empty. The largest class of games on which the core is a well-defined solution is the class of *balanced games*.[g] Superadditivity is neither necessary nor sufficient for a game to have a nonempty core. However, for games with a nonempty core, no partition of the grand coalition can do better. Core allocations are therefore fully efficient.

Remark 6. It can be easily verified that, if i and j are symmetric players, allocations obtained by exchanging x_i and x_j in a core allocation x are also core allocations. Hence, if nonempty, the core contains allocations that give to players i and j equal amounts. Furthermore, the core allocates zero to null players.

Shapley [1971] shows that convex games are balanced and that the core of a convex game is the polytope whose vertices are the marginal contribution vectors as defined by (5). Ichiishi [1981] shows that this is actually a *necessary and sufficient* condition for convexity.

3.4. *Quasi-values and random order values: The Weber set*

The concept of value was introduced by Shapley [1953a] as a measure of what a player may expect from playing a game and the Shapley value belongs to the family of probabilistic values introduced later by Weber [1988]. Consider a collection $q = (q^S \,|\, S \subset N, S \neq \emptyset)$ of $2^n - 1$ nonnegative vectors in $\mathbf{R}^n$ such that for all $S \subset N, q_i^S = 0$ for all $i \notin S$. The resulting object can be written as a $n \times (2^n - 1)$ matrix and we denote by Q_N the set of such matrices. For a given player i, Weber interpretes $(q_i^S \,|\, S \subset N, S \neq \emptyset)$ as a probability distribution over coalitions that may be objective, as the result of some random mechanism, or subjective. The *probabilistic value* associated a probability matrix $q \in Q_N$ is then defined for each player as his expected marginal contribution:

$$\mathrm{PV}_i(N, v, q) = \sum_{S \in \mathcal{C}_i(N)} q_i^S (v(S) - v(S \backslash i)) \quad (i = 1, \ldots, n). \tag{9}$$

A probabilistic value does not necessarily define an efficient payoff vector because probability distributions are unrelated. *Quasi-values* instead are efficient probabilistic value obtained from probability distributions matrices $q \in Q_N$ satisfying

$$\sum_{i \in N} q_i^N = 1 \quad \text{and} \quad \sum_{i \in S} q_i^S = \sum_{i \in N \backslash S} q_i^{S \cup i} \quad \text{for all } S \in \mathcal{C}(N), S \neq N. \tag{10}$$

---

[f]The term "core" was introduced by Gillies [1953, 1959] in connection with von Neumann–Morgenstern stable sets. It was later introduced as an independent solution concept by Shapley. The core has been axiomatized by Peleg [1986] using a reduced game property.
[g]See Bondareva [1963] and Shapley [1967]. For a complete account, see Kannai [1992].

Weber [1988] indeed proves that the probabilistic values defined in (9) satisfy efficiency under (10).[h] By requiring consistency of the probability distributions, quasi-values can be given a normative content. Vasil'ev and van der Laan [2002, Lemma 4.2] prove that (10) is equivalent to the following condition:

$$\sum_{i \in S} \sum_{T: \, S \subset T} q_i^T = 1 \quad \text{for all } S \in \mathcal{C}(N).\tag{11}$$

Let $Q_N^* \subset Q_N$ be the subset of probability distribution matrices satisfying (10) or (11), a polytope whose vertices have been characterized in terms of players' permutations by Vasil'ev [2003, 2007].[i] He proves that the $n!$ vertices of Q_N^* are the matrices $q^\pi (\pi \in \Pi_N)$ defined by

$$(q^\pi)_i^S = \begin{cases} 1 & \text{if } S = \pi^i, \\ 0 & \text{otherwise,} \end{cases}$$

where π^i denotes the set of players preceding i in π, i included.

The probabilistic values associated to the vertices of Q_N^* are then the corresponding marginal contribution vectors:

$$\begin{aligned} \mathrm{PV}_i(N, v, q^\pi) &= \sum_{S \in \mathcal{C}_i(N)} (q^\pi)_i^S (v(S) - v(S \backslash i)) \\ &= v(\pi^i) - v(\pi^i \backslash i) \quad \text{for all } \pi \in \Pi_N. \end{aligned}\tag{12}$$

Random order values are average marginal contribution vectors computed with respect to some probability distribution on players' orderings.

For a given game (N, v), the *random order value* associated to the probability distribution $p \in \Delta(\Pi_N)$ is given by

$$\mathrm{RV}_i(N, v, p) = \sum_{\pi \in \Pi_N} p(\pi) \, \mu_i^\pi(N, v) \quad (i = 1, \ldots, n).$$

The following proposition establishing the equivalence between quasi-values and random order values was proved by Weber [1988].

Proposition 1. *A solution is a quasi-value if and only if it is a random-order value.*

Proof. Consider a game (N, v). We have to show that the sets $\{\mathrm{PV}(N, V, q) \,|\, q \in Q_N^*\}$ and $\{\mathrm{RV}(N, V, p) \,|\, p \in \Delta(\Pi_N)\}$ coincide. Following (12), we know that $Q_N^* = \mathrm{co}\{q^\pi \,|\, \pi \in \Pi_N\}$ and $\mathrm{PV}(N, v, q^\pi) = \mu^\pi(N, v)$ Probabilistic values $\mathrm{PV}(N, V, q)$

---

[h]See also Derks [2005].
[i]See also Vasil'ev and van der Laan [2002].

being linear in q, we have successfully[j]:

$$\{\mathrm{PV}(N,v,q)\,|\,q \in Q_N^*\} = \left\{\mathrm{PV}\left(N,v,\sum_{\pi \in \Pi_N} \eta_\pi q^\pi\right) \,\bigg|\, \eta \in \Delta(\Pi_N)\right\}$$

$$= \left\{\sum_{\pi \in \Pi_N} \eta_\pi \mathrm{PV}(N,V,q^\pi) \,\bigg|\, \eta \in \Delta(\Pi_N)\right\}$$

$$= \mathrm{co}\{\mathrm{PV}(N,v,q^\pi)\,|\,\pi \in \Pi_N\} = \mathrm{co}\{\mu^\pi(N,v)\,|\,\pi \in \Pi_N\}$$

$$= \{\mathrm{RV}(N,v,p)\,|\,p \in \Delta(\Pi_N)\}.$$

In particular, $\mathrm{RV}_i(N,v,p) = \sum_{S \subseteq N} q_i^S(v(S) - v(S\backslash i))$ where the probability distributions (q_i^S) defined by $q_i^S = \sum_{\pi:\,\pi^i=S} p(\pi)$ satisfy (10). $\square$

The set of all quasi-values — or alternatively the set of all random order values — is known as the *Weber set*. Marginal contribution vectors being imputations, the Weber set is a well-defined solution on the class of superadditive games.

The Shapley value is a particular quasi-value. Weber [1988] proves that it is the *unique symmetric* quasi-value. It is defined by probabilities that depend only on coalitions' sizes:

$$q_i^S = \frac{(s-1)!(n-s)!}{n!}$$

i.e.,

$$\mathrm{SV}_i(N,v) = \frac{1}{n!} \sum_{S \subseteq N} (s-1)!(n-s)!(v(S) - v(S\backslash i)). \tag{13}$$

These probabilities correspond to the following two-step random mechanism[k]: first, a coalition size between 1 and n is picked up at random and then each player receives his marginal contribution to a coalition picked up at random among the coalitions of the predetermined size of which he is a member. All sizes have the same probability, namely $1/n$, and the probability of picking up a coalition of size s containing a given player is given by $(s-1)!(n-s)!/(n-1)!$ As a random order value, the Shapley value corresponds to *uniformly* drawn random orders i.e., $p(\pi) = 1/n!$ for all π:

$$\mathrm{SV}_i(N,v) = \frac{1}{n!} \sum_{\pi \in \Pi_N} \mu_i^\pi(N,v) \quad (i=1,\ldots,n). \tag{14}$$

The Shapley value is the average marginal contribution vector and can then be seen as resulting from another two-step random mechanism: first, players are ordered randomly and they then receive their marginal contribution, depending on their

---

[j]The convex hull of a set A, denoted co$[A]$, is the smallest convex set containing A. See Rockafellar [1970].

[k]A random allocation mechanism is "fair" if it treats *ex ante* all players equally.

position in the order that has been picked up. To show that (13) and (14) are equivalent, consider the coalition π^i defined as the subset of players preceding i in π and including player i. Then (13) can be written as

$$\mathrm{SV}_i(N, v) = \frac{1}{n!} \sum_{\pi \in \Pi_N} (v(\pi^i) - v(\pi^i \backslash i)).$$

For a coalition $S \subset N$, there are $(s-1)!(n-s)!$ orderings such that $\pi^i = S$. Hence, we have

$$\mathrm{SV}_i(N, v) = \frac{1}{n!} \sum_{S \in \mathcal{C}_i(N)} (s-1)!(n-s)!(v(S) - v(S \backslash i)) \quad i = 1, \dots, n.$$

The Shapley value defined by (13) or (14) is a well-defined single-valued solution on the class of all games, hence including superadditive games. Covariance follows from (6) and superadditivity ensures that marginal contribution vectors are imputations. Hence, the Shapley value defines an imputation that is not necessarily stable, independently of the core being empty or not. However, as an average of marginal contribution vectors, it defines a core allocation when applied to a convex game. Furthermore, in view of the geometric characterization of the core of a convex game, the Shapley value occupies a central position within the core. It generally differs from the barycenter of the core introduced as a solution concept by Gonzáles-Díaz and Sánchez-Rodríguez [2007]. It also differs from the simple average of core's vertices except for the particular case of convex games with distinct marginal contribution vectors.[1]

3.5. *Dividend distributions: The Harsanyi set*

A distribution of the Harsanyi dividends can be summarized by a matrix λ of dimension $n \times (2^n - 1)$ whose columns are the nonnegative vectors $\lambda^T (T \subset N, T \neq \emptyset)$ that satisfying

$$\sum_{i \in N} \lambda_i^T = 1 \quad \text{and} \quad \lambda_i^T = 0 \quad \text{for all } i \notin T.$$

λ^T specifies how the dividend α_T is allocated within coalition T. In particular, $\lambda_i^{\{i\}} = 1$ for all i and λ^N can be any vector in the unit simplex Δ_n. We denote by M_n the set of all distribution matrices in the case of n players. For a given game (N, v), the *Harsanyi payoff vector* $h(N, v, \lambda)$ derived from a distribution matrix $\lambda \in M_n$ is given by the inner product $h(N, v, \lambda) = \lambda \cdot \alpha(N, v)$:

$$h_i(N, v, \lambda) = \sum_{T \in \mathcal{C}(N)} \lambda_i^T \alpha_T(N, v) = v(i) + \sum_{T \in \mathcal{C}(N) \backslash \{i\}} \lambda_i^T \alpha_T(N, v) \quad (i = 1, \dots, n).$$

$$(15)$$

---

[1]This characterizes what Shapley [1971] calls *strictly* convex games, games with increasing marginal contributions.

It is an allocation:

$$\sum_{i \in N} h_i(N, v, \lambda) = \sum_{i \in N} \sum_{T \in \mathcal{C}(N)} \lambda_i^T \alpha_T(N, v) = \sum_{T \in \mathcal{C}(N)} \alpha_T(N, v) \sum_{i \in N} \lambda_i^T = v(N).$$

We call $h(N, v, \lambda)$ a *Harsanyi value* and the set of all H-payoff vectors obtained by considering all distributions of dividends defines the *Harsanyi set*[m]:

$$H(N, v) = \{x \in \mathbf{R}^n \mid x = h(N, v, \lambda) \text{ for some } \lambda \in M_n\}.$$

Following Remark 2, the Harsanyi set of an inessential game reduces to a single allocation, namely $(v(1), \ldots, v(n))$. The Harsanyi set associated to the unanimity game (N, u_N) is the unit simplex: $H(N, u_N) = \Delta_n$.

For a given subset T, the dividend α_T is allocated between the members of T and, depending on its sign, players in T receive a positive or a negative amount. Hence, the Harsanyi set can alternatively be written as

$$H(N, v) = \sum_{T \in \mathcal{C}(N)} \{x \in \mathbf{R}^n \mid x(T) = \alpha_T, x(N \backslash T) = 0, \text{sign}(x_i) = \text{sign}(\alpha_T)$$

$$\text{for all } i \in N\}. \quad (16)$$

It is obviously a nonempty and convex set. It is covariant. Indeed, if $v''(S) = a\,v'(S) + b(S)$ for some $a > 0$ and $b \in \mathbf{R}^n$, we have

$$\alpha_{\{i\}}(N, v'') = v''(i) = a\,v'(i) + b_i = a\,\alpha_{\{i\}}(N, v') + b_i,$$

$$\alpha_{\{ij\}}(N, v'') = v''(ij) - v''(i) - v''(j)$$

$$= a\,(v'(ij) + b_i + b_j) - (a\,v'(i) + b_i) - (a\,v'(j) + b_j)$$

$$= a\,\alpha_{\{ij\}}(N, v'), \ldots$$

i.e., the additive term affects only the coefficient associated to singletons. Therefore, we have

$$h_i(N, v'', \lambda) = \sum_{T \in \mathcal{C}(N)} \lambda_i^T \alpha_T(N, v'') = a\,\alpha_{\{i\}}(N, v') + b_i + a \sum_{T \in \mathcal{C}(N) \backslash \{i\}} \lambda_i^T \alpha_T(N, v')$$

$$= a\,h_i(N, v', \lambda) + b_i \quad (i = 1, \ldots, n).$$

H-payoff vectors are not necessarily imputations and it may even be that the Harsanyi set contains no imputation at all.[n] However, for almost positive games, H-payoff vectors are imputations, an immediate consequence of (15). The following proposition is due to Derks *et al.* [2000].

Proposition 2. *Marginal contribution vectors are Harsanyi payoff vectors.*

---

[m]The Harsanyi set was introduced as a solution concept by Vasil'ev [1978, 1981] and by Hammer *et al.* [1977], independently. The later used the term *selectope*.
[n]Derks *et al.* [2010] give the example of a game that fails to be superadditive, whose Harsanyi set has no intersection with the imputation set.

Proof. Consider a game (N, v), an arbitrary ordering $\pi \in \Pi_N$ and the distribution matrix $\lambda \in M_n$ defined by

$$\lambda_i^T = \begin{cases} 1 & \text{if } i \in T \text{ and } T \subset \pi^i, \\ 0 & \text{otherwise,} \end{cases}$$

where π^i is the set of players preceding i in π and including i. For any given coalition T, λ gives a positive share *only* to the player in T that has the highest rank in π. Consequently, $\sum_{i \in T} \lambda_i^T = 1$ for all $T \subset N$ and $\lambda \in \Lambda_n$. The corresponding H-payoff vector is then given by

$$h_i(N, v, \lambda) = \sum_{T \in \mathcal{C}(N)} \lambda_i^T \alpha_T = \sum_{T \in \mathcal{C}_i(\pi^i)} \alpha_T = \sum_{T \subset \pi^i} \alpha_T - \sum_{T \subset \pi^i \setminus i} \alpha_T = v(\pi^i) - v(\pi^i \setminus i).$$

Hence, $h(N, v, \lambda)$ is the marginal contribution vector associated to the ordering π. ☐

Hence, following Remark 5, the *Harsanyi imputation set* defined by

$$\text{HI}(N, v) = H(N, v) \cap I(N, v),$$

is a solution set that satisfies the five basic properties when applied to 0-monotonic games.[o]

3.6. *Weighted Shapley values*

The Shapley value relies on symmetry: equal amounts are allocated to symmetric players. Shapley [1953b] derives (13) from the following formula:

$$\text{SV}_i(N, v) = \sum_{T \in \mathcal{C}_i(N)} \frac{1}{t} \alpha_T(N, v) \quad (i = 1, \ldots, n) \tag{17}$$

i.e., the Shapley value is the H-payoff vector associated to the *uniform distribution* of dividends within each coalition: $\lambda_i^T = 1/t$ for all $i \in T$ and $\lambda_i^T = 0$ for all $i \notin T$. Dropping symmetry opens the possibility for symmetric players to be treated differently. Shapley [1953] also introduces an *asymmetric* version of the value obtained by introducing exogenous weights in order to cover asymmetries that are not included in the underlying game. Weighted games are denoted by (N, v, w) where (N, v) is a transferable utility game and $w = (w_1, \ldots, w_n) \in \mathbf{R}_+^n \setminus 0$ are individual weights.

We denote by $\text{SV}(N, v, w)$ the weighted Shapley value associated to the game (N, v, w). It is the Harsanyi payoff vector associated to the dividends' distribution derived from w:

$$\text{SV}_i(N, v, w) = \sum_{T \in \mathcal{C}_i(N)} \frac{w_i}{w(T)} \alpha_T(N, v). \tag{18}$$

---

[o]Notice that for 2-player games, the Harsanyi set coincides with the imputation set. Vasil'ev [1981] provides necessary and sufficient conditions for nonemptiness of the Harsanyi imputation set. See also Derks *et al.* [2010].

That definition is actually valid only if *at most one* of the w_i's is equal to zero. To show this, consider an arbitrary player, say player 1. With positive weights, (18) can be decomposed as follows:

$$\mathrm{SV}_1(N, v, w) = \alpha_{\{1\}} + \sum_{T \in \mathcal{C}_1(N) \backslash \{1\}} \frac{w_1}{w(T)} \alpha_T(N, v)$$

$$= v(1) + \sum_{T \in \mathcal{C}_1(N) \backslash \{1\}} \frac{1}{1 + \sum_{i \in T \backslash 1} (w_i / w_1)} \alpha_T(N, v).$$

Assuming $w_i > 0$ for all $i \neq 1$ and letting $w_1 \to 0$, we obtain a well-defined limit, namely $v(1)$. When there are more than three players and at least two of them are assigned a zero weight, there may be a continuum of values depending on the relative speeds of convergence.

The set of all weighted values is obtained by considering all positive weights and all possible limits of sequences of positive weights. More precisely, considering normalized weights in Δ_n and a sequence of positive weights (w^k) converging to some boundary point $w \in \partial \Delta_n$, the resulting limit is given by

$$\mathrm{SV}_i(N, v, w) = v(i) + \lim_{w^k \to w} \sum_{T \in \mathcal{C}_i(N) \backslash \{i\}} \frac{1}{1 + \sum_{j \in T \backslash i} (w_j^k / w_i^k)} \alpha_T(N, v).$$

It exists and it coincides with $v(i)$ if $w_j^k / w_i^k \to \infty$ for all $j \neq i$. We denote by $\mathrm{WS}(N, v)$ the resulting set of all weighted values.

Applying (18) to the unanimity game (N, u_N), we get

$$\mathrm{SV}_i(N, u_N, w) = \frac{w_i}{w(N)} \quad (i = 1, \dots, n). \tag{19}$$

It is a well-defined expression for all $w \in \mathbf{R}_+^n \backslash 0$ and it equals $1/n$ in the symmetric case.

Weighted Shapley values $\mathrm{SV}(N, v, w)$ can alternatively be obtained as random order values $\mathrm{RV}(N, v, p_w)$ where p_w is a probability distribution on players' orderings depending on w. For an arbitrary players' ordering π, the marginal contributions of player i in the unanimity game (N, u_N) are defined by

$$\mu_i^\pi(N, u_N) = \begin{cases} 1 & \text{if (and only if) } i \text{ comes } last \text{ in } \pi, \\ 0 & \text{otherwise.} \end{cases}$$

Hence, (19) corresponds to the random order values associated to distributions such that $w_i / w(N)$ is the probability that player i comes *last* in an arbitrary ordering, i.e., $p \in \Delta(\Pi_N)$ should satisfy

$$\mathrm{RV}_i(N, u_N, p) = \sum_{\pi^{-i} \in \Pi_{N \backslash i}} p(\pi^{-i}, i) = \frac{w_i}{w(N)} \quad \text{for all } i \in N.$$

Let us assume for a moment that weights are positive and natural numbers, w_i being interpreted as the number of players of type i. We then compute the probability that a given ordering comes out through a sequence of $w(N)$ independent drawings, knowing that each time a player is drawn, he is removed and only the last player of a given type to be removed is placed in the ordering. The number of possible sequences is given by $w(N)!/\prod w_i!$ and they all have the same probability of occurrence. To illustrate the process, let us take $n = 3$ and $w = (1, 2, 3)$. Then, $w(N) = 6$ and there are 60 drawing sequences. For instance, the sequence (3,2,3,3,1,2) leads to the ordering (3,1,2) where player 2 comes last.

Player j comes out last in a given ordering if and only if he is the last to be drawn. This occurs with probability

$$\frac{(w(N) - 1)!}{(w_j - 1)! \prod_{i \neq j} w_i!} \frac{\prod w_i!}{w(N)!} = \frac{w_j}{w(N)}.$$

The probability that player k comes next to last knowing that player j came last is given by:

$$\frac{(w(N \backslash j) - 1)!}{(w_k - 1)! \prod_{i \neq j,k} w_i!} \frac{\prod_{i \neq j} w_i!}{w(N \backslash j)!} = \frac{w_k}{w(N \backslash j)}.$$

Using this argument repeatedly until the second position, the probability that the ordering $\pi = (i_1, \ldots, i_n)$ comes out is given by

$$p_w(\pi) = \frac{w_{i_2}}{w_{i_1} + w_{i_2}} \cdots \frac{w_{i_{n-1}}}{w_{i_1} + \cdots + w_{i_{n-1}}} \frac{w_{i_n}}{w_{i_1} + \cdots + w_{i_n}} = \prod_{k=2}^{n} \frac{w_{i_k}}{\sum_{j=1}^{k} w_{i_j}}$$

or

$$p_w(\pi) = \prod_{k=2}^{n} \frac{1}{1 + \sum_{j=1}^{k-1} (w_{i_j}/w_{i_k})}. \tag{20}$$

This formula is then extended to the case where weights are real numbers.[P] If a player is assigned a zero weight, weighted values are obtained as limit of sequences of positively weighted values. If there is a *single* zero weight player, say player i, the limit distribution is still uniquely defined: player i is *first* with probability 1 and receives his individual worth $v(i)$. When two players or more are assigned a zero weight, a continuum of values may be associated to the same normalized weight system. Considering converging sequences of positive weights, the resulting value may depend on their relative speeds of convergence.

The probability distributions p_w are homogeneous of degree zero in w. Weights may therefore be normalized. For a given set N of players and weights $w \in \Delta_n$, the set of *all* weighted values is obtained from probability distributions in the set

$$F_N(w) = \{p \in \Delta(\Pi_N) \mid p(\pi) = \lim_{w^k \to w} p_{w^k}(\pi)$$

for some converging sequence $(w^k) \subset \text{int}\Delta_n\}.$

[P]I am grateful to Gerard van der Laan for suggesting this procedure. I initially used the sequence of n drawings where, each time a player is drawn, all players of the same type are removed. It leads to a probability distribution where $w_i/\Sigma w_j$ is the probability that player i comes *first*. This is appropriate for cost games and duals of surplus sharing games. The distribution (20) is then obtained by considering the reverse order; see Dehez [2011].

In view of (20), this is a well-defined set: for any positive sequence (w^k) converging to $w \in \Delta_n$, the associated sequence of distributions p_{w^k} converges to a distribution $p_w \in \Delta(\Pi_N)$. The *Shapley set* is then defined as the set of all weighted values:

$$\text{WS}(N, v) = \{x \in \mathbf{R}^n \,|\, x = RV(N, v, p) \text{ for some } p \in F_N(w) \text{ and } w \in \Delta_n\}. \quad (21)$$

Remark 7. There is a one-to-one relationship between the set of *positively weighted values* and the (relative) *interior* of Δ_n: any positively weighted value is associated to a unique *normalized* weight system, and vice versa.

Let us denote by $Z = \{i \in N \,|\, w_i = 0\}$ the set of zero weight players.[a] Given a game (N, v), let (Z, v_Z) be its restriction to Z and define the game $(N \backslash Z, \hat{v})$ by $\hat{v}(S) = v(Z \cup S) - v(Z)$. It is the game that concerns the subset of nonzero weight players, once $v(Z)$ has been distributed to zero weight players. The following proposition establishes that, in order to compute weighted values, positive-weight players and zero weight players can be treated separately.

Proposition 3. *The values of the weighted game (N, v, w) consists of the allocations $x = (x_Z, x_{N \backslash Z})$ where $x_Z \in W(Z, v_Z)$ and $x_{N \backslash Z} = SV(N \backslash Z, \hat{v}, w_{N \backslash Z})$.*

Proof. Inspecting (20), we observe that the distributions in $F_N(w)$ assign a zero probability to orderings in which a nonzero weight player is followed by a zero weight player. Hence, only orderings of the form $\pi = (\pi', \pi'') \in (\Pi_Z \times \Pi_{N \backslash Z})$ do actually matter and the distributions $p_w \in F_N(w)$ are of the form

$$p_w(\pi) = \begin{cases} p_0(\pi')p_{w_{N \backslash Z}}(\pi'') & \text{for all } \pi = (\pi', \pi'') \in \Pi_Z \times \Pi_{N \backslash Z}, \\ 0 & \text{otherwise.} \end{cases}$$

where p_0 is an arbitrary probability distribution on Π_Z. The corresponding allocation is then given by

$$x_i = \sum_{(\pi', \pi'') \in \Pi_Z \times \Pi_{N \backslash Z}} p_0(\pi')p_{w_{N \backslash Z}}(\pi'')\mu_i^{(\pi', \pi'')}(N, v).$$

By definition of the marginal contribution vectors, $\mu_i^{(\pi', \pi'')}$ can be decomposed as follows:

$$\mu_i^{(\pi', \pi'')}(N, v) = (\mu_i^{\pi'}(Z, v_Z), \; \mu_i^{\pi''}(N \backslash Z, \hat{v})).$$

Hence, for a player $i \in Z$, we have

$$x_i = \sum_{\pi' \in \Pi_Z} p_0(\pi') \, \mu_i^{\pi'}(Z, v_Z) \sum_{\pi' \in \Pi_{N \backslash Z}} p_{w_{N \backslash Z}}(\pi'') = \sum_{\pi' \in \Pi_Z} p_0(\pi')\mu_i^{\pi'}(Z, v_Z)$$

i.e., $x_Z = RV(Z, v_Z, p_0)$. Hence, the probability distribution p_0 being arbitrary, we can conclude that $x_Z \in W(Z, v_Z)$.

---

[a]We omit the dependence of Z on w.

Consider now the game $(N \backslash Z, \hat{v})$. For a player $i \in N \backslash Z$ and an arbitrary ordering $(\pi', \pi'') \in (\Pi_Z \times \Pi_{N \backslash Z})$, we have

$$x_i = \sum_{\pi'' \in \Pi_{N \backslash Z}} p_{w_{N \backslash Z}}(\pi'') \mu_i^{\pi''}(N \backslash Z, \hat{v}) \sum_{\pi' \in \Pi_Z} p_0(\pi')$$

$$= \sum_{\pi'' \in \Pi_{N \backslash Z}} p_{w_{N \backslash Z}}(\pi'') \mu_i^{\pi''}(N \backslash Z, \tilde{v})$$

i.e., $x_{N \backslash Z}$ is the weighted value of the game $(N \backslash Z, \hat{v}, w_{N \backslash Z})$. $\qquad \square$

Remark 8. In practice, when more than one player are assigned a zero weight, it would be natural to treat them equally by considering converging sequences such that the ratios of their weights are equal to 1. The distribution is then given by $p_0(\pi) = 1/z!$ for all $\pi \in \Pi_Z$ and the resulting allocation is the symmetric Shapley value of the game (Z, v_Z).[r]

The Shapley set is clearly a nonempty subset of the Weber set and, as a solution, it is covariant. It is however not a convex set in general, as was observed by Monderer *et al.* [1992], except for the 2-player games or for convex games, as we shall see later.

Remark 9. Owen [1968] has been the first to notice that weighted values are not necessarily monotonic with respect to weights: an increase in the weight assigned to a player may indeed result in a decrease of his payoff. Weights being interpreted as measures of players' relative importance (Shapley talks about bargaining abilities), this is an embarrassing fact. It is however no surprise in view of (18), knowing that dividends may be negative.[s] Monotonicity clearly holds for almost positive games. Monderer *et al.* [1992] have shown that it actually holds for (and *only* for) convex games. This can be explained intuitively by the link that exists between a characteristic of convex games and the probability distribution over orderings induced by the weights. Increasing the weight of a player means increasing his probability of arriving late and we know that marginal contributions are increasing with coalition size in convex games. Hence, increasing the weight of a player naturally increases his expected payoff.

3.7. *Relation between solutions*

The question is now to see how the core, the Weber set, the Shapley set and the Harsanyi set are interrelated.

---

[r]See Dehez and Tellone [2013] for an application of the weighted Shapley value with zero weight players.
[s]Owen [1968] suggests to interpret weights as reflecting slowness to reach a decision. An alternative definition of weighted value has been proposed by Haeringer [2006] in which an increase in the weight of a player leads to an increase in his share in positive dividends and a decrease of his share in negative dividends.

Proposition 4. *The following sequence of inclusions holds for 0-monotonic games:*

$$C(N, v) \subset \mathrm{WS}(N, v) \subset W(N, v) \subset \mathrm{HI}(N, v) \subset H(N, v).$$

When applied to convex games, the first three solutions coincide.

Proof. Weber [1988] has shown that the core is a subset of the Weber set and that, when applied to convex games, the two solutions coincide. Actually this coincidence is a *necessary and sufficient* condition for convexity. We have already seen that weighted values are random order values.[t] Monderer *et al.* [1992] have shown that the core is a subset of the set of weighted values. By Proposition 2, random order values are convex combinations of Harsanyi imputation vectors. Hence, the Weber set is a subset of the Harsanyi set. The sequence of inclusions then follows from the convexity of the Harsanyi set.   □

Proposition 5. *All solutions coincide on the set of almost positive games:*

$$C(N, v) = \mathrm{WS}(N, v) = W(N, v) = \mathrm{HI}(N, v) = H(N, v).$$

This is a corollary of the following proposition due to Hammer *et al.* [1977] and Vasil'ev [1978].

Proposition 6. *The core and the Harsanyi set coincide on the set of almost positive games.*

Proof. Let (N, v) be an almost positive (and thereby convex) game. Looking at its 0-normalization, we have

$$v_0(S) = \sum_{T \in \mathcal{C}(N)} \alpha_T(N, v_0) \, u_T(S),$$

where the $\alpha'_T s$ are all nonnegative. For any given $T \subset N$, the core of the game $(N, \alpha_T u_T)$ is given by

$$C(N, \alpha_T u_T) = \{x \in \mathbf{R}^n_+ \,|\, x(T) = \alpha_T \text{ and } x_i = 0 \text{ for all } i \notin T\}.$$

Indeed $u_T(i) = 0$ for all $T \neq \{i\}$, $u_{\{i\}}(i) = 1$ and $u_{\{i\}}(j) = 0$ for all $j \neq i$. The core is additive on the class of convex games. This follows from the following two lemmas.

Lemma 3. *The core is a superadditive solution* [Peleg, 1986].

Lemma 4. *The Weber set is a subadditive solution* [Dragan *et al.*, 1989].

Hence, we have

$$C(N, v_0) = \sum_{T \in \mathcal{C}(N)} C(N, \alpha_T u_T) = \sum_{T \in \mathcal{C}(N)} \{x \in \mathbf{R}^n_+ \,|\, x(T) = \alpha_T \text{ and } x(N \backslash T) = 0\},$$

where the right-hand side is the Harsanyi set of the game (N, v_0).   □

---

[t]In fact, the Shapley set is in general a dimensionally small subset of the Weber set.

Actually, Hammer *et al.* [1977] and Vasil'ev [1981] prove that the core and the Harsanyi set coincide *only if* they apply to almost positive games. Knowing that core allocations (if any) are H-payoff vectors, another way to prove Proposition 6 consists in showing that the reverse inclusion $H(N, v) \subset C(N, v)$ holds for almost positive games. Indeed, using (15), we have

$$\sum_{i \in S} h_i(N, v, \lambda) = \sum_{i \in S} \sum_{T \in \mathcal{C}(N)} \lambda_i^T \alpha_T = \sum_{i \in S} \sum_{T \in \mathcal{C}(S)} \lambda_i^T \alpha_T + \sum_{i \in S} \sum_{\substack{T \subset \mathcal{C}(N) \\ T \not\subset S}} \lambda_i^T \alpha_T$$

$$= \sum_{T \in \mathcal{C}(S)} \alpha_T \sum_{i \in T} \lambda_i^T + \sum_{i \in S} \sum_{\substack{T \subset \mathcal{C}(N) \\ T \not\subset S}} \lambda_i^T \alpha_T$$

$$= v(S) + \sum_{i \in S} \sum_{\substack{T \subset \mathcal{C}(N) \\ T \not\subset S}} \lambda_i^T \alpha_T \geq v(S).$$

At this stage, we can conclude that the core, the Shapley set, the Weber set and the Harsanyi imputation set all satisfy the five basic properties when applied to convex games. In particular, the Shapley set is a convex set in this case.

By Proposition 6, $C(N, v^+) = H(N, v^+)$ and $C(N, v^-) = H(N, v^-)$. Furthermore, Derks *et al.* [2000] proves that $H(N, v) = C(N, v^+) - C(N, v^-)$.

Remark 10. Looking at the Shapley set and assuming convexity, there is a homeomorphism between the *relative interior* of the unit simplex and the relative interior of the core. This homeomorphism cannot be extended to nonnegative weights and boundary core allocations, except if the game is strictly convex.

4. Characterizing Solutions

There are different ways to characterize values and solutions. We will consider two ways: by axioms and by restrictions on dividend distributions.

4.1. *Characterization by axioms*

Given a player set N, consider the following properties applying to a value $\varphi \colon G(N) \to \mathbf{R}^n$:

- *Efficiency:* $\sum \varphi_i(N, v) = v(N)$.
- *Weak positivity:* $v \in G^+(N) \Rightarrow \varphi(N, v) \in \mathbf{R}_+^n$.
- *Strong positivity:* $v \in \mathrm{MG}(N) \Rightarrow \varphi(N, v) \in \mathbf{R}_+^n$.
- *Null player:* i null in $(N, v) \Rightarrow \varphi_i(N, v) = 0$.
- *Additivity:* $\varphi(N, v_1 + v_2) = \varphi(N, v_1) + \varphi(N, v_2)$.

These are usual properties. The following proposition is due to Vasil'ev [1982, 2006].[u] We give here a simple proof.

---

[u] Vasil'ev also requires homogeneity although only additivity is actually needed.

Proposition 7. *For any given player set N, a value ϕ satisfies efficiency, weak positivity, null player and additivity if and only if φ is a Harsanyi value.*

Proof. H-payoffs vectors are efficient and the H-payoff of a null player is obviously zero. Weak positivity follows from the definition of H-payoffs (15). Additivity follows from dividends' additivity.

Consider the unanimity game $(N, \beta u_T)$ where $T \subset N$ and $\beta \in R$. Because unanimity games are positive and players outside T are null players, an application φ satisfying efficiency, weak positivity and null player must be such that

$$\sum \varphi_i(N, \beta u_T) = \beta,$$

$$\varphi_i(N, \beta u_T) \geq 0 \quad \text{for all } i \in N,$$

$$\varphi_i(N, \beta u_T) = 0 \quad \text{for all } i \notin T.$$

Hence, $\varphi(N, \beta u_T)$ is a H-payoff of the game $(N, \beta u_T)$: $\varphi(N, \beta u_T) = h(N, \beta u_T, \lambda)$ for some $\lambda \in M_n$. More precisely, there exists $\lambda \in M_n$ such that $\varphi_i(N, \beta u_T) = \lambda_i^T \beta$ for all $i \in N$. Now consider an arbitrary characteristic function v on N and its decomposition $v = \sum \alpha_T u_T$ in terms of dividends. Following (7), v can be written as $v = v^+ - v^-$ where (N, v^+) and (N, v^-) are positive games that decompose as $v^+ = \sum_{\alpha_T > 0} \alpha_T u_T$ and $v^- = \sum_{\alpha_T < 0} -\alpha_T u_T$. By additivity, we have

$$\varphi(N, v^+) = \varphi(N, v^+ - v^-) + \varphi(N, v^-) \Leftrightarrow \varphi(N, v) = \varphi(N, v^+) - \varphi(N, v^-).$$

As a consequence, $\varphi(N, v)$ is a H-payoff vector of the game (N, v):

$$\varphi_i(N, v) = \sum_{T: \alpha_T > 0} \varphi_i(N, \alpha_T u_T) - \sum_{T: \alpha_T < 0} \varphi_i(N, -\alpha_T u_T)$$

$$= \sum_{T \subseteq N} \lambda_i^T \alpha_T \quad \text{for all } i \in N. \qquad \square$$

Notice that the Harsanyi set, as a solution, is not additive.[v] Indeed, considering for instance the unanimity game (N, u_N), we have

$$H(N, u_N + (-u_N)) = \{0\},$$

$$H(N, u_N) + H(N, -u_N) = \Delta_n - \Delta_n \neq \{0\}.$$

Vasil'ev [1981] provides an axiomatization of the Harsanyi set that requires convexity and a restricted notion of additivity applying to disjoint games.[w]

Weber [1988] proved that strengthening the positivity axiom results in the Weber set.[x]

Proposition 8. *For any given player set N, a value φ satisfies efficiency, strong positivity, null player and additivity if and only if φ is a random order value.*

---

[v] The author is grateful to a referee for pointing out the non-additivity of the Harsanyi set.
[w] See also Vasil'ev and van der Laan [2002].
[x] See also Derks *et al.* [2000].

To obtain weighted Shapley values, a specific axiom is needed. Derks *et al.* [2000] use the following axiom[y]:

Consistency: $i \in S \subset T \Rightarrow \varphi_i(N, \varphi(N, u_T)(S) \, u_S) = \varphi_i(N, u_T)$.

Proposition 9. *For any given player set N, a value ϕ satisfies efficiency, consistency, null player and additivity if and only if φ is a weighted Shapley value.*

It is easy to check that consistency is satisfied by the weighted value when associated to positive weights. Indeed, we have

$$i \in T \Rightarrow \mathrm{SV}(N, u_T, w) = \frac{w_i}{w(T)}$$

and

$$i \in S \subset T \Rightarrow \mathrm{SV}(N, \varphi(N, u_T)(S) \, u_T, w) = \frac{w_i}{w(S)} \frac{w(S)}{w(T)} = \frac{w_i}{w(T)}.$$

4.2. *Characterization by restrictions on dividend distributions*

Harsanyi payoff vectors are defined by distribution matrices in M_n without any further restrictions. A natural question concerns the identification of restrictions on distribution matrices such that the resulting set of H-payoff vectors corresponds to particular solution sets.

Derks *et al.* [2000] suggest to link distributions within a coalition to distributions within its sub-coalitions by requiring that a player's share in the dividend of a coalition does not increase if the coalition is enlarged:

$$i \in S \subset T \Rightarrow \lambda_i^S \geq \lambda_i^T. \tag{22}$$

It means that if a player leaves a coalition that should not reduce the share of those remaining in the coalition. This monotonicity property imposes strong restrictions on distribution matrices. In particular, if the share of a player in a coalition is zero, his share must be equally zero for all larger coalitions.

We denote by $H^m(N, v)$ the subset of Harsanyi payoffs vectors derived from monotonic distribution matrices. The following proposition is due to Derks *et al.* [2000].

Proposition 10. *Distribution matrices associated to random order values are monotonic.*

Proof. Let us fix a player set N and a nonmonotonic distribution matrix $\lambda \in M_n$ i.e., there exist a player i and coalitions S and T in N such that

$$i \in S \varnothing T \quad \text{and} \quad \lambda_i^S < \lambda_i^T.$$

---

[y] Consistency is a weaker version of the axiom of partnership consistency used by Kalai and Samet [1987] to axiomatize the weighted Shapley value.

Consider the game (N, v) defined by $v = u_S + u_{T \setminus S} - u_T$. Following Remark 1, its dividends are all zero except for coalitions S, T and $T \setminus S$:

$$\alpha_S(N, v) = \alpha_{T \setminus S}(N, v) = 1,$$

$$\alpha_T(N, v) = -1.$$

The H-payoff of player i is therefore given by $h_i(N, v, \lambda) = \lambda_i^S - \lambda_i^T < 0$. Hence, (N, v) being positive valued and monotonic, $h(N, v, \lambda)$ cannot be a random order value. □

We therefore have the following inclusion: $W(N, v) \subset H^m(N, v)$. There may however be H-payoff vectors derived from monotonic distribution matrices that are not random order value. To have equality, a stronger monotonicity requirement is needed. Vasil'ev [1988], Dragan [1994] and Derks *et al.* [2006] prove the following proposition.

Proposition 11. *H-payoff vectors are random order values if and only if they are derived from distribution matrices λ satisfying the following inequalities:*

$$\sum_{S:\, S \supset T} (-1)^{s-t} \lambda_i^S \geq 0 \quad \text{for all } T \subset N \quad \text{and} \quad i \in T. \tag{23}$$

Probabilities (q_i^S) and dividend distributions λ are then related by a Möbius transform:

$$q_i^S = \sum_{T:\, T \supset S} (-1)^{t-s} \lambda_i^T \quad \text{for all } S \subset N \quad \text{and} \quad i \in S,$$

$$\lambda_i^S = \sum_{T:\, T \supset S} q_i^T \quad \text{for all } S \subset N \quad \text{and} \quad i \in S,$$

and $H^{\text{sm}}(N, v) = W(N, v)$ where $H^{\text{sm}}(N, v)$ denotes the set of Harsanyi payoff vectors derived from distribution matrices satisfying the strong monotonicity condition (23).

An even stronger restriction consists in assuming that the distribution vectors λ^S are consistent in the Bayesian sense:

$$i \in S \subset T \Rightarrow \lambda_i^S = \frac{\lambda_i^T}{\lambda^T(S)} \quad \text{if } \lambda^T(S) > 0. \tag{24}$$

In the case where $\lambda^T(S) = 0$, λ^S is any distribution on Δ_n satisfying $\lambda_i^S = 0$ for all $i \notin S$. It is easily verified that distribution matrices satisfying (24) also satisfy the monotonicity properties (22) and (23). The following proposition can be found in Derks *et al.* [2000] or Billot and Thisse [2005].

Proposition 12. *The weighted Shapley values associated to the weight system $w \in \mathbf{R}_+^n \setminus 0$ coincides with the set of H-payoff vectors of the form $h(N, v, \lambda)$ where λ is a distribution matrix satisfying (24) for $\lambda^N = w/w(N)$.*

Proof. Let us fix some $w \in \Delta_n$. If $w \gg 0$, the equivalence follows for the distribution matrix λ satisfying (24) with $\lambda^N = w$:

$$h_i(N, v, \lambda) = \sum_{T \subset N} \lambda_i^T \alpha_T = \sum_{T \in \mathcal{C}_i(N)} \frac{\lambda_i^N}{\lambda^N(T)} \alpha_T = \sum_{T \in \mathcal{C}_i(N)} \frac{w_i}{w(T)} \alpha_T = SV_i(N, v, w).$$

Assume now that the set Z of zero weight players is nonempty. We first observe that, for all $S \subset N$ such that $S \cap (N \backslash Z) \neq \emptyset$, (24) implies $\lambda_i^S = 0$ for all $i \in Z \cap S$. Hence, we can treat zero weight and nonzero weight players separately. Consider first nonzero weight players and the distribution matrix $\hat{\lambda}$ derived from the distribution vector $\lambda^{N\backslash Z} = w_{N\backslash Z}$ using (24). Applying the above argument for the player set $N\backslash Z$, we obtain $h(N\backslash Z, \hat{v}, \hat{\lambda}) = WS(N\backslash Z, \hat{v}, w_{N\backslash Z})$ where $(N\backslash Z, \hat{v})$ is the game defined by $\hat{v}(S) = v(Z \cup S) - v(Z)$ introduced within the proof of Proposition 3. Consider now zero weight players and an arbitrary distribution λ^Z on Z. The distribution vectors λ^T for subsets $T \subset Z$ are then obtained applying (24). Considering all distributions on Z generates the Harsanyi set $H(Z, v_Z)$ which coincides with $W(Z, v_Z)$ by Proposition 8.

Hence, $H^b(N, v) = WS(N, v)$ where $H^b(N, v)$ denotes the set of H-payoff vectors derived from distribution matrices satisfying (24). $\qquad\square$

4.3. *Implications of monotonic dividend distributions*

Monotonicity of dividend distribution can be considered as a natural restriction. Can we characterize $H^m(N, v)$, the set of H-payoff vectors derived from distribution matrices satisfying (24)?

We already know that $W(N, v) \subset H^m(N, v)$. Furthermore, the core and the Harsanyi set coincide if (and only if) they apply to almost positive games, in which case $C(N, v) = H^m(N, v) = H(N, v)$: core allocations can be written as H-payoff vectors derived from monotonic dividend distributions.

What about a larger class of games? Billot and Thisse [2005] claim that $H^m(N, v)$ coincides with the core if (and only if) the game is convex. Actually, this is true only for 2- and 3-player games.

Consider a 3-player game and let x be the H-payoff vector corresponding to a monotonic dividend distribution matrix λ. Using (22), individual rationality results from superadditivity:

$$
\begin{aligned}
x_1 - v(1) &= \lambda_1^{12}(v(12) - v(1) - v(2)) + \lambda_1^{13}(v(13) - v(1) - v(3)) \\
&\quad + \lambda_1^{123}(v(123) - v(12) - v(13) - v(23) + v(1) + v(2) + v(3)) \\
&= (\lambda_1^{12} - \lambda_1^{123})(v(12) - v(1) - v(2)) + (\lambda_1^{13} - \lambda_1^{123}) \\
&\quad \times (v(13) - v(1) - v(3)) + \lambda_1^{123}(v(123) - v(23) - v(1)) \\
&\geq \lambda_1^{123}(v(123) - v(23) - v(1)) \geq 0.
\end{aligned}
\tag{25}
$$

Considering 2-player coalitions, superadditivity implies the following inequality:

$$(x_1 + x_2) - v(12) = \lambda_1^{13}(v(13) - v(1) - v(3)) + \lambda_2^{23}(v(23) - v(2) - v(3))$$
$$+ (\lambda_1^{123} + \lambda_2^{123})(v(123) - v(12) - v(13) - v(23)$$
$$+ v(1) + v(2) + v(3))$$
$$\geq \lambda_1^{123}(v(123) - v(12) - v(23) + v(2)) + \lambda_2^{123}$$
$$\times (v(123) - v(12) - v(13) + v(1)),$$

where the last part is nonnegative under convexity. The argument applies identically to all single players and 2-player coalitions, confirming that x is a core allocation under convexity.

For more than three players, convexity is not sufficient to ensure that H-payoff vectors are core allocations under monotonicity. Consider the 4-player convex game defined by $v(S) = s - 1$. Its dividends are given by

$$\alpha_{\{i\}} = 0,$$

$$\alpha_{\{i,j\}} = 1,$$

$$\alpha_{\{i,j,k\}} = -1,$$

$$\alpha_{\{1,2,3,4\}} = 1.$$

The H-payoffs associated to the monotonic matrix given in Table 1 are (0.4, 0.7, 0.8, 1.1), an allocation that does not belong to the core: the coalition $\{1, 2, 3\}$ indeed obtains only 1.9.

For 3-player games, (25) tells us that superadditivity is enough to ensure that monotonic dividend distributions define imputations. For 4-player games, adding convexity ensures individual rationality. To simplify and without loss of generality, let us assume $v(i) = 0$ for all i. Then, by Lemma 1, $v(S) \geq 0$ for all S. The H-payoff of player 1 for an arbitrary distribution matrix λ is given by

$$x_1 = \lambda_1^{12}v(12) + \lambda_1^{13}v(13) + \lambda_1^{14}v(14) + \lambda_1^{123}(v(123) - v(12) - v(13) - v(23))$$
$$+ \lambda_1^{124}(v(124) - v(12) - v(14) - v(24))$$
$$+ \lambda_1^{134}(v(134) - v(13) - v(14) - v(34))$$
$$+ \lambda_1^{1234}(v(1234) - v(123) - v(124) - v(134) - v(234)$$
$$+ v(12) + v(13) + v(14) + v(23) + v(24) + v(34)).$$

Table 1.   Monotonic distribution matrix.

	12	13	14	23	24	34	123	124	134	234	1234
1	0.5	0.4	0.3	0	0	0	0.4	0.3	0.3	0	0.2
2	0.5	0	0	0.5	0.3	0	0.3	0.3	0	0.2	0.2
3	0	0.6	0	0.5	0	0.3	0.3	0	0.3	0.2	0.2
4	0	0	0.7	0	0.7	0.7	0	0.4	0.4	0.6	0.4

Using the monotonicity condition (22), we obtain

$$x_1 \geq \lambda_1^{123}(v(123) - v(13) - v(23)) + \lambda_1^{124}(v(124) - v(12) - v(24))$$
$$+ \lambda_2^{134}(v(123) - v(14) - v(34))$$
$$+ \lambda_1^{1234}(v(1234) - v(123) - v(124) - v(134) - v(234)$$
$$+ v(12) + v(13) + v(14) + v(23) + v(24) + v(34)).$$

Rearranging the above expression, we get

$$x_1 \geq (\lambda_1^{123} - \lambda_1^{1234})(v(123) - v(13) - v(23)) + (\lambda_1^{124} - \lambda_1^{1234})$$
$$\times (v(124) - v(12) - v(24)) + (\lambda_1^{134} - \lambda_1^{1234})$$
$$\times (v(134) - v(14) - v(34)) + \lambda_1^{1234}(v(1234) - v(234)).$$

Combining (22) and convexity, we finally get $x_1 \geq \lambda_1^{1234}(v(1234) - v(234)) \geq 0$.

Convexity is needed for the result to hold. Indeed, consider the following 4-player game and associated dividends.

$$
\begin{aligned}
v(i) &= 0 & \alpha_{\{i\}} &= 0 \\
v(i,j) &= 1 & \alpha_{\{i,j\}} &= 1 \\
v(1,i,j) &= 1 \quad \rightarrow \quad & \alpha_{\{i,j,k\}} &= 1 - 3 = -2 \\
v(2,3,4) &= 2 & \alpha_{\{2,3,4\}} &= 2 - 3 = -1 \\
v(1,2,3,4) &= 2 & \alpha_{\{1,2,3,4\}} &= 2 - 5 + 6 = 3.
\end{aligned}
$$

This game is superadditive but not convex. The distribution matrix given in Table 2, while satisfying the conditions of monotonicity, leads to an allocation that violates individual rationality. Player 1 is indeed allocated a negative amount: $x_1 = 2.1 - 4.2 + 1.65 = -0.45 < 0$.

Table 2. 4-player monotonic distribution matrix.

	12	13	14	23	24	34	123	124	134	234	1234
1	0.7	0.7	0.7	0	0	0	0.7	0.7	0.7	0	0.55
2	0.3	0	0	0.7	0.3	0	0.15	0.15	0	0.15	0.15
3	0	0.3	0	0.3	0	0.3	0.15	0	0.15	0.15	0.15
4	0	0	0.3	0	0.7	0.7	0	0.15	0.15	0.7	0.15

Table 3. 5-player monotonic distribution matrix.

	12	13	14	15	123	124	125	134	135	145	1234	1235	1245	1345	12345
1	0.1	0.1	0.1	0.1	0.1	0.1	0.1	0.1	0.1	0.1	0	0	0	0	0
2	0.9	0	0	0	0.45	0.45	0.45	0	0	0	0.33	0.33	0.33	0	0.25
3	0	0.9	0	0	0.45	0	0	0.45	0.45	0	0.33	0.33	0	0.33	0.25
4	0	0	0.9	0	0	0.45	0	0.45	0	0.45	0.33	0	0.33	0.33	0.25
5	0	0	0	0.9	0	0	0.45	0	0.45	0.45	0	0.33	0.33	0.33	0.25

Beyond four players, convexity does not ensure individual rationality under monotonicity. Indeed, consider the following convex game:

$$
\begin{aligned}
v(i) &= 0 & \alpha_{\{i\}} &= 0 \\
v(i,j) &= 1 & \alpha_{\{i,j\}} &= 1 \\
v(1,i,j) &= 2 & \alpha_{\{1,i,j\}} &= -1 \\
v(i,j,k) &= 3 & \quad\rightarrow\quad \alpha_{\{i,j,k\}} &= 0 \\
v(1,i,j,k) &= 4 & \alpha_{\{1,i,j,k\}} &= 1 \\
v(2,3,4,5) &= 5 & \alpha_{\{2,3,4,5\}} &= -1 \\
v(1,2,3,4,5) &= 6 & \alpha_{\{1,2,3,4,5\}} &= -1.
\end{aligned}
$$

Again here, player 1 is allocated a negative amount (-0.2) on the basis of the monotonic distribution matrix given by Table 3.[z]

4.4. *Graph structures and restrictions on dividend distributions*

Given a game, additional data can be used to place restrictions on dividend distribution. This question has been studied by van den Brink *et al.* [2014] assuming that players are ordered following a *directed graph*. Given a finite set N, a directed graph $D \subset N \times N$ is a set of pairs (i,j) such that $(i,i) \notin D$. The set of directed graphs is denoted by $\mathcal{D}$. Nodes are players and $(i,j) \in D$ means that i "precedes" j. For a given game, the payoff of a player then depends not only upon the characteristic function but also on his position on the graph.[aa]

More specifically, van den Brink *et al.* suggest that, for any given pair of players, the shares in the dividends of all coalitions containing them must be larger or equal for the player that precedes the other:

$$
(i,j) \in G \Rightarrow \lambda_i^S \geq \lambda_j^S \quad \text{for all } S \text{ containing } i \text{ and } j. \tag{26}
$$

Given a game (N,v) and a graph $D \in \mathcal{D}$, this condition leads to a subset of the Harsanyi set that we denote by $H^g(N,v,D)$.

The authors consider only positive games, in which case $H^g(N,v,D)$ is a subset of the core by Proposition 6, that they call *Harsanyi constrained core*. Core allocations are then H-payoff vectors derived from Bayesian consistent distributions and $(i,j) \in D$ is equivalent to $\lambda_i^N \geq \lambda_j^N$. Equivalently, core allocations are weighted values and $(i,j) \in D$ is equivalent to $w_i \geq w_j$. They prove the following two propositions, the second one applying exclusively to positive games.

---

[z] We only reproduce the shares of player 1. Shares can easily be allocated to the other players so as to satisfy monotonicity.

[aa] Another instance of graph-driven restrictions on distributions is given by van den Brink *et al.* [2011] who consider games with communication graphs à la Myerson [1977]. Their idea is to link the dividend distribution to the "power" of players as measured for instance by the size of their neighborhood.

Proposition 13. *For any given game* (N, v) *and graph* $D \in \mathcal{D}$, *the Shapley value* $SV(N, v)$ *is an element of* $H^g(N, v, D)$. *Furthermore, the Shapley value is the unique element of* $H^g(N, v, D)$ *if and only if* D *is the complete directed graph* $\bar{D} = \{(i, j) \in N \times N \mid i \neq j\}$.

Proof. The Shapley value is defined by $\lambda_i^S = 1/s$ for all $i \in S$ and $S \subset N$ and therefore (26) is verified for all graph $D \in \mathcal{D}$. If $D = \bar{D}, \lambda_i^S = \lambda_j^S$ for all $i, j \in S$ and $S \subset N$ and $\lambda_i^S = 1/s$ for all $i \in S$ and $S \subset N$. Hence, $H^g(N, v, \bar{D}) = \{SV(N, v)\}$. Next, consider a graph D such that $(j, k) \notin D$ for some j and $k, j \neq k$, and the unanimity game $(N, u_{\{j,k\}})$. From Remark 1, the allocation $\bar{x} \in \mathbf{R}^n$ defined by $\bar{x}_k = 1$ and $\bar{x}_i = 0$ for all $i \neq k$ belongs to $H^g(N, u_{\{j,k\}}, D)$. It differs from the Shapley value $SV(N, u_{\{j,k\}})$ which allocated $1/2$ to j and k. $\square$

Proposition 14. *For any positive game* $(N, v) \in G^+(N)$ *and graph* $D \in \mathcal{D}$, *we have*

$$H^g(N, v, D) = C(N, v) \quad \text{if and only if} \quad D = \emptyset.$$

Proof. When $D = \emptyset, H^g(N, v, D) = H(N, v)$ and $H(N, v) = C(N, v)$ by Proposition 6. When instead $D \neq \emptyset$, there exist j and k such that $(j, k) \in D$ and the allocation $\bar{x}$ defined in the proof of Proposition 13 does not belong to $H^g(N, u_{\{j,k\}}, D)$. However, it belongs to $C(N, u_{\{j,k\}})$. $\square$

van den Brink *et al.* [2014] also characterize axiomatically the Harsanyi constrained core on the class $G^+(N)$ of positive games. A solution associates a subset $\Phi(N, v, D)$ to any game $(N, v) \in G^+(N)$ and directed graph $D \in \mathcal{D}$. They show that the Harsanyi constrained core is *maximal* among the solutions satisfying efficiency, null player property, additivity, weak positivity, together with the additional property of *structural monotonicity* defined by

For all game $(N, v) \in G^+(N)$ and directed graph $D \in \mathcal{D}$, allocations $x \in \Phi(N, v, D)$ are such that $x_i \geq x_j$ if $(i, j) \in D$ and i is necessary to j in (N, v).

4.5. *An illustration: Liability games*

Liability games have been introduced in Dehez and Ferey [2013].[bb] They cover situations where damage has been caused to a victim by several tortfeasors. The causality question is solved once the damage $v(S)$ that the members of any coalition S *would have caused* is known, their *potential* damage. The problem is to divide the actual damage $v(N)$ between the n tortfeasors. In this framework, the symmetric Shapley value stands as a benchmark from which a judge may deviate if he considers that some tortfeasors are faultier than others. Furthermore, the core has

[bb]The legal aspects are documented in Dehez and Ferey [2016].

an interesting interpretation: core allocations of a liability game are *fair judgments* in the sense that they satisfy the following two (equivalent) conditions:

- no coalition of players contributes *less than its potential damage*

$$x(S) \geq v(S) \quad \text{for all } S \subset N,$$

- no coalition of players contributes *more than its additional damage*

$$x(S) \leq v(N) - v(N \setminus S) \quad \text{for all } S \subset N.$$

Here we will consider the sequential case, usually considered as a difficult one in the legal literature. Following the natural order $1, 2, \ldots, n$, each player is responsible for an additional damage, d_i for player i. The associated game (N, v) is then given by:

$$v(S) = 0 \qquad \text{if } 1 \notin S,$$
$$v(S) = d_1 \qquad \text{if } 1 \in S \text{ and } 2 \notin S,$$
$$v(S) = d_1 + d_2 \quad \text{if } 1, 2 \in S \text{ and } 3 \notin S,$$

and so on. Defining $T_i = \{1, \ldots, i\}$ as the set of successive players, starting with 1 and ending with i, the characteristic function can be written as

$$v(S) = \sum_{i \in N} d_i \, u_{T_i}(S)$$

and Harsanyi dividends are given by:

$$\alpha_T(N, v) = \begin{cases} d_i & \text{if } T = T_i, \\ 0 & \text{otherwise.} \end{cases} \tag{27}$$

Hence, sequential liability games are positive (and thereby convex) and all solutions coincide with the core.[cc] In the 3-player case, the vector of dividends is given by $(d_1, 0, 0, d_2, 0, 0, d_3)$.

The problem is to divide the total damage $v(N) = d_1 + \cdots + d_n$ among the n players. The resulting allocation specifies the compensation that each player must pay to the victim. Using (8), it can be verified that the core of a liability game can be written in terms of the T_i's:

$$C(N, v) = \{x \in \mathbf{R}_+^n \mid x(N) = d(N) \text{ and } x(T_i) \geq v(T_i) \text{ for all } i \in N\}.$$

We observe that the allocation that imposes to the first player to pay the entire damage as well as the allocation $x = d$ that imposes to players to pay each his additional damage are core allocations. Sequential liability games being convex, the core is the polytope whose vertices are the 2^{n-1} distinct marginal contribution vectors.

[cc]Liability games are dual of airport games, known to be positive games.

In the 3-player case, there are four distinct marginal contribution vectors:

$$\mu^{(1,2,3)} = (d_1, d_2, d_3),$$

$$\mu^{(1,3,2)} = \mu^{(3,1,2)} = (d_1, d_2 + d_3, 0),$$

$$\mu^{(2,1,3)} = (d_1 + d_2, 0, d_3),$$

$$\mu^{(2,3,1)} = \mu^{(3,2,1)} = (d_1 + d_2 + d_3, 0, 0).$$

The core and the Shapley set coincide. Hence, fair judgments can be defined equivalently as weighted Shapley values with normalized weights w or Harsanyi payoff vectors with distribution matrix λ satisfying the Bayesian consistency condition (24) such that $\lambda^N = w$. In the context of Tort Law, weights can be used to reflect differences in the degree of misconduct or negligence. Given (27), the H-payoff vector associated to $\lambda^N \gg 0$ defines the following apportionment rule:

$$\mathrm{SV}_i(N, v, w) = \sum_{j=i}^{n} \frac{\lambda_i^N}{\lambda^N(T_j)} d_j. \tag{28}$$

In the 3-player case, for positive weights, (28) reduces to

$$x_1 = d_1 + \lambda_1^{12} d_2 + \lambda_1^{123} d_3 = d_1 + \frac{\lambda_1^{123}}{\lambda^{123}(12)} d_2 + \lambda_1^{123} d_3,$$

$$x_2 = \lambda_2^{12} d_2 + \lambda_2^{123} d_3 = \frac{\lambda_2^{123}}{\lambda^{123}(12)} d_2 + \lambda_2^{123} d_3,$$

$$x_3 = \lambda_3^{123} d_3.$$

This triangular formula shows the one-to-one relationship that exists under convexity between the relative interior of the core and the relative interior of the unit simplex: to each interior core allocation is associated one and only one weight vector in $\lambda^N \in \mathrm{int}\Delta_n$ and vice versa. For boundary core allocations, some players may be exempted and there is indeterminacy if there exists j, $1 < j < n$, such that $\lambda^N(T_i) = 0$ for $i = 1, 2, \ldots, j$. If it is the case, then any distribution λ^S on T_i for $i = 1, 2, \ldots, j$ is possible, for all $S \varnothing N$.

As a matter of illustration, consider the case where $n = 4$. If $\lambda^{1234} = (0, 0, 0, 1)$, we have

$$\lambda^{12} = (a, 1 - a, 0, 0) \quad \text{and} \quad \lambda^{123} = (b, c, 1 - b - c, 0)$$

for some $a, b, c \in [0, 1]$ such that $b + c \leq 1$. The choice $a = 1/2$ and $b = c = 1/3$ corresponds to the Shapley value restricted to the player set $\{1, 2, 3\}$. In the case $n = 3$, if $\lambda^{123}(12) = 0$, we have

$$\lambda^{123} = (0, 0, 1) \quad \text{and} \quad \lambda^{12} = (a, 1 - a, 0)$$

for some $a \in [0, 1]$. It corresponds to the allocation $(d_1 + a d_2, (1 - a) d_2, d_3)$. The choice of $a = 1/2$ corresponds to the Shapley value restricted to the player set $\{1, 2\}$. Instead, the allocations (d_1, d_2, d_3) and $(d_1 + d_2, 0, d_3)$ corresponds to $a = 0$ and

$a = 1$, respectively. The allocation $(d_1 + d_2 + d_3, 0, 0)$ that exempts players 2 and 3 is associated to the weight vector $\lambda^{123} = (1, 0, 0)$. The allocation $(d_1, d_2 + d_3, 0)$ that exempts player 3 is associated to the weight vector $\lambda^{123} = (0, 1, 0)$. In this way, we have covered the four vertices of the core.

The symmetric Shapley value is given by

$$\text{SV}_i(N, v) = \sum_{j=i}^{n} \frac{1}{j} d_j.$$

In the 3-player case, we get the following allocation:

$$x_1 = d_1 + \frac{1}{2} d_2 + \frac{1}{3} d_3,$$

$$x_2 = \frac{1}{2} d_2 + \frac{1}{3} d_3,$$

$$x_3 = \frac{1}{3} d_3.$$

It is important to observe that, *as a rule*, weighted values (or Harsanyi payoffs) are such that no one is liable for damage caused *downstream* in the sequence: what player i contributes depends only on $(d_i, \ldots, d_n)$. This is a characteristic of the core.

5. Concluding Remarks

Other solution concepts could be considered, for instance the *nucleolus* introduced by Schmeidler [1969]. When the core is nonempty, it is a core selection and we know that it is then a particular H-payoff vector. Can it be characterized in terms of dividend distributions? The answer is negative: sequential liability games offer a counterexample. As shown in Dehez and Ferey [2013], the *nucleolus* of a 3-player sequential liability game is given by

$$\varphi(N, v) = \begin{cases} \left(d_1 + \dfrac{d_2}{2} + \dfrac{d_3}{4}, \dfrac{d_2}{2} + \dfrac{d_3}{4}, \dfrac{d_3}{2} \right) & \text{if } d_3 \leq 2d_2, \\[2ex] \left(d_1 + \dfrac{d_2 + d_3}{3}, \dfrac{d_2 + d_3}{3}, \dfrac{d_2 + d_3}{3} \right) & \text{if } d_3 \geq 2d_2. \end{cases}$$

In the first case, it is the average of core's vertices which coincides with the H-payoff associated to $\lambda_1^{12} = 1/2$ and $\lambda_1^{123} = \lambda_2^{123} = 1/4$. In the second case, it is the *equal loss* allocation to which it is not possible to associate an admissible distribution matrix. Furthermore, as an apportionment rule, the nucleolus violates the "downstream" condition: in the second case, what player 3 contributes depends upon damage caused by player 2.

Among the questions that remain open, there is the identification of restrictions on dividend distributions such that the resulting H-payoff vectors are imputations. Following Proposition 11 and individual rationality of random order values for 0-monotonic games, we know that it is the case under strong monotonicity for 0-monotonic games. There may be some weaker restrictions. At this stage, we have only learned that monotonicity is not sufficient even in the case of convex games. Another question concerns the set of H-payoffs resulting from restrictions on dividend distributions associated to graphs, possibly combined with monotonicity and/or convexity.

Appendix A

Proof of Lemma 1. Consider a coalition $S \subset N$ and a player $i \notin S$. By superadditivity, we have

$$v(S \cup i) \geq v(S) + v(i).$$

Hence, if $v(i) \geq 0$ for all $i \in N$, adding a player to a coalition does not decrease its worth. This extends to the addition of any number of players. Positivity of v follows using the above inequality, starting with $S = \{j\}, j = 1, \ldots, n$. □

Proof of Lemma 2. Consider a coalition $T \subset N \backslash i$ not containing j. We already know that $\alpha_T = v(j) = 0$ if $T = \{j\}$. Assume now that $\alpha_S = 0$ for all $S \subset T$ such that $j \in S$. We proceed by induction using (2). Because i is necessary for j in (N, v), we have

$$\alpha_T = v(T) - \sum_{S \subset T} \alpha_S = v(T) - \sum_{S \subset T \backslash j} \alpha_S = v(T) - v(T \backslash j) = 0. \qquad \square$$

Proof of Lemma 3. Given a player set N, consider two set functions $v_1, v_2 \in G(N)$ and an allocation $x \in C(N, v_1) + C(N, v_2)$. Hence there exist x^1 and x^2 such that $x = x^1 + x^2, x^1(S) \geq v_1(S)$ and $x^2(S) \geq v_2(S)$ for all $S \subset N$. Consequently, we have

$$x(S) = x^1(S) + x^2(S) \geq (v_1 + v_2)(S) \quad \text{for all } S \subset N$$

i.e., $x \in C(N, v_1 + v_2)$. □

Proof of Lemma 4. Given a player set N, consider two set functions $v_1, v_2 \in G(N)$ and the corresponding marginal contribution vectors μ^1 and μ^2 as defined by (5). By definition of convex hull [Rockafellar, 1970], we have

$$W(N, v_1) + W(N, v_2) = \text{co}\{\mu^1(\pi) \,|\, \pi \in \Pi_N\} + \text{co}\{\mu^2(\pi) \,|\, \pi \in \Pi_N\}$$

$$= \text{co}(\{\mu^1(\pi) \,|\, \pi \in \Pi_N\} + \{\mu^2(\pi) \,|\, \pi \in \Pi_N\}).$$

The marginal contribution vectors associated to the game $(N, v^1 + v^2)$ are the sum of the marginal contribution vectors. Hence, $W(N, v^1 + v^2) = \text{co}\{\mu^1(\pi) + \mu^2(\pi) \,|\, \pi \in \Pi_N\}$ where

$$\{\mu^1(\pi) + \mu^2(\pi) \,|\, \pi \in \Pi_N\} \subset \{\mu^1(\pi) \,|\, \pi \in \Pi_N\} + \{\mu^2(\pi) \,|\, \pi \in \Pi_N\}.$$

Consequently, by the definition of the convex hull, we have

$$\text{co}\{\mu^1(\pi) + \mu^2(\pi) \,|\, \pi \in \Pi_N\} \subset \text{co}(\{\mu^1(\pi) \,|\, \pi \in \Pi_N\} + \{\mu^2(\pi) \,|\, \pi \in \Pi_N\}). \qquad \square$$

Acknowledgments

The author is grateful to Gerard van der Laan and others for useful comments and suggestions.

References

Ambec, S. and Sprumont, Y. [2002] Sharing a river, *J. Econ. Theory* **107**, 453–462.

Aumann, R. J. and Maschler, M. [1985] Game theoretic analysis of a bankruptcy problem from the Talmud, *J. Econ. Theory* **36**, 195–213.

Billot, A. and Thisse, J. [2005] How to share when context matters: The Möbius value as a generalized solution for cooperative games, *J. Math. Econ.* **41**, 1007–1029.

Bondareva, O. N. [1963] Some applications of linear programming methods to the theory of cooperative games, *Probl. Kybern.* **10**, 119–139 (in Russian).

Dehez, P. [2011] Allocation of fixed costs: Characterization of the (dual) weighted Shapley value, *Int. Game Theory Rev.* **13**, 1–16.

Dehez, P. and Ferey, S. [2013] How to share joint liability. A cooperative game approach, *Math. Social Sci.* **66**, 44–50.

Dehez, P. and Ferey, S. [2016] Multiple causation, apportionment and the Shapley value, *J. Legal Stud.* **45**, 143–171.

Dehez, P. and Tellone, D. [2013] Data games: Sharing public goods with exclusion, *J. Public Econ. Theory* **15**, 654–673.

Derks, J. [2005] A new proof for Weber's characterization of the random order values, *Math. Social Sci.* **49**, 327–334.

Derks, J., Haller, H. and Peters, H. [2000] The selectope for cooperative games, *Int. J. Game Theory* **29**, 23–38.

Derks, J., van der Laan, G. and Vasil'ev, V. [2006] Characterization of the random order values by Harsanyi payoff vectors, *Math. Methods Oper. Res.* **64**, 155–163.

Derks, J., van der Laan, G. and Vasil'ev, V. [2010] On the Harsanyi payoff vectors and Harsanyi imputations, *Theory and Decis.* **68**, 301–310.

Dragan, I. [1994] On the multiweighted shapley value and the random order values, *Proc. 10th Conf. Applied Mathematics*, University of Central Oklahoma, pp. 33–47.

Dragan, I., Potters, J. and Tijs, S. [1989] Superadditivity for solutions of coalitional games, *Libertas Math.* **9**, 101–108.

Gonzáles-Díaz, J. and Sánchez-Rodríguez, E. [2007] A natural selection from the core of a TU game: The core-centre, *Int. J. Game Theory* **36**, 27–46.

Gillies, D. B. [1953] Some theorems on *n*-person games, Ph.D. Thesis, Princeton University.

Gillies, D. B. [1959] Solutions to general nonzero-sum games, in *Contributions to the Theory of Games*, Vol. 4, eds. Tucker, A. W. and Luce, D. R. Annals of Mathematics Study, Vol. 40 (Princeton University Press, NJ, USA), pp. 47–85.

Hart, S. and A. Mas-Colell [1989] Potential, value and consistency, *Econometrica* **57**, 589–614.

Haeringer, G. [2006] A new weight scheme for the Shapley value, *Math. Social Sci.* **52**, 88–98.

Hammer, P. J., Peled, U. N. and Sorensen, S. [1977] Pseudo-Boolean function and game theory I. Core elements and Shapley value, *Cah. Centre Etudes Rech. Opér.* **19**, 156–176.

Harsanyi, J. C. [1959] A bargaining model for the cooperative *n*-person game, in *Contributions to the Theory of Games*, Vol. 4, eds. Tucker, A. W. and Luce, D. R. Annals of Mathematics Study, Vol. 40 (Princeton University Press, NJ, USA), pp. 325–355.

Harsanyi, J. C. [1963] A simplified bargaining model for the *n*-person game, *Int. Econ. Rev.* **4**, 194–220.

Ichiishi, T. [1981] Super-modularity: Applications to convex games and the greedy algorithm for LP, *J. Econ. Theory* **25**, 283–286.

Kalai, E. and Samet, D. [1987] On weighted Shapley values, *Int. J. Game Theory* **16**, 205–222.

Kannai, Y. [1992] The core and balancedness, in *Handbook of Game Theory*, Vol. 1, eds. Aumann, R. J. and Hart, S. (Elsevier, Amsterdam, Netherlands), pp. 355–395.

Littlechild, S. C. and Owen, G. [1973] A simple expression for the Shapley value in a special case, *Manag. Sci.* **20**, 370–372.

Maniquet, F. [2003] A characterization of the Shapley value in queueing problems, *J. Econ. Theory* **109**, 90–103.

Maschler, M., Peleg, B. and Shapley, L.S. [1972] The kernel and bargaining set for convex games, *Int. J. Game Theory* **1**, 73–93.

Monderer, D., Samet, D. and Shapley, L. S. [1992] Weighted values and the core, *Int. J. Game Theory* **21**, 27–39.

Myerson, R. B. [1977] Graphs and cooperation in games, *Math. Oper. Res.* **2**, 225–229.

Owen, G. [1968] A note on the Shapley value, *Manag. Sci.* **14**, 731–732.

Peleg, B. [1986] On the reduced game property and its converse, *Int. J. Game Theory* **15**, 187–200.

Rockafellar, R. T. [1970] *Convex Analysis* (Princeton University Press, NJ, USA).

Schmeidler, D. [1969], The nucleolus of a characteristic function game. *SIAM Journal of Applied Mathematics* **17**, 1163–1170.

Shapley, L. S. [1953a] Additive and non-additive set functions, Ph.D. thesis, Department of Mathematics, Princeton University.

Shapley, L. S. [1953b] A value for *n*-person games, in *Contributions to the Theory of Games*, Vol. 2, eds. Kuhn, H. and Tucker, A. W. Annals of Mathematics Study, Vol. 28 (Princeton University Press, NJ, USA), pp. 307–317. Reprinted in Roth, A. E. (ed.) *The Shapley Value. Essays in Honor of Lloyd Shapley* (Cambridge University Press, Cambridge, UK), pp. 31–40.

Shapley, L. S. [1967] On balanced sets and cores, *Naval Res. Logist. Quart.* **14**, 453–460.

Shapley, L. S. [1971] Cores of convex games, *Int. J. Game Theory* **1**, 11–26.

Shapley, L. S. [1981] Discussion comments on "Equity considerations in traditional full cost allocation practices: An axiomatic approach", in *Joint Cost Allocation, Proc. University of Oklahoma Conf. Costs Allocations*, ed. Moriarity, S. (Centre for Economic and Management Research, University of Oklahoma, OK, USA), pp. 131–136.

van den Brink, R., van der Laan, G. and Pruzhansky, V. [2011] Harsanyi power solutions for graph-restricted games, *Int. J. Game Theory* **40**, 87–110.

van den Brink, R., van der Laan, G. and Vasil'ev, V. A. [2014] Constrained core solutions for totally positive games with ordered players, *Int. J. Game Theory* **43**, 351–368.

Vasil'ev, V. A. [1975] The Shapley value for cooperative games of bounded polynomial variation, *Optim. Vyp* **17**, 5–27 (in Russian).

Vasil'ev, V. A. [1978] Support function of the core of a convex game, *Optim. Vyp* **21**, 30–35 (in Russian).

Vasil'ev, V. A. [1981] On a class of imputations in cooperative games, *Sov. Math. Dokl.* **23**, 53–57.

Vasil'ev, V. A. [1982] On a class of operators in a space of regular set functions, *Optim. Vyp* **28**, 102–111 (in Russian).

Vasil'ev, V. A [1988] Characterization of the cores and generalized NM-solutions for some classes of cooperative games, *Proc. Inst. Math. Novosibirsk Nauk* **10**, 63–89 (in Russian).

Vasil'ev, V. A. [2003] Extreme points of the Weber polyhedron, *Discret. Anal. Issled. Oper.* **10**, 17–55 (in Russian).

Vasil'ev, V. A. [2006] Cores and generalized NM-solutions for some class of cooperative games, in *Russian Contributions to Game Theory and Equilibrium Theory*, eds. Driessen T. *et al.* (Springer, Berlin), pp. 91–149.

Vasil'ev, V. A. [2007] Weber polyhedron and weighted Shapley values, *Int. Game Theory Rev.* **9**, 139–150.

Vasil'ev, V. A. and van der Laan, G. [2002] The Harsanyi set for cooperative TU-games, *Sib. Adv. Math.* **12**, 97–125.

von Neumann, J. and Morgenstern, O. [1944] *Theory of Games and Economic Behavior*, (Princeton University Press, NJ, USA) (3rd and last edition dated 1953).

Weber, R. J. [1988] Probabilistic values for games, in *The Shapley Value. Essays in Honor of Lloyd Shapley*, ed. Roth, A. E. (Cambridge University Press, Cambridge, UK), pp. 101–119.

Chapter 27

An Axiomatization for Two Power Indices for (3,2)-Simple Games

Giulia Bernardi

Dipartimento di Matematica, Politecnico di Milano
P.zza Leonardo da Vinci, 32, Milano
20133, Italy

giulia.bernardi@polimi.it

Josep Freixas*

Departament de Matemàtiques
Escola Politècnica Superior d'Enginyeria de Manresa
Universitat Politècnica de Catalunya
Av. Bases de Manresa, 61–73.
Manresa, 08242, Catalunya, Spain

josep.freixas@upc.edu

The aim of this work is to give a characterization of the Shapley–Shubik and the Banzhaf power indices for (3,2)-simple games. We generalize to the set of (3,2)-simple games the classical axioms for power indices on simple games: transfer, anonymity, null player property and efficiency. However, these four axioms are not enough to uniquely characterize the Shapley–Shubik index for (3,2)-simple games. Thus, we introduce a new axiom to prove the uniqueness of the extension of the Shapley–Shubik power index in this context. Moreover, we provide an analogous characterization for the Banzhaf index for (3,2)-simple games, generalizing the four axioms for simple games and adding another property.

Keywords: Games with abstention; power indices; axioms; voting.

1. Introduction

In classical cooperative theory, simple games are used to model voting situation. However, in a simple game players can vote only either yes or no, while in many real-life voting procedure voters are allowed to have other opinions. For instance, it is possible to take into account the possibility of abstention. A theoretical model

*Corresponding author.

for game with abstention was introduced in Felsenthal and Machover [1997] and extended to voting rules with several levels of approval in input and different levels of output, called (j, k)-games in Freixas and Zwicker [2003]. A voting procedure with abstention can be seen as a (3,2)-simple games: voters can choose among three different options (namely, voting yes, abstaining and voting no) and the outcome is 0 or 1.

The Shapley–Shubik index, presented in Shapley and Shubik [1954], and the Banzhaf index, defined independently in Penrose [1946] and in Banzhaf [1964], are the most well-known and widely accepted ways to measure the power of players in simple games. Two indices analogous to these ones have been defined for (3,2)-simple games to measure the power of players also in this specific voting situation. In particular the Shapley–Shubik power index for (3,2)-simple games was defined by Felsenthal and Machover [1997] and extended to (j, k) games in Freixas [2005b]; while the Banzhaf index for (3,2)-simple games is discussed in Felsenthal and Machover [1998] and for (j, k) games in Freixas [2005a]. Other authors proposed different approaches to the problem of taking into account abstention or different levels of voting approval. One may find different values following the ideas of the Shapley value for different contexts as multichoice games, r-games or bicooperative games. We refer the interested reader to Hsiao and Raghavan [1992], [1993], Bolger [1993], Klijn *et al.* [1999], Grabisch and Lange [2007] and Bilbao *et al.* [2008].

A classical axiomatization of these two power indices for simple games has been provided in Dubey [1975] and in Dubey and Shapley [1979]. The axioms used to characterize the indices are anonymity, transfer, null player, efficiency for the Shapley–Shubik index, and Banzhaf total power for the Banzhaf index. The aim of this work is to provide a characterization for the Shapley–Shubik and the Banzhaf indices for (3,2)-simple games, extending the classical axioms from simple games to (3,2)-simple games. However, as we shall discuss, these classical axioms generalized in the context of (3,2)-simple games are not enough to uniquely characterize the indices on the space of (3,2)-simple games. It is necessary to add another property to describe the behavior of the power index on unanimity games.

The Shapley–Shubik and the Banzhaf indices can be defined on unanimity games and then extended to the family of simple games using the transfer axiom, actually this property holds for any semivalue, as discussed in Carreras *et al.* [2003] and in Bernardi and Lucchetti [2015]. The behavior on unanimity games is crucial in order to uniquely characterize an index. The axioms we propose compare the power of a player in a unanimity game when votes "yes" in a minimal winning tripartition and when he abstains in the same situation, i.e., all other players do not change their vote. We give different conditions for the differences of power in the two situations and use these conditions to deduce the Shapley–Shubik and the Banzhaf index for (3,2)-simple games, respectively.

The chapter is organized as follows. In Sec. 2, we introduce some preliminaries and definitions. In Sec. 3, we discuss axioms. In Sec. 4, we prove the characterization of the Shapley–Shubik index for (3,2)-simple games and the independence of the axioms used. The analogous results for the Banzhaf index for (3,2)-simple games are discussed in Sec. 5. Section 6 concludes the work.

2. Preliminaries and Definitions

A game with abstention is a model of a voting situation in which players have three different possibilities: voting "yes", voting "no", and abstaining. This is a generalization of the standard model of simple games in which players can only vote in support of or against the status quo.

Given a finite set of players N, in simple games the set 2^N represents the set of all coalitions. Actually any coalition $T \in 2^N$ can be seen as a bipartition (T_1, T_2) in which $T_1 = T$ and $T_2 = N \setminus T$. We can view a coalition T as the set of players supporting a decision and the coalition $N \setminus T$ as the set of players against it. Analogously, in the context of (3,2)-simple games we consider the set 3^N of all tripartitions. By tripartition, we denote any element $S = (S_1, S_2, S_3)$, where S_1, S_2, S_3 are mutually disjoint subsets of N such that $S_1 \cup S_2 \cup S_3 = N$ and any S_i can be empty. An element $S = (S_1, S_2, S_3) \in 3^N$ describes a voting situation in which the players in S_i are voting at "level i" of approval. It is supposed that level 1 is the highest, level 2 is the intermediate, and level 3 is the lowest. Hereafter, with the idea of modeling a voting situation, we will say that players in S_1 are voting "yes", players in S_2 are abstaining, players in S_3 are voting "no".

A partial order $\subseteq$ on the set 3^N is defined as follows. If $S, T \in 3^N$, then $S \subseteq T$ means $S_1 \subseteq T_1$ and $S_2 \subseteq T_1 \cup T_2$. We use $S \subset T$ if $S \subseteq T$ and $S \neq T$. In other words, a tripartition S is contained in the tripartition T if players in T are voting as in S or some of them are increasing their level of support. This means that S can be transformed into T by shifting one or more players to higher levels of approval. For instance[a] $(a, b, c) \subseteq (ab, c, \emptyset)$, the second tripartition is obtained by the first one when player b changes from abstaining to voting "yes" and player c switches from voting "no" to abstaining. The tripartition $(\emptyset, \emptyset, N)$ is the minimum of the order $\subseteq$, while the maximum is the tripartition $(N, \emptyset, \emptyset)$.

Definition 1. A (3,2)-simple game is a pair (N, v) in which N is the set of players (or voters) and $v : 3^N \to \{0, 1\}$ is the value function such that:

- it is monotonic, i.e., if $S \subseteq T$ then $v(S) \leq v(T)$;
- $v(\emptyset, \emptyset, N) = 0$ and $v(N, \emptyset, \emptyset) = 1$.

We denote with $\mathfrak{T}^N$ the set of all (3,2)-simple games on the finite set N.

As for simple games, any game $v \in \mathfrak{T}^N$ can be described by the set of *winning tripartitions* $\mathcal{W}(v) = \{S \in 3^N : v(S) = 1\}$ or by the set of *minimal winning tripartitions* $\mathcal{W}^m(v) = \{S \in 3^N : v(S) = 1, v(T) = 0, \forall T \subset S\}$.

---

[a]To simplify the notation we omit the braces to denote the sets in a tripartition, for instance the informal notation (a, b, c) stands for $(\{a\}, \{b\}, \{c\})$.

Definition 2 (Unanimity game). For any tripartition $S \neq (\emptyset, \emptyset, N)$, the *unanimity game* u_S is defined as

$$u_S(T) = \begin{cases} 1 & \text{if } S \subseteq T, \\ 0 & \text{otherwise.} \end{cases}$$

Given a tripartition $S = (S_1, S_2, S_3)$ and a player $a \notin S_3$ we denote with $S_{\downarrow a}$ the tripartition in which player a decreases his or her support of one level

$$S_{\downarrow a} = \begin{cases} (S_1 \setminus \{a\}, S_2 \cup \{a\}, S_3) & \text{if } a \in S_1, \\ (S_1, S_2 \setminus \{a\}, S_3 \cup \{a\}) & \text{if } a \in S_2. \end{cases}$$

Of course, there is also the possibility that player $a \in S_1$ switches from supporting a decision to vote against it

$$S_{\downarrow\downarrow a} = (S_1 \setminus \{a\}, S_2, S_3 \cup \{a\}).$$

In an analogous way, given S and a player $a \notin S_1$, we define the tripartition in which a increases the support

$$S_{\uparrow a} = \begin{cases} (S_1 \cup \{a\}, S_2 \setminus \{a\}, S_3) & \text{if } a \in S_2, \\ (S_1, S_2 \cup \{a\}, S_3 \setminus \{a\}) & \text{if } a \in S_3 \end{cases}$$

and if $a \in S_3$

$$S_{\uparrow\uparrow a} = (S_1 \cup \{a\}, S_2, S_3 \setminus \{a\}).$$

It is possible to characterize some players according to their role in the game. In the following definition, we describe players that do not have influence at all in a voting situation: the outcome of the game does not change whatever they vote.

Definition 3 (Null player). A player $a \in N$ is *null* in the game v if $v(S) = v(S_{\downarrow\downarrow a})$ for any tripartition S such that $a \in S_1$.

Note that a is a null player *if and only if* it holds $v(S) = v(S_{\downarrow a})$ for any tripartition S such that $a \in S_1 \cup S_2$. Note also that the previous definition is equivalent to say that a is a null player if $a \in S_3$ for any $S \in \mathcal{W}^m(v)$.

In the context of simple games, that describe a voting situation, one of the key point is to evaluate the power of players, in order to establish how their vote influences the outcome of the game. For this reason, in simple games, the interesting class of solution concepts is the one of *power indices*. In the following, we define this family on the set of (3,2)-simple games and then focus on the extension of two main power indices.

Definition 4 (Power index). A power index for (3,2)-simple games is a function $\psi : \mathfrak{T}^N \to \mathbb{R}^n$, that assigns to every game v a vector in which the ath component is a measure of the power of player a in the voting system described by v.

As for simple games, there are different power indices for (3,2)-simple games to capture different properties. The well-known Banzhaf [1964] and

Shapley and Shubik [1954] power indices have an equivalent in the context of (3,2)-simple games, as defined in Felsenthal and Machover [1997] and Freixas [2005a].

Definition 5 (Banzhaf index for (3,2)-simple games). For any game $v \in \mathfrak{T}^N$ and any player $a \in N$, define $\eta_a(v)$ as the number of yes–no swings for player a, that is

$$\eta_a(v) = |\{S : a \in S_1 \text{ and } v(S) - v(S_{\downarrow\downarrow a}) = 1\}|.$$

The *Banzhaf index for (3,2)-simple games* is then given by

$$\beta_a(v) = \frac{\eta_a(v)}{3^{n-1}}.$$

The Shapley–Shubik power index for (3,2)-simple games was introduced in Felsenthal and Machover [1997] using the idea of roll-calls. Let $\mathbf{Q}_N$ be the space of all the permutation of N, and let 3^N be the set of all tripartitions of N. The *ternary roll-call space* $\mathbf{R}_N$ is defined as

$$\mathbf{R}_N = \mathbf{Q}_N \times 3^N.$$

Each roll-call $\mathcal{R}$ is given by a queue $q\mathcal{R}$ and a tripartition R, that is $\mathcal{R} = (q\mathcal{R}, R)$ where $q\mathcal{R}$ represents the order in which players are voting and R represents how each one of them is voting. For instance, $q\mathcal{R}(a) = i$ means that a is the ith to vote and $a \in R_1$ means that a is voting "yes". The number of the elements in $\mathbf{R}_N$ is $n!3^n$.

A player a is said to be *pivotal* in $\mathcal{R}$ for the game v (and we write $\mathrm{piv}(\mathcal{R}, v)$) if a is the first player in the queue whose vote fixes the outcome of the voting procedure; this means that after a's vote the outcome is decided, either as winning or losing, no matter what the players after a in $q\mathcal{R}$ are going to vote.

Definition 6 (Shapley–Shubik index for (3,2)-simple games). For any $v \in \mathfrak{T}^N$ and any player $a \in N$, the *Shapley–Shubik index for (3,2)-simple games* is defined as

$$\phi_a(v) = \frac{|\{\mathcal{R} \in \mathbf{R}_N : a = \mathrm{piv}(\mathcal{R}, v)\}|}{3^n n!}. \tag{1}$$

For the purpose of this chapter, we introduce some notation related to roll-calls. Consider the set of roll-calls $\mathbf{R}_N$; for any player $a \in N$, we define the following subsets which form a partition of $\mathbf{R}_N$:

$$\mathcal{R}_a^{\text{yes}} = \{\mathcal{R} : a \in R_1\} = \{\text{roll-calls in which player } a \text{ votes "yes"}\},$$

$$\mathcal{R}_a^{\text{abs}} = \{\mathcal{R} : a \in R_2\} = \{\text{roll-calls in which player } a \text{ abstains}\},$$

$$\mathcal{R}_a^{\text{no}} = \{\mathcal{R} : a \in R_3\} = \{\text{roll-calls in which player } a \text{ votes "no"}\}.$$

Thus,

$$\mathbf{R}_N = \mathcal{R}_a^{\text{yes}} \cup \mathcal{R}_a^{\text{abs}} \cup \mathcal{R}_a^{\text{no}}$$

and $|\mathcal{R}_a^{\text{yes}}| = |\mathcal{R}_a^{\text{abs}}| = |\mathcal{R}_a^{\text{no}}| = n!3^{n-1}$.

Given a player a and a roll-call $\mathcal{R} = (qR, R) \notin \mathcal{R}_a^{no}$, we define the roll-call $\mathcal{R}_{\downarrow a}$ in which players are in the same order as in $\mathcal{R}$, all players in $N \setminus \{a\}$ vote as in $\mathcal{R}$, while a decreases the support of one level

$$\mathcal{R}_{\downarrow a} = (qR, R_{\downarrow a}).$$

Note that if $\mathcal{R} \in \mathcal{R}_a^{yes}$, then $\mathcal{R}_{\downarrow a} \in \mathcal{R}_a^{abs}$; if $\mathcal{R} \in \mathcal{R}_a^{abs}$, then $\mathcal{R}_{\downarrow a} \in \mathcal{R}_a^{no}$.

We also define the roll-call in which a decreases the support of two levels, changing the vote from "yes" to "no": if $\mathcal{R} \in \mathcal{R}_a^{yes}$, then $\mathcal{R}_{\downarrow\downarrow a} \in \mathcal{R}_a^{no}$ is

$$\mathcal{R}_{\downarrow\downarrow a} = (qR, R_{\downarrow\downarrow a})$$

For a roll-call $\mathcal{R} \notin \mathcal{R}_a^{yes}$ we analogously define $\mathcal{R}_{\uparrow a} = (qR, R_{\uparrow a})$ and if $\mathcal{R} \in \mathcal{R}_a^{no}$ we define $\mathcal{R}_{\uparrow\uparrow a} = (qR, R_{\uparrow\uparrow a})$.

Note that, for instance, if $\mathcal{R} \in \mathcal{R}_a^{yes}$ we have: $(\mathcal{R}_{\downarrow a})_{\uparrow a} = \mathcal{R}$, which shows that there is a one-to-one correspondence between $\mathcal{R}_a^{yes}$ and $\mathcal{R}_a^{abs}$ with these changes. In addition, from $(\mathcal{R}_{\downarrow\downarrow a})_{\uparrow\uparrow a} = \mathcal{R}$, the one-to-one correspondence of the three sets $\mathcal{R}_a^{yes}, \mathcal{R}_a^{abs}, \mathcal{R}_a^{no}$ follows.

We also introduce the following sets, for any player $a \in N$ and any game v:

$$Y_{a,v} = \{\mathcal{R} \in \mathcal{R}_a^{yes} : a = \text{piv}(\mathcal{R}, v)\},$$

$$A_{a,v} = \{\mathcal{R} \in \mathcal{R}_a^{abs} : a = \text{piv}(\mathcal{R}, v)\},$$

$$N_{a,v} = \{\mathcal{R} \in \mathcal{R}_a^{no} : a = \text{piv}(\mathcal{R}, v)\}$$

and the following subsets of $A_{a,v}$ and $N_{a,v}$:

$$AY_{a,v} = \{\mathcal{R} \in A_{a,v} : \mathcal{R}_{\uparrow a} \in Y_{a,v}\}, \quad A\overline{Y}_{a,v} = \{\mathcal{R} \in A_{a,v} : \mathcal{R}_{\uparrow a} \notin Y_{a,v}\},$$

$$NY_{a,v} = \{\mathcal{R} \in N_{a,v} : \mathcal{R}_{\uparrow\uparrow a} \in Y_{a,v}\}, \quad N\overline{Y}_{a,v} = \{\mathcal{R} \in N_{a,v} : \mathcal{R}_{\uparrow\uparrow a} \notin Y_{a,v}\}.$$

Thanks to the previous notation, the Shapley–Shubik index as defined in (1) can be written as

$$\phi_a(v) = \frac{1}{3^n n!}[|Y_{a,v}| + |A_{a,v}| + |N_{a,v}|]$$

or as

$$\phi_a(v) = \frac{1}{3^n n!}[|Y_{a,v}| + |AY_{a,v}| + |A\overline{Y}_{a,v}| + |NY_{a,v}| + |N\overline{Y}_{a,v}|].$$

Lastly, given two games $v, w \in \mathfrak{T}^N$, we define the following games:

Disjunction: The game $v \vee w$ is defined as $(v \vee w)(S) = \max\{v(S), w(S)\}$.

Conjunction: The game $v \wedge w$ is defined as $(v \wedge w)(S) = \min\{v(S), w(S)\}$.

Let us make some remarks about these operations:

(1) $W(v \vee w) = W(v) \cup W(w)$ and $W(v \wedge w) = W(v) \cap W(w)$;
(2) if $W^m(v) = \{S_1, \ldots, S_k\}$ then $v = u_{S_1} \vee \cdots \vee u_{S_k}$;
(3) given two unanimity games u_S and u_T, then their conjunction is still a unanimity game and in particular $u_S \wedge u_T = u_Z$ with $Z_1 = S_1 \cup T_1$, $Z_2 = (S_2 \cup T_2) \setminus Z_1$ and $Z_3 = N \setminus (Z_1 \cup Z_2)$.

Table 1.   Shapley–Shubik and Banzhaf indices for (3,2)-simple games with two players.

	W^m	SS index	Bz index
1	$(12, \emptyset, \emptyset)$	$(\frac{1}{2}, \frac{1}{2})$	$(\frac{1}{3}, \frac{1}{3})$
2	$(1, 2, \emptyset)$	$(\frac{2}{3}, \frac{1}{3})$	$(\frac{2}{3}, \frac{1}{3})$
3	$(1, 2, \emptyset)$ and $(2, 1, \emptyset)$	$(\frac{1}{2}, \frac{1}{2})$	$(\frac{2}{3}, \frac{2}{3})$
4	$(1, 2, \emptyset)$ and $(2, \emptyset, 1)$	$(\frac{1}{6}, \frac{5}{6})$	$(\frac{1}{3}, \frac{1}{3})$
5	$(1, \emptyset, 2)$	$(1, 0)$	$(1, 0)$
6	$(1, \emptyset, 2)$ and $(2, \emptyset, 1)$	$(\frac{1}{2}, \frac{1}{2})$	$(\frac{2}{3}, \frac{2}{3})$
7	$(1, \emptyset, 2)$ and $(\emptyset, 12, \emptyset)$	$(\frac{5}{6}, \frac{1}{6})$	$(1, \frac{2}{3})$
8	$(1, \emptyset, 2)$ and $(\emptyset, 2, 1)$	$(\frac{1}{3}, \frac{2}{3})$	$(\frac{1}{3}, \frac{2}{3})$
9	$(\emptyset, 12, \emptyset)$	$(\frac{1}{2}, \frac{1}{2})$	$(\frac{2}{3}, \frac{2}{3})$
10	$(\emptyset, 1, 2)$	$(1, 0)$	$(1, 0)$
11	$(\emptyset, 1, 2)$ and $(\emptyset, 2, 1)$	$(\frac{1}{2}, \frac{1}{2})$	$(\frac{2}{3}, \frac{2}{3})$
12	$(1, \emptyset, 2)$ and $(2, \emptyset, 1)$ and $(\emptyset, 12, \emptyset)$	$(\frac{1}{2}, \frac{1}{2})$	$(\frac{2}{3}, \frac{2}{3})$

Example 1. In Table 1, the Shapley–Shubik and the Banzhaf indices are computed for all the (3,2)-simple games (up to isomorphism) on $N = \{1, 2\}$.

3. Axioms for Power Indices for (3,2)-Simple Games

3.1. Classical axioms for (3, 2)-simple games

In the following, $\psi : \mathfrak{I}^N \to \mathbb{R}^n$ is a power index for (3,2)-simple games.

Anonymity. The index ψ satisfies *anonymity* if for all game $v \in \mathfrak{I}^N$, any permutation π of N and any $a \in N$

$$\psi_a(v) = \psi_{\pi(a)}(\pi v),$$

where $(\pi v)(S) = v(\pi(S))$.

Null Player. The index ψ satisfies the *null player* axiom if given a null player a in the game v, then

$$\psi_a(v) = 0.$$

Transfer. The index ψ satisfies *transfer* if for any $v, w \in \mathfrak{I}^N$

$$\psi(v) + \psi(w) = \psi(v \wedge w) + \psi(v \vee w).$$

Efficiency. The index ψ satisfies *efficiency* if for any $v \in \mathfrak{I}^N$

$$\sum_{a \in N} \psi_a(v) = 1.$$

Banzhaf Total Power. The index ψ satisfies *Banzhaf total power* if for any $v \in \mathfrak{I}^N$

$$\sum_{a \in N} \psi_a(v) = \frac{1}{3^{n-1}} \sum_{a=1}^{n} \sum_{\substack{S \in 3^N \\ a \in S_1}} [v(S) - v(S_{\downarrow \downarrow a})].$$

As for simple games, both Shapley–Shubik and Banzhaf indices for (3,2)-simple games satisfy anonymity, null player and transfer. Moreover, the Shapley–Shubik index for (3,2)-simple games satisfies also efficiency, while the Banzhaf index for (3,2)-simple games satisfies the tautological Banzhaf total power axiom. However, as we will discuss later on, these axioms are not sufficient to characterize the Shapley–Shubik and the Banzhaf indices for (3,2)-simple games.

Lemma 1. *The Shapley–Shubik index for* (3, 2)*-simple games satisfies the anonymity, the null player, the transfer and the efficiency axioms.*

Proof. Let us consider the different properties.

Anonymity. Let π be a permutation of N. Given a roll-call $\mathcal{R} = (qR, R)$, define $\pi(\mathcal{R}) = (\pi(qR), \pi(R))$. This means that if $\pi(a) = b$ then b votes in $\pi(\mathcal{R})$ in the same position and in the same level of approval of a in $\mathcal{R}$.

If a is pivotal in the game v for the roll-call $\mathcal{R}$, then $\pi(a)$ is pivotal in the game πv for the roll-call $\pi(\mathcal{R})$. Then

$$\phi_a(v) = \frac{|\{\mathcal{R} : a = \mathrm{piv}(\mathcal{R}, v)\}|}{3^n n!} = \frac{|\{\pi(\mathcal{R}) : \pi a = \mathrm{piv}(\pi \mathcal{R}, \pi v)\}|}{3^n n!} = \phi_{\pi(a)}(\pi v).$$

So the Shapley–Shubik index for (3,2)-simple games satisfies the anonymity axiom.

Null player. If a is a null player in a game v, there is not a roll-call $\mathcal{R}$ such that $a = \mathrm{piv}(\mathcal{R}, v)$. Then $\phi_a(v) = 0$.

Transfer. Let v and w be two (3,2)-simple games, then consider the following sets of roll-calls:

$$A = \{\mathcal{R} : a \text{ is pivotal in } v \text{ and in } w\},$$

$$B = \{\mathcal{R} : a \text{ is pivotal in } v \text{ but not in } w\},$$

$$C = \{\mathcal{R} : a \text{ is pivotal in } w \text{ but not in } v\}.$$

Note that A and B form a partition of the set of roll-calls for which a is pivotal in v, while A and C form a partition of the set of roll-calls for which a is pivotal in w. Note also that A is the set of roll-calls for which a is pivotal in the game $v \wedge w$, while A, B, C form a partition of the set of roll-calls in which a is pivotal in the game $v \vee w$. For any $a \in N$, we have

$$\phi_a(v) = \frac{|A| + |B|}{3^n n!}, \quad \phi_a(w) = \frac{|A| + |C|}{3^n n!},$$

$$\phi_a(v \vee w) = \frac{|A| + |B| + |C|}{3^n n!}, \quad \phi_a(v \wedge w) = \frac{|A|}{3^n n!}.$$

Thus $\phi(v \vee w) + \phi(v \wedge w) = \phi(v) + \phi(w)$ and the Shapley–Shubik index satisfies the transfer axiom.

Efficiency. In every roll-call there is one and only one player that is pivotal, from the definition of the Shapley–Shubik index in (1), we get that it satisfies efficiency. $\square$

Remark 1. As it is well-known these axioms for simple games are independent and they fully characterized the Shapley–Shubik index for simple games. This is not true for (3,2)-simple games. For instance, let $\bar{\phi}$ be the standard Shapley–Shubik index for simple games. Then consider the index φ for (3,2)-simple games defined as $\varphi(v) = \bar{\phi}(V)$ where V is the simple game associated to the (3,2)-simple game v and defined as

$$V(S) = 1 \quad \text{if and only if } v(S, N \smallsetminus S, \emptyset) = 1.$$

Then φ satisfies the anonymity, null player, transfer and efficiency axioms for (3,2)-simple games since the Shapley–Shubik index satisfies them on simple games, but it is different from ϕ, for instance $\varphi(u_{(a,b,\emptyset)}) = (1,0)$ while $\phi(u_{(a,b,\emptyset)}) = (\frac{2}{3}, \frac{1}{3})$.

Lemma 2. *The Banzhaf index for* (3,2)-*simple games satisfies the anonymity, the null player, the transfer and the Banzhaf total power axioms.*

Proof. Let us consider the different properties.

Anonymity. The Banzhaf index for (3,2)-simple games satisfies anonymity: actually if $a \in N$ and a is yes–no swinger for tripartition S, then it is clear that πa is a yes–no swinger for πS.

Null player. If a is a null player in the game v, then $v(S) = v(S_{\downarrow\downarrow a})$ for all $S \in 3^N$ such that $a \in S_1$. So $\beta_a(v) = 0$.

Transfer. Let v and w be two (3,2)-simple games and V and W be the set of their winning tripartitions, then consider the following sets:

$$A = \{S \in 3^N : a \in S_1, S \in V \smallsetminus W, S_{\downarrow\downarrow a} \notin V\},$$

$$B = \{S \in 3^N : a \in S_1, S \in W \smallsetminus V, S_{\downarrow\downarrow a} \notin W\},$$

$$C = \{S \in 3^N : a \in S_1, S \in V \cap W, S_{\downarrow\downarrow a} \in W \smallsetminus V\},$$

$$D = \{S \in 3^N : a \in S_1, S \in V \cap W, S_{\downarrow\downarrow a} \in V \smallsetminus W\},$$

$$E = \{S \in 3^N : a \in S_1, S \in V \cap W, S_{\downarrow\downarrow a} \notin V \cap W\}.$$

Note that the sets A, C and E form a partition for the set of yes–no swings of a in v; B, D and E form a partition for the set of yes–no swings of a in w. All the five sets form the set of yes–no swings for player a in $v \vee w$, while E is the set of swings for player a in $v \wedge w$. So the Banzhaf index, which counts the number of yes–no swings, satisfies the transfer axiom.

Banzhaf total power. This axiom is trivially satisfied from the definition of the Banzhaf index. □

Remark 2. Again, anonymity, null player, transfer and Banzhaf total power are independent axioms on simple games, but the Banzhaf index for (3,2)-games is not uniquely determined using only these four. For instance, let $\bar{\beta}$ be the standard

Banzhaf index for simple games, consider the index φ for (3,2)-simple games defined as $\varphi(v) = \bar{\beta}(V)$ where V is the simple game associated to the game v and defined as $V(S) = 1$ *if and only if* $v(S, N \setminus S, \emptyset) = 1$. Then φ satisfies the anonymity, null player, transfer and Banzhaf total power axioms since the Banzhaf index satisfies them on simple games, but it is different from β, for instance $\varphi(u_{(a,b,\emptyset)}) = (1,0)$ while $\beta(u_{(a,b,\emptyset)}) = (\frac{2}{3}, \frac{1}{3})$.

As we have seen, these classical axioms are not enough to characterize the Shapley–Shubik and the Banzhaf indices on the space of (3,2)-simple games. We are going to propose two new axioms in order to uniquely determine these two indices.

3.2. *A new axiom for the Shapley–Shubik index for* $(3, 2)$-*simple games*

Let us introduce another index for (3,2)-simple games, that we use to state the new axiom for the Shapley–Shubik index for (3,2)-simple games and that is a generalization of the decisiveness index defined by Carreras [2005].

Definition 7. Given $v \in \mathfrak{T}^N$, let $\mathcal{W}(v)$ be the set of its winning coalitions. Then the *structural decisiveness index* is defined as the map $\delta : \mathfrak{T}^N \to \mathbb{R}$

$$\delta(v) = 3^{-n}|\mathcal{W}(v)|.$$

This index is assumed to provide a measure of the formal *effectiveness* of the game to pass decisions without taking into account either the personality of concrete agents or their preferences as to a particular proposal versus the status quo. For example, let us consider the unanimity game u_S with $S = (S_1, S_2, S_3)$ and $a \in S_1$; then

$$\delta(u_S) = \frac{2^{s_2}}{3^{s_1+s_2}}, \quad \delta(u_{S\downarrow a}) = 2\delta(u_S), \quad \delta(u_{S\downarrow\downarrow a}) = 3\delta(u_S).$$

We now discuss the behaviour of the index on unanimity games.

Yes-abstain loss on unanimity games. An index ψ satisfies the *yes-abstain loss on unanimity games* if for any tripartition $S \in 3^N$ with $a \in S_1$ it holds:

$$\psi_a(u_S) - \psi_a(u_{S\downarrow a}) = \psi_a(u_{S\downarrow a}) - \frac{\delta(u_{S\downarrow\downarrow a})}{s_1 + s_2}. \tag{2}$$

In the following, we denote with $f(u_S) = \frac{\delta(u_{S\downarrow\downarrow a})}{s_1+s_2}$, so that the previous equation can also be written as

$$\psi_a(u_S) + f(u_S) = 2\psi_a(u_{S\downarrow a}).$$

Equation (2) expresses a balance between $\psi_a(u_S)$ and $\psi_a(u_{S\downarrow a})$. If $f(u_S)$ is interpreted as the per-capita decisiveness for non-null players in u_S, then $\psi_a(u_S)$ plus this quantity is two times the value $\psi_a(u_{S\downarrow a})$. Thus, the smaller the per-capita decisiveness is, the closer $\psi_a(u_S)$ is to the double of $\psi_a(u_{S\downarrow a})$.

In order to prove that the Shapley–Shubik index for (3,2)-simple games satisfies the yes-abstain loss on unanimity games, we have to focus on unanimity games. In particular, from now on, we fix a tripartition S with $a \in S_1$ and consider the game u_S. We want to compute $\phi_a(u_S)$ and then compare it with $\phi_a(u_{S\downarrow a})$.

Lemma 3. *Let u_S be a unanimity game, then player $a \in S_1$ is pivotal in the roll-call $\mathcal{R} \in \mathcal{R}_a^{no}$ if and only if a is pivotal in the roll-call $\mathcal{R}_{\uparrow a} \in \mathcal{R}_a^{abs}$.*

Proof. A roll-call is winning in the game u_S *if and only if* all players belonging to S_1 are voting "yes" and all players belonging to S_2 are not voting "no".

If $a \in S_1$ is pivotal by voting "no" in the roll-call $\mathcal{R}$, then the outcome of $\mathcal{R}$ is negative. The roll-call $\mathcal{R}_{\uparrow a}$ represents the same situation of $\mathcal{R}$ with the only difference that player a abstains instead of voting no. But a is still pivotal abstaining and fixing as negative the outcome of the roll-call.

Analogously, if player a is pivotal by abstaining, in the same situation a is also pivotal by voting "no". □

Remark 3. Lemma 3 implies that $|A_{a,u_S}| = |N_{a,u_S}|$, but also $|AY_{a,u_S}| = |NY_{a,u_S}|$ and $|A\overline{Y}_{a,u_S}| = |N\overline{Y}_{a,u_S}|$, because of the one-to-one correspondence among each pair of the sets $\mathcal{R}_a^{yes}$, $\mathcal{R}_a^{abs}$, and $\mathcal{R}_a^{no}$.

Lemma 4. *Let u_S be a unanimity game, if player $a \in S_1$ is pivotal in the roll-call $\mathcal{R} \in \mathcal{R}_a^{yes}$, then a is pivotal in the roll-call $\mathcal{R}_{\downarrow a} \in \mathcal{R}_a^{abs}$ and in the roll-call $\mathcal{R}_{\downarrow\downarrow a} \in \mathcal{R}_a^{no}$.*

Proof. A roll-call is winning in the game u_S *if and only if* all players belonging to S_1 are voting "yes" and all players belonging to S_2 are not voting "no".

If $a \in S_1$ is pivotal in the roll-call $\mathcal{R}$ by voting "yes", this means that after a's vote the outcome is positive and all the other players in S_1 and S_2 voted before a. This also means that after a only some of the players belonging to S_3 are going to vote, but they are null players and cannot be pivotal.

In the roll-call $\mathcal{R}$, a is the last player who has the power to change the outcome of the game, thus a is pivotal also in $\mathcal{R}_{\downarrow\downarrow a}$ voting "no" and in $\mathcal{R}_{\downarrow a}$ abstaining. □

The converse is not true. For instance consider the tripartition $S = (a, b, c)$ and the game u_S. In any roll-call in which a is the first to vote, he is pivotal abstaining or voting "no". On the other hand if a votes "yes" as first player, then b is pivotal: if she votes "no" the outcome is negative, while if she abstains or votes "yes" the outcome is positive.

Remark 4. Note that by definition $|AY_{a,u_S}| \leq |Y_{a,u_S}|$ and $|NY_{a,u_S}| \leq |Y_{a,u_S}|$. Lemma 4 implies that $|Y_{a,u_S}| \leq |AY_{a,u_S}|$ and $|Y_{a,u_S}| \leq |NY_{a,u_S}|$. Thanks to these considerations and Lemma 3 we have $|Y_{a,u_S}| = |AY_{a,u_S}| = |NY_{a,u_S}|$.

Hence, from the previous remarks, the Shapley–Shubik index on the unanimity $(3,2)$-simple game u_S of player $a \in S_1$ is

$$\phi_a(u_S) = \frac{1}{3^n n!}(|Y_{a,u_S}| + |A_{a,u_S}| + |N_{a,u_S}|)$$

$$= \frac{1}{3^n n!}(|Y_{a,u_S}| + |AY_{a,u_S}| + |A\overline{Y}_{a,u_S}| + |NY_{a,u_S}| + |N\overline{Y}_{a,u_S}|)$$

$$= \frac{1}{3^n n!}(3|Y_{a,u_S}| + 2|N\overline{Y}_{a,u_S}|). \tag{3}$$

It is possible to calculate the value $|Y_{a,u_S}|$ thanks to the following lemma.

Lemma 5. *A player $a \in S_1$ is pivotal in the game u_S for the roll-call $\mathcal{R} \in \mathcal{R}_a^{\text{yes}}$ if and only if all players in $(S_1 \setminus \{a\}) \cup S_2$ are before him in qR, they vote "yes" if they belong to S_1, and they vote " yes" or abstain if they belong to S_2.*
    In particular

$$|Y_{a,u_S}| = 2^{s_2} 3^{s_3} \frac{(s_1 + s_2 + s_3)!}{s_1 + s_2}.$$

Proof. A roll-call is winning in the game u_S *if and only if* all players belonging to S_1 are voting "yes" and all players belonging to S_2 are not voting "no". Thus if a player $a \in S_1$ is pivotal in the roll-call $\mathcal{R} \in \mathcal{R}_a^{\text{yes}}$, he is the last player of the set $S_1 \cup S_2$ to vote and he is the last player that has the possibility to influence the outcome and fix it as positive. All the players before him were not pivotal, so they voted "yes" if they belong to S_1, they abstained or voted "yes" if they belong to S_2.
    To prove the second part of the hypothesis, we have to count the number of roll-calls $\mathcal{R} \in Y_{a,u_S}$. Actually, if $j = 0, \ldots, s_3$, player a can vote j positions after that all the players in $S_1 \setminus \{a\} \cup S_2$ voted. This means that j players belonging to S_3 vote before a and $s_3 - j$ players belonging to S_3 vote after a. There are $\binom{s_3}{j}$ different ways to choose the j players, then $(s_1 + s_2 + j - 1)!2^{s_2}3^j$ possibilities for the players before a and $(s_3 - j)!3^{s_3-j}$ for the players after a (see Fig. 1). Hence,

$$|Y_{a,u_S}| = \sum_{j=0}^{s_3} \binom{s_3}{j}(s_1 + s_2 + j - 1)!2^{s_2}3^j(s_3 - j)!3^{s_3-j}$$

$$= 2^{s_2}3^{s_3} \frac{(s_1 + s_2 + s_3)!}{s_1 + s_2}. \qquad \square$$

$$(s_1 + s_2 + j)^{th}$$

players in S_1 and in S_2 not pivotal and j players in S_3		$\downarrow$ $\quad s_3 - j$ players in S_3
	a	can vote anything

$$\underbrace{\binom{s_3}{j}(s_1 + s_2 + j - 1)!2^{s_2}3^j}\qquad\qquad \underbrace{(s_3 - j)!3^{s_3-j}}$$

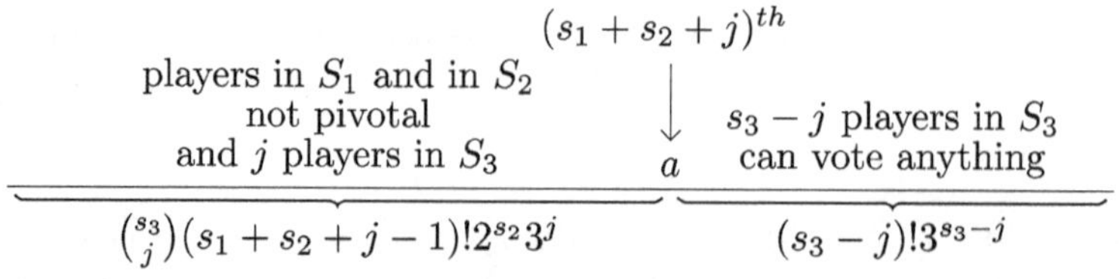

Fig. 1.   Roll-calls in which $a \in S_1$ is pivotal by voting "yes".

We can now discuss how the power of player a changes when a decreases the support of one level. We evaluate the Shapley–Shubik index for (3,2)-simple games of player $a \in S_1$ in the unanimity game $u_{S\downarrow a}$, generated by the tripartition $(S_1 \setminus \{a\}, S_2 \cup \{a\}, S_3)$.

Firstly note that if a is pivotal by abstaining in $u_{S\downarrow a}$, then in the same situation a is pivotal also by voting "yes"; this means that $|A\overline{Y}_{a,u_{S\downarrow a}}| = 0$. Then observe that if $\mathcal{R} \notin A\overline{Y}_{a,u_S}$ and a is pivotal in $\mathcal{R}$ for u_S, then a is pivotal in $\mathcal{R}$ also for $u_{S\downarrow a}$. This means that $Y_{a,u_{S\downarrow a}} = Y_{a,u_S}$ and

$$AY_{a,u_{S\downarrow a}} = AY_{a,u_S}, \quad A\overline{Y}_{a,u_{S\downarrow a}} = \emptyset,$$

$$NY_{a,u_{S\downarrow a}} = NY_{a,u_S}, \quad N\overline{Y}_{a,u_{S\downarrow a}} = N\overline{Y}_{a,u_S}.$$

Hence,

$$\phi_a(u_{S\downarrow a}) = \frac{1}{3^n n!}(3|Y_{a,u_S}| + |N\overline{Y}_{a,u_S}|). \tag{4}$$

Finally, to establish how the index changes when a player switches from voting "yes" to abstaining in a unanimity game, we compare Eqs. (3) and (4) and get the following:

$$\phi_a(u_S) - \phi_a(u_{S\downarrow a}) = \frac{1}{3^n n!}|N\overline{Y}_{a,u_S}|.$$

To compare the difference with something explicitly, we also get the following equivalently

$$2\phi_a(u_{S\downarrow a}) - \phi_a(u_S) = \frac{3}{3^n n!}|Y_{a,u_S}|$$

that can be re-written as

$$\phi_a(u_S) - \phi_a(u_{S\downarrow a}) = \phi_a(u_{S\downarrow a}) - \frac{3}{3^n n!}|Y_{a,u_S}|.$$

From this result and Lemma 5 we obtain the following.

Proposition 1. *The Shapley–Shubik index satisfies the yes-abstain loss on unanimity games.*

4. The Characterization of the Shapley–Shubik Index for (3,2)-Simple Games

We can now state the main theorem to characterize the Shapley–Shubik index for (3,2)-simple games. We also prove the independence of the five axioms we are going to use, so all of them are necessary in order to uniquely characterize the Shapley–Shubik power index for (3,2)-simple games.

Theorem 1 (Shapley–Shubik index for (3,2)-simple games). *Let $\psi : \mathfrak{T}^N \to \mathbb{R}^n$ be an index for $(3,2)$-simple games, then ψ satisfies the anonymity, null player, transfer, efficiency and yes-abstain loss on unanimity games axioms if and only if ψ is the Shapley–Shubik index for $(3,2)$-simple games.*

Proof. In Lemma 1 and in Proposition 1 it is proved that the Shapley–Shubik index for (3,2)-simple games satisfies all the axioms, we just need to prove that only one index satisfies all of them.

So, let ψ be an index that satisfies the hypothesis. We will prove that it is uniquely determined on a game v, using induction on the number of minimal winning tripartitions of v.

First, suppose that $|\mathcal{W}^m(v)| = 1$. Then $v = u_S$ for some tripartition $S \in 3^N$ and $S \neq (\emptyset, \emptyset, N)$. We again use induction on the number of elements in S_2.

$|S_2| = 0$ Then $S = (S_1, \emptyset, N \setminus S_1)$ for some $S_1 \subseteq N$. All players in $S_3 = N \setminus S_1$ are null players, so if $c \in S_3$: $\psi_c(u_S) = 0$, on the other hand all players in S_1 have the same role, thus, thanks to the anonymity and efficiency axioms we have $\psi_a(u_S) = \frac{1}{s_1}$, for any $a \in S_1$.

$|S_2| = t + 1$ Suppose now that the hypothesis is true for any tripartition T such that $|T_2| \leq t$, we want to prove it for a tripartition S such that $|S_2| = t + 1$. Given the tripartition $S = (S_1, S_2, S_3)$, there exist a player $p \in S_2$ and a tripartition $T = (T_1, T_2, T_3)$ such that $T_{\downarrow p} = S$ and $|T_2| = t$. Since ψ satisfies the yes-abstain loss on unanimity games:

$$\psi_p(u_S) = \psi_p(u_{T \downarrow p}) = \frac{1}{2}[\psi_p(u_T) + f(u_T)]$$

then the induction hypothesis and anonymity imply that $\psi_b(u_S)$ is uniquely determined for all players $b \in S_2$. Thanks to anonymity and efficiency:

$$s_1 \psi_a(u_S) + s_2 \psi_b(u_S) = 1,$$

so we can determine $\psi_a(u_S)$ for $a \in S_1$. All players in S_3 are null, so $\psi_c(u_S) = 0$ if $c \in S_3$.

Thus, ψ coincides with the Shapley–Shubik power index for (3,2)-simple games for any unanimity game u_S.

Now, suppose that the hypothesis holds for any game v such that $|\mathcal{W}^m(v)| \leq k - 1$; we need to prove it for v such that $|\mathcal{W}^m(v)| = k$.

If $\mathcal{W}^m(v) = \{S^1, \ldots, S^k\}$, then $v = u_{S^1} \vee u_{S^2} \vee \cdots \vee u_{S^k}$, since ψ satisfies the transfer axiom:

$$\psi(v) = \psi(u_{S^1}) + \psi(u_{S^2} \vee \cdots \vee u_{S^k}) - \psi(u_{S^1} \wedge u_{S^2} \wedge \cdots \wedge u_{S^k}).$$

The conjunction of unanimity games is still a unanimity game, so all games in the right-hand side of the previous equation have a number of minimal winning tripartitions smaller than $|\mathcal{W}^m(v)|$. Using the induction hypothesis, ψ coincides with the Shapley–Shubik index for all of them and this ends the proof. $\square$

Now, we show how these five axioms allow to compute the index for every unanimity game by means of a recursive procedure.

Consider the unanimity game u_S with the set S_3 of people voting no. Then consider the game u_{T^0} where $T^0 = (\emptyset, N \setminus S_3, S_3)$. For every player $a \in S_3$ thanks to null player we have $\phi_a(u_{T^0}) = 0$, for every $a \in N \setminus S_3$ thanks to anonymity and efficiency it holds

$$\phi_a(u_T) = \frac{1}{n - s_3}.$$

Then consider the game u_{T^1} where $T^1 = (p, N \setminus (S_3 \cup p), S_3))$ for some player $p \in S_1$. Thanks to (2) we can compute $\phi_p(u_{T^1})$, then for any $a \neq p \in N \setminus S_3$, $\phi_a(u_{T^1})$ can be computed using efficiency and players in S_3 are still null players.

It is clear that this process can be reiterated until we reach the games u_S and establish the value for all players in S_1 and S_2 using the yes-abstain loss on unanimity games and efficiency.

4.1. *Independence of the axiom for the Shapley–Shubik power index for* (3, 2)-*simple games*

The five axioms for (3,2)-simple games are independent. We are going to give examples of power indices for (3,2)-simple games that satisfy only four of them, as summarized in Table 2.

Not anonymity. Consider the index ψ^1 defined on unanimity games as follows:

- If $s_3 = n - 2$ then for any two players a and b such that $a < b$,
  — if $S = (ab, \emptyset, N \setminus \{a, b\})$, then

  $$\psi_a^1(u_S) = \frac{1}{2} + \varepsilon, \quad \psi_b^1(u_S) = \frac{1}{2} - \varepsilon;$$

  — if $S = (a, b, N \setminus \{a, b\})$, then

  $$\psi_a^1(u_S) = \frac{2}{3} + \frac{\varepsilon}{2}, \quad \psi_b^1(u_S) = \frac{1}{3} - \frac{\varepsilon}{2};$$

  — if $S = (b, a, N \setminus \{a, b\})$, then

  $$\psi_a^1(u_S) = \frac{1}{3} + \frac{\varepsilon}{2}, \quad \psi_b^1(u_S) = \frac{2}{3} - \frac{\varepsilon}{2};$$

Table 2. Independence of the axioms that characterize the Shapley–Shubik index for (3,2)-simple games.

	Anonymity	Null	Transfer	Efficiency	Yes-abs. loss on unan. games
ψ^1	—	✓	✓	✓	✓
ψ^2	✓	—	✓	✓	✓
ψ^3	✓	✓	—	✓	✓
ψ^4	✓	✓	✓	—	✓
φ	✓	✓	✓	✓	—

— if $S = (\emptyset, ab, N \setminus \{a, b\})$, then

$$\psi_a^1(u_S) = \frac{1}{2} + \frac{\varepsilon}{4}, \quad \psi_b^1(u_S) = \frac{1}{2} - \frac{\varepsilon}{4},$$

where $\varepsilon > 0$.

- If $s_3 \neq n - 2$, $\psi^1(u_S) = \phi(u_S)$ where ϕ is the Shapley–Shubik index for (3,2)-simple games.

Then extend ψ^1 to $\mathfrak{T}^N$ using transfer.

It is clear that this index satisfies null player and efficiency. It also satisfies the yes-abstain loss on unanimity games, because it coincides with the Shapley–Shubik index for (3,2)-simple games when $s_3 \neq n - 2$ and if $s_3 = n - 2$ the yes-abstain loss on unanimity games is satisfied by the definition of ψ^1, as it is easy to check. However, ψ^1 does not satisfy anonymity, because for instance

$$\psi_a^1(u_{(ab,\emptyset,N \setminus \{a,b\})}) - \psi_b^1(u_{(ab,\emptyset,N \setminus \{a,b\})}) = 2\varepsilon \neq 0.$$

Not null player. Consider the index ψ^2 defined on unanimity games as follows.

- If $S = (\emptyset, a, N \setminus \{a\})$ for some $a \in N$, then

$$\psi_a^2(u_S) = 1 - \varepsilon, \quad \psi_b^2(u_S) = \frac{\varepsilon}{n - 1}$$

for any $b \neq a$, with $\varepsilon > 0$;
- if $S = (a, \emptyset, N \setminus \{a\})$ for some $a \in N$, then

$$\psi_a^2(u_S) = 1 - 2\varepsilon, \quad \psi_b^2(u_S) = \frac{2\varepsilon}{n - 1}$$

for any $b \neq a$, with $\varepsilon > 0$;
- for any other $S \in 3^N$, $\psi^2(u_S) = \phi(u_S)$ where ϕ is the Shapley–Shubik index for (3,2)-simple games.

Then extend ψ^2 to $\mathfrak{T}^N$ using transfer.

It is clear that this index satisfies anonymity and efficiency. It also satisfies the yes-abstain loss on unanimity games since it coincides with the Shapley–Shubik index on unanimity games such that $s_3 \neq n - 1$ and if $s_3 = n - 1$ and $a \in S_1$

$$2\psi^2(u_{S \downarrow a}) - \psi^2(u_S) = 1 = f(u_{(a,\emptyset,N \setminus \{a\})}).$$

However, ψ^2 does not satisfy null player: any $b \neq a$ is a null player in the game $u_{(\emptyset,a,N \setminus \{a\})}$ but $\psi_b^2(u_{(\emptyset,a,N \setminus \{a\})}) = \frac{\varepsilon}{n-1} \neq 0$.

Not transfer. Consider the index ψ^3 defined as $\psi^3(u_S) = \phi(u_S)$ for any unanimity game u_S and for any other game v

$$\psi_a^3(v) = \begin{cases} 0 & \text{if } a \text{ is a null player,} \\ \dfrac{1}{k} & \text{otherwise,} \end{cases}$$

where $k = |\{p \in N : p \text{ is not a null player in } v\}|$.

The index ψ^3 satisfies the null player, the anonymity, and the efficiency axioms; it also satisfies the yes-abstain loss on unanimity games, since it coincides with the Shapley–Shubik index on unanimity games. From the definition of $\psi^3(v)$, it is clear that this index does not satisfy the transfer axiom.

Not efficiency. Consider the index ψ^4 defined on unanimity game as

$$\psi_a^4(u_S) = \begin{cases} 1 & \text{if } a \in S_1, \\ \dfrac{1}{2} + \dfrac{2^{s_2-1}}{3^{s_1+s_2-1}(s_1+s_2)} & \text{if } a \in S_2, \\ 0 & \text{if } a \in S_3 \end{cases}$$

and extended to $\mathfrak{T}^N$ using transfer. Then ψ^4 satisfies anonymity, null player and transfer. It also satisfies the yes-abstain loss on unanimity games since for any tripartition S with $a \in S_1$:

$$2\psi_a^4(u_{S\downarrow a}) = 2\left[\frac{1}{2} + \frac{2^{s_2-1}}{3^{s_1+s_2-1}(s_1+s_2)}\right]$$

$$= 1 + \frac{2^{s_2}}{3^{s_1+s_2-1}(s_1+s_2)} = \psi_a^4(u_S) + f(u_S).$$

However, ψ^4 does not satisfy efficiency, for instance $\psi^4(u_{(N,\emptyset,\emptyset)}) = (1,\ldots,1)$ so that $\sum_{a \in N} \psi^4(u_{(N,\emptyset,\emptyset)}) = n \neq 1$.

Not yes-abstain loss on unanimity games. As we examined in Remark 1, the index φ, that is Shapley–Shubik index for simple games computed on the simple games associated to the game with abstention, satisfies null player, anonymity, transfer, and efficiency, but does not satisfy the yes-abstain loss on unanimity games.

5. A Similar Approach for the Banzhaf Index for (3,2)-Simple Games

We now consider the Banzhaf index for (3,2)-simple games and characterize this index, as we did in the previous sections with the Shapley–Shubik index for (3,2)-simple games. Let us start with a preliminary lemma that shows how simple it is to compute the Banzhaf index for (3,2)-simple games on unanimity games. This lemma will be used in the following to prove the new axioms we are going to introduce in order to characterize the Banzhaf index for (3,2)-simple games.

Lemma 6. *Let $S \neq (\emptyset, \emptyset, N)$ be a tripartition of N, then the Banzhaf index for* $(3,2)$-*simple games on the unanimity games u_S is*

$$\beta_p(u_S) = \begin{cases} 3\delta(u_S) & \text{if } p \in S_1, \\ \dfrac{3}{2}\delta(u_S) & \text{if } p \in S_2, \\ 0 & \text{if } p \in S_3. \end{cases}$$

From the previous lemma, it follows, in particular, that if $a \in S_1$ and $b \in S_2$, it holds

$$\beta_a(u_S) = 2\beta_b(u_S).$$

Proof. We have to compute $\eta_a(u_S)$ for any player $a \in N$. Assume first that $a \in S_1$. Remember that $u_S(T) = 1$ *if and only if* $S_1 \subseteq T_1$ and $S_2 \subseteq T_1 \cup T_2$. Note that if $a \in S_1 \cap T_1$, the condition $u_S(T) = 1$ implies $u_S(T_{\downarrow\downarrow a}) = 0$. Moreover, if $a \in S_1$, the conditions $a \in T_1$ and $u_S(T) = 1$ are equivalent to $S \subseteq T$. Hence,

$$\eta_a(u_S) = |\{T \in 3^N : a \in T_1, u_S(T) - u_S(T_{\downarrow\downarrow a}) = 1\}|$$

$$= |\{T \in 3^N : a \in T_1, u_S(T) = 1\}|$$

$$= |\{T \in 3^N : S \subseteq T\}| = 2^{s_2} 3^{s_3}.$$

Suppose now $a \in S_2$. Analogously we have

$$\eta_a(u_S) = |\{T \in 3^N : a \in T_1 \text{ and } S \subseteq T\}| = 2^{s_2 - 1} 3^{s_3}.$$

Finally, assume $a \in S_3$. Players in S_3 are null, so $\beta_a(u_S) = 0$.

Since the Banzhaf index for (3,2)-simple games is given by

$$\beta_a(u_S) = \frac{\eta_a(u_S)}{3^{n-1}}$$

and $n = s_1 + s_2 + s_3$, we have the hypothesis.   □

In Dubey and Shapley [1979] characterization of the Banzhaf power index for simple games, the Banzhaf total power axiom is introduced in order to replace efficiency, that is used for the Shapley–Shubik power index. However, the Banzhaf total power axiom is not a convincing axiom; some subsequent axiomatic characterization of the Banzhaf power index avoided this axiom, see for instance Laurelle and Valenciano [2001], Lehrer [1988] and Albizuri [2001].

We want to follow the classical approach and use the same set of axioms to characterize the Shapley–Shubik and the Banzhaf indices for (3,2)-simple games. However, we will replace the Banzhaf total power with a weaker condition that refers only to unanimity games.

Total power on unanimity games. An index ψ satisfies the total power on unanimity games if for any tripartition $S \neq (\emptyset, \emptyset, N)$

$$\sum_{a \in N} \psi_a(u_S) = \frac{3}{2}(2s_1 + s_2)\delta(u_S).$$

Lemma 7. *The Banzhaf index for* $(3, 2)$*-simple games satisfies the total power on unanimity games axiom.*

Proof. The hypothesis follows from anonymity and Lemma 6.   □

As we have previously done for the Shapley–Shubik index on (3,2)-simple games, it is necessary to add another axiom in order to uniquely characterize the Banzhaf index on (3,2)-simple games. The new axiom defined in Eq. (2), describes what a player is losing when passing from voting "yes" to abstaining; we introduce the following axiom to describe a different idea: the power of player a does not change in the game u_S and in $u_{S\downarrow a}$.

Yes-abstain null loss on unanimity games. An index ψ satisfies the yes-abstain null loss axiom if for any unanimity game u_S and $a \in S_1$ it holds

$$\psi_a(u_S) = \psi_a(u_{S\downarrow a}).$$

Note that while now we demand for ψ_a the equality for games (u_S) and $u_{S\downarrow a}$, it turned out that this relation was almost the double for the Shapley–Shubik power index for (3,2)-simple games.

We immediately have the following result.

Proposition 2. *Consider the unanimity game u_S with $a \in S_1$. Then the Banzhaf index on (3, 2)-simple games satisfies the yes-abstain null loss axiom and in particular*

$$\beta_a(u_S) = \beta_a(u_{S\downarrow a}).$$

5.1. *Characterization of the Banzhaf index for (3, 2)-simple games*

Theorem 2. *Let $\psi : \mathfrak{T}^N \to \mathbb{R}^n$ be an index for (3, 2)-simple games, then ψ satisfies anonymity, null player, transfer, total power on unanimity games and yes-abstain null loss if and only if ψ is the Banzhaf index for (3, 2)-simple games.*

Proof. We already proved that the Banzhaf index for (3,2)-simple games satisfies all these properties, so we just need to prove that if ψ is a power index that satisfies the hypothesis, then it is uniquely determined. We use induction on the number of minimal winning tripartitions of the game v.

Suppose that $|\mathcal{W}^m(v)| = 1$, then $v = u_S$ for some tripartition S. So we start proving that ψ coincides with the Banzhaf index on unanimity games.

We again use induction, this time on the cardinality of S_2.

$|S_2| = 0$. Then $S = (S_1, \emptyset, N \setminus S_1)$ for some $S_1 \subseteq N$. Then players in $S_3 = N \setminus S_1$ are null, so $\psi_c(u_S) = 0$ for all $c \in S_3$. Players in S_1 are symmetric and thanks to anonymity and total power on unanimity game, if $a \in S_1$

$$\beta_a\big(u_{(S_1, \emptyset, N \setminus S_1)}\big) = \frac{1}{s_1} \frac{2 s_1 2^{-1}}{3^{s_1 - 1}} = \frac{1}{3^{s_1 - 1}}.$$

So, ψ is uniquely determined on unanimity games with $s_2 = 0$ and it coincides with the Banzhaf index for (3,2)-simple games.

$|S_2| = t+1$. Suppose now that the hypothesis is true for any tripartition T such that $|T_2| \leq t$, we want to prove this for a tripartition S such that $|S_2| = t+1$. Given a tripartition $S = (S_1, S_2, S_3)$ such that $|S_2| = t+1$, there exist a player $p \in S_2$ and a tripartition $T = (T_1, T_2, T_3)$ such that $T_{\downarrow p} = S$ and $T_2 = t$. Since ψ satisfies yes-abstain null loss:

$$\psi_p(u_S) = \psi_p(u_{T \downarrow p}) = \psi_p(u_T),$$

then the induction hypothesis and anonymity imply that $\psi_b(u_S)$ is uniquely determined for all players $b \in S_2$.

Using again anonymity and the total power on unanimity game:

$$s_1 \psi_a(u_s) + s_2 \psi_b(u_S) = (2s_1 + s_2) \frac{2^{s_2 - 1}}{3^{s_1 + s_2 - 1}}$$

with $a \in S_1$ and $b \in S_2$. So, we can determine $\psi_a(u_S)$ for $a \in S_1$. Thanks to null player we have that $\psi_c(u_S) = 0$ if $c \in S_3$. Thus, ψ coincides with the Banzhaf index for (3,2)-simple games for any unanimity game u_S.

We suppose that the hypothesis holds for any game v such that $|\mathcal{W}^m(v)| \leq k-1$, and prove it for v such that $|\mathcal{W}^m(v)| = k$.

If $\mathcal{W}^m(v) = \{S^1, \ldots, S^k\}$, then $v = u_{S^1} \vee u_{S^2} \vee \cdots \vee u_{S^k}$, since ψ satisfies the transfer axiom:

$$\psi(v) = \psi(u_{S^1}) + \psi(u_{S^2} \vee \cdots \vee u_{S^k}) - \psi(u_{S^1} \wedge u_{S^2} \wedge \cdots \wedge u_{S^k}).$$

The conjunction of unanimity games is still a unanimity game, so all games on the right-hand side of the previous equation have a number of minimal winning tripartitions smaller than $|\mathcal{W}^m(v)|$. Using the induction hypothesis, ψ coincides with the Banzhaf index for (3,2)-simple games for all of them and this ends the proof. □

5.2. *Independence of the axiom for the Banzhaf index for (3, 2)-simple games*

The five axioms for (3,2)-simple games used in Theorem 2 are independent. We are going to give examples of power indices on (3,2)-simple games that satisfy only four of them, as summarized in Table 3.

Table 3.   Independence of the axioms that characterize the Banzhaf index for (3,2)-simple games.

	Anonymity	Null	Transfer	Total power	Yes-abs. null loss
γ^1	—	✓	✓	✓	✓
γ^2	✓	—	✓	✓	✓
γ^3	✓	✓	—	✓	✓
γ^4	✓	✓	✓	—	✓
γ^5	✓	✓	✓	✓	—

Not anonymity. Consider the index γ^1 defined on unanimity games as follows:

- If $s_3 = n - 2$ then for any two players a and b such that $a < b$,
  — if $S = (ab, \emptyset, N \setminus \{a, b\})$, then
  $$\gamma_a^1(u_S) = \frac{1}{3} + \varepsilon, \quad \gamma_b^1(u_S) = \frac{1}{3} - \varepsilon;$$
  — if $S = (a, b, N \setminus \{a, b\})$, then
  $$\gamma_a^1(u_S) = \frac{2}{3} + \varepsilon, \quad \gamma_b^1(u_S) = \frac{1}{3} - \varepsilon;$$
  — if $S = (b, a, N \setminus \{a, b\})$, then
  $$\gamma_a^1(u_S) = \frac{1}{3} + \varepsilon, \quad \gamma_b^1(u_S) = \frac{2}{3} - \varepsilon;$$
  — if $S = (\emptyset, ab, N \setminus \{a, b\})$, then
  $$\gamma_a^1(u_S) = \frac{2}{3} + \varepsilon, \quad \gamma_b^1(u_S) = \frac{2}{3} - \varepsilon;$$
  where $\varepsilon > 0$.
- If $s_3 \neq n - 2$, $\gamma^1(u_S) = \beta(u_S)$ where β is the Banzhaf index for (3,2)-simple games.

Then extend γ^1 to $\mathfrak{T}^N$ using transfer.

  This index satisfies null player, total power on unanimity games and yes-abstain null loss. However, γ^1 does not satisfy anonymity, because for instance

$$\gamma_a^1(u_{(ab,\emptyset,N \setminus \{a,b\})}) \neq \gamma_b^1(u_{(ab,\emptyset,N \setminus \{a,b\})}).$$

Not null player. Consider the index γ^2 defined on unanimity games as follows:

- If $S = (\emptyset, a, N \setminus \{a\})$ or $S = (a, \emptyset, N \setminus \{a\})$ for some $a \in N$, then
  $$\gamma_a^2(u_S) = 1 - \varepsilon, \quad \gamma_b^2(u_S) = \frac{\varepsilon}{n - 1}$$
  for any $b \neq a$ and with $\varepsilon > 0$;
- for any other $S \in 3^N$, $\gamma^2(u_S) = \beta(u_S)$ where β is the Banzhaf index for (3,2)-simple games.

  Then extend γ^2 to $\mathfrak{T}^N$ using transfer.
  This index satisfies anonymity, total power on unanimity games and yes-abstain null loss. However, γ^2 does not satisfy null player: any $b \neq a$ is a null player in the game $u_{(\emptyset,a,N \setminus \{a\})}$ but $\gamma_b^2(u_{(\emptyset,a,N \setminus \{a\})}) = \frac{\varepsilon}{n-1} \neq 0$.

Not transfer. Consider the index γ^3 defined as $\gamma^3(u_S) = \beta(u_S)$ for any unanimity game u_S and for any other game v

$$\gamma_a^3(v) = \begin{cases} 0 & \text{if } a \text{ is a null player,} \\ \dfrac{1}{k} & \text{otherwise,} \end{cases}$$

where $k = |\{p \in N : p \text{ is not a null player in } v\}|$.

The index γ^3 satisfies the null player and the anonymity axioms. It also satisfies the total power on unanimity games and the yes-abstain null power, since it coincides with the Banzhaf index on unanimity games. From the definition of $\gamma^3(v)$ it is clear that this index does not satisfy the transfer axiom.

Not total power on unanimity games. Consider the index γ^4 defined on unanimity game as

$$\gamma_a^4(u_S) = \begin{cases} \dfrac{1}{s_1 + s_2} & \text{if } a \in S_1 \cup S_2, \\ 0 & \text{if } a \in S_3 \end{cases}$$

and extended to $\mathfrak{T}^N$ using transfer.

The index γ^4 satisfies anonymity, null player, and yes-abstain null loss. However, γ^4 satisfies efficiency instead of total power on unanimity game.

Not yes-abstain null loss. Consider the index γ^5 defined on unanimity game as

$$\gamma_a^5(u_S) = \begin{cases} \dfrac{2s_1 + s_2}{s_1 + s_2} \dfrac{2^{s_2 - 1}}{3^{s_1 + s_2 - 1}} & \text{if } a \in S_1 \cup S_2, \\ 0 & \text{if } a \in S_3 \end{cases}$$

and extended $\mathfrak{T}^N$ using transfer.

The index γ^4 satisfies anonymity, null player, and total power on unanimity games. However, it does not satisfy yes-abstain null loss:

$$\gamma_a^5(u_S) = \frac{2s_1 + s_2}{s_1 + s_2} \frac{2^{s_2 - 1}}{3^{s_1 + s_2 - 1}} \neq \frac{2s_1 + s_2 - 1}{s_1 + s_2} \frac{2^{s_2}}{3^{s_1 + s_2 - 1}} = \gamma_a^5(u_{S \downarrow a}).$$

6. Conclusion

In this work, we provide an axiomatization for the Shapley–Shubik index for (3,2)-simple games. The definition of this index in Felsenthal and Machover's seminal work is given as the expected probability under the discrete uniform distribution of each player of being pivotal in the roll-call space. Thus, there was a bargaining interpretation of this index but there was not, in that original paper, an axiomatic characterization of the index and the list of the properties that it satisfies.

Our work focus on the classical axiomatization of the Shapley–Shubik and the Banzhaf indices for simple games, due to Dubey [1975] and Dubey and Shapley [1979], respectively, and generalize these approaches to the family of (3,2)-simple games. The characterizations we give for the two indices have a very similar structure and they can be of future interest in order to define new indices for (3,2)-simple games analogous to probabilistic values and semivalues defined for cooperative and simple games, as in Dubey *et al.* [1981] and Monderer and Samet [2002].

Moreover, it can be of future interest to study other axiomatizations of power indices for (3,2)-simple games, not related to the behavior on unanimity games, as done, only for the Banzhaf index for (3,2)-simple games, in Freixas and Lucchetti [2016] who use the axiom of individual block effect and in Bernardi [2018], who uses the average gain–loss balance. Following this line of investigation, it could be interesting also to use different properties extending to (3,2)-simple games other axiomatizations for cooperative games such as Casajus [2012].

Finally, the axiom we introduce, together with anonimity and efficiency, allows to compute the Shapley–Shubik index for (3,2)-simple games on unanimity games using a recursive formula. Thus, using the transfer property it is possible to compute the value for any game, without explicitly use the definition and count the number of roll-calls in which a player is pivotal. Further research could be addressed to compute the Shapley–Shubik index (3,2)-simple games for any (3,2)-simple game by means of a direct formula.

Acknowledgments

Josep Freixas acknowledges the Spanish Ministry of Economy and Competitiveness (MINECO) and the European Union (FEDER funds) under grant MTM2015-66818-P (MINECO/FEDER). The authors are thankful for the valuable comments and suggestions to improve the quality of this work.

References

Albizuri, M. J. [2001] An axiomatization of the modified Banzhaf Coleman index, *Int. J. Game Theory* **33**(2), 167–176.

Banzhaf, J. F. [1964] Weighted voting doesn't work: A mathematical analysis, *Rutgers L. Rev.* **19**, 317–343.

Bernardi, G. [2018] A new axiomatization of the Banzhaf index for (3,2)-simple games, *Group Decision Negot.* **27**(1), 165–177.

Bernardi, G. and Lucchetti, R. [2015] Generating semivalues via unanimity games, *J. Optim. Theory Appl.* **166**(3), 1051–1062.

Bilbao, J. M., Fernández, J. R., Jiménez, N. and López, J. J [2008] The Shapley value for bicooperative games, *Ann. Operat. Res.* **158**(1), 99–115.

Bolger, E. [1993] A value for games with *n* players and *r* alternatives, *Int. J. Game Theory*, **22**(4), 319–334.

Carreras, F. [2005] A decisiveness index for simple games, *Europ. J. Operat. Res.* **163**(2), 370–387.

Carreras, F., Freixas, J. and Puente, M. A. [2003] Semivalues as power indices, *Europ. J. Operat. Res.* **149**(3), 676–687.

Casajus, A. [2012] Amalgamating players, symmetry, and the Banzhaf value, *Int. J. Game Theory* **41**(3), 497–515.

Dubey, P. [1975] On the uniqueness of the Shapley value, *Int. J. Game Theory* **4**(3), 131–139.

Dubey, P. and Shapley, L. S. [1979] Mathematical properties of the Banzhaf power index, *Math. Operat. Res.* **4**(2), 99–131.

Dubey, P., Neyman, A. and Weber, R. J. [1981] Value theory without efficiency, *Math. Operat. Res.* **6**(1), 122–128.

Felsenthal, D. S. and Machover, M. [1997] Ternary voting games, *Int. J. Game Theory* **26**(3), 335–351.

Felsenthal, D. S. and Machover, M. [1998] *The Measurement of Voting Power: Theory and Practice, Problems and Paradoxes* (Edward Elgan Publishing).

Freixas, J. [2005a] Banzhaf measures for games with several levels of approval in the input and output, *Ann. Operat. Res.* **137**(1), 45–66.

Freixas, J. [2005b] The Shapley–Shubik power index for games with several levels of approval in the input and output, *Decision Support Syst.* **39**(2), 185–195.

Freixas, J. and Lucchetti, R. [2016] Power in voting rules with abstention: An axiomatization of a two components power index, *Ann. Operat. Res.* **244**(2), 455–474.

Freixas, J. and Zwicker, W. S. [2003] Weighted voting, abstention, and multiple levels of approval, *Social Choice Welf.* **21**(3), 399–431.

Grabisch, M. and Lange, F. [2007] Games on lattices, multichoice games and the Shapley value: A new approach, *Math. Methods Operat. Res.* **65**(1), 153–167.

Hsiao, C. and Raghavan, T. [1992] Monotonicity and dummy free property for multi-choice cooperative games, *Int. J. Game Theory* **21**(3), 301–312.

Hsiao, C. and Raghavan, T. [1993] Shapley value for multichoice cooperative games, I *Games Econ. Behav.* **5**(2), 240–256.

Klijn, F., Slikker, M. and Zarzuelo, J. [1999] Characterizations of a multi-choice value, *Int. J. Game Theory* **28**(4), 521–532.

Laruelle, A. and Valenciano, F. [2001] Shapley–Shubik and Banzhaf indices revisited, *Math. Operat. Res.* **26**(1), 89–104.

Lehrer, E. [1988] An axiomatization of the Banzhaf value, *Int. J. Game Theory* **17**(2), 89–99.

Monderer D. and Samet D. [2002] Variations on the Shapley value, in *Handbook of Game Theory with Economic Applications* Vol. 3, (Elsevier, North-Holland), pp. 2055–2076.

Penrose, L. S. [1946] The elementary statistics of majority voting, *J. Royal Statis. Soc.* **109**(1), 53–57.

Shapley, L. S. and Shubik, M. [1954] A method for evaluating the distribution of power in a committee system, *Amer. Polit. Sci. Rev.* **48**(3), 787–792.

Chapter 28

Indices of Criticality in Simple Games

Marco Dall'Aglio*

Department of Economics and Finance, LUISS University
Viale Romania 32, 00197 Roma, Italy
mdallaglio@luiss.it

Vito Fragnelli

Department of Sciences and Innovative Technologies (DISIT)
University of Eastern Piedmont
Viale T. Michel 11, 15121 Alessandria, Italy
vito.fragnelli@uniupo.it

Stefano Moretti

Université Paris Dauphine, PSL Research University
CNRS, LAMSADE, 75775 Paris, France
stefano.moretti@dauphine.fr

Power indices in simple games measure the relevance of a player through her ability in being critical, i.e. essential for a coalition to win. We introduce new indices that measure the power of a player in being decisive through the collaboration of other players. We study the behavior of these criticality indices to compare the power of different players within a single voting situation, and that of the same player with varying weight across different voting situations. In both cases we establish monotonicity results in line with those of Turnovec [1998]. Finally, we examine which properties characterizing the indices of Shapley–Shubik and Banzhaf are shared by these new indices.

Keywords: Simple games; power indices; criticality; weighted majority games; monotonicity.

1. Introduction

In the theory of simple games, power indices based on marginal contributions [Straffin, 1977; Dubey *et al.*, 1981] are used to convert information about winning

---

*Corresponding author.

and losing subsets (i.e., coalitions) of the player set into a personal attribution (of power) to each of the players, keeping into account the criticality of players over all possible coalitions (a player i is considered critical for a coalition S containing i, if S is winning and S without i is not). However, the notion of criticality itself is not sufficiently rich to encompass all the interaction possibilities among players and, consequently, the indices based on this notion are not able to distinguish the role of players across quite different simple games. Consider for instance two simple games with only two players i and j, and such that in the first game only the grand coalition $\{i,j\}$ is winning, whereas in the second game all coalitions, except the empty set, are winning. The Shapley and Shubik [1954] power index (or the Banzhaf [1965] one) yields the same power to both players and in both games, and this may be basically seen as the consequence of the fact that each player is critical only once in both situations (precisely, each player is only critical for coalition $\{i,j\}$ in the first game, and only for the corresponding singleton coalition in the second one; see Example 1 for more details). On the other hand, without any further assumption about the probability to form coalitions of different size, the two situations are quite different for the following reason: Suppose that the grand coalition $\{i,j\}$ forms; in the first game, the grand coalition may be threatened by both players, while in the second game, this is not true anymore, and each player becomes critical only if the other leaves the grand coalition. Of course, an answer to the question whether a player is more powerful in one game than in the other one is highly context-specific. Nevertheless, we can argue that the information contained in the characteristic function of a simple game cannot be adequately represented by a single attribute which exclusively relies on the notion of criticality.

In order to introduce a more general concept of criticality, in a previous paper, see Dall'Aglio *et al.* [2016], we analyzed those situations in which a governing majority does not correspond to a minimal winning coalition, i.e., it includes some parties, called non-critical, in the sense that, differently from the critical ones, they may recede from the majority, with no effect on the government. Our main interest focused on the matter that a non-critical party may become critical if other non-critical parties leave the majority; in this way, we introduced different orders of criticality. More precisely, we defined a party to be critical of order k if it is necessary that at least $k-1$ other non-critical parties leave the majority without affecting it, before that its defection makes the coalition a losing one. It should be clear that the order of criticality of a party may be different in the various possible majorities it may belong to. In view of this, we want to associate to each party a unique value that aggregates the different orders of criticality that it has in the different winning coalitions.

The concept of criticality of the first order is the basis of the first power indices due to Shapley and Shubik [1954] and Banzhaf [1965]. Other elements may influence the power of an agent; Myerson [1977] proposed to use an undirected graph, called *communication structure*, in order to represent the relationships among the agents, while Owen [1977] introduced the *a priori unions*, or *coalition structures*, that account for existing agreements, not necessarily binding, among the agents. In the following years, Deegan and Packel [1978] and Holler [1982] defined two new indices that consider only the *minimal winning coalitions*, i.e., those coalitions in which each agent is critical; Johnston [1978] used a similar model considering the *quasi-minimal winning coalitions*, i.e., those coalitions in which at least one agent is critical. Another important contribution is due to Kalai and Samet [1987] that added a *weight* to the elements characterizing each agent, besides the issue of criticality. Winter [1989] extended Owen's idea of *a priori* unions by requiring that the different unions may join only according to a predefined scheme, that he called *levels structure*. In the years that followed, some papers dealt with situations in which only some coalitions are feasible, not only for communication reasons but also for possible incompatibilities among the agents; we refer to Algaba *et al.* [2003; 2004], Bilbao *et al.* [1998], Bilbao and Edelman [2000], Gilles *et al.* [1992], Van den Brink and Gilles [1996], Van den Brink [1997]. Recently, in a monodimensional voting space, Amer and Carreras [2001] considered the ideological distance of the agents and Fragnelli *et al.* [2009] considered the issue of *contiguity*, while Chessa and Fragnelli [2011] introduced the idea of *connectedness* in a possibly multidimensional voting space.

More in general, given a monotonic simple game, for each order of criticality, first, we introduce an appropriate criticality index aimed at measuring the number of times a player is critical of that order (and given the probabilities that various coalitions will form). Second, we propose a collective index aimed at aggregating the indices corresponding to different order of criticality into a unique value for each player. In order to compare our collective index with classical power indices, we analyze some monotonicity properties for weighted majority games, [Turnovec, 1998] and we prove that within a given game, the behavior of a collective index is coherent with the one of classical power indices, while a disaggregate representation of the criticality indices provides a more refined picture of the variation of each player's power across different games.

The chapter is organized as follows. In the next section, we provide some preliminary notions and notations. Section 3 is devoted to the definition of alternative indices of criticality and to the analysis of their basic properties. In Sec. 4, we study some monotonicity properties of a collective index in weighted majority games. Section 5 reconsiders, the criticality indices in the light of the properties that characterize the Shapley–Shubik and Banzhaf indices. Section 6 presents some computational examples for the indices introduced in the previous sections. Section 7 concludes with some open problems.

2. Preliminaries

A *cooperative game with transferable utility* (TU-game) is a pair (N, v), where $N = \{1, 2, \ldots, n\}$ denotes the *finite set of players* and $v : 2^N \to \mathbb{R}$ is the *characteristic function*, with $v(\varnothing) = 0$. $v(S)$ is the worth of coalition $S \subseteq N$, i.e., what players in S may obtain standing alone.

A TU-game (N, v) is *simple* when $v : 2^N \to \{0, 1\}$, with $S \subseteq T \Rightarrow v(S) \le v(T)$[a] and $v(N) = 1$. If $v(S) = 0$ then S is a *losing* coalition, while if $v(S) = 1$ then S is a *winning* coalition. Given a winning coalition S, if $S \setminus \{i\}$ is losing then $i \in N$ is a *critical player* for S. When a coalition S contains at least one critical player for it, S is a *quasi-minimal winning* coalition; when all the players of S are critical, it is a *minimal winning* coalition. A simple game may be defined also assigning the set of winning coalitions or the set of minimal winning coalitions.

A particular class of simple games is represented by the *weighted majority games* (*w.m.g*). A vector of weights $(w_1, w_2, \ldots, w_n)$ is associated to the players that leads to the following definition of the characteristic function of the corresponding weighted majority game (N, w):

$$w(S) = \begin{cases} 1 & \text{if } \sum_{i \in S} w_i \ge q \\ 0 & \text{otherwise} \end{cases}, \quad S \subseteq N,$$

where q is the *majority quota*. A *weighted majority situation* (*w.m.s.*) is often denoted as $[q; w_1, w_2, \ldots, w_n]$. Usually, we ask that the game is *proper* or *N-proper*, i.e., if S is winning then $N \setminus S$ is losing; for this aim it is sufficient to choose $q > \frac{1}{2} \sum_{i \in N} w_i$. Note that a simple game may not correspond to any weighted majority situation.

Given a TU-game (N, v), an *allocation* is a n-dimensional vector $(x_i)_{i \in N} \in \mathbb{R}^N$ assigning to player $i \in N$ the amount x_i; an allocation $(x_i)_{i \in N}$ is *efficient* if $x(N) = \sum_{i \in N} x_i = v(N)$. A *solution* is a function ψ that assigns an allocation $\psi(v)$ to every TU-game (N, v) belonging to a given class of games $\mathcal{G}$ with player set N.

For simple games, and in particular for weighted majority games, a solution is often called a *power index*, as each component x_i may be interpreted as the percentage of power assigned to player $i \in N$. In the literature, several power indices were introduced; among others, we recall the following definitions.

The Shapley and Shubik [1954] index, ϕ, is the natural version for simple games of the Shapley [1953] value. It is defined as the average of the marginal contributions of player i with respect to all the possible orderings and it can be written as:

$$\phi_i(v) = \sum_{S \subseteq N, S \ni i} \frac{(s-1)!(n-s)!}{n!} m_i(S), \quad i \in N,$$

where n and s denote the cardinalities of the set of players N and of the coalition S, respectively and $m_i(S) = v(S) - v(S \setminus \{i\})$ denotes the marginal contribution of player $i \in N$ to coalition $S \subseteq N, S \ni i$.

---

[a]This property is called *monotonicity*.

The Banzhaf [1965] index, β, is similar to the Shapley–Shubik index, but it considers the marginal contributions of a player to all possible coalitions, independently from the order of the players; first, we define:

$$\beta_i(v) = \frac{1}{2^{n-1}} \sum_{S \subseteq N, S \ni i} m_i(S), \quad i \in N.$$

Consider the following properties:

- Anonymity

  Let π be a permutation of the set of players N; a solution ψ satisfies anonymity if $\psi_i(v) = \psi_{\pi(i)}(\pi v)$ for each $i \in N$, where πv is the game in which $\pi v(S) = v(\pi(S))$ for each $S \subseteq N$.

- Null player

  A player $i \in N$ is a null player if $v(S \cup \{i\}) = v(S)$ for each $S \subseteq N \backslash \{i\}$; a solution ψ satisfies null player property if $\psi_i = 0$ for each null player $i \in N$.

- Transfer

  A solution ψ satisfies the transfer property if, given two games v and w with the same set of players N, then $\psi(v) + \psi(w) = \psi(v \vee w) + \psi(v \wedge w)$, where $v \vee w = \max\{v, w\}$ and $v \wedge w = \min\{v, w\}$, with $(v \vee w)(S) = \max\{v(S), w(S)\}$ and $(v \wedge w)(S) = \min\{v(S), w(S)\}$ for each $S \subseteq N$.

- Banzhaf total power

  A solution ψ satisfies the Banzhaf total power property if $\sum_{i \in N} \psi_i = \sum_{i \in N} \frac{1}{2^{n-1}} \sum_{S \subseteq N, S \ni i} m_i(S)$.

The Shapley–Shubik index is the unique efficient solution that satisfies the properties of Anonymity, Null player and Transfer (see Dubey [1975]), while the Banzhaf index is characterized by the properties of Anonymity, Null player, Transfer and Banzhaf total power (see Dubey and Shapley [1979]).

Given a simple game (v, N), we reconsider the definition of criticality given in Dall'Aglio *et al.* [2016].

Definition 1. Let $k \geq 0$ be an integer, let $M \subseteq N$, with $|M| \geq k+1$, be a winning coalition. We say that player i is *critical of the order $k+1$ for coalition M, via coalition K*, with $|K| = k$ if and only if $K \subseteq M \backslash \{i\}$ is a set of minimal cardinality such that

$$v(M \backslash K) - v(M \backslash (K \cup \{i\})) = 1. \tag{1}$$

The meaning of the definition is that K is a coalition of minimal cardinality such that $M \backslash K$ is still a winning coalition, while $M \backslash (K \cup \{i\})$ becomes a losing one.[b]

In the above definition, coalition K is often omitted and we simply say that i is critical of a certain order for coalition M.

---

[b]In other terms, $v(M \backslash T) = 0$ or $v(M \backslash (T \cup \{i\})) = 1$ for any $T \subset M \backslash \{i\}$ with $|T| < k$.

Notice also that, when $k = 0$, $K = \varnothing$, thus, a player is critical of order 1 if and only if it is critical in the usual sense.

Example 1 (Following the discussion in the Introduction). Consider two simple games (N, u) and (N, w) with $N = \{1, 2\}$ and such that $u(\{1, 2\}) = 1$, $u(\{1\}) = u(\{2\}) = 0$ and $w(\{1, 2\}) = w(\{1\}) = w(\{2\}) = 1$. Both players 1 and 2 are critical of the first order for coalition $\{1, 2\}$ in game u, while each player $i \in N$ is critical of the first order for coalition $\{i\}$ in game w. Moreover, each player $i \in N$ is critical of the second order for coalition $\{1, 2\}$ via the other player $j \neq i$ in game w.

Example 2. Consider a simple game (N, v) with $N = \{1, 2, 3, 4, 5, 6\}$ as the set of players and the following minimal winning coalitions (any superset of them is winning):

$$\{1, 2, 3\} \quad \{1, 2, 4\} \quad \{1, 2, 5\} \quad \{1, 3, 4\} \quad \{1, 3, 5\}.$$

If we consider the winning coalition $M = \{1, 2, 3, 4, 5\}$, player 1 is critical of the first order, i.e., critical in the usual sense. Player $i = 2, 3$ is critical of the second order (consider $K = \{2, 3\}\backslash\{i\}$), while player $i = 4, 5$ is critical of the third order (consider $K = \{2, 4, 5\}\backslash\{i\}$ or $K = \{3, 4, 5\}\backslash\{i\}$). If we consider $M = N$ the order of criticality for the first five players does not change, while player 6 is never critical: He cannot have any role in changing the status of a coalition.

The following results hold: (i) Let $M \subseteq N$ be a winning coalition, then the players in M may be partitioned according to their order of criticality, including the subset of those players that are never critical (Corollary 1 in Dall'Aglio *et al.* [2016]); (ii) Let $i \in M$ be a player critical of the order $k + 1, k \geq 1$ for coalition M, via coalition $K \subset M$; if a player $j \in K$ leaves the coalition, then i is a player critical of the order k for coalition $M\backslash\{j\}$, via coalition $K\backslash\{j\}$ (Proposition 3 in Dall'Aglio *et al.* [2016]); (iii) If player $i \in M$ is critical of the order $k + 1$ for coalition M, via coalition $K \subset M\backslash\{i\}$, then in K there are at most $h - 1$ players critical of order not greater than $h, h = 1, \ldots, k$, otherwise i should be critical of order at most k.

3. Indices of Criticality

The aim of this section is to introduce an index for measuring the criticality of an agent. The starting point is the formula in Shapley and Shubik [1954], that is based on the number of swings for each agent considering all the possible orderings of the agents. The motivation of this choice is to have an index that coincides with the Shapley–Shubik index when the criticality order is one.

The Shapley–Shubik index in a weighted majority game v can be defined as the average number of times in which a player is critical with respect to a coalition when players enter the coalition in a random order. More formally, for any $i \in N$, let

$$\phi_i(v) = \frac{1}{n!} \sum_{\pi \in \Pi} \sigma_i(\pi), \tag{2}$$

where π is any permutation of N, Π is the class of all permutations and

$$\sigma_i(\pi) = \begin{cases} 1 & \text{if player } i \text{ is critical in } P_\pi^i \cup \{i\}, \\ 0 & \text{otherwise.} \end{cases}$$

and P_π^i is the set of players in N that precede i in the order π.

We may, therefore, introduce an index which measures for any player i his power in being k order critical, with $k = 1, 2, \ldots, n$.

$$\phi_{i,k}(v) = \frac{1}{n!} \sum_{\pi \in \Pi} \sigma_{i,k}(\pi) \tag{3}$$

with

$$\sigma_{i,k}(\pi) = \begin{cases} 1 & \text{if } i \text{ is critical of order } k \text{ in } P_\pi \cup \{i\}, \\ 0 & \text{otherwise.} \end{cases}$$

With simple combinatorial arguments, we may write these indices in a way that is familiar to anybody working with the Shapley–Shubik index:

$$\phi_{i,k}(v) = \sum_{S \not\ni i} \frac{|S|!(n - |S| - 1)!}{n!} dc_k(i, S \cup \{i\}), \tag{4}$$

where

$$dc_k(i, M) = \begin{cases} 1 & \text{if } i \text{ is critical of order } k \text{ in the winning coalition } M, \\ 0 & \text{otherwise.} \end{cases} \tag{5}$$

We may therefore associate a whole distribution of indices of criticality to a player:

$$\Phi_i(v) = (\phi_{i,1}(v)\phi_{i,2}(v), \ldots, \phi_{i,n}(v)).$$

Example 1 (Continued). It is immediate to verify that: $\Phi_1(u) = \Phi_2(u) = \left(\frac{1}{2}, 0\right)$ and $\Phi_1(w) = \Phi_2(w) = \left(\frac{1}{2}, \frac{1}{2}\right)$.

Example 2 (Continued). With the aid of Mathematica, we computed the distribution of indices for every player:

$$\Phi_1(v) = \left(\frac{17}{30}, 0, 0, 0, 0, 0\right), \quad \Phi_2(v) = \Phi_3(v) = \left(\frac{3}{20}, \frac{3}{10}, 0, 0, 0, 0\right)$$

$$\Phi_4(v) = \Phi_5(v) = \left(\frac{1}{15}, \frac{3}{20}, \frac{1}{5}, 0, 0, 0\right), \quad \Phi_6(v) = (0, 0, 0, 0, 0, 0).$$

Recalling the efficiency property for the Shapley–Shubik index, we have

$$\sum_{i \in N} \phi_{i,1}(v) = \sum_{i \in N} \phi_i(v) = v(N) = 1. \tag{6}$$

Some bounds can be established.

Proposition 1. *The following inequalities hold*

$$1 \le \sum_{k=1}^{n} \sum_{i\in N} \phi_{i,k}(v) \le n \tag{7}$$

and these are sharp, i.e., they can be attained.

Proof. Clearly, by (6)

$$\sum_{k=1}^{n} \sum_{i\in N} \phi_{i,k}(v) \ge \sum_{i\in N} \phi_{1,k}(v) = 1.$$

To verify that the bound can be attained, consider the following game

$$\underline{v}(S) = \begin{cases} 1 & \text{if } S = N, \\ 0 & \text{otherwise.} \end{cases}$$

Clearly, the players are symmetrical and no player can be critical beyond the first order. Thus

$$\Phi_i(\underline{v}) = \left(\frac{1}{n}, 0, \dots, 0\right).$$

To prove the second inequality, note that no player can be critical of more than one order. Therefore, for any permutation $\pi \in \Pi$,

$$\sum_{k=1}^{n} \sum_{i\in N} \sigma_{i,k}(\pi) \le n. \tag{8}$$

Thus, recalling the definition (3),

$$\sum_{k=1}^{n} \sum_{i\in N} \phi_{i,k}(v) = \sum_{k=1}^{n} \sum_{i\in N} \frac{1}{n!} \sum_{\pi\in\Pi} \sigma_{i,k}(\pi) = \frac{1}{n!} \sum_{\pi\in\Pi} \sum_{k=1}^{n} \sum_{i\in N} \sigma_{i,k}(\pi) \le n$$

the inequality deriving from (8).

The following game achieves the upper bound

$$\overline{v}(S) = \begin{cases} 0 & \text{if } S = \varnothing, \\ 1 & \text{otherwise.} \end{cases}$$

Now, in each permutation π of N, player $\pi(1)$ is critical of the first order, player $\pi(2)$ is critical of the second order, player $\pi(3)$ is critical of order 3, and so on. Therefore,

$$\Phi_i(v) = \left(\frac{1}{n}, \dots, \frac{1}{n}\right)$$

and the upper bound is reached. $\qquad\qquad\qquad\qquad\qquad\qquad\qquad\square$

In Example 2, player 6 exhibits null values because it is always non-critical. To point out the role of non-critical players, we may introduce a suitable index:

$$\phi_{i,NC}(v) = \sum_{S \not\ni i} \frac{|S|!(n - |S| - 1)!}{n!} dc_{NC}(i, S \cup \{i\}),$$

where

$$dc_{NC}(i, M) = \begin{cases} 1 & \text{if } i \text{ is never critical in the winning coalition } M, \\ 0 & \text{otherwise.} \end{cases}$$

Example 2 (Continued). Again, using Mathematica, we obtain:

$$\phi_{1,NC}(v) = \phi_{2,NC}(v) = \phi_{3,NC}(v) = \phi_{4,NC}(v) = \phi_{5,NC}(v) = 0 \quad \phi_{6,NC}(v) = \frac{23}{60}.$$

The Shapley–Shubik index of power is naturally paired with the (non-normalized) Bahnzaf index. This index counts the proportion of times that a certain player is critical with respect to any coalition that includes him. It is thus natural to define a Bahnzaf index of criticality of any order:

$$\beta_{i,k}(v) = \sum_{S \not\ni i} \frac{dc_k(i, S \cup \{i\})}{2^{n-1}}. \tag{9}$$

Also in this case we have a distribution of indices:

$$B_i(v) = (\beta_{i,1}(v)\beta_{i,2}(v), \dots, \beta_{i,n}(v))$$

and an index of non-criticality

$$\beta_{i,NC}(v) = \sum_{S \not\ni i} \frac{dc_{NC}(i, S \cup \{i\})}{2^{n-1}}.$$

Now, we return to the introductory examples.

Example 1 (Continued). Since the weights in (4) and (9) coincide when $n = 2$, then we have $B_1(u) = B_2(u) = \left(\frac{1}{2}, 0\right)$ and $B_1(w) = B_2(w) = \left(\frac{1}{2}, \frac{1}{2}\right)$.

Example 2 (Continued). Here, the values differ from the Shapley-derived:

$$B_1(v) = \left(\frac{5}{8}, 0, 0, 0, 0, 0\right), \quad B_2(v) = B_3(v) = \left(\frac{1}{4}, \frac{3}{16}, 0, 0, 0, 0\right)$$

$$B_4(v) = B_5(v) = \left(\frac{1}{8}, \frac{3}{16}, \frac{1}{16}, 0, 0, 0\right), \quad B_6(v) = (0, 0, 0, 0, 0, 0)$$

and

$$B_{1,NC}(v) = B_{2,NC}(v) = B_{3,NC}(v) = B_{4,NC}(v) = B_{5,NC}(v) = 0, \quad B_{6,NC}(v) = \frac{5}{16}.$$

The indices of power provide an easy-to-use and easy-to-understand tool to compare the capability of players in being decisive, i.e. critical in forming a winning coalition. When we extend these notions by including higher order of criticality, we need to compare distributions (i.e., vectors) of indices. In what follows, we provide numeric indices that synthesize the whole distribution and allow the comparison of power among players.

Definition 2. The Collective Shapley–Shubik (CSS) power index through all orders of criticality for player $i \in N$ is defined by

$$\bar{\Phi}_i(v) = \frac{\sum_{h=1}^{n} \phi_{i,h}(v) h^{-1}}{\sum_{h=1}^{n} h^{-1}}. \tag{10}$$

Similarly, the Collective Banzhaf (CB) power index through all orders of criticality for the same player is

$$\bar{B}_i(v) = \frac{\sum_{h=1}^{n} \beta_{i,h}(v) h^{-1}}{\sum_{h=1}^{n} h^{-1}}. \tag{11}$$

Those two indices range between 0 and 1, but no result concerning their total power – something similar to the efficiency property for Shapley–Shubik index, or the Bahnzaf total power, is currently known.[c] Their importance lies in their ability to compare the power of different players. Let us denote as $\succeq_{CSS}$ ($\succeq_{CB}$, respectively) the order established among the players by the Collective Shapley–Shubik (Collective Bahnzaf, respectively) index.

Example 1 (Continued). By relation (10), it is easy to calculate the collective Shapley index for the two players in the games u and w, we have: $\bar{\Phi}_1(u) = \bar{\Phi}_2(u) = \frac{1}{3}$ and $\bar{\Phi}_1(w) = \bar{\Phi}_2(w) = \frac{1}{2}$. Notice that the collective Shapley index emphasizes the different role of the two players over the games u and w (while the Shapley–Shubik index is the same in the two games). Identical conclusions hold for CB.

Example 2 (Continued). Clearly

$$\bar{\Phi}_1(v) = \frac{34}{147}, \quad \bar{\Phi}_2(v) = \bar{\Phi}_3(v) = \frac{6}{49},$$

$$\bar{\Phi}_4(v) = \bar{\Phi}_5(v) = \frac{25}{294}, \quad \bar{\Phi}_6(v) = 0$$

and

$$\bar{B}_1(v) = \frac{25}{98}, \quad \bar{B}_2(v) = \bar{B}_3(v) = \frac{55}{392},$$

$$\bar{B}_4(v) = \bar{B}_5(v) = \frac{115}{1176}, \quad \bar{B}_6(v) = 0.$$

Clearly,

$$1 \succ_{CSS} 2 \sim_{CSS} 3 \succ_{CSS} 4 \sim_{CSS} 5 \succ_{CSS} 6,$$

and the same order applies through the Collective Bahnzaf index.

---

[c]The lower bound is obvious, all the quantities being non-negative; as both $\phi_{i,h}(v)$ and $\beta_{i,h}(v)$ are never larger than 1, for any i and h, then the numerators in both (10) and (11) are not greater than the denominators, so the upper bound holds.

4. Monotonicity Properties for Weighted Majority Games

As we said in the preliminaries, a particular class of simple games is represented by the weighted majority games.

When it comes to measuring power in committees or parliaments, most often a party's power is not perfectly proportional to its size. However, if no impediment occurs, the rankings in power and in size should agree within the same assembly and across different assemblies, when the weights clearly show the advantage of a party over the others. These situations are captured by the following properties.

Definition 3. Let $[q; w_1, \ldots, w_n]$ be a w.m.s. and let v_m be the corresponding game. A power index $\psi : \mathcal{W}(N) \to \mathbb{R}_+^n$ is *locally monotone* whenever $w_i > w_j$ implies

$$\psi_i(v_m) \geq \psi_j(v_m).$$

Consider now two weighted majority situations $[q; w_1, \ldots, w_n]$ and $[q; w_1', \ldots, w_n']$, with

$$\sum_{i \in N} w_i = \sum_{i \in N} w_i' = C. \tag{12}$$

Let v_m and v_m' be the corresponding games. The power index ψ is *globally monotone* if

$$w_{i^*} > w_{i^*}' \quad \text{and} \quad w_j \leq w_j' \quad \text{for all } j \neq i^* \tag{13}$$

implies

$$\psi_{i^*}(v_m) \geq \psi_{i^*}(v_m').$$

Turnovec [1998] shows that the (first order) Shapley–Shubik power index is both locally and globally monotonic. The same properties apply to the (non-normalized) Banzhaf power index. We now extend these properties to the Collective indices just introduced.

Adapting Turnovec [1998], we define the sets

$$C_{iks}(v_m) := \{S \subset N : i \text{ is critical of order } k \text{ with respect to } S \text{ and } |S| = s\}$$

and

$$\mathfrak{C}_{iks}(v_m) := \bigcup_{\ell \leq k} C_{i\ell s}(v_m).$$

We can now write the Shapley–Shubik index of criticality of any order k as

$$\phi_{i,k}(v_m) = \sum_{s=1}^{n} \frac{(s-1)!(n-s)!}{n!} |C_{iks}(v_m)|,$$

where, as usual, $|\cdot|$ denotes the cardinality of a set.

A similar expression is found for the CSS index

$$\bar{\Phi}_i(v_m) = \sum_{s=1}^{n} \frac{(s-1)!(n-s)!}{n!} \left(\frac{\sum_{h=1}^{n} |C_{ihs}(v_m)| h^{-1}}{\sum_{h=1}^{n} h^{-1}} \right). \tag{14}$$

Simpler expressions hold for the Bahnzaf-type indices

$$\beta_{i,k}(v_m) = \sum_{s=1}^{n} \frac{|C_{iks}(v_m)|}{2^{n-1}}$$

and

$$\bar{B}_i(v_m) = \sum_{s=1}^{n} \frac{1}{2^{n-1}} \left(\frac{\sum_{h=1}^{n} |C_{ihs}(v_m)| h^{-1}}{\sum_{h=1}^{n} h^{-1}} \right). \tag{15}$$

We begin by showing the local monotonicity of the indices.

Lemma 1. *Consider a w.m.s.* $[q; w_1, \ldots, w_n]$, *and the associated game* v_m. *Then,* $w_i > w_j$ *implies*

$$|\mathcal{C}_{iks}(v_m)| \geq |\mathcal{C}_{jks}(v_m)| \quad \text{for any } k, s = 1, \ldots, n. \tag{16}$$

Proof. Assume $w_i > w_j$. We now show that to any $S \in C_{jks}(v_m)$ there corresponds a set $S' \in C_{i\ell s}(v_m)$ with $\ell \leq k$. Since no pair of sets S' coincide, this proves (16).

Since $S \in C_{jks}(v_m)$, then $|S| = s$, and there exists $K \subset S \setminus \{j\}$, $|K| = k - 1$, such that

$$v_m(S \setminus K) = 1, \quad v_m(S \setminus (K \cup \{j\})) = 0 \tag{17}$$

and the K is the set of minimal cardinality satisfying (17). Denoting $\tilde{K} = S \setminus (K \cup \{j\})$, we may restate (17) as

$$v_m(\tilde{K} \cup \{j\}) = 1, \quad v_m(\tilde{K}) = 0. \tag{18}$$

We now consider several cases.

Case 1. $i \in K \subset S$. We show that i is critical of order k or lower for the same coalition S by showing that an analogous of (17), or, equivalently, (18), holds for S, i and $K' = K \cup \{j\} \setminus \{i\}$. We do not need to check whether this is a set of minimal cardinality, as long as we do not care about the exact order of criticality for player i.

Now $S = K' \cup \tilde{K} \cup \{i\}$,

$$v_m(S \setminus K') = v_m(\tilde{K} \cup \{i\}) \geq v_m(\tilde{K} \cup \{j\}) = 1,$$

by (18), with the inequality justified by the assumption $w_i \geq w_j$. Moreover, by (18) again,

$$v_m(S \setminus (K' \cup \{i\})) = v_m(\tilde{K}) = 0.$$

Case 2. $i \in S \setminus K$. We show that i, S and K satisfy an analogous of condition (18). This time $S = K \cup \tilde{K}' \cup \{i\}$ with $\tilde{K}' = \tilde{K} \cup \{j\} \setminus \{i\}$. Now (18) and the assumption imply,

$$v_m(S\backslash K) = v_m(\tilde{K}' \cup \{i\}) = v_m(\tilde{K} \cup \{j\}) = 1$$

and

$$v_m(S\backslash(K \cup \{i\})) = v_m(\tilde{K}') = v_m(\tilde{K} \cup \{j\}\backslash\{i\}) \leq v_m(\tilde{K}) = 0.$$

Case 3. $i \notin S$. We now show that i is critical of order k or lower with respect to coalition $S' = S \cup \{i\}\backslash\{j\}$ of size s. Here $S' = K \cup \tilde{K} \cup \{i\}$, while conditions (18) imply:

$$v_m(S'\backslash K) = v_m(\tilde{K} \cup \{i\}) \geq v_m(\tilde{K} \cup \{j\}) = 1$$

and

$$v_m(S'\backslash(K \cup \{i\})) = v_m(\tilde{K}) = 0.$$

Once again, player i is critical of order k or lower for a coalition of cardinality s.

$\square$

This leads to the following:

Theorem 1. *Both CSS and CB are locally monotonic.*

Proof. If $w_i > w_j$, Lemma 1 shows that

$$\frac{\sum_{h=1}^{n} |C_{ihs}(v_m)| h^{-1}}{\sum_{h=1}^{n} h^{-1}} \geq \frac{\sum_{h=1}^{n} |C_{jhs}(v_m)| h^{-1}}{\sum_{h=1}^{n} h^{-1}} \quad \text{for any } s = 1, \ldots, n,$$

and, therefore, that $\bar{\Phi}_i(v_m) \geq \bar{\Phi}_j(v_m)$ and $\bar{B}_i(v_m) \geq \bar{B}_j(v_m)$.

$\square$

To verify the global monotonicity of the same indices, we need to compare the two w.m.s. $[q; w_1, \ldots, w_n]$ and $[q; w'_1, \ldots, w'_n]$, and the corresponding games, v_m and v'_m under (12) and (13). The first lemma is, in its essence, a reworking of a result by Turnovec.

Lemma 2 (Turnovec [1998], Lemma 5). *Suppose (12) and (13) hold. Then for any $P \subset N\backslash\{i^*\}$ we have*

$$v_m(P \cup \{i^*\}) \geq v'_m(P \cup \{i^*\}) \quad \text{and} \quad v_m(P) \leq v'_m(P). \tag{19}$$

Proof. By (12), we can write

$$w_{i^*} = C - \sum_{j \neq i^*} w_j, \quad w'_{i^*} = C - \sum_{j \neq i^*} w'_j$$

and by (13) we have

$$w_{i^*} - w'_{i^*} = \sum_{j \neq i^*} (w'_j - w_j) \geq \sum_{j \in P} (w'_j - w_j) \geq 0.$$

Therefore,

$$w_{i^*} + \sum_{j \in P} w_j \geq w'_{i^*} + \sum_{j \in P} w'_j$$

and

$$\sum_{j \in P} w_j \leq \sum_{j \in P} w'_j$$

which leads to (19).   □

To establish global monotonicity we need another lemma.

Lemma 3. *Assumptions* (12) *and* (13) *imply*

$$|\mathfrak{C}_{i^* ks}(v_m)| \geq |\mathfrak{C}_{i^* ks}(v'_m)| \quad \text{for any } k, s = 1, \ldots, n. \tag{20}$$

Proof. We will show that, if $S \in C_{i^* ks}(v'_m)$, then $S \in \mathfrak{C}_{i^* \ell s}(v_m)$ for some $\ell \leq k$. In fact, for such a coalition S we have that there exists $K \subset S \setminus \{i^*\}$, $|K| = k - 1$ such that

$$v'_m(S \setminus K) = 1 \quad \text{and} \quad v'_m(S \setminus (K \cup \{i^*\})) = 0 \tag{21}$$

and any $T \subset S \setminus \{i^*\}$ with cardinality lower than $k - 1$ does not satisfy both equations.

A straightforward application of Lemma 2 yields

$$v_m(S \setminus K) \geq v'_m(S \setminus K) = 1$$

and

$$v_m(S \setminus (K \cup \{i^*\})) \leq v'_m(S \setminus (K \cup \{i^*\})) = 0$$

and, therefore, concludes the proof.   □

This leads to the following.

Theorem 2. *Both* CSS *and* CB *are globally monotonic.*

Proof. If assumptions (12) and (13) hold, Lemma 3 shows that

$$\frac{\sum_{h=1}^{n} |C_{i^* hs}(v_m)| h^{-1}}{\sum_{h=1}^{n} h^{-1}} \geq \frac{\sum_{h=1}^{n} |C_{i^* hs}(v'_m)| h^{-1}}{\sum_{h=1}^{n} h^{-1}} \quad \text{for any } s = 1, \ldots, n,$$

and, therefore, that $\bar{\Phi}_i^*(v_m) \geq \bar{\Phi}_i^*(v'_m)$ and $\bar{B}_i^*(v_m) \geq \bar{B}_i^*(v'_m)$.   □

Remark 1. Turnovec [1998] shows (Theorem 4) that, under suitable symmetry assumptions, global monotonicity implies local monotonicity for power indices. Whether this fact applies here as well remains an open question.

5. First Steps Toward a Characterization

We may ask which among the properties that characterize the indices of Shapley–Shubik and Banzhaf may be used for characterizing the criticality indices. We begin with an easy result.

Proposition 2. *The criticality indices $\Phi, \bar{\Phi}, B$ and $\bar{B}$ satisfy anonymity and null player properties.*

Proof. It is straigthforward, after noticing that relations (4) and (9) satisfy the two properties.   □

As previously noted, no property concerning the efficiency or the total power is known for these indices. Also the transfer property is not satisfied, as the following example shows.

Example 3. Consider the following example where $N = \{1, 2, 3\}$, v is the simple game whose unique minimal winning coalition is $\{1, 2\}$ and w is the simple game whose minimal winning coalitions are $\{1, 3\}, \{2, 3\}$. In this case, we have:

index/game	v	w	$v \vee w$	$v \wedge w$
Φ_1	$\left(\frac{1}{2}, 0, 0\right)$	$\left(\frac{1}{6}, \frac{1}{3}, 0\right)$	$\left(\frac{1}{3}, \frac{1}{3}, 0\right)$	$\left(\frac{1}{3}, 0, 0\right)$
Φ_2	$\left(\frac{1}{2}, 0, 0\right)$	$\left(\frac{1}{6}, \frac{1}{3}, 0\right)$	$\left(\frac{1}{3}, \frac{1}{3}, 0\right)$	$\left(\frac{1}{3}, 0, 0\right)$
Φ_3	$(0, 0, 0)$	$\left(\frac{2}{3}, 0, 0\right)$	$\left(\frac{1}{3}, \frac{1}{3}, 0\right)$	$\left(\frac{1}{3}, 0, 0\right)$
$\bar{\Phi}$	$\left(\frac{3}{11}, \frac{3}{11}, 0\right)$	$\left(\frac{2}{11}, \frac{2}{11}, \frac{4}{11}\right)$	$\left(\frac{3}{11}, \frac{3}{11}, \frac{3}{11}\right)$	$\left(\frac{2}{11}, \frac{2}{11}, \frac{2}{11}\right)$
B_1	$\left(\frac{1}{2}, 0, 0\right)$	$\left(\frac{1}{4}, \frac{1}{4}, 0\right)$	$\left(\frac{1}{2}, \frac{1}{4}, 0\right)$	$\left(\frac{1}{4}, 0, 0\right)$
B_2	$\left(\frac{1}{2}, 0, 0\right)$	$\left(\frac{1}{4}, \frac{1}{4}, 0\right)$	$\left(\frac{1}{2}, \frac{1}{4}, 0\right)$	$\left(\frac{1}{4}, 0, 0\right)$
B_3	$(0, 0, 0)$	$\left(\frac{3}{4}, 0, 0\right)$	$\left(\frac{1}{2}, \frac{1}{4}, 0\right)$	$\left(\frac{1}{4}, 0, 0\right)$
$\bar{B}$	$\left(\frac{3}{11}, \frac{3}{11}, 0\right)$	$\left(\frac{9}{44}, \frac{9}{44}, \frac{9}{22}\right)$	$\left(\frac{15}{44}, \frac{15}{44}, \frac{15}{44}\right)$	$\left(\frac{3}{22}, \frac{3}{22}, \frac{3}{22}\right)$

It is easy to check that the transfer property does not hold, as in $v \vee w$ and $v \wedge w$ the players are symmetric, while in v and w they are not.

More generally, the following result holds.

Proposition 3. *The transfer property is not satisfied by any weighted average of the Shapley–Shubik (Banzhaf, respectively) indices of criticalities of all orders, with strictly positive weights.*

Proof. Referring to Example 3, notice that $\Phi_3(v) + \Phi_3(w) = \left(\frac{2}{3}, 0, 0\right)$ and $\Phi_3(v \vee w) + \Phi_3(w \wedge w) = \left(\frac{2}{3}, \frac{1}{3}, 0\right)$ and no set of posistive weights can equate the averages of the two distributions.   □

In view of this, we need other properties for characterizing the criticality indices.

6. Examples

Consider Example 3 in Dall'Aglio *et al.* [2016]. In the weighted majority situation $[51; 44, 3, 3, 3, 3, 3, 3]$, party 1 is critical of order 1, while the other six parties are critical of order 4 and there are no parties critical of order 2 or 3. If we account the ordering (we write the weights of the parties instead of their numbers):

$$44, 3, 3, 3, 3, 3, 3$$

the fourth party is critical of order 1, the fifth party is critical of order 2 via the fourth party (or the second, or the third), the sixth party is critical of order 3 via the fourth and the fifth parties (or any pair of parties from the second to the fifth), the seventh party is critical of order 4 via the fourth, the fifth and the sixth parties (or any triple of parties from the second to the sixth). The same reasoning applies to all the permutations of the parties with weight 3.

Accounting the ordering:

$$3, 3, 3, 3, 44, 3, 3$$

the fifth party is critical of order 1, the sixth party is critical of order 3, the seventh party is critical of order 4. Again, the same reasoning applies to all the permutations of the parties with weight 3.

If we consider all the possible orderings we have Table 1, where the Roman number below each party after the critical one denotes its order of criticality.

In terms of distributions, we have

$$\Phi_1 = \left(\frac{24}{42}, 0, 0, 0, 0, 0, 0\right), \quad \Phi_2 = \cdots = \Phi_7 = \left(\frac{3}{42}, \frac{4}{42}, \frac{5}{42}, \frac{6}{42}, 0, 0, 0\right),$$

Table 1.   Orders of criticality for each ordering of the parties in $[51; 44, 3, 3, 3, 3, 3, 3]$.

44	3	3	3	3	3	3
			I	II	III	IV
3	44	3	3	3	3	3
			I	II	III	IV
3	3	44	3	3	3	3
			I	II	III	IV
3	3	3	44	3	3	3
				I	II	III	IV
3	3	3	3	44	3	3
					I	III	IV
3	3	3	3	3	44	3
						I	IV
3	3	3	3	3	3	44
							I

$$\bar{\Phi} = \left(\frac{80}{363}, \frac{3}{40}, \dots, \frac{3}{40} \right),$$

$$B_1 = \left(\frac{21}{32}, 0, 0, 0, 0, 0, 0 \right), \quad B_2 = \dots = B_7 = \left(\frac{5}{32}, \frac{5}{32}, \frac{5}{64}, \frac{1}{64}, 0, 0, 0 \right),$$

$$\bar{B} = \left(\frac{245}{968}, \frac{1}{10}, \dots, \frac{1}{10} \right).$$

Now, consider Example 1 in Dall'Aglio *et al.* [2016]. In the weighted majority situation [51; 40, 8, 5, 5, 5], party 1 is critical of order 1, while the other four parties are critical of order 2. If we consider all the possible orderings we have Table 2.

In terms of distributions, we have

$$\Phi_1 = \left(\frac{1}{2}, 0, 0, 0, 0 \right), \quad \Phi_2 = \left(\frac{1}{4}, \frac{1}{5}, 0, 0, 0 \right), \quad \Phi_3 = \Phi_4 = \Phi_5 = \left(\frac{1}{12}, \frac{3}{10}, 0, 0, 0 \right),$$

$$\bar{\Phi} = \left(\frac{30}{137}, \frac{21}{137}, \frac{14}{137}, \frac{14}{137}, \frac{14}{137} \right),$$

$$B_1 = \left(\frac{1}{2}, 0, 0, 0, 0 \right), \quad B_2 = \left(\frac{3}{8}, \frac{1}{16}, 0, 0, 0 \right), \quad B_3 = B_4 = B_5 = \left(\frac{1}{8}, \frac{3}{16}, 0, 0, 0 \right),$$

$$\bar{B} = \left(\frac{30}{137}, \frac{195}{1096}, \frac{105}{1096}, \frac{105}{1096}, \frac{105}{1096} \right).$$

Now, consider the weighted majority situation [12; 10, 2, 2, 1, 1], party 1 is critical of order 1, while the other four parties are critical of order 3 and there are no parties critical of order 2. If we consider all the possible orderings we have Table 3.

In terms of distributions, we have:

$$\Phi_1 = \left(\frac{7}{10}, 0, 0, 0, 0 \right), \quad \Phi_2 = \Phi_3 = \left(\frac{7}{60}, \frac{11}{60}, \frac{1}{5}, 0, 0 \right),$$

$$\Phi_4 = \Phi_5 = \left(\frac{1}{30}, \frac{1}{10}, \frac{1}{5}, 0, 0 \right),$$

$$\Phi_{NC} = \left(0, 0, 0, \frac{7}{60}, \frac{7}{60} \right), \quad \bar{\Phi} = \left(\frac{42}{137}, \frac{33}{274}, \frac{33}{274}, \frac{9}{137}, \frac{9}{137} \right),$$

$$B_1 = \left(\frac{13}{16}, 0, 0, 0, 0 \right), \quad B_2 = B_3 = \left(\frac{3}{16}, \frac{1}{4}, \frac{1}{16}, 0, 0 \right),$$

$$B_4 = B_5 = \left(\frac{1}{16}, \frac{1}{8}, \frac{1}{16}, 0, 0 \right),$$

$$B_{NC} = \left(0, 0, 0, \frac{3}{16}, \frac{3}{16} \right), \quad \bar{B} = \left(\frac{195}{548}, \frac{20}{137}, \frac{20}{137}, \frac{35}{548}, \frac{35}{548} \right).$$

Table 2.   Orders of criticality for each ordering of
the parties in [51; 40, 8, 5, 5, 5].

40	8	5 I	5 II	5 II
40	5	8 I	5 II	5 II
40	5	5	8 I	5 II
40	5	5	5 I	8 II
8	40	5 I	5 II	5 II
5	40	8 I	5 II	5 II
5	40	5	8 I	5 II
5	40	5	5 I	8 II
8	5	40 I	5 II	5 II
5	8	40 I	5 II	5 II
5	5	40	8 I	5 II
5	5	40	5 I	8 II
8	5	5	40 I	5 II
5	8	5	40 I	5 II
5	5	8	40 I	5 II
5	5	5	40 I	8 II
8	5	5	5	40 I
5	8	5	5	40 I
5	5	8	5	40 I
5	5	5	8	40 I

Table 3.   Orders of criticality for each ordering of the parties in $[12; 10, 2, 2, 1, 1]$.

10	2 (I)	2 (II)	1 (0)	1 (III)	2	2	10 (I)	1 (0)	1 (III)
10	2 (I)	1 (0)	2 (II)	1 (III)	2	1	10 (I)	2 (II)	1 (III)
10	2 (I)	1 (0)	1 (II)	2 (III)	2	1	10 (I)	1 (II)	2 (III)
10	1 (I)	2 (II)	2 (III)*	1	1	2	10 (I)	2 (II)	1 (III)
10	1 (I)	2 (II)	1 (III)*	2	1	2	10 (I)	1 (II)	2 (III)
10	1 (I)	1 (II)	2 (III)*	2	1	1	10 (I)	2 (II)	2 (III)
2	10 (I)	2 (II)	1 (0)	1 (III)	2	2	1	10 (I)	1 (III)
2	10 (I)	1 (0)	2 (II)	1 (III)	2	1	2	10 (I)	1 (III)
2	10 (I)	1 (0)	1 (II)	2 (III)	2	1	1	10 (I)	2 (III)
1	10 (I)	2 (II)	2 (III)*	1	2	2	10 (I)	1 (III)
1	10 (I)	2 (II)	1 (III)*	1	2	1	10 (I)	2 (III)
1	10 (I)	1 (II)	2 (III)*	1	1	2	10 (I)	2 (III)
2	2	1	1	10 (I)					
2	1	2	1	10 (I)					
2	1	1	2	10 (I)					
1	2	2	1	10 (I)					
1	2	1	2	10 (I)					
1	1	2	2	10 (I)					

7. Conclusions

In this chapter, we have introduced a new notion of criticality index that generalizes the classical one of power index to different orders of criticality. It turns out that the representation of the players' interaction in terms of the n-dimensional vector containing all criticality indices (where n is the size of the player set) is a good compromise between the exponentially high complexity associated with the characteristic function, and the rough sketch provided by a personal attribution of power to each player. A first attempt to aggregate such criticality indices into a collective one is also presented and studied with respect to several monotonicity properties in weighted majority games. Notice that the majority of the results dealing with the Shapley index and the Banzhaf index in this chapter, may be easily adapted to every semivalue Dubey *et al.* [1981].

An open question for future research concerns a better understanding of the nature of power associated to each index of criticality: how such a power may be carried out by means of the bargaining abilities of players? which trade-off exists between the different orders of criticality and between the corresponding indices?

Another interesting issue is related to an alternative interpretation of the indices of criticality in terms of the likelihood that a coalition will form. For instance, following Example 1, according to the index of criticality of second order, the formation of the grand coalition seems to be more likely in game u, where players have a null critical index of second order, than in game w, where such an index is one-half for both players. In other words, it is not clear whether the indices of criticality may be used to compare the possibility that a coalition will form over different games with equivalent distribution of (first order) criticality.

Acknowledgments

The authors gratefully acknowledge the useful comments and suggestions that allowed for improving this work.

References

Algaba, E., Bilbao, J. M., Brink van den, R. and Jiménez-Losada, A. [2003] Axiomatizations of the restricted Shapley value for cooperative games on antimatroids, *Math. Methods Operat. Res.* **57**, 49–65.

Algaba, E., Bilbao, J. M., Brink van den, R. and Jiménez-Losada, A. [2004] An axiomatization of the Banzhaf value for cooperative games on antimatroids, *Math. Methods Operat. Res.* **59**, 147–166.

Amer, R. and Carreras, F. [2001] Power, cooperation indices and coalition structures, in *Power Indices and Coalition Formatio*, eds. Holler, M. J. and Owen, G. (Kluwer Academic Publishers, Dordretch), pp. 153–173.

Banzhaf, J. F. [1965] Weighted voting doesn't work: A mathematical analysis, *Rutgers Law Rev.* **19**, 317–343.

Bilbao, J. M., Jiménez, A. and Lopez, J. J. [1998] The Banzhaf power index on convex geometries, *Math. Social Sci.* **36**, 157–173.

Bilbao, J. M. and Edelman, P. H. [2000] The Shapley value on convex geometries, *Discrete Appl. Math.* **103**, 33–40.

Brink van den, R. [1997] An axiomatization of the disjunctive permission value for games with a permission structure, *Int. J. Game Theory* **26**, 27–43.

Brink van den, R. and Gilles, R. P. [1996] Axiomatizations of the conjunctive permission value for games with permission structures, *Games Econ. Behav.* **12**, 113–126.

Chessa, M. and Fragnelli, V. [2011] Embedding classical indices in the FP family, *AUCO Czech Econ. Rev.* **5**, 289–305.

Dall'Aglio, M., Fragnelli, V. and Moretti, S. [2016] Orders of criticality in voting games, *Operat. Res. Deci.* **26**, 53–67.

Deegan, J. and Packel, E. W. [1978] A new index of power for simple n-person games, *Int. J. Game Theory* **7**, 113–123.

Dubey, P. [1975] On the uniqueness of the Shapley value, *Int. J. Game Theory* **4**, 131–139.

Dubey, P., Neyman, A. and Weber, R. J. [1981] Value theory without efficiency, *Math. Operat. Res.* **6**, 122–128.

Dubey, P. and Shapley, L. S. [1979] Mathematical properties of the Banzhaf power index, *Math. Operat. Res.* **4**, 99–131.

Fragnelli, V., Ottone, S. and Sattanino, R. [2009] A new family of power indices for voting games, *Homo Oeconom.* **26**, 381–394.

Gilles, R. P., Owen, G. and Brink van den, R. [1992] Games with permission structures: The conjunctive approach, *Int. J. Game Theory* **20**, 277–293.

Holler, M. J. [1982] Forming coalitions and measuring voting power, *Political Stud.* **30**, 262–271.

Johnston, R. J. [1978] On the measurement of power: Some reactions to laver, *Environ. Plan. A* **10**, 907–914.

Kalai, E. and Samet, D. [1987] On weighted Shapley values, *Int. J. Game Theory* **16**, 205–222.

Myerson, R. [1977] Graphs and cooperation in games, *Math. Operat. Res.* **2**, 225–229.

Owen, G. [1977] Values of games with *a priori* unions, *Lect. Notes Econ. Math. Syst.* **141**, 76–88.

Shapley, L. S. and Shubik, M. [1954] A method for evaluating the distribution of power in a committee system, *Amer. Polit. Sci. Rev.* **48**, 787–792.

Shapley, L. S. [1953] A value for n-person games, in *Contributions to the Theory of Games* II. eds. Kuhn, H. W. and Tucker, A. W. (Princeton University Press, Princeton), pp. 307–317.

Straffin, P. D. [1977] Homogeneity, independence, and power indices, *Public Choice* **30**, 107–118.

Turnovec, F. [1998] Monotonicity and power indices, in *Trends in Multicriteria Decision Making*, Lecture Notes in Economics and Mathematical Systems, eds. Stewart, T. J. and van den Honert, R. C. Vol. 465 (Springer-Verlag), pp. 199–214.

Winter, E. [1989] A value for cooperative games with levels structure of cooperation, *Int. J. Game Theory* **18**, 227–240.

Index

CPSIA information can be obtained
at www.ICGtesting.com
Printed in the USA
BVHW012306131019
560925BV00001B/1/P